D0758558

EARTHQUAKE ENGINEERING

CONTRIBUTING AUTHOR

BRUCE A. BOLT

Professor of Seismology and Director of the Seismographic Stations
University of California
Berkeley, California

DONALD E. HUDSON

Division of Engineering and Applied Science
California Institute of Technology
Pasadena, California

H. BOLTON SEED

Department of Civil Engineering
University of California
Berkeley, California

JOSEPH PENZIEN

Professor of Civil Engineering and
Director, Earthquake Engineering Research Center
University of California
Berkeley, California

JOHN A. BLUME

President, John A. Blume & Associates, Engineers
San Francisco, California

M. G. BONILLA

Geologist, U.S. Geological Survey
Menlo Park, California

J. G. BOUWKAMP and DIXON REA

Department of Civil Engineering
University of California
Berkeley, California

ROBERT L. WIEGEL

Department of Civil Engineering
University of California
Berkeley, California

NATHAN M. NEWMARK

University of Illinois
Urbana, Illinois

T. Y. LIN

Department of Civil Engineering
University of California
Berkeley, California

GEORGE W. HOUSNE

Division of Engineering and Applied Scienc
California Institute of Technolog
Pasadena, Californi

KARL V. STEINBRUGG

Pacific Fire Rating Bure
San Francisco, Californ

RAY W. CLOU

Department of Civil Engineer
University of Califo
Berkeley, Calif

HENRY J. DEGE

President, H. J. Degenkolb &
San Francisco,

JOHN

Standard Oil Company
San Francis

EARTHQUAKE ENGINEERING

ROBERT L. WIEGEL

Coordinating Editor

PRENTICE-HALL, INC., Englewood Cliffs, N. J.

Current printing (last digit):
19 18 17 16 15 14 13 12

13-222646-4

Library of Congress Catalog Card Number 75-76876

Printed in the United States of America

PRENTICE-HALL INTERNATIONAL, INC., *London*
PRENTICE-HALL OF AUSTRALIA, PTY. LTD., *Sydney*
PRENTICE-HALL OF CANADA, LTD., *Toronto*
PRENTICE-HALL OF INDIA PRIVATE LTD., *New Delhi*
PRENTICE-HALL OF JAPAN, INC., *Tokyo*

Contents

Preface

The Alaska Earthquake of March 27, 1964, was the strongest earthquake ever recorded on the North American Continent. Loss of life, although substantial, was not nearly as great as that resulting from a number of other earthquakes; for example, nearly 11,000 persons lost their lives in the earthquake in northeastern Iran on August 31, 1968. Property damage from the Alaska Earthquake, however, was extensive—of the order of $300 million. The extent of physical suffering and mental anguish of the survivors cannot be estimated, but the enormity of it spurs man to improve his ability to plan better located installations and to design and build structures more resistant to earthquakes and associated phenomena.

The Alaska Earthquake of 1964 stimulated a wide interest in earthquake engineering among many practicing engineers, with a number of them expressing a desire to learn more about earthquakes and what could be done to lessen the loss of life and decrease property damage in the future.

A short course was organized and given in early September 1965 at the University of California, Berkeley. The course was structured to present the state of seismology, geology, and engineering as they applied to earthquake engineering. Lectures were given on causes of earthquakes, seismic waves, faulting, strong ground motion and its measurement, structural damage caused by earthquakes, the "design earthquake," dynamic testing of models and prototype structures, soil problems and soil behavior, linear and nonlinear structural response (both deterministic and probabilistic), soil-pile foundation interaction, and tsunamis. These were followed by a series of lectures on the design of earthquake-resistant structures, including earth structures, poured-in-place concrete structures, prestressed and precast concrete structures, and steel frame structures.

The interest in the course was so great that it appeared desirable to prepare a book which was largely modeled on the lectures. Thus, sixteen writers have contributed their efforts in putting together this comprehensive book with the hope that the information presented in it may serve as a "plateau" from which students and practitioners can build.

ROBERT L. WIEGEL
Berkeley, California

Chapter 1

Elastic Waves in the Vicinity of the Earthquake Source

BRUCE A. BOLT

Professor of Seismology and Director of the Seismographic Stations, University of California, Berkeley, California

1.1 INTRODUCTION

Principal attention has been given in seismology to the properties of seismic waves at places far from the region where an earthquake is felt. One reason is that elastic wave theory is simpler if the dimensions of the source can be ignored; another is that, independent of the generating mechanism, many important properties of the Earth along the path of propagation can be inferred from the observed elastic waves; and yet another is that very few seismograms showing motion near to the source are available.

The dynamics of wave motion near to the earthquake source will be emphasized in this chapter because of its relevance to earthquake engineering. It should not be overlooked, however, that earthquakes are known to cause damage at considerable distances from the source region, through the vibrations caused by elastic waves traveling in the Earth or the effects of water waves (tsunamis) in the ocean. In the case of the California earthquake of April 18, 1906, which ruptured the San Andreas fault, there was considerable building damage in the town of Los Banos (compare intensity zone IX–X southwest of Merced in Fig. 1.7), which is situated in the Central Valley of California some 60 km from the nearest point of the fault trace. As a second example, slight structural damage has been reported in Sacramento, California, following only moderate earthquakes that have occurred over 150 km away to the east of the Sierra Nevada. At these distances, the wave train carrying the seismic energy has the form that is predicted quite closely by the elastic wave theory outlined in this chapter.

The earthquake mechanism becomes an important ingredient of the wave dynamics near to the source. The ground shaking, for example, will be somewhat different if it is produced by progressive rupture on a nearby fault than if it is generated by a nearby explosive source. Ground motion from dip-slip faulting will be different to ground motion from strike-slip faulting. At all distances from the source, however, mechanical properties of the elastic basement rocks, such as incompressibility, rigidity, and density play a role, as does the effect of layering of the rocks and physical properties of the soil.

In this chapter we consider the behavior of both traveling and standing seismic waves in a layered deformable media. The present stage of continuum elasticity theory allows the use of high-speed computers to determine response functions for reasonably realistic models of the soil and basement rocks. It is unfortunate that more studies of this kind, of special interest to earthquake engineers, have not been made and published in suitable journals.

It must be emphasized that the discussion in this chapter concerns the behavior of *elastic waves*. The shaking of surface soils and rocks from incident elastic waves from earthquakes sometimes leads to severe *nonelastic behavior* including differential slumping and soil liquefaction. The study of these important problems of earthquake engineering goes beyond elastic wave theory.

Several specialized textbooks in the English language give detailed accounts of elastic wave theory with general seismological applications. Only those parts of this theory are outlined in the following sections that are required for either an analysis of ground vibrations or the critique of the hypotheses of earthquake source mechanism made in Chapter 2. Certain assumptions upon which the mathematical theory rests are emphasized. The reader who wishes to follow more rigorously the mathematical formulation might consult the following references:

Brekhovskikh, L. M., *Waves in Layered Media* (trans.), New York: Academic Press, 1960.

Bullen, K. E., "Seismic Wave Transmission," *Encyclopedia of Physics*, **47**, 75–118. Berlin: Springer-Verlag, 1956.

Bullen, K. E., *An Introduction to the Theory of Seismology*, 3rd ed., New York: Cambridge University Press, 1963.

Ewing, W. M., W. S. Jardetzky, and F. Press, *Elastic Waves in Layered Media*, New York: McGraw-Hill, 1957.

White, J.E., *Seismic Waves*, New York: McGraw-Hill, 1965.

1.2 THE ELASTIC WAVE MODEL

The basic physical model in seismology is that of a perfectly elastic medium in which the infinitesimal strain approximation of elasticity theory is adopted. Although this representation is quite adequate for many geophysical problems, and although it is often a successful first approximation for problems in earthquake engineering, we must keep in mind that near to the earthquake source higher powers of the components of strain may become significant. Anisotropy, imperfections in elasticity, and gross geological inhomogeneities also modify the ground response predicted by the simpler theory. From a mathematical point of view, there is the additional complication that near to the source the presence of fractures and faults introduces discontinuities into the elastic field functions. Some modifications to deal with some of these special problems will be mentioned in the course of setting down the main results for the perfectly elastic model.

We work in the notation of Cartesian tensors in which the Cartesian coordinates of a point A of the medium are given by x_i ($i = 1, 2, 3$). On the occurrence of an earthquake, let A be displaced an amount u_i; then a neighboring particle B, initially at $x_i + dx_i$, is displaced an amount

$$u_i + \frac{\partial u_i}{\partial x_j}\,dx_j \tag{1.1}$$

where the convention is that a repeated suffix stands for the summation over that suffix.

A measure of the deformation is

$$\begin{aligned} d(AB)^2 &= \left(\frac{\partial u_j}{\partial x_i} + \frac{\partial u_i}{\partial x_j}\right) dx_i\,dx_j + \frac{\partial u_i}{\partial x_j}\frac{\partial u_i}{\partial x_k}\,dx_j\,dx_k \\ &= 2e_{ij}\,dx_i\,dx_j + \text{second-order terms}, \end{aligned}$$

where

$$e_{ij} = \frac{1}{2}\left(\frac{\partial u_j}{\partial x_i} + \frac{\partial u_i}{\partial x_j}\right). \tag{1.2}$$

For infinitesimal strain, the second-order terms are neglected. Because of its symmetry the (infinitesimal) strain tensor e_{ij} has only six independent components of which e_{11}, e_{22}, e_{33} correspond to *extensions* parallel to the Cartesian axes and e_{23}, e_{31}, e_{12} measure the angular deformation or *shear strain.*

The increase in volume of a unit cube of the medium through rarefaction is $e_{11} + e_{22} + e_{33} = e_{ii}$, to first order. In the limit, as the volume becomes vanishingly small, e_{ii} is a measure of the *dilatation* θ or negative compression. On this infinitesimal strain theory, the deformation of the ground is a function only of the derivatives $\partial u_i/\partial x_j$. Now, using Eq. 1.2, we may write

$$\frac{\partial u_i}{\partial x_j} = e_{ij} - w_{ij},$$

where

$$w_{ij} = \frac{1}{2}\left(\frac{\partial u_j}{\partial x_i} - \frac{\partial u_i}{\partial x_j}\right). \tag{1.3}$$

The ground deformation is thus the resultant of strain e_{ij} and *rotation* w_{ij}.

In earthquakes, effects attributable to the three quantities θ, e_{ij}, and w_{ij} may be expected. Substantial changes of level in wells have been observed following earthquakes* (compare Blanchard and Byerly, 1935; Eaton and Takasaki, 1959). The well can be thought of as a *dilatometer* that has magnified the compression $-\theta$ by movement of the entrapped water. More than three decades ago H. Benioff (1935) designed a linear *strainmeter* that provided subsequently precise measurements of linear strain $\partial u_i/\partial x_i$. In one form the strainmeter consists of a quartz tube (aligned horizontally or vertically) about 50–100 m long and fixed at one end to the rock. The variations in distance between the other end of the tube and a neighboring point in the rock are measured by a sensitive electromagnetic or capacitance transducer. Strains following earthquakes as small as 10^{-10} can be measured by such devices.

Deformations associated with earthquakes in the form of ground tilts produce oscillations (*seiches*) in lakes and reservoirs and are measured by *tiltmeters* such as long-period pendulums or long tubes containing liquid. Attempts have been made to record the components of *rotation* w_{ij}, but so far unsuccessfully.

Translation and deformation of an elastic body are conceived as arising from the applications of two types of forces: body forces and surface tractions. The *stress* at a point A of the medium arises from the surface tractions on small plane interfaces surrounding the point. As the area of the interfaces becomes vanishingly small the ratio of the tractions to the area is called the stress at A. These pressures may be summarized by a symmetrical *stress tensor* p_{ij}. If the reference axes are selected so that the shear components p_{23}, p_{31}, p_{12} are all zero, the components p_{11}, p_{22}, p_{33} are the *principal stresses* at A.

For a hydrostatic pressure p, the three principal stresses will each equal minus p. More generally, let P be the mean of the principal stresses, i.e.,

$$P = \tfrac{1}{3}p_{ii}.$$

Then the applied stresses can be treated as the resultant of the mean stress P together with deviations P_{ij} from it. In symbols,

*After the great 1964 Alaskan earthquake, oscillations of the water level in wells were observed worldwide.

$$p_{ij} = P\delta_{ij} + P_{ij}, \tag{1.4}$$

where P_{ij} is the *stress deviation tensor* and δ_{ij} is the Kronecker delta ($\delta_{ij} = 1$ if $i = j$; $\delta_{ij} = 0$ if $i \neq j$). In the corresponding physical representation, the *strain deviator* E_{ij} is given by

$$e_{ij} = \tfrac{1}{3}\theta\delta_{ij} + E_{ij}. \tag{1.5}$$

For many seismological purposes, stress is taken to be related to strain through Hooke's relation, which is both linear and time invariant. Neither restriction probably holds exactly in the neighborhood of an earthquake source, particularly in the case of the greatest earthquakes. Observations of many earthquakes do, however, indicate (cf. Section 1.6) that the Hookian assumption is a reasonable approximation in most circumstances in rocks beyond the immediate region of energy release. The effect of the time dependence is considered in Section 1.4.

To allow for the most general anisotropy, we take

$$p_{lm} = c_{ijlm}e_{ij} \tag{1.6}$$

where c_{ijlm} is a set of 81 coefficients. The number of independent coefficients is reduced to 21 by the symmetry of the stress and strain tensors and by certain thermodynamical conditions. Further reduction depends upon the conditions of strain symmetry. An *isotropic medium* has complete elastic symmetry. Its elastic behavior is described completely by two coefficients λ and μ, where μ is the *rigidity modulus* of the medium and $\lambda = k - \frac{2}{3}\mu$; k is the *incompressibility* or *bulk modulus*. The full stress–strain relations for a perfectly elastic isotropic medium are

$$p_{ij} = (k - \tfrac{2}{3}\mu)\theta\delta_{ij} + 2\mu e_{ij}. \tag{1.7}$$

Substitution in Eq. 1.7 from Eqs. 1.4 and 1.5 yields

$$P_{ij} = 2\mu E_{ij}. \tag{1.8}$$

It follows that μ is the ratio of the shearing stress to the change in angle produced by the shearing stress; thus, for a perfect fluid, the rigidity μ is zero. We also note that the sum of the symmetric elements of Eq. 1.7 gives

$$p_{ii} = 3k\theta = 3ke_{ii}. \tag{1.9}$$

From the ratio of the transverse strain (contraction) to the longitudinal strain (extension) of an elastic cylinder, subject to uniform tension over its plane ends and free from lateral traction, a useful relation between k and μ may be derived that is called Poisson's ratio σ:

$$\sigma = \frac{3k - 2\mu}{2(3k + \mu)},$$

where

$$-1 < \sigma < \tfrac{1}{2}.$$

For a fluid $\sigma = 0.5$, for granitic rocks $\sigma \approx 0.21$, and for the sedimentary column of the Earth's crust (average density about 2.4 gm/cm^3) $\sigma \approx \frac{1}{3}$. The assumption $\sigma = 0.25$, called Poisson's relation, is sometimes used as an approximation to simplify the mathematical development.

The work done by the forces that deform a perfectly elastic medium is stored in the form of *strain energy*. On removal of the forces, say at the time of an earthquake, this energy will be released by such processes as heat dissipation and seismic wave propagation (kinetic energy). Under certain assumptions it may be shown that the elastic energy per unit volume stored through the working of surface tractions and body forces is

$$W = \tfrac{1}{2}p_{ij}e_{ij},$$

or, from the stress–strain relations,

$$W = \tfrac{1}{2}c_{ijkl}e_{ij}e_{kl},$$

in the general anisotropic case.

For isotropic media, substitution from Eqs. 1.5 and 1.7 gives

$$W = \tfrac{1}{2}k\theta^2 + \mu E_{ij}E_{ij}. \tag{1.10}$$

The expression Eq. 1.10 provides a measure of the strain energy per unit volume stored in the rocks of the Earth; it will be used in the discussion of earthquake mechanism in Chapter 2.

We now set down some parts of the theory of elastic waves that are necessary to analyze the ground vibrations during an earthquake. Consider an element, with volume dV and surfaces dS, of a continuous perfectly elastic medium of density ρ. Suppose that the equilibrium state at time t is perturbed by an earthquake that changes the body force per unit mass by X_i and the stress by p_{ij}. The element will suffer an acceleration, $\partial^2 u_i/\partial t^2$ approximately, so that, by Newton's law,

$$\int_V \rho\frac{\partial^2 u_i}{\partial t^2}\,dV = \int_S \nu_j p_{ji}\,dS + \int_V \rho X_i\,dV \tag{1.11}$$

where the ν_j are the unit normals to the volume element.

On the assumption of continuous and single-valued functions p_{ji} and their derivatives, Gauss's divergence theorem yields

$$\rho\frac{\partial^2 u_i}{\partial t^2} = \frac{\partial p_{ji}}{\partial x_j} + \rho X_i, \qquad i, j = 1, 2, 3, \tag{1.12}$$

because the region of integration is arbitrary.

The three partial differential equations (Eq. 1.12) are the basic equations of elastic wave theory. For a nonhomogeneous but isotropic medium, substitution from Eq. 1.7 gives

$$\rho\frac{\partial^2 u_i}{\partial t^2} = \frac{\partial}{\partial x_i}\left[\left(k - \frac{2}{3}\mu\right)\theta\right] + \frac{\partial}{\partial x_j}\left[\mu\left(\frac{\partial u_j}{\partial x_i} + \frac{\partial u_i}{\partial x_j}\right)\right] + \rho X_i. \tag{1.13}$$

It is commonplace for the rocks and soils of the crust of the Earth to vary in composition markedly within tectonically active regions. Although variations in *chemical* constitution do not always correspond to significant changes in *elastic* properties, geological circumstances do occur that make it necessary to consider the influence of the spatial variation in k and μ on a propagating disturbance. Because of the difficulties in solving Eq. 1.13, however, we first develop the theory of wave propagation in a homogeneous medium and discuss the necessary modification in Section 1.4.

It can be shown that the usual body forces, e.g., gravity, produce insignificant effects on short-period seismic waves. Therefore, omitting the X_i term and the gradients of k and μ, Eq. 1.13 becomes

$$\rho\frac{\partial^2 u_i}{\partial t^2} = \left(k + \frac{1}{3}\mu\right)\frac{\partial\theta}{\partial x_i} + \mu\nabla^2 u_i. \tag{1.14}$$

Following a method used by Helmholtz in electromagnetic theory, we may analyze the time variations of the vector displacements u_i by putting

$$u_i = \frac{\partial\varphi}{\partial x_i} + (\text{curl}\,\boldsymbol{\psi})_i,$$

where φ is a scalar and ψ a vector potential. Then $\text{div}\,\boldsymbol{u} = \nabla^2\varphi = \theta$.

Substitution in the equations of motion Eq. 1.12 demonstrates that they are satisfied if

$$\rho\frac{\partial^2\varphi}{\partial t^2} = \left(k + \frac{4}{3}\mu\right)\nabla^2\varphi \tag{1.14a}$$

and

$$\rho\frac{\partial^2\psi}{\partial t^2} = \mu\nabla^2\psi. \tag{1.14b}$$

The expression Eq. 1.14a is the wave equation for waves of dilatation with velocity

$$\alpha = \sqrt{\frac{k + \frac{4}{3}\mu}{\rho}}, \tag{1.15}$$

and Eq. 1.14b is the equation for shear waves with velocity

$$\beta = \sqrt{\frac{\mu}{\rho}}. \tag{1.16}$$

We note that in a fluid (i.e., $\mu = 0$) there are no shear waves. In a gas, $\alpha = (k/\rho)^{1/2}$, which, using the usual adiabatic relation between pressure and density, becomes $\alpha = (\gamma p/\rho)^{1/2}$, where γ is the gas constant. Elastic dilatational waves are a type of acoustic wave. Earthquakes are known frequently to generate audible sounds in open spaces; the sounds are compared sometimes to thunder or artillery fire.

In seismology, the dilatational waves are called *P* (primary) waves and the shear waves are called *S* (secondary) waves.* From an earthquake source within an elastic medium, both types of body waves will propagate outward into the medium. If the source can be approximated by a point or small sphere, the wave fronts will be spherical in a homogeneous isotropic medium; at large distances from a source of arbitrary shape the wave fronts are effectively plane, so that the *P* motion is *longitudinal*

*Often thought of as "push" and "shake" waves, respectively.

and the S motion is *transverse* to the direction of propagation.

It may be demonstrated further that S plane waves may be plane–polarized. In the Earth, polarization is observed. Relative to the Earth's surface, S waves that cause particles of the medium to move in a vertical plane containing the direction of propagation are denoted by SV; horizontally polarized waves are called SH waves.

When an elastic body wave encounters an interface or boundary that separates rock of different elastic properties it will, like sound and electromagnetic waves, undergo reflection and refraction. There is the special complication in the elastic wave case that conversion between mode types occurs: either an incident P or SV wave can yield, in general, reflected P and SV waves *and* refracted P and SV waves; an incident SH wave yields only reflected and refracted SH waves.

Boundaries between layers and between the country rock and massive intrusions also give rise to *diffracted* waves and *scattering* of the waves; in this way, seismic wave energy appears in regions which on simple ray theory would be "in shadow" (compare Section 1.5). Diffracted and scattered waves, although the reason for some of the complications observed in recorded ground motion after earthquakes, usually do not carry sufficient energy to compete with the effects of the direct P and S waves (see Stokes, 1849; Hudson and Knopoff, 1966). "Head" waves such as P_n and S_n are exceptional in this regard, (see Section 1.3).

However, interfaces between elastic layers permit the existence of *additional classes of seismic waves*, which are of first importance to earthquake engineering. Because we are concerned primarily with effects at the ground surface we shall now consider the elastic waves that propagate only along the free surface of the Earth. (Mechanical coupling between the surface rocks and the atmosphere is quite unimportant here.)

Consider a simple crustal model that consists of an elastic half-space $x_3 > 0$ with physical constants k, μ, ρ. We consider the two-dimensional case in which the waves are independent of the x_1 coordinate. Let elastic waves propagate in the vertical plane containing the x_2, x_3 axes. These waves may be described by *scalar potentials* φ and ψ that satisfy two-dimensional forms of the wave equations (Eqs. 1.14a and 1.14b). It can be shown that an elastic traveling wave may exist that has a motion restricted near to the free surface. It can be described by the potentials φ and ψ, and, at $x_3 = 0$, satisfies the boundary conditions that all components of stress vanish (i.e., $p_{32} = p_{33} = 0$). The surface wave criterion that motion becomes vanishingly small as $x_3 \longrightarrow \infty$ is satisfied by taking the wave amplitude to decay exponentially with distance from the surface.

The restrictions on the forms of the potentials lead to the equation

$$4\left(1 - \frac{c^2}{\alpha^2}\right)^{1/2}\left(1 - \frac{c^2}{\beta^2}\right)^{1/2} = \left(2 - \frac{c^2}{\beta^2}\right)^2, \quad (1.17)$$

which was first obtained by Lord Rayleigh in 1885. A real root c ($c < \beta < \alpha$) of Eq. 1.17 can always be found. For example, when the medium satisfies Poisson's relation, $\sigma = 0.25$, a value near that of many rocks, the speed of Rayleigh waves is found from Eq. 1.17 to be $c_R \approx 0.53\alpha \approx 0.92\beta$.

A further result of the theory is that the motion of the particles near the surface of the ground will be elliptical. For the above elementary model, the vertical axis of the ellipse is about $1\frac{1}{2}$ times the horizontal axis and the particle motion is retrograde, i.e., particles move in an anticlockwise sense in a vertical plane when viewed along the x_1 axis in the sense that x_1 is increasing.

The theory of P, S, and Rayleigh waves outlined above provides a framework for the detail that follows. The central role of these three types of seismic motion is well attested to by laboratory experiments and observations of earthquake motion. A seismogram selected to

Fig. 1.1. Seismogram from a long-period seismograph showing the vertical component of elastic wave motion recorded at Oroville, California, during part of one day. The third trace from the bottom is from a magnitude 5 earthquake in Alaska. The time between breaks in the trace is 1 min. The maximum recorded amplitudes for P, S, and Rayleigh waves correspond to ground displacements of 1.3, 0.8, and 4.4 μ, respectively (1 μ = 10^{-4} cm).

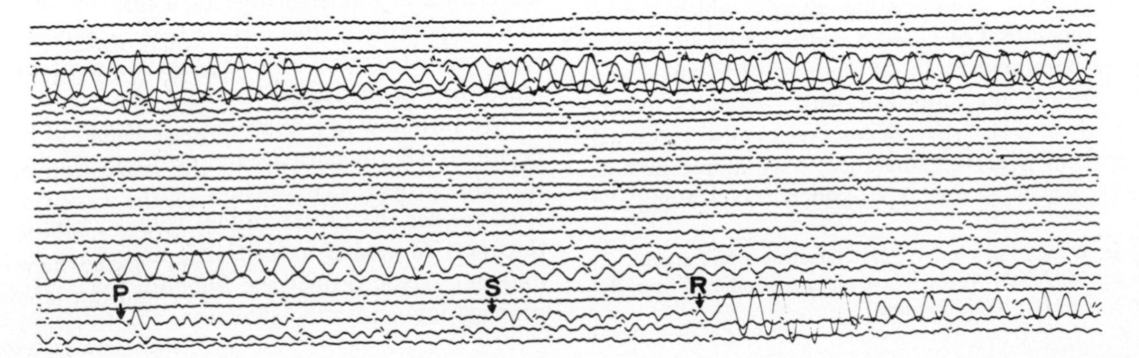

illustrate the characteristics and dominance of these waves is reproduced in Fig. 1.1. Because the seismogram was written by a vertical component seismograph, only waves with vertical components of motion (P, SV, and Rayleigh waves) are recorded. It is noteworthy that the order of the onsets of the three phases accords with theory ($c_R < \beta < \alpha$) and that there is comparative quiescence between the three wave onsets. Such intervals of quiescence diminish as the distance from the earthquake source decreases until in the focal region the ground never ceases to vibrate throughout the earthquake and it becomes difficult to separate the P, S, and Rayleigh wave trains (compare Fig. 1.8). It is noteworthy that in Fig. 1.1 the Rayleigh waves have the largest amplitudes at this distance (~3000 km); we deal with the question of seismic wave amplitudes when considering attenuation in Section 1.4.

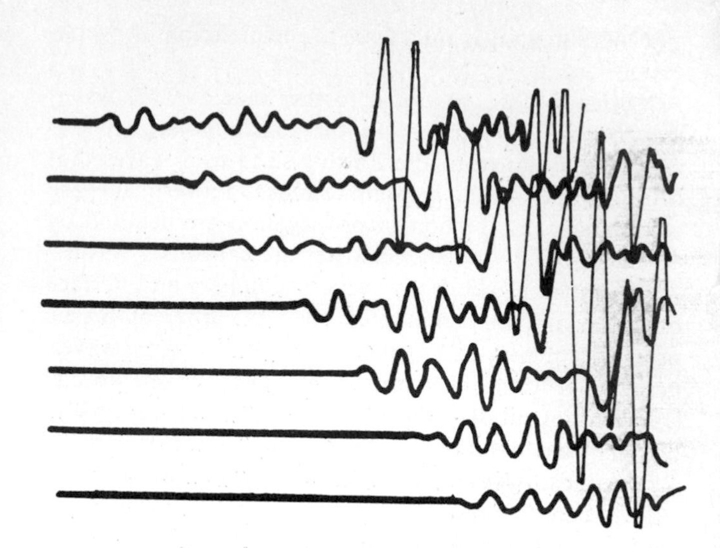

Fig. 1.2. Model experiment showing variation in the wave trains with distance from the source when there is a distinct horizontal layer above the basement medium. The first recorded wave is the direct P wave (through the brass layer); the later arriving waves with the large amplitudes are the head waves, which have traveled in part along the interface between the brass and aluminum.

1.3 LAYERED MEDIA

In the previous section we defined the wave types that are possible in a perfectly elastic, homogeneous, isotropic solid with a plane-free boundary. This is the simplest model for the surface regions of the Earth. The theory shows that an earthquake might generate dilatational (or P) waves, shear (or S) waves, and Rayleigh waves. The first two types would travel through the body of the solid; at the free surface, or any interface, reflections of one type give rise, in general, to waves of both types. The Rayleigh waves have no particle motion normal to the plane defined by the vertical and the direction of propagation; they are confined to propagation along the surface with an exponential decay of amplitude with depth.

The presence of layers of distinct elastic media complicates the ground motion through complication of both the body and surface waves. For both P and S waves, distinct wave packets may appear on the record, each corresponding to a favorable ray path through a particular layer. For example, consider the case of a single surficial layer over basement material.* One set of P and S waves will travel, between the source and a receiver on the surface, through the layer at the appropriate velocities α_1 and β_1, say. The receiver will also record, in general, considerable energy which corresponds to a second set of P and S waves which are refracted into the basement material at the critical angle and travel just below the interface with velocities α_2, β_2 ($\alpha_1 < \alpha_2$; $\beta_1 < \beta_2$), say. These critically refracted waves,† beyond a certain distance, may arrive first at the receiver; waves of this type, which travel along the boundary between the crust and mantle of the Earth (the Mohorovičić discontinuity), are called P_n and S_n body waves.

The direct and diffracted P wave onsets are illustrated in Fig. 1.2 by a seismic model experiment. The source and receiver were both placed on the free edge of a strip of brass bonded to one edge of a rectangular aluminum plate. Each trace from the top corresponds to an increase in the distance D between the transducers. The time interval between the arrival of the direct P and the diffracted P waves can be seen to grow shorter as D is increased. (Some of the motion following the first P arrivals arises from waves *reflected* from the interface.)

It has long been observed that surface waves with strong horizontal surface motion transverse to the direction of wave propagation occur following P and S body waves. Since Rayleigh waves do not possess such an SH component, the simple homogeneous half-space model is seriously inadequate with respect to horizontally polarized transverse shear motions. Surface waves containing SH motion, it turns out, can arise by complicating the mathematical model in one of a number of ways. The most important of these ways from a geological standpoint are structural layering, rock heterogeneity, and anisotropy of the rocks.

We turn now to a quantitative analysis of the effect of layering on surface waves and postpone the treatment of the other properties to Section 1.4.

Consider the LOHS model with a layer of uniform thickness H welded to an elastic half-space $x_3 > 0$ with different elastic properties. Suppose that SH waves are propagating across the structure; transmission of waves

*Hereafter we call this model LOHS (layer-over-half-space).

†Often called "head" waves. Wave energy is radiated from the interface into the layer by diffraction.

along the layer will be governed by the S wave equation (like Eq. 1.14b) and the boundary conditions at the free surface $x_3 = -H$ and the interface $x_3 = 0$. Set suffixes 1, 2 on the parameters μ, ρ, β in the layer and half-space, respectively.

It may be shown that surface waves that satisfy these conditions exist if

$$\tan\left[\kappa H\left(\frac{c^2}{\beta_1^2} - 1\right)^{1/2}\right] = \frac{\mu_2}{\mu_1}\frac{\left(1 - \frac{c^2}{\beta_2^2}\right)^{1/2}}{\left(\frac{c^2}{\beta_1^2} - 1\right)^{1/2}}. \quad (1.18)$$

In this equation, derived by A. E. H. Love, c is the phase velocity and $\kappa = 2\pi/\lambda$ where λ is the wavelength. Love-type waves also exist, in general, when more than one layer is present and in other circumstances (see Section 1.4). In these cases the equations corresponding to Eq. 1.18 are more complicated.

The formulation for Love waves (and for Rayleigh waves), when a large number of plane parallel layers needs to be taken into account, is usually in the form of *transfer layer matrices.* A medium that is vertically inhomogeneous in this way can be approximated by a series of homogenous layers; numerical solutions are obtained by high-speed computers (see Haskell, 1953; Dorman and Prentiss, 1960). It is interesting that the appropriate matrix methods were used in the theory of mechanical structures and similar engineering fields (compare Pestel and Leckie, 1963) long before they were introduced with such success by Haskell into seismology.

Equation 1.18 serves the purpose of illustrating certain properties possessed in general by surface waves that are important in understanding observed ground motion. First, we notice that in Eq. 1.18, c is a function of wave number κ. In other words, the velocity of an harmonic wave component depends on its wavelength; long waves have a velocity that approaches the S-wave velocity β_2 in the lower medium and for short waves $c \longrightarrow \beta_1$. Such behavior characterizes a *dispersive* wave train; a transient pulse emitted at the earthquake source will disperse into a wave train as it propagates outward. Because layering is the rule rather than the exception in the crust, both Love and Rayleigh waves are found to be strongly dispersive in general. This property contributes to the marked persistence of surface waves compared with body waves. A clear example of a dispersed Rayleigh wave train may be found following the letter R in Fig. 1.1; in this case, ground motion from the earthquake surface waves continues for more than 8 min.

Second, because the tangent in Eq. 1.18 is a many-valued function, for a given wavelength there will be multiple roots c. The value of c when $\kappa H(c^2/\beta_1^2 - 1)^{1/2} < \pi/2$ corresponds to the fundamental mode of propagation; the next value, when $\pi/2 < \kappa H(c^2/\beta_1^2 - 1)^{1/2} < 3\pi/2$, corresponds to the first higher Love mode, and so on. This family of modes of the traveling waves can be thought of as the counterpart of the overtones of the normal modes of vibration of the layered system (standing waves).

An analysis of the motion in terms of standing waves has received intermittent study, with a recent revival of interest because of the availability of high-speed computers (see Sezawa and Kanai, 1937; Haskell, 1960, 1962; Phinney, 1964; Gupta, 1966). The vibration of a building structure was treated along these lines by Kanai and Yoshizawa (1963).

The standing waves (or normal modes) of a given type (say, SH motion) arise from the interference of the system of traveling (incident, reflected, and refracted) waves of that type. The amplitude of the displacement vector at the ground surface is in general a function of frequency.

The theoretical results can be illustrated by the simple case of a P wave incident normally from below on a single surface layer over basement material, i.e., the LOHS model defined for Love waves. Let the incident wave be expressed as $A \exp[i\kappa(-x_3 - ct)]$. Then it may be shown that the vertical displacement at the free surface is

$$S_z = \frac{2A}{\cos\kappa H + i(\rho_1\alpha_1/\rho_2\alpha_2)\sin\kappa H}e^{-i\kappa ct}, \quad (1.19)$$

where the wavelength $\lambda = 2\pi/\kappa$. Incident waves with wavelengths that are integral submultiples of twice the layer thickness ($m\lambda = 2H$, $m = 1, 2, \ldots$) are passed unaffected by the layer; the layered medium acts as a wave filter (compare Fig. 1.4).

When $H = 0$, Eq. 1.19 reduces to the well-known result $S_z = 2Ae^{-i\kappa ct}$; i.e., for a half-space, the vertical amplitude at the free surface is double that of the normally incident P wave.

Let us calculate the relative amplification between two points on the free surface, one on the half-space and the other on the LOHS model. The *relative amplification factor* is

$$\left|\left\{\cos\kappa H + i\frac{\rho_1\alpha_1}{\rho_2\alpha_2}\sin\kappa H\right\}^{-1}\right| = \left\{\left(\frac{\rho_1\alpha_1}{\rho_2\alpha_2}\right)^2 + \left[1 - \left(\frac{\rho_1\alpha_1}{\rho_2\alpha_2}\right)^2\right]\cos^2\kappa H\right\}^{-1/2}.$$

The maximum amplification factor is therefore $\rho_2\alpha_2/\rho_1\alpha_1$ ($\rho_1 < \rho_2$) and the minimum is unity. The maximum amplification occurs when $2\kappa H = (2m + 1)\pi$, $m = 0, 1, 2 \ldots$. In terms of the period T, the condition for the maximum is $T = 2H/(2m + 1)\alpha_1$; the graph of the amplification factor against wave period thus will show a series of peaks and troughs. Similar results hold for shear waves vertically incident from below on a single surface layer.

Consider, as a numerical example, a layer of alluvium, with a thickness of 100 m, a P velocity of 570 m/sec, and a density of 1.70 gm/cm³. Suppose the bedrock to be shale with $\alpha_2 = 2200$ m/sec and $\rho_2 = 2.2$ gm/cm³. Then the above results indicate that the maximum amplification of P waves at the surface of the alluvium would be about 5 compared with the amplitudes on outcrops of shale. The peak of maximum amplification that has the longest period ($m = 0$) occurs at $T = 0.35$ sec.

The range of validity of this type of elastic wave analysis is not well understood at the present time. Near to the source of a large earthquake and in the case of special soil conditions the theoretical assumptions, as pointed out in Section 1.4, may be violated to a greater or lesser extent (e.g., slumping). Near to the focal region not much experimental evidence is available on the relative modal energies; the higher modes may be quite important for some engineering response problems.

For traveling waves, because of the interference between the various harmonic components of the dispersed seismic pulse, the ground motion at a particular point will be a series of wave groups. The velocity U of a wave group (measured at, say, a wave peak or trough) is related to the phase velocity c by the relation

$$U = c - \lambda \frac{dc}{d\lambda}.$$

The discussion of the theory has been in terms of crustal models with plane parallel layering. Until very recently, no general treatment of surface waves for curved or dipping boundaries had been discovered. Some progress on this problem is now being made. A promising method of approximation for the problem of the transmission of Rayleigh waves past a step change in elevation has been published by Mal and Knopoff (1965). Numerical approximations such as finite element analysis may best yield the surface response when the underground structures have wedge-like geometrics.

Geological circumstances sometimes occur in which there are a number of layers with strong contrasts in elastic properties as, e.g., in clays and alluvium. Special problems of wave motion then can arise that are relevant to engineering practice.

There may be reduction of rigidity μ with depth leading to subsurface channels in which the seismic P and S velocities are less than in the overburden. Appreciable seismic energy may be trapped in such low velocity layers; certain modes of vibration may be suppressed while others may be enhanced. The surface ground motion above such a region may have quite a different frequency response spectrum from that above a neighboring region below which the low velocity layer is absent.

It is not easy to predict without quantitative treatment the ground response at the free surface of superimposed layers with markedly different properties. An illustration comes from some recent work (Mooney and Bolt, 1966) on the dispersion of Rayleigh surface waves. As noted above, the relative persistence of surface waves at moderate distances from the source gives them their importance in cases of damage due to continued shaking.

In this research Mooney and Bolt computed the characteristics of the dispersion at the surface of the LOHS single-layer structure. Specific numerical results for a number of geophysical models included the variation of surface particle motion with frequency for the fundamental mode and the first and second higher modes.

Let us consider the results for a layer of unconsolidated alluvium, with a thickness of 100 m, over bedrock resembling Pierre shale in elastic properties (McDonal *et al.*, 1958). In the alluvium, $\alpha_1 = 0.57$ km/sec and $\beta_1 = 0.2$ km/sec; for the shale, $\alpha_2 = 2.2$ km/sec and $\beta_2 = 0.8$ km/sec. These empirical velocities correspond to an unusually high Poisson's ratio σ of about 0.4 in each case. The density of the alluvium was taken as 1.70 gm/cm^3 and of the shale 2.20 gm/cm^3.

The complicated dispersion pattern is shown in Fig. 1.3 with the curves for phase velocity c and group velocity U shown on the upper part of the diagram. The group velocity curve for the fundamental mode exhibits two steeply rising limbs separated by an inflection. The first and second higher Rayleigh modes have no less than three group velocity minima each. Such points of inflection or extremes of the group velocity curve are known to be associated with relative concentrations of seismic energy. Thus, in this case, the dominant fundamental wave group has a frequency near 1 cps and travels slowly—about 100 m/sec. It will consequently arrive at a given place much later than the major part of the dispersed wave train.

The shape of the curves drawn on Fig. 1.3 is valid for layer thicknesses other than 100 m. The diagram is drawn so that the only effect of increasing the layer thickness by a factor is to multiply the period scale (horizontal axis) by the same factor. It follows that for a 25-m layer of alluvium the predominant group for the fundamental wave motion has a frequency of 4 cps.

The lower part of the figure shows the ratio, at the surface, of the horizontal ground displacement S_x to the vertical displacement S_z. It will be seen that the particle motion is strongly *frequency dependent*, unlike the constant elliptical form for Rayleigh waves in a half-space (compare Eq. 1.17). Points where the curves cross the horizontal axis correspond to purely *vertical* surface displacement, while points where the curve approaches infinity yield purely horizontal displacement. Positive values of the ratio S_x/S_z correspond to retrograde particle motion, and negative values correspond to progressive or prograde motion.

In the case illustrated in Fig. 1.3, there are extensive frequency ranges for each of the fundamental and two higher modes, in which the ground displacements are mainly horizontal. (See Fig. 1.8 for an example of recorded ground displacement.) For the fundamental it is of interest that near the frequency of the group velocity minimum (~ 1 cps) the ground motion is all vertical, while for lower frequencies the motion becomes predom-

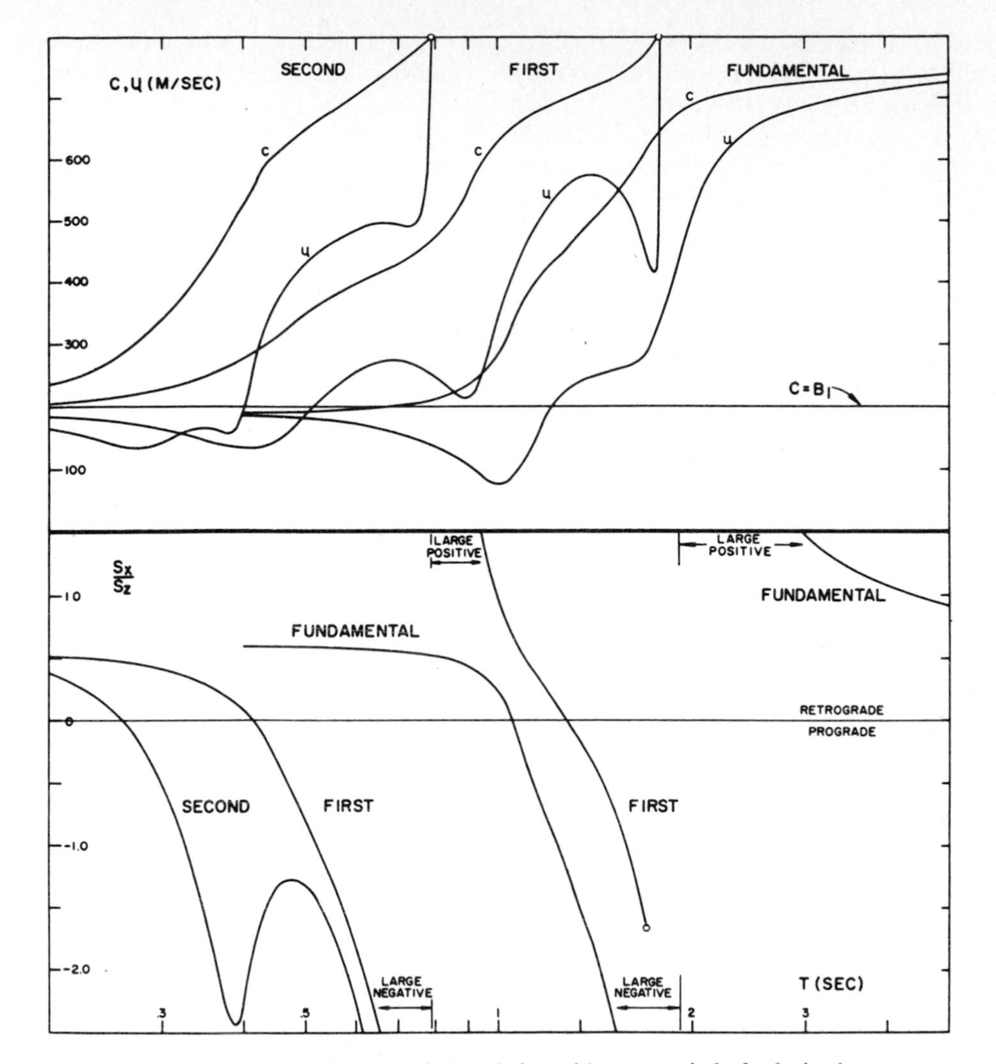

Fig. 1.3. Graphs showing the theoretical variation with wave period of velocity (upper section) and surface particle motion (lower section) for Rayleigh waves traveling through 100-m thick alluvium over shale.

inantly horizontal. A surprise in this computation is the presence of *prograde motion* in the fundamental mode; Fig. 1.3 shows that it is confined to the period range between 1 and 2 sec. Normally observed Rayleigh wave motion is retrograde elliptical in agreement with the theory of Section 1.2, although some workers in seismic prospecting have given accounts of measuring prograde motion.

The ground displacement-frequency spectrum for this crustal model provides an opportunity to explain a possibility for the design of structures to minimize shaking, at least at moderate distances from probable earthquake sources. We noted in Fig. 1.3 the almost complete absence of horizontal motion at periods of 1 sec and just above (at least for the graver mode). A seismic survey of the basement material in some areas with predominantly horizontal layering would lead to enough data for the surface wave (and, for that matter, the body wave) response characteristics to be calculated from theory. The calculated ground response might show certain frequency bands where horizontal motion is minimal but in the possible range of the shear resonance frequency of a proposed structure. Deliberate design selection might then lead to mitigation of damage from horizontal motion.

1.4 ANISOTROPY, VISCOSITY, AND INHOMOGENEITY

Most minerals have significant elastic anisotropy; e.g., a *cubic* crystal such as rock salt has three distinct elastic constants. The usual country rock can be regarded as assemblages of various minerals with more or less random directions of crystal axes; for the wavelength of seismic waves of interest, say from 10 m to 10 km, isotropy is thus generally a close approximation. The same situation holds for soils and alluvium. Geological conditions are known, however, where such physical conditions as sedimentary deposition or tectonic folding establish directional elastic properties. A field example of anisotropy in the rock near Halifax harbor has been discussed by Macpherson (1960).

If anisotropy is a significant property of the elastic medium, the observed wave motion may be modified in a number of ways. For a homogeneous anisotropic medium, the equation of motion Eq. 1.12 can be written, using Eq. 1.6, and dropping the body force term, as

$$\rho\frac{\partial^2 u_i}{\partial t^2} = c_{ijkl}\frac{\partial^2 u_i}{\partial x_j \partial x_l}. \tag{1.20}$$

Equation 1.20 does not permit, in general, separation of the dilatational from the shear motion as was done with Eq. 1.14; hence, there are no purely P or S body waves in an anisotropic medium for an arbitrary direction.

Let the displacements u_i have the form

$$u_j = B_j \exp\left[i(\kappa_m x_m - \omega t)\right],$$

where B_j and κ_m are constants.

By substitution in Eq. 1.20 the condition for a propagating disturbance is

$$\det\left[M_{jk}\right] = 0, \tag{1.21}$$

where

$$M_{jk} = c_{ijkl}\kappa_i\kappa_l - \rho\omega^2\delta_{jk}.$$

For any given frequency ω and for a given direction of propagation, the determinantal sextic polynomial, Eq. 1.21, will have real roots that yield the wave velocities; the velocities depend upon the numerical values of the elastic parameters and the density of the anisotropic medium. It can be shown, further, that in general, propagating plane waves are now neither purely transverse nor purely longitudinal but certain directions may exist along which pure P and S waves can travel. For further analysis of seismological interest, specific symmetries of the media need to be considered. (see Rudski, 1911)

A number of special studies have been published on elastic waves in an anisotropic half-space (e.g., Stoneley, 1955; Buchwald, 1961; Kraut, 1963). We list a few central results: surface waves of the Rayleigh type are generally either dissipative or are not propagated at all. If the free surface is a plane of symmetry, nondissipative Rayleigh waves may exist but, in this case, the particle motion will be elliptical, in a vertical plane *inclined to the instantaneous direction of propagation*. In anisotropic surface rocks or soils there will usually not be a complete distinction between Love and Rayleigh waves; surface waves will have particle motion with components both transverse to and along the direction of propagation.

The theory of Section 1.2 also needs modification when the viscous properties of the rocks and soils are considered. One way of dealing with deviations from perfect elasticity, mathematically, is by the addition of terms to the relation in Eq. 1.8.

A model with features that resemble "elastic afterworking" or "creep" is

$$P_{ij} + \tau\frac{dP_{ij}}{dt} = 2\mu E_{ij} + 2\nu\frac{dE_{ij}}{dt}, \tag{1.22}$$

where τ and ν are the viscous parameters. The application of a sudden constant stress causes an initial strain $E_{ij} = \tau P_{ij}/2\nu$, which increases exponentially with time to $P_{ij}/2\mu$. For disturbances with slowly varying stresses, the main terms are P_{ij} and $2\mu E_{ij}$ and the response is perfectly elastic, while for rapidly varying stresses the time-derivative terms dominate and the behavior is like a perfectly elastic solid with rigidity ν/τ. Intermediate cases will lead to damping of the disturbance.

Another treatment of wave motion in linear viscoelastic solids may be developed as follows (see Bland, 1960). Suppose that all dependent variables such as stress and strain vary sinusoidally with time. We introduce the complex variables p_{ij}^*, e_{ij}^* which are functions of space coordinates only, such that

$$p_{ij} = \mathscr{R}(p_{ij}^* \exp i\omega t)$$

$$e_{ij} = \mathscr{R}(e_{ij}^* \exp i\omega t),$$

where $\mathscr{R}$ denotes the real part. Now in place of k and μ, we define the complex dilatational and deviatoric moduli, $K(\omega)$ and $M(\omega)$, such that

$$p_{ii}^* = 3Ke_{ii}^* \quad \text{and} \quad P_{ij}^* = 2ME_{ij}^*,$$

where, as usual, summation holds over the repeated suffix. There is exact correspondence with the Eqs. 1.9 and 1.8 in Section 1.2.

The equation of motion corresponding to Eq. 1.14 is

$$-\rho\omega^2 u_i^* = \left(K + \frac{1}{3}M\right)\frac{\partial^2 u_j^*}{\partial x_j \partial x_i} + M\nabla^2 u_i^*. \tag{1.23}$$

Because of the correspondence, separation into waves of P and S type may be achieved as in the case of perfect elasticity.

A spherically symmetric solution for dilatational waves is

$$\operatorname{div} \boldsymbol{u}^* = \frac{A}{r}\exp\left(\pm\frac{i\omega r}{\alpha^*}\right),$$

where A is a real constant and $\alpha^* = \sqrt{(K + \frac{4}{3}M)/\rho}$, i.e., for diverging waves

$$\operatorname{div} \boldsymbol{u} = \frac{A}{r}\mathscr{R}\left\{\exp i\omega\left(t - \frac{r}{\alpha^*}\right)\right\}.$$

The velocity of the spherical P waves is then $\mathscr{R}(\alpha^*)$ and the amplitude at radius r is

$$\frac{A}{r}\exp\left(\omega\mathscr{I}\frac{r}{\alpha^*}\right). \tag{1.24}$$

It should be noted that because the complex elastic moduli are frequency dependent, both the velocity and the attenuation factor are functions of frequency ω so that P body waves may be, in general, *dispersive* as well as dissipative. The same property holds for S waves for which the velocity is $(\mathscr{R}\sqrt{\rho/M})^{-1}$ and the attenuation factor $-\omega\mathscr{I}\sqrt{\rho/M}$.

From Eq. 1.24 it is apparent that the amplitude of ground motion generated by earthquakes will decrease with distance in two ways:

1. *Geometrical spreading*. For spherical waves, such as P and S, at large distances compared with the dimension of the source, there will be a reduction in amplitude due to geometrical spreading of the wave fronts proportional to r. For surface (cylindrical) waves the corresponding reduction factor can be shown to be $\sqrt{r}$, so that far from the source the surface waves predominate in the complete wave train.

At distances close to the earthquake source, the simple spreading factors are probably less effective in accounting for observed amplitudes than local crustal heterogeneity and the kinematics of the source. Strong velocity variations in the crustal layers as well as discontinuities between structures will focus the elastic waves so that the wave intensities will vary in some complex way on the surface; the variation will be different for P, S, Rayleigh waves, etc. In addition, the length of a fault source will influence the wave amplitudes nearby as will the mechanism of faulting; dip-slip rupture will tend to enhance P and SV motion over SH motion while strike-slip faulting should lead to relatively large SH amplitudes locally. These problems are discussed further in Section 1.5.

2. *Frictional attenuation*. The frictional dissipation will introduce an exponential decay; the higher frequencies will have greater damping.

Empirical attenuation factors for rocks have been estimated in the laboratory and a very few in the field (e.g., Gutenberg, 1945; McDonal *et al.*, 1958). The method is to compute the frequency spectrum of waves recorded by seismographs at various distances from the source. Practical difficulties in field experiments arise from the variation in ground response at each recording station, from the spacial configuration of the earthquake source, possibly from its extension in time through aftershock activity, and from the imprecisions in estimating focal depth.

The available experimental evidence indicates (see Knopoff, 1964) that the (positive) attenuation factor $a(\omega)$ in the wave amplitude term $\exp\{-a(\omega)r\}$ of Eq. 1.24 is approximately a linear function of frequency. A specific dimensionless attenuation factor Q, often used to measure the damping, is defined by

$$a(\omega) = \frac{\omega}{2cQ},$$

where c is the phase velocity. The higher the value of Q, the smaller the damping.

For the Late Cretaceous Pierre Shale in the Rocky Mountain area, McDonal *et al.* (1958) found $Q = 23$ for P waves and $Q = 10$ for S waves for $20 < \omega < 125$ cps. Laboratory experiments on shales and sandstones yield Q values of the same order. For more loessial materials the Q values might be expected to be lower. Water content of soils is an important factor in wave attenuation and ground amplification (compare, e.g., Goodman and Appuhn, 1966). Unfortunately we lack experimental results for earthquakes.

No strong-motion accelerometer, e.g., recorded the ground motion in Anchorage from the great 1964 Prince William Sound earthquake. However, engineering reports of damage to high-rise buildings rather than single-story ones may reflect in part the predominance of longer periods in the elastic waves reaching Anchorage. Anchorage is some 150 km from the epicenter. At such a distance, attenuation would lead to a relative amplitude enhancement of 2–3 for seismic waves of period 1 sec over those of period 0.5 sec, for an average Q of 20.

Variations from place to place in the elastic parameters and density of the crustal rocks also will modify the wave forms. Elastic inhomogeneity may be severe over a short distance, say, at the boundary between an alluvial valley and rocky uplands. However, in general, k and μ are only slowly varying functions of space coordinates in the Earth's crust.

Mathematical work for elastic wave propagation through heterogeneous media has been done for a number of special cases (e.g., Newlands, 1950; Jeffreys and Hudson, 1965). The basic equation of motion is Eq. 1.13 in Section 1.2. The presence of the spatial derivations of the elastic constants means that once again there is no clear decomposition, in general, into P and S motion; a dilatational movement will be accompanied by a shear movement and vice versa. Dispersive effects can be expected for all phases.

The practical importance of effects due to heterogeneity is not clear; they depend upon the amount that μ and k vary within a wavelength. In the near zone, wave frequencies of most engineering interest might range between 5 and 0.5 cps while velocities of the most energetic waves might be as low as 100 m/sec. Strong changes in μ and k in a few tens of meters therefore might lead to significant modification of the waves.

As has been mentioned previously, if the inhomogeneity is a function of depth only, the variation has been successfully represented by means of plane-parallel homogeneous layers; the stack of layers gives a piecewise approximation to the continuous variation. If the inhomogeneity is horizontal not much can be done at present in a quantitative way by these methods; mathematical models in which the continuum is replaced by lumped mass elements appear more suitable. This case seems to be of considerable importance to earthquake engineering because of the field evidence for amplitude variation of the elastic waves from earthquakes as they pass from one geological province into another.

Model seismic studies are useful here. Figure 1.4

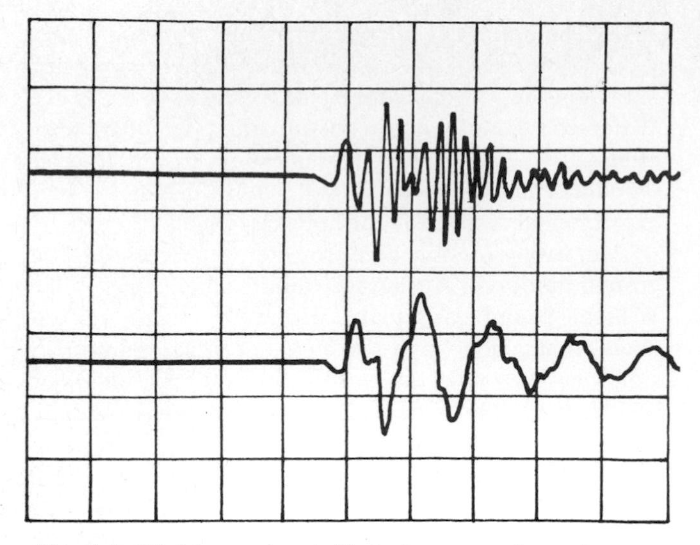

Fig. 1.4. Model experiment illustrating wave forms (to same scale) at the ground surface on opposite sides of a vertical steplike discontinuity separating a rectangular layer of less rigid material from the basement rock. The upper trace is the signal on the basement.

shows the wave forms recorded at two places, A and B, say, on the edge EE' of a square sheet of brass with a side of 2 ft. The generating transducer was placed on the opposite edge of the plate and on the line of symmetry OO' between A and B. The plate was homogeneous except for a bonded rectangle of Plexiglas inserted in the recording edge EE', so that the steplike discontinuity was along OO'. The length and width of the insert were 10′ and $\frac{1}{4}$ in. $AB = \frac{1}{2}$ in.; and the pulse rate was 10 μsec/cm.

The rectangular insert might be taken to model the boundary between one geological structure and another. The recorders at A and B indicate the occurrence of significantly different ground motion on the two sides of the boundary. In this case, the brass side OE' which models the bedrock has shorter periods than the Plexiglas side which models the soil.

1.5 THE EXTENDED SOURCE

Chapter 2 discusses the evidence on the nature of the earthquake source. Although there are at least two competing hypotheses on the cause of tectonic earthquakes, at the present time the most thoroughly worked out theory, at least for shallow earthquakes, is that an earthquake is caused by the sudden release of strain energy (which has accumulated slowly in the rocks) by rupture of a fault. In what follows we shall adopt this point of view, although a number (but not all) of the same general conclusions on wave motion follow from competing hypotheses of the cause of earthquakes.

The elastic strain energy hypothesis holds that there is a minimum rock volume around the focal region of any earthquake that must be strained to the rupture point. Calculations based on Eq. 1.10, made by Bullen (1953) and Tsuboi (1956), show that the minimum critically strained region just prior to the largest earthquakes has a volume of the order of 10^5 km^3. (The shape of the region need not, of course, be spherical.)

In a few earthquakes the length of a fault break has been measured (see Tocher, 1963). These direct geological measurements suggest that the linear extent of a seismic source may extend up to hundreds of kilometers. (There was clear field evidence after the 1906 San Francisco earthquake of rupture of the San Andreas fault from San Juan Bautista to Point Arena, a distance of 190 mi.)

Further evidence that the size of the source region may be quite large comes from the study of aftershocks (see Chapter 2, Section 2.2.2). If the aftershocks are taken as the result of local readjustment of stress by visco-elastic flow* following the main stress drop at the time of the principal earthquake, then their extent will give an upper bound to the extent of the previously strained region. In major earthquakes, aftershocks are found to extend over hundreds of kilometers (see Fig. 2.4).

An example of an aftershock distribution following a moderate earthquake near Parkfield in the coast ranges of central California is given in Fig. 1.5. More than 100 aftershocks with magnitudes greater than one were recorded in the first 3 hours. These Parkfield earthquakes, which occurred in June 1966, are particularly interesting in that they were associated with surface cracks along some 30 km of the San Andreas fault (compare Chapter 2, Section 2.3.2).

The elastic rebound theory of earthquake generation of H.F. Reid (see Lawson, 1908, Vol. 1, pp. 147–151) states (compare Chapter 2, Section 2.4.2) that the sudden displacement or "fling" is initiated at a point of rupture somewhere along the fault.† The rupture then propagates along the fault plane; the greater the extent of fault rupture, in general, the larger and more persistent the overall wave motion from the earthquake. On this view, the *first waves* are generated within some small region on the fault plane: This is the *focus* of the earthquake, and the point above it on the Earth's surface is the *epicenter*. Because P waves have the maximum possible velocity in an elastic medium, the waves to reach first any point on the Earth's surface will be (except in extraordinary circumstances) the P waves from the focus (see Nakano, 1923). The observed times of travel of the first arrivals are, in fact, used to determine (with a precision now perhaps of the order of a few kilometers) the position of the earthquake focus.

*See the discussion related to Eq. 1.22.

†As Reid remarked, the common experience in tearing paper is that the tear begins at one place—not at several.

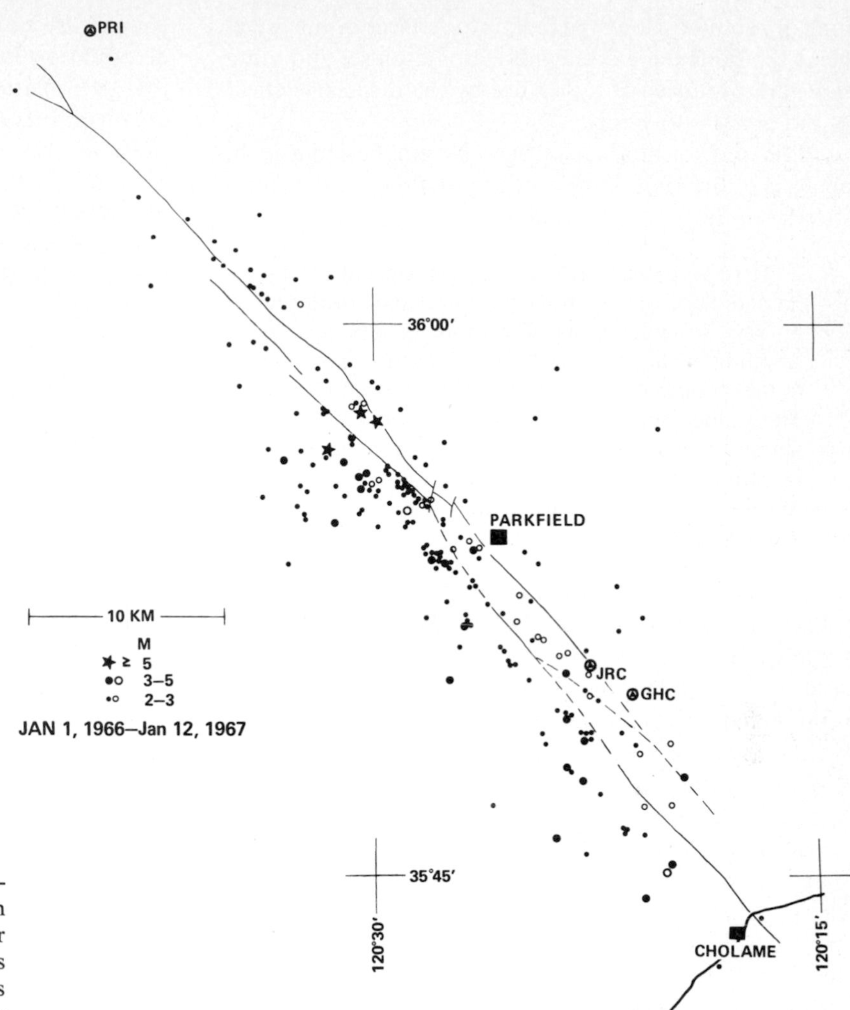

Fig. 1.5. Map showing the distribution of earthquake epicenters and their relation to the San Andreas fault in the sequence of June 1966 near Parkfield, California. Open circles are locations with special near-station control. Mapped faults (not cracking) in the area are shown. (Courtesy T. V. McEvilly.)

In the case of the great 1906 earthquake, e.g., it has been verified (Bolt, 1968) that, within reasonable bounds of precision, the first onsets of P and S waves arrived at points around the world at times consistent with a single initiation of rupture. This estimated focus agrees with an initial rupture in the crust, on the San Andreas fault, in the vicinity of San Francisco Bay. The rupture then proceeded bilaterally but not necessarily smoothly.

There is some indirect evidence from seismograms that the focal region within which P and S waves are first generated is relatively small, perhaps of order of a few kilometers in linear dimension. The presumption is based on theoretical considerations that suggest that the linear dimension of the focal region is linearly proportional to the duration of the first body wave oscillations (see Bullen, 1963, p. 274; Honda, 1962).

It is usually found that earthquakes that cause damage to structures and loss of life have their foci within, or just below, the crust of the Earth, i.e., no deeper than 60 km. (The deepest recorded earthquakes have foci 700 km or so below the surface.) The depth of focus affects the spectrum of elastic waves in a number of ways. Consider the case of fundamental mode surface waves, e.g., whose particle motion decays exponentially with depth (Section 1.2). By a reciprocal theorem of normal mode theory, it follows that because a deep source is on a node of these waves it will not excite displacement at the surface. This is in agreement with observations that indicate that the deeper the earthquake the less prominent the *surface wave* train.

In a theoretical investigation of a simple buried source Nakano (1925) obtained the result that, for a focal depth H, fully developed Rayleigh waves are not to be expected on the surface at epicentral distances less than $c_R H/(\alpha^2 - c_R^2)^{1/2}$. For $c_R = 0.5\alpha$ (see Section 1.2) and a focal depth of 15 km, this distance is 10 km. Later theoretical and numerical work of Pekeris and Lifson (1957) indicates that the amplitude of the Rayleigh surface wave is insignificant compared to P and S waves out to distances that are as much as 5 times the focal depth. At greater epicentral distances the Rayleigh waves become very prominent.

Let us now face the problem of the actual mechanical generation of the seismic waves. One theory is that the fling or rebound along the fault is the source of the waves. Near the fault there is a unidirectional pulse, and vibrations are produced at a distance by the dispersive effects of the transmitting rock.

Reid put an alternative view by emphasizing in his analysis of the 1906 San Francisco earthquake the role of friction in the wave generation:

> It is probable that the whole movement at any point did not take place at once, but that it proceeded by very irregular steps. The more or less sudden stopping of the movement and the friction gave rise to the vibrations which are propagated to a distance. The sudden starting of the motion would produce vibrations just as would its sudden stopping; and vibrations are set up by the friction of the moving rock, exactly as the vibrations of a violin string are caused by the friction of the bow.

In Reid's view, the elastic waves are generated at numerous places on the fault plane. Perhaps both mechanisms are significant. Haskell (1964, 1966) has recently studied the model "in which the fault displacement is represented as a coherent wave only over segments of the fault and the radiation from adjacent segments is assumed to be statistically independent, or incoherent, so that the energies are additive." The geophysical situation may be that the rupture begins suddenly and then spreads with periods of acceleration and retardation along the relatively weakly welded fault zone. The ground motion at any point in general will be affected by these variations. In the limiting case, a series of separate ruptures may occur in succession along a fault; the sudden cessation of each rupture would give rise to wave pulses, called *stopping phases* (compare Savage, 1966). Evidence that the great 1964 Alaskan earthquake occurred as a series of separate shocks, separated by time intervals of the order of 10 sec and spread over a distance of 250 km, has been found by Wyss and Brune (1967) in the *P* wave portion of seismograms.

The repetition of onsets in the *P* wave train might be expected to add to the complexity of motion in the later arriving ground vibrations.

The exact nature of the source of the waves is of considerable importance in evaluating earthquake hazards near to faults, and the question will be taken up again in Chapter 2, Section 2.4.

A number of plausible ways of modeling stress fields that might account for the observed variation in strain during the fault rupture have been suggested. Perhaps the simplest scheme is one involving torques from two couples at right angles. Let us take moments of the couples to be equal and opposite, one pair of forces being parallel to the strike* of the fault, FOF', and the other in a direction normal to it, NON'. These couples are statically equivalent to zero but will produce deformation in the elastic crust. It will be seen at once that deformation will be in one sense, say, dilatation, in the quadrants FON' and NOF and in the other (compression) in the quadrants NOF' and FON'. Quadrantal patterns of this type are commonly observed in *the direction of the first motion of the P waves* recorded by seismographs at stations around the focal region. (If there is a dip-slip component to the fault displacement or if the force system causing the displacement is more complex, then the distribution of first motions will be more complicated.) In general, the spatial variation in properties of a wave type around the source defines the *radiation pattern* for that mode. Experimental radiation patterns for *P*, *S*, and surface waves have been studied for many earthquakes, and there is an extensive literature (see Byerly, 1955; Brune, 1961; Stauder, 1962; Ben-Menahem and Toksoz, 1963).

As an illustration of observational work on source radiation patterns, Fig. 1.6 shows the recorded *P* and *S* waves at the Berkeley Seismographic Station from a number of small, shallow focus earthquakes near Antioch, California. These earthquakes that occurred as part of a sequence in September 1965 possess an almost exact similarity in *S* wave motion. The measurements have been used to infer the stationarity of focal mechanism throughout the sequence of earthquakes (McEvilly and Casaday, 1967). We shall return to a discussion of observed source radiation effects in Chapter 2, Section 2.4.2.

Ground vibrations in the near field no doubt are influenced by the source radiation pattern,* but due to the lack of strong motion records no detailed experimental evidence can be cited. In principle, the above theory would predict that if the earthquake source was elastic rebound along a transcurrent fault under a double-couple system the trace of the fault would be a *nodal* line for *dilatational* (*P*) motion and a *loop* line (maximum response) for *shear* (*S*) motion. In practice, some vibrations of the surface of the ground could always be expected, however; first because a principal part of the seismic waves originates in a finite volume at some depth below the surface and second because irregularities in rock structure cause wave scattering and diffraction and third because the fault zone near the surface usually consists of thoroughly crushed and weakened gouge material that might be expected to lead to special mechanical conditions.

Powerful mathematical techniques have been applied in recent years to the analysis of more detailed models of faulting (see Knopoff and Gilbert, 1959). In 1961 and 1962, Ben-Menahem was successful in determining the

*We restrict ourselves, for the sake of simplicity, to strike-slip, or transcurrent, faulting.

*Formulas for the displacements of elastic waves generated by a general system of forces at a point in an isotropic homogeneous elastic medium are given by Stokes (1849, Eqs. 1.39 and 1.40) without any approximation depending upon the distance from the source.

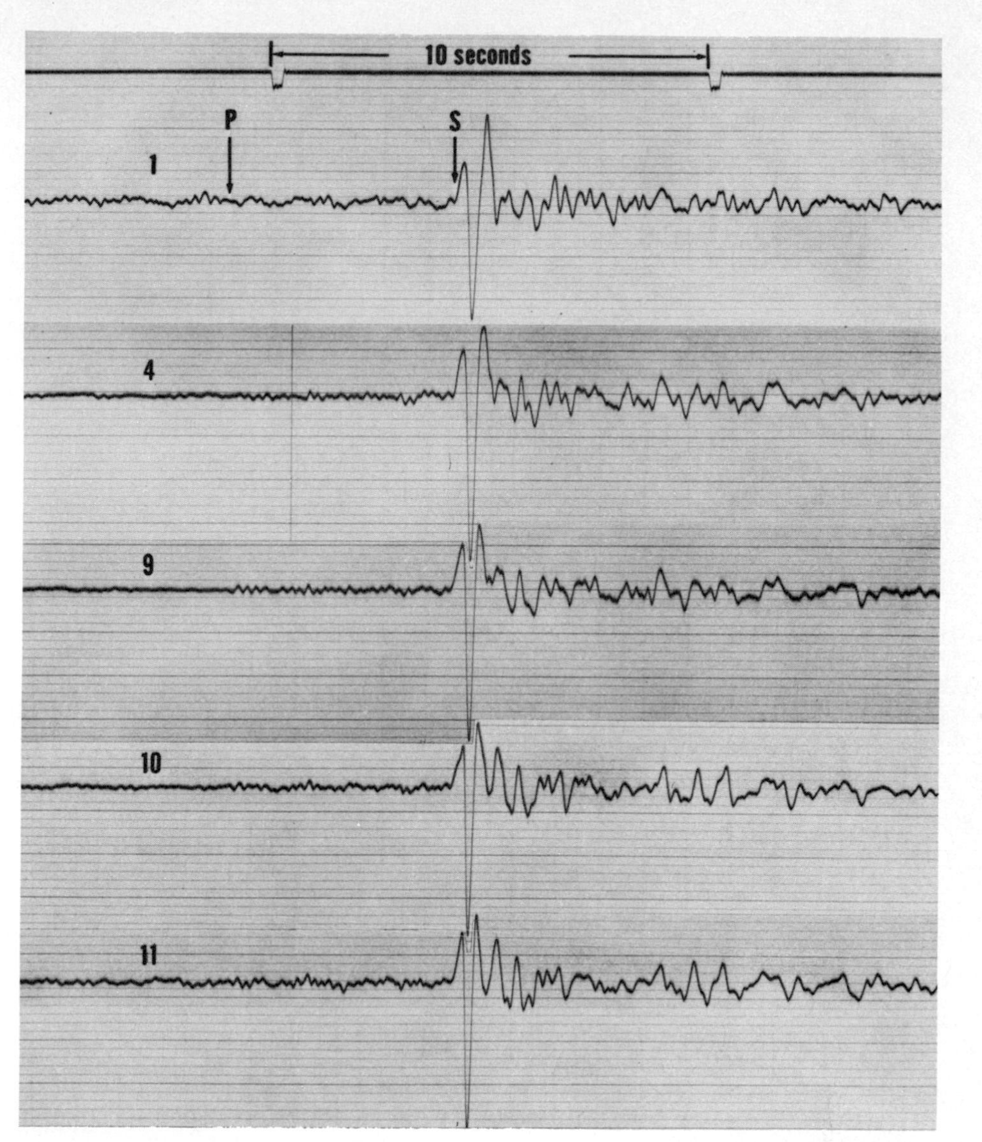

Fig. 1.6. A comparison of seismograms showing horizontal ground displacement (S waves) at Berkeley, California, from a number of earthquakes some 40 km away. Trace 1 is from a foreshock, magnitude 2.1, of the Antioch earthquake on September 10, 1965; traces 4, 9, 10, and 11 are from aftershocks, magnitudes 2.6, 2.8, 2.4, 2.3, respectively.

radiation pattern of both seismic body and surface waves from a finite source moving unilaterally. For a vertical strike-slip model he found that the extent of the source significantly affects the wave pattern *when the wavelength is of the order of the fault size;* a method was given to determine the rupture velocity (in the unilateral, constant velocity case). The speed of fault rupture has been seldom observed directly in the field, but Ben-Menahem and others have found indirectly from seismograms of large earthquakes values of order 3 km/sec. Filson and McEvilly (1967) found that a constant rupture velocity of 2.2 km/sec enables them to explain the frequency spectrum of Love waves recorded at Berkeley after the 1966 Parkfield earthquakes in California.

Various generalizations of the problem of one-dimensional rupture have now been published. *Bilateral* rupture was treated by Knopoff and Gilbert (1959) and by Hirasawa and Stauder (1965). Savage (1966) has analyzed the form of body waves radiated when rupture starts at one focus of an ellipse and then spreads out radially on the fault plane. When the rupture reaches the surface of the Earth a pulselike wave is generated, called the *breakout* phase; when the progression of the rupture ceases there is a stopping phase.

More complete experimental data are necessary to test these mathematical models, including especially detailed field recordings of ground motions along a fault ruptured in a large shallow-focus earthquake. There is nothing that advances seismological knowledge as much as more observations, i.e., a large earthquake—always supposing that there are first-rate seismologists and first-rate recording instruments available for its study!

1.6 EARTHQUAKE GROUND MOTION AND EARTHQUAKE MAGNITUDE

Sufficient theory has been mentioned in the previous sections to demonstrate the difficulties in the way of prediction of detailed ground motion within the region

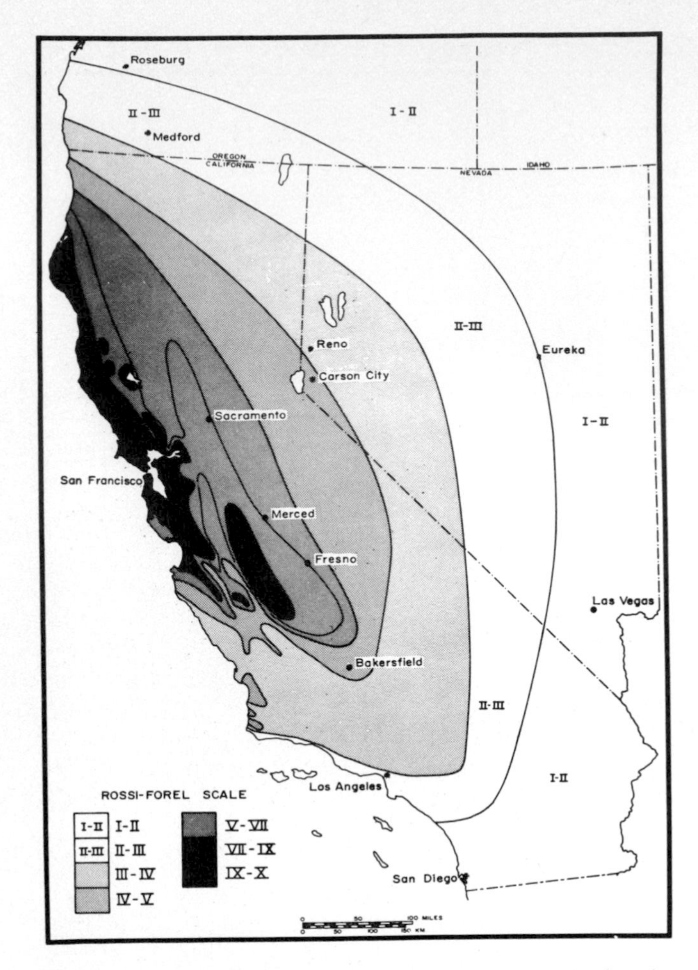

Fig. 1.7. Isoseismals showing the seismic intensity following the great 1906 California earthquake. The intensity scale used is that produced by Rossi and Forel which has now largely been superseded by the modified Mercalli scale (with 12 rather than 10 degrees): Rossi–Forel V–VI is about equivalent to modified Mercalli V. The map has been redrawn from the original given in the *Report of the State Earthquake Investigation Commission* (Lawson, 1908).

of damage (the *meizoseismal area*). Among other parameters, the motion will depend upon the soil, upon the underground geological structures at the observing site, upon the source mechanism and extent, and upon the distance from the source. In regions where the earthquake source mechanism is known not to vary from earthquake to earthquake, similar wave motion might be expected to be produced at a given place by each earthquake (see Fig. 1.6) of the same magnitude.

The test of this hypothesis and the application of the elastic wave theory depends upon the observation of ground motions* in the near zone following local earthquakes. Measurements of this kind are made in two main ways:

1. Collection of field reports on earthquake intensity.
2. Instrumental recording of the kinematics of the ground by strong-motion seismographs.

Seismologists for a long time have used felt reports and estimates of damage in the analysis of earthquake motion. After the disastrous earthquake in Calabria, Italy, in 1783 a special commission gathered information on damage to structures and hence inferred that the *amount of damage depended on the nature of the site* (buildings on the alluvial plains were more affected than those on rocky hills). Such observed effects came to be rated in terms of *intensity scales*. In the United States the modified Mercalli scale is used (Wood and Neumann, 1931). This scale has 12 degrees (see Table 4.9): e.g., degree I, "Felt only in especially favorable circumstances" and degree VIII "Much damage to ordinary substantial buildings; fall of chimneys."

The use of the intensity scale is illustrated in Fig. 1.7, which shows the distribution of seismic intensity associated with the great 1906 California earthquake; the boundary lines are called *isoseismal curves*.

Such intensity maps have great value, not only in terms of relative local propensity for damage but also in throwing light on the extent and energy of the source (see Gutenberg and Richter, 1942, 1956) and similar data needed for the general theory of earthquakes. In Fig. 1.7 the elongated isoseismals reflect in a general way the great length of the fault break (see Section 1.5), and the average decrease in intensity reflects the shallow rupture depth and the attenuation of the waves through frictional dissipation and geometrical spreading (see Section 1.3).

The exceptional pockets of higher or lower intensity (relative to the mean trend) that appear in Fig. 1.7 are of considerable engineering importance. The damage at Los Banos in the pocket with intensity IX–X west of Merced has been mentioned already; the pocket at the southern end of San Francisco Bay and the Y-shaped region north of the Bay (which contains Santa Rosa) provide two additional examples of high intensity in regions where the basement material is alluvium or recent fluvial deposits. Many field studies of earthquakes have pointed to the same phenomenon (see Richter, 1958).

It must be stressed, however, that intensity data do not give directly a measure of ground motion; the modified Mercalli scale is a pastiche of psychological, engineering, and geological criteria. Because ground accelerations must be inferred from this kind of material there is much room for disagreement. Hershberger (1956) and others have derived numerical relations between intensities and ground accelerations, but the matter needs much more study.

Procedure 2 above, i.e., instrumental recording, is much more quantitative but more expensive. *Strong-*

*The utility of the elastic wave theory to predict ground motion following an underground nuclear explosion has recently been examined by Power (1966). A 5-kiloton nuclear explosion in a salt dome at a depth of 825 m near Hattiesburg, Mississippi, caused damage to dwellings. Power finds that observed ground motion was "approximately what would be expected (theoretically) from the known geologic structure."

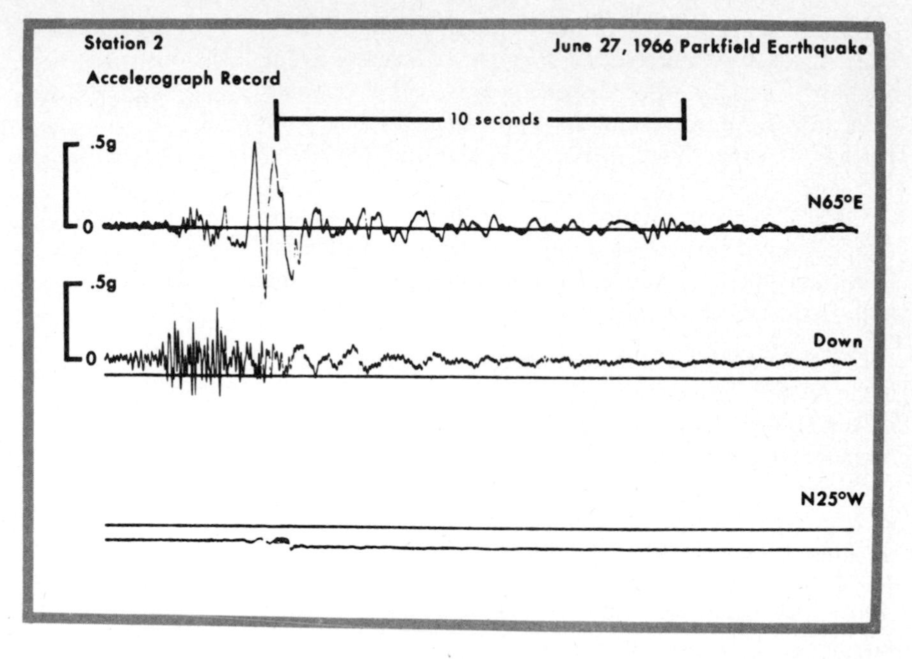

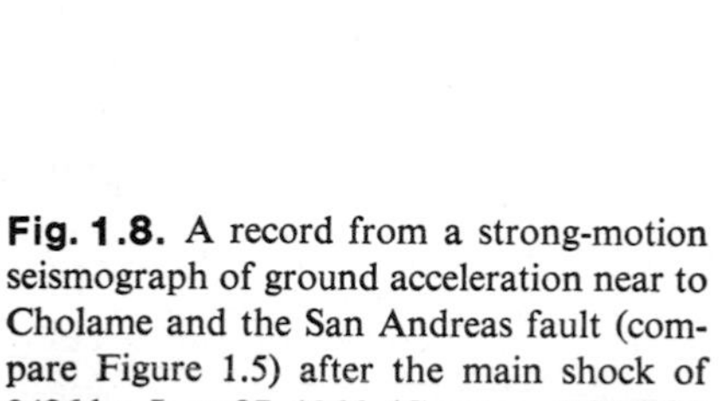

Fig. 1.8. A record from a strong-motion seismograph of ground acceleration near to Cholame and the San Andreas fault (compare Figure 1.5) after the main shock of 0426 hr, June 27, 1966. (Courtesy USCGS.)

*motion seismographs** measure the local ground motion directly and so provide numerical estimates of the frequency spectra of acceleration, velocity, and displacement of the ground at the recording station. It is necessary to stress that any seismogram shows only the *response* of the recording instrument to the ground motion. This response will depend crucially upon the free period of the pendulum of the seismograph. Usually, strong-motion seismographs are specially designed so that *within a certain range of wave frequencies* the seismogram shows a wave form proportional to the *ground acceleration*. The record is then called an *accelerogram*. Outside of the designed frequency range, the seismogram no longer yields a direct measurement of ground acceleration.

Generally speaking, considerable portions of most available strong motion accelerograms can be interpreted in terms of the *P*, *S*, and surface wave phases of elastic wave theory, although because the strong-motion instruments are usually triggered† into operation by the first strong pulse, an unknown amount of the body phases may be missing. Most available strong-motion seismograms were recorded, however, at distances of tens of kilometers from the earthquake focus; as has been pointed out earlier in this discussion, it is at shorter distances that this type of description might be found seriously inadequate. An accelerogram written during the June 27, 1966, Parkfield earthquake is reproduced in Fig. 1.8.

The upper trace is the horizontal component of ground acceleration as measured by the seismograph in a direction N65°E, which is, at this station, about perpendicular to the strike of the fault; the second trace is the vertical component of acceleration. (The orthogonal horizontal component did not record.) The recording station is within the fault zone some 300 ft from a slippage crack that appeared across Highway 46 intersecting the San Andreas fault (compare Fig. 1.5). The station was on alluvium.

These records are probably the closest yet obtained to an earthquake source. A number of wave properties are worthy of notice.

1. *The vertical and horizontal motions are quite dissimilar.* The vertical motion, composed of waves of *P*, *SV*, and Rayleigh type, contains initially relatively more short period motion. The horizontal record, because of the orientation of the pendulum and the position of the recorder relative to the earthquake focus, should be predominantly of shear-type (*SH*) body waves and Love waves. The horizontal ground acceleration reached 0.5 g, i.e., about 450 cm/sec^2.

2. There is an onset of high frequency motion which is common to both components about 1.8 sec after the beginning of the record; this might be interpreted as the arrival of the *S* body waves (*SV* and *SH*) from the focus. Let us assume that (a) the beginning of the record does *not* correspond to the first *P* wave arrival and (b) that the dominant waves originated at the focus of the main shock, which has been located (McEvilly, 1966) some 30 km away in a direction NW along the fault (see Fig. 1.5). The measured time interval between the arrival of the *S* body waves and the long-period large-amplitude motion on the horizontal component is about 1.5 sec. The latter wave motion, therefore, could well,

**Seismoscopes* (conical pendulums) are also used; these give only the horizontal particle motion of the ground at one frequency.

†Torsional pendulums are continuously operated at the Berkeley Seismographic Station, which have a magnification of only about four times the ground displacement for 1 cps waves.

correspond to the arrival of shear-type surface waves that have traveled with a velocity of order 2.5 km/sec. From studies of surface wave dispersion this is a reasonable value for Love waves in the fundamental mode with periods of order 2 sec (see also Filson and McEvilly, 1967).

3. The vertical component of acceleration indicates that a train of surface waves of Rayleigh type with relatively long periods and small amplitudes followed the high-frequency crescendo.

Another aspect of the interpretation of seismic waves is the relation between the ground motion and the *energy* of the source. In principle, the amplitudes of ground-surface movements estimated from seismograms should provide an estimate of the seismic energy released by an earthquake. Because of the problem of local crustal response and other reasons this is difficult in practice (see e.g., Niazi, 1964).

The principle is, however, the basis of a scheme introduced into seismology by Richter (1935) whereby an earthquake can be alloted a number M on a *magnitude scale*. For Southern California earthquakes, Richter defined the magnitude* as the logarithm (to base 10) of the maximum amplitude (in microns) traced on a standard (Wood–Anderson) seismograph at distance 100 km from the epicenter. The conveniences of describing the "strength" of an earthquake by just one number (the magnitude) proved so great that Gutenberg and others have extended the usage of magnitude so that any earthquake at any distance from a calibrated seismograph is nowadays assigned a magnitude.

It should be noticed that *magnitude*, unlike *intensity*, is an instrumental measure. The magnitude scale does not have upper or lower bounds: with more sensitive instruments than were available in 1935, earthquakes of magnitude less than -2.0 are now recorded by some observatories; at the other extreme, the great Alaskan earthquake of March 28, 1964, in Prince William Sound had a magnitude estimated as 8.6 by the Berkeley Seismographic Station.

A number of empirical relations have been set down connecting the energy E with M, usually of the form

$$\log E - \log E_0 = aM. \tag{1.25}$$

In a recent determination, Gutenberg and Richter (1956) give as an average $a = 1.5$, $E_0 = 2.5 \times 10^{11}$ ergs. The formula has been checked against the available seismic energy liberated by underground nuclear explosions; for any particular earthquake and observatory the calculated seismic energy may be uncertain by a factor of ten or more. Equation 1.25 entails that a one unit increase in M is equivalent to a 32-fold increase in energy (see Chapter 2, Section 2.2.3).

*A detailed discussion is given in Richter's book (1958).

1.7 SUMMARY

The elastic waves predicted by the infinitesimal strain theory of a perfectly elastic, Hookian, isotropic medium with homogeneous layers can account, in general, for the principal part of the observed motion on rock or firm ground following earthquakes. Certain anomalous behavior can be explained if deviations from the above restrictions are allowed; particularly important under certain circumstances are structural and elastic heterogeneity, viscosity, non-linear elastic behavior and anisotropy of the soil and rock (see Idriss and Seed, 1968). Distortion of wave fronts, scattering, diffraction, and mode conversion occurs at structural interfaces, and the wave motion bears the key signature imparted by the nature and extent of the source; much of the complication of the ground vibrations are from these phenomena.

More precise weighing of the agreement (or lack of it) between elastic wave theory and observed motion is hardly possible at present for three main reasons:

1. Despite the sophistication of modern experimental methods in seismology, very few records exist of detailed ground motion near to the epicenter of an earthquake. Tests of the theory for the near displacement field must await such recording. (Work from large underground explosions indicates that the restricted theory makes accurate predictions of ground motion, within a few hundred meters of the source, *if the properties of the ground are known*.)

2. Subsurface structures in the medium through which the waves are traveling are usually unknown. This ignorance precludes the full theoretical calculation of wave forms necessary to compare with relevant observations. Although practical algorithms have not yet been deduced for the ground motion above complex wedge and other structures, theory has progressed to a stage where reasonably adequate approximate results could be computed for many geological structures. The adequacy of the restricted elastic wave theory to predict structural variations, of course, is well attested to by the experience of seismic prospectors in the oil industry.

3. Near to the earthquake focus the spectrum and duration of the ground disturbance are strongly dependent upon the source dynamics. To the extent that source mechanism remains uncertain (compare Chapter 2), seismologists are unable to make firm predictions concerning ground motion in earthquake country.

ACKNOWLEDGMENTS

My thanks go to Dr. P. Byerly, Dr. T. V. McEvilly, and Dr. A. K. Mal for their helpful comments.

I am indebted to Dr. D. Sutton for providing Figs. 1.2 and 1.4. The modeling work was carried out in the California Research Corporation Model Laboratory at the University of Claifornia, Berkeley. This Laboratory was equipped by a grant from the California Research Corporation.

REFERENCES

Benioff, H. (1935). "A Linear Strain Seismograph," *Bull. Seism. Soc. Am.*, **25,** 283.

Ben-Menahem, A. (1961). "Radiation of Seismic Surface-Waves from Finite Moving Sources," *Bull. Seism. Soc. Am.*, **51,** 401.

Ben-Menahem, A. (1962). "Radiation of Seismic Body Waves from a Finite Moving Source in the Earth," *J. Geophys. Res.*, **67,** 345.

Ben-Menahem, A. and M. N. Toksoz (1963). "Source-Mechanism from Spectra of Long-Period Seismic Surface Waves. 3. The Alaska Earthquake of July 10, 1958," *Bull. Seism. Soc. Am.*, **53,** 905.

Blanchard, F. B. and P. Byerly (1935). "A Study of a Well Gauge as a Seismograph," *Bull. Seism. Soc. Am.*, **25,** 313.

Bland, D. R. (1960). *The Theory of Linear Viscoelasticity*, New York: Pergamon Press.

Bolt, B. A. (1968). "The Focus of the 1906 California Earthquake," *Bull. Seism. Soc. Am.*, **58,** 457.

Brune, J. N. (1961). "Radiation Pattern of Rayleigh Waves from the Southeast Alaska Earthquake of July 10, 1958," *Publ. Dom. Obs. Ottawa*, **24,** 373.

Buchwald, V. T. (1961). "Rayleigh Waves in Transversely Anisotropic Media," *Quart. J. Mech. Appl, Math.*, **14,** 293.

Bullen, K. E. (1953). "On Strain Energy and Strength in the Earth's Upper Mantle," *Trans. Am. Geophys. Union*, **34,** 107.

Byerly, P. (1955). "Nature of Faulting as Deduced from Seismograms," *Geol. Soc. Am.*, Special Paper 62, p. 61.

Dorman, J. and D. Prentiss (1960). "Particle Amplitude Profiles for Rayleigh Waves in a Heterogeneous Earth," *J. Geophys. Res.*, **65,** 3805.

Eaton, J. P. and K. J. Takasaki (1959). "Seismological Interpretation of Earthquake-Induced Water-Level Fluctuations in Wells," *Bull. Seism. Soc. Am.*, **49,** 227.

Filson, J. and T. V. McEvilly (1967). "Love Wave Spectra and the Mechanism of the 1966 Parkfield Sequence," *Bull. Seism. Soc. Am.*, **57,** 1245.

Goodman, R. E and R. A. Appuhn (1966). "Model Experiments on the Earthquake Response of Soil-Filled Basins," *Bull. Geol. Soc. Am.*, **77,** 1315.

Gupta, I. N. (1966). "Standing Waves in a Layered Half-Space," *Bull. Seism. Soc. Am.*, **56,** 1153.

Gutenberg, B. (1945). "Amplitudes of *P*, *PP*, and *S* and Magnitudes of Shallow Earthquakes," *Bull. Seism. Soc. Am.*, **35,** 57.

Gutenberg, B. and C. F. Richter (1942). "Earthquake Magnitude, Intensity, Energy and Acceleration," *Bull. Seism. Soc. Am.*, **32,** 163; second paper (1956), ibid., **46,** 105.

Haskell, N. A. (1953). "The Dispersion of Surface Waves on Multilayered Media," *Bull. Seism. Soc. Am.*, **43,** 17.

Haskell, N. A. (1960). "Crustal Reflection of Plane *SH* Waves," *J. Geophys. Res.*, **65,** 4147.

Haskell, N. A. (1962). "Crustal Reflection of Plane *P* and *SV* Waves," *J. Geophys. Res.*, **67,** 4751.

Haskell, N. A. (1964). "Total Energy and Energy Spectral Density of Elastic Wave Radiation from Propagating Faults, Part I," *Bull. Seism. Soc. Am.*, **54,** 1811.

Haskell, N. A. (1966). "Total Energy and Energy Spectral Density of Elastic Wave Radiation from Propagating Faults, Part II," *Bull. Seism. Soc. Am.*, **56,** 125.

Hershberger, John (1956). "A Comparison of Earthquake Accelerations with Intensity Ratings," *Bull. Seism. Soc. Am.*, **46,** 317.

Hirasawa, T. and W. Stauder (1965). "On the Seismic Body Waves from a Finite Moving Source," *Bull. Seism. Soc. Am.*, **55,** 237.

Honda, H. (1962). "Earthquake Mechanism and Seismic Waves," *Geophys. Notes*, Geophys. Inst., Tokyo, **15,** 1.

Hudson, J. A. and L. Knopoff (1966). "Signal Generated Seismic Noise," *Geophys. J., R. Astro. Soc.*, **11,** 19.

Idriss, I. M. and H. B. Seed (1968). "An Analysis of Ground Motions during the 1957 San Francisco Earthquake," *Bull. Seism. Soc. Am.*, 58, 2131.

Jeffreys, H. and J. A. Hudson (1965). "On Very Long Love Waves," *Geophys. J., R. Astro. Soc.*, **10,** 175.

Kanai, K. and S. Yoshizawa (1963). "Some New Problems of Seismic Vibrations of a Structure, Part I," *Bull. Earthquake Res. Inst.* (*Tokyo*), **41,** 825.

Knopoff, L. (1958). "Energy Release in Earthquakes," *Geophys. J., R. Astro. Soc.*, **1,** 44.

Knopoff, L. (1964). "Q," *Rev. Geophys.*, **2,** 625.

Knopoff, L. and F. A. Gilbert (1959). "Radiation from a Strike-Slip Fault," *Bull. Seism. Soc. Am.*, **49,** 163.

Kraut, F. A. (1963). "Advances in the Theory of Anisotropic Elastic Wave Propagation," *Rev. Geophys.*, **1,** 401.

Lawson, A. C. (1908). *California State Earthquake Commission Report*, Carnegie Institution of Washington.

MacPherson, J.D. (1960). "Anisotropy of the Rock in the Approaches to Halifax-Harbour, Nova Scotia," *Bull. Seism. Soc. Am.*, **50,** 575.

Mal, A. K. and L. Knopoff (1965). "Transmission of Rayleigh Waves Past a Step Change in Elevation," *Bull. Seism. Soc. Am.*, **55,** 319.

McDonal, F. J., F. A. Angona, R. L. Mills, R. L. Sengsbush, R. G. Van Nostrand, and J. E. White (1958). "Attenuation of Shear and Compressional Waves in Pierre Shale," *Geophysics*, **23**, 421.

McEvilly, T. V. (1966). "Preliminary Seismic Data, June–July, 1966; Parkfield Earthquakes of June 27–29, 1966, Monterey and San Luis Obispo Counties, California," *Bull. Seism. Soc. Am.*, **56**, 967.

McEvilly, T. V. and K. B. Casaday (1967). "The Earthquake Sequence of September, 1965, near Antioch, California, *Bull. Seism. Soc. Am.*, **57**, 113.

Mooney, H. M. and B. A. Bolt (1966). "Dispersive Characteristics of the First Three Rayleigh Modes for a Single Surface Layer," *Bull. Seism. Soc. Am.*, **56**, 43.

Nakano, H. (1923). "Notes on the Nature of the Forces Which Give Rise to the Earthquake Motions," *Seism. Bull. Cent. Meteor. Obs. Japan*, **1**, 92.

Nakano, H. (1925). "On Rayleigh Waves," *Jap. J. Astr. and Geophys.*, **2**, 233.

Newlands, M. (1950). "Rayleigh Waves in a Two-Layer Heterogeneous Medium," *Mon. Not. R. Astro. Soc., Geophys. Supp.*, **6**, 109.

Niazi, M. (1964). "Partition of Seismic Energy," Ph.D. thesis, University of California, Berkeley.

Pekeris C. C. and H. Lifson (1957). "Motion of the Surface of a Uniform Elastic Half-Space Produced by a Buried Pulse," *J. Acoust. Soc. Am.*, **29**, 1233.

Pestel, E. C. and F. A. Leckie (1963). *Matrix Methods in Elastomechanics*, New York: McGraw-Hill.

Phinney, R. A. (1964). "Structure of the Earth's Crust from Spectral Behavior of Long-Period Body Waves," *J. Geophys. Res.*, **69**, 2997.

Power, D. V. (1966). "A Survey of Complaints of Seismic-Related Damage to Surface Structures Following the Salmon Underground Nuclear Detonation," *Bull. Seism. Soc. Am.*, **56**, 1413.

Richter, C. F. (1935). "An Instrumental Earthquake Magnitude Scale," *Bull. Seism. Soc. Am.*, **25**, 1.

Richter, C. F. (1958). *Elementary Seismology*, San Francisco: W. H. Freeman.

Rudski, M. P. (1911). "Parametrische Darstellung der elastischen Welle in anisotropen Medien," *Anz. Akad. Wiss. Krakau*, **5**, 503.

Savage, J. C. (1966). "Radiation from a Realistic Model of Faulting," *Bull. Seism. Soc. Am.*, **56**, 577.

Sezawa, K. and K. Kanai (1937). "On the Free Vibrations of a Surface Layer Due to an Obliquely Incident Disturbance," *Bull. Earthquake Res. Inst. (Tokyo)*, **15**, 375.

Stauder, W. (1962). "The Focal Mechanism of Earthquakes," *Ad. Geophys.*, **9**, 1.

Stokes, J. G. (1849). "On the Dynamical Theory of Diffraction," *Trans. Camb. Phil. Soc.*, **9**, 243.

Stoneley, R. (1955). "The Propagation of Surface Elastic Waves in a Cubic Crystal," *Proc. Roy. Soc. (London)*, **A, 232, 447**.

Tocher, D. (1958). "Earthquake Energy and Ground Breakage," *Bull. Seism. Soc. Am.*, **48**, 147.

Tsuboi, C. (1956). "Earthquake Energy, Earthquake Volume, Aftershock Area, and Strength of the Earth's Crust," *J. Physics of Earth (Tokyo)*, **4**, 63.

Wood, H. O. and F. Neumann (1931). "Modified Mercalli Intensity Scale of 1931," *Bull. Seism. Soc. Am.*, **23**, 277.

Wyss, M. and J. N. Brune (1967). "The Alaska Eathquake of 28 March 1964: A Complex Multiple Rupture," *Bull. Seism. Soc. Am.*, **57**, 1017.

Chapter 2

Causes of Earthquakes

BRUCE A. BOLT

Professor of Seismology and Director of the Seismographic Stations, University of California, Berkeley, California

2.1 INTRODUCTION

Like many fundamental problems in science, the genesis of earthquakes is controversial. In Newtonian mechanics, acceptance of the equation $F = ma$ moves one beyond the enigmatic notions of force and mass to the generally unquestioned results on mechanics; similarly, in seismology, the equations of deformable media separate the realms of largely qualitative speculation on earthquakes from those of quantitative analysis. In many ways, seismology has developed into one of the most precise disciplines: seismographs are now in routine operation that record earth motion with amplitudes as small as 10^{-10} cm (or about the size of a carbon dioxide molecule); seismology provides far more information on the crust and the deep interior of the Earth than any other geophysical discipline.

Of those portions of seismology of direct interest to earthquake engineers, the *mechanism of earthquakes* is of central interest and, as we shall demonstrate, mechanism cannot be separated from earthquake causation. It must be said at the outset that no fully worked out description of earthquake mechanism is yet available; there may be a number of mechanisms depending primarily on the depth. Nevertheless, a much sharper description of this problem can now be offered than when G. Hartwig (1887) wrote that "the causes of earthquakes are still hidden in obscurity, and probably will ever remain so, as these violent convulsions originate at depths far below the realms of human observations."

In order to maintain relevance to earthquake engineering, we limit the discussion in the main to the class of earthquakes in which the energy release is both near the Earth's surface and large enough to damage structures. We do not concern ourselves directly with the relatively weak earthquakes caused by volcanic action or by rock falls, rock bursts in mines, etc.

Shallow-focus earthquakes* are related, by a chain of cause and effect, to the forces that bring about deformation of the Earth as a whole. The underlying causes are clearly part of those tectonic processes that produce mountains, rift valleys, midoceanic ridges, and ocean trenches at the Earth's surface. These global processes are beginning to be better understood in terms of sea-floor spreading (Isacks *et al.*, 1968). It is perhaps sufficient here to state that the two main hypotheses (see Lyttleton, 1963; Wilson, 1965) for deformation of the Earth's surface (such as mountain building) are (1) radial contraction (or expansion) of the whole Earth and (2) slow convective motion of the material within the Earth's mantle.† It is possible that either or both processes may occur at any geological epoch; neither theory has been developed to the stage where it can predict all of the present pattern of stresses and temperature in the Earth's crust and upper mantle. Volcanism is no longer regarded as the general prime cause of large, shallow earthquakes, nor are differences in the amount of isostatic compensation between mountains and plains. Observed gravity differences indicate that compensation usually holds to first order over large seismic areas; however, in some regions vertical faulting may arise from gravitational forces. This mechanism is discussed by Banerji (1957) in connection with the Himalayas.

In my view it is necessary to be content at the present time with an analysis that takes the regional strain (deformation) patterns as empirical data. Shallow earthquakes give information on these patterns and, ultimately, on the global tectonic processes. This restricted approach does not satisfy some seismologists. For example, in 1907 Omori considered that there was a causal relation between the occurrence of the large earthquakes along the Pacific Coast of North and South America; he went so far as to predict from the spatial pattern of previous earthquakes that "the probable position of the next great shock on the Pacific side of America . . . would be to the south of the equator (Chile and Peru)." He took the large earthquake near Valparaiso on August 17, 1906, to be confirmation of his prediction. References to other attempts to discover global interrelations between earthquakes can be found in the review article by Lomnitz (1966).

Another line of attack on the problem of the general cause of earthquakes has been to work from very general mechanical principles such as, say, the laws of thermodynamics. The globe is regarded as a heat engine which, by doing work, produces earthquakes. An interesting description of an "earthquake machine" based on thermodynamical theory is given by Matuzawa (1964).

If the causes of earthquakes were fully understood in a quantitative way, prediction of the time and place of damaging earthquakes might become feasible in some probability sense.* There is a danger here: earthquake engineers must be alert to oppose the view that successful prediction would eliminate (or even substantially mitigate) public hazard. Forewarning of an earthquake will not prevent damage, obviously, to poorly built or already weakened structures.

2.2 EARTHQUAKE OCCURRENCE

2.2.1 Global Seismicity

The present pattern of the distribution of earthquakes over the Earth has been known since early in this century. This result was the first to flow from the establishment of a worldwide system of seismographic stations. The most substantial catalog of world seismicity based on this instrumental record is that of the *International Seismological Summary*.† This catalog lists earthquake hypocenters, origin times, recorded travel times, and certain other data for earthquakes recorded globally since 1918. Other important international lists of earthquakes now appearing are the *Bulletin Mensuel du Bureau Central Seismologique* (produced at Strasbourg) and the catalogs of the U.S. Coast and Geodetic Survey (Washington, D.C.). At the present time, the latter organization mails cards giving preliminary determination of epicenters to interested parties a few days after the occurrence of earthquakes listed on the card.

An illustration of the completeness of the present earthquake catalogs is given by Fig 2.1, on which are

*A number of seismological terms such as "focus" are defined in Chapter 1, Section 1.5.

†A solid region of the Earth, some 2900 km thick, which lies between the base of the crust and the Earth's liquid core.

*For a recent review of developments in forecasting earthquakes, see Press and Brace (1966).

†In 1964 this summary was replaced by the *Bulletin of the International Seismological Center* (*BISC*), Edinburgh.

The detailed depth distribution for 30 years of data is tabulated by Gutenberg and Richter (1954). On the average, the frequency of occurrence falls off rapidly below 200 km depth, but some earthquake foci (as deep as 700 km) have been located by instrumental methods. Over 75 % of the average annual seismic energy is released by earthquakes with foci less than about 60 km deep. Such earthquakes are called shallow-focus earthquakes by Gutenberg and Richter. This division is largely arbitrary and is unsatisfactory in discussions of earthquake genesis because it is unrelated to either Earth structure or the *in situ* mechanical properties of the rocks (see Section 2.4.1). We will therefore use the description "shallow-focus earthquake" henceforth to distinguish earthquakes with depths that place the foci within the Earth's crust.

The frequency of occurrence of earthquakes in a given geographical region is observed to be significantly increased immediately following a large near-surface earthquake. Such a clustering of earthquakes in time and space would appear to be a cardinal aspect of earthquake mechanism; it is dealt with separately in Section 2.2.2.

As the above description indicates, the study of the global morphology of earthquakes brings to light certain key properties that are germane to any attempt to explain earthquake genesis by a general theory. In order to have a ready place of reference for the development that follows it is convenient to list here the most important of these observational properties (about which there is little or no controversy):

1. Earthquakes are global but their present geographical distribution is structured (see Fig. 2.1) with extensive aseismic regions and belts of high seismicity.

2. Earthquakes have a very great range in the amount of energy released (see Section 2.2.3).

3. Earthquakes that release relatively moderate-to-large amounts of seismic energy may occur under both continents and oceans (see Fig. 2.1). [The causes of most tsunamis (long ocean waves) are large submarine earthquakes.]

4. Earthquakes sometimes *cluster* strongly in both space and time (e.g., aftershock sequences) (see Figs. 2.1 and 2.4).

5. Earthquake foci vary in depth from near-surface to depths of about 700 km. The frequency distribution of earthquakes as a function of focal depth is neither uniform with depth nor with geographical region (see Fig. 2.1).

6. Where field observations are possible (i.e., in continental or near-continental regions) earthquakes of shallow focus and moderate-to-large energy release are usually accompanied by pronounced ground deformation at the surface. This deformation often takes the form of fault rupture (see Chapter 3), surface uplift or subsidence.

It is not possible here to develop each of these six points in detail. The main point of the above summary is to emphasize the great range of properties associated with seismicity. An argument might be developed that this diversity entails a variety of causes for earthquakes.* Rather than pursue this controversial question we restrict our attention in the following sections to the class of earthquakes that is damaging to the works of man. If the matter of damage produced by tsunamis is left aside, this class excludes all but the moderate-to-large-energy continental (or near continental) earthquakes of shallow focus. In Sections 2.2.2 and 2.2.3, the properties of this class of earthquakes are gone into in more detail in order to lay an empirical basis for the theoretical development of Section 4.

2.2.2 Fine Structure; Earthquake Sequences

Patterns in the occurrence of earthquakes are of the first importance in any theory of mechanism. It has long

*The bulk of the present evidence inferred from waves recorded on seismograms can be accommodated by supposing that there is only one general type of generating mechanism (see Section 2.4.1).

Fig. 2.2. Display of historic fault breaks and epicenters of the larger earthquakes (up to 1965) in the western United States. (Courtesy of Ryall, Slemmons and Gedney, 1966.)

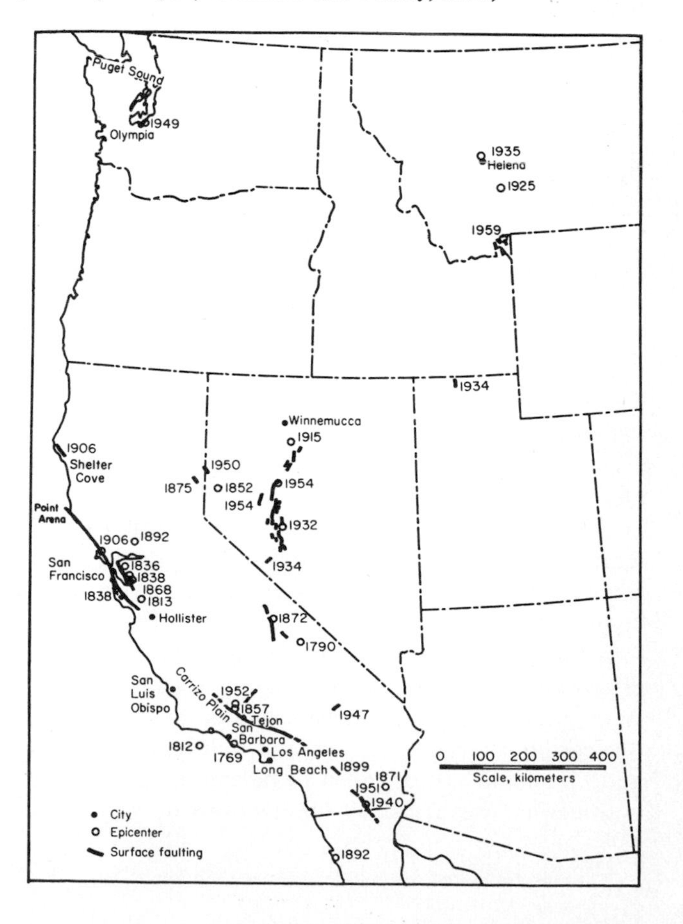

been recognized that in the great majority of cases, a shallow-focus earthquake above a magnitude of 4.0 to 5.0 is followed by a sequence of other, usually smaller, earthquakes in the same vicinity. Furthermore, many (but not all) shallow-focus earthquakes occur in regions where present geological faulting is common and, indeed, some earthquakes have been accompanied by perceptible surface faulting. We now turn our attention to these two properties, taking examples largely from recent work in California.

In California, the continuous historical record goes back to about 1800, the period of the development of the Spanish missions. A useful summary plot of the major known earthquakes* in the western United States (due to Ryall, Slemmons, and Gedney, 1966) is shown in Fig. 2.2. An estimate of the fault break in each earthquake is also shown; uncertainties are high for early shocks.

The basic reference for California is the compendium of Townley and Allen (1939), which runs from 1769 to 1928. Earthquakes were reported in 1800 that damaged Mission San Juan Bautista; 1812 was seismically so active that it was called "the year of earthquakes." In 1836 and 1838 there are records of large earthquakes near San Francisco, and in 1857 one near Fort Tejon (central California) that left rents or cracks in the earth. These reports are the first indication of earthquake-associated rupture on what is now called the San Andreas fault. On October 31, 1868, a large earthquake occurred on the Hayward fault that branches from the San Andreas fault in the vicinity of Hollister (see Fig. 2.3). This earthquake was severely felt in the region about San Francisco Bay with damage to buildings, particularly in Hayward, on the east side of the bay, and in San Francisco. Reports are incomplete but observations were reported of ground breakage along the Hayward fault from near Oakland to 20 mi southeast of San Leandro. In his description, Lawson (1908, Vol. 1, Part II, p. 435) writes that "in some places where it crossed the lower ground, the crack showed faulting or displacement of 8 to 10 inches, but from the accounts given it is not clear in what direction the faulting took place. The statements indicate a slight downthrow on the southwest side. In other places a displacement of 3 feet is said to have been observed."

The great California earthquake of April 18, 1906, marked a turning point in the study of earthquake mechanism. For many seismologists it has come to be the "type" shallow-focus earthquake; it led directly to the formulation by H. F. Reid of the elastic rebound theory of earthquake genesis. Other seismologists have taken the view that the geological circumstances that gave rise to this earthquake were freakish and that inferences based upon it should not be taken to be universal. This view is somewhat weakened by observations of very similar fault ruptures accompanying some more recent earthquakes, e.g., in Turkey, New Zealand, and Iran.

Because of its location and time of occurrence the earthquake could be studied in great detail. In the report of the California Earthquake Commission (1908) the cause of the earthquake is given as "the sudden rupture of the Earth's crust [with right-lateral offset] along a line or lines extending from the vicinity of Point Delgada to a point in San Benito County near San Juan; a distance in a nearly straight course, of about 270 miles. For a distance of 190 miles from Point Arena to San Juan, the fissure formed by this rupture is known to be practically continuous. Beyond Point Arena it passes out to sea so that its continuity with the similar crack near Point Delgada is open to doubt." Recent oceanographic evidence (Currey and Nason, 1967) indicates that the San Andreas fault continues as a submarine feature from Point Arena to Point Delgada (Shelter Cove). This finding adds some weight to the view that the rupture was continuous from San Juan Bautista to near Cape Mendocino. Additional analysis of recorded seismic waves (Bolt, 1968) yields a focus within the crust on the San Andreas fault close to the San Francisco peninsula. It should be noted that this entails bilateral rupture to the north and south along the San Andreas fault whereas most cases of faulting studied since 1906 show unilateral rupture. Further discussion of the properties of the 1906 earthquake will be taken up in following sections.

The few great earthquakes in California are separated by continuous seismic activity of a lesser kind. In recent decades there have been sufficient seismographic stations in the state to keep track of the position and magnitude of most earthquakes with an energy release that is perceptible to man in the vicinity. In southern California a regional earthquake catalog is provided by the Seismological Laboratory of the California Institute of Technology; in northern California similar catalogs are available from the Seismographic Station of the University of California, Berkeley. A general catalog, which contains the important felt reports, entitled *United States Earthquakes*, is issued by ESSA, U.S. Coast and Geodetic Survey.

Detailed regional seismicity studies throw light on the extent of the tectonic forces that provide the earthquake energy and also, indirectly, on the rock properties and physical conditions that pertain within the crust and upper mantle in the region studied. This can be illustrated by a study of Fig. 2.3. On a fault map of central California have been plotted all the epicenters of earthquakes located by the Seismographic Station, University of California, Berkeley, for the period January 1962 through June 1965. (An analysis for southern California has been given by Allen *et al.*, 1965).

The earthquakes generally occur in clusters closely associated with the main fault system. There are, however scattered earthquakes that are not clearly associated

*That is, estimated magnitudes greater than about 7.

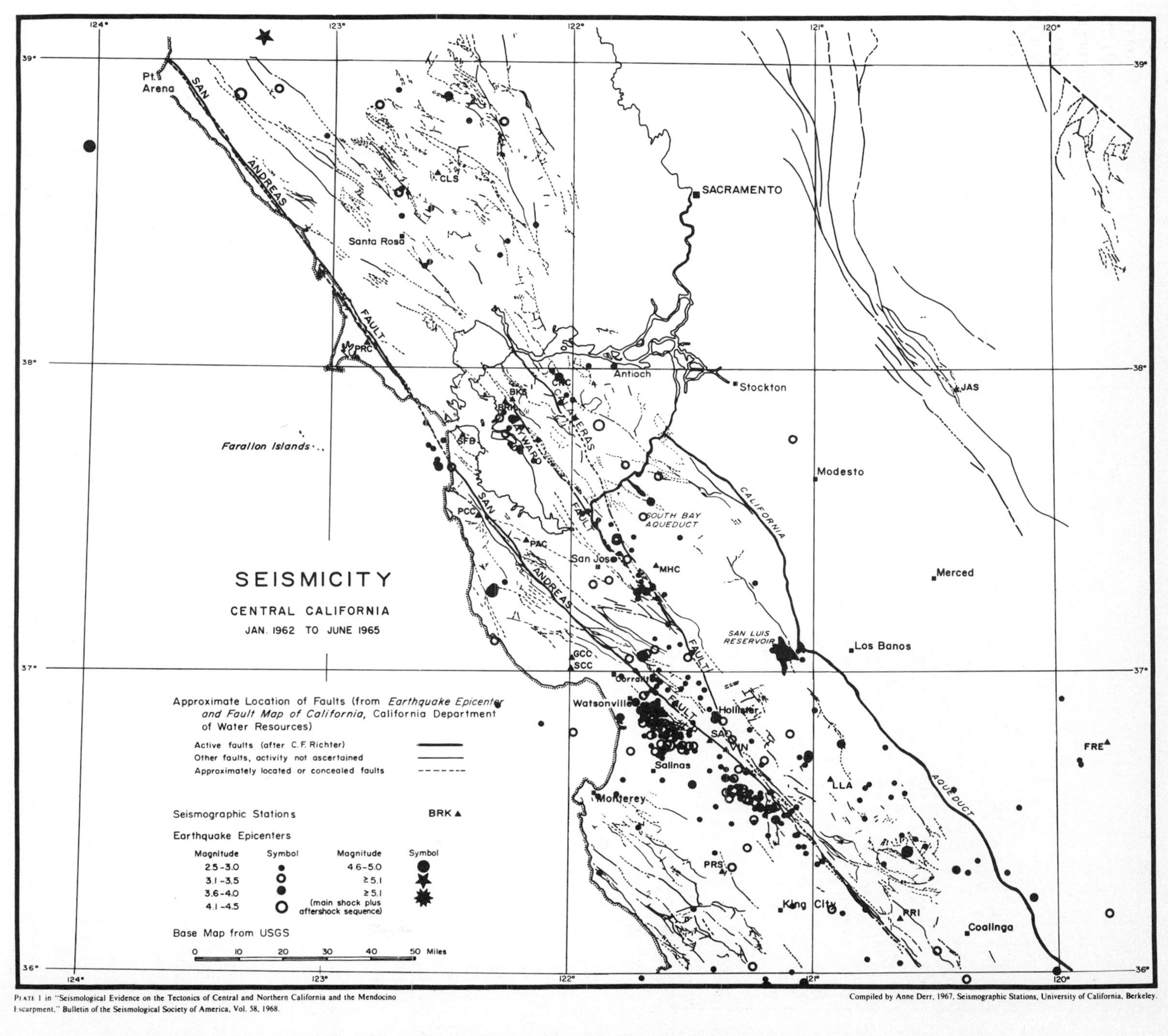

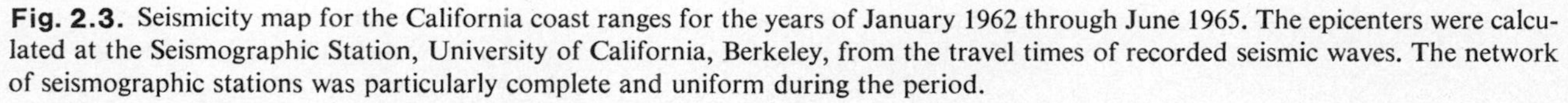

Fig. 2.3. Seismicity map for the California coast ranges for the years of January 1962 through June 1965. The epicenters were calculated at the Seismographic Station, University of California, Berkeley, from the travel times of recorded seismic waves. The network of seismographic stations was particularly complete and uniform during the period.

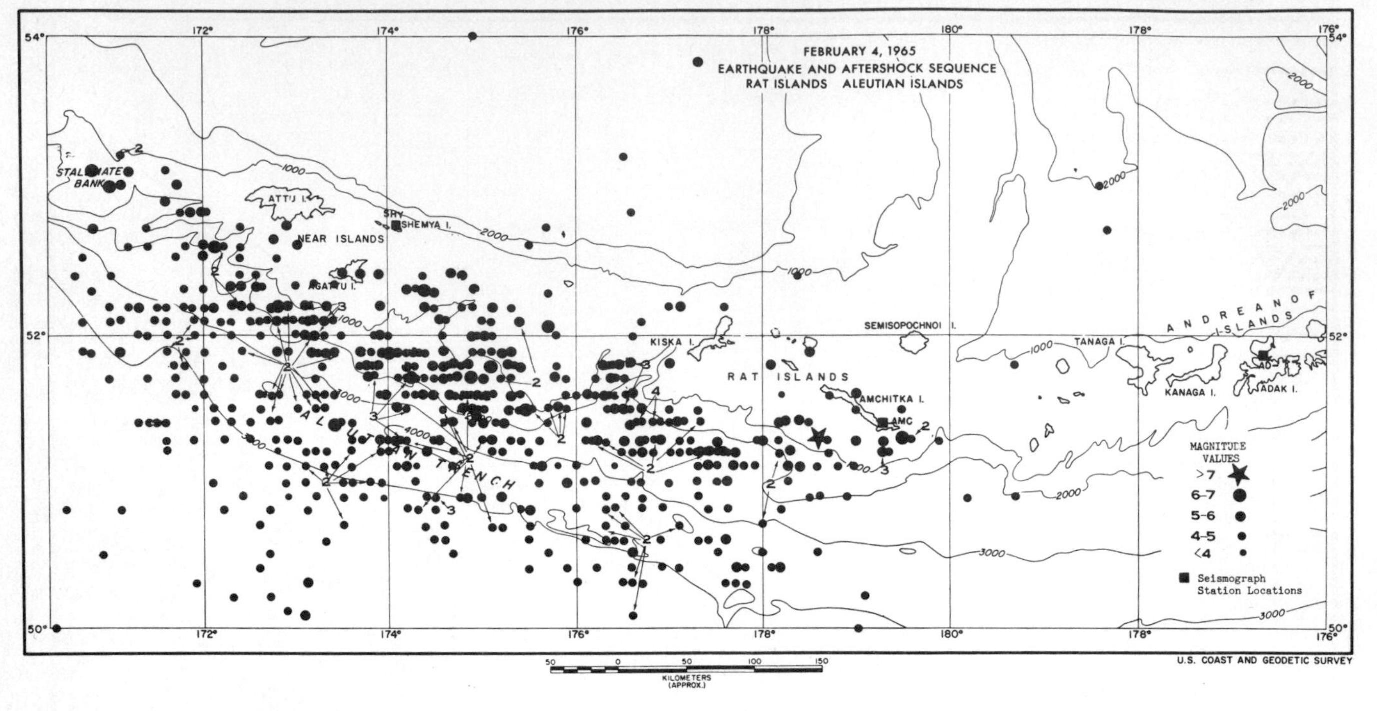

Fig. 2.4. The particularly dense sequence of aftershocks that followed an energetic earthquake in the Rat Islands on February 4, 1965. Approximately 750 earthquakes were located in an interval of 24 days following the main shock. (Courtesy of U.S. Coast and Geodetic Survey.)

in this way. It will be seen at once that no earthquakes were located in this period on the San Andreas fault from Point Arena to near Corralitos with the exception of a cluster where the fault becomes submarine off San Francisco. The quiescence along this whole section has been normal since the 1906 fault rupture and even before (Byerly, 1937). The present activity off San Francisco is evidently associated with a magnitude 5.6 earthquake that occurred in the same vicinity on March 22, 1957, and caused some damage on San Francisco Peninsula.

From Corralitos southward there is more or less continuous seismicity of a moderate kind until, northeast of King City, the activity along the San Andreas fault again diminishes markedly. It is of interest that no fault slippage (see Section 2.3.2) has been discovered on the San Andreas north of Corralitos although a number of cases have been observed near Hollister and on the Hayward fault. Byerly (1937) has inferred from the seismicity pattern that "the regions of greatest fault displacement at the time of great shocks are distinct from those of more or less continuous activity of a minor sort. Probably the frictional resistance to movement (due perhaps to compression transverse to the faults) is much greater along the faults in those regions which show little activity throughout long periods." A test of this hypothesis in California must await further great earthquakes.

Almost all California earthquakes of magnitude 4 or

Fig. 2.5. Two sequences of earthquakes that occurred in August and September of 1963 near Watsonville, California. The broken line indicates the position of the San Andreas fault.

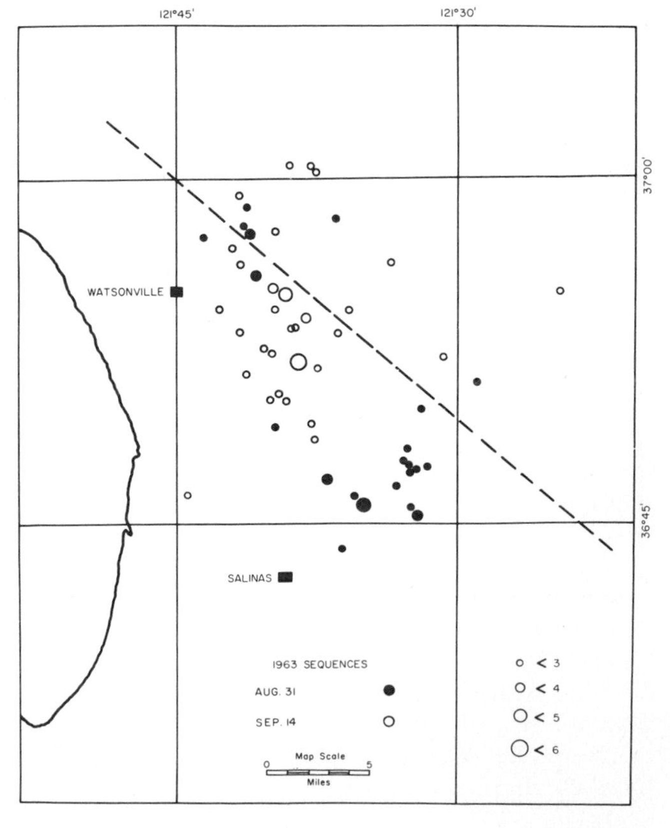

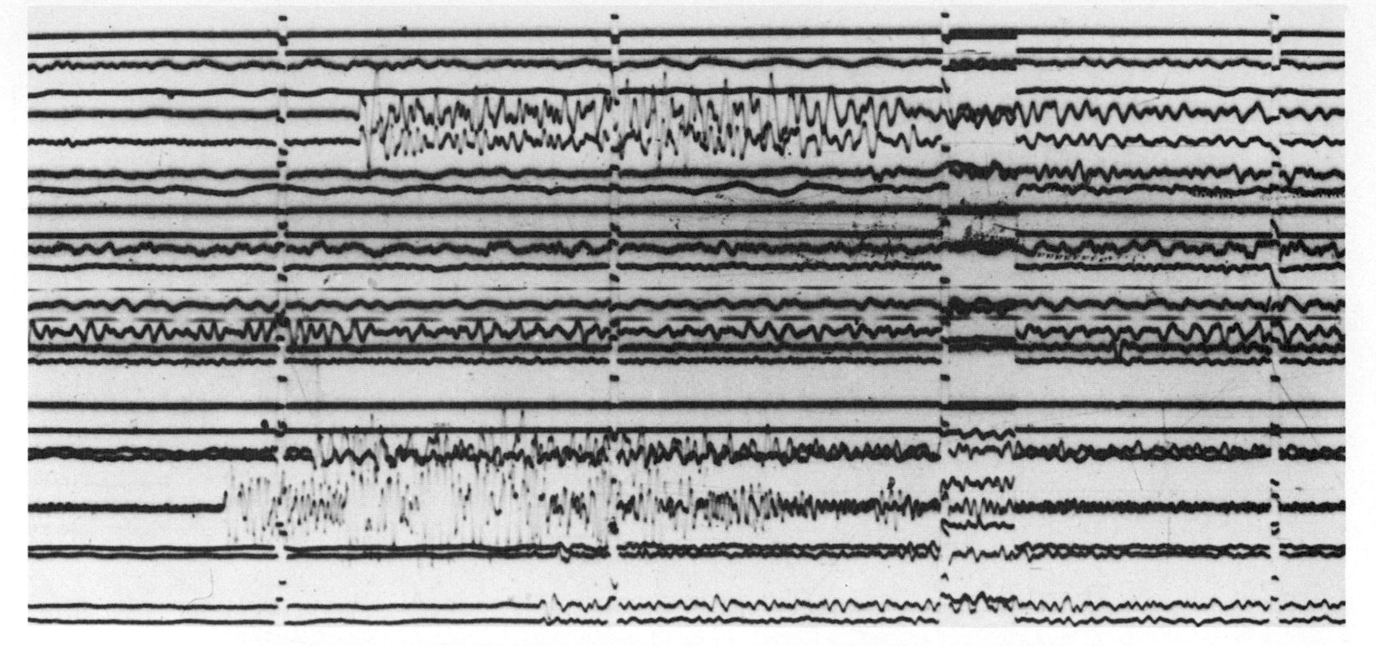

Fig. 2.6. The superposition of two separate film recordings made on the Develocorder at the Berkeley Seismographic Station. Each film has 13 traces; the interval between trace offsets is 10 sec. The films show the recording of *P* waves from different aftershocks (magnitude $2\frac{1}{2}$, 18 hr apart) of the 1964 sequence near Corralitos, California.

more are followed in the same region by a sequence of smaller shocks. This feature is typical of shallow-focus earthquakes around the world. The largest aftershock sequences extend over hundreds of miles along the Earth's surface and over many months in time (e.g., Fig. 2.4). The smallest sequences extend only over a few tens of miles and over several days. Obviously, any causative theory of earthquake genesis must include an explanation of the aftershocks.

In contrast with the extensive sequence shown in Fig. 2.4 we consider two short sequences that occurred in the Salinas–Watsonville region of central California in August–September 1963. Figure 2.5 shows the extent of the epicenters of the two sequences spread over an area of dimensions about 20 mi. Some 100 earthquakes with magnitudes greater than 1 were recorded in 30 days.

The depths were mainly in the range of 4 to 7 km, which was, at the time, something of a surprise because a depth of 15 km had become widely accepted for all California earthquakes. Since 1962, when a telemetry array of standardized seismographic stations went into operation, focal depths and coordinates of most earthquakes along the San Andreas fault in the Coastal Ranges region can be determined within a few kilometers. The majority of focal depths are less than 6 km, with a few as deep as 12 km. A number of special studies of aftershock sequences have been made since that of the Salinas–Watsonville sequence. McEvilly and Casaday (1967) have classified them into two types: One type, like the Watsonville sequence and the 1966 Parkfield sequence (see Chapter 1, Fig. 1.5), has a relatively large area of aftershock activity; the second type, like the Corralitos sequence of 1964, has a very restricted region of activity. In the Corralitos sequence, about 40 shocks with magnitude greater than unity occurred in 26 days before the background activity returned to normal. These shocks were concentrated in a region within the crust, some 10 km deep and 5 km across. Any theory of mechanism must account for the two types.

For the second type, rock deformations must be such that energy can be released suddenly in discrete quanta for 26 days from a volume of linear dimensions 5 km.

No sophisticated analysis is necessary to demonstrate that aftershocks are sometimes repeatedly generated in the same small volume of rock. Figure 2.6 shows the superposition of two film recordings on the Berkeley telemetry system of separate aftershocks of the Corralitos sequence. (Traces correspond to different stations of the seismographic network.) At each station of the network, the *P* onsets from the two shocks have arrived at identical times!

Consistent patterns in the seismic waves generated by earthquakes (see Chapter 1) have been noticed by seismologists for many decades. Because such patterns are, to a large extent, a consequence of the mechanism of the source that generated the waves, careful measurements of the patterns place constraints on the earthquake mechanism. Polarity, amplitude, and frequency of body waves such as *P*, *S*, and surface waves (see Chapter 1, Section 1.2) have all been used in these investigations. It will be sufficient here to limit the discussion to the use of the direction of first motion (polarity) of the first arriving *P* waves at any station; a review of the subject has been published by Stauder (1962). For a simple, completely

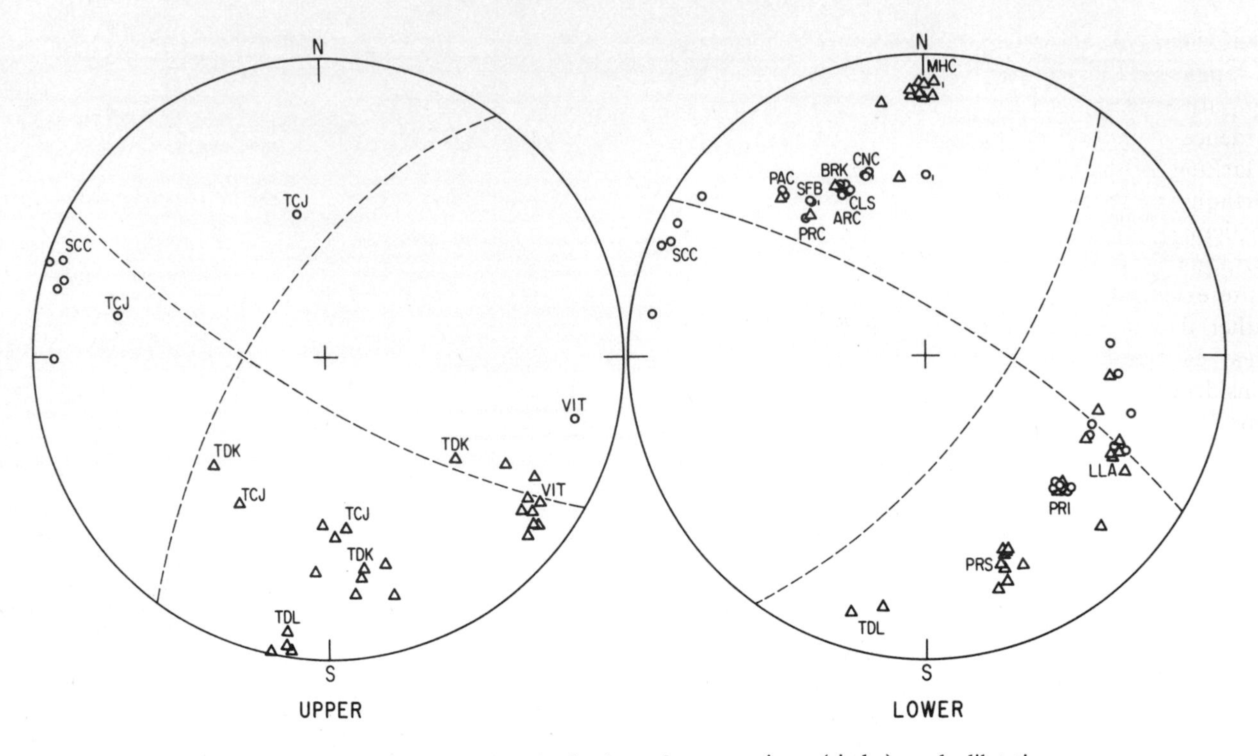

Fig. 2.7. Diagrams showing the distribution of compressions (circles) and dilatations (triangles) in the first motion of the *P* wave from the aftershocks of the September 14, 1963, sequence near Watsonville, California. The left diagram shows the projection of the *P* rays upon the upper half of the focal sphere; the right diagram is the projection upon the lower hemisphere. The dashed lines are the computed nodal lines.

symmetrical source such as an outward-moving, spherical, pressure pulse, the first motion of the *P* wave at any place would be the same; it would be recorded at the ground surface as an upward movement, corresponding to a *compression*. In contrast to this simple source the great majority of earthquakes (both shallow and deep focus), however, are observed to generate *P* motion of both polarities, corresponding to both dilatations and compressions. Consider the first motions from the aftershocks of the 1963 Watsonville sequence. In a special study, Udias (1965) plotted in Fig. 2.7 the measured polarities at all the near stations (each designated by three letters) using a common projection that takes account of the actual ray paths.

Allowing for measurement errors (the earthquakes were of small magnitude), the plot clearly shows a consistent quadrantal pattern in the dilatations and compressions. These are shown separated by dashed lines that represent the nodal planes. Furthermore, the pattern was closely identical with that obtained for the main shock. The first conclusion is that for the great majority of the earthquakes *the mechanism of energy release remained similar to that of the main shock throughout the sequence*. This result has been found to hold in many earthquake sequences not only in California but in other seismic regions of the world.

A further important result was obtained from the polarity diagram of Fig. 2.7. The strike of one of the nodal planes was N54°W. *This orientation differs by only 7° with the strike of the San Andreas fault trace in the region (N47°W)*. Thus, adopting this nodal plane as the fault plane, the motion is consistent with right-handed strike slip along the plane striking N54°W and with a dip of 60°NE.

A considerable number of earthquakes in northern California have now been studied in this way, from Parkfield, north of the Corrizo Plain, to Cape Mendocino, near Shelter Cove (see Fig. 2.2). In a majority of cases a clear quadrantal pattern of first motions of *P* waves is recorded with one nodal line lying within a few degrees of the strike of the major fault in the vicinity. All these observations are consistent with right-lateral transcurrent motion (i.e., oceanic side north, relatively). A point of possible engineering importance (supported by field geology) is that there is little dip-slip (or vertical) component shown in these solutions; the presumption is that the dominant strong motion of the ground from large earthquakes along the San Andreas fault in central and northern California will be horizontal. The tectonic implications of the focal mechnism solutions of earth quakes in the region considered above are discussed in detail by Bolt, Lomnitz, and McEvilly (1968).

Not all shallow-focus earthquakes can be fitted into such a complete picture as the one sketched above for

California. Not all damaging shallow-focus earthquakes in Japan and Europe (particularly Italy and Yugoslavia), e.g., are accompanied by observable faulting. Similarly, evidence of motion restricted sharply to a surface fault is lacking for the 1811 New Madrid and 1886 Charleston earthquakes in the United States. A further point of particular importance for the larger shallow-focus earthquakes is that the source of the waves may be quite extended in space (compare Section 1.5). Indeed, rather than a smoothly rupturing source, the cause of great earthquakes may, perhaps, be more realistically considered as a series of multiple events in space and time.

2.2.3 Energy Release

Estimates of the energy budget involved in earthquakes are important for both theoretical and practical purposes. From the standpoint of practice, the frequency spectrum of energy released in earthquakes gives a bound on the hazard to be expected from earthquakes of different size; it also has been valuable in work on the detection of underground nuclear explosions and their engineering applications.

On the cosmological scale, the maximum amount of energy released each year during historical time by earthquake sources is a small fraction of the total geothermal energy released which is of order 10^{28} ergs/year. By contrast, the annual release of seismic energy, as estimated from seismicity catalogs and the energy-magnitude relation (see Eq. 2.2 below) is of the order of 10^{25}–10^{26} ergs. [It should be noted that slow viscous deformation of the Earth, such as fault creep (see Section 2.3.2), will also dissipate mechanical energy of perhaps appreciable amount.]

The energy released in earthquakes is estimated in two main ways. The first method uses measurements of the seismic wave motion either recorded on seismograms or estimated from reported intensity of shaking; the second uses geodetic surveys before and after the earthquake to estimate the strain energy release (assuming elastic strain energy as the earthquake source). Other possible energy sources are gravitational (potential) energy (see isostasy in Section 2.1) and chemical energy. The first can be shown to be small for most sizable earthquakes: the chemical source has been considered mainly for deep-focus earthquakes.

The P seismic wave energy at distance r, assuming spherical symmetry and perfectly elastic media, can be shown to be

$$4\pi r^2 \rho\alpha \int \left\{\frac{\partial u}{\partial t}\right\}^2 dt, \tag{2.1}$$

where ρ is the density, α is the P wave velocity, and u is the displacement of the ground. There are similar expressions involving the kinetic energy for other body waves and for surface waves. By the use of integrals of this kind, the recorded motion at a seismographic station can be used, taking into account the frictional attenuation of the waves along their path, to deduce the total seismic energy released at the earthquake source. The numerical analysis has been carried through in surprisingly few cases, partly because of the difficulty in handling photographic seismograms and partly because the instrumental response of only a few seismographs has been known until recently. Jeffreys, Gutenberg and Richter, and DeNoyer and Båth, in particular, have worked on this problem; Båth (1966) has recently published a comprehensive review.

Estimates of the above kind, together with estimates of energy released from large explosions, can be used to calibrate the magnitude-energy formula, see Chapter 1, Eq. 1.25. The 1956 empirical relation of Gutenberg and Richter is

$$\log E_s = 11.8 + 1.5\,M, \tag{2.2}$$

where E_s is the energy (in ergs) radiated and M is the magnitude of the earthquake determined by a standard method from the measured amplitudes on a seismogram. This formula gives 10^{24} ergs for a magnitude 8–8.5 earthquake and 10^{21} ergs for magnitude 6. Studies by other seismologists have given somewhat different constants; nevertheless, formulas such as Eq. 2.2 enable the order of total seismic energy E_s to be calculated for any earthquake. This energy may then be compared to estimates of total energy E_T released at the source, which will include heat generation and other nonelastic effects. Let η be the efficiency of conversion of source energy released to seismic energy. Then

$$E_s = \eta E_T \qquad \eta < 1. \tag{2.3}$$

Little is known of the numerical value of η. Its variability will presumably depend upon the earthquake mechanism; the fault rupture theory of earthquakes, because it entails an elongated source, is consistent with η being approximately independent of earthquake magnitude (see Section 2.5).

A variant of the above procedure makes use of intensities measured in the meizoseismal zone (i.e., the zone of damage). This scheme uses damage and felt reports instead of seismographs (which are often not available) to measure earthquake wave motion (see Gutenberg and Richter, 1942). In time, strong-motion accelerometers may provide the intensity measurements.

Because $a^2 \propto (2\pi/T)^2 E$, where T is the wave period and E the wave energy and $\log a$ is approximately a linear function of intensity I, an empirical relation between energy and I (see Section 1.6) is

$$\frac{E_A}{E_B} = \left(\frac{T_A}{T_B}\right)^2 10^{k(I_A - I_B)}, \tag{2.4}$$

where k is a constant. The absolute energy level depends upon calibration at a particular place. Equation 2.4 was used by Byerly and DeNoyer (1958) in conjunction with Eq. 2.15, Section 2.5, to infer the depth* (h) of the San Andreas fault break. (They found h to be 10 km, approximately, on the assumption that a lubricated surface existed at a depth h.)

The second common method of estimation of energy release is based upon the extent of deformation of the crustal rocks that occurred during the earthquake; this method assumes that the energy in earthquakes is derived, in one way or another, from the stored elastic strain energy.

It may be shown easily that adequate strain energy is available to produce even the largest earthquakes. From Eq. 1.10 in Chapter 1 the strain energy in shear is

$$W = \iiint \mu E_{ij} E_{ij}\, dV, \tag{2.5}$$

where E_{ij} is the strain deviation tensor. The expression shows at once that the very large range of energy observed in earthquakes (a factor of 10^{12} or so) can be accommodated in terms of variations in the volume of strained rock affected by a given rupture. For shear energy release due to fault rupture,

$$W = \int_{-\infty}^{\infty} \int_0^L \int_0^h \mu E_{ij} E_{ij}\, dx\, dy\, dz, \tag{2.6}$$

where the y axis is taken along the strike of the fault of length L. To obtain an order of magnitude, let μ and the strain be constant throughout the volume.† Then Eq. 2.6 shows that the shear strain energy is, approximately, $wLh\mu\cdot(\text{strain})^2$, where w is the effective breadth of the strained region. For an earthquake like that of 1906, $L = 400$ km, $h = 10$ km, and $w = 2 \times 10$ km, so that the shear strain energy is $W = 2.4 \times 10^{31} \times (\text{strain})^2$ ergs. The shear strain along the San Andreas fault was of order of 5 m/10 km, i.e., 5×10^{-4} (see also Eq. 2.13). Therefore, $W = E \simeq 10^{25}$ ergs.

Bullen (1953) and Tsuboi (1956) have used Eq. 2.5 to estimate the modulus of elastic resilience (strain energy density at the point of fracture) and the minimum volume of strained rock to yield an earthquake of a given size, taking the likely strength of rocks in the Earth's crust into account. Bullen concludes that an upper bound on W is of order 10^{25} ergs, using a figure of 10^9 dynes/cm² for the strength of crustal rocks.

Byerly and DeNoyer (1958) make allowance for a variation of rigidity away from the fault. They derive the formula, consistent with Reid's elastic rebound theory (see Section 2.4.2),

$$E_T = \frac{\mu_0 h D^2 L \alpha}{3\pi}, \tag{2.7}$$

*Sometimes called fault width.

†In all calculations in this chapter, μ is taken as 3×10^{11} dynes/cm².

where D is the fault offset, α gives the rate of decrease of the strain from the fault, and μ_0 is the rigidity in the fault zone. They calculate E_T for the 1906 earthquake as 0.9×10^{23} ergs and $E_T = 9.6 \times 10^{21}$ for the 1940 Imperial Valley (El Centro) earthquake, magnitude about 7.

The dilatational strain energy term has been omitted in Eq. 2.5 above. Because empirical values are lacking, theoretical models of earthquake sources have been used to indicate the relative size of the two terms. Explosion-type sources require significant dilatational energy. The fault-rupture theory does *not* preclude it. For example, Niazi (1964) has investigated a spherical focal model in which the release of shear energy at the time of rupture sets up a laminar displacement field. This field is taken, in line with geodetic measurements, to decrease monotonically away from the point where rupture along the fault starts. His conclusions are that (1) the rupture leads to a redistribution of the strain energy with only about one-half of the stored energy released, (2) one-half of the released shear energy is converted into dilatational energy retained in the focal region, and (3) at most about one-third of the original stored strain energy is released in the form of seismic waves (i.e., in Eq. 2.3, $\eta = \frac{2}{3}$).

2.3 THE STRAIN FIELD

2.3.1 Geodetic Surveys

Considerable deformation of the Earth's surface is often observed after a large shallow-focus earthquake. Some of the deformation is of a secondary nature, such as landslides, soil compaction, earth lurches, and slumping. Other effects can be fairly clearly associated with the earthquake mechanism itself. There may be widespread changes in ground elevation as well as horizontal or shear displacements.

The Niigata earthquake of June 16, 1964 (magnitude 7.5), near the west coast of Honshu, Japan, provides an example. (The soil liquefaction in Niigata city brought about by the shaking is a notable feature of this earthquake.) A small island, Awashima, is situated about 5 km from the epicenter and about 60 km from Niigata. Regional leveling surveys and other geodetic measurements showed that Awashima Island was uplifted (during a period of a few months that included the shock) by an amount ranging from 80 to 160 cm with a tilt along a NNE striking axis downwards on the northwesterly side; adjacent to the island, soundings delineated a zone of subsidence amounting to 4 m. Fresh submarine dip-slip faulting was detected by echo sounding.

Many other examples of precise measurements of earthquake-associated deformation in the Japanese islands are described in the literature (e.g., Tsuboi, 1939; Richter, 1958; Matuzawa, 1964). The Tango earthquake of March 7, 1928 (magnitude 8), was associated with substantial left-lateral displacement of the Gomura fault; the displacements have been used in studies of earthquake mechanism (Kasahara, 1958).

Detailed surveying also has been carried out in a few other places after earthquakes, notably recently in connection with the great 1960 Chilean earthquake and the great 1964 Alaskan shock. Earthquake-associated deformations clearly provide basic material for theories of earthquake genesis. Geodetic measurements, on the one hand, can suggest appropriate models for earthquake generation and, on the other, provide quantitative tests for the validity of a particular hypothesis. The geodetic evidence is not always clear-cut. Some damaging earthquakes have been followed by no observable large-scale primary faulting but by substantial changes in the local ground elevation. Evison (1963) gives as an example of this type the Napier earthquake of February 2, 1931, in New Zealand, in which a region 100 km long and up to 16 km wide was raised 1–2 m relative to sea level. Such vertical movement often can be explained either in terms of a change of volume of rock at depth (see Section 2.4.1) or of normal or reversed slip on a dipping fault.

However, many cases of postearthquake deformation have been documented that do have faulting as an essential or primary feature. Ten examples of this type in California and Nevada have been discussed by Tocher (1958). The relation between fault rupture length and magnitude derived by Tocher is

$$\log L = 1.02M - 5.77, \tag{2.8}$$

where L is in kilometers. More recently, a list of 42 earthquakes with measured fault displacement has been compiled by King and Knopoff (1968). (In some cases listed, however, the extent of faulting had been inferred and not directly observed.) The shocks occurred in North and South America, Japan, Taiwan, New Zealand, Mongolia, Turkey, Bulgaria, and East Africa; the magnitudes varied from 8.5 to 5.5. [A 3.5 magnitude earthquake in the Imperial Valley, California, also is an interesting likely case (Brune and Allen, 1967).] The lengths of the ruptured faults varied from 600 km to a few kilometers. The maximum fault displacements listed were 6.55 m horizontally (Lituya Bay shock, 1958, July 10) and 13.3 m vertically* (Montana, August 18, 1959). The correlation between fault length L and maximum vertical or horizontal offsets D (both in centimeters) can be expressed as (King and Knopoff, 1968)

$$\log LD^2 = 1.90M - 2.65. \tag{2.9}$$

In a few cases, the quality and repetition of geodetic triangulation has been sufficiently high to enable estimates of the form of the strain field in a seismically active region to be determined numerically. Three sets of measurements of the U.S. Coast and Geodetic Survey have been particularly important. Triangulation points were occupied before and after the faulting in the 1906 California earthquake, the Imperial Valley (California) earthquake of 1940, and the Fairview Peak (Nevada) shock of 1954. The first two earthquakes were predominantly transcurrent (horizontal) while the third had substantial vertical offset.

Consider first the 1906 earthquake. By 1907, three geodetic surveys of the region traversed by the 1906 fault break had been made by the U.S. Coast and Geodetic Survey: (I) 1851–65, (II) 1874–1892, and (III) 1906–1907. These surveys can be interpreted as showing (1) that only minor changes occurred in relative *elevation* in the region, (2) that significant *horizontal* displacements of the land occurred with the main movement *parallel* to the San Andreas fault in the right-lateral sense, (3) that the displacements were a maximum near to the fault between surveys II and III, and (4) that the total relative displacement of distant points on opposite sides of the fault between I and III amounted to about 3.2 m or 11 ft with the Farallon Islands (see Fig. 2.3) moving north. These results formed the basis of Reid's elastic rebound theory for the cause of earthquakes.

Byerly and DeNoyer (1958) found that the following curves closely fitted the measured horizontal displacements v (in the direction y of the strike) that occurred between the pre-earthquake and postearthquake measurements for the three earthquakes mentioned above:

$$\text{1906 earthquake: } v = 194 \cot^{-1} (0.325 \times 10^{-5})x \tag{2.10}$$

$$\text{1940 earthquake: } v = 144 \cot^{-1} (0.308 \times 10^{-5})x \tag{2.11}$$

$$\text{1954 earthquake: } v = 118 \cot^{-1} (0.085 \times 10^{-5})x. \tag{2.12}$$

In these formulas, x is the normal horizontal distance in kilometers from the fault plane and v is in centimeters. It should be noted that it is an assumption that all the offset occurred at the time of the earthquake; a portion may have occurred by creep after the initial rupture (see Section 2.3.2). If all of v occurred as fling, the two halves of the curves, represented by Eqs. 2.10–2.12 on each side of the fault, separated on the fault and each became a straight line.

From Eq. 1.2, the shear strain in the vicinity of the San Andreas fault, e.g., just before the earthquake was, using Eq. 2.10 above and ignoring signs,

$$\begin{aligned} e_{12} &= \frac{1}{2}\frac{\partial v}{\partial x} \\ &= \frac{3.2 \times 10^{-4}}{1 + 0.11 \times 10^{-10}x^2}. \end{aligned} \tag{2.13}$$

An estimate of the shear-strain energy relieved by the displacements can be obtained by using Eqs. 2.10–2.12 together with Eq. 1.10. In this procedure the assump-

*Direct field measurement gave a maximum of about 6 m (M.G. Bonilla, personal communication).

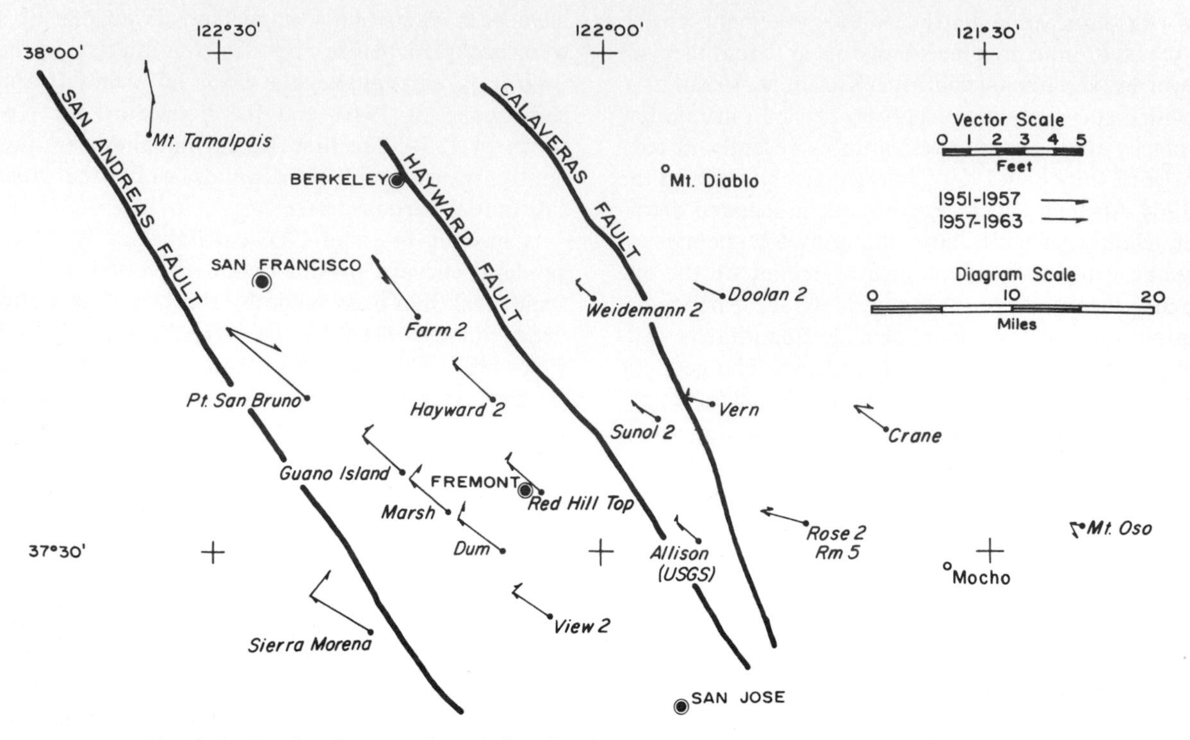

Fig. 2.8. Results of recent triangulations by the U.S. Coast and Geodetic Survey in the San Francisco Bay area. Three surveys, 6 years apart, provided the vector displacements shown by scaled arrows.

tion is made that linear elasticity conditions prevail, for the most part.

Equation 1.10 contains a shear-strain term and a term that depends upon pure compression. The partition of energy between these terms has not yet been estimated from geodetic observations. It is at least reasonable to suppose that most of the change in strain in an earthquake like the 1906 California earthquake is in the shear term. In earthquakes like the 1931 Napier shock, release of dilatational energy may have been important. Even more precise and frequent geodetic leveling and triangulation measurements than hitherto are needed in earthquake regions if definitive observations on the strain redistribution are to be obtained.

Finally, it should be mentioned that the ongoing program of geodetic surveys in California continues to show right-lateral horizontal deformation along the San Andreas fault system (see, e.g., Whitten and Clair, 1960; Pope, Stearn, and Whitten, 1966). Whitten (1956) has computed from the geodetic measurements an equivalent slip of 8 cm/year along the fault zone in the Imperial Valley.

An illustration of recent triangulation measurements in the vicinity of San Francisco Bay is given in Fig. 2.8 (personal communication, C. A. Whitten). Two survey intervals of 6 years are shown. The data relate mainly to the Hayward fault. The displacement vectors show a significant relative right-lateral movement of about $\frac{1}{2}$ ft across this fault in a zone about 15 miles wide. Most displacement took place in the first 6-year period.

An interesting reduction concerning the movement of the Farallon Islands (see Fig. 2.3) recently has been made available by C. A. Whitten (personal letter, June 16, 1967):

> Satisfactory measurements are available from the 1880 era, 1906 after the earthquake, 1922, 1948 and 1957. In the last two sets of measurements, observations were made from the island to the mainland, in addition to the repeat observations from the mainland to the island. All of these observations, when adjusted simultaneously, indicate an annual change in latitude northward of 4.5 cm and westward in longitude of 1.1 cm. Without any attempt to control the direction of movement, these differences give a vector that is W76°N, or approximately parallel to the San Andreas fault. This movement is relative to three points considered to be "fixed": Mt. Diablo, Mt. Sonoma, and Mt. Ross.

The overall conclusion is that the crust of coastal California continues to be deformed horizontally with the oceanic margin moving north relative to the continent. The system of forces that led to shear strain along the San Andreas fault system is being maintained. Similar, if not so detailed, evidence for a direct link between earthquakes and present-day regional tectonic forces can be found in a number of other seismic areas of the world (see Isacks *et al.*, 1968).

2.3.2 Fault Slippage

Deformation of the Earth's crust from tectonic forces involves not only elastic strain (see Section 1.2) but also plastic deformation and "rigid body" translations and rotations. Along fault zones in particular, displacement may occur by slow, secular, differential slippage as well as by sudden rupture. This slow slippage (or fault "creep") has important theoretical (as well as practical engineering and land surveying) consequences. The hypothesis of fault slippage was first proposed in the course of a theoretical argument on the relation between faults and shallow earthquakes by G. D. Louderback (1942). He held that the typical geologic section in fault zones showed a succession of crushed and sheared rock, gouge, breccia, and clays. Mechanically, such materials would behave under pressure more as a viscous, plastic solid than as a brittle, elastic one. "Such material cannot offer the resistance to deformation required to build up elastic strain capable of suddenly giving rise to a strong earthquake. Progressive displacement would be much more likely to take place by slow creep." Since this remarkable prediction a number of instances of fault slip ("creep") in California have been discovered, particularly on the San Andreas and Hayward faults.

The first case reported involves overthrusting on a fault in the Buena Vista Hills, Kern Country; at the last published report (Wilt, 1958), the average movement was about 0.8 in./year. The second case is at the now-famous winery straddling the San Andreas fault in the Coast Ranges of California near Hollister at which progressive building deformation was described by Steinbrugge and Zacher (1960). Measurements of alignments of walls and floor slabs made for more than $7\frac{1}{2}$ years from 1956 gave an average relative displacement across the line of slippage of 8.5 cm (or 3.3 in.). One-sixth of the total slippage occurred as sudden fault movement in the nearby earthquakes of 1960 (magnitude 5.0) and 1961 (two shocks, magnitudes 5.6 and 5.5). Intervals of fault creep begin abruptly, but usually not at the time of local earthquakes. Dr. Tocher has informed me recently that the movement has continued through 1968 at about the same rate. In the same vicinity, extensive offsets in structures have been discovered recently in the town of Hollister (see Fig. 2.3). Fault movements at the Baldwin Hills reservoir from 1951 to 1964 were found by Hudson and Scott (1965) to have some mechanical similarities with the two earlier cases. If growth of a crack in the concrete drainage chamber under the reservoir is assumed to give a measure of fault movement in the above period, the displacement rate was of the order of 0.1 in./year.

Evidence of slippage at four localities on the Hayward fault was published in 1966 in a number of studies (Cluff and Steinbrugge, 1966; Bonilla, 1966; Blanchard and Laverty, 1966; Radbruch and Lennert, 1966; Bolt and Marion, 1966; Pope, Stearn and Whitten, 1966). The observations all indicated right-lateral motion; they suggested that the rate of slippage was not uniform with time, but when it was averaged over a number of years the mean rate was of about 0.1 in./year, or a little greater. At several localities, direct evidence for slippage could be traced back for several decades. Measurements made by M. G. Bonilla (1966) of deformation of a railroad where it crosses the Hayward fault are shown in Fig. 2.9. A differential transformer, used as a displacement meter across a crack in a culvert under the Memorial Stadium, University of California, Berkeley (Bolt and Marion, 1966), has shown the occurrence of right-lateral offsets through 1966 to 1968, particularly in the winter (rainy) months.

The extent of the observations at the present time gives credence to the possibility that fault slippage is a characteristic property of certain stages of fault development. Because certain sections of the San Andreas and the Hayward faults have been associated with major earthquakes in historical times (see Fig. 2.2), and small earthquakes continue to occur along particular segments of the faults (see Fig. 2.3), a common cause for the slippage and the earthquakes is likely.

A model of the slippage is proposed in the following paragraphs, from which certain consequences follow. As Louderback pointed out, it is evident from the geological evidence that elastic strain deformation across faults will not occur uniformly from the ground surface into the crust. The gouge of the fault zone may extend to depths of some kilometers. The increased pressure and temperature with depth as well as changed hydrological conditions will lead to changed mechanical properties; ultimately the downward plunging fault plane will encounter the crystalline rocks of the deeper crust. This fault model will have, across the fault zone, a variable coefficient of friction which increases, apart from local irregularities, with depth.

At the surface, the slicken-sided nature of the gouge material and the presence of much ground water (as indicated by springs and sag ponds) would ensure that, at many places along the fault, the friction coefficient is very small. In the basement rocks, on the other hand, the fault rupture, between earthquakes, would be effectively welded by the large lithostatic forces normal to the fault plane. Regional deformation in the basement would occur, therefore, as elastic strain. The surficial, low-rigidity material near to the fault might respond in a number of ways to the movements beneath and to the

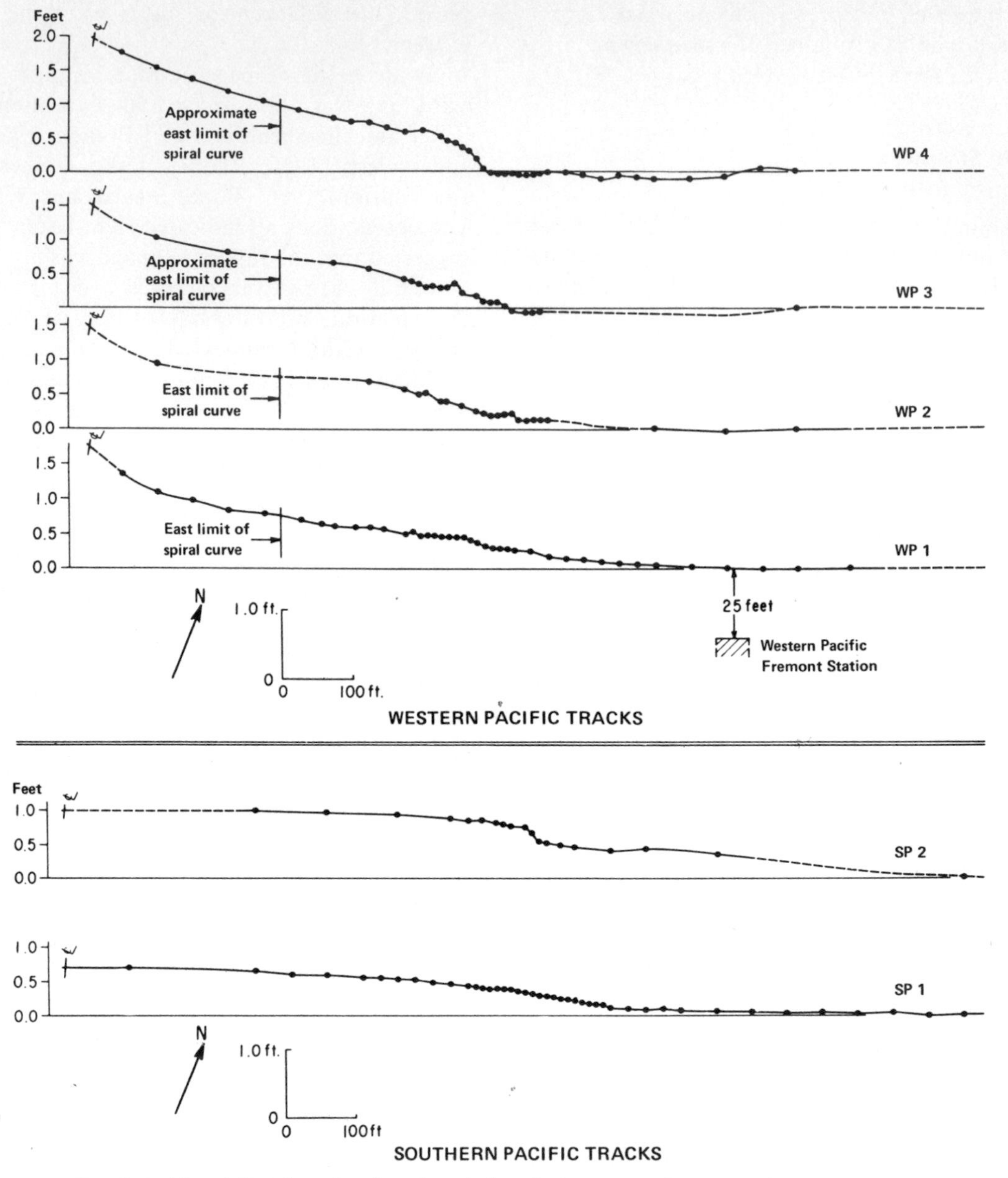

Fig. 2.9. Theodolite data showing the relative displacement of six sets of railroad tracks; the southerly rail of each set is shown. The deformation of the Western Pacific and Southern Pacific tracks was measured by M. G. Bonilla (1966). He interprets the deformation as arising from right-lateral slippage on the Hayward fault in California.

side, depending upon the balance of such mechanical properties as viscosity and cohesive strength as well as elastic properties.

One mode is slippage along the discontinuity provided by the lubricated fault plane. The stronger country rock, some distance from the fault, on each side of the fault would be transported in opposite directions by the basement. On any given side of the fault, the gouge material would tend, under certain conditions, to be carried along by the adjacent stronger surface rocks. Variations in rock type and structure in the fault zone would entail variations in slippage rate along the trace.

This model predicts that observed slippage on a transcurrent fault may well be a mark of the accumulation of strain energy at depth. It also suggests that on portions of a fault where slippage occurs, there may be reduced *surface offset* at the time of an earthquake that accompanies rupture on the fault at depth. On the portion where slippage is small or nonexistent, surface offsets should be a maximum and of the order of the throw at depth.

The Parkfield earthquakes of June 1966 on the San Andreas fault in central California present interesting further evidence on fault slippage associated with earthquakes. The first motions of the P seismic waves of the main shock were consistent with right-lateral transcurrent faulting. In 1966 Allen and Smith pointed out that surficial (right-lateral) displacements had been taking place along the Quaternary trace of the fault just south of Parkfield for several years *before* the earthquakes. (There also were reports of surface cracking along the fault at the time of the 1934 Parkfield earthquakes.) The amount of fault slippage during the earthquake is not directly known, but many postearthquake measurements show that fault slippage occurred at a number of places along the fault in the following days. In particular, steady right-lateral offset of a white line on a highway near Cholame (about 30 km southeast from the epicenter along the fault) took place; the offset had amounted to about 3 in. some 40 hr after the main shock. Allen and Smith remark that "unlike the fault break farther north, the relatively new pavement here apparently was not cracked prior to the June 27 event, although very old fence lines nearby had been offset more than 12 inches along the fault sometime in the years prior to this shock."

2.4 THEORIES OF EARTHQUAKE MECHANISM

2.4.1 The Dilatational Source Theory

The notion of an explosive source for earth tremors goes back at least to Greek science. Because Athens was close to the Aegean volcanoes there appeared to be a natural link between earthquakes and volcanic phenomena.* Strabo, e.g., recognized that earthquakes occurred more frequently along the sea coasts, but he wrote of subterranean winds that ignite combustible materials.

In the eighteenth century the most commonly encountered explanation of earthquakes was still in terms of subterranean fires. (There is the well-known example in Voltaire's *Candide*.) Clear statements offering mechanical energy as an alternative to chemical energy did not appear until the nineteenth century. The Irish engineer Robert Mallet wrote that earthquakes are generated by "either the sudden flexure and constraint of the elastic materials forming a portion of the Earth's crust, or by the sudden relief of this constraint by withdrawal of the force, or by their giving way and becoming fractured."

As the mechanical properties of materials as well as elasticity theory became more thoroughly understood, Mallet's thesis was more refined and the explosive earthquake source became eclipsed. Recently, however, it has been reexamined, particularly for deep-focus earthquakes, in new forms.

This renewal of interest was brought about by two main developments. In the first place (as has been brought out by Griggs and Handin (1960), Brace (1964), and others working on the mechanics of rocks) at relatively shallow depths in the Earth's crust Coulomb friction of any dry plane surface becomes very large. Shearing displacement could occur only if the shear stress exceeds the force of friction which is a constant (the coefficient of friction) μ_{Dry} times the overburden pressure. At a depth of 5 km, the lithostatic pressure is already of the order of 10^9 dynes/cm^2, which is about equal to the estimated strength of granite and similar rocks at the pressure and temperature (500°C) appropriate for that depth. In other words, shearing forces of the magnitude necessary to bring about strike-slip frictional sliding could not be attained below a few kilometers. Phase transformations of various kinds, such as solid–liquid, were considered as possible alternatives by Griggs and Handin.

It has been long recognized by geologists that frictional forces get in the way of simple explanations of the mode of emplacement of great overthrusted blocks in the Alps and elsewhere. Hubbert and Rubey (1959) have shown the key importance of interstitial water as a lubricant. The effective friction coefficient μ_E (not to be confused with the rigidity) can be greatly reduced by the pressure of the water; in symbols, $\mu_E = \mu_{\text{Dry}}(1 - \lambda)$ where $\lambda \longrightarrow 1$ under certain conditions of permeability. The presence of water as springs and so on is often evident in fault zones; it would seem that down to depths of 10 km or so lubrication of the fault planes by ground water and connate water provides a physically adequate mechanism for sudden rupture.* (Variations in μ_E along the fault could well produce checks in the rupture and local grinding, see Chapter 1, Section 1.5.) One possible mode for accumulating and holding fluid in shear zones, related to the dilatancy theory of Osborne Reynolds, has been discussed by Frank (1965) and Orowan (1968).

At somewhat greater depths lubrication by means of local melting (particularly in volcanic regions) along a zone of weakness may be important. Jeffreys (1942) has computed from the heat conduction equation that, with dry friction, sudden fling would melt the rocks near the fault at quite shallow depths. (The absence of geological evidence for fused fault contacts might be taken as evidence for effective lubrication.)

A further important source of lubrication lately has been discovered by Raleigh and Paterson (1965). The laboratory experiments of these workers were particularly germane to seismology because the rock behavior was observed not only at high pressures (3.5–5 kbar, i.e., corresponding to a depth in the crust of about 20

*It may be more than a coincidence that proponents of some form of a dilatational source theory are to be found in Japan and New Zealand.

*Some verfication of the importance of water lubrication on faults as a cause for earthquakes has come from a series of earthquakes near Denver, Colorado. These shocks, in 1962–1967, followed the pumping of waste liquids into the rock strata down a bore 4 km deep at the Rocky Mountain Arsenal. Previously, earthquakes were almost unknown in the region.

km) but also at an appropriate temperature (500–700°C). They found that the hydrous mineral serpentinite became dehydrated and that "brittle" fracture occurred (i.e., a sudden stress drop), evidently facilitated by the fluid pressure. Griggs has suggested that this earthquake mechanism might operate down to depths of a 100 km or so in the Earth. It might be expected that further sources of lubrication will be found as rock mechanics experiments come to represent more realistic geophysical conditions.

The second development that renewed interest in an explosive source mechanism was the seismological dilemma of distinguishing sharply between small, shallow earthquakes and underground explosions. It was argued (e.g., Evison, 1963) that this difficulty implied that the earthquake mechanism was similar to an explosion. (What it actually implied was that the existing seismographic observatories were not equipped or geographically situated to record seismic waves adequately enough to show the discriminatory properties. For example, it is now quite clear that, with proper recording care, underground explosions will produce compressional first arrivals in the *P* phase at all azimuths and distances; earthquakes of comparable magnitude will produce patterns of dilatations and compressions.)* Benioff (1963) suggested on the basis of observed wave forms from three earthquakes that a class of deep earthquakes may arise when a phase transition occurs through a region of rock, causing a sudden change in volume. Seismologists previously had generally discounted such a view on the grounds that deep focus shocks generate clear shear (*S*) waves as well as compressional (*P*) waves. However, one of the early difficulties in distinguishing underground nuclear explosions from earthquakes of comparable depth was the discovery that explosions also generate significant *S* and Love surface waves (although the relative amplitudes and frequency spectra of the wave trains differ for the two sources).

Evison (1963) put forward chemical phase transitions as generators of even shallow focus shocks. He recognized, however, that the transient nature of the seismic source presents an important constraint: "For phase transitions to be capable of generating earthquakes, the transformation would need to be almost simultaneous throughout a substantial volume. This appears to raise difficulties. In recent experiments on the rate at which polymorphic transitions take place at high temperatures, however, there is a hint that suitable rates might not be out of the question."

The phase-change mechanism has been further studied by Dennis and Walker (1965), who argue that the sudden release of chemically stored energy into seismic energy is not possible under equilibrium thermodynamical conditions; the rate of change of energy of the system will be that of conduction or convection. However, they consider that metastable states may be attained by rocks in the mantle under the action of slow convection: "A transition, once initiated, would proceed spontaneously to completion if the pressure-temperature environment of reaction differed substantially from that of equilibrium." With the explosive source hypothesis in mind, Randall (1964) recently developed a mathematical model for a spherical dilatational source. He found that at a depth of 60 km a volume change of 1% was sufficient to generate a seismic wave energy E_s per unit source volume of 3×10^{22} ergs/km^3. A mechanism for aftershocks on this hypothesis has not been worked out.

As was stated in Section 2.1, we are not justified in a work on earthquake engineering to pursue exhaustively the causes of deeper earthquakes. Mechanism models for them are not lacking, however. Recent analyses of E. Orowan (1960) have been in terms of creep instability. Griggs has investigated shear instability resulting from (molecular) thermal activation. Along these lines Alec Riechel has suggested (personal communication) the possible relevance of the Wigner release energy that is found to be of importance in nuclear reactor dynamics; at least at low temperatures, lattice defects are known to store up to 100 cal/gm.

In conclusion, we note that even for deep-focus earthquakes there is considerable *seismological* evidence that is difficult to explain by the hypothesis of a dilatational source. Detailed studies on earthquakes in the Hindu Kush provide a strong illustration. The Hindu Kush in Afghanistan is the site of moderate-to-large earthquakes in a restricted region some 220 km below the Earth's surface. Stevens (1966) has summarized the results of relevant studies; focal mechanisms were worked out to satisfy recorded polarity of *P* waves, polarization of *S* waves, and the radiation patterns of surface waves. The conclusion was that more than 50 earthquakes studied yielded results that were consistent with a double-couple mechanism. Such a mechanism is indistinguishable from ones determined for a number of shallow-focus earthquakes but is inconsistent with an explosive (or implosive) source.

2.4.2 The Elastic Rebound Theory

The elastic rebound theory of earthquake generation is credited to H. F. Reid (1911).* Like the dilatational source theory, this hypothesis has historical antecedents. Faulting plays an essential part in the elastic rebound

*A comparison between the recorded polarities of an underwater chemical explosion, Chase V, and a contemporaneous nearby earthquake is given by Bolt and Lomnitz (1967).

*In a footnote Reid states: "This conception of the causes of the 1906 earthquake was first stated by Professor A. L. Lawson in the *Report of the State Earthquake Investigation Commission* [see Lawson, 1908]." Reid developed the theory in the second volume.

Fig. 2.10. Aerial panorama of part of Marin County, California, showing the San Andreas rift zone from Bolinas Lagoon (foreground) to Bodega Head. In the great 1906 California earthquake horizontal ground offset along the fault ranged from about 15–20 ft in this region. Olema is at the head of the long finger of water (Tomales Bay) in the upper part of the photograph.

theory; an intimate relation between faults and shallow earthquakes must be accepted before the theory can be formulated.

In Greece and Italy earthquakes are not usually accompanied by observable surface faulting while in Turkey and Palestine they are. It comes as little surprise, therefore, that one of the earliest statements of a connection between earthquakes and transcurrent faulting may be found in the Bible in Zechariah, Chapter 14, Verse 4: "The Mount of Olives shall cleave in the midst thereof toward the east and toward the west, and there shall be a very great valley; and half of the mountain shall remove toward the north and half of it towards the south."

As was mentioned in Section 2.4.1, reasons for a causal connection between shallow earthquakes and faults started to appear when knowledge of the behavior of brittle materials became available. In 1859, Robert Mallet wrote on rupture in iron and applied these results to earthquake mechanism. Noting fault rupture extending for 40 mi after the 1891 Mino-Owari earthquake in Japan, Professor B. Koto made explicit his view* that "it can be confidently asserted that the sudden [faulting] was the actual cause [and not the effect] of the earthquake."

A few years later, on the basis of his field work after the great Assam earthquake of June 12, 1897, R. D. Oldham concluded that mechanical forces and fault thrusting were the cause of the shock.*

*Quoted in Richter (1958).

The hypothesis of Lawson and Reid on the cause of the 1906 San Francisco earthquake therefore was in harmony with the trend of thought on earthquake mechanism. Reid had the advantage of firm geodetic survey data, and he made the important step of testing the proposed mechanism by means of a model laboratory experiment using jelly.

Reid supposed that in the 1906 earthquake there had been sudden slippage along the fault that allowed the strained rocks immediately to the west of the trace to move northwest and the rocks to the east to move southeast. The maximum measured strike-slip displacement of adjacent points on the San Andreas fault after the 1906 earthquake reached 21 ft near Olema in the Point Reyes region (see Fig. 2.10). The sudden slip at one point increased the existing stress at adjacent points near the fault plane. This augmentation of stress resulted in rupture, and the break was propagated along the fault. A graphic account of the process has been given by Benioff (1964). In Reid's words:

*Oldham (1926) later called on a deep source with change in volume to explain both the earthquake waves and the surface ruptures for both the 1897 earthquake and the great 1906 California earthquake.

> It is impossible for rock to rupture without first being subjected to elastic strains greater than it can endure. We conclude that the crust in many parts of the Earth is being *slowly* displaced and the difference between displacements in neighboring regions sets up elastic strains, which may become larger than the rock can endure. A rupture then takes place and the strained rock rebounds under its own elastic stresses, until the strain is largely or wholly relieved. In the majority of cases, the elastic rebounds on opposite sides of the fault are in opposite directions.

The earthquake energy therefore arises in the elastic rebound theory from the elastic strain energy stored slowly in the deformed rocks (see Section 2.2.3).

In the primitive form given above the rebound theory [compare the complete statements of Reid (1911, p. 436)] gives no explanation of aftershocks.* If the plastic properties of rocks are introduced, however, the difficulty disappears (see, e.g., Benioff, 1955). Rock mechanics studies are important here. An often-quoted, early high-pressure experiment was that of Bridgman (1952). At a certain stress level, he found that deformation of a sample of rock proceeded in the form of jerks with successive small stress drops. The phenomenon, suggestive of aftershocks, is open to a different interpretation however. More recent experiments on the deformation of rocks have been carried out at more realistic conditions. Griggs and Handin (1960), e.g., deformed a number of rocks at 500°C and 5 kbars confining pressure. The important result emerged that generally deformation was by slow plastic yield with continuous rehealing and without loss of cohesion or sudden release of strain energy. In one test only a specimen of Eureka quartzite broke as a brittle material.

The aftershock mechanism can be explained as follows: After elastic rebound on a lubricated fault there is a readjustment and repartition of the remaining dilatational and shear strain in the region. Plastic creep occurs perhaps with strain hardening and water movement. The stress grows on a section of the fault (possibly weakened by the preceding fracture) until slip again occurs locally. In this way readjustment of the stress field continues by flow; every so often a fracture occurs until the stress everywhere is below the threshold necessary for rupture. (As a corollary, foreshocks may be included in the same mechanism.)

One of the attractions of the modified elastic-rebound theory of shallow earthquakes is that, in principle, it leads to predictions that can be directly checked. In contrast, no crucial tests on the validity of some of the competing theories of earthquake mechanism seem to be possible, e.g., certain forms of the chemical phase-change hypothesis.

*Incidentally, felt reports of aftershocks of the 1906 earthquake continued in central and northern California for some 12 months.

It is of particular interest that Reid used his theory to forecast crudely the time of the next great earthquake near San Francisco. (The theory, of course, predicted that the *place* would be the San Andreas or associated, fault.)

The argument was briefly as follows: From the U.S. Coast and Geodetic Survey surveys Reid inferred that during 50 years (the approximate interval between the first and third surveys) the relative displacement across the San Andreas fault system had reached 3.2 m in the vicinity of San Francisco. After the rebound on the fault on April 18, 1906, a maximum relative displacement D_{max} of some 6.5 m was observed. Therefore, $6.5/3.2 \times 50 \approx 100$ years would elapse before the next great earthquake. (If a displacement of less than 6.5 m is taken as more representative, the interoccurrence term is reduced.) The argument assumes that the regional strain proceeds uniformly and that the various constraints are not altered by a great earthquake (by, e.g., fault creep or by weakening of the country rock adjacent to the fault).

The elastic rebound theory accounts, in a straightforward way, for the observed radiation pattern of seismic waves. As described in Section 2.2, the first motion of *P* waves, e.g., shows in general a pattern of dilatation and compression. This property led Byerly to develop his "fault-plane method" of studying earthquake mechanism (see Byerly, 1926, 1955). There are observational difficulties arising from the uneven distribution of seismographic stations over the Earth's surface; also, the recorded waves are generally only those radiated from the lower half of the seismic source. Under optimum conditions for polarity studies, what is striking, however, is the great consistency of the polarity patterns for shocks throughout an extensive region of given tectonic type (compare, e.g., the fault plane solutions of Sykes, 1967; Bolt, Lomnitz, and McEvilly, 1968).

The elastic rebound involved in either strike-slip or dip-slip faulting can be accounted for, mechanically, in terms of an equivalent pair (a "double-dipole" source) of force couples with the force pairs at right angles (compare Section 1.5). Other force systems have been considered as possible sources (see Stauder, 1964) and the predicted wave motion compared with that recorded from earthquakes. The majority of recent comparisons, using *P*, *S*, and surface waves recorded with fidelity by standarized seismographs around the world, show that the mechanism of most of the earthquakes studied is compatible with conditions near the focus equivalent to a double couple arrangement of forces.*

If the two couples have equal moment, then the overall force system is statically equivalent to zero. However, because the tractions making up the couples occur in an elastic body, a complicated stress system can

*Investigations of special worth on the 1964 Alaskan earthquake sequence are reported by Stauder and Bollinger (1966).

occur in the crustal rocks in the focal region (see Chapter 1, Section 1.2).

The main criticisms of the modified elastic rebound theory have already been treated. The most telling objections have come from work in rock mechanics that show the difficulty with dry stick-slip rupture in rock under pressure. These objections have now been partly met by imaginative extensions of the early experiments to include higher temperatures and hydrous minerals (see Section 2.4.1). One almost insuperable difficulty remains in drawing inferences on the cause of earthquakes from laboratory work in tectonophysics. This problem, recognized by the investigators themselves, arises from the very different rates of deformation in the Earth and in the laboratory.

It should be mentioned that it remains likely that novel earthquake mechanisms will be discovered with further investigations of rock deformation. An example comes from the experimental work of Paterson and Weiss (1966) on folding of rocks with strong layering or foliations. In slates, phyllites, and schists, prominent kinking and folding by flexural slip on the folded surfaces was achieved at confining pressures equivalent to 10–15 km down in the crust *without* lubrication (see Fig. 2.11).

An interesting feature of the compression was the sudden release of energy marked by a noise and jump of the pressure apparatus just at the initiation of the kink bands. Paterson and Weiss conclude that this suggests "the possibility of kinking being an earthquake focal mechanism if circumstances are such that it occurs on a sufficiently large scale and fast enough . . . it can also occur under conditions of deep burial giving a shearing displacement where fracture faulting cannot be expected." Because the generation of elastic waves would be similar to that in Reid's theory it would probably be difficult to distinguish the two mechanisms by purely seismological methods.

In summary, on the elastic rebound theory, faulting is the cause of shallow focus earthquakes—not a consequence of them. Relative offsets along faults are the sum of displacements during earthquakes along the fault together with displacements arising from slow creep* (compare Section 2.3.2). The initiation and maintenance of faults will depend upon the contemporary system of tectonic stresses in the Earth's crust. In the course of time the stress patterns change as they respond to changing mechanical, chemical, and thermodynamical conditions in the Earth's interior. As a consequence, the seismicity

*Movements of either kind, in historical times, are usually taken to define an "active fault."

Fig. 2.11. Model experiment using a deck of cards carried out by L. E. Weiss to illustrate flexural slip folding by conjugate kinking in a finite body compressed along the foliations.

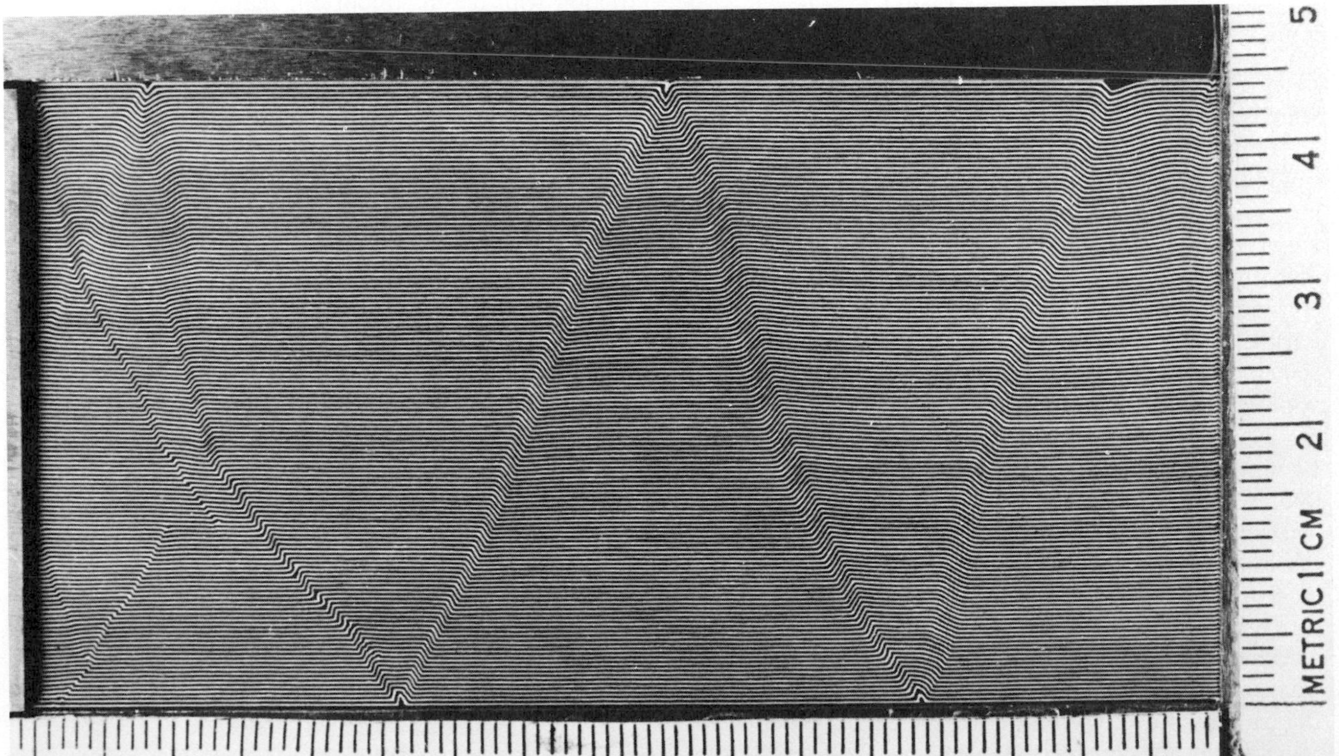

will change, with earthquakes occurring where they have not in the past; in terms of human history these changes may not be of much significance. An extension of the analysis is necessary to include the complexities observed in faulted regions. In some regions, such as along some parts of the San Andreas fault in California, there is clearly a major (or "primary") fault *that is, however, surrounded by a multitude of smaller secondary faults* (see Fig. 2.3). In other places a system of faults is not associated with a "major" fault.

On Reid's theory, earthquakes in any region, in principle, should be explicable in terms of strain release by elastic rebound along the faults in the region. As a corollary, the pattern of secondary faulting might be calculated theoretically. Some attention has been given to these questions. For example, Chinnery (1966) has shown that after rupture on a major fault the stress at the ends of the rupture will rise above pre-earthquake levels. In this model, secondary faulting may occur at the ends of a shear fault to reduce the concentration of stress there.

2.5 MECHANISM PARAMETERS

As was pointed out in Sections 1.5 and 2.4.2, the *kinematics* of sudden fault-slip have not yet been directly observed. A value for the velocity of rupture (ranging from about 2 to 3.5 km/sec) has been inferred for a few earthquakes from seismograms. The time variation of the slippage at any point along the fault is even less well determined. One plausible form for the fling is

$$v = \frac{D}{2}\left(1 + \frac{nt}{3}\right)e^{-nt} \qquad (2.14)$$

which corresponds to a critically damped oscillation.

This dynamical model represents the slippage as a single unidirectional heave with period $T = 2\pi/n$; the maximum acceleration $\frac{1}{6}Dn^2$ occurs at $t = 0$. (In previous sections we have considered the possibility of less uniform motion.) As examples of the use of (2.14) we consider the 1906 California earthquake and the 1966 Parkfield earthquake. For the 1906 shock, take $D =$ 500 cm; then the maximum acceleration is $3\,g/T^2$, approximately. The period T is not known directly but for $T < 1.5$ sec the maximum acceleration will exceed 1 g approximately. Damage and felt reports along the fault would seem to indicate that the peak acceleration was not as great as 1 g; a range of $2 < T < 6$ sec for the duration of the heave might be inferred.

For the Parkfield earthquake, the peak acceleration along the fault at Cholame (30 km southeast of the epicenter) was perhaps 0.5 g and the period of the heave was perhaps of the order of 1 sec. The formula for the maximum acceleration based on Eq. 2.14 then yields $D \simeq 80$ cm; by comparison, measured fault offsets after the earthquake (see Section 2.3.2) were of order 10 cm.

The presence of the squared power of T, apart from the various simplifications, makes Eq. 2.14 difficult to use.

Quantitative predictions on the *dynamics* of the ground motion during shallow-focus earthquakes may be based on measurements of ground deformation for particular earthquakes together with a theory of the earthquake genesis. Consider the empirical formulas, Eqs. 2.10–2.12, for the horizontal displacements adjacent to a fault given by Byerly and DeNoyer (1958). The observations were found to fit the form

$$v = \frac{D}{\pi}\cot^{-1}\alpha x \qquad (2.15)$$

where the fling displacement v is in centimeters and x is in kilometers.

Recently, the theory of dislocations in elastic solids (Nabarro, 1951; Steketee, 1958) has been applied to the earthquake problem. In this theory the basic model is one of a dislocation line surrounded by a stress field; the line of rupture is replaced by a moving dislocation. The applicability of this theory is supported by the field geodetic measurements; dislocation theory has been used (e.g., Teisseyre, 1961) to derive Eq. 2.15 for surface deformation.

The elastostatical problem of the energy release in a shear crack was first treated by Starr (1928) and, in 1958, Knopoff extended the work to a two-dimensional model of strike-slip faulting. Weertman (1964) derived formulas for the fault offset (length of the Burgers vector of the dislocation) in terms of the frictional stress on the fault and discussed the distribution of friction with depth. Important mathematical extensions have been made by Chinnery (1964) and Maruyama (1964), among others. Some equations have resulted from this theoretical work that allow the estimation of key parameters for quantitative comparison of earthquakes. Some of these are now summarized. It should be remembered that many of the derivations contain simplifying assumptions of only approximate validity such as constancy of rigidity μ, neglect of viscosity, and the vanishing of stresses on the crack boundary. End effects on the fault (possibly important for small and moderate earthquakes) are usually ignored. A critical comparison between the theoretical surface displacement fields produced by sloping dislocation surfaces and the observed ground deformation in the 1954 Fairview Peak (Nevada) earthquake, the 1959 Hebgen Lake (Montana) earthquake, and the great 1964 Alaskan earthquake has been given by Savage and Hastie (1966).

The total energy lost, by linear elasticity theory, when a rupture is introduced into a strained elastic half-space (so that there is no discontinuity at the lower edge of the crack) is

$$E_T = \gamma\pi\mu D^2 L, \qquad (2.16)$$

where μ is the rigidity, L is the rupture length, and γ is a constant which depends upon the model. If the shear stress does not drop to zero after the earthquake there is a further multiplicative factor (which is a function of the ratio of the pre- and postearthquake stress) on the right side of Eq. 2.16. Knopoff (1958) found $\gamma = 1/16$; Niazi (1964) gave for his model $\gamma \approx 1/70$. For the California earthquake, we take the values: $\gamma = 1/16$, $L = 400$ km, and $D = 500$ cm. Then Eq. 2.16 yields $E = 5 \times 10^{23}$ ergs. Now this earthquake has an estimated magnitude of $8\frac{1}{4}$ (see, e.g., Bolt, 1967), which, by Eq. 2.2, corresponds to $E_s \simeq 10^{24}$ ergs. The calculation suggests that the efficiency factor η (see Eq. 2.3) is close to unity, which is mechanically somewhat doubtful.

The introduction of the crack in the Earth's crust will lead to a sudden reduction of strain Δe. For the strike-slip fault model of Knopoff (1958), $\Delta e = D/2h$, where h is the fault depth. Therefore, from Eq. 1.8 in Chapter 1, the stress drop at the time of the earthquake is given by

$$\Delta p = 2\mu\Delta e = \frac{\mu D}{h}. \tag{2.17}$$

Substitution in Eq. 2.17 of the 1906 earthquake values gives a stress drop of order 100 bars when $h = 10$ km. This cannot be considered a large pressure drop; it is equivalent to the hydrostatic pressure at a depth of only about 1/2 km or 1700 ft in the crust. Stress drops for a number of earthquakes have been estimated (see, e.g, Aki, 1967); the values range from the order of 100 bars to as low as 1 bar (the 1966 Parkfield main shock). In 1955, Benioff demonstrated that the average elastic stress in the country rock near the White Wolf fault in California before the 1952 earthquake sequence was only of order 25 bars; this provided an upper limit to the stress available for that sequence.

A further parameter specifying a cardinal property of earthquake models is the moment $\mathscr{M}$ of the equivalent double-couple force system (see Section 2.4.2). The expression derived from the dislocation formulation for transcurrent faulting (Maruyama, 1961; Haskell, 1963) is simply

$$\mathscr{M} = \mu DLh. \tag{2.18}$$

We again venture a numerical calculation for the 1906 earthquake. In this case $\mathscr{M} = 6 \times 10^{27}$ dynes/cm. Several properties contribute to the worth of $\mathscr{M}$: First, like E_T and ΔP, its computation depends on the static field measurement of D and L and not upon the dynamics of the faulting (variations in rupture velocity along the fault, i.e., "multiple earthquakes," might well be expected (see Reid's view in Section 1.5); second, theory indicates that the ground motion of an earthquake caused by transcurrent or dip-slip faulting should be directly proportional to the moment (and not its logarithm, say). Against these benefits, h cannot be estimated directly and indeed if, as is likely, the brittle portion of the crust gradually gives way with depth to more plastic conditions, the physical meaning of h becomes obscure. Also, as has been mentioned in Section 2.3.2, the surface offset may vary considerably along the fault trace;* if slow fault slippage occurs, D should be put equal to the maximum measured offset rather than the average. There is also the difficulty that a substantial fraction of the value of D (which is usually measured days after the earthquake) may not have occurred during the earthquake but by slippage afterward. Such creep may be mechanically more likely in some regions than in others, depending on the regional lateral pressures across the fault zone (see Section 2.2.2).

ACKNOWLEDGMENTS

I should like to thank Dr. Perry Byerly, Dr. Don Hudson, Dr. M. Niazi, and Fr. L. Drake, S.J., for their criticisms. It should be mentioned that we continue to hold different views on some issues raised in the text. Mr. George Mitchell kindly assisted with the preparation of the photographic material.

REFERENCES

Aki, K. (1967). "Scaling Law of Seismic Spectrum," *J. Geophys. Res.*, **72**, 1217.

Allen, C. R. (1965). "Transcurrent Faults in Continental Areas," in "A Symposium on Continental Drift," *Phil. Trans. Roy. Soc.*, **258**, 82.

Allen, C. R., P. St. Amand, C. F. Richter, and J. M. Nordquist (1965). "Relationship Between Seismicity and Geologic Structure in the Southern California Region," *Bull. Seism. Soc. Am.*, **55**, 753.

Banerji, S. K., (1957). *Earthquakes in the Himalayan Region*, Calcutta: *Indian Association of Cultivation of Science.*

Bath, M. (1966). "Earthquake Energy and Magnitude," in *Physics and Chemistry of the Earth*, **7**, New York Pergamon Press.

Benioff, H. (1951). "Earthquakes and Rock Creep, 1, Creep Characteristics of Rocks and the Origin of Aftershocks," *Bull. Seism. Soc. Am.*, **41**, 31.

Benioff, H. (1963). "Source Wave Forms of Three Earthquakes," *Bull. Seism. Soc. Am.*, **53**, 893.

Benioff, H. (1964). "Earthquake Source Mechanisms," *Science*, **143**, 1399.

Blanchard, F. B., and G. L. Laverty (1966). "Development in the Claremont Water Tunnel at an Intersection with the Hayward Fault," *Bull. Seism. Soc. Am.*, **56**, 291.

Bolt, B.A. (1959). "Seismic Travel-Times in Australia," *Proc. Roy. Soc. N.S.W.*, **91**, 64.

*Often, of course, D is not directly measurable at all.

Bolt, B.A., and W.C. Marion (1966). "Instrumental Measurement of Slippage on the Hayward Fault," *Bull. Seism. Soc. Am.*, **56,** 305.

Bolt, B. A. (1968). "The Focus of the 1906 California Earthquake," *Bull. Seism. Soc. Am.*, **68,** 457.

Bolt, B. A., C. Lomnitz, and T. V. McEvilly (1968). "Seismological Evidence on the Tectonics of Central and Northern California and the Mendocino Escarpment," *Bull. Seism. Soc. Am.*, **58,** 1725.

Bonilla, M. G. (1966). "Deformation of Railroad Tracks by Slippage on the Hayward Fault in the Niles District of Fremont, California," *Bull. Seism. Soc. Am.*, **56,** 281.

Brace, W. F. (1964). "Brittle Fracture of Rocks," in *State of Stress in the Earth's Crust*, W. R. Judd, ed. New York: Elsevier.

Bridgman, P.W. (1952). *Studies in Large Plastic Flow and Fracture*, New York: McGraw-Hill.

Brune, J. N., and C. R. Allen (1967). "A Low-Stress Drop, Low-Magnitude Earthquake with Surface Faulting: The Imperial, California Earthquake of March 4, 1966," *Bull. Seism. Soc. Am.*, **57,** 501.

Bullen, K. E. (1953). "On Strain Energy and Strength in the Earth's Upper Mantle," *Trans. Am. Geophys. Union*, **34,** 107.

Byerly, P. (1926). "The Montana Earthquakes of June 28, 1925," *Bull. Seism. Soc. Am.*, **16,** 209.

Byerly, P. (1937). "Earthquakes Off the Coast of Northern California," *Bull. Seism. Soc. Am.*, **27,** 73.

Byerly, P. (1955). "Nature of Faulting as Deduced from Seismograms," *Geol. Soc. Am.*, Special Paper **62,** 75.

Byerly, P., and J. DeNoyer (1958). "Energy in Earthquakes as Computed from Geodetic Observations," in *Contributions in Geophysics: In Honor of Beno Gutenberg*, New York: Pergamon Press.

Byerly, P. (1960). "Release of Energy at the Source of an Earthquake," *Publ. Dom. Obs., Ottawa*, **24,** 303.

Chinnery, M. A. (1966). "Secondary Faulting, I. Theoretical Aspects," *Can. J. Earth Sci.*, **3,** 163.

Chinnery, M. A. (1964). "The Strength of the Earth's Crust Under Horizontal Shear Stress," *J. Geophys. Res.*, **69,** 2085.

Cluff, L. S., and K. V. Steinbrugge (1966). "Hayward Fault Slippage in the Irvington-Niles Districts of Fremont, California," *Bull. Seism. Soc. Am.*, **56,** 257.

Currey, J. R., and R. D. Nason (1967). "San Andreas Fault North of Point Arena, California," *Bull. Geol. Soc. Am.*, **78,** 413.

De Montessus de Ballore, F. (1915). "La sismologia en la Biblia," *Bol. del Servicio Sismol. de Chile*, **11.**

De Montessus de Ballore, F. (1924). *La géologie sismologique*, Paris: Armand Colin.

Dennis, J. G., and C. T. Walker (1965). "Earthquakes Resulting from Metastable Phase Transitions," *Tectonophysics*, **2,** 401.

Evison, F. F. (1963). "Earthquakes and Faults," *Bull. Seism. Soc. Am.*, **53,** 873.

Frank, F. C. (1965). "On Dilatancy in Relation to Seismic Sources," *Rev. Geophys.*, **3,** 485.

Griggs, D., and J. Handin (1960). "Observations on Fracture and Hypothesis of Earthquakes," *Geol. Soc. Am. Memoirs*, **79,** 347.

Gutenberg, B. and C. F. Richter (1942). "Earthquake Magnitude, Intensity, Energy and Acceleration," *Bull. Seism. Soc. Am.*, **32,** 163.

Gutenberg, B., and C. F. Richter (1954). *Seismicity of the Earth*, 2nd ed. Princeton, New Jersey: Princeton University Press.

Hartwig, G. (1887). *Volcanoes and Earthquakes*, London: Longmans, Green.

Haskell, N. A. (1963). "Radiation Pattern of Rayleigh Waves from a Fault of Arbitrary Dip and Direction of Motion in a Homogeneous Medium," *Bull. Seism. Soc. Am.*, **53,** 619.

Honda, H. (1962). "Earthquake Mechanism and Seismic Waves," *Geophys. Notes*, Geophys. Inst., Tokyo, **15,** 1.

Hubbert, M. K., and W. W. Rubey (1959). "Roles of Fluid Pressure in Mechanics of Overthrust Faulting, I," *Bull. Geol. Soc. Am.*, **70,** 115.

Hudson, D. E., and R. F. Scott (1965). "Fault Motions at the Baldwin Hills Reservoir Site," *Bull. Seism. Soc. Am.*, **55,** 165.

Isacks, B., J. Oliver, and L.R. Sykes (1968). "Seismology and the New Global Tectonics," *J. Geophy. Res.*, **73,** 5855.

Jeffreys, H. (1942). "On the Mechanics of Faulting," *Geol. Mag.*, **79,** 291.

Kasahara, K. (1958). "Fault Origin Models of Earthquakes, with Special References to the Tango Earthquake, 1927," *J. Phys. Earth*, **6,** 15.

Kasahara, K. (1958). "Physical Conditions of Earthquake Faults as Deduced from Geodetic Data," *Bull. Earthquake Res. Inst. (Tokyo)*, **36,** 455.

King, Chi-Yu, and L. Knopoff (1968). "Stress Drop in Earthquakes," *Bull. Seism. Soc. Am.*, **58,** 249.

Knopoff, L. (1958). "Energy Release in Earthquakes," *Geophys. J.*, **1,** 44.

Lawson, A.C. (1908). *California State Earthquake Commission Report*, Carnegie Institution of Washington.

Lee, S. P. (ed.) (1956). *Chronological Tabulation of Chinese Earthquake Records*, Peking: Chinese Academy of Sciences.

Lomnitz, C. (1966). "Statistical Prediction of Earthquakes," *Rev. Geophys*, **4,** 377.

Lomnitz, C., and B. A. Bolt (1967). "Evidence on Crustal Structure in California from the Chase V. Explosion and the Chico Earthquake of May 24, 1966," *Bull. Seism. Soc. Am.*, **57,** 1093.

Louderback, G. D. (1942). "Faults and Earthquakes," *Bull. Seism. Soc. Am.*, **32,** 305.

Lyttleton, R. A. (1963). "On the Origin of Mountains," *Proc. Roy. Soc., London A*, **275,** 1.

Maruyama, T. (1963). "On the Force Equivalent of Dynamic Elastic Dislocation with Reference to the Earthquake Mechanism," *Bull. Earthquake Res. Inst.*, **41,** 467.

Maruyama, T. (1964). "Statical Elastic Dislocations in an Infinite and Semi-Infinite Medium," *Bull. Earthquake Res. Inst.*, **42,** 289.

Matuzawa, T. (1964). *Study of Earthquakes*, Tokyo: Uno Shoten.

McEvilly, T. V., and K. B. Casaday (1967). "The Earthquake Sequence of September 1965 near Antioch, California," *Bull. Seism. Soc. Am.*, **57,** 113.

Nabarro; F. R. N. (1951). "The Synthesis of Elastic Dislocation Fields," *Phil. Mag.*, **42,** 1224.

Niazi, M. (1964). "Partition of Energy in the Focal Region of Earthquakes," *Bull. Seism. Soc. Am.*, **54,** 2175.

Niazi, M. (1964). "Seismicity of Northern California and Western Nevada," *Bull. Seism. Soc. Am.*, **54,** 845.

Oldam, R. D. (1926). "The Cutch Earthquake of 16 June 1819 with a Revision of the Great Earthquake of 12 June 1897," *Mem. Geol. Survey, India*, **46,** 48.

Orowan, E. (1960). "Mechanism of Seismic Faulting, Rock Deformation," *Geol. Soc. Am. Memoir*, **79,** 323.

Orowan, E. (1966). "Dilatancy and the Seismic Focal Mechanism," *Rev. of Geophys.* **4,** 395.

Paterson, M. S., and L. E. Weiss (1966). "Experimental Deformation and Folding in Phyllite," *Bull. Geol. Soc. Am.*, **77,** 343.

Pope, A. J., J. L. Stearn, and C. A. Whitten (1966). "Surveys for Crustal Movement Along the Hayward Fault," *Bull. Seism. Soc. Am.*, **56,** 317.

Press, F., and W. F. Brace (1966). "Earthquake Prediction," *Science*, **152,** 1575.

Radbruch, D. H., and B. J. Lennert (1966). "Damage to Culvert Under Memorial Stadium, University of California, Berkeley, Caused by Slippage in the Hayward Fault Zone," *Bull. Seism. Soc. Am.*, **56,** 295.

Raleigh, C. B., and M. S. Paterson (1965). "Experimental Deformation of Serpentinite and Its Tectonic Implications," *J. Geophys. Res.*, **70,** 3965.

Randall, M. J. (1964). "Seismic Energy Generated by a Sudden Volume Change," *Bull. Seism. Soc. Am.*, **54,** 1291.

Reid, H. F. (1911). "The Elastic-Rebound Theory of Earthquakes," *Bull. Dept. of Geol.*, **6,** 413.

Richter, C. (1958). *Elementary Seismology*, San Francisco: Freeman.

Ryall, A., D. B. Slemmons, and L. D. Gedney (1966). "Seismicity, Tectonics, and Surface Faulting in the Western United States during Historic Time," *Bull. Seism. Soc. Am.*, **56,** 1105.

Savage, J. C., and L. M. Hastie (1966). "Surface Deformation Associated with Dip-Slip Faulting," *J. Geophys. Res.*, **71,** 4897.

Starr, A. T. (1928). "Slip in a Crystal and Rupture in a Solid Due to Shear," *Proc. Camb. Phil. Soc.*, **24,** 489.

Stauder (S. J.), W. (1962). "The Focal Mechanism of Earthquakes," in *Advances in Geophysics*, Vol. 9, New York: Academic Press.

Stauder (S. J.), W. and G. A. Bollinger (1966). "The *S* Wave Project for Focal Mechanism Studies—The Alaska Earthquake Sequence of 1964," Scientific Report No. 1 of Contract AF 19(628)–5100.

Steinbrugge, K. V. and E. G. Zacher (1960). "Creep on the San Andreas Fault; Fault Creep and Property Damage," *Bull. Seism. Soc. Am.*, **50,** 389.

Steketee, J. A. (1958). "On Volterra's Dislocations in a Semi-Infinite Medium," *Can. J. Phys.*, **36,** 192.

Stevens, A. E. (1966). "*S*-wave Focal Mechanism Studies of the Hindu Kush Earthquake of July 6, 1962," *Can. J. Earth Sci.*, **3,** 367.

Sykes, L. R. (1967). "Mechanism of Earthquakes and Nature of Faulting on the Mid-Oceanic Ridges," *J. Geophys. Res.*, **72,** 2131.

Teisseyre, R. (1961). "Dynamic and Time Relations of the Dislocation Theory of Earthquakes," *Acta Geophys. Polonica*, **9,** 3.

Teisseyre, R. (1964). "The Method of the Continuous Dislocation Field and Its Application to the Fold Theory," *Bull. Seism. Soc. Am.*, **54,** 1059.

Tocher, D. (1958). "Earthquake Energy and Ground Breakage," *Bull. Seism. Soc. Am.*, **48,** 147.

Tocher, D. (1960). "Creep on the San Andreas Fault; Creep Rate and Related Measurements at Vineyard, California," *Bull. Seism. Soc. Am.*, **50,** 396.

Townley, S. D., and M. W. Allen (1939). "Descriptive Catalog of Earthquakes of the Pacific Coast of the United States 1769 to 1928," *Bull. Seism. Soc. Am.*, **29,** 1.

Tsuboi, C. (1939). "Deformation of the Earth's Crust as Disclosed by Geodetic Measurements," *Beitr. Z. Geophys. Ergeb. Kosmischen Physik*, **4,** 106.

Udias, A. (1965). "A Study of the Aftershocks and Focal Mechanism of the Salinas-Watsonville Earthquakes of August 31 and September 14, 1963," *Bull. Seism. Soc. Am.*, **55,** 85.

Wertmann, J. (1964). "Continuum Distribution of Dislocations on Faults with Finite Friction," *Bull. Seism. Soc. Am.*, **54,** 1035.

Whitten, C. A. (1956). "Crustal Movement in California and Nevada," *Trans. Am. Geophys. Union*, **37,** 393.

Whitten, C. A., and C. N. Claire (1960). "Analysis of Geodetic Measurements Along the San Andreas Fault," *Bull. Seism. Soc. Am.*, **50,** 404.

Willis, B. (1928). "Earthquakes in the Holy Land," *Bull. Seism. Soc. Am.*, **18,** 73.

Wilson, J.T. (1965). "A New Class of Faults and their bearing on Continental Drift, "*Nature*, 4995, 343.

Wilt, J. W. (1958). "Measured Movement Along the Surface Trace of an Active Thrust Fault in the Buena Vista Hills, Kern Country, California," *Bull. Seism. Soc. Am.*, **48,** 169.

Chapter 3

Surface Faulting and Related Effects

M. G. BONILLA

Geologist, U.S. Geological Survey
Menlo Park, California

3.1 INTRODUCTION

Faults and faulting are important to engineers because (1) they can severely damage or destroy structures by shearing, compression, extension, and rotation caused by tilting or bending; (2) earthquakes may occur along them; and (3) past faulting may have greatly affected the physical properties of foundation materials by decreasing their strength, changing their permeability, or bringing together rock units with very different physical properties.

This chapter is based primarily on experience in North America and discusses only items 1 and 2 above, with emphasis on item 1. Item 2 is treated more fully in other chapters of this book, particularly the chapters by Bruce Bolt and G. W. Housner.

Fault rupture and the ground deformation closely associated with it can have extremely serious consequences even though the area directly affected is small compared to the area affected by shaking, landsliding, compaction, and liquefaction. Buildings, bridges, dams, tunnels, canals, and pipelines have been severely damaged by fault rupture; damage of this kind is described in reports by Lawson and others (1908), Ambraseys (1960), Duke (1960), and the California Department of Water Resources (1967).

Not all faults are important to engineering. Some have displacements of only a few inches and lengths of a few to a few hundred feet. Their effects on the physical properties of the rock may have been minor; furthermore, many once-active faults are now healed and as sound as the surrounding rock. Most faults are not now the site of earthquakes. Many faults are very ancient, and the absence of movement for hundreds of millions of years can be demonstrated for some of them.

3.2 GLOSSARY

A short glossary defining selected geological terms used in this report is given below. The meanings apply to the terms as used in this chapter and are not intended as general definitions. Geological terms not in the glossary can be found in a standard dictionary or in the *Glossary of Geology and Related Sciences* (American Geological Institute, 1960).

Dip: The angle that a stratum, joint, fault, or other structural plane makes with a horizontal plane.

Dip slip: The component of the slip parallel with the dip of the fault.

Fault: A fracture or fracture zone along which the two sides have been displaced relative to one another parallel to the fracture. The displacement may range from a few inches to many miles.

Fault creep: Apparently continuous displacement along a fault at a low but varying rate, usually not accompanied by felt earthquakes (see also tectonic creep). As used in this chapter, fault creep is not necessarily tectonic in origin; it may result from artificial withdrawal of fluids or solids.

Fault displacement: Relative movement of the two sides of fault, measured in any specified direction.

Fault sag: A narrow tectonic depression common in strike-slip fault zones. Fault sags are generally closed depressions less than a few hundred feet wide and approximately parallel to the fault zone; those that contain water are called sag ponds.

Fault scarp: A cliff or steep slope formed by displacement of the ground surface.

Fracture: A general term for discontinuities in rock; includes faults, joints, and other breaks.

Graben: A fault block, generally long and narrow, that has been dropped down relative to the adjacent blocks by movement along the bounding faults. The same form of the word is used for both the singular and plural.

Landslide: The downward and outward movement of slope-forming materials, such as rock, soil, artificial fill, or combinations of these materials (Varnes, 1958, p. 20); the topographic feature and the deposit resulting from such movement.

Left slip: Strike-slip displacement in which the block across the fault from an observer has moved to the left; also called sinistral strike slip.

Normal fault: A fault in which the block above an inclined fault surface has moved downward relative to the block below the fault surface; also includes vertical faults with vertical slip.

Oblique slip: A combination of strike slip and normal or reverse slip.

Reverse fault: A fault in which the block above an inclined fault surface has moved upward relative to the block below the fault surface.

Right-normal slip: Fault displacement consisting of nearly equal components of right slip and normal slip; also called dextral normal.

Right slip: Strike-slip displacement in which the block across the fault from an observer has moved to the right; also called dextral strike slip.

Slip: The relative displacement of points on opposite sides of a fault, measured on the fault surface.

Strike: The direction or bearing of a horizontal line in the plane of an inclined or vertical stratum, joint, fault, or other structural plane.

Strike slip: The component of the slip parallel with the strike of the fault; the horizontal component of slip.

Strike-slip fault: A fault in which the slip is approximately in the direction of the strike of the fault; also called wrench or transcurrent fault. The historic displacements on strike-slip faults discussed in this chapter have, in places along those faults, included a vertical component that has generally been less than one-quarter of the horizontal component.

Tectonic: Of, pertaining to, or designating the rock structure and external forms resulting from deep-seated crustal and subcrustal forces in the earth.

Tectonic creep: Fault creep of tectonic origin; also called slippage.

3.3 SURFACE MANIFESTATIONS OF FAULTING

Surface manifestations of faulting and closely related processes include sudden rupture and displacement, creep, warping, tilting, and gross changes in land level. The first of these is of greatest importance for most engineering structures and consequently is treated more fully than the others.

3.3.1 Sudden Rupture and Displacement

Sudden rupture and displacement occurs with normal, reverse, strike-slip, or oblique-slip faulting (see glossary). The historic record of surface faulting in the continental United States and adjacent parts of Mexico, as currently known, is summarized in chronological order in Table 3.1.

3.3.2 Length of Surface Ruptures

The length of surface ruptures given in column 3 of the table is the distance between the ends of continuous or nearly continuous breaks that formed at the surface in the listed earthquakes. This length can be substantially less than the length inferred from the distribution of aftershocks, from dislocation theory, or from other indirect means. The longest surface rupture (partly submarine) on record occurred along 270 mi of the San Andreas fault in 1906. The length of subsurface faulting that occurred in the 1964 Alaskan earthquake is estimated at about 370 mi by Savage and Hastie (1966, Table 1) and about 450 mi by Housner (Chapter 4, this volume).

3.3.3 Fault Displacements

Maximum recorded surface displacements accompanying earthquakes have ranged from 0.05 ft of strike slip in the Imperial, California, earthquake of 1966 (Brune and Allen, 1967) through 35 ft of vertical displacement in the Assam earthquake of 1897 (Oldham, 1899, p. 145) to possibly as much as 42 ft of vertical displacement in the Yakutat Bay, Alaska, earthquake of 1899 (Bonilla, 1967, pp. 9–10). The largest measured strike slip, 29 ft, occurred in the Gobi–Altai earthquake of 1957 (Florensov and Solonenko, 1965, p. 288). From measurements of offset stream channels, Wallace (1968) has suggested that strike slip on the San Andreas fault may have been 30 ft in 1857.

The vertical displacements for normal faults given in columns 4, 5, and 6 of Table 3.1 are the scarp heights except where otherwise specified. The scarp height generally is more critical for engineering purposes than the vertical component of fault displacement; furthermore, many published reports give only scarp heights. Scarps produced by normal faulting commonly are of greater height than the vertical component of fault displacement, chiefly because gravity graben form along the fault (Gilbert, 1890, p. 354; Slemmons, 1957, pp. 367–375). This is shown in Fig. 3.1, a diagrammatic cross section of a typical graben formed by gravity settling of part of the hanging wall of a normal fault. The vertical component of fault displacement, equal to the vertical distance from *A* to *B*, is less than the scarp height *AC*. In order to avoid having to accommodate the full scarp height, a structure across the main fault would have to bridge the graben (Fig. 3.2). Because the width (Fig. 3.1, *CD*) of the graben is generally more than 10 ft and can be as much as 300 ft (Witkind, 1964, p. 45), structures may bear on the graben and have to accommodate the full scarp height.

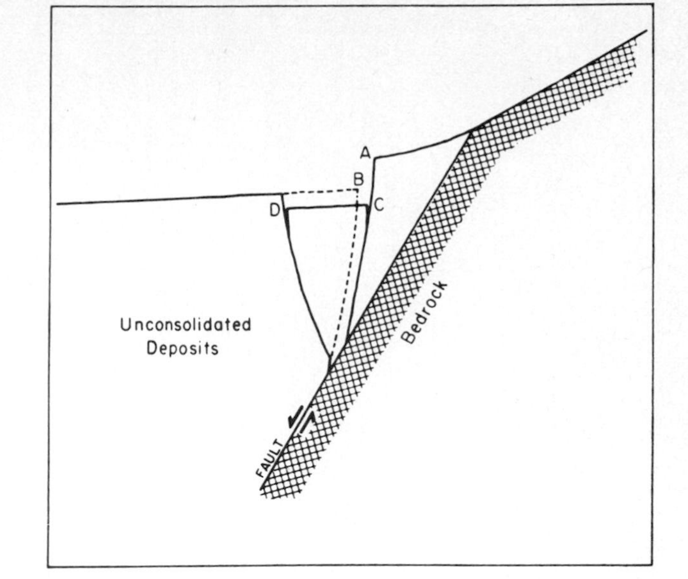

Fig. 3.1. Cross section of a gravity graben associated with a normal fault. The relative movement of the fault is shown by arrows.

In addition to the effects of graben formation, scarp heights may be increased by minor landsliding and other erosional processes that cause a gradual uphill retreat of the brow (Fig. 3.1, point *A*) of the scarp. Scarp heights are not given in the table for specific points where erosional processes are known to have substantially increased them as, e.g., parts of the Fairview Peak scarps formed in 1954 (Slemmons, 1957, pp. 373–375).

Although the scarp heights of normal faults are commonly greater than the vertical component of fault movement at the surface, they can be substantially less than inferred fault displacement at depth. The maximum scarp produced in the 1959 Montana earthquake, e.g., was 20 ft (Witkind, 1964, p. 37; Myers and Hamilton, 1964, p. 81), but the subsurface displacement based on dislocation theory was more than 40 ft (Brune and Allen, 1967, p. 510).

Small to moderate amounts of vertical movement have accompanied some strike-slip faulting. According to dislocation theory, a systematic quadrantal distribution of very small elevations and depressions is expected for strike-slip faulting that is not infinitely long (Chin-

Table 3.1. HISTORIC SURFACE FAULTING IN THE CONTINENTAL UNITED STATES AND ADJACENT PARTS OF MEXICO

Fault (name or location), date, and type of displacement* (See notes at end of table)	Magnitude (Richter) of associated earthquake	Length of surface rupture (miles)	Displacement (feet), main fault (maximum)	Displacements (feet) at indicated distances (miles) from center of main fault zone		Distances (maximum) from center of main zone to outer limits of			Remarks	Principal references
				Branch faulting	Secondary faulting	Main zone	Branch faulting	Secondary faulting		
1	2	3	4	5	6	7	8	9	10	11
1. New Madrid, Missouri; 1811–1812; N(?)			6(?) V						Fault whose scarp bounds Reelfoot Lake shows vertical separation of 40 ft in Eocene beds 160 ft below the surface. Uplift as well as subsidence occurred in this earthquake. See text.	Fuller, 1912; Fisk, 1944; U.S. Army Corps of Engineers, 1950
2. Hayward, California; 1836; Rs(?)		38(?)								Louderback, 1947
3. San Andreas, California; 1838; Rs(?)		35(?)								Louderback, 1947
4. San Andreas, California; 1857; Rs		200±	Large, possibly 30							Lawson *et al.*, 1908; Wood, 1955; Allen *et al.*, 1965; Brown and Vedder, 1967; Wallace, 1968
5. Hayward, California; 1868; Rs(?)		30±	3Rs(?), 1V	Displacement unknown	1.5V at 1.4		0.8± mi	1.8 mi	Given length includes a 23-mi southern segment and a probable segment 0.3 mi long, 7 mi to the north.	Lawson *et al.*, 1908; Radbruch, 1967
6. Owens Valley, California; 1872; RN and LN(?)	8.3 (estimated)	60+	23N; 16–20Rs		18V at 1.6+; 4N at 8; 2.5N at 8; 15V at $1\frac{1}{2}$	0.5 mi		8 mi	Displacements given for secondary faults at 8 mi are scarp heights; net displacements were $1\frac{1}{2}$ and 1 ft.	Knopf and Kirk, 1918; Whitney, 1888; Hobbs, 1910; Bateman, 1961; Bonilla, 1967 and unpublished data
7. Mohawk Valley, California; 1875; N(?)									May have been landsliding rather than faulting.	Turner, 1891, 1896, 1897; Gianella, 1957; Bonilla, 1967
8. Sonora, Mexico; 1887; N		35+	26N			500± ft			Possible secondary faulting at maximum distance of 13 mi from main fault but contemporaneity is doubtful.	Aguilera, 1920; Goodfellow, 1888; Richter, 1958
9. San Jacinto, California; 1899; Rs(?)		2								Danes, 1907; Allen *et al.*, 1965
10. Yakutat Bay, Alaska; 1899; N(?) and Ls(?)	8.5–8.6	Unknown	29–42N (?)	See remarks	See remarks		See remarks	See remarks	Maximum uplift 47 ft. Inferred principal faults under water. Uplift, warping, and possible faulting in area at least 30 by 15 mi and probably much greater. Secondary (?) faulting produced scarps as much as 8 ft high, 21 mi from the inferred principal faults.	Tarr and Martin, 1906, 1912; Martin, 1907; Richter, 1958; Bonilla, 1967

Fault (name or location), date, and type of displacement* (See notes at end of table)	Magnitude (Richter) of associated earthquake	Length of surface rupture (miles)	Displacement (feet), main fault (maximum)	Displacements (feet) at indicated distances (miles) from center of main fault zone		Distances (maximum) from center of main zone to outer limits of			Remarks	Principal references
				Branch faulting	Secondary faulting	Main zone	Branch faulting	Secondary faulting		
1	2	3	4	5	6	7	8	9	10	11
11. Gold King, Nevada; 1903(?); N(?)		3+							Possibly 12 mi long. Fault marked by open crack 3 to 5 ft wide. No data available on vertical or horizontal components of displacement. Movement also occurred on this fault in 1954.	Slemmons *et al.*, 1959
12. San Andreas, California; 1906; Rs	8.3	270	20Rs; 3V		2V at 1.5; 0.5Rs at 1.3; 4Ls at 0.3; 1V at 0.2; 4Rs and 2.5V at 0.6	200 ft		1.5 mi	Small cracks in bedrock as much as 10 miles from fault. A tunnel perpendicular to the fault was offset and deformed along nearly a mile of its length; at 4000 ft from the fault the displacement was 14 in.	Lawson *et al.*, 1908; Bonilla, 1967
13. Shelter Cove (San Andreas?), California; 1906; Rs, RN(?)		2+	(?) Rs; 4(?)V	Displacement unknown	Displacement unknown				Right-slip movement indicated by appearance of trace. May be the San Andreas fault itself or a branch or secondary fault 1.5 to 7 mi east of the San Andreas.	Lawson *et al.*, 1908; Curray and Nason, 1967
14. Pleasant Valley, Nevada; 1915; N	7.6	20 to 40	15N	None(?)	3V at 2.5	500 ft		2.5 mi	Northern 5 mi of fault is *en echelon* to principal segment, partly overlaps it, and is $2\frac{1}{2}$ mi perpendicularly from it.	Jones, 1915; Page, 1935; Muller, Ferguson, and Roberts, 1951; Ferguson, Roberts, and Muller, 1952
15. Cedar Mountain, Nevada; 1932; RN	7.3	38	2.8Rs; 4V	See remarks	See remarks				Discontinuous traces scattered over a belt 4 to 9 mi wide and 38 mi long.	Gianella and Callaghan, 1934
16. Excelsior Mountains, Nevada; 1934; N	6.5	0.9	0.4N; slight Ls	None	None					Callaghan and Gianella, 1935
17. Hansel Valley (Kosmo), Utah; 1934; N	6.6	5+	1.7N							Neumann, 1936; Ryall *et al.*, 1966; Eppley, 1965
18. San Jacinto, Mexico; 1934; Rs(?)	7.1								Faulting inferred from aerial photos taken in 1935.	Kovach *et al.*, 1962; Biehler *et al.*, 1964
19. Imperial (El Centro), California; 1940; Rs	7.1	40+	19Rs; 4V	0.08Rs and 0.17V at 0.5	None	300 ft	0.5 mi			Ulrich, 1941; Richter, 1958; J. P. Buwalda, unpublished field notes
20. Vacherie, Louisiana; 1943; N		1	0.7N	None	None				In the Red River fault zone and on the flank of a salt dome. Evidence at surface of an earlier fracture; drilling indicates a vertical separation of $3\frac{1}{2}$ ft at depth. See text.	Fisk, 1944; U. S. Army Corps of Engineers, 1950

Table 3.1. Historic surface faulting in the continental United States and adjacent parts of Mexico (cont.)

Fault (name or location), date, and type of displacement* (See notes at end of table)	Magnitude (Richter) of associated earthquake	Length of surface rupture (miles)	Displacement (feet), main fault (maximum)	Displacements (feet) at indicated distances (miles) from center of main fault zone		Distances (maximum) from center of main zone to outer limits of			Remarks	Principal references
				Branch faulting	Secondary faulting	Main zone	Branch faulting	Secondary faulting		
1	2	3	4	5	6	7	8	9	10	11
21. Manix, California; 1947; Ls	6.4	1	0.25Ls	None	None				Surface faulting may be secondary to concealed right-slip rupture.	Richter, 1958; Allen *et al.*, 1965
22. N. of Bakersfield, California; 1949; N(?)	No quake	2							May be related to subsidence.	Hill, 1954; Allen *et al.*, 1965
23. Fort Sage, California; 1950; N	5.6	5.5	0.6–2N	None	0.25V at 0.25	0.1 mi		0.25 mi	The given distance from the center of the main zone to its outer limits is one-half the perpendicular distance between overlapping *en echelon* segments.	Gianella, 1957
24. Superstition Hills, California; 1951; Rs	5.6	2±							Strike-slip indicated by *en echelon* fractures but amount of displacement unknown.	Dibblee, 1954; Allen *et al.*, 1965
25. White Wolf, California; 1952; LRv and N	7.7	33 (discontinuous)	2.5Ls; 4VRv; 4VN	1Ls at 1.1	0.3N at 8	0.5 mi	1.7 mi	8 mi	Ten feet of shortening measured across main fault zone at one locality. Shaking or regional readjustment of strain produced 0.5 ft vertical faulting for 400 ft along Garlock fault, 20 mi from White Wolf fault.	Buwalda and St. Amand, 1955; Dibblee, 1955; Kupfer *et al.*, 1955; Richter, 1958, pp. 83–84; Whitten, 1955
26. Rainbow Mountain, Nevada; 1954, July; N	6.6	11	1N	None	0.15V at 0.3	0.2 mi		0.3 mi		Tocher, 1956
27. Rainbow Mountain, Nevada; 1954, August; N	6.8	19	2.5N	None	?V at 0.3			0.3 mi	Partly overlaps the July 1954 Rainbow Mountain ruptures and increased the displacement on some of them.	Tocher, 1956
28. Fairview Peak, Nevada; 1954, December; RN	7.1	36	14Rs; 12N	?V at 1.6	3N at 2; 1.5Rs at 2.5; 1.7Rs at 0.6; 1.5N at 3±; 0.5N at 4±	0.5 mi	1.6 mi	4± mi	Produced scarps 16 to 23 ft high. Movement occurred along part of this zone of faulting in 1903 (Gold King fault). Maximum oblique slip was 16 ft.	Slemmons, 1957; Romney, 1957; Steinbrugge and Moran, 1957
29. Dixie Valley, Nevada; 1954, December; N	6.8	38	7 + N (15′ scarp)	None(?)	2N at 1.4; 0.5N at 2.4; 0.2N at 1.5; 0.2N at 2	3000 ft		2.5 mi		Same
30. San Miguel, Mexico; 1956; RN	6.8	12+	3N; 2.6Rs	None	0.75N at 0.4	450 ft		0.5 mi		Shor and Roberts, 1958
31. Fairweather, Alaska; 1958; Rs	8.0	115–124	21.5Rs; 6V		5N at 0.4			0.6 mi	Vertical displacement recorded along 0.25 mi of the fault. Vertical displacement was 3.5 ft where horizontal displacement was 21.5 ft, indicating oblique slip of 21.8 ft.	Tocher, 1960a; Tocher and Miller, 1959

Fault (name or location), date, and type of displacement* (See notes at end of table)	Magnitude (Richter) of associated earthquake	Length of surface rupture (miles)	Displacement (feet), main fault (maximum)	Displacements (feet) at indicated distances (miles) from center of main fault zone		Distances (maximum) from center of main zone to outer limits of			Remarks	Principal references
				Branch faulting	Secondary faulting	Main zone	Branch faulting	Secondary faulting		
1	2	3	4	5	6	7	8	9	10	11
32. Hebgen Lake, Montana; 1959: N	7.1	15±	20N	3N at 3	2.75N at 4.5±; 1N at 4±; 1N at 7.5±; 1V at 8.5±; 0.7V at 8±; 3N at 8	500 ft	3 mi	8.5 mi		Myers and Hamilton, 1964; Witkind, 1964
33. Patton Bay, Alaska; 1964; Rv	8.4	39+	20–23 VRv; 1.4Ls(?); 26± dip slip	None	None	1500 ft			In addition to faulting of 8 ft at one place, distortion of 1 part vertical in 56 parts horizontal occurred within 800 ft of the fault. Magnitude given is for main shock, whose epicenter was more than 75 mi from the surface faulting. Four aftershocks within 50 mi of the faulting had magnitudes ranging from 6.2 to 6.6. Simultaneous faulting occurred 6 mi away (see Hanning Bay fault).	Plafker, 1965; Plafker, 1967
34. Hanning Bay, Alaska; 1964; Rv		4	16VRv	None	None	650 ft			For magnitude see Patton Bay fault, which occurred simultaneously 6 mi away.	Same
35. Imperial, California; 1966, March; Rs	3.6	6	0.05Rs	None	None					Brune and Allen, 1967
36. San Andreas, Parkfield, California; 1966, June; Rs	5.5	23	0.58Rs 0.16V		0.08Rs at 0.85	10 ft			Displacement given includes tectonic creep that occurred within 50 days following main shock. Initial strike-slip displacement unknown at this locality; at another locality strike-slip displacement totaled about 1.8 in. 10 hr after the shock and 4.7 in. 37 days later.	Brown and Vedder, 1967; Wallace and Roth, 1967
37. Buena Vista Hills, California; continuing fault creep; Rv	No quake								Fault creep has been occurring on this reverse fault, without felt earthquakes, for more than 30 years. Total dip-slip displacement 1.6 ft between 1933 and 1958. See text for other localities where creep has occurred.	Koch, 1933; Wilt, 1958

*Abbreviations for type of displacement: Rs, right-slip; Ls, left-slip; N, normal slip (includes vertical faults); RN, right-normal slip; LN, left-normal slip; Rv, reverse (both high angle and low angle); LRv, left-reverse slip; V, vertical (either normal or reverse); VN, vertical displacement on normal fault; VRv, vertical displacement on reverse fault. Query (?) indicates uncertainty as to type, quantity, or identification. Blank spaces in table indicate no reliable data available.

Fig. 3.2. Ranch buildings astride a gravity graben that formed in the 1959 Montana earthquake. The view is from the top of the main scarp looking toward the opposing scarp about 2 ft high. *A* and *D* correspond to *A* and *D* on Fig. 3.1. Part of the collapsed concrete block wall can be seen under the building. Photo by J. R. Stacy, U.S. Geological Survey.

nery, 1961), but the observed vertical movements in strike-slip faulting in California and Alaska have not been described as systematic. The maximum vertical displacements reported were 3 ft for the 1906 faulting (Lawson *et al.*, 1908, p. 147), 4 ft for the 1940 faulting (Buwalda and Richter, 1941), and 6 ft for the 1958 faulting (Tocher, 1960a, p. 276).

The maximum horizontal and vertical movements given in Table 3.1 generally have not been at the same point on the fault.

3.3.4 Map Pattern of Faults

Fault ruptures may consist of a single narrow main break, but commonly they are much more complex (Fig. 3.3) and are accompanied by subsidiary breaks. The following description of the 1906 California faulting (Lawson *et al.*, 1908, p. 53) makes this point:

> The width of the zone of surface rupturing varied usually from a few feet up to 50 feet or more. Not uncommonly there were auxiliary cracks either branching from the main fault-trace obliquely for a few hundred feet or yards, or lying subparallel to it and not, so far as disturbance of the soil indicated, directly connected with it. Where these auxiliary cracks were features of the fault-trace, the zone of surface disturbance which included them frequently had a width of several hundred feet. The displacement appears thus not always to have been confined to a single line of rupture, but to have been distributed over a zone of varying width. Generally, however, the greater part of the dislocation within this zone was confined to the main line of rupture, usually marked by a narrow ridge of heaved and torn sod.

For descriptive purposes it is convenient to classify surface ruptures into three categories or zones. The subsidiary faults can be subdivided into branch faults and secondary faults, the main fault constituting the third category. This classification is illustrated in Fig. 3.4, which shows some of the surface faulting that accompanied the Fairview Peak, Nevada, earthquake of 1954. The main fault and closely associated faults which, at a map scale of 1:250,000, form a band of varied width, constitute zone I. For this classification the fault with the greatest displacement, length, and continuity at the surface is considered the main fault for a particular episode of faulting. Some of the main surface faults (e.g., Patton Bay, Alaska, fault of 1964) actually may be subsidiary to a concealed principal fault that is more directly related to the earthquake-generating process. Zone II contains the branch faults; these diverge from and extend well beyond the main zone of faults. They either join the main fault at the surface or can reasonably

Fig. 3.3. Part of the main zone of faulting along the San Andreas fault in 1906. The main trace passes through the center of the photo, and another strand passes through the notch in the skyline to the right of the photo center. The dashed line is drawn to left of the fault traces; the dotted line indicates the inferred position of the traces. The maximum horizontal distance between the lines (on the ridge crest) is 90 ft. Photo by G. K. Gilbert, U.S. Geological Survey.

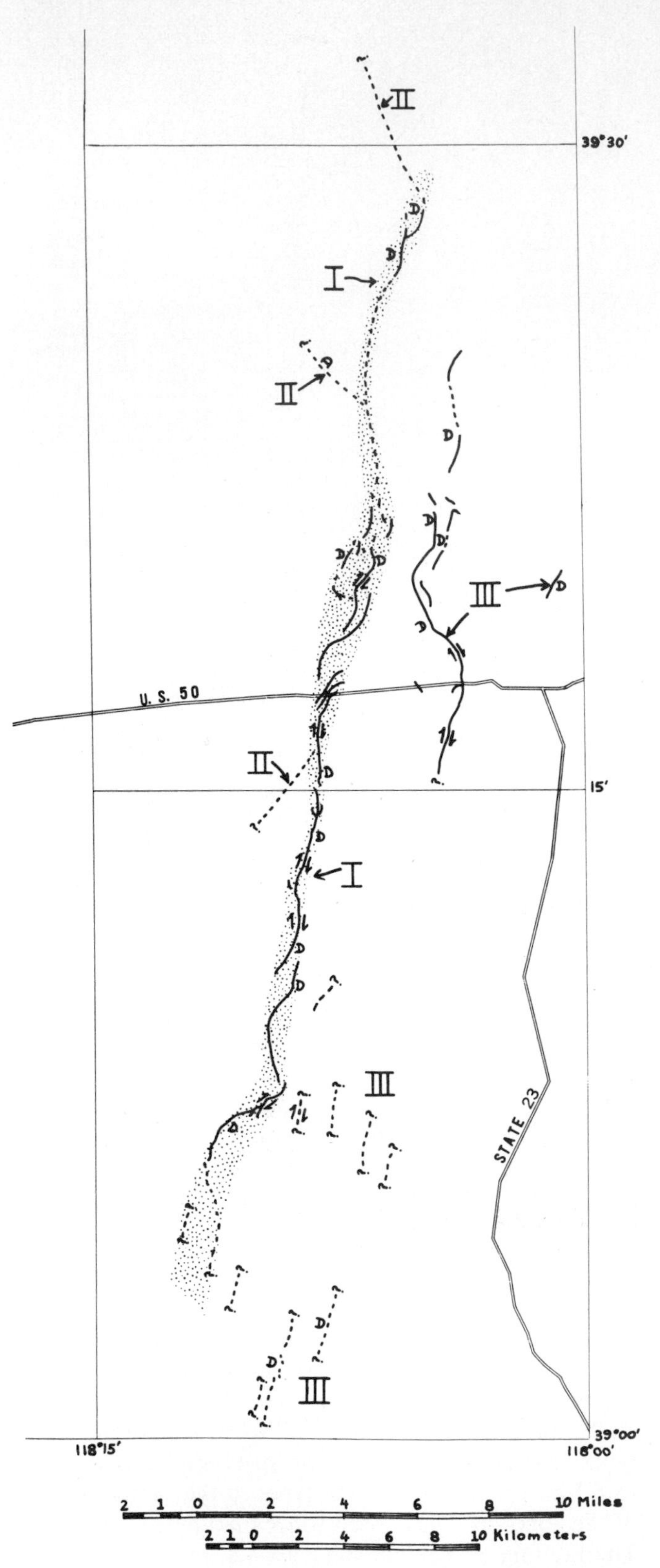

Fig. 3.4. Map of part of the Fairview Peak, Nevada, 1954, faulting, showing main fault zone (I), branch faults (II), and secondary faults (III). The dashed lines indicate faults seen from a distance or interpreted from aerial photographs; the query (?) indicates that the end of break was not determined; *D* indicates the downthrown side; the single-barbed arrow indicates the relative horizontal displacement. Modified after Slemmons, 1957.

be inferred to do so underground. The distinction between the main fault and branch faults is, of necessity, somewhat arbitrary and often difficult to make. The secondary faults that make up zone III have no surface connection with the main fault.

Although the concept of zones is useful, it is not applicable to all historic surface faulting. In the Cedar Mountain, Nevada, faulting of 1932, e.g., the surface ruptures were widely scattered and there was no single continuous main fault. Another example is the Yakutat Bay, Alaska, faulting of 1899 in which several large faults were postulated but no main fault has yet been identified.

3.3.5 Subsidiary Faults

At least half of the historic faulting events in North America have included subsidiary faulting, and the proportion is probably even greater because in only about one-sixth of these events is there good evidence that it did not occur. The importance of subsidiary faulting is indicated in Fig. 3.5, which shows the cumulative lengths of the main and subsidiary surface faults for 15 events. For some of these events the cumulative length of the subsidiary faulting was less than 5%, and for others more than 95%, of the length of the main fault.

The displacements on subsidiary faults can be substantial, even at some distance from the main fault, as shown in Fig. 3.6—a figure based on the data in Table 3.1. The occurrence of displacements of one to a few feet at distances as great as 8.5 mi from the main fault is worthy of note. The distances given in columns 5 through 9 of Table 3.1 were measured at right angles to the trend of the main fault from its approximate centerline. The distances given in columns 5 and 6 are to points where the displacement was actually measured or estimated by the investigator; the corresponding distances in columns 8 and 9 are generally greater, because they were measured to the most distant parts of the branch or secondary ruptures.

The data of Fig. 3.6 have been replotted on Fig. 3.7 with the displacements on the faults expressed as percentages of the displacements on the corresponding main fault. A curve that includes all but three of the data points below it has been sketched on the graph. The curve crosses the 20% line at a distance of 3 or 4 mi from the main fault and decreases at a low rate beyond that, but of course the curve could be drawn in other ways also. The part of the curve to the left of the 1-mi line represents an inference as to the maximum displacement that might occur. The three points above the sketched curve may not be in the correct positions with regard to the amount of displacement during the respective earthquakes. The upper pair were not measured until many years after the 1872 Owens Valley, California,

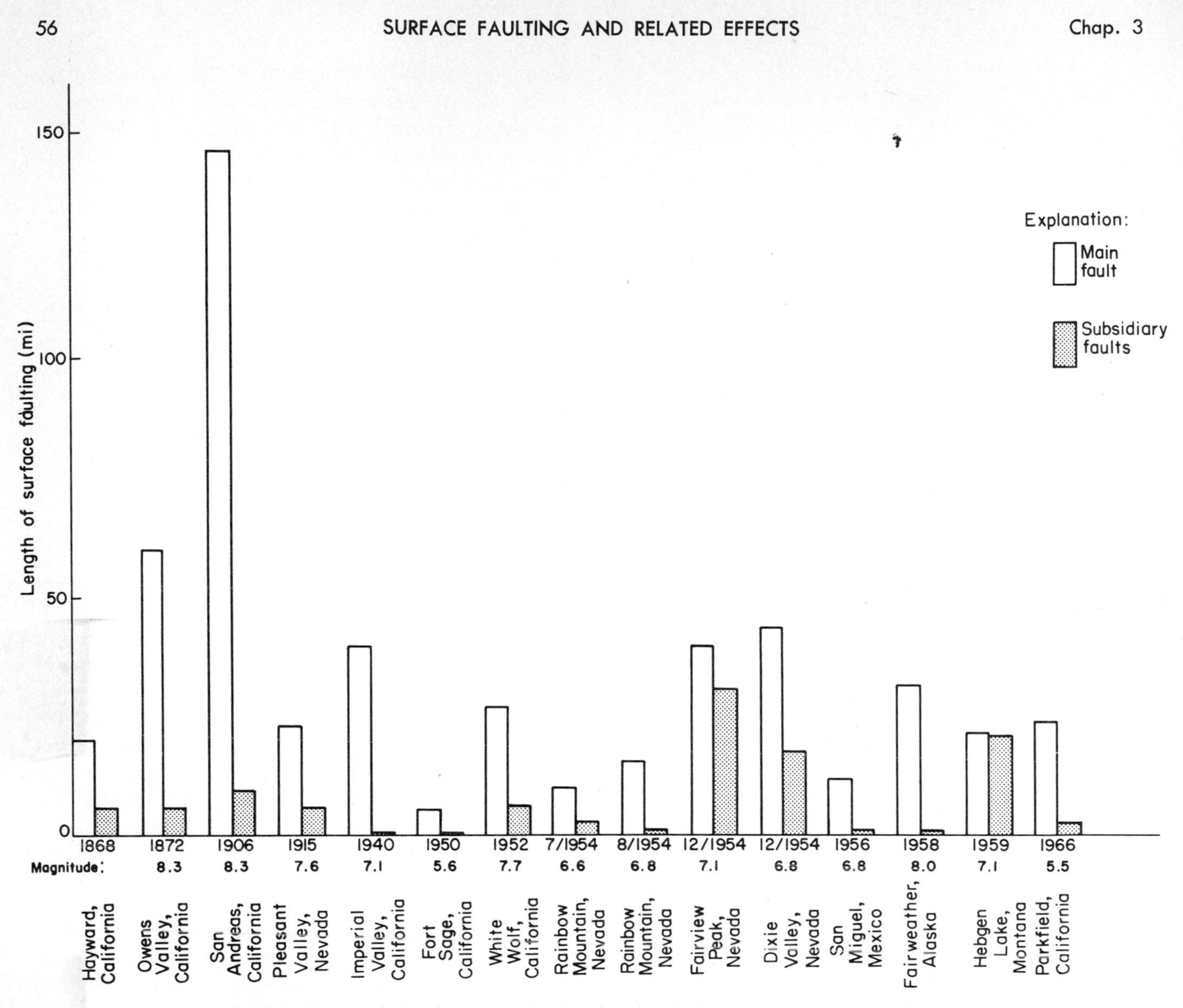

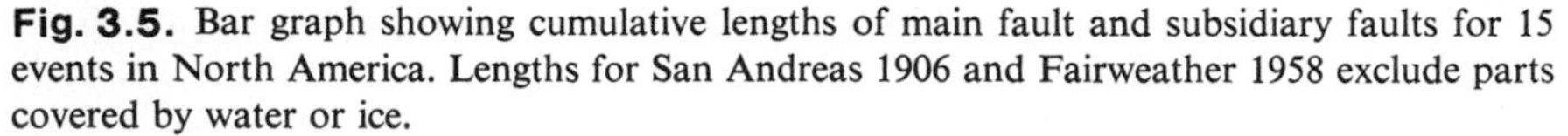
Fig. 3.5. Bar graph showing cumulative lengths of main fault and subsidiary faults for 15 events in North America. Lengths for San Andreas 1906 and Fairweather 1958 exclude parts covered by water or ice.

earthquake, and the lower one is based on the accounts of residents who experienced the 1868 Hayward, California, earthquake.

3.3.6 Width of Zones of Faulting

The maximum distances from the centerline of the main zone of faulting to the outer edges of the main, branch, and secondary zones of faulting are plotted against earthquake magnitude (Richter) in Fig. 3.8. The correlation between magnitude and distance to the outer edges of the zones is very poor. The figure serves to illustrate, however, that the maximum widths of the three zones differ among the four types of faults in this sample and that the zones of strike-slip faults are the narrowest. For each type of fault shown in the figure the maximum distance to the outer edge of the three zones is indicated by roman numerals—I for the main zone, II for the zone of branch faults, and III for the zone of secondary faults. The maximum distance to the outer edge of zone I is less than 0.06 mi for strike-slip faults but between 0.5 and 0.6 mi for the other types; for zone II, 0.5 mi for strike-slip faults and 1.6 to 3 mi for the other types; for zone III, 1.5 mi for strike-slip faults and 8 to 8.5 mi for the other three types in the sample. Some of the zones may have been wider than indicated above. The faulting at Yakutat Bay, Alaska, in 1899 occurred over a broad area, but the main fault has not been identified. Secondary faulting is reported to have occurred 1.8 mi from the Hayward, California, strike-slip fault in 1868 (Lawson *et al.*, 1908, pp. 435 and 444; Radbruch, 1967) but is not shown on the figure because the magnitude is unknown.

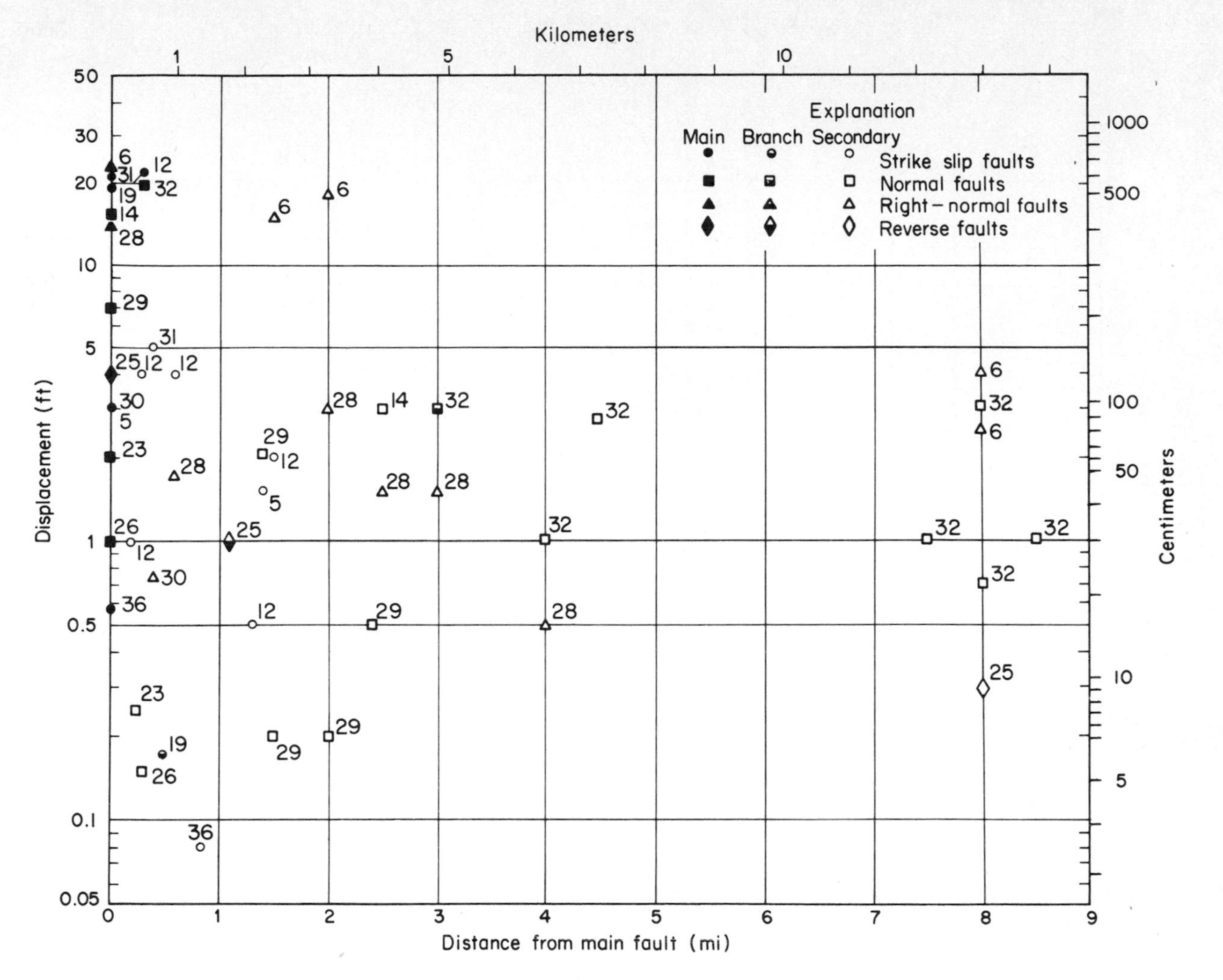

Fig. 3.6. Fault displacement as related to the distance from the main fault. The numbers beside the symbols refer to events listed in Table 3.1.

Fig. 3.7. Fault displacement (in percent of displacement on main fault) as related to distance from the main fault. The numbers beside the symbols refer to events listed in Table 3.1.

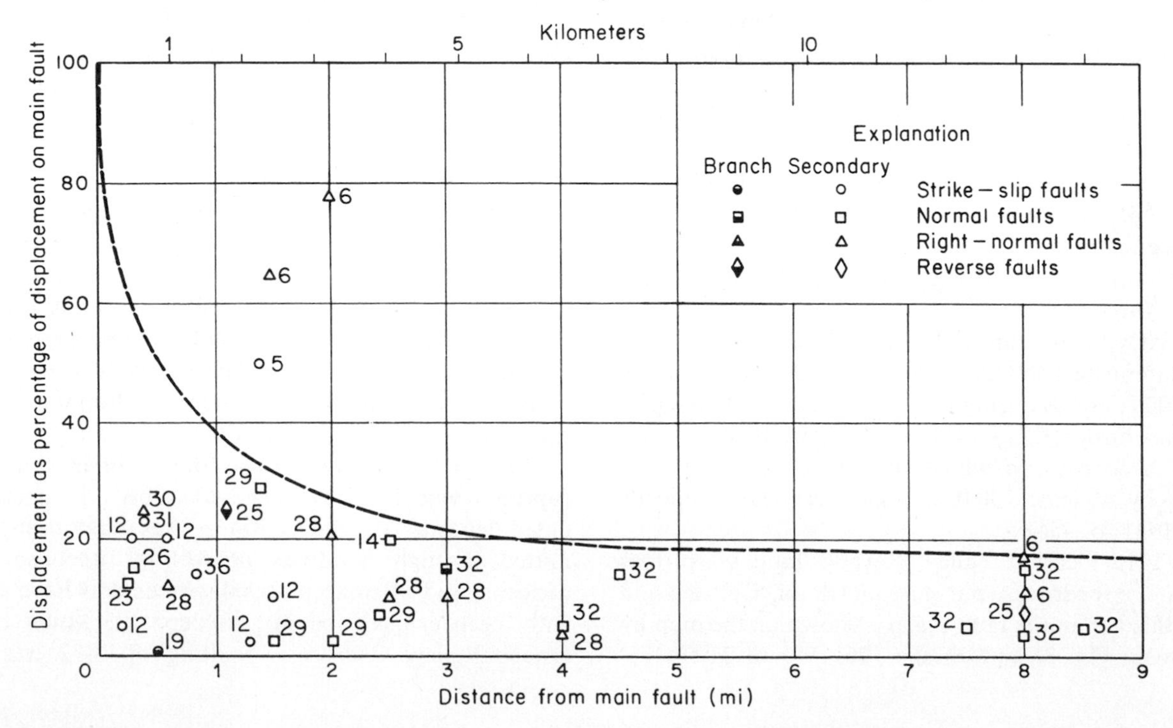

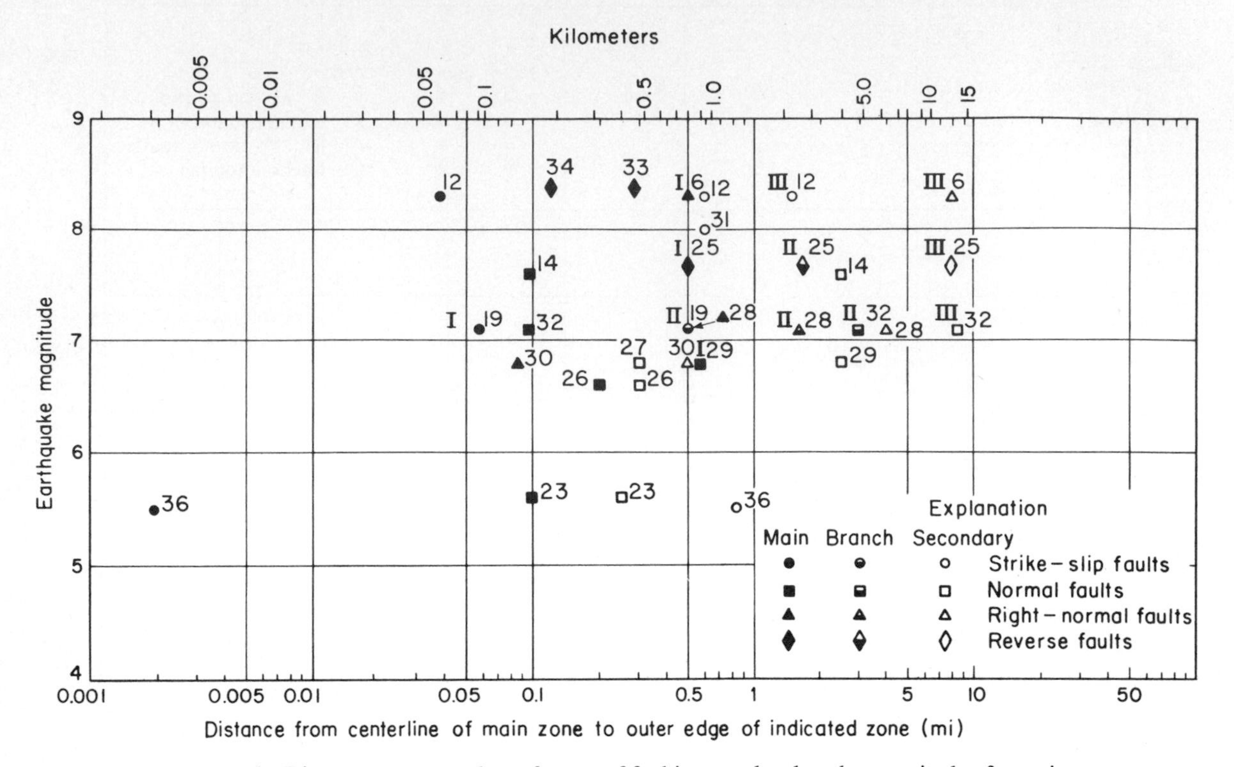

Fig. 3.8. Distances to outer edges of zones of faulting as related to the magnitude of associated earthquakes. The numbers beside the symbols refer to events listed in Table 3.1.

3.3.7 Absorption of Ruptures in Rock and Soil

Fault ruptures can be absorbed (i.e., die out or become indistinguishable) in short distances in rock or soil, but they also can be transmitted through thick deposits of unconsolidated sediments. For example, a part of the White Wolf, California, faulting of 1952 displaced railroad tunnels, but near the surface it seems to have been locally absorbed. Buwalda and St. Amand (1955, p. 48) state: "We have the dilemma that the faults indicated at the tunnels show displacements of at least several feet while the moletracks which are presumably their continuation on the hill above show relatively small offsets both horizontally and vertically." Kupfer and others (1955, p. 74) suggest that fractures conspicuous in rigid concrete might go unnoticed or be distributed and absorbed in the fractured and weathered bedrock and soil near the surface.

Similarly, strike-slip fault displacement of 8 ft in a tunnel diminished to less than 3 ft at the ground surface about 500 ft above during the Idu, Japan, earthquake of 1930 (Suyehiro, 1932, pp. 32–37; Richter, 1958, p. 580). The volcanic rock in which the tunnel was driven is overlain by at least 130 ft of sandy clay lake deposits (Nasu, 1931, p. 456).

The 1915 Pleasant Valley, Nevada, fault scarp does not cross a bedrock spur just north of Cottonwood Creek; this gap in the fault scarp is shown on the map by Page (1935, Fig. 3) and on the 1961 Mount Tobin 15-min topographic map. Just north of the spur the scarp is about 10 ft high but dies out rapidly as it ascends the spur, then reappears to the south; possibly the displacement was taken up by bending rather than distinct faulting. The Red Canyon fault that accompanied the 1959 Montana earthquake accommodated the displacement locally by warping rather than by the usual high single scarp (Myers and Hamilton, 1964, p. 83).

One of the best examples of local absorption of faulting occurred on the Patton Bay, Alaska, fault in 1964 and is shown in Fig. 3.9. Reverse faulting produced a scarp $8\frac{1}{2}$ ft high (*A*, Fig. 3.9) in the gravel-covered bedrock at beach level, but no comparable scarp could be found where the principal trace of the fault cut the top of the sea cliff (*B*, Fig. 3.9). Thus more than 8 ft of displacement was absorbed in rock between points *A* and *B*—a distance of about 700 ft. Scarps behind and parallel to the sea cliff suggest incipient landsliding (Plafker, 1967, p. G13), but only minor sloughing occurred during the earthquake; evidently the faulting was distributed and taken up along the numerous joints and minor faults in the rock.

From the foregoing, one might infer that fault ruptures would generally be absorbed by unconsolidated deposits; on the contrary, they have been transmitted through hundreds of feet of unconsolidated deposits, and in some places displacements have apparently been exaggerated in soft deposits. Much of the Owens Valley, California, faulting of 1872 was near

Fig. 3.9. Patton Bay, Alaska, faulting of 1964 produced a scarp $8\frac{1}{2}$ ft high at *A* but no comparable scarp at *B*. The cliff is about 500 ft high, and points *A* and *B* are about 700 ft apart; note the helicopter in foreground.

the center of the valley—an area underlain by 500 ft or more of unconsolidated to semiconsolidated alluvium and lake deposits. The water table there was less than 10 ft below the surface in 1909 (Lee, 1912, pp. 72–74) and probably just as shallow in 1872. Similarly, the Imperial Valley, California, faulting of 1940 that involved strike slip of more than 15 ft at the surface was propagated upward through many feet of poorly consolidated deposits. Logs of old water wells in Holtville and El Centro, about 5 mi east and west, respectively, of the 1940 fault, describe the sediments to depths in excess of 700 ft as clay, sand, and soil (Hutchins, 1914, pp. 213–222). A log of a boring in El Centro indicates a depth of 6 ft to groundwater in 1946. This log shows loam and clay to a depth of 100 ft (Duke and Leeds, 1962, Station Data Sheet 64). The sedimentary deposits through which the faulting was propagated are believed to be similar, in thickness and kind, to those described in the logs.

3.3.8 Extension and Compression

In addition to the shearing displacements that have been discussed above, surface faulting is often accompanied by extension or compression approximately perpendicular to the fault. An example of extension is shown by the fracture (Fig. 3.10) which formed 15 ft from the Patton Bay, Alaska, fault scarp. Numerous fractures of this kind formed in bedrock on the upthrown sides of the Patton Bay and Hanning Bay faults; they were as much as 0.4 ft wide and 200 ft long and were found as much as 1000 ft from the fault scarps (Plafker, 1967, pp. G7–G13 and G34–G35). Open fractures of much larger size commonly accompany normal faulting, because of a change in dip of the fault near the ground surface. An example is a fracture 9 ft wide on the Pleasant Valley, Nevada, fault (Jones, 1915, p. 203 and Fig. 10); another example is an open fracture about 10 ft wide on the Fairview Peak, Nevada, fault (Slemmons, 1957, Fig. 15). Open fractures are sometimes associated with strike-slip faulting, usually at or close to the main fault trace, and are arranged in an *en echelon* pattern. Some of the fractures close immediately, as was the case when a cow fell into a wide fracture along the San Andreas fault in 1906 and was entombed when the crack closed again (Lawson *et al.*, 1908, p. 72).

Fault sags, which are common along strike-slip faults, probably are caused in part by extension transverse to the fault, permitting the settlement of blocks bounded by faults. Many fault sags along the San Andreas fault were deepened in 1906, generally only a few inches but locally as much as 2 ft (Lawson *et al.*, 1908, pp. 32–33, 67, 69, 72–73). Fault sags are found 0.5 to 1 mi from the 1906 trace (Lawson *et al.*, 1908, pp. 33, 75; Higgins, 1961, p. 57).

Compression transverse to the fault occurred, at least locally, in the 1906 California earthquake. For example, at a place where a road perpendicular to the fault was severed and displaced 8 ft, board fences on each side of the road were broken, the boards overlapped, and the adjacent telephone wires sagged, indicating compression perpendicular to the fault (Lawson and others, 1908, p. 102).

Damage to structures by extension and compression can and has occurred on strike-slip faults without net

Fig. 3.10. Fracture produced in bedrock near the 1964 Patton Bay, Alaska, fault scarp. The fracture 1.6 in. wide follows a pre-existing tight mineralized joint.

extension or shortening normal to the fault. This occurs where a structure crosses the fault obliquely and the ends of the structure are brought closer together or pulled farther apart as the walls of the fault move. Structures crossing a right-slip fault obliquely from right to left (observer looking along the fault) will be lengthened, and structures crossing from left to right will be shortened; the reverse is true for left-slip faults (Reid, 1910, pp. 33–34). Many examples of this effect were noted in 1906 where the San Francisco aqueduct crossed the fault and was pulled apart or telescoped (Lawson and others, 1908).

3.3.9 Tilting, Warping, and Level Changes

Tilting, warping, and changes in elevation can seriously affect canals and shoreline facilities of various kinds by changing their relation to water level. The movements may be restricted to local areas adjacent to a fault or they may affect thousands of square miles. Figure 3.11 shows an example of a large shift in the shoreline as a result of tilting and subsidence of the Hebgen Lake basin in the 1959 Montana earthquake. The tilting extended 5 mi or more from the Hebgen fault scarp. In places the tilting ended not against a fault scarp but against a zone of warping in which 9 ft of vertical change occurred in a horizontal distance of about 650 ft (1 part in 72) without recognized faulting (Myers and Hamilton, 1964, pp. 81–82 Plate 2).

The Yakutat Bay, Alaska, earthquake of 1899 was accompanied by both widespread elevation and local depression of the shoreline. The maximum uplift relative to sea level was more than 47 ft; in places substantial warping occurred, reaching 1 part vertically to 360 parts horizontally between points 2400 ft apart on the west shore of Disenchantment Bay (Tarr and Martin, 1912, Plate 14). Even steeper warping (1 in 56) occurred in the 1964 Alaskan earthquake within about 800 ft of the Patton Bay fault, and similar warping occurred near the Hanning Bay fault (Plafker, 1967, Fig. 2, pp. G7, G35, and Plate 1, Section A-A′), producing only the open fractures described in Section 3.3.8.

Fig. 3.11. Emergence of shoreline as a result of tilting of the Hebgen Lake basin in the 1959 Montana earthquake. Photo by J. R. Stacy, U.S. Geological Survey.

Regional tectonic movements accompanying large earthquakes have produced changes in level (uplift or depression) over very large areas. The 1899 Yakutat Bay uplift and subsidence has already been mentioned. Similar movements affected possibly 110,000 mi² in the 1964 Alaskan earthquake, producing uplifts of as much as 38 ft and downwarps of more than 7 ft (Plafker, 1967, pp. G2–G4). A somewhat lesser area was affected in the 1960 Chilean earthquake where uplift of $2\frac{1}{2}$ m and subsidence of 2 m has been reported (Saint-Amand, 1963, p. 350); recent work by Plafker (written communication, 1968) shows that the maximum uplift and subsidence were 5.7 and 2.7 m, respectively and that more than 75,000 mi² were affected. The New Madrid, Missouri, earthquakes of 1811–1812 were accompanied by uplift and depression that produced Reelfoot Lake and enlarged St. Francis Lake; Reelfoot Lake is 8–10 mi long, 2–3 mi wide, and at least 20 ft deep (Fuller, 1912, p. 73).

3.3.10 Fault Creep

A description of the process of fault creep or slippage and some of its theoretical and practical consequences is given in Section 2.3, and only a few additional comments will be made here. As noted in the glossary, fault creep as used in this chapter is not necessarily limited to tectonic movements. Withdrawal of petroleum, water, sulfur, salt, or other substances can result in surface subsidence accompanied by extensional and compressional movements on faults. Where this occurs in an area that may be tectonically active it is sometimes extremely difficult to separate natural and artificial causes of fault creep. Examples of such areas are along the Casa Loma and San Jacinto faults, California (Fett, Hamilton, and Fleming, 1967); in parts of the city of Hollister, California, adjacent to the Calaveras fault (Rogers and Nason, 1967, p. 102); along the Buena Vista Hills, California, fault (Allen *et al.*, 1965, pp. 765–766; Whitten, 1961, pp. 318–319; Whitten, 1966, pp. 72–76); and near the Baldwin Hills Reservoir, California, faulting of 1963 (Kresse, 1966). Thus a general term such as fault creep is useful for those situations where the relative importance of tectonic creep and artificially induced creep is unknown.

In addition to the areas of fault creep cited in Chapter 2, several other areas can be mentioned. Tectonic creep has occurred on the San Andreas fault between the winery at Vineyard (Tocher, 1960b) and the Parkfield-Cholame area, a distance of about 90 mi (Brown and Wallace, 1968), as well as north of the winery (Tocher, 1966) and near San Juan Bautista (Rogers and Nason, 1967). Additional areas of fault creep or probable creep also have been found on the Hayward fault as far north as Richmond, the Calaveras fault near Gilroy, and the Pleasanton fault near Pleasanton, California (Radbruch, 1968; Gibson and Wollenberg, 1968).

Movements suggestive of tectonic creep occurred at least locally prior to the 1959 Montana earthquake. At the Madison Fork Ranch, 8.5 mi from the Red Canyon fault, several prequake scarps showed new movements ranging from a few inches to 1 ft. A lodge built across the projection of one of the scarps was being slowly deformed before the earthquake (Myers and Hamilton, 1964, p. 60), which strongly suggests that tectonic creep was active across this normal fault. In addition to discrete faulting, local warping in this vicinity affected a stream, ditch, and the local runoff pattern.

Fault creep also has occurred at various locations in Texas, where movements on faults have damaged roads, buildings, pipelines, and other structures (Bryan, 1933, p. 439; Sheets, 1947, p. 216; Bell and Brill, 1938; Weaver and Sheets, 1962; Wiggins, 1954, p. 308). Some of these movements are undoubtedly related to the withdrawal of fluids or to secondary effects related to the presence of salt domes, but some probably are tectonic (Weaver and Sheets, 1962, p. 254; Russell, 1957, p. 69).

Tectonic creep at a rate probably greater than normal is known to have directly followed some faulting and is inferred in other instances. Postquake creep in the Parkfield-Cholame, California, area continued for many weeks after the June 27, 1966, earthquake. Measurements started shortly after the earthquake showed that in 2 weeks about 2 in. of creep occurred at a rapidly decreasing rate and then continued slowly (Wallace and Roth, 1967, Fig. 25). A similar pattern of postquake creep at a rapidly decreasing rate occurred on the Tanna, Japan, fault in 1930 and 1931. However the postquake creep there was less than 0.5 mm (0.02 in.) in the first 2 weeks after the earthquake (Takahasi, 1931, Fig. 10). Postquake creep probably accounts for the following: the fault movement noted after the 1962 Iran earthquake (Ambraseys, 1965a, pp. V-7, V-10); several inches of progressive overlapping of the boards of a broken fence on the Hayward, California, fault in 1868 (Lawson *et al.*, 1908, p. 442); small movements on the 1940 Imperial, California, fault (Richter, 1958, pp. 74–75); and movements at four locations on the White Wolf, California, fault in 1952 (Buwalda and Saint Amand, 1955, pp. 46, 48, 49; Kupfer *et al.*, 1955, p. 68). The postquake creep in the Parkfield-Cholame area exceeded the amount of the initial rupture, but in the other cases it was only a fraction of the initial rupture.

The foregoing examples show that fault creep is a widespread phenomenon that should be considered, along with the possibility of sudden rupture, in planning engineering structures on or near faults. Although not spectacular, it is persistent and capable of causing damage to some kinds of structures. Long-term rates of tectonic creep have ranged from about 0.1 in./year on the Hayward fault (see Chapter 2) to about 0.8 in./year for a long segment of the San Andreas fault (Brown and Wallace, 1968). Most reported fault creep has been concentrated in single narrow zones of a few tens of feet or less in width, but some seems to be distributed in zones as much as 500 ft wide (Brown and Wallace, 1968) or is in parallel overlapping *en echelon* zones more than 175 ft apart (Radbruch, 1968, p. 50; Nason, 1968, p. 87).

3.4 GEOGRAPHIC DISTRIBUTION OF HISTORIC SURFACE FAULTING

In North America nearly all of the historic faulting has been in the western part of the continent, as shown in Figs. 3.12 and 3.13 and in Fig. 2.2. The faults shown on Figs. 3.12 and 3.13 can be identified by the numbers, which are keyed to Table 3.1.

Faulting at the surface has not been unequivocally

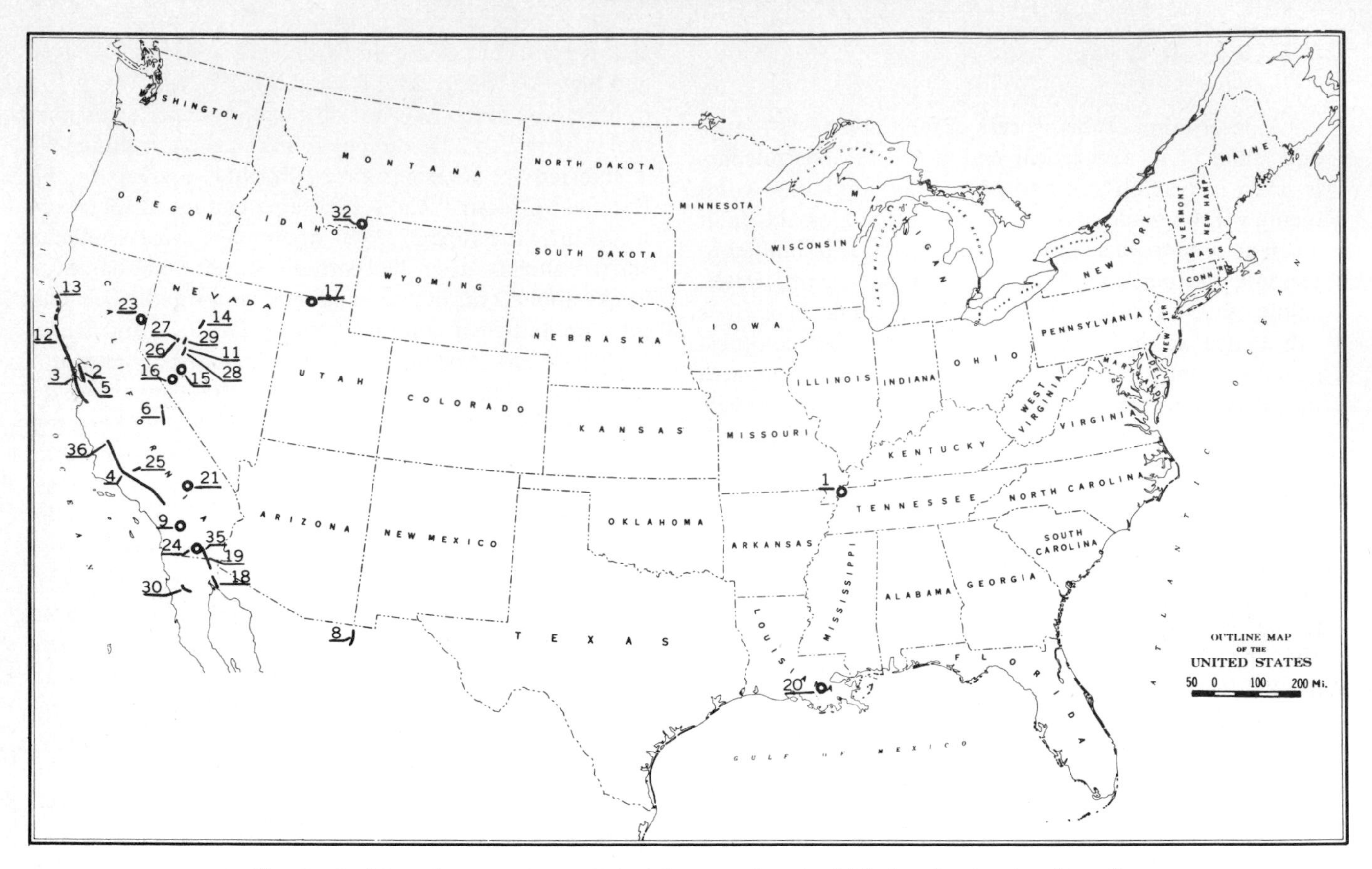

Fig. 3.12. Map of conterminous United States and part of Mexico showing location of historic surface faulting. Numbers identify faults (see Table 3.1).

established for the great New Madrid, Missouri, earthquakes of 1811–1812 (number 1 in table 3.1), but the available evidence strongly suggests that it did occur. Historic accounts mention the formation of both barriers and waterfalls across the Mississippi River near New Madrid; one of the waterfalls was estimated to be 6 ft high (Fuller, 1912, pp. 58, 59 and 62). Reelfoot Lake, which formed in the earthquake, is bounded on its southwest side by a fault, one side of which was uplifted while the other side subsided (Fuller, 1912, p. 75; Fisk, 1944, p. 25 and Fig. 33; U.S. Army Corps of Engineers, 1950, pp. 6–11). This fault extends below the surficial sediments, and borings show a vertical separation of 40 ft in Eocene beds 160 ft below the surface (U.S. Army Corps of Engineers, 1950, Fig. 4). Other areas that sank or rose during the earthquake also may be bounded by faults, but no definite information about them is available.

Faults that are expressed in the present topography are found in several parts of the lower Mississippi Valley (Fisk, 1944; U.S. Army Corps of Engineers, 1950; Veatch, 1906), and faulting of a Pleistocene terrace in the nearby southern part of Illinois has been reported by Ross (1963). This is a seismic region that experienced other great earthquakes prior to 1811 (Fuller, 1912, pp. 12–13) and has had many small to moderate earthquakes since then (Heinrich, 1941; U.S. Army Corps of Engineers, 1950, pp. A9–A17; Wollard, 1958; Heyl and Brock, 1961, p. D-4).

The tectonic origin of the Vacherie, Louisiana, faulting of 1943 (number 20 in Table 3.1) remains in doubt. This fault movement was accompanied by a small earthquake felt locally. The nearest seismograph, which was 50 mi away and designed to record large distant shocks, did not record the earthquake. The initial displacement was 3 in. but it increased to about 8 in. in the next 24 hr (U.S. Army Corps of Engineers, 1950, pp. A34–A37; Fisk, 1944, p. 33). The area is on the flank of a salt dome and is also in the Red River fault zone (Fisk, 1944, p. 33). A well being drilled nearby encountered a strong flow of water under 2000 lb of pressure at a depth of 8800 ft shortly before the surface faulting occurred. This flow suggested a possible cause-and-effect relation, but prior movement also had occurred on this fault. Drilling revealed 3.5 ft of vertical separation of upper Pleistocene sediments at a depth of 55 ft (Fisk, 1944, Plate 17).

Several other episodes of surface faulting or probable surface faulting are not shown on the map or table because little is known of them. Most of these have been on the San Andreas fault system in California and include episodes at Dos Palmas (near Salton) in 1868 (Townley and Allen, 1939, p. 50); near Chittenden, in

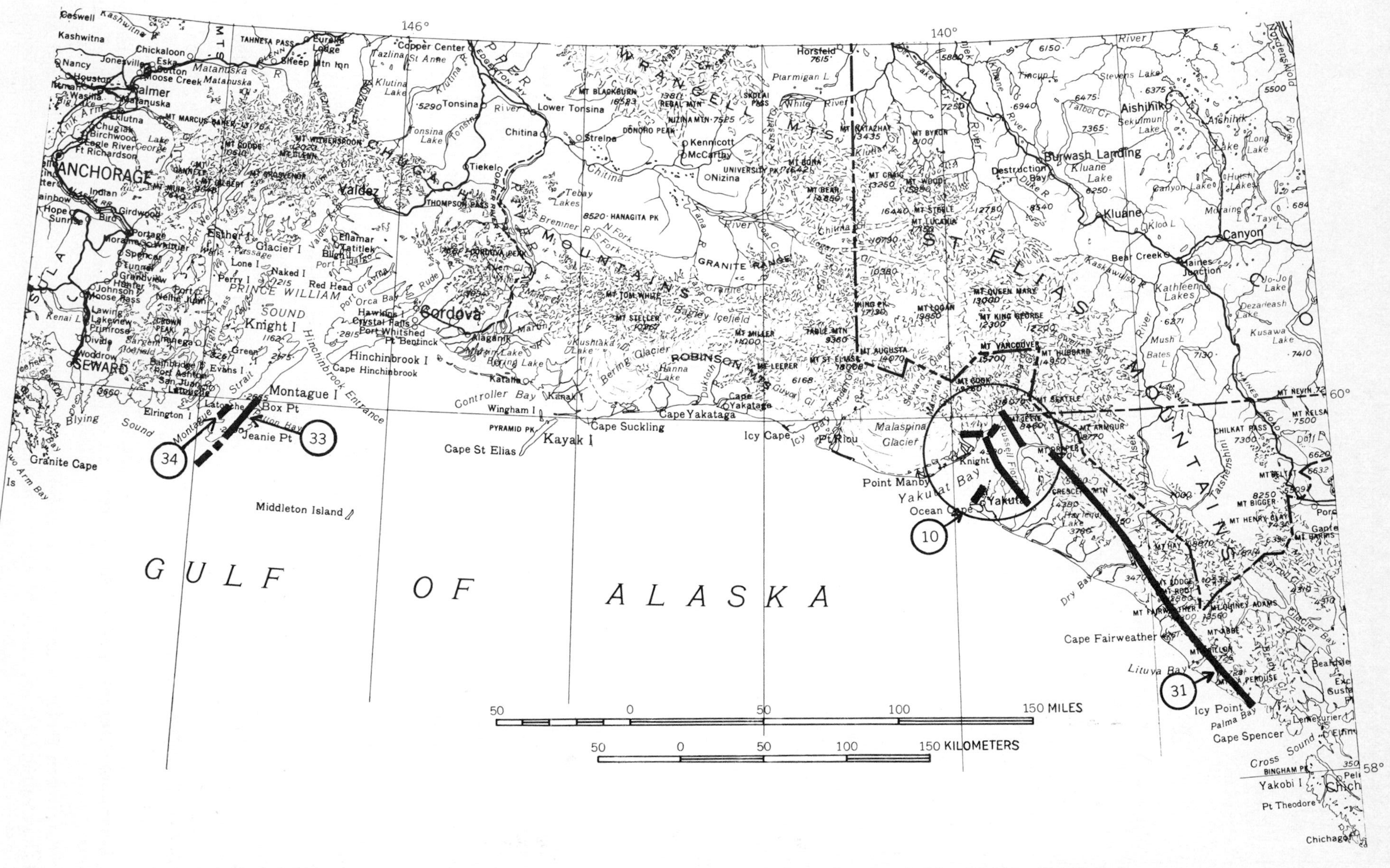

Fig. 3.13. Map of part of Alaska showing the location and approximate extent of historic surface faulting. Numbers identify faults (see Table 3.1).

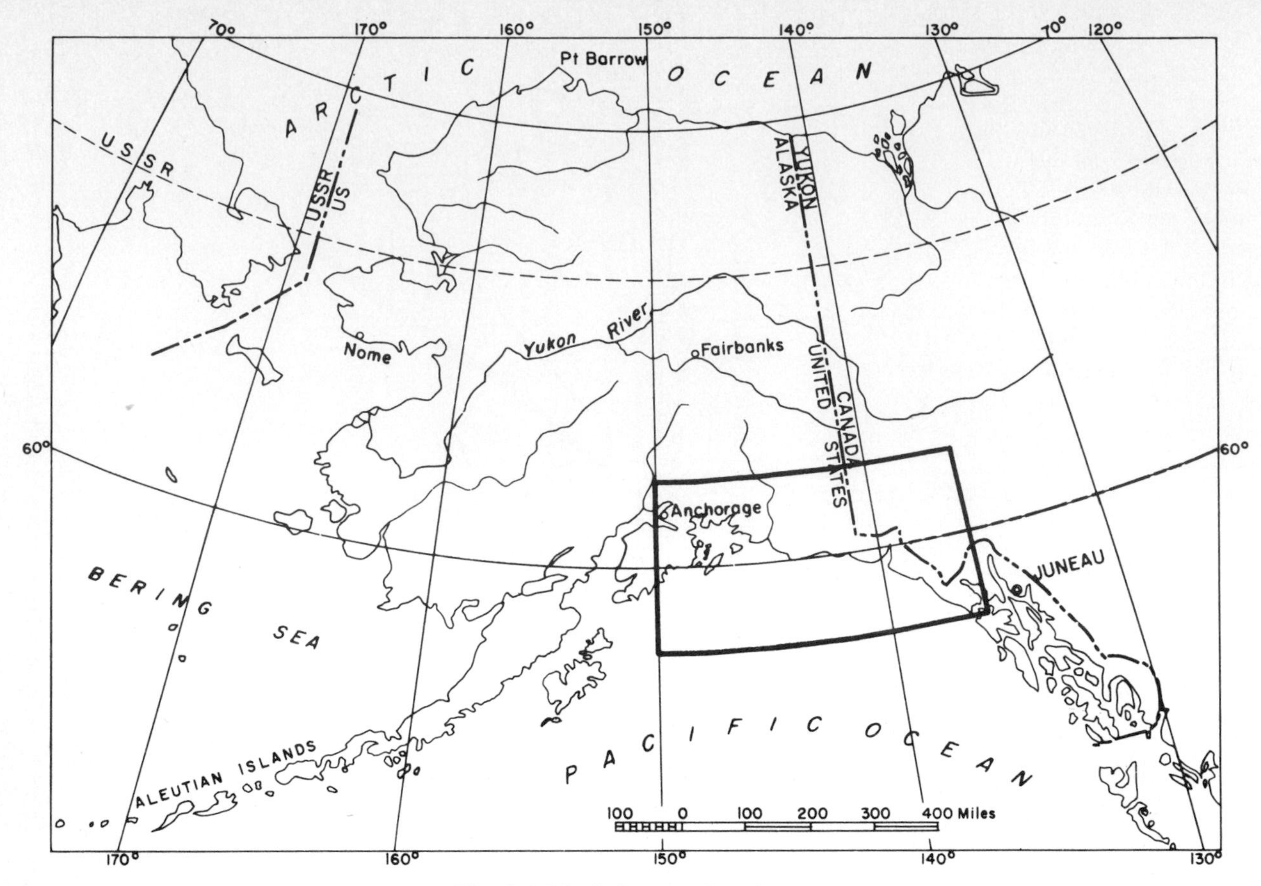

Fig. 3.13A. Index map for Fig. 3.13.

1890 (Lawson and others, 1908, p. 449); near Parkfield, in 1901, 1922, and 1934 (McEvilly, 1966, p. 970; Brown and Vedder, 1967, pp. 9–10); and possibly near Vineyard in 1961. Faulting or possible faulting also has been reported on the Calaveras fault in 1861 near Dublin, California (Radbruch, 1968, pp. 52–53); in 1852 on the Big Pine fault, California (Vedder and Brown, 1968, p. 256); in 1875 in the Mohawk Valley, California (Gianella, 1957, p. 177); and in 1869 on the Olinghouse fault zone, Nevada (Slemmons, 1967, Table 1 and Fig. 2).

Historic surface faulting has occurred in many places outside North America, but a detailed treatment of it is beyond the scope of this chapter. Faulting has occurred at least once in the following places: Argentina, Bulgaria, Greece, Hawaii, Japan, India, Iran, Kenya, Mongolia, New Zealand, Pakistan, Peru, Sudan, Sumatra, Taiwan, Turkey, and perhaps Yugoslavia. Nearly all of it occurred in the seismically active areas that are apparent in Fig. 2.1. A summary of most of this faulting is given by Richter (1958).

3.5 FAULTING AND EARTHQUAKES

3.5.1 Earthquake Intensity Near the Fault

Most American geologists and seismologists believe that shallow earthquakes are caused by elastic rebound (see Chapter 2) occurring at faults. This theory leads to the conclusion that shaking effects should be great near the fault and decrease away from it; experience shows this to be true in a general way. It does not necessarily follow, however, that the intensity of shaking rises to a high peak right at the fault. This idea was developed by Louderback (1942), who pointed out that the source of the earthquake waves, at least of strong earthquakes, was likely to be some miles beneath the surface and that the energy reaching the surface would be about the same in a moderately wide zone along the fault. Housner (Chapter 4) suggests that the accelerations decrease at a slow rate for about the same distance from the fault as the vertical dimension of the fault rupture. Accelerometer records for one earthquake (Parkfield, California, 1966) support this suggestion; they show very little diminution of maximum acceleration within 4 mi of the fault and rapid decrease beyond that. The records show 0.5 g at 270 ft from the fault, 0.46 g at 3.3 mi, 0.4 g at 4 mi, and 0.28 g at 5.7 mi (Cloud and Perez, 1967, Fig. 10). The vertical extent of the faulting has been inferred to be on the order of 6 mi or less (McEvilly, Bakun, and Casaday, 1967, p. 1240) for this earthquake. The lack of markedly greater shaking damage to structures adjacent to the surface trace of faults has been reported by several investigators (Jones, 1915, p. 195; Gianella and Callaghan, 1934, p. 367; Louderback, 1942, pp. 316–319; Steinbrugge and Cloud, 1962, p. 231; and Ambraseys, 1963, p. 735); note, however, the qualifications regarding the Hebgen Lake, Montana, earthquake given by Steinbrugge (Chapter 9 of this volume).

The examples cited in the reports listed above indicate that neither multidirectional shaking nor "fling" were effective agents of destruction adjacent to the faults. Fling is the rapid displacement of rock masses to positions of no (or greatly reduced) elastic strain, as postulated in the elastic rebound theory. The displacement should be essentially parallel to the fault. Housner (1965, pp. III-104) concluded from a theoretical analysis that a maximum acceleration on the order of 0.5 g would be produced near the fault, using a differential fault displacement equal to the 1906 San Andreas displacement (about 20 ft) in the analysis. The 0.5 g acceleration that was measured 270 ft from the 1966 Parkfield, California, rupture (fault displacement less than 1 ft at the surface) thus was unexpected; moreover, the displacement pulse was nearly perpendicular rather than parallel to the fault. Seismoscope records from three earthquakes in 1960 and 1961 on the San Andreas fault also show maximum motion at a high angle to the fault (Cloud, 1967, p. 1446). At least locally, displacement pulses directed at a high angle to the fault also occurred in 1906 (Lawson *et al.*, 1908, p. 192). Some short-duration but damaging earthquakes such as the ones at Port Hueneme, California (Housner and Hudson, 1958), Agadir, Morocco (American Iron and Steel Institute, 1962, pp. 81–82) and Skopje, Yugoslavia (Berg, 1964, p. 33; Ambraseys, 1965b, p. S23) have included unidirectional pulses possibly related to fling. Neither the importance nor even the existence of fling is universally accepted, and more facts are clearly needed about it. (See also Chapters 1 and 2 in this volume.)

3.5.2 Nonseismic Faults

Louderback (1942, p. 328) reasoned that some active faults, because of their small size or shallowness or because they cut incompetent rocks (even though long and deep), may produce only slight or unfelt earthquakes. Paterson (1958, p. 473), on the basis of laboratory work on marble, suggested that earthquakes would not be associated with faults that develop in calcite rocks. The discovery of fault creep at many places since publication of the reports by Louderback and Paterson leaves no doubt that fault displacement can occur at the surface without felt earthquakes. However, the Imperial, California, 1966 earthquake (magnitude 3.6, modified Mercalli intensity V) had its source "within the soft sedimentary section" (Brune and Allen, 1967, p. 512), and the Hebgen Lake, Montana, faults of 1959 cut through formations containing a large proportion of calcite rocks (Witkind, Hadley, and Nelson, 1964, Plate 5, p. 201). Furthermore, most of the fault segments affected by tectonic creep have had moderate-to-strong earthquakes (e.g., the Hayward fault, 1868) along them in the historic past. Some historic fault movement has occurred on faults on which no historic earthquakes have been reported (see Section 3.3.10 and numbers 22 and 37 in Table 3.1), but the tectonic origin of those movements is uncertain. In the present state of knowledge, it does not seem prudent to conclude that a given active fault, because of the kinds of rocks it cuts near the surface or the occurrence of tectonic creep along it, will not produce damaging earthquakes.

3.5.3 Relation of Fault Displacement and Length at the Surface to Earthquake Magnitude

Figures 3.14 and 3.15 show the relation between the maximum displacement on the main fault at the ground surface and the magnitude of the associated earthquake. The displacement generally increases as the magnitude increases but with considerable scatter of individual points. A line of best fit (*A* in Fig. 3.14) for all the points has been obtained by the method of least squares, yielding the equation

$$\log D = 0.57\,M - 3.39 \qquad (3.1)$$

in which D is the maximum displacement in feet and M is the Richter magnitude. The line of best fit for strike-slip faults alone (not shown on graphs) is almost the same as the line for all the faults; the line for normal faults has a somewhat higher slope than the line for all faults. Inasmuch as only a small number of points are presently available for each of the various types of faults and the best-fit lines are not greatly different, all types have been combined in the calculations.

Another line (*B* in Fig. 3.14) that includes the largest displacements for all the faults has been drawn parallel to the line of best fit. Its equation is

$$\log D = 0.57\,M - 2.67 \qquad (3.2)$$

Still another line (*C* in Fig. 3.14) corresponding to line *B* has been drawn on the other side of line *A*, making the separation between lines *A* and *C* the same as between *A* and *B*. The equation for line *C* is

$$\log D = 0.57\,M - 4.11 \qquad (3.3)$$

Line *C* bounds all but one of the smallest displacements. The excluded point (number 21) represents the Manix, California, faulting of 1947. Richter has suggested that the surface faulting at Manix was secondary to a concealed main rupture (Richter, 1958, pp. 517–518; Allen *et al.*, 1965, p. 768). A larger displacement would shift point 21 closer to, or perhaps to the other side of, line *C*.

Lines *A*, *B*, and *C* on Figs. 3.14 and 3.15 and the corresponding Eqs. 3.1, 3.2, and 3.3 can be used to esti-

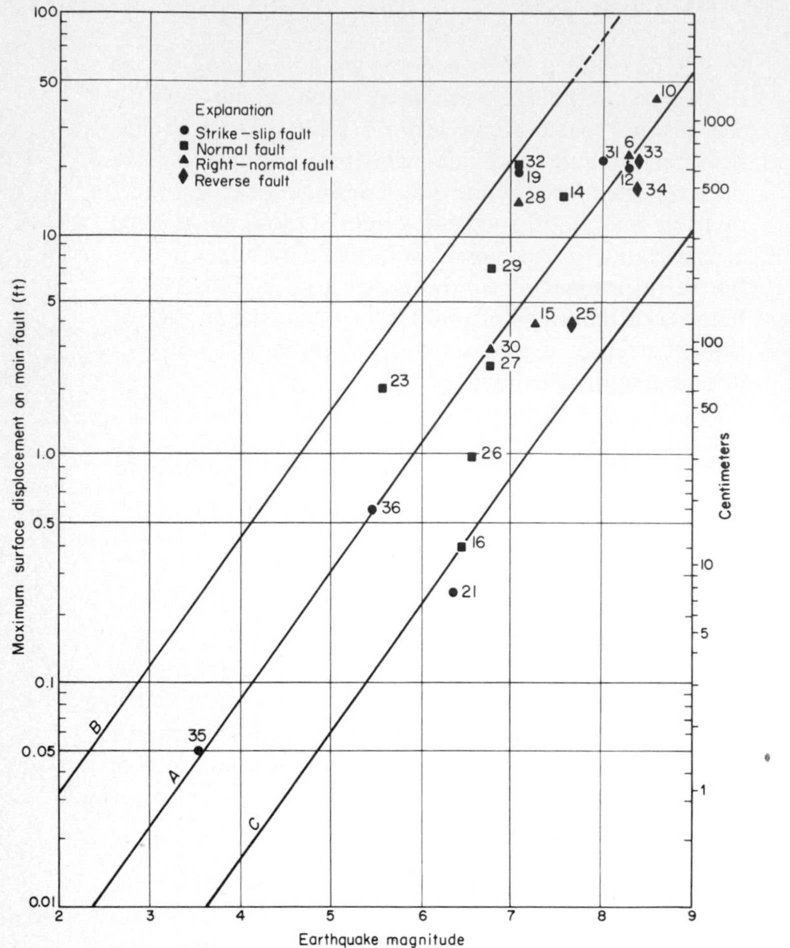

Fig. 3.14. Maximum displacement on main fault at the surface as related to earthquake magnitude (logarithmic plot).

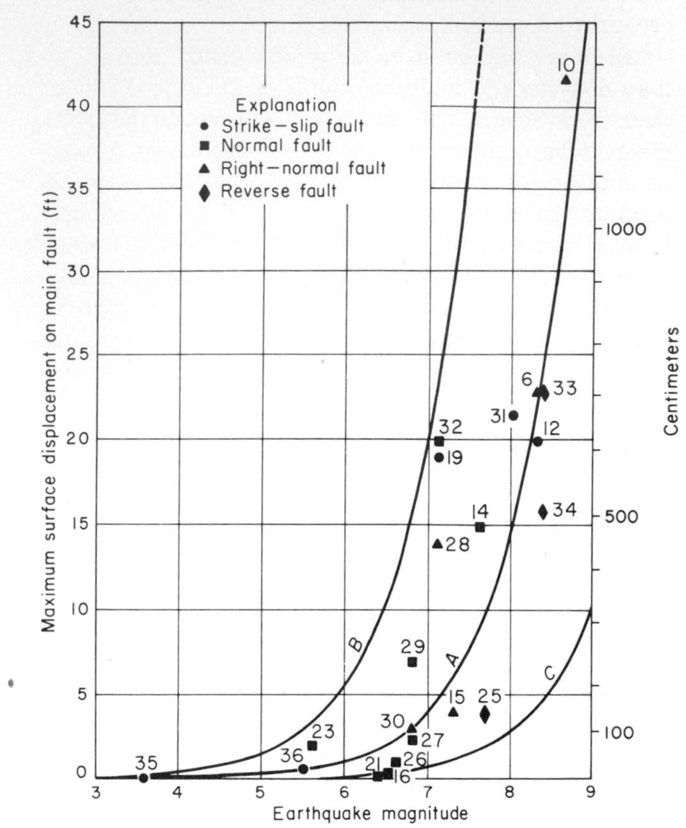

Fig. 3.15. Maximum displacement on main fault at the surface as related to earthquake magnitude (arithmetic plot).

Fig. 3.16. Length of surface rupture on main fault as related to earthquake magnitude.

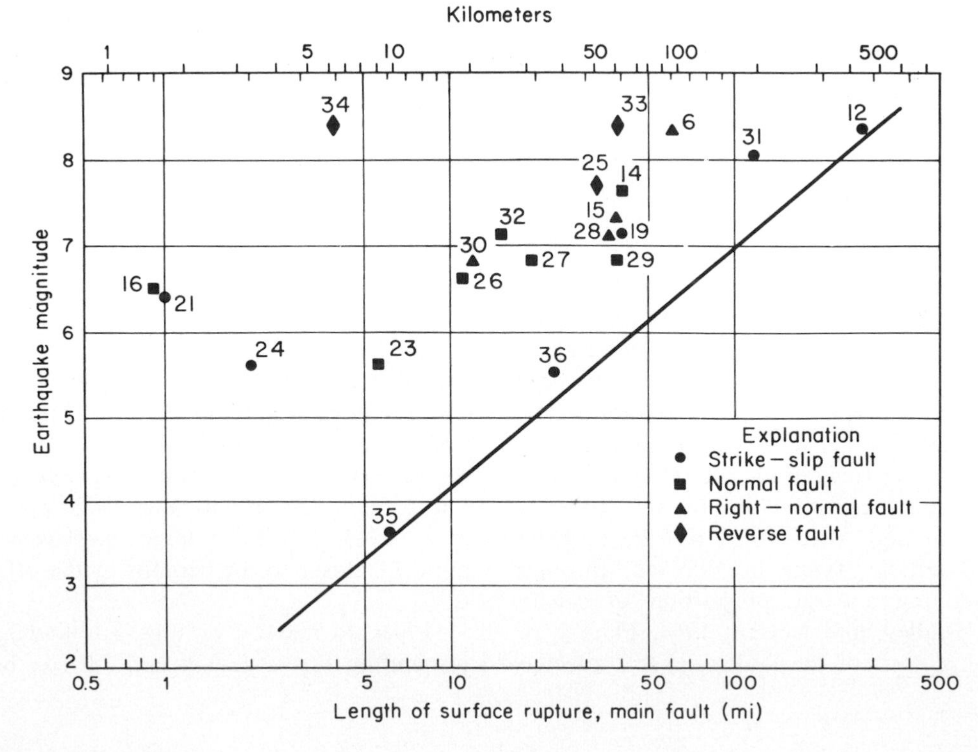

mate fault displacement at the ground surface that may accompany an earthquake of a given magnitude in the continental United States. Whether line *A*, *B*, or *C* is used depends upon the degree of risk that can be tolerated. For high magnitudes, line *B* indicates displacements substantially larger than any that have been recorded to date, and therefore the line is dashed for magnitudes greater than 7.5. With this exception, the lines permit realistic estimates of fault displacement.

Figure 3.16 shows the relation between earthquake magnitude and the length of surface rupture on the main fault. The points show considerable scatter and a least-squares fit was not made, but a line has been drawn that bounds all of the data points on the graph. The line can be used as an aid in estimating the maximum length of faulting that may occur in an event of a given magnitude. The position of this line is strongly influenced by two small earthquakes accompanied by surface faulting (numbers 35 and 36) that occurred in 1966. Its position may have to be altered if future small earthquakes show even longer surface ruptures.

Figure 3.17 shows in a logarithmic plot the relation between the maximum surface displacement and the length of the surface rupture along the main fault in a given event. The general increase of maximum displacement with length of rupture is apparent. The line of best fit, obtained by the method of least squares, has the equation

$$\log D = 0.86 \log L - 0.46 \tag{3.4}$$

where D is maximum displacement in feet and L is length of surface rupture in miles. This graph can be used as an aid in roughly estimating the maximum displacement that may occur on a fault of known length.

Fig. 3.17. Relation of maximum surface displacement to length of surface rupture on main fault. Numbers beside symbols refer to events listed in Table 3.1.

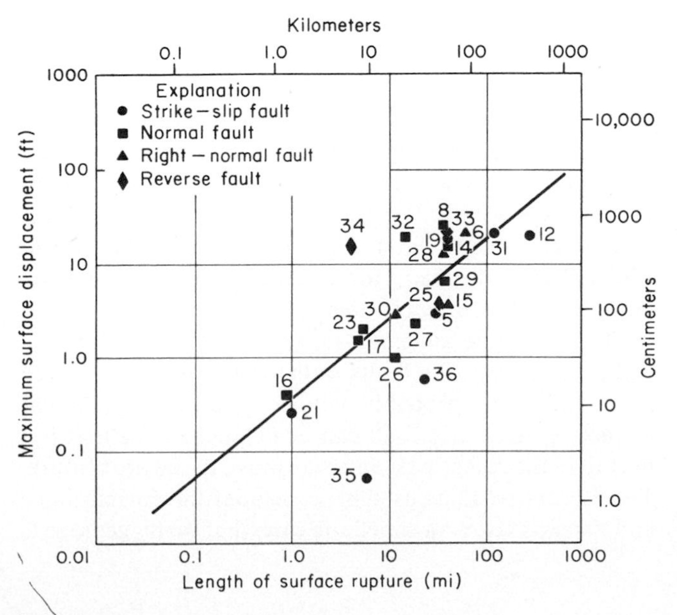

Surface faulting in a particular earthquake generally extends over just a part of the total length of the preexisting fault. The length of faulting accompanying historic earthquakes in southern California, according to Albee and Smith (1966, p. 20), has commonly been one-half to one-fifth the total length of the fault system on which it occurred, but there is a wide range in this ratio.

Numerous studies have been made relating earthquake magnitude to the product of the length and displacement on the fault. The earliest of these was by Tocher (1958), using California and Nevada faulting; followed by Iida (1959, 1965), using Japanese and worldwide faulting; and Slemmons (1966), using California and Basin-and-Range Province faulting. Slemmons' formula is

$$M = 3.68 + 0.41 \log LD \tag{3.5}$$

in which L is surface fault length and D is average surface displacement, both in centimeters. Slemmons (1966, p. 83) found that the average displacement in 15 examples ranged from 18 to 60% of the maximum displacement, and the median was 40%. The formulas of Tocher, Iida, and Slemmons are of the same form. King and Knopoff (1968, p. 253) have introduced the square of the displacement to give

$$\log LD^2 = 1.90\, M - 2.65 \tag{3.6a}$$

which can be written as

$$M = 1.4 + \frac{\log LD^2}{1.9} \tag{3.6b}$$

in which L and D are in centimeters. The King–Knopoff formula given above is based on data from 42 worldwide events. The King–Knopoff form of the equation seems to best express the relation between magnitude, fault length, and displacement. However, the coefficients in their equations are based partly on fault dimensions that have been theoretically derived (e.g., 40 ft for Montana, 1959, whereas the maximum scarp height was only 20 ft; see Section 3.3.3 above).

The existing fomulas and curves relating earthquake magnitude to fault length and fault displacement can serve as general guides, but without doubt all of them will need to be changed in the future as more data become available. All the formulas have some weaknesses. Among these are (1) the mixing of data from different kinds of faults, (2) the uncertain relation between fault displacement in unconsolidated materials and displacement in rock, (3) insufficient data to evaluate postquake fault creep as a component of fault displacement, (4) inferences regarding the relation between maximum and average fault displacement, both at the surface and at depth, and (5) inferences regarding length of faulting at depth. When the formulas and curves are used, the uncertainties in them need to be seriously considered.

3.6 FUTURE FAULTING

Anticipation of future surface faulting involves questions of whether, where, how much, what kind, and when. The answers currently available are not very satisfactory, but much research is underway. These questions are closely related to earthquake prediction, and a better understanding of the processes involved should progressively increase the quality of the answers. Some approaches to these answers are outlined below.

3.6.1 Location of Future Faulting; the Active Fault

One of the best guides to the location of future faulting is the location of past faulting. Geologic evidence shows that most faults have had repeated movement on them, and some are known to have been active for millions of years. The San Andreas fault, e.g., has developed a zone of shearing that is a mile or more wide in some places because of repeated movement.

One would expect that once faults and fault zones were well established, the stresses would tend to be relieved along them rather than in the more sound rock nearby, and historic ruptures support this inference. All of the main faults listed in Table 3.1 (except possibly Sonora, 1887) followed preexisting faults for all or nearly all of their extent. Many of these faults were known prior to the rupturing, and all (except possibly Sonora) could have been identified beforehand; however, extensive investigation would have been necessary in some cases, and some would probably have been considered inactive. The White Wolf fault,which had surface displacements in 1952, had been classed as "dead" on the fault map of California (*Seism. Soc. Am.*, 1922); by 1930 it might have been classed as active, however, because Hoots (1930, p. 315) inferred that large post-Pliocene movement had occurred on it. At least one-third of the branch and secondary faulting listed in Table 3.1 also occurred on preexisting faults that could have been identified as faults by simple geologic investigations; whether the rest could have been identified is problematical. A possible instance of some new faulting occurred in 1959; inconclusive evidence suggests that part of the Hebgen Lake, Montana, faulting of 1959 may have advanced locally into previously unfaulted rock (Myers and Hamilton, 1964, p. 85).

Geologic and historic evidence suggests that most or all of the later movements along parts of some faults have been concentrated in narrow zones. For example, in most (but not all) segments of the San Andreas fault one or two narrower strands within the broad fault zone can be readily identified as being the locus of the more recent movements. Wallace (1968, Fig. 8 and discussion) has presented evidence that 40 or more displacements have occurred in a strand less than 100 ft wide on one segment of the fault; however, he also notes that particular strands die out and the displacement is taken up by nearby *en echelon* strands. The 1906 ruptures were *en echelon* in at least two places, but unfortunately the width of the *en echelon* zone is not known (Lawson and others, 1908, p. 68; Taber, 1906, p. 307 and Fig. 2).

The 1966 Parkfield, California, faulting closely followed two strands within the fault zone that showed recent movement and had been identified prior to the earthquake; moreover, some of the 1966 ruptures were about 25 ft from ruptures formed in the 1934 earthquake (Brown and Vedder, 1967, pp. 4, 10). A similar example is provided by the 1966 Imperial faulting that occurred exactly where the 1940 ruptures had occurred (Brune and Allen, 1967, pp. 501, 502). The Borrego Mountain, California, faulting of April 9, 1968, closely followed earlier lines of faulting, but it also crossed areas that had no surface evidence of previous faulting (Allen *et al.*, 1968).

The examples given above were on strike-slip faults; repetition of breaks on two normal faults show more discordance between older and newer breaks. Most of the August 23, 1954, Rainbow Mountain, Nevada, faulting coincided with or extended the July 6, 1954, faulting, but some of the new ruptures were subparallel to the older ones (Tocher, 1956). Part of the December 1954 faulting north of Fairview Peak, Nevada, on the Gold King fault coincided with ruptures formed about 1903, but over most of its length it did not. Study of a somewhat generalized map of the 1903 breaks shows that the 1954 faulting crisscrossed the earlier faulting but was more than 200 ft from it in many places and more than 1000 ft in some places (Slemmons *et al.*, 1959, pp. 262–263, map).

Geologic evidence shows that some faults that were dormant for a long time have been reactivated. An example of this is the Hurricane fault in southwestern Utah, which has had several episodes of movement separated by periods of quiescence. Some of the quiescent periods were long enough to permit erosion of the faulted terrane and emplacement of lava flows. Its latest rejuvenation extends into the present and some earthquakes have occurred on the fault in historic time (Gardner, 1941; Averitt, 1964). Little is known of the rate at which faults, dormant for a long period (in geologic terms), have been reactivated.

The available evidence suggests that future faulting will occur on existing faults rather than on newly formed faults and that it probably will occur on those faults that are active. An active fault can be defined as one that has moved in the recent past and may move in the near future. The "recent past" as used here includes the current hour and extends back an indefinite time that many geologists would take to include at least the Holocene Epoch (about 10,000 years). The "near future" as used above includes a length of time on the order of the useful life of engineering structures or the time span considered in long-range plans for the future. The determination of whether a fault is "active," as defined above, involves geology, geophysics, geodesy, and engineering. Some criteria currently in use are (1) the occurrence of earthquakes that can be related to the fault with reasonable assurance; (2) one or more episodes of surface rupture (including tectonic creep) or acute bending in the recent past as defined above; (3) instrumental evidence of elastic or inelastic strain; and (4) structural coupling to another fault (or other tectonic feature such as a monocline) that is active. At present some active faults may not be identifiable, but the ability to identify them should improve with time.

3.6.2 Amount of Future Faulting

At present, estimates of the type and amount of future faulting that may occur are based on the historic and geologic record. The type (i.e., normal, reverse, etc.) of faulting that has occurred in the past is generally assumed to be the type that will occur in the future. The length and displacement of prehistoric faulting sometimes can be estimated from the displacement of the ground surface—the height and length of scarps and the amount of offset of streams. The general rates of older displacements often can be estimated by measuring the displacement of formations of known age. With this geologic information as background, useful estimates of future displacement can be made by using various empirical curves and formulas (such as those in Section 3.5.3) that interrelate length, displacement, and earthquake magnitude. Conversely, estimates of magnitudes of prehistoric earthquakes can be made if the length and displacement of prehistoric ruptures can be determined by geologic investigations. Research currently underway may someday permit estimates of future displacement based on measurements of strain on the fault system.

3.6.3 Likelihood of Future Faulting

The most difficult question to answer regarding active faults is if and when surface faulting will occur. Surface faulting is more commonly associated with the larger- than the smaller-magnitude earthquakes, although this imbalance is probably much less than the present record indicates. The surface displacements that accompany small earthquakes are small, and the evidence is likely to be quickly obliterated, especially for strike-slip displacements in areas that have few artificial structures; moreover, many earthquakes of moderate to small size in the western United States were never investigated in the field by geologists or seismologists. Furthermore, the recognition of the process of tectonic creep introduces the possibility that surface rupture on faults may accompany shallow-focus earthquakes of any magnitude.

The question of when surface faulting will occur cannot be satisfactorily answered at present on either a long-term or short-term basis. The problem is closely allied to earthquake prediction, and current research toward that goal probably will produce useful results. Various methods have been used to estimate the time between large earthquakes. By using strain rates, Reid estimated a return period of about 100 years after 1906 for the next great earthquake on the San Andreas fault in central California (see Chapter 2). By using rates of slip, Wallace (1968) has tentatively suggested a 700-year recurrence interval for another part of the San Andreas fault; he also suggested that the recurrence intervals on different parts of the fault may differ by several orders of magnitude. Another approach involves the use of the statistical relationship between the frequency of earthquakes and their magnitude. A study by Allen and others (1965) of the southern California region, using earthquake statistics for the 29-year period between 1934 and 1963, yielded a recurrence interval of 52 years for large earthquakes in that region; the historic record of large earthquakes since 1800 suggests that the calculated rate may be approximately correct. The same study, however, when applied to small areas within southern California yielded results that the authors believed were unrealistic. Recurrence curves for much of the western United States have been published by Ryall, Slemmons, and Gedney (1966). Other approaches involve measuring changes in various properties of the rock in the vicinity of the fault, but no practical method of estimating time of occurrence is yet available.

ACKNOWLEDGMENTS

A review of historic surface faulting that forms much of the basis for this chapter was sponsored by the U.S. Atomic Energy Commission, Division of Reactor Development and Technology. The writer also acknowledges with thanks the ideas and information supplied by various colleagues, particularly E. H. Pampeyan, George Plafker, D. H. Radbruch, and Julius Schlocker—the sources of other ideas and data are cited in the text. Jane M. Buchanan compiled some of the data and prepared many of the illustrations. Publication of this chapter was authorized by the Director, U.S. Geological Survey.

REFERENCES

Aguilera, J. G. (1920). "The Sonora Earthquake of 1887," *Seism. Soc. Am. Bull.*, **10** (1), 31–44.

Albee, A. L., and J. L. Smith (1966). "Earthquake Characteristics and Fault Activity in Southern California," in *Engineering Geology in Southern California*, pp. 9–34, Glendale, Los Angeles Section of Association of Engineering Geologists.

Allen, C. R., A. Grantz, J. N. Brune, M. M. Clark, R. V. Sharp, T. G. Theodore, E. W. Wolfe, and M. Wyss (1968). "The Borrego Mountain, California, Earthquake of 9 April 1968—A Preliminary Report," *Seism. Soc. Am. Bull.*, **58** (3), 1183–1186.

Allen, C. R., P. St. Amand, C. F. Richter, and J. M. Nordquist (1965). "Relationship Between Seismicity and Geologic Structure in the Southern California Region," *Seism. Soc. Am. Bull.*, **55** (4), 753–797.

Ambraseys, N. N. (1960). *On the Seismic Behavior of Earth Dams, Proceedings of the Second World Conference on Earthquake Engineering*, Vol. 1, pp. 331–358, Tokyo and Kyoto, Japan.

Ambraseys, N. N. (1963). "The Buyin–Zara (Iran) Earthquake of September 1962—A Field Report, *Seis. Soc. Am. Bull.*, **53** (4), 705–740.

Ambraseys, N. N. (1965a). *An Earthquake Engineering Study of the Buyin–Zahra Earthquake of September 1, 1962, in Iran, Proceedings of the Third World Conference on Earthquake Engineering*, Vol. 3, pp. V7–V26, New Zealand.

Ambraseys, N. N. (1965b). *An Earthquake Engineering Viewpoint of the Skopje Earthquake, 26th July, 1963, Proceedings of the Third World Conference on Earthquake Engineering*, Vol. 3, pp. S22–S38, New Zealand.

American Geological Institute (1960). *Glossary of Geology and Related Sciences*, 2nd ed., Washington, D.C.

American Iron and Steel Institute (1962). *The Agadir, Morocco, Earthquake*, New York: American Iron and Steel Institute.

Averitt, P. (1964). "Table of Post-Cretaceous Geologic Events Along the Hurricane Fault, Near Cedar City, Iron County, Utah," *Geol. Soc. Am. Bull.*, **75** (9), 901–908.

Bateman, P. C. (1961). "Willard D. Johnson and the Strike-Slip Component of Fault Movement in the Owens Valley, California, Earthquake of 1872," *Seism. Soc. Am. Bull.*, **51**(4), 483–493.

Bell, D. E., and V. A. Brill (1938). "Active Faulting in Lavaca County, Texas," *Am. Assoc. Petrol. Geol. Bull.*, **22**(1), 104–106.

Berg, G. V. (1964). *The Skopje, Yugoslavia, Earthquake July 26, 1963*, New York: American Iron and Steel Institute.

Biehler, S., R. L. Kovach, and C. R. Allen (1964). "Geophysical Framework of Northern End of Gulf of California Structural Province," in T. H. van Andel and G. B. Shor, Jr. (eds.), *Marine Geology of the Gulf of California—A Symposium*, pp. 146–143. American Association of Petroleum Geologists.

Bonilla, M. G. (1967). "Historic Surface Faulting in Continental United States and Adjacent Parts of Mexico," *U. S. Geological Survey Open-File Report;* also *U.S. Atomic Energy Commission Report TID-24124.*

Brown, R. D., Jr., and J. G. Vedder (1967). "Surface Tectonic Fractures Along the San Andreas Fault, California," in R. D. Brown, Jr., *et al.*, *The Parkfield–Cholame, California, Earthquakes of June–August 1966—Surface Geologic Effects, Water-Resources Aspects, and Preliminary Seismic Data*, U. S. Geological Survey Professional Paper 579, pp. 2–23.

Brown, R. D., Jr., and R. E. Wallace (1968). "Current and Historic Fault Movement Along the San Andreas Fault Between Paicines and Camp Dix, California," in W. R. Dickinson and A. Grantz (eds.), *Proceedings of the Conference on Geologic Problems of the San Andreas Fault System, September 14–16, 1967, Stanford University*, pp. 22–41, Stanford, California: Stanford University Press.

Brune, J. N., and C. R. Allen (1967). "A Low-Stress-Drop, Low-Magnitude Earthquake with Surface Faulting—The Imperial, California, Earthquake of March 4, 1966," *Seism. Soc. Am. Bull.*, **57**(3), 501–514.

Bryan, F. (1933). "Recent Movements on a Fault of the Balcones System, McLennan County, Texas," *Am. Assoc. Petrol. Geol. Bull.*, **17**(4), 439–442.

Buwalda, J. P., and C. F. Richter (1941). "Imperial Valley Earthquake of May 18, 1940," *Geol. Soc. Am. Bull.*, **52**(12), 1944–1945.

Buwalda, J. P., and P. St. Amand (1955). "Geological Effects of the Arvin–Tehachapi Earthquake," *Calif. Div. Mines Bull.*, **171**, 41–56.

California Department of Water Resources (1967). "Earthquake Damage to Hydraulic Structures in California," *California Department of Water Resources Bulletin 116-3.*

Callaghan, E., and V. P. Gianella (1935). "The Earthquake of January 30, 1934, at Excelsior Mountains, Nevada," *Seism. Soc. Am. Bull.*, **25**(2), 161–168.

Chinnery, M. A. (1961). "The Deformation of the Ground Around Surface Faults," *Seism. Soc. Am. Bull.*, **51**(3), 355–372.

Cloud, W. K. (1967). "Seismoscope Results from Three Earthquakes in the Hollister, California, Area," *Seism. Soc. Am. Bull.*, **57**(6), 1445–1448.

Cloud, W. K., and V. Perez (1967). "Accelerograms—Parkfield Earthquake," *Seism. Soc. Am. Bull.*, **57**(6), 1179–1192.

Curray, J. R., and R. D. Nason (1967). "San Andreas Fault North of Point Arena, California," *Geol. Soc. Am. Bull.*, **78**(3), 413–418.

Danes, J. V. (1907). "Das Erdbeben von San Jacinto am 25 Dezember 1899," *Geog. Gesell. Wien Mitt.*, **50**, 339–347.

Dibblee, T. W., Jr. (1954). "Geology of the Imperial Valley Region, California," Chap. 2 in R. H. Jahns (ed.), *Geology of Southern California*, pp. 21–28, California Division of Mines Bulletin 170.

Dibblee, T. W., Jr. (1955). *Geology of the Southeastern Margin of the San Joaquin Valley, California*, pp. 23–24, California Division of Mines Bulletin 171.

Duke, C. M. (1960). *Foundations and Earth Structures in Earthquakes, Proceedings of the Second World Conference on Earthquake Engineering*, Vol. 1, pp. 435–455, Tokyo and Kyoto, Japan.

Duke, C. M., and D. J. Leeds (1962). *Site Characteristics of Southern California Strong-Motion Earthquake Stations*, University of California at Los Angeles, Dept. Eng. Rept. No. 62-55.

Eppley, R. A. (1965). "Stronger Earthquakes of the United States (Exclusive of California and Western Nevada)," Part I in *Earthquake History of the United States*, rev. ed. (through 1963): U.S. Coast and Geodetic Survey No. 41-1.

Ferguson, H. G., R. J. Roberts, and S. W. Muller (1952) *Geologic Map of the Golconda Quadrangle, Nevada*, U.S. Geological Survey Geologic Quadrangle Map [GQ-15], scale 1:125,000.

Fett, J. D., D. H. Hamilton, and F. A. Fleming (1967). "Continuing Surface Displacements Along the Casa Loma and San Jacinto Faults in San Jacinto Valley, Riverside County, California," *Engr. Geol.*, **4** (1), 22–32.

Fisk, H. N. (1944). *Geological Investigation of the Alluvial Valley of the Lower Mississippi River*, U.S. Mississippi River Commission Report.

Florensov, N. A., and V. P. Solonenko (eds.) (1963). "Gobi–Altayskoye Zemletryasenie," *Iz. Akad. Nauk SSSR.;* also 1965, *The Gobi–Altai Earthquake*, U.S. Department of Commerce (Eng. trans.).

Fuller, M. L. (1912). *The New Madrid Earthquake*, U. S. Geological Survey Bull. 494.

Gardner, L. S. (1941). "The Hurricane Fault in Southwestern Utah and Northwestern Arizona," *Am. J. Sci.*, **239**(4), 241–260.

Gianella, V. P. (1957). "Earthquake and Faulting, Fort Sage Mountains, California, December 1950," *Seism. Soc. Am. Bull.*, **47**(3), 173–177.

Gianella, V. P., and E. Callaghan (1934). "The Cedar Mountain, Nevada, Earthquake of December 20, 1932," *Seism. Soc. Am. Bull.*, **24** (4), 345–384.

Gibson, W. M., and H. A. Wollenberg (1968). "Investigations for Ground Stability in the Vicinity of the Calaveras Fault, Livermore and Amador Valleys, Alameda County, California," *Geol. Soc. Am. Bull.*, **79** (5), 627–638.

Gilbert, G. K. (1890). *Lake Bonneville*. U.S. Geological Survey Monograph 1.

Goodfellow, G. E. (1888). "The Sonora Earthquake," *Science*, **11,** 162–166.

Heinrich, R. R. (1941). "A Contribution to the Seismic History of Missouri," *Seism. Soc. Am. Bull.*, **31** (3), 187–224.

Heyl, A. V., Jr., and M. R. Brock (1961). "Structural Framework of the Illinois-Kentucky Mining District and Its Relation to Mineral Deposits," in *Geological Survey Research, 1961*, pp. D3–D6, U.S. Geological Survey Professional Paper 424-D.

Higgins, C. G. (1961). "San Andreas Fault North of San Francisco, California," *Geol. Soc. Am. Bull.*, **72** (1), 51–68.

Hill, M. L. (1954). "Tectonics of Faulting in Southern California" in R. H. Jahns (ed)., *Geology of Southern California*, pp. 5–13, California Division of Mines Bulletin 170.

Hobbs, W. H. (1910). "The Earthquake of 1872 in the Owens Valley, California," *Beitr. Geophys.*, **10** (3), 352–385.

Hoots, H. W. (1930). *Geology and Oil Resources Along the Southern Border of San Joaquin Valley, California*, pp. 243–338, U.S. Geological Survey Bulletin 812-D.

Housner, G. W. (1965). *Intensity of Earthquake Ground Shaking Near the Causative Fault, Proceedings of the Third World Conference on Earthquake Engineering, New Zealand, 1965*, Vol. 3, pp. III 94–III 115.

Housner, G. W., and D. E. Hudson (1958) "The Port Hueneme [California] Earthquake of March 18, 1957," *Seism. Soc. Am. Bull.*, **48** (2), 163–168.

Hutchins, W. A. (1914). *Report on Investigation of Wells in Imperial Valley, 1914*, pp. 212–228, California Department of Engineering, Fourth Biennial Report.

Iida, Kumizi (1959). "Earthquake Energy and Earthquake Fault," *Nagoya Univ., J. Earth Sci.*, **7** (2), 98–107.

Iida, K. (1965). "Earthquake Magnitude, Earthquake Fault, and Source Dimensions," *Nagoya Univ., J. Earth Sci.*, **13** (2), 115–132.

Jones, J. C. (1915). "The Pleasant Valley, Nevada, Earthquake of October 2, 1915," *Seism. Soc. Am. Bull.*, **5** (4), 190–205.

King, Chi-Yu, and L. Knopoff (1968). "Stress Drop in Earthquakes," *Seism. Soc. Am. Bull.*, **58** (1), 249–257.

Knopf, Adolph, and E. Kirk (1918). *A Geological Reconnaissance of the Inyo Range and the Eastern Slope of the Southern Sierra Nevada, California, with a Section on the Stratigraphy of the Inyo Range by Edwin Kirk*, U.S. Geological Survey Professional Paper 110.

Koch, T. W. (1933). "Analysis and Effects of Current Movement on an Active Thrust Fault in Buena Vista Hills Oil Field, Kern County, California," *Am. Assoc. Petrol. Geol. Bull.*, **17** (6), 694–712.

Kovach, R. L., C. R. Allen, and F. Press (1962). "Geophysical Investigations in the Colorado Delta Region," *J. Geophys. Res.*, **67** (7), 2845–2871.

Kreese, F. C. (1966). "Baldwin Hills Reservoir Failure of 1963," in *Engineering Geology in Southern California*, pp. 93–103. Glendale, Los Angeles Section of Association of Engineering Geologists.

Kupfer, D. H., S. Muessig, G. I. Smith, and G. N. White (1955). *Arvin-Tehachapi Earthquake Damage Along the Southern Pacific Railroad Near Bealville, California*, California Division of Mines Bulletin 171, pp. 67–74.

Lawson, A. C., *et al.* (1908). *The California Earthquake of April 18, 1906—Report of the State Earthquake Investigation*

Commission, Vol. 1, Part 1, pp. 1–254; Part 2, pp. 255–451. Carnegie Institution of Washington, Publication 87.

Lee, C. H. (1912). *An Intensive Study of the Water Resources of a Part of Owens Valley, California*, U.S. Geological Survey, Water-Supply Paper 294.

Louderback, G. D. (1942). "Faults and Earthquakes," *Seism. Soc. Am. Bull.*, **32** (4), 305–330.

Louderback, G. D. (1947). "Central California Earthquakes of the 1830's," *Seism. Soc. Am. Bull.*, **37** (1), 33–74.

Martin, L. (1907). "Possible Oblique Minor Faulting in Alaska," *Econ. Geol.*, **2**, 576–579.

McEvilly, T. V. (1966). "Preliminary Seismic Data, June–July, 1966," in "Parkfield Earthquakes of June 27–29, 1966, Monterey and San Luis Obispo Counties, California—Preliminary Report," *Seism. Soc. Am. Bull.*, **56** (4), 967–971.

McEvilly, T. V., W. H. Bakun, and K. B. Casaday (1967). "The Parkfield, California, Earthquakes of 1966," *Seism. Soc. Am. Bull.*, **57** (6), 1221–1244.

Muller, S. W., H. G. Ferguson, and R. J. Roberts (1951) *Geology of the Mount Tobin Quadrangle, Nevada*, U.S. Geological Survey Geologic Quadrangle Map [GQ-7], scale 1:125,000.

Myers, W. B., and W. Hamilton (1964). "Deformation Accompanying the Hebgen Lake, Montana, Earthquake of August 17, 1959," in *The Hebgen Lake, Montana, Earthquake of August 17, 1959*, pp. 55–98. U.S. Geological Survey Professional Paper 435.

Nason, R. D. (1968). *Fault Slippage at Hayward, California*, pp. 86–87, Geological Society of America, Cordilleran Section—Seismological Society of America—Paleontological Society, Pacific Coast Section, 64th Annual Meeting, Tucson, Arizona, 1968, Program.

Nasu, N. (1931). *Comparative Studies of Earthquake Motions Above-Ground and in a Tunnel* (*Part I*), Vol. 9, pp. 454–472. Tokyo University Earthquake Research Institute Bulletin.

Neumann, F. (1936). "The Utah Earthquake of March 12, 1934," in *United States Earthquakes, 1934*, pp. 43–48. U.S. Coast and Geodetic Survey Serial 593.

Oldham, R. D. (1899). "Report on the Great Earthquake of 12th June, 1897," *India Geol. Survey Mem.*, **29**, 1–379.

Page, B. M. (1935). "Basin-Range Faulting of 1915 in Pleasant Valley, Nevada," *J. Geol.*, **43** (7), 690–707.

Paterson, M. S. (1958). "Experimental Deformation and Faulting in Wombeyan Marble," *Geol. Soc. Am. Bull.*, **69** (4), 465–475.

Plafker, G. (1965). "Tectonic Deformation Associated with the 1964 Alaska Earthquake," *Science*, **148** (3678), 1675–1687.

Plafker, G. (1967). *Surface Faults on Montague Island Associated with the 1964 Alaska Earthquake*, pp. G1–G42. U.S. Geological Survey Professional Paper 543-G.

Radbruch, D. H. (1967). *Approximate Location of Fault Traces and Historic Surface Ruptures Within the Hayward Fault Zone Between San Pablo and Warm Springs, California*, U.S. Geological Survey Miscellaneous Geological Investigation Map I-522, scale 1:62,500.

Radbruch, D. H. (1968). "New Evidence of Historic Fault Activity in Alameda, Contra Costa and Santa Clara Counties, California," in W. R. Dickinson and A. Grantz (eds.), *Proceedings of the Conference on Geologic Problems of the San Andreas Fault System, September 14–16, 1967, Stanford University*, pp. 46–54, Stanford, California: Stanford University Press.

Reid, H. F. (1910). "The Mechanics of the Earthquake," Vol. 2 in *The California Earthquake of April 18, 1906, Report of the State Earthquake Investigation Commission*, Carnegie Institution of Washington, Publication 87.

Richter, C. F. (1958). *Elementary Seismology*, San Francisco: W. H. Freeman.

Rogers, T. H., and R. D. Nason (1967). "Active Faulting in the Hollister Area" in *Guidebook, Gabilan Range and Adjacent San Andreas Fault*, pp. 102–104, American Association of Petroleum Geologists, Pacific Section, and Society of Economic Paleontologists, Pacific Section, Annual Field Trip.

Romney, C. F. (1957). "Seismic Waves from the Dixie Valley–Fairview Peak Earthquakes [Nevada]," *Seism. Soc. Am. Bull.*, **47** (4), 301–319.

Ross, C. A. (1963). *Faulting in Southernmost Illinois*, p. 229, Geological Society of America Special Paper 73.

Russell, W. L. (1957). "Faulting and Superficial Structures in East-Central Texas," *Gulf Coast Assoc. Geol. Soc. Trans.*, **7**, 65–72.

Ryall, A., D. B. Slemmons, and L. D. Gedney (1966). "Seismicity, Tectonism, and Surface Faulting in the Western United States During Historic Time," *Seism. Soc. Am. Bull.*, **56** (5), 1105–1135.

Saint-Amand, P. (1963). *The Great Earthquakes of May 1960 in Chile*, pp. 337–363, Smithsonian Institution, Washington Publication 4550.

Savage, J. C., and L. M. Hastie (1966). "Surface Deformation Associated with Dip-Slip Faulting," *J. Geophys. Res.*, **71** (20), 4897–4904.

Seismological Society of America (1922). *Fault Map of the State of California*, scale 1:506,880.

Sheets, M. M. (1947). "Diastrophism During Historic Time in the Gulf Coastal Plain," *Am. Assoc. Petrol. Geol. Bull.*, **31** (2), 201–226.

Shor, G. G., Jr., and E. E. Roberts (1958). "San Miguel, Baja California Norte [Mexico], Earthquakes of February, 1956—A Field Report," *Seism. Soc. Am. Bull.*, **48** (2), 101–116.

Slemmons, D. B. (1957). "Geological Effects of the Dixie Valley–Fairview Peak, Nevada, Earthquakes of December 16, 1954," *Seism. Soc. Am. Bull.*, **47** (4), 353–375.

Slemmons, D. B. (1966). *Long-Term Strain Release from Earthquake History and Late Cenozoic Faulting in the Basin-and-Range Province*, pp. 82–84, *Proceedings of the*

Second United States–Japan Conference on Research Related to Earthquake Prediction.

Slemmons, D. B. (1967). "Pliocene and Quaternary Crustal Movements of the Basin-and-Range Province, USA," *Osaka City Univ. J. Geosci.*, **10**, 91–103.

Slemmons, D. B., K. V. Steinbrugge, D. Tocher, G. B. Oakeshott, and V. P. Gianella (1959). "Wonder, Nevada, Earthquake of 1903," *Seism. Soc. Am. Bull.*, **49** (3), 251–265.

Steinbrugge, K. V., and W. K. Cloud (1962). "Epicentral Intensities and Damage in the Hebgen Lake, Montana, Earthquake of August 17, 1959," *Seism. Soc. Am. Bull.*, **52** (2), 181–234.

Steinbrugge, K. V., and D. F. Moran (1957). "Engineering Aspects of the Dixie Valley–Fairview Peak Earthquakes," *Seism. Soc. Am. Bull.*, **47** (4), 335–348.

Suyehiro, K. (1932). "Engineering Seismology—Notes on American Lectures," *Am. Soc. Civil Engr. Proc.*, **58** (4), 1–43.

Taber, S. (1906). "Some Local Effects of the San Francisco Earthquake," *J. Geology*, **14** (4), 303–315.

Takahasi, R. (1931). "Results of the Precise Levellings Executed in the Tanna Railway Tunnel and the Movement Along the Slicken-Side that Appeared in the Tunnel," *Tokyo Univ., Earthquake Res. Inst. Bull.*, **9** (4), 435–453.

Tarr, R. S., and L. Martin (1906). "Recent Changes of Level in the Yakutat Bay Region, Alaska," *Geol. Soc. Am. Bull.*, **17** (1), 29–64.

Tarr, R. S., and L. Martin (1912). *The Earthquakes at Yakutat Bay, Alaska, in September 1899, with a Preface by G. K. Gilbert*, U. S. Geological Survey Professional Paper 69.

Tocher, D. (1956). "Movement on the Rainbow Mountain Fault [Nevada]," in "The Fallon–Stillwater Earthquakes of July 6, 1954, and August 23, 1954," *Seism. Soc. Am. Bull.*, **46** (1), 10–14.

Tocher, D. (1958). "Earthquake Energy and Ground Breakage," *Seism. Soc. Am. Bull.*, **48,** 147–153.

Tocher, D. (1960a). "The Alaska Earthquake of July 10, 1958, Movement on the Fairweather Fault and Field Investigation of Southern Epicentral Region," *Seism. Soc. Am. Bull.*, **50** (2), 267–292.

Tocher, D. (1960b). *Movement on Faults, Proceedings of the Second World Conference on Earthquake Engineering*, Vol. 1, pp. 551–564, Japan, 1960.

Tocher, D. (1966). *Fault Creep in San Benito County, California*, p. 72, Geological Society of America, Cordilleran Section—Seismological Society of America—Paleontological Society, Pacific Coast Section, 62d Annual Meeting, Reno, Nevada. Program.

Tocher, D. and D. J. Miller (1959). "Field Observations on Effects of Alaska Earthquake of 10 July 1958," *Science*, **129** (3346) 394–395.

Townley, S. D., and M. W. Allen (1939). "Descriptive Catalog of Earthquakes of the Pacific Coast of the United States, 1769 to 1928," *Seism. Soc. Am. Bull.*, **29** (1), 1–297.

Turner, H. W. (1891). "Mohawk Lake Beds [Plumas County, California]," *Philos. Soc. Wash. Bull.*, **11,** 385–409.

Turner, H. W. (1896). "Further Contributions to the Geology of the Sierra Nevada," *U.S. Geol. Survey Ann. Rept.*, **17** (1), 521–740.

Turner, H. W. (1897). *Description of the Downieville Quadrangle, California*, U.S. Geological Survey Geological Atlas, Folio 37.

Ulrich, F. P. (1941). "The Imperial Valley Earthquakes of 1940," *Seism. Soc. Am. Bull.*, **31** (2), 13–31.

U.S. Army Corps of Engineers (1950). *Geological Investigation of Faulting in the Lower Mississippi Valley*, U.S. Army Corps of Engineers Waterways Experiment Station Technical Memorandum 3-311.

Varnes, D. J. (1958). "Landslide Types and Processes," Chap. 3 in E. B. Eckel (ed.), *Landslides and Engineering Practice*, pp. 20–47, National Research Council, Highway Research Board Special Report 29, NAS-NRC Publication 544.

Veatch, A. C. (1906). *Geology and Underground Water Resources of Northern Louisiana and Southern Arkansas*, U.S. Geological Survey Professional Paper 46.

Vedder, J. G., and R. D. Brown, Jr. (1968). "Structural and Stratigraphic Relations Along the Nacimiento Fault in the Southern Santa Lucia Range and San Rafael Mountains, California," in W. R. Dickinson and A. Grantz (eds.), *Proceedings of the Conference on Geologic Problems of the San Andreas Fault System, September 14–16, 1967, Stanford University*, pp. 242–259, Stanford, California: Stanford University Press.

Wallace, R. E. (1968). "Notes on Stream Channels Offset by the San Andreas Fault, Southern Coast Ranges, California," in W. R. Dickinson and A. Grantz (eds.), *Proceedings of the Conference on Geologic Problems of the San Andreas Fault System, September 14–16, 1967, Stanford University*, pp. 6–21, Stanford, California: Stanford University Press.

Wallace, R. E., and E. F. Roth (1967). "Rates and Patterns of Progressive Deformations," in R. D. Brown, Jr., *et al.*, *The Parkfield–Cholame, California, Earthquakes of June–August 1966—Surface Geologic Effects, Water-Resources Aspects, and Preliminary Seismic Data*, pp. 23–40, U.S. Geological Survey Professional Paper 579.

Weaver, P., and M. M. Sheets (1962). "Active Faults, Subsidence, and Foundation Problems in the Houston, Texas, Area," in E. H. Rainwater, and R. P. Zingula, *Geology of the Gulf Coast and Central Texas and Guidebooks of Excursions*, pp. 254–265, Geological Society of America and Associated Societies, Annual Meeting, Houston, Texas.

Whitney, J. D. (1888). "The Owens Valley Earthquake," in W. A. Goodyear, *Inyo County*, pp. 288–309, California Mining Bureau Eighth Annual Report State Mineralogist.

Whitten, C. A. (1955). *Measurements of Earth Movements in California*, pp. 75–80, California Division of Mines Bulletin 171.

Whitten, C. A. (1961). "Measurement of Small Movements in the Earth's Crust," *Acad. Sci. Fennicae Ann. Ser. A, Geol. Geograph. Suomalainen Tiedeakatemia*, **3** (61), 315–320.

Whitten, C. A. (1966). "Crustal Movements from Triangulation Measurements," in *ESSA Symposium on Earthquake Prediction, Rockville, Maryland, 1966*, pp. 72–76, U.S. Environmental Science Services Administration, Washington, U.S. Government Printing Office.

Wiggins, P. N. (1954). "Geology of Ham Gossett Oil Field, Kaufman County, Texas," *Am. Assoc. Petrol. Geol. Bull.*, **38** (2), 306–318.

Wilt, J. W. (1958). "Measured Movement Along the Surface Trace of an Active Thrust Fault in the Buena Vista Hills, Kern County, California," *Seism. Soc. Am. Bull.*, **48** (2), 169–176.

Witkind, I. J. (1964). "Reactivated Faults North of Hebgen Lake," in *The Hebgen Lake, Montana, Earthquake of August 17, 1959*, pp. 37–50, U.S. Geological Survey Professional Paper 435.

Witkind, I. J., J. B. Hadley, and W. H. Nelson (1964). "Pre-Tertiary Stratigraphy and Structure of the Hebgen Lake Area," in *The Hebgen Lake, Montana, Earthquake of August 17, 1959*, pp. 199–207, U.S. Geological Survey Professional Paper 435.

Wollard, G. P. (1958). "Areas of Tectonic Activity in the United States as Indicated by Earthquake Epicenters," *Am. Geophys. Union Trans.*, **39** (6), 1135–1150.

Wood, H. O. (1955). "The 1857 Earthquake in California," *Seism. Soc. Am. Bull.*, **45** (1), 47–67.

Chapter 4

Strong Ground Motion

G. W. HOUSNER

California Institute of Technology
Pasadena, California

The shaking of the surface of the ground during an earthquake is produced by the passage of stress waves (Richter, 1958). These seismic waves emanate from a region of the Earth's crust where a stress failure has resulted in a sudden change in the equilibrium stress state. The size of earthquakes and the frequency of occurrence depend on the state of stress in the Earth's crust. The relatively frequent occurrence of earthquakes in some regions, e.g., in California, results in the release of relatively large amounts of strain energy over the years (Allen *et al.*, 1965, Ryall *et al.*, 1966), and this implies that there is a process that is putting strain energy into the Earth's crust at approximately the same average rate as energy is being released by earthquakes. A common plausible view is that a viscous flow in the Earth's interior is the active process responsible for straining the Earth's crust, primarily in the horizontal plane (see Chapter 2). Of course, it is possible that after the active process has ceased earthquakes will still continue as part of a slow relaxation process during which a nonuniform state of stress in the Earth's crust approaches stationary equilibrium. Actually, the change of stress resulting from a stress failure also involves an increase of stress in surrounding regions which, at depth, presumably gradually relaxes. Thus, the slow changes in strain observable at the surface of the Earth may indicate an increase of strain energy by an active tectonic process or it may indicate a redistribution of strain energy by a relaxation process.

From a broad point of view, shallow focus earthquakes are manifestations of a failure process taking place in the Earth's crust. Shearing stress failures take place along fault planes whose orientation can be deduced from seismograms. Where such fault planes intersect the surface of the Earth, evidence of past fault displacements is usually visible and indications of fracturing and grinding of rock during stress failures are often evident (Richter, 1953). The intersection of the fault plane with the local horizon plane defines an intersection line useful for describing the relative motion of the two sides of the fault during a stress failure. At any point on the fault plane, the relative displacement of the two sides of the fault defines a line whose length and direction are the amplitude and direction of the fault displacement. The component of fault displacement parallel to the intersection line is called strike-slip displacement, and the component normal to this line is called dip-slip displacement.

During the 1906 San Francisco earthquake, the maximum relative fault displacement was 21 ft (strike slip). Analyses of surface fault displacements and associated surface strains indicate that during a large earthquake the stress drop across a fault in basement granite is of the order of 500–1000 psi (see Eq. 2.17 in Chapter 2). This value is very low compared to the hydrostatic compression due to the weight of the rock above, which thus rules out the possibility of ordinary Coulomb friction sliding during the fault displacements.

4.1 MAGNITUDE OF AN EARTHQUAKE

A large earthquake is associated with a stress failure (slip) over a large fault area, with a large release of strain energy in the form of seismic waves and with a large surface area subjected to strong ground shaking. It is important for engineering purposes to be able to describe in a quantitative way the size of the earthquake. In 1935, C. F. Richter of the California Institute of Technology defined the magnitude of an earthquake for shallow shocks as

$$M = \log_{10} \frac{A}{A_0}$$

where M is the magnitude of the earthquake, A is the maximum amplitude recorded by a Wood–Anderson seismograph at a distance of 100 km from the center of the disturbance, and A_0 is an amplitude of one thousandth of a millimeter. The Wood–Anderson instrument has a natural period of 0.8 sec, almost critical damping, and nominal static magnification of 2800. In practice, the recordings must be made at distances that are large compared to the dimensions of the slipped fault area. The recordings are then extrapolated to a distance of 100 km from the center of the shock. For best results an average value of M is determined from a number of recordings from different seismological stations (Gutenberg and Richter, 1956).

Earthquakes of magnitude 5.0 or greater generate ground motions sufficiently severe to be potentially damaging to structures. For magnitudes less than approximately 5.0 the ground motion is unlikely to be damaging because of its very short duration and moderate acceleration.

The use of the magnitude scale is a convenient way of classifying earthquakes according to size. However, due to nonuniformity of the Earth's crust, different fault orientations, etc., M is not a precise measure of the size of an earthquake.

4.2 NATURE OF FAULT SLIP

An idealized earthquake may be thought of as being generated by means of a penny-shaped crack in an elastic solid. If the two sides of the crack are displaced relative to each other by shear stresses applied to the faces of the crack, the relative displacement of the center points being 0′-0′ as shown in Fig. 4.1a, a certain amount of strain energy is stored. If the stresses on the faces of the crack are suddenly released, stress waves are originated and the material returns to its unstressed condition. The time-history of the motion of a point on the centerline perpendicular to the crack and at some distance from it might appear as shown in Fig. 4.1b. This motion represents a simple type of earthquake generated by a relatively small slipped fault. Such motion was recorded during the Port Hueneme earthquake of March 18, 1957 (Housner and Hudson, 1958).

Most earthquakes of engineering significance are generated by slip over relatively large areas which, presumably, have a nonuniform distribution of prestress and, moreover, the material through which the seismic waves travel is nonuniform. As a result, the motion of a point on the Earth's surface will be much more complex than the motion shown in Fig. 4.1b.

For most smaller earthquakes, say $M < 6.0$, the vertical and horizontal dimensions of the slipped fault area are presumably approximately the same, but for large earthquakes the length of the slipped area may be measured in hundreds of miles whereas the perpendicular dimension is thought to be at most some 10–20 mi

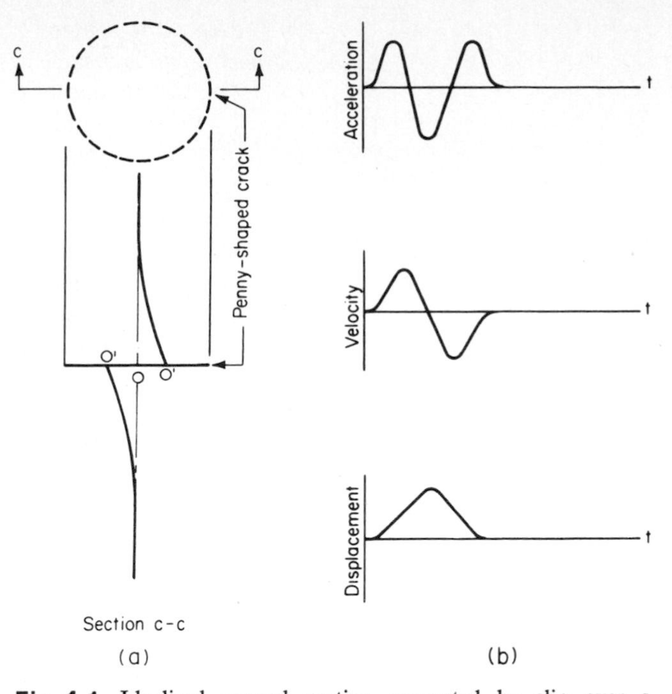

Fig. 4.1. Idealized ground motion generated by slip over a prestressed penny-shaped crack in an elastic solid. During slip, the release of shear stress causes the two faces of the crack to return to the unstressed position, e.g., points 0′ return to point 0.

in extent. It is possible to construct an idealized relation between M and the length L of slipped fault as given in Table 4.1. This table may be compared with the inferred lengths of slipped fault for some actual earthquakes that had surface fault displacement shown in Table 4.2 (Tocher, 1958; see Eqs. 2.8 and 2.9 in Chapter 2).

Table 4.1 shows that L for a magnitude 8.0 earthquake is about 7.5 times as long as for a magnitude 7.0 earthquake, i.e., the $M = 8.0$ event is essentially $7\frac{1}{2}$ $M = 7.0$ events laid end to end. This is a significant observation because it emphasizes the fact that the energy

Table 4.1. IDEALIZED RELATION BETWEEN MAGNITUDE AND LENGTH OF SLIPPED FAULT

Magnitude	Length (miles)
8.8	1000
8.5	530
8.0	190
7.5	70
7.0	25
6.5	9
6.0	5
5.5	3.4
5.0	2.1
4.5	1.3
4.0	0.83
3.0	0.33
2.0	0.14(735′)
1.0	0.05(270′)
0	0.018(100′)

Table 4.2. INFERRED LENGTHS OF SLIPPED FAULT OF ACTUAL EARTHQUAKES

Earthquake	Magnitude	Length (miles)
Chile, 1960	8.5–8.6	600±
Alaska, 1964	8.4	450±
San Francisco, 1906	8.2	250±
El Centro, 1940	7.1	40±
Baja California, 1956	6.8	15±

released in the form of seismic waves does not originate at a point source, as is the case in an underground nuclear detonation, but originates in a volume of rock that is greater for large earthquakes than for small shocks. If, as appears to be the case, the shear stress released by slip on the fault is approximately the same for larger earthquakes, it follows that the slipped area is directly related to the magnitude. It is this fact that makes the magnitude meaningful to engineers. It indicates approximately the size of the earthquake source and, hence, the approximate area affected by ground shaking.

4.3 EPICENTER

In the early days of the subject seismologists considered an earthquake to originate in a relatively small volume of rock at depth. The center of this volume was called the hypocenter, or sometimes the focus, and the point on the surface of the ground directly above the hypocenter was called the epicenter (Richter, 1958). The instrumentally determined epicenter can be located by means of the expression

$$(t_s - t_c) = \left(\frac{1}{v_s} - \frac{1}{v_c}\right) d$$

where t_s and t_c are the travel times to the recording seismograph of the shear curves and the compression waves respectively, v_s and v_c are the velocities of the shear and the compression waves respectively, and d is the distance traveled by the waves, i.e., $d = v_s t_s$, $d = v_c t_c$. The quantity $(t_s - t_c)$ is determined from the seismogram as the difference in the initial arrival times of the shear and the compression waves, and therefore, the wave velocities v_s and v_c being known, the distance d to the epicenter can be determined. Approximate values of the velocities in granite are $v_c = 19{,}000$ ft/sec, $v_s = 10{,}000$ ft/sec, and Rayleigh surface wave velocity $v_r = 9000$ ft/sec. The epicenter is located by determining its distance from several different seismograph stations, preferably surrounding the epicenter. The accuracy with which the epicenter can be located depends largely on how accurately the wave velocities are known and how well surrounded the epicenter is by seismographic stations. That the location of the epicenter is subject to error is emphasized by Perry Byerly's definition: "An epicenter is a mark made on a map by a man who calls himself a seismologist." Actually, it appears that the locations of epicenters provided by seismologists have a probable error of perhaps 5 mi under favorable circumstances and perhaps 20 mi under unfavorable circumstances. Note that a probable error of 5 mi means that on the average one-half of the cases will have errors greater than 5 mi.

It is important to recognize that the instrumentally determined epicenter and hypocenter do not indicate the center of energy release—rather they indicate the point where fault slip began. In the case of the smaller earthquakes, $M < 6$, the center of energy release and the point where slip begins are not far apart, but in the case of large earthquakes these points may be hundreds of miles apart. For example, the Chilean earthquake of 1960 had a fault slip about 600 mi in length and the epicenter was at the northern end of the slip some 300 mi from the center of energy release (Housner, 1963). The engineer should not ask how far the epicenter was from the damaged city but how far distant was the causative fault.

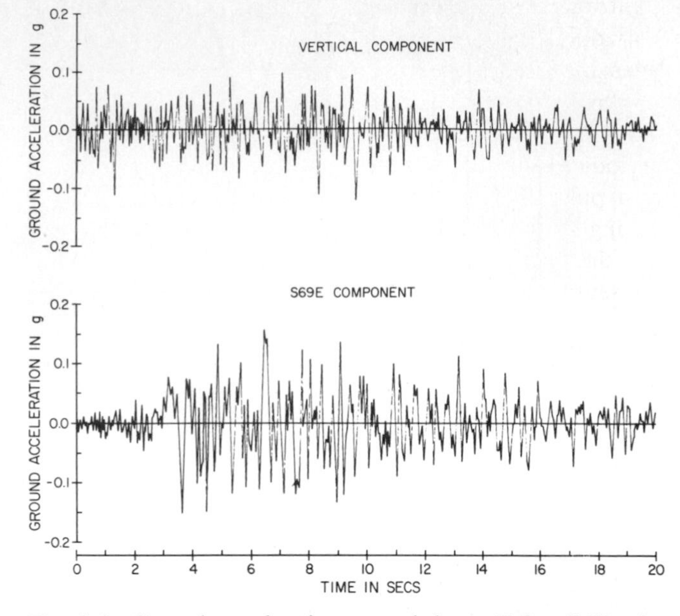

Fig. 4.2. Ground acceleration recorded at Taft, California, approximately 25 mi from the causative fault of the magnitude 7.7 Arvin–Tehachapi earthquake of July 21, 1952. Recorded on approximately 25 ft of alluvium overlying sedimentary rock. Component S69E.

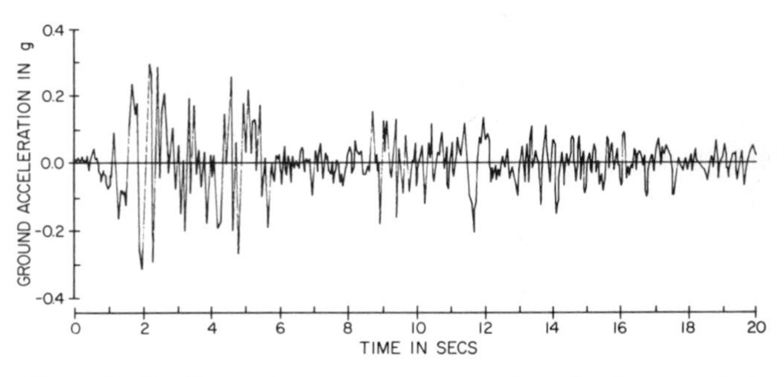

Fig. 4.3. North–south component of ground acceleration recorded at El Centro approximately 4 mi from the causative fault of the magnitude 7.1, May 18, 1940, earthquake. Recorded on very deep (5000+ ft) alluvium with ground water close to the surface.

4.4 INTENSITY

In the absence of any instrumental recordings of the ground motion, seismologists describe the severity of the ground shaking by assigning Modified Mercalli intensity numbers (see Table 4.9). The MM scale ranges from I, ground motion not felt by anyone, to XII, damage total. The MM number is a shorthand description of the effect of the ground shaking (Hodgson, 1964). It is not a precise engineering measure of the severity of ground shaking (see Chapter 1 for examples).

4.5 EARTHQUAKE GROUND MOTION

The basic data of earthquake engineering are the recordings of ground accelerations during earthquakes. (Hudson, 1963; Halverson, 1965). A knowledge of the ground motion is essential to an understanding of the earthquake behavior of structures. A network of strong-motion accelerographs is maintained by the Seismological Field Survey of the U.S. Coast and Geodetic Survey, and descriptions of the recordings and of the earthquakes are given in an annual publication (Coast and Geodetic Survey Annual). Most strong-motion recordings in the United States have been made on alluvium with only a few (Helena, Montana 1935; Taft, California 1952; Golden Gate Park, San Francisco 1957) recorded on sedimentary rock. Figure 4.2 shows one horizontal and one vertical component of recorded ground acceleration at Taft, California, during the magnitude 7.7 earthquake of July 21, 1952. The accelerograph was approximately 25 mi from the causative fault. Figure 4.3 shows one component of ground acceleration recorded at El Centro, California, on deep alluvium during the magnitude 7.1 earthquake of May 18, 1940. During the El Centro earthquake a surface fault slip developed over a length of 40+ mi, and the maximum relative fault displacement at the surface was 15 ft. The recording accelerometer was approximately 4 mi from the surface fault trace, rather closer to the northern end than to the center. The 1952 earthquake originated on the White Wolf fault, which is perpendicular to the San Andreas fault, and the fault motion was mainly dip slip. The 1940 El Centro earthquake originated on the Imperial fault, a southerly extension of the San Andreas fault system, and the fault motion was strike slip.

Figure 4.4 shows the recorded ground acceleration and the computed velocity and displacement at Taft,

California. For most recorded earthquakes, the two horizontal components of ground motion have nearly the same intensity.

An earthquake accelerogram is a random function that may be thought of as being composed of a non-periodic sequence of acceleration pulses. It is the area of a pulse that is a measure of its effectiveness in producing structural vibrations. The amplitude of the pulse, i.e., the maximum acceleration, is often used to indicate the severity of the ground motion. This is satisfactory if

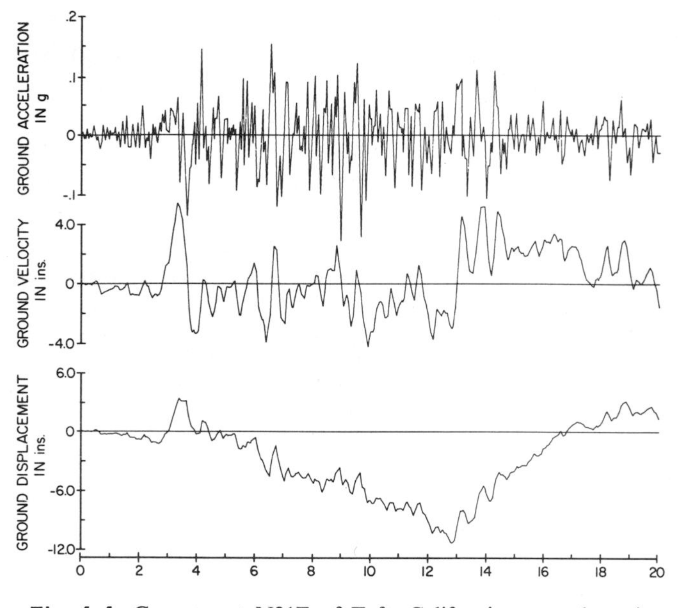

Fig. 4.4. Component N21E of Taft, California, ground motion July 21, 1952, showing recorded acceleration and integrated velocity and displacement.

the pulse durations are similar in all earthquakes, but it is not a reliable measure if the pulses have different durations. The intensity of the strong phase of shaking is characterized by the size and shape of the pulses, i.e., the maximum accelerations and the number of zero crossings per second. The effect of the ground shaking also depends upon the number of pulses, i.e., the duration of strong shaking. The most significant engineering characteristics of the ground motion, as discussed later, are exhibited by the Response Spectrum or by the Fourier Spectrum of the ground acceleration (Housner, 1959; Hudson, 1956; U.S. Atomic Energy Commission, 1963). In most earthquakes the vertical ground acceleration is from one-third to two-thirds as intense as the horizontal acceleration and has higher frequency components, i.e., it has approximately 50% more zero crossings per second.

The amplitude of the ground accelerations decreases with distance from the causative fault. The higher frequency components attenuate more rapidly than the lower frequency components so that the spectrum at a distance will be relatively depressed in the high frequency end (Housner, 1959).

4.6 IDEALIZED GROUND MOTION

Although there is not a rich supply of recorded ground accelerations of destructive earthquakes, there are enough recordings to indicate general trends related to magnitude, distance from fault, etc. Idealized earthquake ground motions can then be described that will portray the general characteristics of the earthquake problem, in a probability sense. The recorded ground accelerations have been obtained mostly on relatively firm deep alluvium, so that the idealized earthquakes described in the following paragraphs are representative of ground motions on such ground. Actual ground motions, of course, may deviate from the idealized motions.

4.6.1 Maximum Ground Acceleration

Although the maximum ground accelerations decrease with distance from the causative fault, the rate of decrease is relatively small over a distance comparable to the vertical dimension of the slipped fault. Table 4.3 lists idealized maximum ground accelerations in the vicinity of the causative fault for earthquakes of various magnitudes. These values are, in general, on the high side, and it can be expected that many actual earthquakes will have somewhat smaller values than are indicated in Table 4.3. For example, the maximum acceleration recorded during the $M = 7.1$ El Centro 1940 shock was 33% g. This is the largest acceleration recorded in the United States during an earthquake of large magnitude. Several times relatively small earthquakes have had higher accelerations than those shown in Table 4.3; an example is the Parkfield, California, earthquake of June 27, 1966, on the San Andreas fault (Housner, 1967; Hudson, 1967). This $M = 5.6$ shock was associated with evidence of surface slip extending over a length of about 20 mi and a shallow depth of slipped fault. Two hundred feet from the fault the recorded

Table 4.3. MAXIMUM GROUND ACCELERATIONS AND DURATIONS OF STRONG PHASE OF SHAKING

Magnitude	Maximum acceleration (% g)	Duration (sec)
5.0	9	2
5.5	15	6
6.0	22	12
6.5	29	18
7.0	37	24
7.5	45	30
8.0	50	34
8.5	50	37

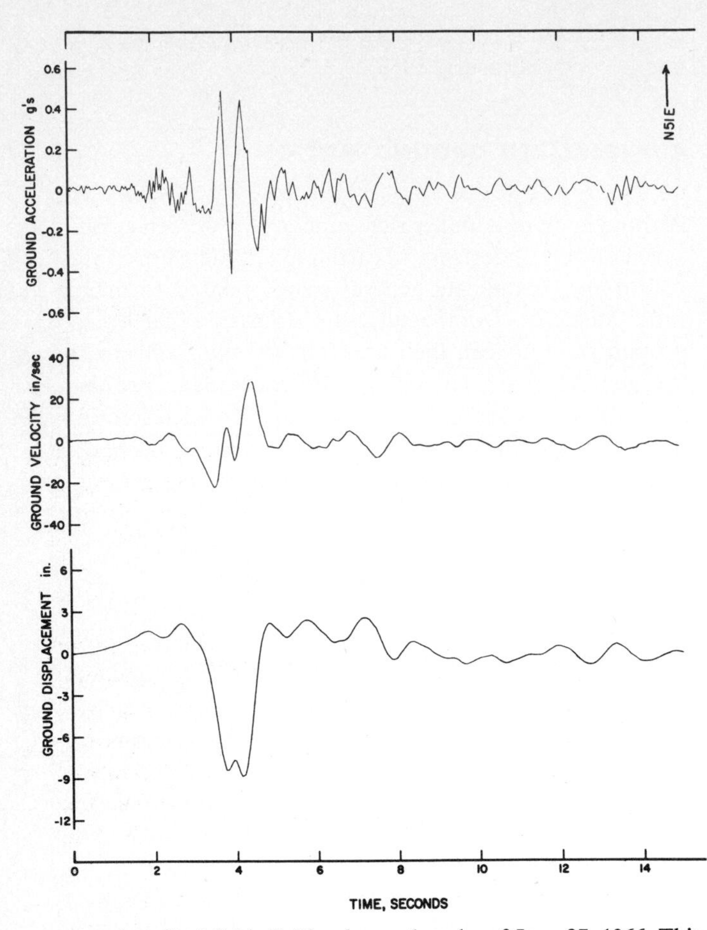

Fig. 4.5. Parkfield, California, earthquake of June 27, 1966. This magnitude 5.6 shallow earthquake originated on the San Andreas fault. The acceleration shown was recorded on alluvium of 100–200 ft depth overlying sedimentary rock, and the accelerograph was 200 ft from the fault trace that ran in the N39W direction.

maximum ground acceleration was 50% g and the duration of the strong phase of shaking was 1.5 sec, as shown in Fig. 4.5. The area of the maximum acceleration pulse was 2.0 ft/sec, which agrees with estimates that have been made for the largest expected earthquake pulse (Housner, 1965). The fact that the fault slip extended to the surface of the ground and that the accelerograph was only 200 ft from the fault means that this motion was recorded under nontypical conditions. Also, the long, shallow fault slip was not typical of $M = 5.6$ shocks. A more typical condition, as indicated by Table 4.1, is a diameter of fault slip of some 4 mi at a depth of perhaps 4–5 mi. It is thought that the Parkfield accelerogram may be indicative of the first $1\frac{1}{2}$ sec of ground motion of a very large earthquake. The Parkfield ground motion did remarkably little damage, presumably because of its very short duration (Cloud, 1967).

4.6.2 Duration of Ground Shaking

The duration of the strong phase of the idealized ground acceleration is given in Table 4.3 (Housner, 1965). In addition to the durations noted, there is an attenuated shaking that follows for a relatively long time. The table gives the duration of the strong phase in the vicinity of the fault; however, at greater distances from the fault the duration of shaking is longer and the intensity of shaking is less.

4.6.3 Spatial Distribution of Ground Shaking

The intensity of ground shaking is greatest in the vicinity of the causative fault and decreases with distance from the fault. For a small earthquake the idealized intensity distribution would be bell-shaped, but for a large earthquake having a slipped length of fault of several hundred miles the idealized intensity distribution would be elongated, as shown in Fig. 4.6. Idealized intensity distributions can be constructed as given in Table 4.4 which indicates the areas affected by ground shaking of various intensities. The accelerations in

Table 4.4. AREA IN 1000 mi² COVERED BY GROUND ACCELERATION (% g)

	M						
Acceleration	5.0	5.5	6.0	6.5	7.0	7.5	8.0
$\geq$ 5	0.4	1.6	3.6	6.8	13	28	56
$\geq$ 10		0.6	1.6	3.6	7.6	14	32
$\geq$ 15			0.6	2.0	4.4	9.6	21
$\geq$ 20				0.9	2.5	6.0	14
$\geq$ 25					1.3	4.0	10
$\geq$ 30					0.25	2.0	6.4
$\geq$ 35						0.6	4.0
$\geq$ 40							1.2

Fig. 4.6. Idealized contour lines of equal intensity of ground shaking in the vicinity of a slipped length of fault of 120 mi.

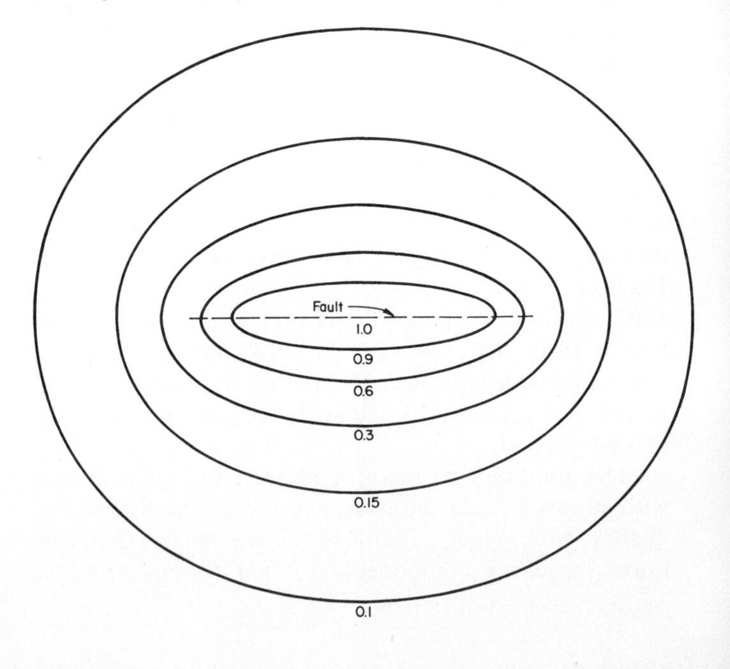

Table 4.4 are based on the maximum accelerations in Table 4.3 and, hence, are on the high side.

4.7 FREQUENCY OF OCCURRENCE

Assuming that the frequency distribution of larger California earthquakes, for which statistical data is rather sparse, has the same form as the frequency distribution of world earthquakes, the expectation of earthquakes in California (150,000 square miles) can be expressed as in Table 4.5. Data on frequencies of occurrence are given in several publications (Gutenberg and Richter, 1965; Coast and Geodetic Survey, 1958 and 1961; see also Chapter 5).

Table 4.5. EXPECTATION OF EARTHQUAKES IN CALIFORNIA (150,000 mi²)*

Magnitude	Number per 100 years
4.75–5.25	250
5.25–5.75	140
5.75–6.25	78
6.25–6.75	40
6.75–7.25	19
7.25–7.75	7.6
7.75–8.25	2.1
8.25–8.75	0.6

*Associated with large earthquakes are clusters of smaller earthquakes that may produce repetitive shaking of the same location. To discount this for probability calculations, some of the earthquakes having $M < 6.0$ have been omitted.

4.8 FREQUENCY OF GROUND SHAKING

When an earthquake of magnitude 7, e.g., occurs in California, according to Table 4.4 it will cover approximately 13,000 mi² with ground shaking of 5% g or greater, and it will cover some 5400 mi² with maximum ground shaking between 5 and 10% g, and so forth. As an illustration, let it be assumed that the occurrence of earthquakes in California is random in time and in space, thus disregarding the fact that all portions of California are not equally seismic. For a site in California there is thus a certain probability of occurrence of, say, a magnitude 7 shock and a certain probability that it will be located at such a distance that the site experiences ground shaking between 5 and 10% g. A simplified calculation of probabilities can be made on the basis of the following analogous problem: Given a large area A, let a small area Δa be placed at random in it. After a number of Δa areas have been placed at random so that $\Sigma \Delta a = a$, the probability that a certain point in A has not been covered by any of the $\Sigma \Delta a$ can be written $p(0, a)$, that is the probability of zero hits when a total area $\Sigma \Delta a = a$ has been dropped. The probability of zero hits after $a + \Delta a$ has been dropped is the product of the independent probabilities:

$$p(0, a + \Delta a) = p(0, a)p(0, \Delta a)$$

Noting that

$$p(0, \Delta a) = 1 - p(1, \Delta a) = 1 - \frac{\Delta a}{A}$$

the equation can be written in the form

$$\frac{p(0, a + \Delta a) - p(0, a)}{\Delta a} + \frac{1}{A}p(0, a) = 0$$

Letting $\Delta a \longrightarrow 0$, the foregoing equation approaches in the limit

$$\frac{dp}{da} + \frac{1}{A}p = 0$$

The solution of this equation is $p(0, a) = e^{-a/A}$ which is the probability that a point in A is covered by $\Sigma \Delta a$. The probability that the point is not missed, i.e., the probability that it is hit is

$$p(h, a) = 1 - e^{-a/A} \tag{4.1}$$

This equation may be used, e.g., to calculate the probability that a certain site in California will experience maximum ground shaking between 5 and 10% g in a 50-year period, if the frequency of occurrence of earthquakes is as shown in Table 4.5 and the spatial distribution of ground shaking is as shown in Table 4.4. The computed probabilities are given in Table 4.6. There is,

Table 4.6. PERCENT PROBABILITY OF ACCELERATION AT A LOCATION IN CALIFORNIA

Acceleration (% g)	In period of years			
	10	25	50	100
≥ 5	65	92	99	99
≥ 10	37	70	88	98
≥ 15	19	41	64	87
≥ 20	10	23	40	63
≥ 25	5	12	22	37
≥ 30	2.5	5.5	10	19
≥ 35	1.0	2.5	4.4	8.7

of course, a possibility that the site is covered more than once by the specified ground acceleration and the probability of being covered n times is

$$p(n, a) = \left(\frac{a}{A}\right)^n \frac{1}{n!} e^{-a/A} \tag{4.2}$$

It should be noted that the probabilities given in Table 4.6 are too high for some of the less seismic regions of California and too low for some of the highly seismic regions.

4.9 UPPER BOUND FOR INTENSITY OF GROUND SHAKING

It is of some engineering value to know that there is an upper bound for the intensity of ground shaking. For example, if there were no upper bound, cities would be destroyed at intervals because it would be only a matter of time until completely destructive ground motions would occur. In such a situation the building code requirements undoubtedly would be established solely on the basis of initial cost and the cost of repair of damage. Hazard to life and limb would not play a large role because it would be impossible to design structures to be completely safe. On the other hand, if there is an upper bound for ground shaking that is not inordinately high, the building code requirements can be based on two conditions: (1) no loss of life, i.e., no collapse in the event of the strongest ground shaking, and (2) an economical balance between damage loss and initial cost. The present building code requirements in California, it is generally agreed, are aimed at ensuring that no injury or loss of life occurs in the event of severe ground shaking and that no damage happens in the event of moderate ground shaking that has a high probability of occurring during the life of the structure. This, in effect, is assuming that there is an upper bound for ground shaking. Therefore, it is of interest to review the evidence for an upper bound for United States earthquakes.

Observations of damage caused by large earthquakes that had surface fault displacements show that the intensity of ground shaking is approximately the same within a few miles of the fault. That is, the intensity does not rise to an especially high value right at the fault, in a cusp shape. This has been specifically pointed out for the following earthquakes: El Centro, California (Richter, 1958), May 18, 1940; Hebgen Lake, Montana (Steinbrugge and Cloud, 1962), August 17, 1959; Pleasant Valley, Nevada (Jones, 1915), October 2, 1915; and Buyin-Zara, Iran (Ambraseys, 1963), September 1963. During the El Centro earthquake, ground acceleration recorded at a point 4 mi from the fault trace had a maximum acceleration of 33% g. Although this motion was not necessarily the most severe in the region affected by the earthquake, observations of damage did not reveal any other locations obviously shaken more severely. Therefore, it may be concluded that an earthquake associated with fault slip over a length of some 40+ mi, in a region covered by relatively deep alluvium, will have a maximum acceleration pulse of about 33% g with pulse area about 1.6 ft/sec, as listed in Table 4.7.

Table 4.7. MAXIMUM RECORDED ACCELERATION PULSES

Amplitude (% g)	Duration (sec)	Area (ft/sec)
50	0.25	2.0
33	0.23	1.6
23	0.29	1.4
30	0.18	1.1
27	0.14	0.8
31	0.1	0.6

If the recorded maximum accelerations of all past United States earthquakes are plotted versus magnitude and distance and an extrapolation is made, an upper bound of approximately 50% g is indicated (Housner, 1965). This reflects the fact that a magnitude 8.0 earthquake with 190 mi of slipped fault is essentially just $7\frac{1}{2}$ magnitude 7.0 earthquakes laid end to end. Since the intensity of ground shaking at a point near the fault depends almost entirely upon the stress released by the fault within 30 mi of the point under consideration, it can be expected that there will be little difference in shaking near the midpoint of the slipped fault whether it is 60 or 160 mi in length.

Basically, the intensity of ground shaking must be governed by the energy released by the fault slip, i.e., it depends upon the stress drop and upon the area of slip. Since there is an upper bound for the failing stress of rock, there is an upper bound for the stress drop and, hence, an upper bound for the energy released per unit area of slipped fault. Observations of surface strains resulting from large earthquakes with surface fault displacements indicate that the stress drop is of the order of 500 to 1000 psi. This corresponds approximately to a release of shear strain at the fault of $\gamma = 0.0002$. If such strain were released instantaneously over the entire area of fault slip, essentially a step function shear wave would be generated initially having an amplitude $\gamma = 0.0002$ and traveling perpendicular to the fault at a velocity of $c = 10{,}000$ ft/sec. The passage of the wave front would generate an acceleration pulse whose area would be $\gamma c = 2.0$ ft/sec. This indicates an upper bound for the area of an acceleration pulse of approximately 2.0 ft/sec, which corresponds to a pulse whose amplitude is 50% g and whose base is 0.25 sec (Housner, 1965). It should be noted that it is the area of the pulse that is significant for engineering purposes rather than the amplitude. The largest pulses from a number of the strongest recorded ground motions are listed in Table 4.7. A list of stronger recorded earthquakes is given in Table 4.8.

4.10 INFLUENCE OF GROUND

The foregoing discussion of ground motions was based on typical conditions. It did not take into consideration such special conditions as vibrations or lurching of very soft soils, landslides, gross movement of rock, etc. It is known that under certain circumstances special soil conditions can have a strong influence on the surface ground motions. A good example of this is the ground motion recorded at Mexico City in May 1962 (Zeevaert, 1964). The center of the city is on an old lake bed that has several hundred feet of very soft soil of volcanic ash origin overlying the rock. This clearly behaved like a bowl of jelly during the earthquake, and the spectrum of

Table 4.8. EARTHQUAKES WITH STRONGER RECORDED GROUND ACCELERATIONS

Date and location	Horizontal distance to slipped fault (miles)	Magnitude	Component	Maximum acceleration (% g)
1. June 27, 1966	200 ft	5.6	N65°E	50
Parkfield, California			S25°E	—
2. May 18, 1940	4	7.0	N-S	33
El Centro, California			E-W	23
3. April 13, 1949	10	7.1	S80°W	31
Olympia, Washington			S10°E	18
4. December 30, 1934	35	6.5	N-S	13
El Centro, California			E-W	12
5. June 30, 1941	10	5.9	S45°E	24
Santa Barbara, California			N45°E	23
6. March 9, 1949	10	5.3	S01°W	23
Hollister, California			N89°W	11
7. March 10, 1933	10	6.3	S82°E	19
Vernon, California			N08°E	13
8. July, 21, 1952	25	7.7	S69°E	18
Taft, California			N21°E	17
9. April 29, 1965	35	6.5	S04°E	18
Olympia, Washington			S86°W	18
10. October 31, 1935	5	6.0	E-W	16
Helena, Montana			N-S	14
11. September 11, 1938	25	5.5	N45°E	8.2
Ferndale, California			S45°E	16
12. October 3, 1941	20	6.4	N45°E	13
Ferndale, California			S45°E	12
13. March 22, 1957	8	5.3	S80°E	13
San Francisco,			N10°E	9.5
Golden Gate Park				
14. October 2, 1933	20	5.3	S82°E	12
Vernon, California			N08°E	8.5
15. March 22, 1957	10	5.3	S09°E	10
San Francisco			S81°W	6
State Building				
16. April 13, 1949	35	7.1	N88°W	7.5
Seattle, Washington			S02°W	5.8
17. February 9, 1941	60	6.6	N45°E	7.5
Ferndale, California			S45°E	4
18. October 2, 1933	15	5.3	N39°E	6.5
Los Angeles			N51°W	6
Survey Terminal				
19. May 10, 1933	15	6.25	N51°W	6.5
Los Angeles			N39°E	4
Subway Terminal				
20. March 22, 1957	11	5.3	N81°E	5
San Francisco,			N09°W	5
Alexander Building				
21. March 22, 1957	12	5.3	N45°E	5
San Francisco,			N45°W	4
Southern Pacific Building				
22. March 22, 1957	17	5.3	N26°E	5
Oakland, California			S64°E	4

the ground acceleration had a large peak at 2.5 sec period, the fundamental period of vibration of the body of soft soil. Similar soil conditions and similar jellylike oscillations have been reported in Japan.

The factors that can influence the surface ground motion are

1. The nature of the source mechanism; the dimensions and orientation of the slipped area of fault; the stress drop; the nature of the fault movement, its amplitude, direction, time, and history.

2. The travel path of the seismic waves; the physical properties of the rock; discontinuities; layering; etc.

3. Local geology; physical properties of soil layers and sedimentary rock; vertical and horizontal dimensions of bodies of soils and rock; orientations of bedding planes; etc.

In the case of Mexico City, item 3 has a dominant influence, but recorded earthquakes in the United States do not exhibit such strong effects of local geology. For example, strong ground motions recorded at El Centro, California, at Taft, California, and at Olympia, Washington, have similar spectra despite the fact that the local geology differs greatly at these three sites. None of these earthquakes had a predominant peak on the spectrum like that of Mexico City. It also is found that when two earthquakes are recorded at the same site, e.g., at Olympia, Washington (1949, 1965), the differences between their spectra are as great as the differences between the Olympia and El Centro spectra. These earthquakes, of course, were recorded on soils that were firm compared to that at Mexico City.

If the basement granitic rock is overlain by a reasonably thick, uniform layer of alluvium, and a plane wave travels vertically upward, theory says that the amplitude of the wave at the surface of the alluvium should be twice the amplitude at the bottom of the alluvium layer. If there are multiple layers over the basement rock the amplification may be greater or less than the single layer produces, depending upon the properties of the layers. Actually, strong motions have never been recorded on basement rock so that it is not known what amplification of motion does in fact occur. There is some evidence to indicate that the local geology has a more noticeable influence on very weak ground motions than on motions of potentially destructive intensities. Further research is needed to establish the conditions under which the influence of local geology is of engineering significance.

Fig. 4.7. Artificially generated accelerogram. This random function was generated on a digital computer. Its spectrum is similar to the spectrum of recorded earthquake ground motions. A set of such accelerograms were prepared at the California Institute of Technology for use in computing the response of structures for design purposes.

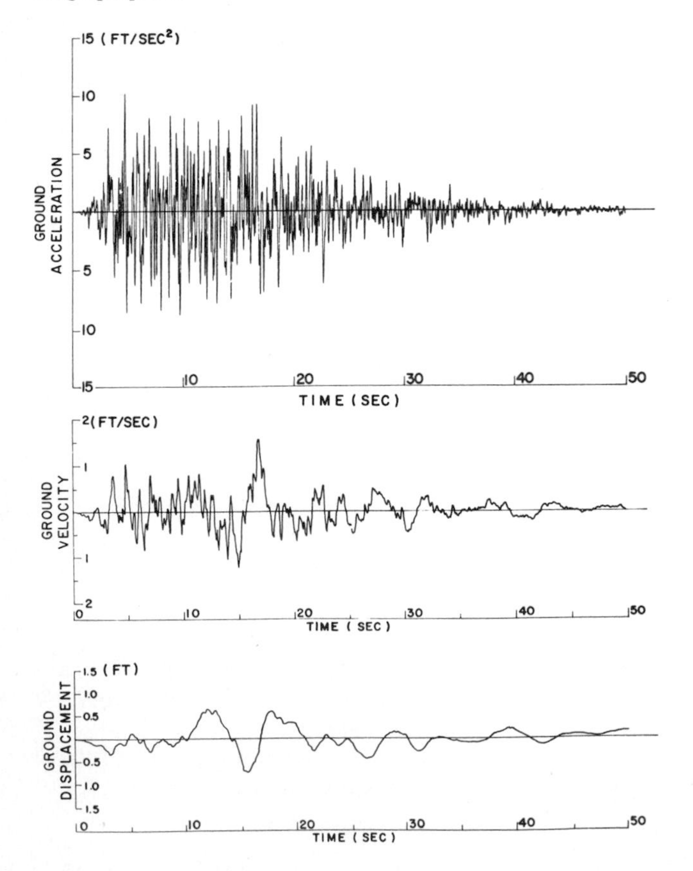

4.11 ARTIFICIALLY GENERATED GROUND MOTIONS

Recorded earthquake ground accelerations have properties similar to nonstationary random functions, but, as yet, not enough ground motions have been recorded to determine precisely the statistical properties. Because of the incomplete sample of earthquake recordings, artificially generated accelerograms are used in some studies (Housner and Jennings, 1964; Amin and Ang, 1966). These are generated by specified stochastic processes. They have the advantage that accelerograms of any duration can be generated, and the statistical properties of the accelerograms are known. An example of an artificial accelerogram is shown in Fig. 4.7.

4.12 FOURIER SPECTRUM

A standard method of exhibiting the frequency content of a function, such as an accelerogram, is by means of the Fourier amplitude spectrum. The simple linear oscillator of mass m and stiffness k, shown in Fig. 4.8a, has the equation of motion $m\ddot{y} + ky = -m\ddot{z}$, where y is the relative displacement and $\ddot{z}$ is the acceleration of the base. The vibratory response is given by the well-known expression for the relative displacement at time t:

$$y(t, \omega) = \frac{1}{\omega}\int_0^t \ddot{z}(\tau) \sin \omega(t - \tau)\, d\tau \qquad (4.3)$$

where $\omega^2 = k/m = (2\pi/T)^2$, and T is the natural period of vibration. This expression is also the vibratory response of the oscillator when the base is fixed and there is a force $(-m\ddot{z})$ applied to the mass, as shown in Fig. 4.8b. For this problem the total energy in the oscillator is $E = \frac{1}{2}m\dot{y}^2 + \frac{1}{2}ky^2$, the sum of the kinetic and the strain energies. Using Eq. 4.3 and some algebraic manipulation, the total energy can be written in the following form:

$$E(t, \omega) = \frac{1}{2}m\left[\left(\int_0^t \ddot{z} \sin \omega\tau\, d\tau\right)^2 + \left(\int_0^t \ddot{z} \cos \omega\tau\, d\tau\right)^2\right] \qquad (4.4)$$

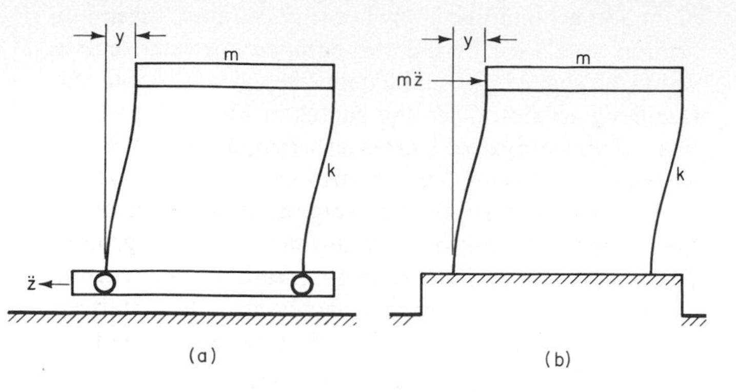

Fig. 4.8. Single-degree-of-freedom oscillator (a) subjected to base acceleration $\ddot{z}$, (b) subjected to applied force $(m\ddot{z})$. The response y is the same in both cases.

The square root of twice the energy per unit mass is

$$\sqrt{\frac{2E(t, \omega)}{m}} = \left[\left(\int_0^t \ddot{z} \sin \omega\tau \, d\tau\right)^2 + \left(\int_0^t \ddot{z} \cos \omega\tau \, d\tau\right)^2\right]^{1/2} \tag{4.5}$$

If the duration of $\ddot{z}$ is from $t = 0$ to $t = t_1$, the square root of twice the energy per unit mass at time t_1 is

$$\sqrt{\frac{2E(t_1, \omega)}{m}} = \left[\left(\int_0^{t_1} \ddot{z} \sin \omega\tau \, d\tau\right)^2 + \left(\int_0^{t_1} \ddot{z} \cos \omega\tau \, d\tau\right)^2\right]^{1/2} \tag{4.6}$$

The right-hand side of Eq. 4.6 is a function of $\omega = 2\pi/T = 2\pi f$. When it is evaluated and plotted as a function of ω or T or f, it is called the Fourier amplitude spectrum. It is customary to plot earthquake spectra as functions of the period T. A Fourier spectrum of an earthquake accelerogram (Hudson, 1963) is shown in Fig. 4.9. It is a measure of the final energy in the oscillator as a function of period. The peaks on the spectrum curve represent periods at which relatively large amounts of energy were put into the system.

Fig. 4.9. Fourier spectrum of ground acceleration component S69E recorded at Taft, California, July 21, 1952. The Fourier spectrum (dotted line) is similar to the zero-damped, velocity-response spectrum (solid line).

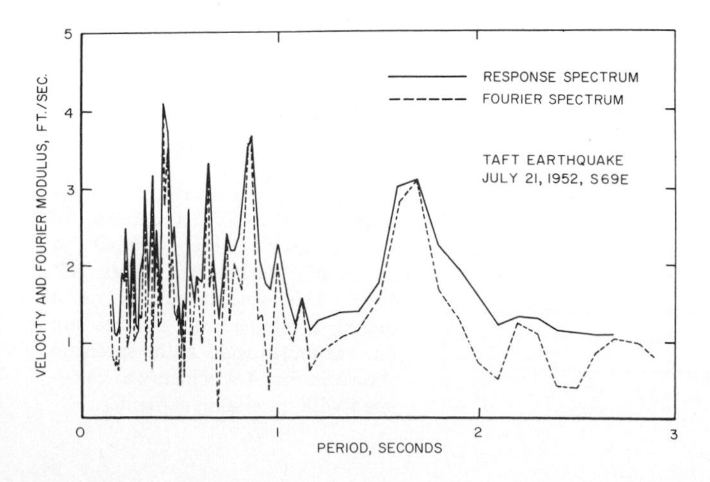

4.13 RESPONSE SPECTRUM

The integrals in Eq. 4.6 are evaluated at the end of the accelerogram, at $t = t_1$, and the Fourier spectrum is thus a measure of the energy $E(t_1, \omega)$ at the end of the excitation. The maximum value of the energy $E(t_m, \omega)$ will likely occur at some time $t_m < t_1$. From an engineering point of view the maximum energy is of more interest than the final energy because it is a measure of the maximum displacement and, hence, the maximum base shear and the maximum stress. If Eq. 4.4 is evaluated at the time t_m, which gives a maximum value of the energy $E(t_m, \omega)$, and this is plotted as a function of period or frequency, the curve is called the Energy Response Spectrum. The quantity $\sqrt{2E(t_m, \omega)/m}$ is the maximum (possible) velocity of the oscillator, and when it is plotted as a function of period or frequency it is called a Maximum Velocity Response Spectrum. This has also been plotted in Fig. 4.9, in which it is seen that its shape is similar to that of the Fourier Spectrum but its amplitude is somewhat larger.

If the oscillator in Fig. 4.8 has viscous damping its response is given by

$$y(t, \omega, n) = \frac{1}{\omega_n} \int_0^t \ddot{z}(\tau) e^{-n\omega_n(t-\tau)} \sin \omega_n(t - \tau) \, d\tau \tag{4.7}$$

where n is the fraction of critical damping and $\omega_n = \omega\sqrt{1 - n^2}$. For $n < 0.2$ the value of ω_n is, for practical purposes, the same as the value of ω. The maximum value of the displacement occurs at t_m, and it is $|y(t_m, \omega, n)|$. This is called the Displacement Response Spectrum S_d. It is customary to plot S_d as a function of period, for several values of n. The maximum velocity is $|\dot{y}(t_m, \omega, n)|$, and this is called the Velocity Response Spectrum S_v. The maximum relative acceleration is not of interest, but the maximum absolute acceleration $(\ddot{y} + \ddot{z})_{max}$ is of particular interest. The maximum force exerted on the mass m is $ky(t_m, \omega, n)$, and hence the maximum absolute acceleration is $(k/m)y(t_m, \omega, n)$. Therefore, the absolute Acceleration Spectrum is $S_a = (k/m)S_d = \omega^2 S_d = (2\pi/T)^2 S_d$.

The so-called pseudovelocity spectrum is defined to be S_{pv}, where

$$\left(\frac{T}{2\pi}\right)S_a = S_{pv} = \left(\frac{2\pi}{T}\right)S_d$$

The physical significance of S_{pv} may be explained as follows: The maximum displacement corresponds to a condition of zero kinetic energy and maximum strain energy $\frac{1}{2}kS_d^2$. If this energy were in the form of kinetic energy $\frac{1}{2}m(\dot{y})^2 = \frac{1}{2}kS_d^2$, the maximum relative velocity would be

$$\dot{y} = \sqrt{\frac{k}{m}} S_d = \left(\frac{2\pi}{T}\right) S_d = S_{pv}$$

For most earthquake ground motions, S_{pv} and S_v do not differ much. However, when $T > t_1$ there may be a marked difference. It is obvious that in the limit as T becomes very large, $S_{pv} \longrightarrow 0$ while $S_v \longrightarrow (\dot{z})_{\text{max}}$.

Figures 4.10–4.16 give examples of computed velocity and acceleration response spectra (Alford *et al.*, 1951; Housner *et al.*, 1953). It is seen that the response spectra express the maximum effect of the ground acceleration on a structure with a single period of vibration.

Figure 4.17 shows the velocity response spectrum for ground motion recorded in Mexico City (Jennings, 1962). The large peak represents the effect of the shaking of the soft ground like a bowl of jelly. Without the amplification at 2.5 sec period the damped spectrum curves would be shaped approximately like $S_v = C(1 - e^{-\alpha T})$, becoming asymptotic to the curves in Fig. 4.17 for $T >$ 3 sec. This comparison gives an estimate of the actual amplification around $T = 2.5$ sec.

Figure 4.18 is an idealized representation that shows the effects of magnitude and distance on the spectrum. The curves are smoothed representations of the zero-damped spectrum curves (backbone curves). They show that a small, close earthquake may have a stronger effect on short period structures than a large, distant earthquake, whereas for long-period structures just the opposite is true. A rule of thumb for estimating the

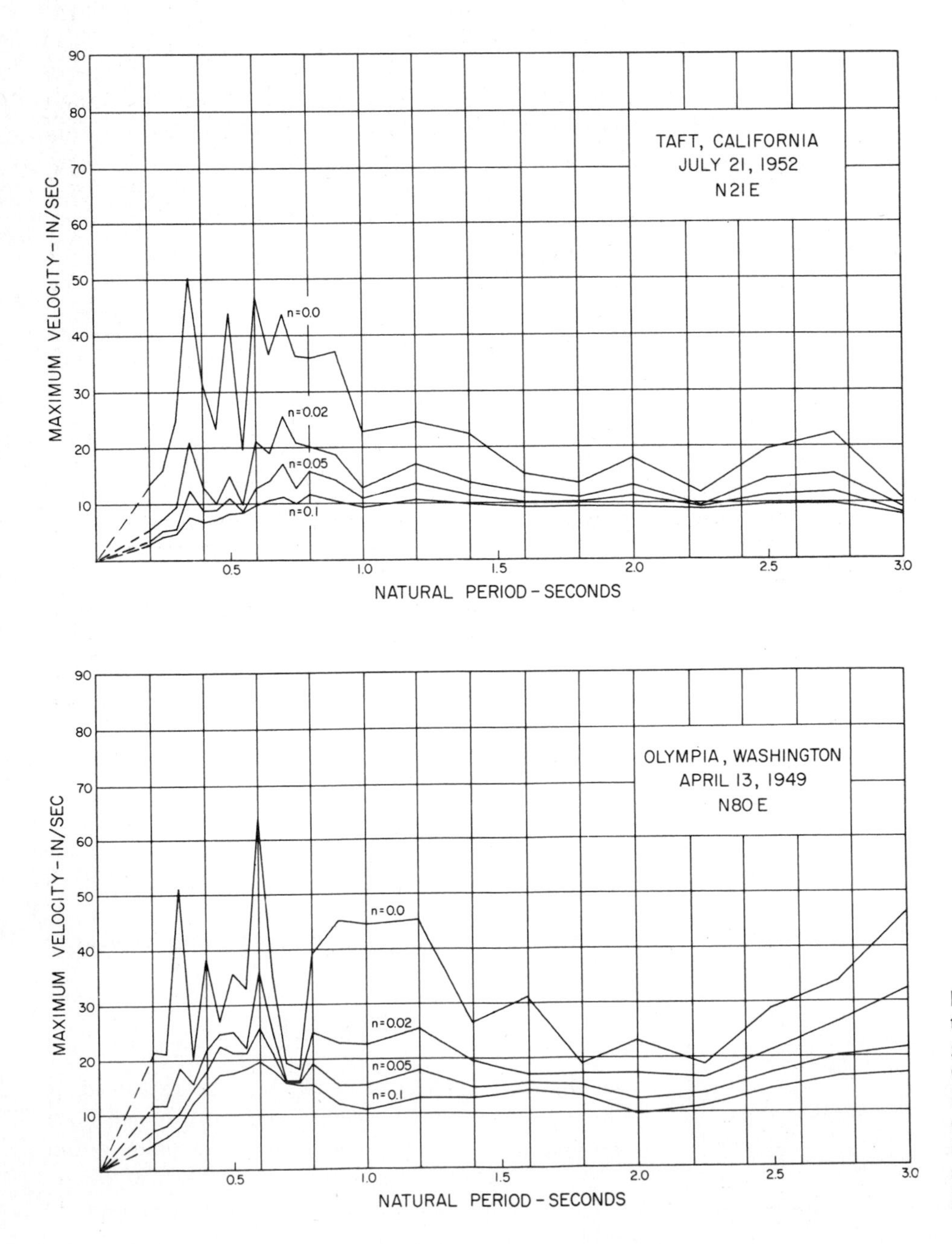

Fig. 4.10. Velocity response spectrum S_v of N21E component of ground acceleration recorded at Taft, California, July 21, 1952.

Fig. 4.11. Velocity response spectrum S_v of N80E component of ground acceleration recorded at Olympia, Washington, April 13, 1949. This was calculated on a digital computer, and it does not have as much detail as the spectrum shown in Fig. 4.9, which was calculated with an analog computer.

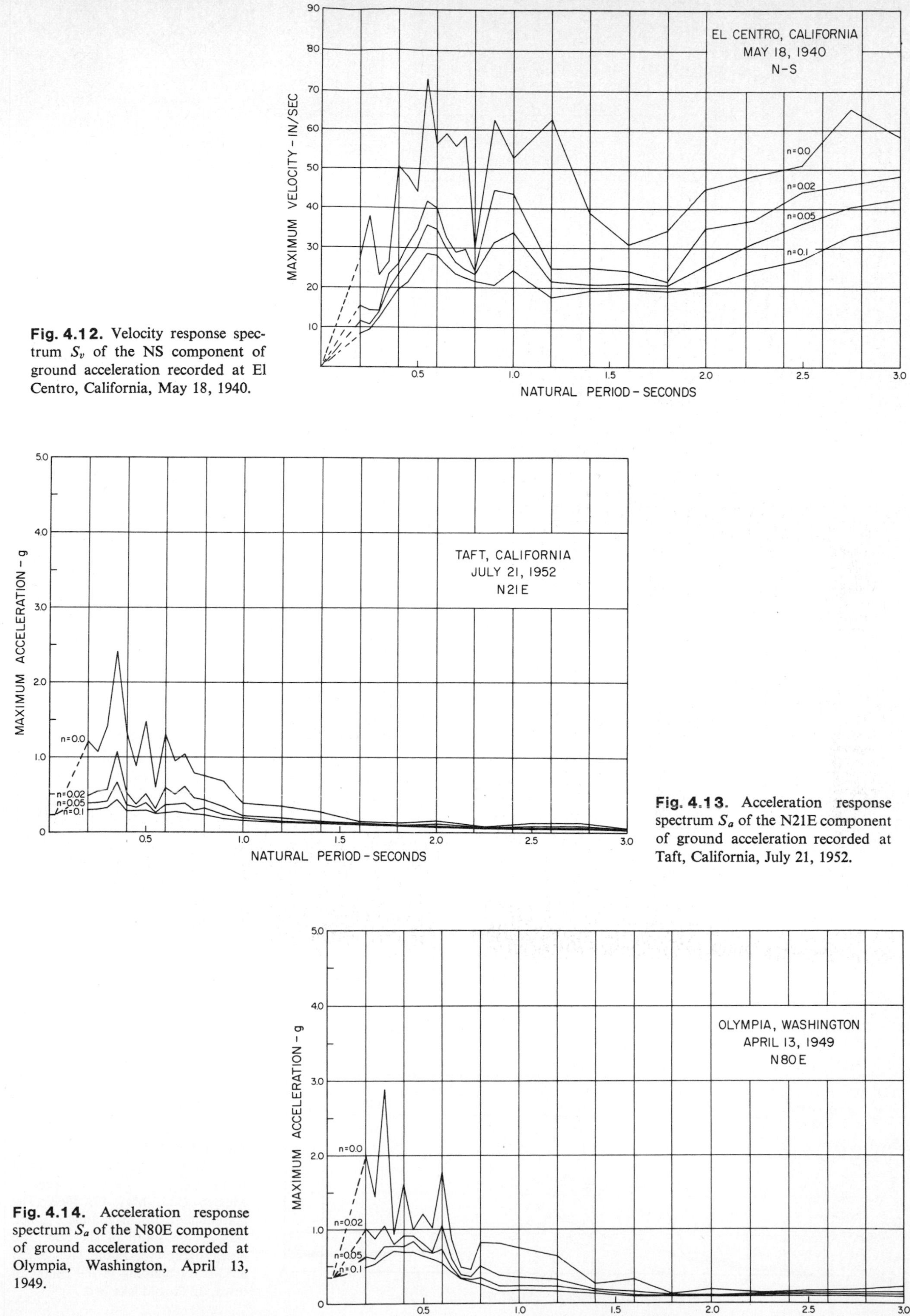

Fig. 4.12. Velocity response spectrum S_v of the NS component of ground acceleration recorded at El Centro, California, May 18, 1940.

Fig. 4.13. Acceleration response spectrum S_a of the N21E component of ground acceleration recorded at Taft, California, July 21, 1952.

Fig. 4.14. Acceleration response spectrum S_a of the N80E component of ground acceleration recorded at Olympia, Washington, April 13, 1949.

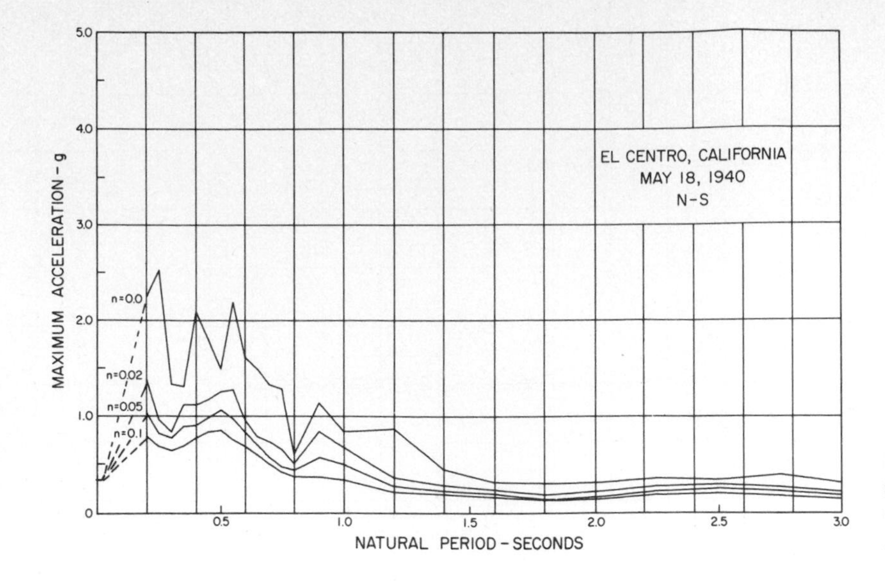

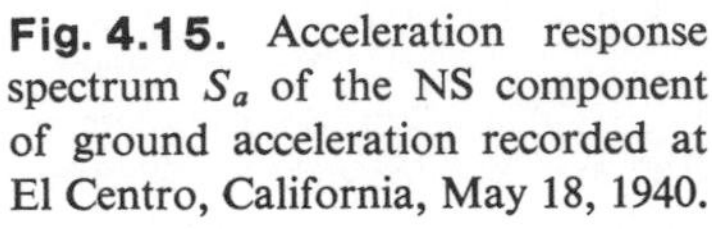
Fig. 4.15. Acceleration response spectrum S_a of the NS component of ground acceleration recorded at El Centro, California, May 18, 1940.

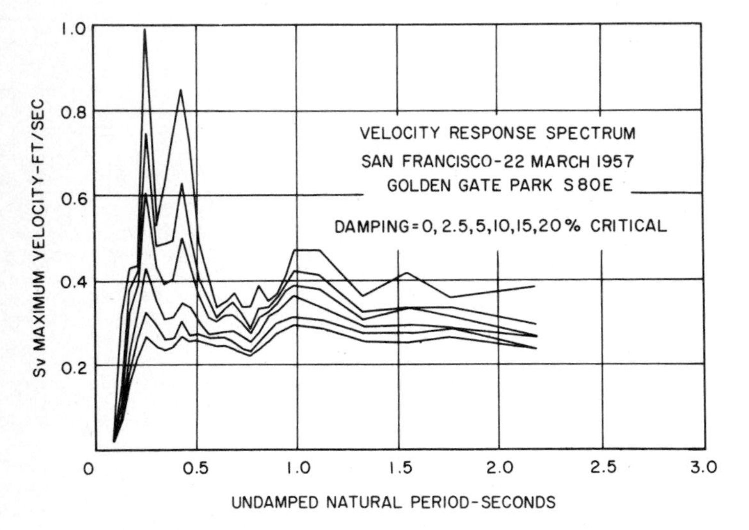

Fig. 4.16. Velocity response spectrum S_v of the S80E component of ground acceleration recorded in Golden Gate Park, San Francisco, March 22, 1957. The epicenter of this magnitude 5.3 shock was 8 mi south of the accelerograph.

fundamental period of vibration of buildings in the United States is $T = 0.1\,N$ sec, where N is the number of stories. The periods of exceptionally flexible or exceptionally stiff buildings may differ substantially from the value given by this formula (Housner and Brady, 1963).

It should be noted that ground accelerations recorded in Lima, Peru, and in Santiago, Chile, had higher frequencies than those recorded in the western United States. For example, the motion recorded in Lima on October 17, 1966, is shown in Fig. 4.19. The velocity response spectrum is shown in Fig. 4.20. Although the maximum acceleration was 40% g, the area of this acceleration pulse was not large and, hence, the velocity response spectrum ordinates are not especially large. This relatively high acceleration caused only moderate damage in Lima, which emphasizes the fact that the maximum acceleration by itself is not a satisfactory measure of severity of shaking when dissimilar ground motions are being compared.

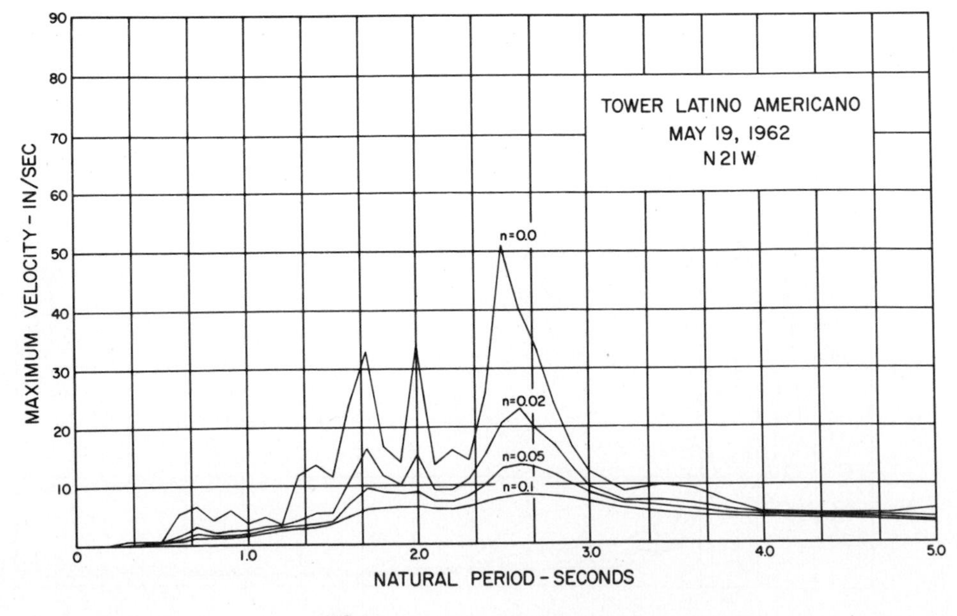

Fig. 4.17. Velocity response spectrum S_v of the N21W component of the ground acceleration recorded in Mexico City, May 19, 1962. The epicenter of this magnitude 7 shock was 200 mi south of the city. The accelerograph was in the basement of a 43-story building in the center of town, on the old lake bed.

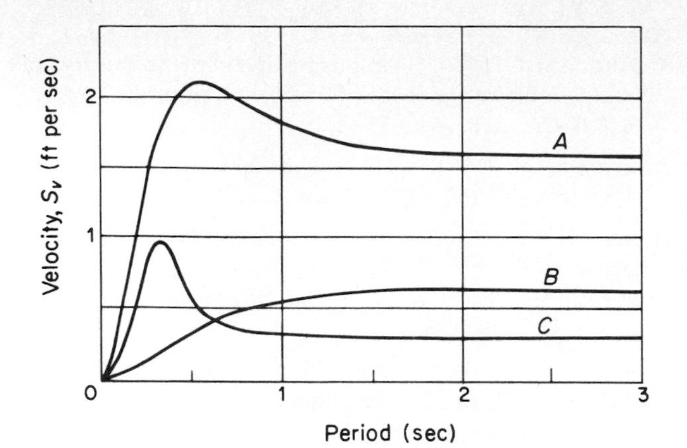

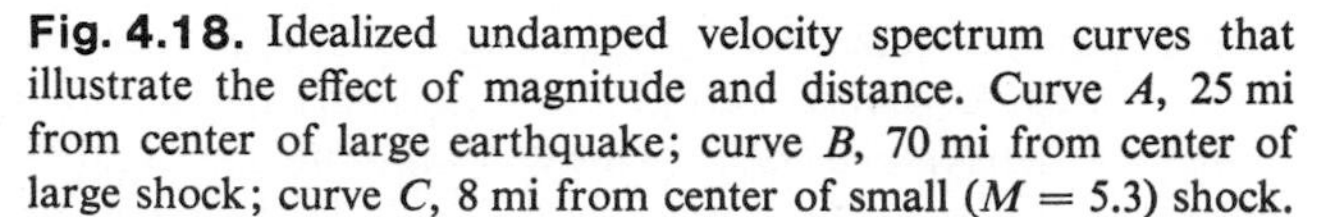

Fig. 4.18. Idealized undamped velocity spectrum curves that illustrate the effect of magnitude and distance. Curve A, 25 mi from center of large earthquake; curve B, 70 mi from center of large shock; curve C, 8 mi from center of small ($M = 5.3$) shock.

4.14 SPECTRUM INTENSITY

The severity of the vibrations experienced by a simple oscillator can be determined from the velocity response spectrum value $S_{pv}(T, n)$ corresponding to the period T and damping n of the oscillator. An average measure of the severity of the ground shaking as regards its effect on elastic structures is given by the so-called spectrum intensity SI, defined by (Housner, 1952, 1954).

$$SI_n = \int_{0.1}^{2.5} S_{pv}(T, n)\, dT \tag{4.8}$$

Another measure is the root mean square (RMS) of the strong phase of the ground acceleration (Housner and Jennings, 1964):

$$\text{RMS} = \left[\frac{1}{t_1}\int_0^{t_1} (\ddot{z})^2\, dt\right]^{1/2} \tag{4.9}$$

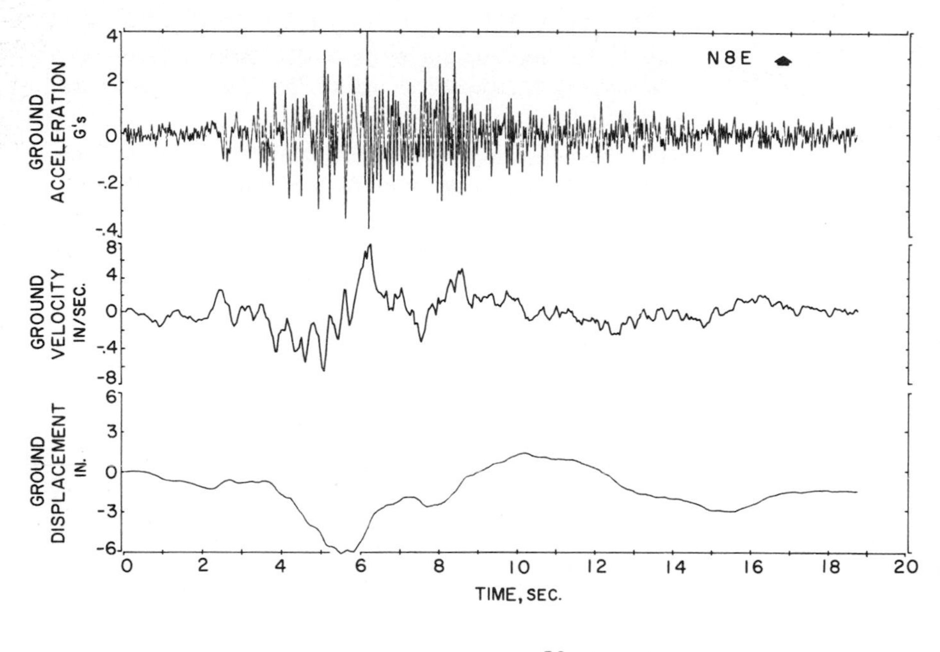

Fig. 4.19. Ground motion recorded on alluvium in Lima, Peru, during the earthquake of October 17, 1966. This motion was noteworthy for the peak acceleration of 40% g and for the high frequencies as compared to United States earthquakes.

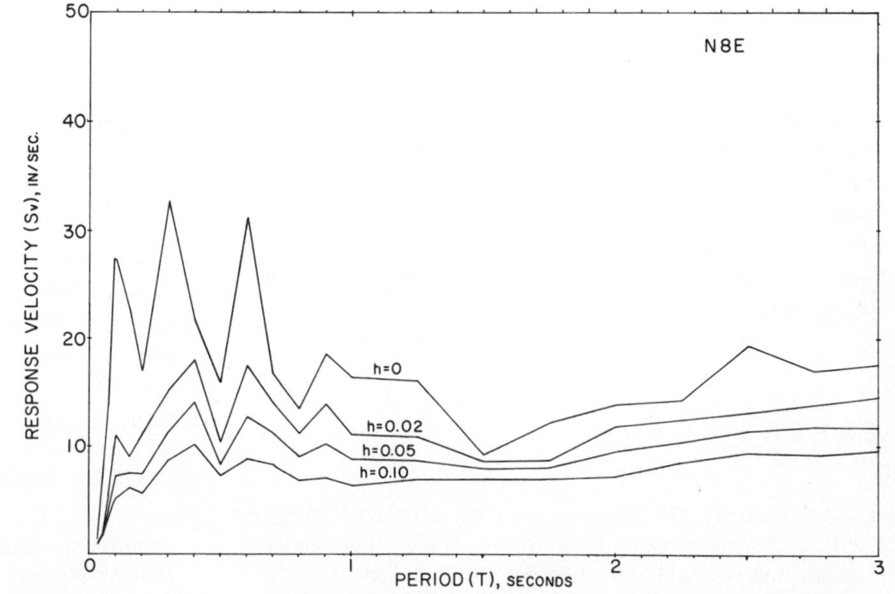

Fig. 4.20. Velocity response spectrum S_v for the ground acceleration recorded in Lima, Peru, October 17, 1966. This spectrum is less than half that of El Centro, NS, 1940, despite the fact that its maximum acceleration was greater.

Table 4.9. ABRIDGED MODIFIED-MERCALLI INTENSITY SCALE*

I.	Detected only by sensitive instruments.
II.	Felt by a few persons at rest, especially on upper floors; delicate suspended objects may swing.
III.	Felt noticeably indoors, but not always recognized as a quake; standing autos rock slightly, vibration like passing truck.
IV.	Felt indoors by many, outdoors by a few; at night some awaken; dishes, windows, doors disturbed; motor cars rock noticeably.
V.	Felt by most people; some breakage of dishes, windows, and plaster; disturbance of tall objects.
VI.	Felt by all; many are frightened and run outdoors; falling plaster and chimneys; damage small.
VII.	Everybody runs outdoors; damage to buildings varies, depending on quality of construction; noticed by drivers of autos.
VIII.	Panel walls thrown out of frames; fall of walls, monuments, chimneys; sand and mud ejected; drivers of autos disturbed.
IX.	Buildings shifted off foundations, cracked, thrown out of plumb; ground cracked; underground pipes broken.
X.	Most masonry and frame structures destroyed; ground cracked; rails bent; landslides.
XI.	New structures remain standing; bridges destroyed; fissures in ground; pipes broken; landslides; rails bent.
XII.	Damage total; waves seen on ground surface; lines of sight and level distorted; objects thrown up into air.

*This scale is a subjective measure of the effect of the ground shaking and is not an engineering measure of the ground acceleration.

A measure of the overall effectiveness of the ground motion is given by the root-square (RS) value

$$\text{RS} = \left[\int_0^{t_1} (\ddot{z})^2\, dt\right]^{1/2} \tag{4.10}$$

The foregoing measures of the intensity of ground shaking are most meaningful for elastic structures. They are indicative of the damage potential of the ground shaking only under restricted conditions. They should be reasonably good measures of the damage potential for brittle structures; however, since the degree of damage to ductile structures depends on the duration of the motion and the number of stress reversals, as well as the amplitude of vibrations, the spectrum intensity should be used to compare the damage potential only between ground motions of approximately the same duration. As yet a really good measure of the damage potential of ground motions of various durations, maximum accelerations, and frequency content is not available. This reflects the fact that there is not a good understanding of the damage process.

REFERENCES

Alford, J. L., G. W. Housner, and R. R. Martel (August 1951). *Spectrum Analysis of Strong-Motion Earthquakes*, California Institute of Technology (revised August 1964).

Allen, C. R., P. St.-Amand, C. F. Richter, and J. M. Nordquist (1965). "Relationship Between Seismicity and Geologic Structure in the Southern California Region," *Bull. Seism. Soc. Am.*, **55** (4).

Ambraseys, N. (1963). "The Buyin-Zara (Iran) Earthquake of September 1962," *Bull. Seism. Soc. Am.*, **53** (4).

Amin, M., and A. Ang (1966). *Non-Stationary Stochastic Model of Earthquake Motions, Proceedings of the Engineering Mechanics Division Specialty Conference*, Washington, D.C.: ASCE.

Cloud, W. K. (1967). "Intensity Map and Structural Damage, Parkfield, California, Earthquake of June 27, 1966," *Bull. Seism. Soc. Am.*, **57** (6).

Coast and Geodetic Survey (1958 and 1961). *Earthquake History of the United States, Part I, Exclusive of California and Western Nevada (1958); Part II, California and Western Nevada (1961)*, Washington, D.C.: U.S. Government Printing Office.

Coast and Geodetic Survey (Annual). *United States Earthquakes 1928–1966*, Washington, D.C.: U.S. Government Printing Office.

Gutenberg, B., and C. F. Richter (1956). "Earthquake Magnitude, Intensity, Energy, and Acceleration," *Bull. Seism. Soc. Am.*, **46** (2).

Gutenberg, B., and C. F. Richter (1965). *Seismicity of the Earth*, New York: Stechert-Hafner (reprint).

Halverson, H. T. (1965). *The Strong Motion Accelerograph, Proceedings of the Third World Conference on Earthquake Engineering*, Vol. I, New Zealand.

Hodgson, J. H. *Earthquakes and Earth Structure*, Englewood Cliffs, New Jersey: Prentice-Hall, 1964.

Housner, G. W. (1952). *Spectrum Intensities of Strong-Motion Earthquakes, Proceedings of the Symposium on Earthquakes and Blast Effects on Structures*, Earthquake Engineering Research Institute.

Housner, G. W. (October 1959). "Behavior of Structures During Earthquakes," *Proc. ASCE*, **85**, EM4.

Housner, G., ed. (1963). "An Engineering Report on the Chilean Earthquakes of May 1960," *Bull. Seism. Soc. Am.*, **53** (2).

Housner, G. W. (1965). *Intensity of Ground Shaking Near the Causative Fault, Proceedings of the Third World Conference on Earthquake Engineering*, Vol. I, New Zealand.

Housner, G. W., and A. G. Brady (August 1963). "Natural Periods of Vibration of Buildings," *J. Eng. Mech. Div., ASCE*, EM4.

Housner, G. W., and D. E. Hudson (1958). "The Port Hueneme Earthquake of March 18, 1957," *Bull. Seism. Soc. Am.*, **48** (2).

Housner, G. W., and P. C. Jennings (February 1964). "Generation of Artificial Earthquakes," *J. Eng. Mech. Div., ASCE*, **90** (EMI), Proceedings Paper 3806.

Housner, G. W., R. R. Martel, and J. L. Alford (1953). "Spectrum Analysis of Strong-Motion Earthquakes," *Bull. Seism. Soc. Am.*, **43** (2).

Housner, G. W., and M. D. Trifunac (1967). "Analysis of Accelerograms—Parkfield Earthquake," *Bull. Seism. Soc. Am.*, **57** (6).

Hudson, D. E. (1956). *The Response Spectrum Technique, Proceedings of the First World Conference on Earthquake Engineering*, Berkeley, California.

Hudson, D. E. (1963). "Some Problems in the Application of Spectrum Techniques to Strong-Motion Earthquake Analysis," *Bull. Seism. Soc. Am.*, **53** (2).

Hudson, D. E. (1963). "The Measurement of Ground Motion of Destructive Earthquakes," *Bull. Seism. Soc. Am.*, **53** (2).

Hudson, D. E., and W. K. Cloud (1967). "An Analysis of Seismoscope Data from the Parkfield Earthquake of June 27, 1966," *Bull. Seism. Soc. Am.*, **57** (6).

Jennings, P. C. (1962). *Velocity Spectra of the Mexican Earthquakes of 11 May and 19 May 1962*, California Institute of Technology.

Jones, J. C. (1915). "The Pleasant Valley, Nevada Earthquake of October 2, 1915," *Bull. Seism. Soc. Am.*, **5** (4).

Richter, C. F. (1958). *Elementary Seismology*, New York: Freeman.

Ryall, A., D. B. Slemmons, and L. D. Gedney (1966). "Seismicity, Tectonism, and Surface Faulting in the Western United States During Historic Time," *Bull. Seism. Soc. Am.*, **56** (5).

Steinbrugge, K. V., and W. K. Cloud (1962). "The Earthquake at Hebgen Lake, Montana, August 17, 1959—Epicentral Intensities and Damage," *Bull. Seism. Soc. Am.*, **52** (2).

Tocher, D. (1958). "Earthquake Energy and Ground Breakage," *Bull. Seism. Soc. Am.*, **48** (2).

U.S. Atomic Energy Commission (1963). *Nuclear Reactors and Earthquakes*, TID 7024, Washington, D. C.: Office of Technical Services.

Zeevaert, L. (1964). "Strong Ground Motions Recorded During Earthquakes of May 11th and 19th 1962 in Mexico City," *Bull. Seism. Soc. Am.*, **54** (1).

Chapter 5

Design Spectrum

G. W. HOUSNER

California Institute of Technology
Pasadena, California

5.1 INTRODUCTION

Basically the behavior of a structure during an earthquake is a vibration problem. The seismic motions of the ground cause the structure to vibrate and the amplitude and distribution of the dynamic deformations, and their duration, are of concern to the engineer. The chief objective of earthquake code requirements is that the structure should not be a hazard to life and limb in the event of strong ground shaking. Some damage may be sustained, but not to the extent of being hazardous. During moderate ground shaking that has a significant probability of occurrence during the life of the structure, the vibrations may be in the elastic range with no damaging amplitudes, but during strong ground shaking members may undergo plastic strains and there may be some cracking. Calculations can be made of the earthquake-induced vibrations of structures, and these will indicate the general nature and amplitude of the deformations that can be expected during earthquake ground shaking.

The actual earthquake design criteria must be based on the following considerations: the probability of occurrence of strong ground shaking; the characteristics of the ground motion; the nature of the structural deformations; the behavior of building materials when subjected to transient oscillatory strains; the nature of the building damage that might be sustained; and the cost of repairing the damage as compared to the cost of providing additional earthquake resistance. The earthquake design criteria must specify the desired strength of structures so that there is approximately a uniform factor of safety for different structures and for different parts of the same structure. This is usually done by means of a design spectrum (Housner, 1959).

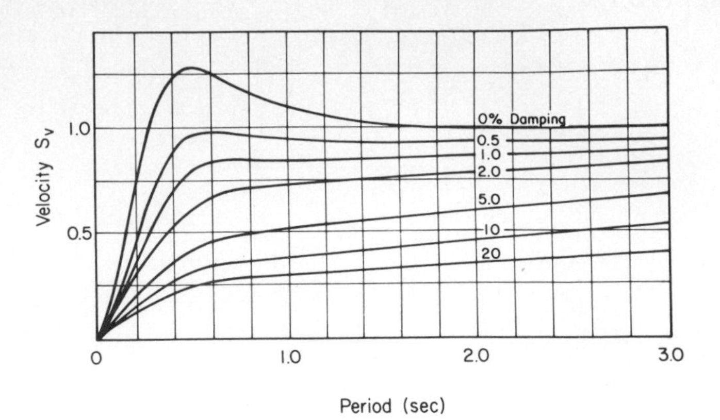

Fig. 5.1. Design spectrum giving velocity S_v as a function of period and damping; arbitrary scale.

5.1.1 Design Spectrum

The recorded ground accelerations and the response spectra of past earthquakes provide a basis for the rational design of structures to resist earthquakes. Although the individual spectra differ from each other, certain standard characteristics are exhibited by ground motions recorded under certain conditions. For example, the velocity response spectra of large United States earthquakes recorded not too far from the causative fault have approximately the characteristics of a "white noise" input for the period range 0.4–5.0$^+$ sec. For periods less than 0.4 sec, the spectrum curves dip down to low values at $T = 0.1$ sec. On the other hand, the velocity response spectra of smaller earthquakes ($M \leq 5.5$) are relatively stronger in the period range 0.2–0.5 sec and exhibit a characteristic peakedness in this region; this reflects the fact that the duration of shaking is short and, hence, the long period components are weak. Very small earthquakes, such as those of magnitude 3, exhibit an even greater peakedness. It should also be noted that ground motions of larger earthquakes that have been recorded in Lima, Peru, and Santiago, Chile, contain higher frequencies than do United States earthquakes of the same magnitudes recorded on similar ground conditions and that for the same maximum acceleration the velocity response spectrum curves have relatively smaller ordinates and are indicative of "white-noise" input over a period range of about 0.25–5.0 sec (see Chapter 4).

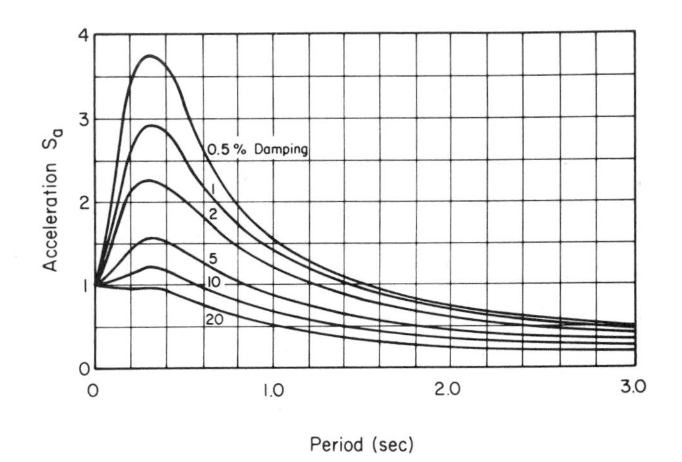

Fig. 5.2. Design spectrum giving acceleration S_a as a function of period and damping; arbitrary scale.

Idealized velocity spectrum curves based on the four strongest recorded ground motions of larger United States earthquakes (Housner, 1959) are shown in Fig. 5.1. Corresponding acceleration spectra and displacement spectra are shown in Figs. 5.2 and 5.3. Although, in an average sense, the shapes of these curves are consistent with ground motions recorded in El Centro and Taft, California, and Olympia, Washington, the shape would not be consistent with ground motions recorded in Mexico City or in Lima, Peru. Neither are the shapes appropriate for earthquakes of small magnitude.

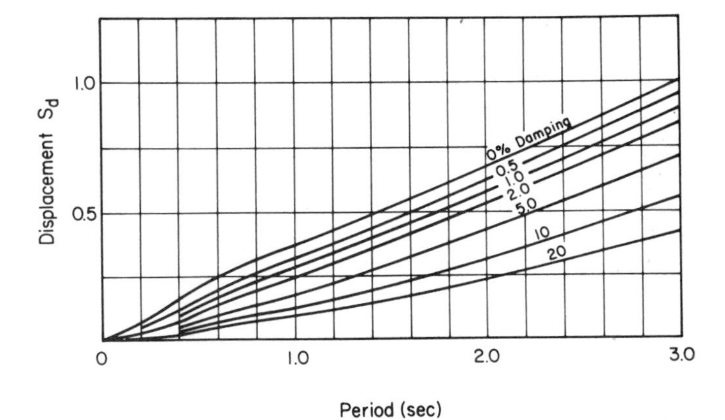

Fig. 5.3. Design spectrum giving displacement S_d as a function of period and damping; arbitrary scale.

It has been proposed that idealized spectrum curves be used as "design spectrum" curves (Housner, 1959; U.S. Atomic Energy Commission, 1963). They have been replotted in Fig. 5.4 on special log paper that permits values of S_a, S_{pv}, and S_d to be read from the same graph. In recent years, various other proposals have been made for constructing design spectrum curves, and in this connection it should be kept in mind that design spec-

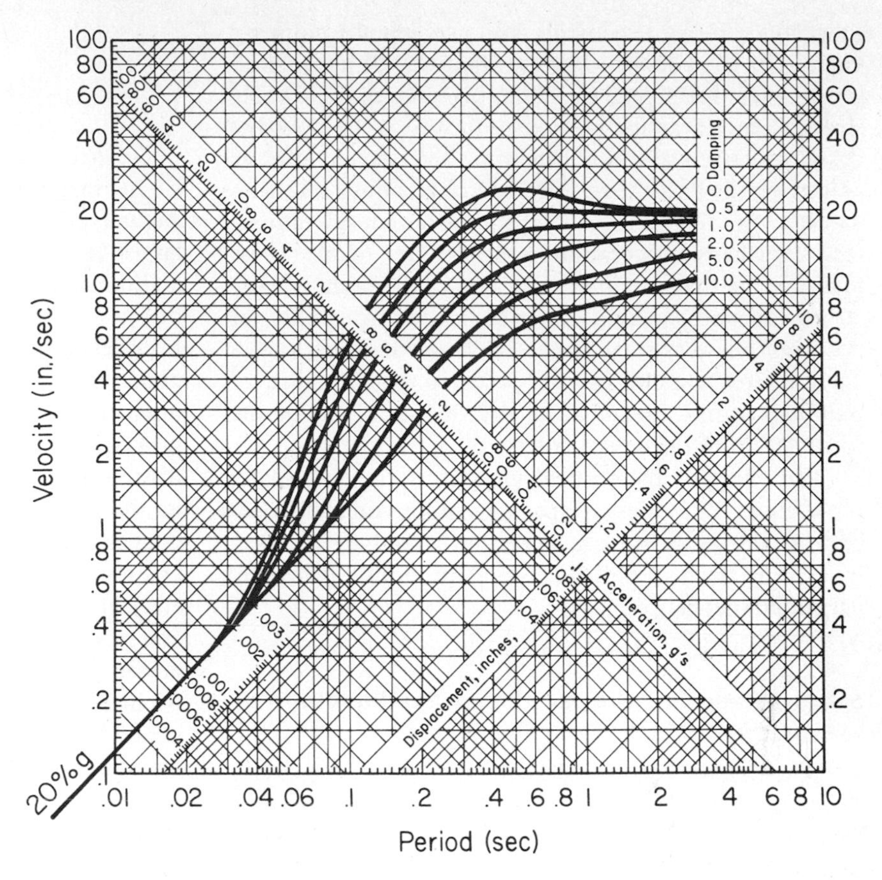

Fig. 5.4. Combined plot of design spectrum giving S_a, S_v, and S_d as a function of period and damping; scaled to 20% g acceleration at zero period.

trum curves must be consistent with ground motions actually recorded in the past. Recorded ground motions form the experimental basis of engineering seismology.

It should be noted that a calculated response spectrum is not the same as a specified design spectrum. The response spectrum is a convenient way of describing a particular earthquake ground motion by plotting the maximum response velocity, response acceleration, or response displacement of an oscillator, and it should be noted that this is an instantaneous value that may not be approached again during the earthquake. For example, the time-history of motion of an oscillator having a period $T = 1.0$ sec may show a maximum acceleration of 25% g, which is the response spectrum value, whereas the second and third highest accelerations might be 20%g and 15% g and there might be numerous peaks of 10% g. This raises the question of whether the design should be based on the instantaneous value of 25% g or whether it should be based on the numerous 10% g values. This question is particularly significant in the short period range ($T < 0.5$ sec), because here the calculated response spectrum is dominated by the single largest acceleration pulse. For example, the Lima, Peru, accelerogram had a peak acceleration of 40% g and, hence, the acceleration response spectrum has a value of 40% g at $T = 0$. This acceleration was associated with a very narrow pulse whose area was less than one-third that of the 33% g pulse on the El Centro, 1940, accelerogram. Such a pulse would not be expected to be very damaging and, in fact, the damage in Lima was moderate. Also, ground motions having short durations, such as in the Parkfield, 1966, earthquake, are not as damaging as might be inferred from the response spectrum (Cloud, 1967).

The design spectrum is not a specification of a particular earthquake ground motion; it is a specification of the strengths of structures. This is analogous to the building code requirements for wind forces that, in effect, specify the strengths of buildings and do not describe actual wind pressures. It should be noted that the design spectrum by itself specifies only the relative strengths of structures of different periods. In order to specify the actual strengths of structures it is necessary to prescribe the damping and the allowable design stresses. The actual earthquake forces used in the design of a structure will depend strongly on the damping that the structure is assumed to have. Using a large design spectrum with a large value of damping may give a smaller design force than using a small spectrum with a small value of damping. The actual damping that structures may have when vibrating strongly is not well known, so this must be estimated. Furthermore, a decision must be made as to the allowable design stresses to be used: Should ordinary code values be used, or ordinary code values plus one-third increase for transient loading, or yieldpoint stresses, etc.? It would not be proper to specify a design spectrum without also taking into account the damping values and the allowable stresses that will be used. Similarly, when specifying the damping and the allowable stresses, consideration should be given as to how the design spectrum was established: on the basis of average values, on the basis of an envelope of response spectrum values, on the basis of theoretical considerations, etc. A further consideration is the amount of overstress and damage that would be tolerated in the event of very strong ground shaking for which the probability of occurrence is very small. In all of these considerations, a fixed reference point is the observed performance of buildings during earthquakes, and this should guide the formulation of the design criteria.

The effect of the vertical ground motion is usually represented by a design spectrum approximately one-half to two-thirds as large as the horizontal design spectrum. The vertical and horizontal motions, of course, act simultaneously.

5.1.2 Estimating the Maximum Earthquake

When establishing the design spectrum it is necessary to have some idea of the ground shaking that might be experienced at the location under consideration. For this purpose the earthquake history (Coast and Geodetic Survey, 1958, 1961; annual) of the region should be examined for information on seismicity, and the geology of the region should be studied for evidence of recent faulting or other indications of tectonic activity. These factors by themselves are usually difficult to evaluate. They are best evaluated in the highly seismic regions of California, where recent faulting is relatively well defined, where there is a relatively full earthquake history, and where measurements of ground motions have been made. A relatively good evaluation also can be made for Florida, where there is no evidence of recent faulting and no history of any significant earthquakes. For regions between these two extremes it is not so easy to establish the meaning of the geologic and seismic evidence and, in such cases, it is usually helpful to compare the evidence with the corresponding evidence in California and Florida and then decide just where, in between these two extremes, the seismicity of the region lies.

A key item in the design of some important structures is the largest earthquake that might occur near the site under consideration. This must be considered, e.g., in planning a nuclear reactor power plant or a major dam. Unfortunately, in most of the United States the geologic evidence is not convincing and the seismic history is fragmentary. The essential difficulty can be explained as follows. The frequencies of earthquakes are reasonably well described by an equation of the form (Gutenberg and Richter, 1966)

$$N = AN_0 e^{-M/B} \tag{5.1}$$

$$n = -\frac{dN}{dM} = \frac{1}{B} AN_0 e^{-M/B} \tag{5.1a}$$

where N is the number per year of shallow earthquakes having magnitude equal to or greater than M, in area A. Actually, the historical record is usually very meager for larger earthquakes because they occur so infrequently and the data is also incomplete for small shocks because these are more difficult to record and tedious to tabulate. It appears, however, that an equation of this form is a good representation down to at least $M = 3$. The annual number of shocks per unit area N_0 having $M \geq 0$ is a measure of the average seismicity of the region. The distribution parameter B describes seismic severity, that is, the relative frequencies of large versus small earthquakes. For a 100,000 mi^2 area of southern California and northern Mexico, which is highly seismic, the parameters have the values (Allen *et al.*, 1965) $N_0 = 1.7/\text{mi}^2$ and $B = 0.48$ as determined on the basis of a 29-year period, 1934–1963. For the entire world (Gutenberg and Richter, 1966), $AN_0 = (2.5)10^7$ and $B = 0.48$, as determined for a 43-year period, 1904–1946. The values for the world are well defined since they are based on a record that contains many large shocks, but the values for southern California are not so well defined for large shocks, because the number of very large earthquakes in this smaller area is relatively low.

For purposes of plotting, it is customary to put Eq. 5.1 a in the form

$$\log_{10} n = a - \frac{M}{b}$$
$$a = \log_{10}\left(\frac{1}{B} AN_0\right) \qquad b = 2.3B \tag{5.2}$$

where (ndM) is the number of shocks having magnitudes between M and $M + dM$. This equation, plotted as a straight line in Fig. 5.5 is a histogram of earthquake frequency. The data for a 43-year period of world earthquakes (Gutenberg and Richter, 1966) fit this line closely for $B = 0.48$, except that above $M = 8$ the observed frequency of earthquakes drops off, as shown in Fig. 5.5, and goes to zero at about $M = 8.7$, which therefore may be taken as a practical upper bound. When the earthquake history of smaller regions is examined, the data again are fitted well by Eq. 5.2, with values of B lying in the range from 0.4 to 0.6. It should be noted that the seismicity is not necessarily constant over the years, and the value of N_0 may fluctuate with time, e.g., as shown in Fig. 5.6.

When smaller areas are considered, the historical record of earthquakes is often not sufficient to define the curve for larger magnitudes as the data may not include any of the larger earthquakes. For a region like California or southern California, where large earthquakes are known to have occurred, it is customary to take $M = 8.5$ as the upper bound. The frequency distribution for southern California is shown in Fig. 5.8, with the dotted line showing the assumed drop-off at large magnitudes. For a less seismic region, such as the state of Oregon or the eastern part of the United States, it is not known what is the correct shape of the frequency distribution curve at large magnitudes. If the tectonic processes were the same in two regions, differing only in rate of straining, the shapes of frequency curves should be the same, but the tectonic processes are not the same throughout the United States. In many regions of the United States the available geologic and seismic evidence indicates that large-magnitude earthquakes are not expected. However, the large New Madrid, Missouri, earthquakes of 1811–1812 and the Charleston, South Carolina, earthquake of 1886 occurred in regions of otherwise low seismicity, which indicates that the possibility of large-magnitude shocks cannot be discounted just because a region has had relatively low seismicity during recent times.

From a practical viewpoint, two different approaches may be used to deal with the question of the maximum earthquake. One, in effect, assumes different shapes for the frequency distribution curves for regions of different seismicity, whereas the other approach takes all regions to have frequency distribution curves of the same shape. The first method is based on a seismic probability map such as that shown in Fig. 5.7. The maximum intensity of shaking in the various zones and the approximate cor-

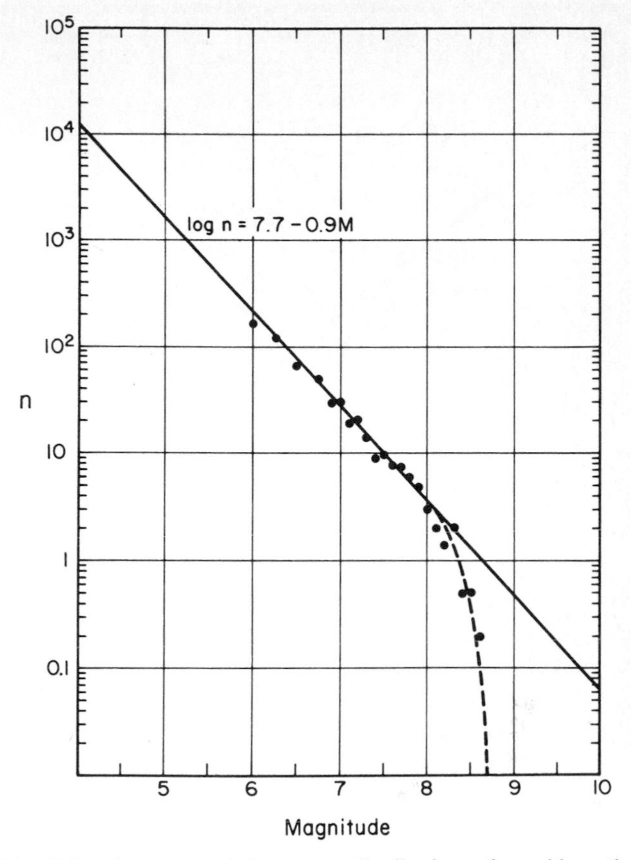

Fig. 5.5. Mean annual frequency distribution of world earthquakes, 1904–1946; $n dM$ is the mean annual number of shocks having magnitudes lying between M and $M + dM$.

Fig. 5.6. Annual number of shocks having magnitudes 3 or greater in a southern California region (Allen *et al.*, 1965). The mean annual number is 220, and deviations about the mean are large. The statistical fluctuations are progressively larger for earthquakes of magnitude greater than 3 during this period.

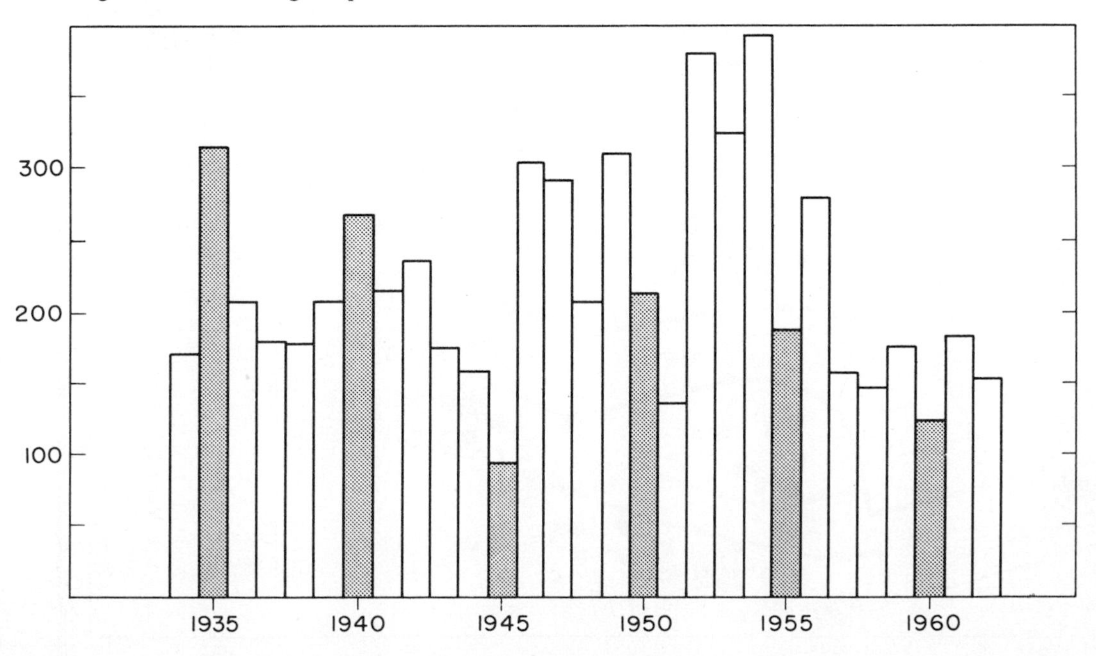

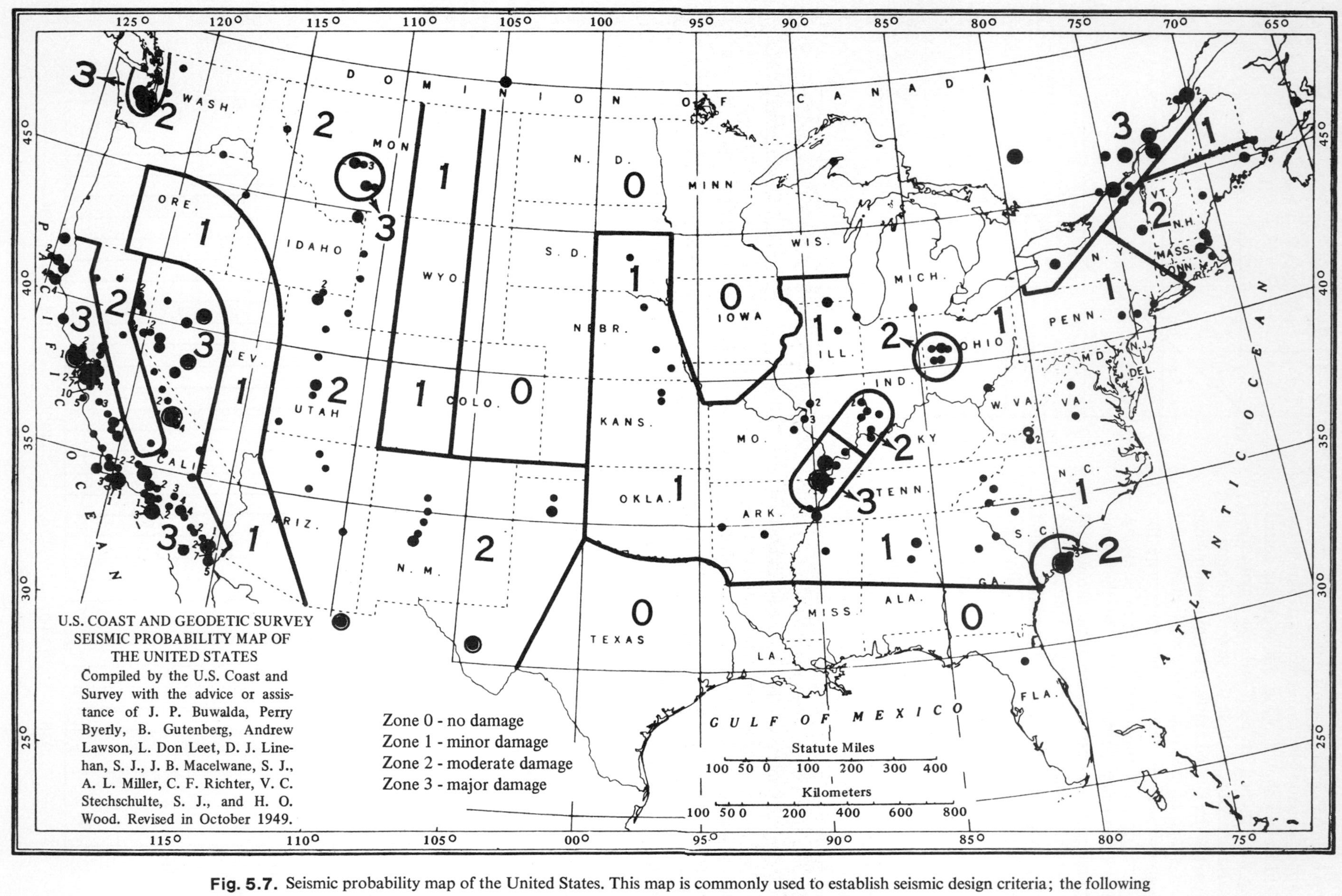

Fig. 5.7. Seismic probability map of the United States. This map is commonly used to establish seismic design criteria; the following maximum ground accelerations are associated with the zones: Zone 3, 33 % g; Zone 2, 16 % g; Zone 1, 8 % g; Zone 0, 4 % g. In Zone 3 close to a major active fault the maximum ground acceleration is estimated to be approximately 50 % g.

responding magnitudes (Housner, 1965) are usually taken as shown in Table 5.1.

Table 5.1. MAXIMUM ZONAL ACCELERATIONS

	Maximum acceleration, %g	M
Zone 3 (near a great fault)	50	8.5
Zone 3 (not near a great fault)	33	7.0
Zone 2	16	5.75
Zone 1	8	4.75
Zone 0	4	4.25

The second approach assumes that earthquakes of all magnitudes are possible in any region of the world. This approach assumes the frequency distribution to be the same in all regions:

$$p(\geq M) = e^{-M/B} \qquad (0 < M < \infty) \tag{5.3}$$

$$p(\geq M; \leq M + dM) = Be^{-M/B}\, dM \tag{5.3a}$$

With this ideal frequency distribution there is no upper bound for magnitude. As a practical matter, however, a lower bound must be set for meaningful probability. For example, in California it seems most unlikely that an earthquake having $M > 8.5$ will occur; hence, when making probability calculations by means of Eq. 5.3, the probability for $M > 8.5$ can be considered to be negligible. This, then, specifies a "negligible probability" to be applied when considering regions of lower seismicity than California. For example, the expectation of shocks having magnitudes greater than M during Y years, per unit area, is

$$E = YN_0e^{-M/B}$$

and if this is less than some specified value E_1 corresponding to M_1, the hazard of earthquakes having magnitudes greater than M_1 is considered to be negligible. Let $M_1 = 8.5$ for California, which has seismicity N_{01}, $B_1 = 2.1$, and $A_1 = 150{,}000$ mi². A second region might have seismicity N_{02}, distribution parameter B_2, and an area A_2. The corresponding upper bound for M_2 is determined by the condition $E_2 = E_1$, which gives

$$M_2 = \frac{B_2}{B_1}M_1 - B_2 \log_e\left(\frac{N_{01}}{N_{02}}\right) \tag{5.4}$$

This is equivalent to taking the frequency distribution curve of all regions to have the shape of the southern California curve shown in Fig. 5.8; for less seismic regions this curve is moved to the left until it agrees with the observed seismicity N_{02} of the region. If both regions have $B = 0.48$, the upper bounds for magnitude are as shown in Table 5.2. These are also shown in Fig. 5.9 as dotted lines. The values given in Table 5.2 are not predictions of maximum earthquakes but are a consistent set, in a probability sense, of upper bounds for magnitudes. When making such probability calculations it must be kept in mind that a certain minimum area is required for an earthquake, e.g., a magnitude 8.5 shock would require an area about that of California.

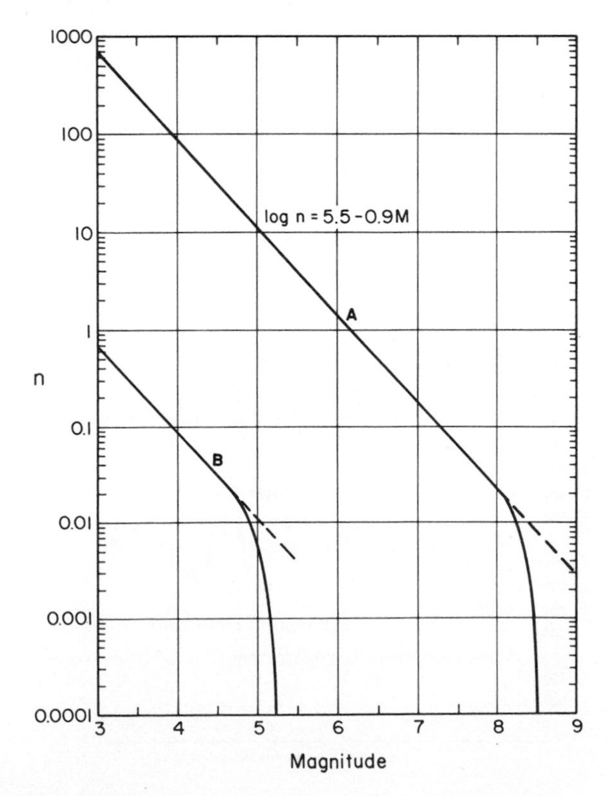

Fig. 5.8. Mean annual frequency distribution of southern California earthquakes, curve A. The distribution curve is based on data for the years 1934–1963 and the estimated upper bound for magnitude is $M = 8.5$. Curve B shows the same distribution curve, moved to the left, to represent a region with a seismicity 0.001 times that of southern California.

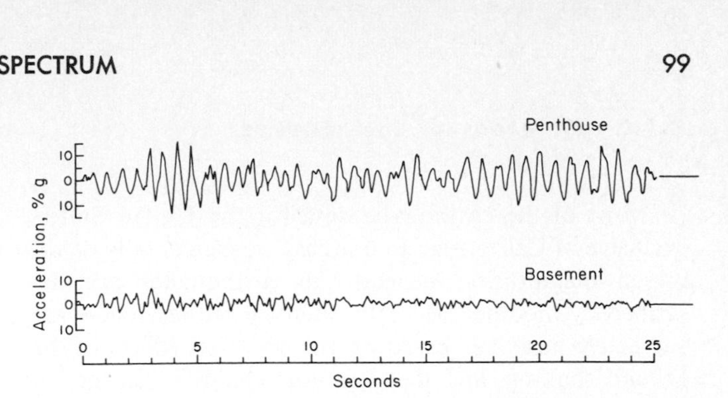

Fig. 5.9. Earthquake-induced vibrations of the 10-story, reinforced concrete Hollywood Storage Building during the July 21, 1952, Tehachapi earthquake (epicentral distance, 65 mi). The exterior concrete walls of the building had relatively few window openings and, hence, were relatively stiff.

Table 5.2. UPPER BOUND M_2 FOR MAGNITUDE AS DETERMINED BY SEISMICITY RATIO

N_{01}/N_{02}	1	2	4	5	10	20	30	100	1000
M_2	8.5	8.15	7.8	7.7	7.4	7.1	6.9	6.3	5.2

5.1.3 Felt Areas of Earthquakes

Most of the earthquake data for the United States, exclusive of California, do not have instrumentally determined magnitudes; instead the earthquakes are described by modified Mercalli intensity ratings. These are subjective measures based on the observed effects of the ground shaking, and they are not reliable indicators of the magnitude of the earthquake. Usually the total area over which the earthquake was felt is also given, and this is a more reliable indicator of the magnitude than is the Modified Mercalli intensity at the epicenter. Because of geologic conditions, the felt area in the eastern United States is greater for a given magnitude than in the western United States. The following empirical formulas are idealized relations between the magnitude M and the felt area A in square miles. The reported values of A have, of course, considerable scatter about these curves.

1. Western United States:

$$M = 2.3 \log_{10} (A + 3000) - 5.1$$

2. Rocky Mountain region and central region:

$$M = 2.3 \log_{10} (A + 14{,}000) - 6.6$$

3. Eastern region:

$$M = 2.3 \log_{10} (A + 34{,}000) - 7.5$$

When the estimated magnitudes for a given region have been computed by means of the foregoing relations, the distribution of magnitudes presumably should be statistically consistent with Eq. 5.1.

5.2 MEASURED BUILDING BEHAVIOR

Knowledge of the true dynamic properties of actual structures is essential for earthquake design. To obtain such information requires that measurements be made at a time when the structure is excited into strong vibration by an earthquake. It is for this reason that the City of Los Angeles Building Code requires that each new large building be instrumented with three strong-motion accelerographs, one in the basement, one at midheight; and one on the top floor. Some information about the dynamic properties of structures can be obtained without waiting for an earthquake. The fundamental periods of vibration of buildings are determined by recording the small vibrations induced by wind and microtremors. Also periods, mode shapes, and damping are determined by forcing the building to vibrate by means of a shaking machine. The potential value of information about the true dynamic properties of structures would justify a much greater effort than is currently being made to measure the true dynamic behavior of structures.

5.2.1 Recorded Earthquake Response of Buildings

Accelerographs installed in the basements and in the upper parts of buildings are used to record the motions induced by earthquakes. Figure 5.9 shows the motion recorded in the Hollywood Storage Building during the earthquake of July 21, 1952. This magnitude 7.7 shock centered approximately 65 mi north of the building, and it caused some damage to multistory buildings in the Los Angeles area but did not damage one- and two-story buildings (short period structures). The Hollywood Storage Building was a 10-story, reinforced concrete structure founded on moderately firm alluvium. The motion recorded on the roof of the building shows strongly the natural period of vibration of the fundamental mode of the structure, whereas the motion recorded in the basememt shows no influence of the building vibration. The response spectrum of the basement motion did not show any peak corresponding to the natural period of vibration of the structure, and this indicates that the vibratory forces exerted by the building did not appreciably deform the soil.

Figure 5.10 shows the motion recorded in a Japanese building during the Niigata earthquake of June, 1964. The six-story reinforced concrete building, shown in Fig. 5.11, had a three-story penthouse and the accelerometer was located in the second story of the penthouse. Also shown in Fig. 5.10 is the computed motion of the second floor of the penthouse (Osawa and Murakami, 1966), using the basement record as input. In the calculations the building was represented by a lumped-mass dynamical system whose stiffness and damping were adjusted to give the best agreement with the recorded penthouse motion.

Fig. 5.10. Earthquake-induced vibrations, east–west of the Akita Prefecture Building during the Niigata, Japan, earthquake of June 16, 1964. The penthouse had a maximum acceleration of approximately 20 % g. The computed response was based on a lumped mass model with 5 % damping (Osawa, 1966). See Fig. 5.11 for locations of accelerographs.

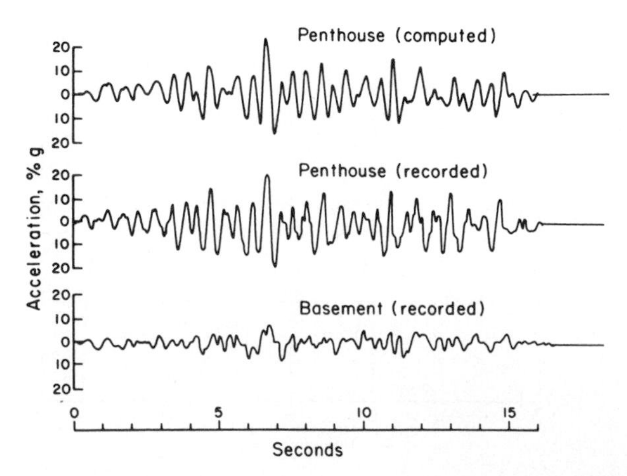

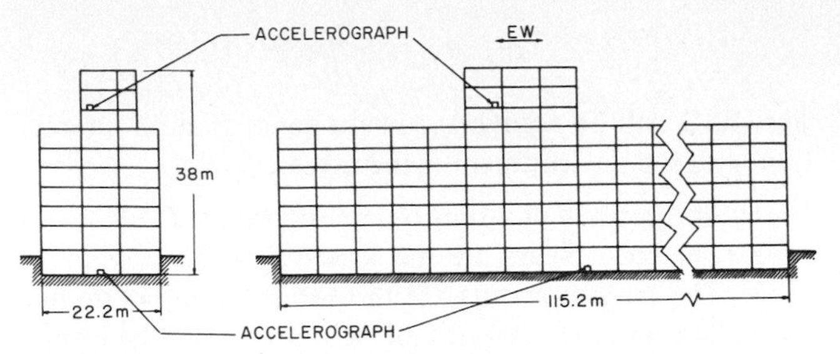

Fig. 5.11. Akita Prefecture Building. This modern six-story plus penthouse, reinforced concrete frame building survived, with no significant damage, the shaking shown in Fig. 5.10.

5.2.2 Measured Dynamic Properties of Buildings

Measurements of the vibrations of structures are made to build up an understanding of the behavior of real structures during earthquakes. One way of obtaining such information is to install a shaking machine in the upper part of the building and force the building into vibration. This permits the mode shapes, the natural periods of vibration, and the damping in the mode to be measured. Figure 5.12 shows the results of such a vibration test made on the library building at the California Institute of Technology (Kuroiwa, 1967). This slender eight-story building had reinforced concrete shear walls so that the deformation was primarily that of a cantilever beam in bending. As shown in Fig. 5.12, the damping was very small. The building was founded on a mat-type footing on a moderately firm layer of alluvium 1000 ft deep. Measurements showed that foundation rocking was negligible. On the other hand, the vibrating building did transmit oscillations to the surrounding ground, and sensitive seismographs recorded this induced ground motion several miles from the building; it was also recorded at the Caltech Seismological Laboratory, which is founded on granite at a distance of 3 mi.

Fig. 5.12. Shaking-test response curves of slender, eight-story reinforced concrete, shear-wall Millikan Library Building at California Institute of Technology. Steady-state sinusoidal motion was induced by means of two vibration generators having counter-rotating masses and a precise speed control. The vibration generators were mounted on the eighth floor, and the measurements, shown by the dots in the diagram, were made on the eighth floor. The response curves define the natural frequency and damping of the first mode. Measurements made simultaneously on different floors define the mode shape. Six different masses were employed having resultant force equal to 432, 834, 1495, 1890, 2620, and 3290 lb. Damping increased with increasing force amplitude ranging from 1.0 to 1.5% of critical.

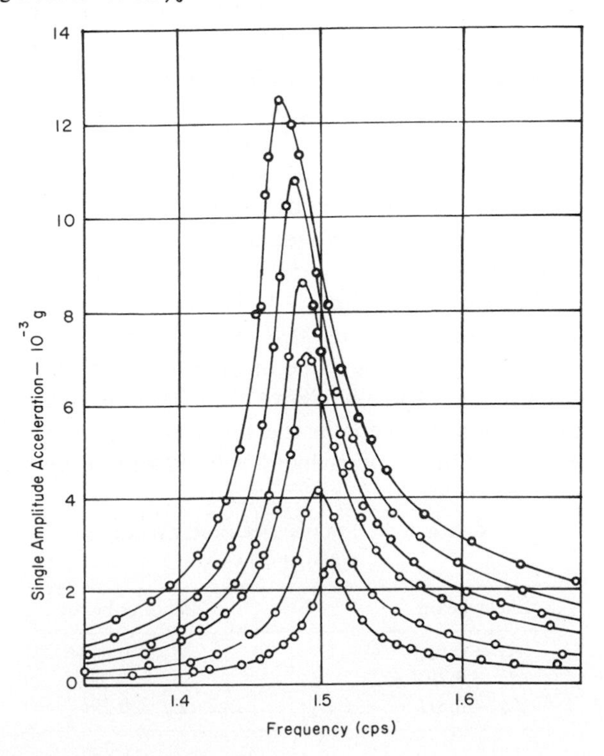

5.3 EARTHQUAKE ANALYSIS AND DESIGN

When making an analysis to determine the earthquake forces for design, several different approaches are used. The simplest is to use the Uniform Building Code earthquake provisions. This is a simplified approach that does not take into account all of the significant dynamic properties of structures. Important structures, such as 50-story buildings, nuclear power plants, large dams, long suspension bridges, and offshore oil drilling platforms, should be given a more thorough dynamic analysis. This is done by specifying the design spectrum, the damping to be used, and the allowable design stresses. The natural periods and modes of vibration are then computed, and the amplitude of each mode is determined from the design spectrum. Commonly, only a few of the lower modes are used and the computed modal stresses and strains are combined by taking the square root of the sum of the squares. A somewhat more conservative approach is to take the sum of the lowest mode plus the square root of the sum of the squares of the higher modes. An upper bound, of course, is given by taking the sum of the absolute values of the modal contributions. For complex structures it is not always clear how the modal contributions are best combined, and in this case the following analysis is made. The response of each of the modes to a prescribed accelerogram is computed, and the modal responses are added to obtain the complete time history of the response. The input acceleration is scaled so that its response spectrum has an ordinate equal to or greater than the ordinate of the design spectrum in the neighborhood of the significant periods. This should be done for several different inputs to see how great a statistical fluctuation there might be.

Sometimes computations are made of the response of

a structure to a single recorded earthquake accelerogram as a step in the design process. There are several difficulties in this approach that in essence specifies an accelerogram as the design criterion. First, it is not known whether the selected accelerogram is appropriate to use. Sometimes the El Centro, 1940, accelerogram is used even though this accelerogram has special features associated with being recorded close to a fault that had a large surface displacement during the earthquake. Second, there is a question as to the appropriate allowable design stress to use when the peak response stress is to govern the design. This would not necessarily be the same as the allowable stress used with a design spectrum. Third, errors in assumed damping and other quantities can be cumulative when a large number of modes are involved.

5.3.1 Equivalent Simple Oscillator

When using the design spectrum the response of each mode of the structure is determined by means of an equivalent simple oscillator (ESO). Since the response spectrum is the computed maximum response of a simple oscillator, the design spectrum also applies to a simple oscillator. A more complex vibrating system, such as the distributed mass beam shown in Fig. 5.13a, which is vibrating freely in a particular mode of vibration

$$y_i = A_i f_i(x) \sin \frac{2\pi}{T_i} t$$

will have an ESO, corresponding to this modal vibration, whose motion is described by

Fig. 5.13. The slender cantilever structure shown in (a) is vibrating in the ith mode. The equivalent simple oscillator shown in (b) has the same period of vibration, the same energy of vibration, and the same base shear and moment.

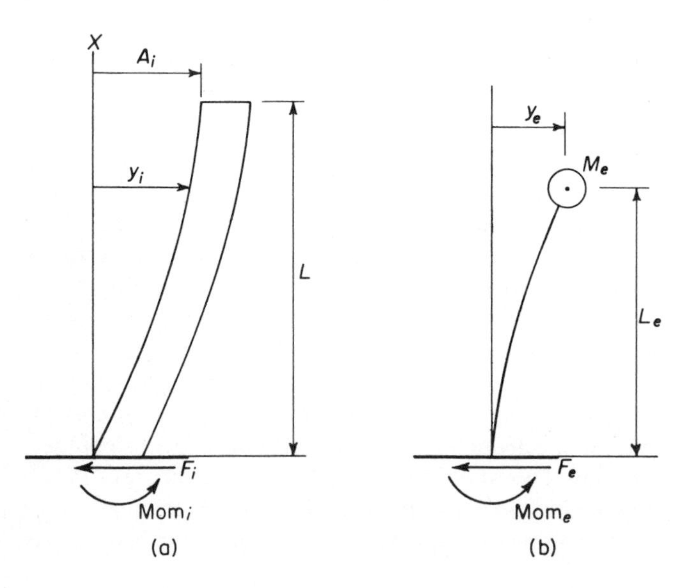

$$y_e = A_e \sin \frac{2\pi}{T_e} t$$

This ESO will be equivalent in the sense that its interaction with the environment is the same.

1. The periods of vibration are the same: $T_e = T_i$.
2. The base shears are the same: $F_e = F_i$.
3. The base moments are the same: $\text{mom}_e = \text{mom}_i$.
4. The kinetic energies of vibration are the same: $\text{KE}_e = \text{KE}_i$

With the condition that $f_i(l) = 1$ so that A_i is the amplitude of vibration at $x = l$, the foregoing four conditions give the following relations for determining the ESO:

$$M_e = \frac{\left[\int f_i(x)\,dm\right]^2}{\int f_i^2(x)\,dm} \qquad A_e = A_i \frac{\int f_i^2(x)\,dm}{\int f_i(x)\,dm}$$

$$L_e = \frac{\int x f_i(x)\,dm}{\int f_i(x)\,dm} \qquad k_e = M_e\left(\frac{2\pi}{T_i}\right)^2 \tag{5.5}$$

The ESO has the same damping as the mode, and its maximum response is determined from the design spectrum curves. The maximum response of the mode then can be determined by means of Eq. 5.4. The distribution of lateral forces is

$$p(x) = mA_i\left(\frac{2\pi}{T_i}\right)^2 f_i(x)$$

The ESO and the design spectrum can be used to exhibit the general dynamic properties of the system, i.e., how the base shear and base moment depend on the proportions of the structure, the mass, the stiffness, and the period. For example, the shape of the mode of vibration has an influence on the properties of the ESO and, hence, on the dynamic forces exerted on the structure. A uniform cantilever beam with shearing deformation only has a fundamental mode that is outwardly convex, whereas a uniform cantilver beam with bending deformation only has a fundamental mode that is outwardly concave. For these two cases, the ESO has the properties shown in Table 5.3. The base shear and the base moment are given by $M_e S_a$ and $M_e L_e S_a$ and, hence, if both beams have the same mass and period of vibration the bending beam will have a smaller base shear and a smaller base moment in the ratios of 0.78 and 0.76, respectively. On the other hand, a building that deforms

Table 5.3. EQUIVALENT PROPERTIES OF VIBRATING BEAMS

Shear beam	Bending beam
$M_e = 0.81M$	$M_e = 0.63M$
$A_e = 0.79A_i$	$A_e = 0.59A_i$
$L_e = 0.81l$	$L_e = 0.79l$

primarily in bending may have a shorter period of vibration than does a similarly proportioned frame building that deforms primarily in shear. In this case, the design spectrum value S_a may be appreciably larger for the shorter period building and its base shear and base moment may exceed those of the longer period building.

5.3.2 Application of Spectrum Techniques to Fluid Oscillations

The design spectrum and the equivalent simple oscillator can be used to exhibit the earthquake behavior of a more complicated system, such as a tank containing fluid with a free surface. Experience shows that during an earthquake oil storage tanks and water storage tanks that are not properly designed to resist earthquake forces may collapse and release their contents. For example, during the 1964 Alaska earthquake, seven oil company tanks in Anchorage collapsed, one of which released 750,000 gal of aviation fuel; and during the 1933 Long Beach earthquake a very large steel water storage tank collapsed. The nature of the earthquake forces involved can be explained by means of the equivalent simple oscillator and the spectrum curves.

In Fig. 5.14a a tank is shown whose fluid contents have been excited into oscillation by the earthquake ground motion. As the tank walls are moved back and forth by the ground, the fluid near the walls will be accelerated back and forth essentially with the walls, whereas

Fig. 5.14. Typical large oil-storage tank with (a) earthquake-induced oscillation of fluid surface and (b) an equivalent mechanical system.

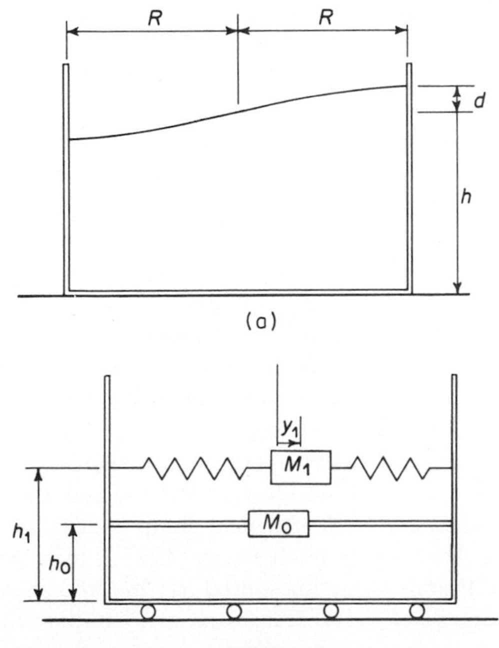

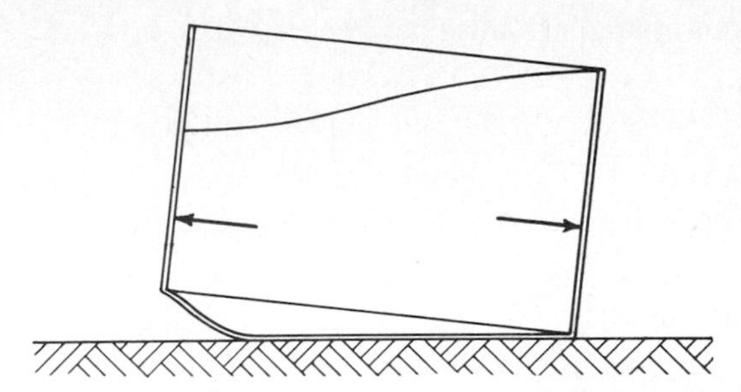

Fig. 5.15. Tank partly lifting up under the action of fluid oscillation.

the fluid near the center of the tank will not be so accelerated, but will tend to oscillate back and forth with the natural period of the sloshing fluid. The fluid pressures on the wall of the tank will reflect these two actions, and one component of the pressure will vary in time directly proportional to the ground acceleration and another component will vary in time with the sloshing of the fluid. As shown in Fig. 5.14b, the action of the dynamic forces exerted by the fluid can be represented by a fixed mass M_0 that moves with the tank walls and an ESO with mass M_1 that has the same period of vibration as does the first mode of the fluid oscillation. If the equivalent system shown in Fig. 5.14b is subjected to earthquake ground motion, the horizontal forces exerted by M_0 and M_1 will be equivalent to the fluid forces that would be exerted on the walls of the tank shown in Fig. 5.15a. The higher modes of fluid oscillation are usually neglected as they are not strongly excited. The properties of the equivalent dynamic system are (Housner, 1967; U.S. Atomic Energy Commission, 1963)

$$M_0 = M\frac{\tanh 1.7R/h}{1.7R/h} \qquad h_0 = \frac{3}{8}h \tag{5.6}$$

$$M_1 = M(0.6)\frac{\tanh 1.8h/R}{1.8h/R} \tag{5.7}$$

$$h_1 = h\left(1 - \frac{\cosh 1.8h/R - 1}{(1.8h/R)\sinh 1.8h/R}\right) \tag{5.8}$$

$$T_1 = 2\pi\sqrt{\frac{h}{g}} \div \sqrt{(1.8h/R)\tanh(1.8h/R)} \tag{5.9}$$

In the expression for the period of vibration T_1, the symbol g represents the acceleration of gravity. The fluid also exerts a net moment on the floor of the tank that was not taken into account when calculating h_0 and h_1. If this is taken into account, the correct total moment on the tank walls and floor is given by taking

$$h_0' = \frac{3}{8}h\left[1 + \frac{4}{3}\left(\frac{M}{M_0} - 1\right)\right]$$

h_1' = same as Eqs. 5.8 and 5.13 except that the hyperbolic cosine term is followed by (−2) instead of (−1)

The amplitude of the earthquake-excited wave corresponding to a maximum displacement A_1 of the mass

M_1 for a circular tank is

$$d = \frac{0.6A}{(g/R)(T/2\pi)^2 - (5A_1/6g)(2\pi/T)^2} \qquad (5.10)$$

A typical oil storage tank containing a fluid weighing 42 lb/ft³ might have the following properties:

$$R = 22.5 \text{ ft} \qquad M_1 = (4.7)10^5 \div g$$
$$h = 28.6 \text{ ft} \qquad T_1 = 3.9 \text{ sec}$$
$$M = (1.9)10^6 \div g \qquad h_0 = 11.0 \text{ ft}$$
$$M_0 = (1.2)10^6 \div g \qquad h_1 = 18.2 \text{ ft}$$

The maximum force exerted on the tank walls by M_0 and M_1 due to one horizontal component of ground acceleration $\ddot{z}_x$, is given by

$$F_x = M_0(\ddot{z}_x)_{\max} + M_1 S_a \sin(\omega t + \phi)$$

where S_a is the acceleration spectrum value. Actually, the fluid oscillation in a circular tank is a degenerate system in which the x and y components combine into a single wave whose amplitude is approximately $\sqrt{2}$ times the single component amplitude. The maximum force on the walls will occur when the sloshing force is synchronous with and parallel to the force exerted by M_0. This would give for the maximum force,

$$F = M_0(\ddot{z}_x^2 + \ddot{z}_y^2)_{\max}^{1/2} + \sqrt{2}M_1 S_a$$

An upper bound for the force is

$$F_u = \sqrt{2}M_0(\ddot{z}_x)_{\max} + \sqrt{2}M_1 S_a$$

Since the x and y components of acceleration are only weakly correlated, a more reasonable value for the maximum force is

$$F = M_0(\ddot{z}_x)_{\max} + \sqrt{2}M_1 S_a$$

For the tank under consideration, and taking $(\ddot{z}_x)_{\max} =$ 0.33 g and $S_a = 14\%$ g, the force has the value

$$F = (40)10^4 + (9.2)10^4 = (49.2)10^4$$

The total overturning moment exerted on the tank wall is

$$(44)10^5 + (16)10^5 = (6)10^6 \text{ ft-lb}$$

The first term represents a force that oscillates with the ground acceleration, and it changes direction some 5 to 10 times/sec, whereas the second term represents a more slowly varying force that has a period of about 4 sec. The overturning moment is resisted by the weight of the tank wall and roof which, at 60,000 lb, has a stabilizing moment of (13.5)10⁵ ft-lb. This stabilizing moment is insufficient, and the tank thus tends to overturn and to raise up the bottom plate, as shown in Fig. 5.15, until the downward fluid pressure on the uplifted portion provides the extra stabilizing force needed to resist the overturning moment. In the uplifted position the tank tends to stand on its toe and the vertical force may buckle the tank wall and may collapse the tank. To avoid this, the design of the bottom plate should be such as to minimize the uplift.

The mass M_0 represents the mass of fluid that effectively moves with the wall of the tank, and the dynamic force produced by it varies in time directly as the ground acceleration varies. This variation is different from the quasi-periodic variation of the force produced by M_1, the ESO of the first mode, or by M_2, the ESO of the second mode, etc. Actually, the planar vibrations of systems always can be decomposed into a set of ESO's with masses M_1, M_2, etc., plus a rigid body mass M_0. For ordinary buildings the M_0 is relatively small and is usually neglected. On the other hand, an earth dam will have a relatively large M_0 in addition to the various modal M_i. Effectively, for an elastic structure the M_0 represents the contribution of all the higher modes whose natural frequencies are above the significant frequency components in the ground acceleration or, in other words, M_0 represents the effect of the higher modes whose natural periods are so short that in this range the acceleration spectrum curve has essentially a constant value equal to the maximum ground acceleration.

5.3.3 Design Spectrum for Inelastic Vibrations

The dynamic forces computed on the basis of the ESO and the design spectrum are relatively large compared to the forces specified in the building codes. For example, the ESO of a 20-story shear-deformation building will have approximately $M_1 = 0.8M$ and $T = 2$ sec and will have a base shear $0.8M_1S_a$. For the design spectrum shown in Fig. 5.16, having $S_a = 0.33g$ at $T = 0$, the base shear is $0.8M_1 \times 0.16g = 0.128W$ for 4% damping. This is about six times larger than the base shear specified by the Uniform Building Code. Figure 5.16 shows a comparison between the design spectrum curves of Fig. 5.4 and the Uniform Building Code equivalent spectrum curve for steel-frame buildings. Since a steel-frame building will have a damping in the range of 2 to 5% of critical, it can be expected that a building designed according to the Code requirements will be strained far beyond the yield point in the event of strong ground shaking. It thus is necessary to consider the nonlinear vibrations of structures strained beyond the proportional limit.

Nonlinear vibrations are much more complicated to analyze than are linear vibrations. Fortunately, studies show that the spectrum curves determined for linearly elastic systems still have significance for nonlinear hysteretic vibrations. A rule of thumb is that the maximum earthquake-induced displacement of a hysteretic nonlinear system is roughly equal to the maximum displacement the system would attain if it remained elastic and had an additional 4% of critical damping because of the yielding (Hudson, 1965; Jennings, 1965; 1968). Another rule of thumb sometimes used is that the maximum displacement of the yielding structure is approximately the same as the maximum displacement of the structure if

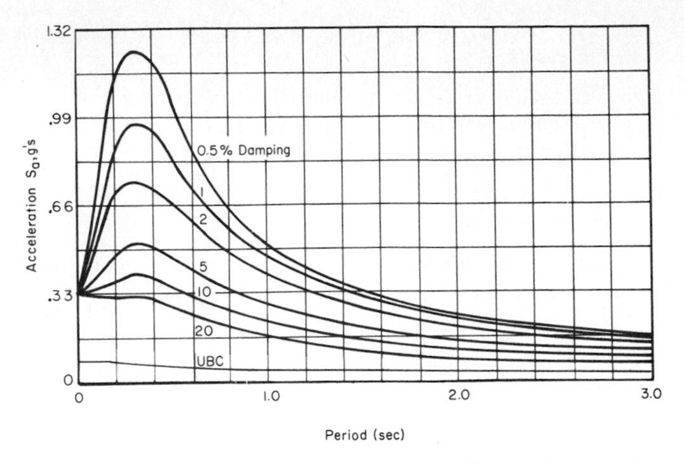

Fig. 5.16. Uniform Building Code earthquake requirements for multistory steel-frame buildings expressed as a design spectrum and compared with a design spectrum for 33 % g acceleration at zero period.

it remained elastic (Clough *et al.*, 1965; Veletsos *et al.*, 1965). The maximum equivalent damping that can be provided by yielding in steady-state vibrations is approximately 16%, and for randomly excited vibrations it is about 4% at moderate amplitudes. Both of the foregoing rules of thumb are applicable in the period range of about 0.4–4.0 sec, and they are applicable only if the plastic displacement is not too large. When using these methods, the maximum displacement is determined by means of the ESO and the spectrum curves, and then the corresponding deformations and strains are determined. During very strong ground shaking a steel-frame building designed according to the Uniform Building Code will experience displacement much greater than the yield-point displacement. How objectionable this is from the point of view of hazard to life and limb, or costliness of damage, can be determined only by special considerations of the use of the structure, type of construction, materials of construction, the probability of occurrence, etc.

When a structure is designed so that it will yield when subjected to strong ground shaking, the design process must differ somewhat from that for a structure that is to remain elastic. For example, in the case of an elasto-plastic structure, it is not pertinent to speak of a design stress because it is known that the maximum stress will be the yield-point stress. In this case the important design consideration is the maximum displacement and the strains it produces. If large plastic displacements are expected, it is important to know how near to collapse the structure is under the combined action of earthquake ground motion and gravity. Studies of the earthquake response of a one-degree-of-freedom elasto-plastic structure with gravity acting (Fig. 5.17) show that the collapse depends strongly on the duration of shaking and on the intensity of ground shaking as compared to the yield-point force (Husid, 1967). It was found that on the average the time needed to collapse was related to the height of the simple structure, the intensity of ground shaking, and the strength of the structure by the following equation:

$$\frac{t}{t_0} = 300\frac{h}{h_0}\frac{(C_y)^2}{I^2} \tag{5.11}$$

where t = time to collapse
t_0 = 100 sec
h = height of simple structure
h_0 = 15 ft
C_y = fraction of g corresponding to yielding
I = intensity of ground acceleration—equal to 2.9 for the strong phase of El Centro, 1940; 1.9 for the strong phase of El Centro, 1934; 2.5 for Olympia, Washington, 1949; and 2.1 for Taft, 1952 (Housner and Jennings, 1964).

Equation (5.11) was derived for an idealized structure so that it cannot be applied directly to actual structures. However it does indicate the kind of behavior to be expected from other structures, e.g., a multistory building whose first story has elasto-plastic columns and whose upper stories form a rigid box. If the first story, $h = 15$ ft, is designed to reach yield point at a shear force equal to 10% of the weight, $C_y = 0.1$, and it is subjected to ground motion having the intensity of $I = 3$, the probable time to collapse is $t = 33$ sec. On the other hand, if the ground motion has a very short duration, say 1.5 sec, like the Parkfield, 1966, earthquake, a very high intensity would be required to produce collapse. According to Eq. 5.11, if identical structures are subjected to $I_1 = 3$, $t_1 = 33$ sec, and to I_2, $t_2 = 1.5$ sec,

$$\frac{t_1}{t_2} = \frac{I_2^2}{I_1^2}$$

and, hence,

$$\frac{I_2}{I_1} = 4.7$$

Thus, on the average, it would take about five times the intensity of ground shaking to make the structure collapse in 1.5 sec as it would to make it fail in 33 sec. This

Fig. 5.17. Vibrating frame under the action of gravity with yield moments at top and bottom of columns.

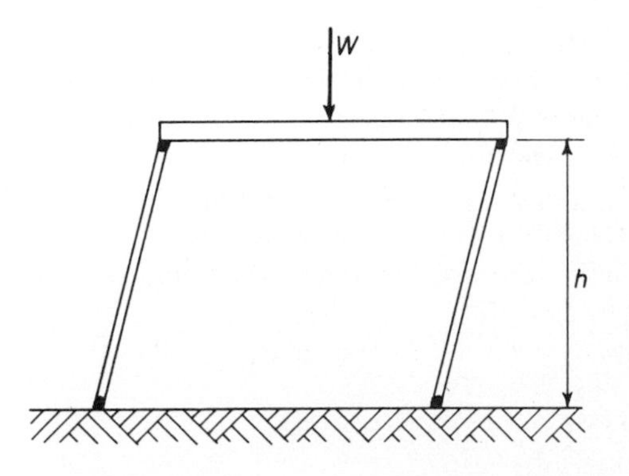

shows why the potential for producing failure of non-brittle structures is low for earthquakes like Parkfield, 1967, even though the maximum acceleration is relatively high and the response spectrum is large.

When using spectrum curves to estimate the total displacement (elastic plus plastic) for the design of structures, it must be kept in mind that there is a critical displacement for which gravity produces a moment just equal to the plastic restoring moment so that the structure will collapse under its own weight when displaced beyond the critical displacement. Consistent factors of safety against collapse can be determined by means of Eq. 5.11.

Some structures are designed so that the north–south shaking is resisted by north–south walls and the east–west shaking is resisted by east–west walls. Other structures are designed so that the north–south and east–west shaking are both resisted by the same structural element. The simplest example of this is a mass supported on a single steel cantilever column. In this case, the east–west and the north–south vibrations of the mass both stress the same column. For elastic vibrations the resultant stress is the significant item, but for elasto-plastic vibrations the resultant plastic displacement is the important consideration. In this case the north–south yielding is affected by the east–west forces and vice versa, and the behavior of the structure differs from that which would be computed if each component of motion were analyzed separately (Nigam, 1967). Because of the interaction, yielding begins sooner than would be indicated if only one component of motion were analyzed at a time. This earlier onset of yielding somewhat reduces the amplitude of motion.

REFERENCES

Allen, C. R. *et al.* (1965). "Relationship Between Seismicity and Geologic Structure in the Southern California Region," *Bull. Seism. Soc. Am.*, **55** (4).

Cloud, W. K. (1967). "Intensity Map and Structural Damage, Parkfield, California, Earthquake of June 27, 1966," *Bull. Seism. Soc. Am.*, **57** (6).

Clough, R. W., K. L. Benuska, and E. L. Wilson (1965). *Inelastic Response of Tall Buildings, Proceedings of the Third World Conference on Earthquake Engineering*, Vol. II, New Zealand.

Coast and Geodetic Survey (1958, 1961). *Earthquake History of the United States, Part I, Exclusive of California and Western Nevada (1958); Part II, California and Western Nevada (1961)*, Washington, D.C.: U.S. Government Printing Office.

Coast and Geodetic Survey (annual). *United States Earthquakes 1928–1966*, Washington, D.C.: U.S. Government Printing Office.

Gutenberg, B., and C. F. Richter (1966). *Seismicity of the Earth*, New York: Stechert-Hafner.

Housner, G. W. (October 1959). "Behavior of Structures During Earthquakes," *Proc. ASCE*, **85** (EM4).

Housner, G. W. (1965). *Intensity of Ground Shaking Near the Causative Fault, Proceedings of the Third World Conference on Earthquake Engineering*, Vol. I, New Zealand.

Housner, G. W. (January 1967). "Dynamic Pressures on Accelerated Fluid Containers," *Bull. Seism. Soc. Am.*, **47** (1).

Housner, G.W., and P. C. Jennings (February 1964). "Generation of Artificial Earthquakes," *J. Engr. Mech. Div., ASCE*, **90** (EM1), Proc. Paper 3806.

Hudson, D. E. (1965). *Equivalent Viscous Friction for Hysteretic Systems with Earthquake-Like Excitation, Proceedings of the Third World Conference on Earthquake Engineering*, Vol. II, New Zealand.

Husid, R. (1967). *Gravity Effects on the Earthquake Response of Yielding Structures*, Pasadena, California: Report of Earthquake Engineering Research Laboratory, California Institute of Technology.

Jennings, P. C. (1965). *Response of Yielding Structures to Statistically Generated Ground Motion, Proceedings of the Third World Conference on Earthquake Engineering*, Vol. II, New Zealand.

Jennings, P. C. (February 1968). "Equivalent Viscous Damping for Yielding Structures," *J. Engr. Mech. Div., ASCE*, **94** (EM1).

Kuroiwa, J. H. (1967). *Vibration Test of a Multistory Building*, Pasadena, California: Report of Earthquake Engineering Research Laboratory, California Institute of Technology.

Nigam, N. C. (1967). *Inelastic Interactions in the Dynamic Response of Structures*, Pasadena, California: Report of Earthquake Engineering Laboratory, California Institute of Technology.

Osawa, Y., and M. Murakami (1966). "Response Analysis of Tall Buildings to Strong Earthquake Motions," *Bull. Earthquake Res. Inst., Tokyo*, **44.**

U.S. Atomic Energy Commission (1963). *Nuclear Reactors and Earthquakes*, TID-7024, Washington, D.C.: Office of Technical Services.

Veletsos, A. S., N. M. Newmark, and C. V. Chelapati (1965). *Deformation Spectra for Elastic and Elasto-Plastic Systems Subjected to Ground Shock and Earthquake Motions, Proceedings of the Third World Conference on Earthquake Engineering*, Vol. II, New Zealand.

Chapter 6

Ground Motion Measurements

DONALD E. HUDSON

Professor of Mechanical Engineering and Applied Mechanics
California Institute of Technology
Pasadena, California

6.1 INTRODUCTION

Any study of earthquake engineering that is to have a sound scientific foundation must be based on accurate knowledge of the motions of the ground during destructive earthquakes. Such knowledge can be obtained only by actual measurements in the epicentral regions of strong earthquakes.

The number of destructive earthquakes for which such measurements are available unfortunately is very small. It is perhaps not generally realized how slender our stock of accurate information really is in this respect. For example, not a single measurement of strong ground motion was obtained for any of the following recent destructive earthquakes: Mexico (1957), Chile (1960), Agadir (1960), Iran (1962), Skopje (1963), Alaska (1964), and Turkey (1966). Among recent major earthquakes, it is only for Niigata (1964) that important ground accelerograph records were obtained. The available strong motion records are thus mainly limited to the several dozen accelerograms collected over the past 30 years by the U.S. Coast and Geodetic Survey network in the Pacific Coast states of the United States.

It is well to emphasize that typical seismological observatories with their sensitive seismographs are not intended to make measurements in the epicentral regions of strong earthquakes and cannot be adapted to do so effectively. Thus, although there are at present some thousand operating seismological stations distributed throughout the world, they cannot be expected to contribute directly to the special problem of the measurement of destructive ground motion.

6.2 SEISMOLOGY AND EARTHQUAKE ENGINEERING

The instruments used by seismologists have been carefully designed for the specific research interests of geophysicists. This has resulted in devices that are unsuitable for direct engineering application for the following reasons: (1) The seismologist desires to record small earthquakes occurring at any point in the earth. This requires sensitive instruments of high magnification. A strong earthquake near the instrument will usually displace the reading off scale or may even damage the instrument. The engineer requires a rugged device that will accurately record the heaviest shocks in the near vicinity of the instrument. (2) Since a study of the internal constitution of the Earth has been a primary objective of seismologists, his instruments are founded if possible on solid bedrock to eliminate the effects of local geological and soil conditions. The engineer wishes to know the ground motion at the sites at which engineering structures are located, often on thick alluvium or at locations otherwise quite unsuitable for seismological observatories. (3) For seismological investigations, accurate absolute timing of wave arrival times is essential, whereas the measurement of true ground motion is often of little importance. The engineer does not need absolute time, but true ground motion must be accurately known if the effects of the earthquake on structures are to be determined.

Thus it will be seen that the fundamentally different objectives of the engineer will require a basically different instrumentation than that needed for seismological studies. Such instruments must be designed, developed, installed, and operated by earthquake engineers who will be thoroughly familiar with the ultimate practical objectives of earthquake-resistant design. It is not implied that the engineer does not derive immense benefits from the activities of the seismologist. It is from the seismologist that the engineer learns of the distribution of seismicity in time and space over the Earth and thus can evaluate at any given site the probability of occurrence of earthquakes of various sizes.

6.3 TYPES OF EARTHQUAKE GROUND MOTION

Earthquake ground motions are of four types that should be carefully distinguished: (1) The earthquake may trigger landslides or similar local surficial movements that may destroy structures by simply removing their foundations. (2) The earthquake ground shaking may result in a large-scale soil and subsoil consolidation or settling, which may damage structures through excessive foundation deformation. (3) Sudden fault displacements may occur at the surface of the ground. (4) The earthquake ground accelerations may induce inertia forces in a structure sufficient to damage it. The first three effects may almost be called "static" effects. Although they are initiated by the earthquake ground vibrations, the large-scale earth motions themselves occur relatively slowly and do not set up appreciable inertia forces in structures.

It is the dynamic type of ground shaking that is to be the major concern of this chapter. Because of the increasing importance of the slower ground motions, however, a brief discussion will be included of the special instrumentation problems that they involve.

A closely related subject is that of nonseismic fault motions that have now been observed and measured in at least four areas in California. The Buena Vista overthrust fault (Wilt, 1958), the San Andreas Fault near Hollister (Steinbrugge and Zacher, 1960), the Hayward Fault (Cluff and Steinbrugge, 1966), and the Baldwin Hills Reservoir faults (Hudson and Scott, 1965), are all conspicuous examples of slow, nonseismic fault motions that have caused considerable structural damage. Recorded displacement measurements made along the San Andreas Fault and the Hayward Fault already have yielded some very interesting results, without as yet clearly revealing the full relationship between such local fault motions and earthquakes.

The measurement problems posed by local surficial movements such as landslides and subsidences are particularly troublesome. Because of the very local nature of such phenomena, the selection of instrument sites involves a considerable uncertainty, and the chances of developing useful information from any reasonable number of instruments seems at present to be small. There remains the possibility that certain areas known to be undergoing such local motions with some regularity might be instrumented to give basic data on such features as the dynamics of landslide initiation. For example, certain unstable areas in the Palos Verdes Hills and at Pacific Palisades in Southern California might be instrumented with slow-speed displacement recording devices that would have a good probability of picking up interesting information within a few years.

Although the following discussion will deal mainly with the dynamic forces associated with ground shaking, it should be kept in mind that the other types of ground motion may in some cases be an even more important source of damage. In the 1964 Alaskan earthquake, e.g., massive landslides accounted for a large fraction of the total damage (Hansen, 1965), and in Niigata in 1964 large-scale consolidation and compaction of soils were responsible for a major part of the destruction (International Institute of Seismology and Earthquake Engineering, 1965).* Such local surficial motions are, however, of limited areal extent, and the risks associated with them can usually be assessed by geological explorations and soils studies. Ground shaking, on the other hand, may involve very large areas and represents an ever-present danger even under the best foundation conditions.

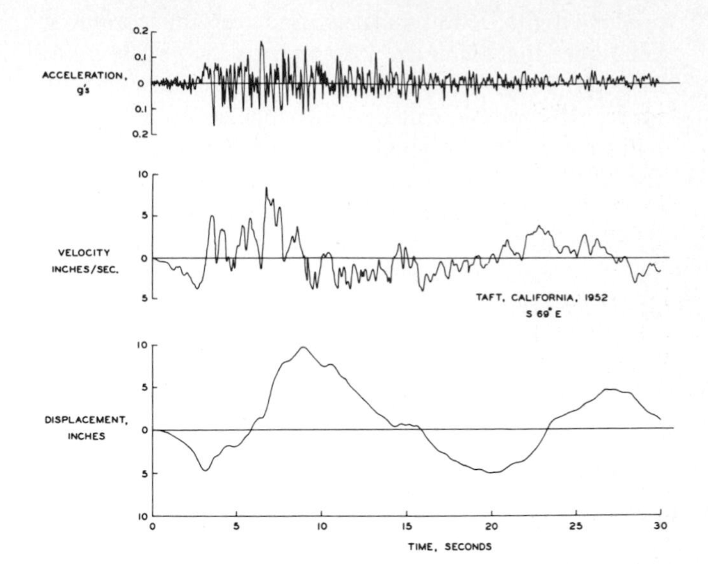

Fig. 6.2. Ground motion for the Taft earthquake of 1952. (From Hudson, 1965.)

6.4 STRUCTURAL RESPONSE DETERMINATIONS

In Fig. 6.1, two equivalent dynamic systems are shown. At the left is indicated schematically a four-story building whose foundation has acquired a horizontal acceleration $\ddot{y}(t)$ as a result of earthquake ground shaking. The right figure indicates that this situation is equivalent to a fixed base building with lateral forces applied to each floor having a magnitude equal to the product of the mass of the floor and the ground acceleration. The systems are equivalent in the sense that the same dynamical equations of motion describe each system. Such a replacement of ground acceleration by lateral inertia forces proportional to the acceleration is a generally valid procedure for all structures and leads to the conclusion that it is the ground acceleration that must be known if the equivalent forces acting on a structure are to be determined.

*See Chapter 10 in this book.

Fig. 6.1. Equivalent dynamic systems for a multistory building. (From Hudson, 1965.)

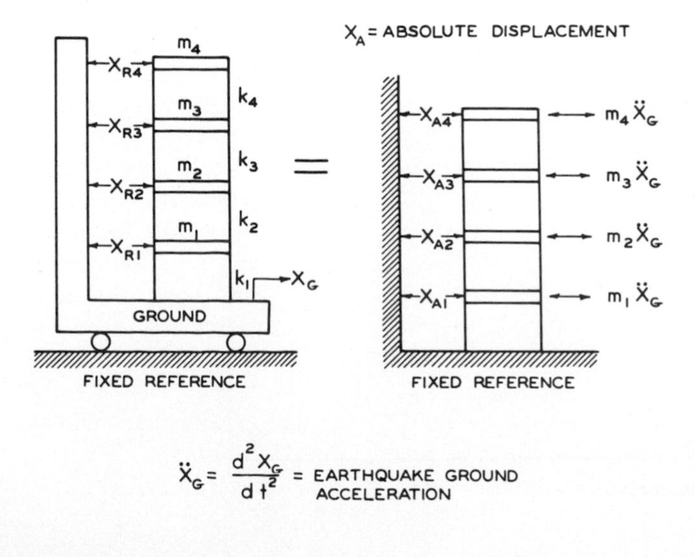

In principle it should be immaterial whether the displacements, velocities, or accelerations of the ground are measured, since there is a simple mathematical relationship between them. To obtain displacement from acceleration requires two integrations, whereas to obtain the acceleration from the displacement requires two differentiations. In practice, however, integrations can be carried out much more accurately than differentiations, since it is easier to determine the way in which the area under a complicated curve varies with time than it would be to measure the slope of the curve with the required accuracy. The state of affairs for a typical earthquake may be seen in Fig. 6.2, which gives the ground acceleration, velocity, and displacement for the Taft, California, earthquake of 1952 (Berg and Housner, 1961). It will be seen readily that to start from the displacement curve and to determine its slopes with a sufficient accuracy to produce the relatively complicated velocity curve and then to repeat the process to obtain the very complicated acceleration curve would be practically a very difficult undertaking. The inevitable loss of accuracy in the differentiation process cannot be avoided no matter how the process is carried out, be it by electrical, mechanical, graphical, or numerical techniques. We thus reach our first conclusion: The basic measurement for earthquake engineering applications should be of the ground acceleration vs time. Once an accurate ground acceleration curve is available, the velocity and displacement curves can be obtained with a satisfactory accuracy, as has been done in Fig. 6.2. The calculation of these velocity and displacement curves from the accelerogram is not a trivial matter, however, and a number of studies of the best way of carrying out the integrations have been made (Hudson, 1962; Berg, 1963; Brady, 1966; Amin and Ang, 1966; Schiff and Bogdanoff, 1967). At present, most methods employ digital computations and certain correction techniques that minimize instrumental errors (Berg and Housner, 1961; Brady, 1966). The largest error in most double integration calculations appears to be in the digitization of the original accelerogram and in the location of the true base line of zero acceleration (Brady, 1966). The digitization errors are partly the consequence of inadequacies in the original analog photographic trace and partly personal and instrumental errors in the digitizing process.

Although a complete analysis of structural response requires the acceleration-time record, there are some reasons for wanting the velocity and displacement curves with a reasonable accuracy. A knowledge of the maximum velocity attained during the ground motion is useful because of its approximate correlation with structural damage in an overall sense. It has been shown by several investigators that for blast loading the single best descriptive number related to structural damage is the maximum ground velocity (Edwards and Northwood, 1959; Neumann, 1958). It also appears that this maximum ground velocity can be correlated in an approximate way with the Modified Mercalli Intensity scale commonly used to give a rough measure of the damage associated with strong earthquakes (Neumann, 1959). These correlations are not surprising in view of the direct relationships between velocity and energy.

The ground displacement curve also has a practical interest because of the direct relationship between ground displacements and the strains to which large structures such as dams and underground pipelines might be subjected.

Figure 6.2 shows that the prominent frequency components that appear in the records are different for the acceleration, velocity, and displacement curves. In the acceleration curves the very short periods are clearly evident, and in the displacement curves the long periods are predominant. Although this same frequency information also is revealed by spectrum curves calculated from the acceleration-time record, the velocity and displacement curves have the merit of clearly emphasizing these various frequency regimes.

Because of the interest in the longer period motions and the difficulties of the numerical integration of accelerograms, the United States Coast and Geodetic Survey (USCGS) has included in a number of its strong-motion instruments longer period transducer elements that for certain frequency ranges will give a record of displacement vs time. It can be shown that if such a transducer element is to faithfully reproduce displacements, it must have a natural period somewhat longer than the displacement component to be measured. The displacement meters in various USCGS installations have periods ranging from about 2 to 10 sec, and they are thus rather limited in their usefulness in producing true displacement data at the longer periods of interest. Modern data processing techniques should make it possible to extend the range of quantitative usefulness of these displacement devices, but so far little has been done in this direction. The design and operation of the longer period displacement devices such as the 10-sec transducer pose many special problems of sensitivity, stability, and instrument leveling in the Earth's gravitational field. It is not entirely clear at the present time just how the overall accuracy of the longer period transducers compares with that obtained by modern integration techniques, and addi-

Fig. 6.3. Fourier spectrum curve for Taft earthquake. (From Hudson, 1965.)

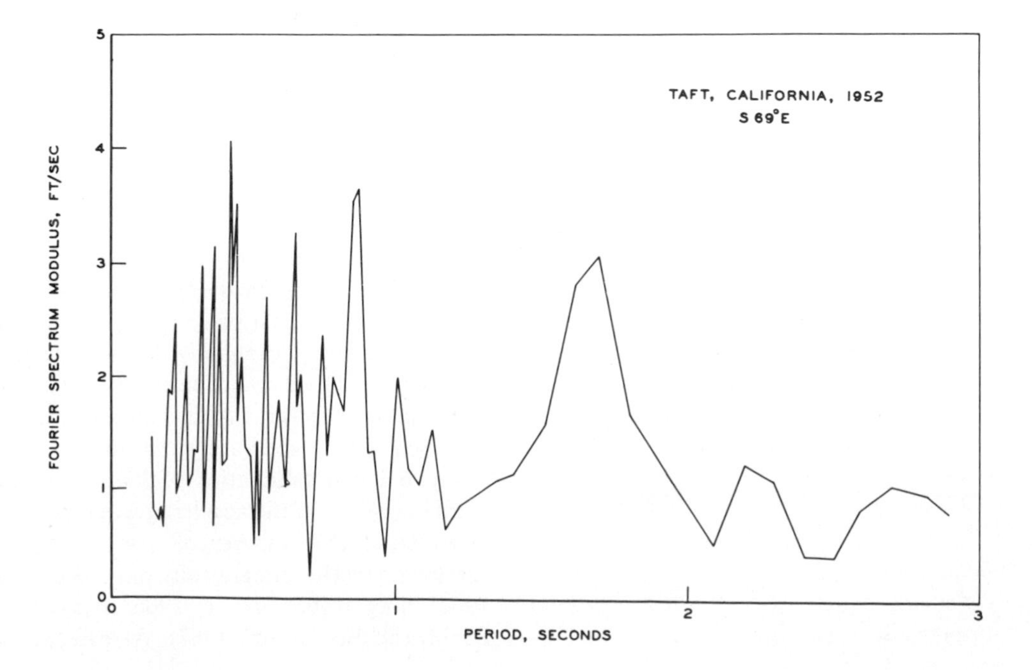

tional studies are needed to resolve these questions. The possibility exists also that the 2-sec transducers could be considered as accelerometer elements for the longer period waves, and an analysis of past data along these lines would be well worth carrying out.

The complexity of the typical acceleration curve of Fig. 6.2 suggests that a wide range of periods is involved in the ground motion. This is illustrated more specifically by the Fourier spectrum plot of Fig. 6.3 for the same earthquake (Hudson, 1962). The individual sharp peaks in the low period range are to a certain extent fortuitous aspects of a particular earthquake and are also partly the result of the data processing techniques. The general trend of such curves is clear, however, and indicates that a significant amount of energy might be introduced into a structure almost uniformly over a band extending from 0.1 to 3 or 4 sec. This is also the range of periods that is likely to be covered by typical engineering structures. Below 0.1 sec a structure becomes so rigid that dynamic design is not ordinarily a significant factor. Above 3 or 4 sec a structure is so flexible that conditions of excessive deformation rather than strength considerations will likely limit design, and other loadings such as wind forces may play an important role. It is thus concluded that an accelerograph suitable for earthquake ground motion measurements should record accurately over a period range of from 0.1 to at least 3 or 4 sec, with the proviso that it also is highly desirable if velocity and displacement data can be obtained for even longer periods up to some 10 sec.

6.5 ACCELEROGRAPH DESIGN PRINCIPLES

We next consider how to design an instrument that will measure the absolute ground acceleration over the required period range in the absence of any fixed reference point. The solution to this problem is indicated in Fig. 6.4 (Beckwith and Buck, 1961; Bradley and Eller, 1961). The crosshatched line in the upper left-hand diagram represents the ground whose motion is to be determined. We attach to the ground a mass m by means of a linear spring k and a viscous damping element c. This constitutes the instrument, the output of which is the relative displacement between the mass and the ground, which can be measured without any fixed reference point. It can readily be shown from the equations of motion of the spring-mass system that at any particular period of ground motion the instrument output is proportional to the ground acceleration. Unfortunately, however, the proportionality factor depends on the period of the ground motion, and the object of the instrument design is to make the response as independent of period as possible. The typical Fourier spectrum of Fig. 6.3 indicates that the ground acceleration wave form includes many period components covering the whole period

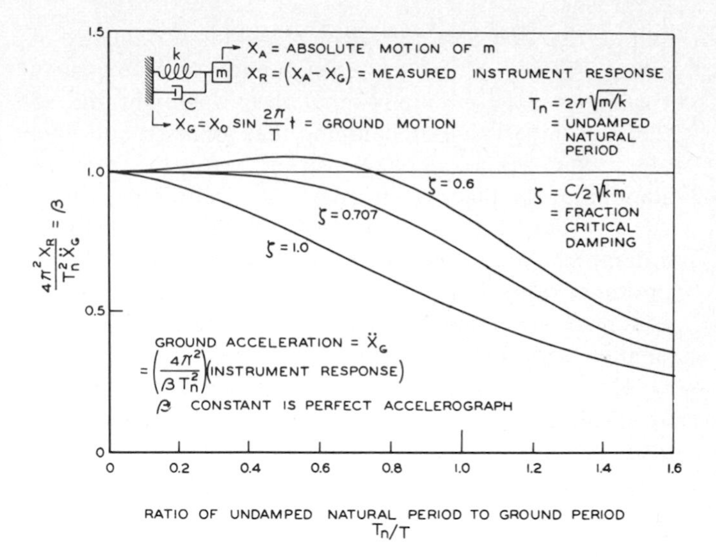

Fig. 6.4. Accelerograph design principles. (From Hudson, 1965.)

range of structural interest. To reproduce the wave form accurately the instrument must record each component with the same factor of proportionality.

The instrument response curve of Fig. 6.4 shows how well this can be done in practice. If the curve were a horizontal straight line, the instrument response would be independent of period. It will be seen that by a judicious selection of damping around a value of from 0.6 to 0.7 of critical it is possible to achieve an instrument characteristic that is approximately constant over a band of periods from zero to a period nearly equal to the natural period of the instrument spring-mass system. It thus appears that the design of an accelerometer transducer must satisfy two main conditions: (1) the natural period of the instrument should be smaller than the smallest period to be measured and (2) the damping should be from 0.6 to 0.7 of critical damping. Under these conditions the instrument will give a faithful record of the smallest period component, and all of the longer periods will be even more accurately measured.

Another difficulty presents itself in the design of instruments of the above type. There may be a phase shift between the ground motion and the relative displacements measured by the transducer. If the phase shift is the same for waves of all periods, then the resulting signal would simply be shifted a little in time, which would be of no consequence for structural response calculations. If the phase shift is different for different periods, however, component waves will add up in the output signal to give a signal which is shaped differently from the input signal. It may be shown that if the phase shift can be made to be a linear function of the frequency, then the resulting output wave will have the same shape as the input wave, with a small, constant shift of phase. This constant phase shift simply moves the time scale a little, which is not important for the present purpose. Fortunately, it develops that a transducer element having a damping of about 70% of critical possesses a phase-shift-frequency curve that is a good approximation to a straight line, so that the same value of damping that gives an optimum amplitude response curve is also the best value from the standpoint of phase shift (Beckwith and Buck, 1961; Bradley and Eller, 1961). Thus it may be concluded that a damping value of from 0.6 to 0.7 of critical will be an optimum value that will produce a satisfactory amplitude and phase response for periods that are less than the smallest period to be measured.

Keeping in mind that the shortest periods of interest in earthquake ground motions are about 0.1 sec, it is evident that the accelerograph period should be somewhat less than 0.1 sec if possible. This is the consideration that has fixed the usual period setting of the standard USCGS Strong Motion Accelerograph in a range from 0.05 to 0.08 sec. The disadvantage of reducing the instrument period still further to give increased accuracy at shorter ground periods is that the sensitivity of the instrument would be reduced. The deflection per unit acceleration input of the instrument is proportional to the square of the period, so that a shorter period instrument would have a smaller deflection for a given ground acceleration.

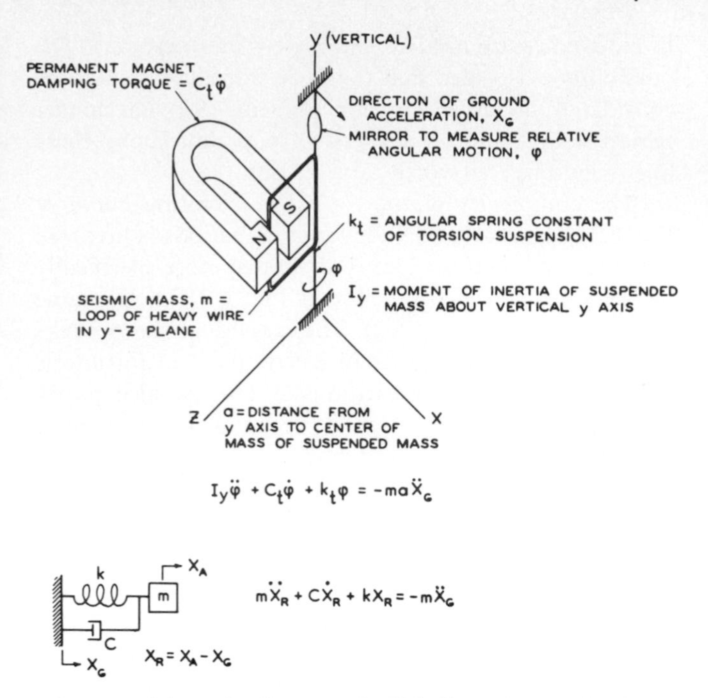

Fig. 6.5. Schematic diagram of USCGS accelerograph. (From Hudson, 1965.)

6.5.1 Accelerograph Design Details

One way in which the spring-mass-damping system of the theoretical accelerograph of Fig. 6.4 appears in practice may be seen by referring to the diagrammatic sketch of the horizontal transducer element from the USCGS accelerograph shown in Fig. 6.5 (Heck, McComb, and Ulrich, 1936). In this instrument the moving mass is a rectangular loop of wire that swings like a door about a vertical torsion suspension axis along one side. The elastic restoring force is thus a torque in the torsion suspension rather than a rectilinear spring as in Fig. 6.4. As the loop of wire rotates, it moves in the magnetic field set up by a permanent magnet, and the induced eddy currents in the loop set up viscous damping forces. The advantage of a torsion rather than a rectilinear arrangement is that the angular displacements can be very easily amplified and recorded optically by mounting a mirror on the torsion suspension. It will be recognized that this transducer element is a form of the Wood–Anderson torsion seismometer that has been used for many years by seismologists for local earthquake recording (Anderson and Wood, 1925; Benioff, 1955).

A modification of this same element is used in the recently designed AR240 and RFT250 accelerographs. In these instruments the single loop of wire forming the seismic mass is replaced by a rectangular many-turn coil of wire. By changing the external resistance in the coil circuit, the damping can be set to any desired value. An additional advantage of the coil system is that an external electrical signal can be introduced easily into the transducer element for calibration purposes.

As an example of a very different way of accomplishing the same end result, Fig. 6.6 shows the basic transducer element of the Japanese SMAC accelerograph (Takahasi, 1956). The seismic mass is supported on the end of a rigid bar, which rotates about a flexure hinge. For the very small motions involved, the system performs essentially rectilinear motion. The seismic mass itself forms the piston of an air damping system, which is provided with sufficiently small clearances so that critical damping is achieved. The air damper has the advantage of relative independence of damping with temperature. The rectilinear motion of the mass is magnified 16 times by a mechanical lever system, and the final record is scribed on a waxed paper by a sapphire stylus. Although the record amplitude is small, the line is very fine so that considerable magnification of the record itself can be made if desired. In order to reduce the battery power requirements, the record paper is driven by a mechanical spring motor that is hand wound.

The same general results attained in the above instruments are achieved by different means in the MO2 strong-motion accelerograph designed in New Zealand (Duflou and Skinner, 1965). The basic transducer element in this device is a small mass mounted on four wires, crossed in pairs, which constrain the mass to rotate about a single axis and which supply the restoring spring force. Damping is provided by attaching to the mass a paddle that moves in a silicone oil. Mirrors are attached to the masses, and a lamp-pinhole arrangement forms recording traces on a 35-mm film.

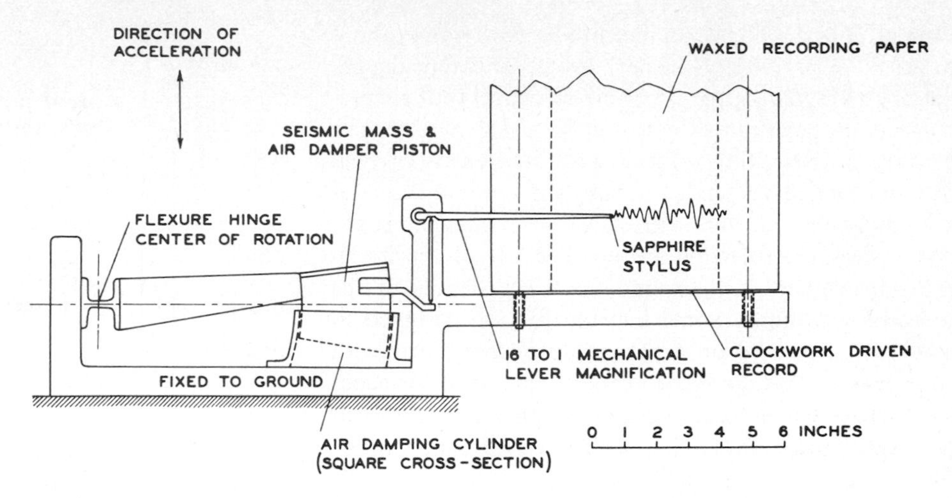

Fig. 6.6. Schematic diagram of SMAC accelerograph. (From Hudson, 1965.)

The RMT-280 magnetic-tape recording strong-motion accelerograph that appeared in 1968 uses transducer elements of the type shown in Fig. 6.5. Instead of the optical recording system, however, the rotating element has attached to it a small vane whose motion alters the air gaps in a magnetic variable-reluctance system. By suitable electronic circuits, motion is recorded as a frequency-modulated signal on a standard $\frac{1}{4}$-in. magnetic tape. Having the record in the form of a magnetic tape offers, of course, many advantages in data processing.

6.5.2 Sensitivity Requirements

The first consideration involved in the required instrument sensitivity is that the largest possible earthquake ground acceleration should stay on scale. It is also required that the scale should be of such a size that acceleration-time data can be read from the record with an accuracy suitable for response calculations.

The most severe earthquake ground motion so far recorded is that of the El Centro earthquake of May 18, 1940, shown in Fig. 6.7 (Neumann, 1942), which was recorded some 4 mi from the surface fault break. Higher accelerations have been recorded, however, during the Parkfield earthquake of June 27, 1966, by an accelerograph located virtually on the causative fault (Cloud and Perez, 1967; Housner and Trifunac, 1967). The Parkfield earthquake consisted of essentially a single displacement pulse, with a corresponding horizontal peak acceleration of just over 0.5 g. Detailed studies have indicated that the maximum ground acceleration caused by ground shaking in firm alluvium is not likely to be greater than approximately 0.5 g, and the duration of

Fig. 6.7. Accelerogram for El Centro earthquake of 1940. (From Hudson, 1965.)

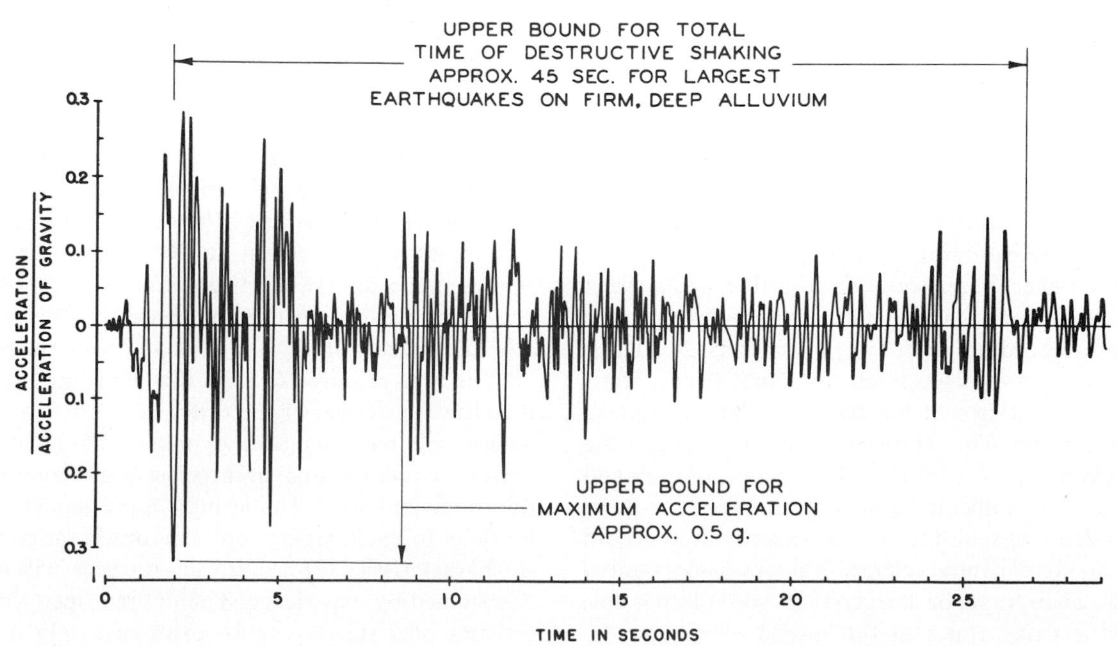

heavy ground shaking is not likely to be greater than about 45 sec (Cloud, 1963; Housner, 1965). Theoretical considerations involving the probable mechanism of generation and propagation of earthquake waves also suggest that the above limiting values are approximately correct. Current practice has tended toward a maximum acceleration limit of 1 g, in the interests of ensuring against the loss of large acceleration peaks. The actual sensitivity of the instrument will depend upon the details of the recording technique. Some instruments are designed to write a small record on a narrow recording paper with a fine line so that the record can be magnified for accurate reading. Other accelerographs employ a wide recording paper producing a large size record that can be directly measured.

A desirable feature would be a provision for two different sensitivities simultaneously recorded, as was done optically in some of the original USCGS accelerographs. The same result also can be obtained by providing the instrument with a nonlinear scale, so that sensitivity decreases as the acceleration level increases. This will prevent the loss of acceleration peaks at the expense of an increased difficulty of reading the record.

The time duration of heavy shaking determines the length of record that must be provided for each earthquake. Some accelerographs are arranged so that a fixed time interval of several minutes is automatically recorded for each earthquake, while others are designed to record for a fixed time after the earthquake ground motion has subsided below the initial triggering level. In any event, it is very desirable that a supply of recording material sufficient for a number of individual earthquakes be available, and after each earthquake the instrument should automatically return itself to readiness for the next shock.

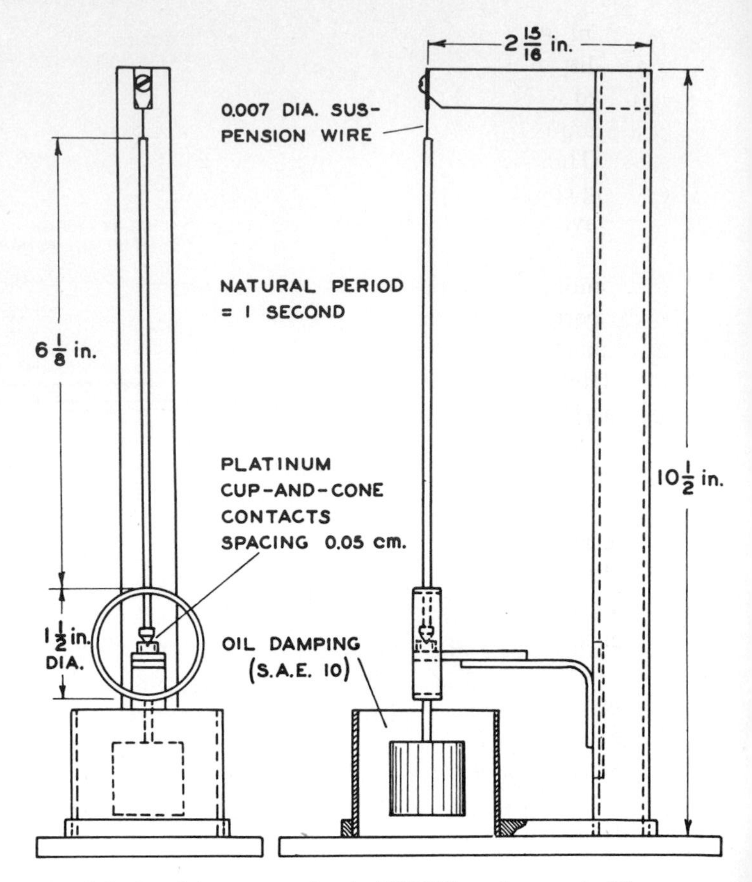

Fig. 6.8. Pendulum starter for the USCGS accelerograph. (From Hudson, 1963.)

6.5.3 Recording Speed and Starting System

The recording speed must be such that the complicated wave forms can be measured with an accuracy adequate for spectrum analysis and for response calculations. A speed of 1 cm/sec has been standard for several strong-motion accelerographs. It is now recognized that a somewhat higher speed would be desirable, and a speed of 2 cm/sec would simplify data analysis procedures.

At these recording speeds, a continuous 24 hr/day recording is completely impracticable and, hence, some type of inertia starting switch activated by the earthquake itself is necessary. This starting device is perhaps the most critical component of the whole accelerograph and is the part most difficult to specify and design. Since the very beginnings of the ground acceleration record may contain significant acceleration peaks, it is essential that the accelerograph be triggered as soon as possible and that the delay times in the inertia element, relay systems, and motor drive be as small as possible. On the other hand, if the starter operates at too low values of acceleration-time excitation, it may be set off by extraneous nonseismic vibrations or by a series of small, nondestructive earthquakes, with the danger that the recording medium supply might be exhausted before a strong earthquake occurred.

The horizontal pendulum starter developed by the USCGS and used successfully for the past 25 years is shown schematically in the diagram of Fig. 6.8 (Heck, McComb, and Ulrich, 1936). The pendulum has a period of 1 sec, and the damping is approximately 30% of critical damping. A displacement of the platinum electrical contacts of some 0.05 cm in any horizontal direction will start the recording cycle. It was found that more reliable operation was obtained with a break-contact start using a holding relay rather than a make-contact type.

The time required to start the recording process with the above system is of the order of 0.2 sec. Any starting device will have its own dynamic characteristics, and various combinations of acceleration magnitude, time duration, and wave shape may cause sufficient relative motions to cause operation. The optimum combination of characteristics for a given seismic area will need to be determined by experience. The desired operating characteristics of a starter can be expressed only in terms of transient response, and this is difficult to do in any generality. Figure 6.9 gives curves that show the time required to close the pendulum contacts as a function of peak amplitude and time duration of a single half-sine pulse (Hudson, 1963). Another way of expressing these starting conditions is to say that if a constant amplitude sine-wave train of given amplitude and period starts at time $t = 0$ with the starter at rest, the curves will show the combination of peak acceleration amplitude and sine-wave period that will ensure starter operation with the first one-half cycle of acceleration. These curves are calculated for the particular characteristics given above for the USCGS starter. This particular half-sine pulse is only a very approximate model for the actual state of affairs, since the operation of the starter likely would be preceded by a gradually increasing series of alternating acceleration peaks.

Although the starting time of the USCGS starter is not as short as might be desired for some purposes, the evidence from past accelerograms indicates that it has been adequate for Pacific Coast strong-motion earthquakes. The major acceleration peaks in practically all cases have been preceded by a series of smaller peaks of a size sufficient to start the accelerograph, and it does not appear that significant information has been lost because of starting delays. The newer accelerographs now in service have starting delays of the order of 0.1 sec.

In the epicentral region, the arriving earthquake waves emerge almost vertically and hence the longitudinal P waves will have a motion that is predominately vertical while the shear waves will correspond to a horizontal motion. Since the longitudinal waves travel faster than the shear waves, the first arrivals would be expected to be vertical. An examination of past strong-motion accelerograms does show that the vertical components are often of an appreciable magnitude when the horizontal starter operates. There thus would be a considerable advantage in using a vertical starter.

After a considerable amount of experimentation, the USCGS abandoned vertical starters because of long-term stability problems. A vertical starter is used, however, in the standard Japanese SMAC accelerograph. With careful adjustment, this vertical starter has given satisfactory service.

A different approach to the starting problem has been made in a recent New Zealand design (Duflou and Skinner, 1965). In the MO2 accelerograph the starter system moves vertically and consists of a coil moving in a magnetic field that generates a voltage pulse. This velocity type device thus avoids the troubles of contact systems and should solve the stability and drift problem for the vertical starter. Such a starter should offer definite advantages over any of the standard devices used in the past.

Another way of reducing the total system starting time is illustrated in the USSR Type UAR accelerograph (Shi-Yuan, Kirnos, and Solovyev, 1961). In this instrument the recording drum is wound up against a spring, and is held in a cocked position by an electrically-operated brake. The starting pendulum releases the brake, and the high initial torque quickly starts the recording drum. The main difficulty in this device would be the provision of automatic resetting.

A number of attempts have been made to produce an

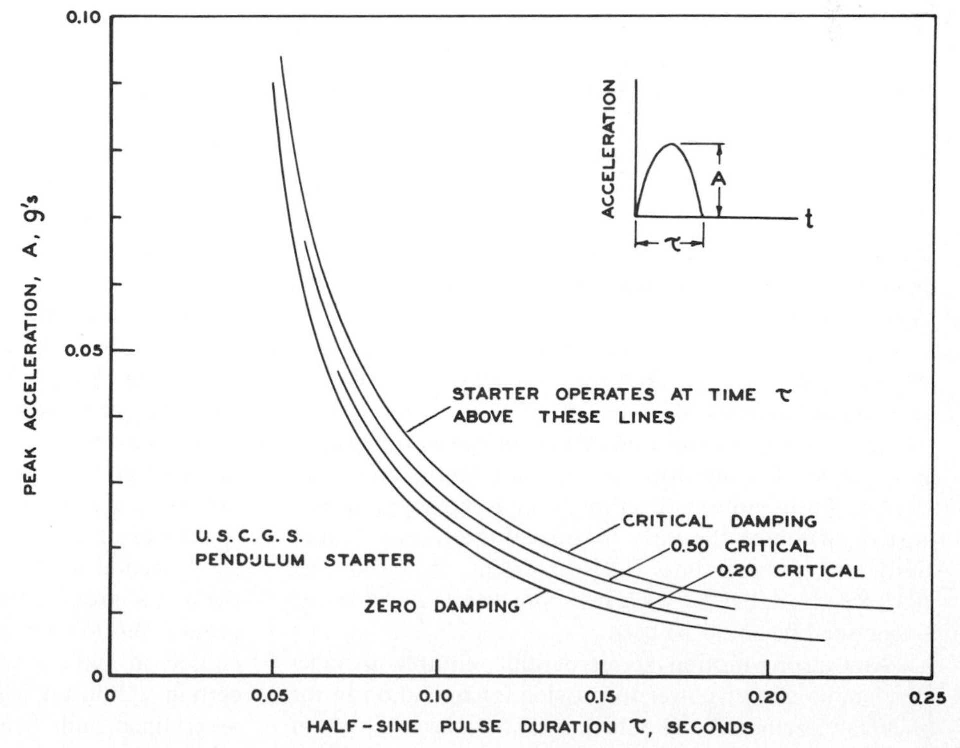

Fig. 6.9. Transient characteristics of the USCGS starter. (From Hudson, 1963.)

instrument with a short memory, which would preserve intact the whole initial portion of the accelerogram. This could be accomplished, e.g., by recording on a continuous tape loop, which would be also continuously erased. The starter and timer would then have only the function of arresting the erasing process and of stopping the recorder after one complete cycle of the loop. Although this would seem to be relatively simple to work out in practice, it does not seem as yet to have been used in the field for strong-motion measurements, although experimental devices have been built (Fremd, 1963). Another idea records the data continuously as a photoluminescent trace that gradually fades out. The starter then serves the function of pressing an auxiliary recording paper against the trace at the appropriate time, thus producing a permanent record (Borisevich, Goldfarb, and Mosyagina, 1961). Such memory systems would inevitably result in a more complicated and hence more expensive and potentially less reliable device. The small amount of increased information would probably not justify much compromise with overall instrument reliability.

6.5.4 Timing and Power Requirements

Although an absolute time scale is not needed for strong-motion work, it is necessary to have accurate time marks on the record so that accurate data processing such as spectrum determinations and response calculations can be carried out. Most standard accelerographs contain an arrangement for making one or two marks per second on the record with a relative timing accuracy of the order of 1%. It also may be advantageous to have interconnections between closely located instruments so that common timing is provided. For example, several accelerographs in the basement and upper floors of a building should be connected together for common time marks. Interconnections also should be provided so that the starting pendulum that first starts will simultaneously start all instruments. Currently available accelerographs are designed so that such starting and timing interconnections can be made easily.

If several instruments located within a few miles can be interconnected so that the first starter to operate will simultaneously start all of the accelerographs, it would be possible to take advantage of wave propagation times to start some instruments prior to the arrival of the first large motion. This might be a more practicable way of obtaining the early portion of the record than to design a memory into the instrument. A radio link device is available for such interconnections, or a leased telephone line could be used.

Any strong-motion accelerograph requires an independent source of power to provide for operation in the event an earthquake should knock out the local power system. This is often arranged by using a storage battery as the main power source that is continuously charged by a trickle charger energized from the local power supply. Since the Japanese SMAC accelerograph derives its main power from a mechanical spring-wound motor, it requires only relatively long-lived dry cells for the starting circuits. Some recent accelerographs employ a lead dioxide rechargeable battery, which has a long service life and can be either periodically replaced and recharged in the laboratory or continuously charged with a trickle charger.

6.5.5 Summary of Existing Accelerographs

Table 6.1 summarizes the characteristics of a number of standard accelerographs that have been used for strong-motion recording (Halverson, 1965). The photographs of Figs. 6.10 and 6.11 will show two of the recently developed strong-motion accelerographs that are already installed in many locations. A review of the table will indicate the extent to which the above requirements have been met by existing instruments and will suggest possible future developments. Although several satisfactory devices are at present available on a commercial basis, there is a general feeling that these instruments are too costly for widespread application. Optically recording accelerographs are commercially available in the $1400–2400 price range. The cost reflects the so far rather limited market and the fact that such instruments in the past have been built in small lots. If means could be found to acquire the numbers of accelerographs that are really needed to more adequately instrument areas of strong seismicity, the costs could undoubtedly be reduced.

6.6 SIMPLIFIED INSTRUMENTATION

Because of the relatively high cost of recording accelerographs of the type discussed above, there always has been a keen interest in the development of a simpler type of device which, although it might yield limited results, could be much more widely distributed. Such a device can be produced by adopting a different point of view in instrument design. The design philosophy behind the recording accelerographs described above produces a record from which the true motion can be derived. Once this true ground motion is known, any desired features can be analyzed, and the effects of such a motion on structures of any kind can be theoretically determined.

A second technique makes no attempt to determine the actual ground motion but measures the effect of the ground motion on a particular system. The system is chosen in such a way that from its known behavior certain significant features of the ground motion can be ascertained and, hence, the behavior of other systems

Table 6.1. CHARACTERISTICS OF CURRENT STRONG-MOTION ACCELEROGRAPHS—1968*

Characteristic	USCGS Standard	Akashi SMAC B/B2	Teledyne AR-240	Teledyne RFT-250	Teledyne RMT-280	New Zealand MO2	USSR UAR
Period, sec	0.043–0.085	0.10/0.14	0.055–0.065	0.05	0.05	0.03	0.05
Sensitivity, mm/0.1 g	5.5–19.7	4.0/6.5	5.0–7.5	1.9	±200 cps FM deviation/ ±1 g	1.5 horizontal 2.2 vertical	1.6
Recording range, g's	0.01–1.0	0.01–1.0/ 0.006–0.5	0.01–1.0	0.01–1.0	0.01–1.0	0.01–1.0	0.025–1.0
Damping, % critical	60	100	55–65	60	60	60	70
Damping mechanism	Magnetic	Air piston	Electro-magnetic	Electro-magnetic	Electro-magnetic	Oil paddle	Electro-magnetic
Recording speed, cm/sec	1	1	2	1	$3\frac{3}{4}$ in./sec	1.5	Approximately 1
Recording medium	Photo paper	Waxed paper	Photo paper	70 mm (Type II perforated)	$\frac{1}{4}$ in. magnetic tape	35 mm film (unperforated)	Photo paper
Recording drive	DC motor	Hand-wound spring	DC motor	DC motor	DC motor	Precision speed DC motor	Spring
Recording duration	$1\frac{1}{4}$ min	3 min	7 sec after last strong motion	5 sec after last strong motion	7 sec after last strong motion	47 (70) sec	60 sec
Repeat cycles	5	5	To end of 150 ft paper roll	To end of 100 ft film	To end of 1100 ft tape cartridge	9 (5)	1
Time marking	2/sec	1 or 5/sec	2/sec at ±1%	2/sec ±2%	2/sec ±2%	Trace interrupt 5 and 50 cps 0.1%	None
Starter type	Horizontal pendulum	Vertical pendulum	Horizontal pendulum	Horizontal pendulum (inverted)	Horizontal pendulum (inverted)	Vertical pendulum	Horizontal pendulum
Pendulum period, sec	1	0.3	1	1	1	0.15	—
Pendulum damping, % critical	30	—	100	Adjustable	Adjustable	Low	—
Damping type	Oil	—	Electro-magnetic	Electro-magnetic	Electro-magnetic	—	—
Starter control	Closed circuit relay	—	Closed circuit relay	Open circuit relay	Open circuit relay	Generator type, no contact	—
Overall time delay, sec	Approximately 0.2	—	0.1–0.15	0.1	0.1	0.1	0.05 to uniform speed
Power supply	12 VDC external storage battery 115 VAC trickle charger	12 VDC internal dry cells	12 VDC external storage battery 115 VAC trickle charger	12 VDC (internal chargeable lead dioxide batteries)	12 VDC (internal chargeable lead dioxide batteries)	12 VDC external	100 VDC and 6 VDC dry cells
Size, in.	13 × 20 × 45	15 × 21 × 21	14 × 16 × 16	$8\frac{3}{4} \times 10\frac{1}{2} \times 19\frac{1}{2}$	9 × 15 × 19	7 × 7 × 17	Approximately 12 × 18 × 24
Weight (including cover), lb	135	220	60	30	42	20	—
Manufacturer or supplier	USCGS	Akashi Seisakusho, Ltd., Tokyo	Earth Sciences, Teledyne, Inc., Pasadena	Earth Sciences, Teledyne, Inc., Pasadena	Earth Sciences, Teledyne, Inc., Pasadena	Victoria Engineering, Ltd., New Zealand	Earth Physics Institute, USSR Academy of Science

*All instruments have three-component seismometers, two horizontal and one vertical.

subjected to the same ground motion can be at least approximately determined (Hudson, 1959; Cloud and Hudson, 1961).

Perhaps the first systematic attempt to measure earthquake ground motions by their effects on very simple structures was made by Galitzin (1911), who suggested that a series of rectangular blocks of various proportions could be calibrated in terms of which blocks would be overturned by a particular ground motion, an idea which was elaborated on by a number of later investigators.

Fig. 6.10. Photograph of the AR240 accelerograph. (From Halverson, 1965.)

The first practical attempt to develop an instrument along the above lines was made by Suyehiro (1926). The "Suyehiro Vibration Analyzer" consisted of a series of cantilever beams of various natural periods, all having the same damping. The motion of these beams caused by the earthquake was recorded on a rotating drum. The instrument can be regarded as a series of dynamic models of structures, covering the range of natural periods likely to be encountered in actual buildings. Although the actual ground motion cannot be uniquely determined by such a device, something even more useful to the engineer is produced since the effects of the ground motion on typical structures are directly indicated.

Typical of later application of similar principles is the device developed by Medvedev as a means of attaching quantitative significance to earthquake intensity scales. Medvedev's (1966) "CBM Seismometer" consists of a conical pendulum of 0.25-sec period whose mass is free to move in any horizontal direction. The pendulum mass is in the form of a copper disk that moves in the air gap of permanent magnets that give the system about 8% critical damping. The motions of the pendulum are recorded on a smoked glass. A relationship has been established between the pendulum displacements and the grades of the earthquake intensity scale (Savarensky and Kirnos, 1955).

6.7 THE RESPONSE SPECTRUM

6.7.1 Response Spectrum Theory

In order to understand more completely the full possibilities of simplified instruments of the above type and how they are to be compared with recording accelerographs, it is useful to introduce the idea of a response spectrum.

To define the response spectrum, suppose that a given ground acceleration is applied to the base of a single-degree-of-freedom system. The behavior of the system as measured, e.g., by its maximum displacement will depend upon the exciting force and upon the natural period and damping of the system. For a given excitation and a particular value of damping, the maximum displacement of the single-degree-of-freedom system could be plotted vs the natural period of the system. A family of such curves, for various values of damping, would then form the response spectrum. Given the response spectrum, the maximum motion of any particular single-degree-of-freedom structure of known period and damping can be directly determined.

The response spectrum reveals directly the aspects of the earthquake ground motion that are of primary concern to the structural engineer and the preparation of

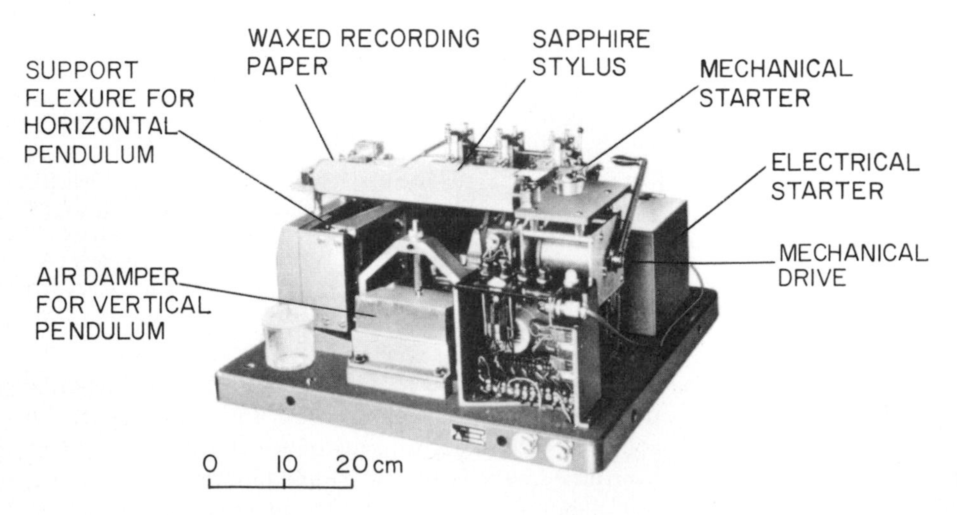

Fig. 6.11. Photograph of the SMAC Accelerograph. (From Hudson, 1963.)

such response spectrum curves is one of the main uses of recorded earthquake accelerograms (Hudson, 1956).

6.7.2 Calculation of Response Spectrum Curves

Response spectrum curves are in practice determined either by analog computation or by the use of high-speed digital computers. The main computation of United States earthquake response spectrum curves was based on a passive-type electric analog computer system operating repetitively at a 10 cps rate (Caughey, Hudson and Powell, 1960). The recent spectrum calculations made for Japanese earthquakes are computed on an electromechanical analog computer having a fixed-frequency responding element and a variable-frequency exciting function generator (Muto *et al.*, 1960). With the widespread availability of high-speed digital computers, it is now becoming customary to make response spectrum calculations by this means (Hudson, 1962; Brady, 1966). In any event, the spectrum calculations, although simple in principle, are either laborious or expensive to carry out with the desired accuracy considering the present form in which the input data are available. A great advantage over the analog traces provided by all existing strong-motion accelerographs would be provided by a tape-recording accelerograph. Such a magnetic tape record then could be easily digitized by an electronic analog to digital converter or could be used directly as the electric input for analog response calculation. Even greater advantages ultimately may be realized by a direct digital recording accelerograph.

6.8 SEISMOSCOPES

6.8.1 The USCGS Seismoscope

It will be recognized that Suyehiro's vibration analyzer described above is one way of measuring the response spectrum directly without the necessity of first measuring the true ground motion vs time and then carrying out a calculation to get the response spectrum. By plotting the maximum deflections of each of the cantilever beam elements in the vibration analyzer vs its natural period, the response spectrum for the one damping value could be obtained. By providing similar instruments having various damping values, the whole set of response spectrum curves could be obtained.

In the same way, it will be seen that Medvedev's seismometer gives directly the one point on the response spectrum curve corresponding to the natural period and damping of the element.

The USCGS seismoscope, a device similar in many respects to Medvedev's seismometer, is designed with the point of view of obtaining one significant value on the response spectrum curve in the simplest possible way (Cloud and Hudson, 1961, Hudson and Cloud, 1967). After an examination of many response spectrum curves calculated from recorded accelerograms of Pacific Coast earthquakes, it was decided that the single spectrum point that would give the maximum information would be a period of 0.75 sec and 10% of critical damping. This decision was based on the fact that at periods above the value the velocity response spectrum curves tended to become constant independent of period, whereas at lower periods the response dropped off markedly. The 10% damping was sufficient to ensure a relatively smoothly varying spectrum curve without local peaks, and yet it was not so high that the response was reduced to a value difficult to measure. This seemed to be the single point around which the most informative extrapolations could be carried out, and in addition the values of period and

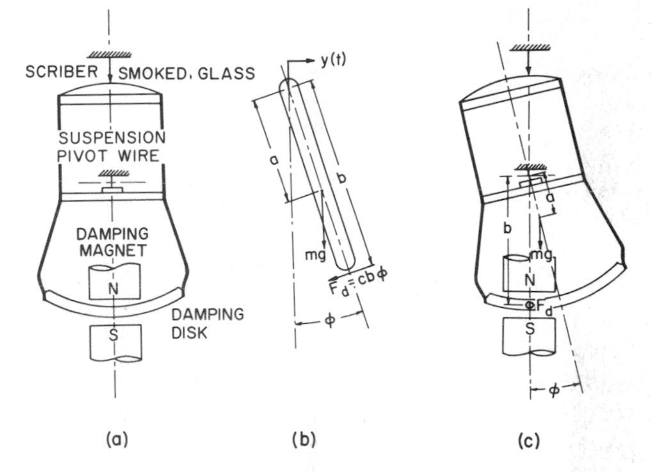

Fig. 6.12. Schematic diagram of USCGS seismoscope. (From Cloud and Hudson, 1961.)

damping were sufficiently close to those of many modern multistory structures so that the device preserved the useful physical concept of serving as a direct dynamic model of a structure. Figure 6.12 shows a schematic diagram of the final model of the seismoscope developed by the USCGS and the California Institute of Technology (Cloud and Hudson, 1961). This instrument was tested in the field by mounting it beside one of the USCGS recording accelerographs and recording simultaneously the same earthquake on each instrument. This has now been done for several earthquakes, the most notable results being obtained from the Parkfield earthquake of June 27, 1966 (Hudson and Cloud, 1967). During the Parkfield earthquake, accelerograph–seismoscope combinations at four different stations gave excellent simultaneous records from which direct comparisons could be made. At the right of Fig. 6.13 is a digital-computer calculated and plotted response of the relative displacement of a single-degree-of-freedom system having the

Fig. 6.13. Comparison of seismoscope results with calculations from measured accelerograms. (From Hudson and Cloud, 1967.)

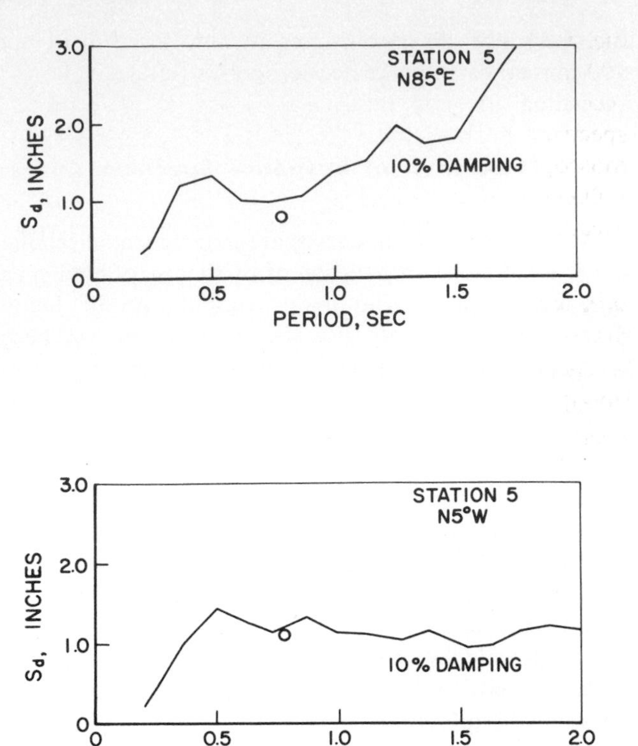

Fig. 6.14. Comparison of seismoscope results and computed response spectrum. (From Cloud and Hudson, 1961.)

period and damping of the seismoscope as computed from the ground acceleration-time curve as measured by the recording accelerograph. At the left of Fig. 6.13 is a photograph of the actual trace recorded on the smoked glass plate of the seismoscope to the same scale. It will be seen that many details of the response are faithfully reproduced. Figure 6.14 shows a comparison for two horizontal components of the complete response spectrum as calculated from the acceleration-time curve, with the single point obtained from the seismoscope (Cloud and Hudson, 1961). The difference between the point and the curve is believed to be consistent with the accuracy to be expected from the seismoscope. When it is considered that relatively complicated instruments and calculations are required to obtain the spectrum curve, it is remarkable that a seismoscope of such a simple form can produce so much information.

6.8.2 Multielement Seismoscopes

A logical extension of the above ideas suggests the idea of several seismoscopes having various periods and damping so that a number of spectrum points can be obtained. One such instrument, developed by Nazarov, contains 12 elements in one instrument (Nazarov, 1959; Savarensky and Kirnos, 1955). Nine of these elements measure horizontal motions, and three of them, vertical motions. The elements all have about the same damping and cover a period range of from 0.08 to 1.2 sec. Records are made on a smoked glass plate, as in the seismoscopes described above.

A similar idea has been developed by Krishna and Chandrasekaran (1965), who have modified the USCGS seismoscopes so that they can be installed in sets of six having periods of 0.40, 0.75, and 1.25 sec at damping values of 5 and 10% of critical damping (Krishna and Chandrasekaran, 1965).

Of course, it should be remembered that even if the exact response spectrum curve could be completely defined, the full information available in a true ground acceleration-time curve would not be at hand. The response spectrum curves, e.g., give only maximum response magnitudes and do not preserve the time differences at which these maxima occur. It thus is not possible to solve exactly from response spectrum curves alone the problem of the response of multidegree-of-freedom systems. In view of the many uncertainties involved in all earthquake response calculations, however, a determination of the response spectrum may be considered as sufficient for many practical applications.

The big advantage of the seismoscope, of course, is the elimination of time recording, which much simplifies the device. Since it is possible to acquire about 25 seismoscopes for the price of one recording accelerograph, the attractive possibilities of the simple device for increasing coverage becomes very evident. It would seem that the optimum use of the intruments would involve one recording accelerograph surrounded in an area of 100 mi² or so by several dozen seismoscopes. From the recording accelerograph the details of the response spectrum curves could be learned, and from the seismoscopes the way in which the magnitude levels are influenced by local geology and soil conditions could be ascertained.

6.9 STRONG-MOTION INSTRUMENT NETWORKS

A strong-motion accelerograph must be located within 30 mi of the epicenter of a strong earthquake if the most useful information is to be obtained. Considering the seismic areas involved in the world, it obviously would be impossible to completely instrument such areas with an adequate network of strong-motion accelerographs. Past practice has been to concentrate such instruments near important cities or sites of major engineering works such as dams or power plants. The optimum location of available instruments to produce a maximum information is clearly a matter of many compromises that deserves more special study than it has received in the past (Vanmarcke and Cornell, 1967).

Figure 6.15 shows the distribution (1966) of strong-motion accelerographs in the United States Pacific Coast network maintained by the USCGS. The dotted circle shows roughly the area of useful information covered by a single instrument. Although there are many notable gaps in the coverage, the record of recovered data from Pacific Coast earthquakes has been very good since the first accelerogram obtained in the Long Beach earthquake of 1933, when the system was initiated. A notable recent exception is the Alaskan earthquake of 1964, for which no strong-motion records were obtained, as there were no instruments in that area prior to the earthquake. Since the earthquake, 15 recording accelerographs have been installed in Alaska. The accelerograph network is in general concentrated in the Los Angeles and San Francisco regions, where the largest centers of population and structural investment are located.

Fig. 6.15. Strong-motion accelerograph network. (From Hudson, 1965.)

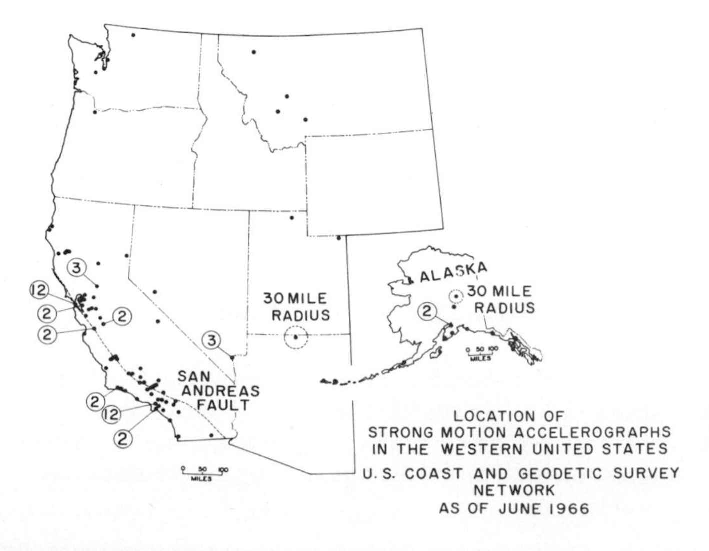

In the 2 years since Fig. 6.15 was prepared, the accelerograph network in the Pacific Coast states has been significantly extended to about 200 recording accelerographs plus 375 seismoscopes, including 100 seismoscopes in Alaska.

Among the recent additions to the network, of special interest are several arrays of instruments extending across the San Andreas fault in regions where major fault motions are to be expected. One of these arrays, installed as a cooperative effort by the California Department of Water Resources and the USCGS, consists of five recording accelerographs and 16 seismoscopes arranged in a line about 15 mi long at right angles to the San Andreas fault. The Parkfield earthquake of June 27, 1966, gave excellent records on all of these instruments, one of the recording accelerographs being within 300 ft of the surface trace of the fault (Cloud and Perez, 1967; Housner and Trifunac, 1967; Hudson and Cloud, 1967).

The strong-motion accelerograph program in the United States is operated by the Seismological Field Survey of the USCGS. Information on the network of strong-motion instruments, as well as copies of instrumental records, are contained in the annual publication *United States Earthquakes*, issued beginning in 1928 by the U.S. Department of Commerce, Coast and Geodetic Survey. In addition to the U.S. Pacific Coast network, the USCGS has installed single accelerographs in the Canal Zone and in Colombia, El Salvador, Guatemala, Peru, Ecuador, Costa Rica, Chile, and Venezuela.

The Japanese strong-motion accelerograph network has expanded very rapidly since the formation of the Strong-Motion Acceleration Committee in 1951. As of 1968 there were about 300 recording accelerographs installed in Japan, making that country the most completely instrumented region for strong ground motion in the world (Earthquake Research Institute, 1966). The strong-motion records obtained from these instruments are prepared by the Strong-Motion Earthquake Observation Committee and published by the Earthquake Research Institute of the University of Tokyo.

An extensive strong-motion accelerograph network also is being built up in Mexico, which now has some 30 accelerographs installed. Smaller networks already exist in New Zealand, Canada, and India, and a number of other countries have one or several instruments in operation.

At present, no seismic region in the world can be said to be satisfactorily covered by strong-motion accelerographs. This fact was recognized at the 1964 UNESCO Intergovernmental Meeting on Seismology and Earthquake Engineering, which recommended that provision should be made for increased numbers and improved distribution of strong-motion accelerographs. UNESCO established a special working group having representatives from Japan, the USSR, and the United States to define suitable characteristics for such instruments. The instruments described in Table 6.1 all meet the general specifications set up by this working group.

In addition to the measurement of earthquake ground motion, the standard strong-motion accelerographs are useful to measure the earthquake response in tall buildings. By simultaneous measurements of ground motion and structural response during actual earthquakes, a great deal of information on the dynamic characteristics of structures under earthquake excitation can be derived. One opportunity to carry out calculations of this kind occurred during the San Francisco earthquake of 1957 (Hudson, 1960). A similar situation existed in Akita, Japan, during the Niigata earthquake of 1964, when excellent accelerograph records were simultaneously obtained of the ground motion and structural response of a modern structure (Osawa and Murakami, 1966). In 1966 an amendment to the City of Los Angeles Building Code required that all multistoried buildings over a certain height contain three recording accelerographs in the basement, roof, and at an intermediate location.

6.10 FUTURE ACCELEROGRAPH DEVELOPMENT

Although existing instruments have been reasonably satisfactory, there are many obvious improvements that should be made, some of which have been suggested above. It must be concluded that none of the existing field devices really exploits modern instrumentation developments to any significant degree. A major drawback is the form of analog record, which has been a major stumbling block to the introduction of modern data processing techniques. There would be a great advantage to a record that would directly produce an electric signal, such as a magnetic tape, since the data then could be easily transformed into any desired form. At present a great deal of routine work is required to produce electric signals for analog computer studies or to produce digital data for digital computation. There is no basic reason why the instrument output itself should not be in a digital form as this would permit a maximum flexibility in data processing.

It is also clear that a good deal would be gained by the kind of cooperative program that would make it possible to increase the manufacture lot size of accelerographs. The high cost of past accelerographs has been in large part caused by the fact that many of them have been built on an almost individual basis. The number of such devices that can be economically justified for the seismic regions of the world is such that it should be possible to significantly reduce production costs.

Compared with the impressive network of teleseismic seismological stations distributed throughout the world and the detailed studies of seismologists and geophysicists, the strong-motion earthquake measurement program must be considered to be in a very rudimentary state. The potential importance of the subject makes it imperative that a much more energetic approach be made to the problem. It is hoped that the near future will see rapid and comprehensive developments in all phases of the subject.

6.11 FAULT SLIP MEASUREMENTS

As mentioned above, nonseismic fault slip is becoming of direct interest to structural engineers because of the damage potential for engineering works of all kinds. Two instrumental investigations have involved the recording of displacements across cracks in structures undergoing such slow fault slip motions. The first of these was along the San Andreas fault near Hollister (Tocher, 1960). Three creep recorders were designed for this application. The first involved a measurement of a relative displacement between two concrete piers set in the ground 3 ft apart. A 20-to-one mechanical amplification linkage with a recording drum speed of 116 mm/day was used. A second instrument made use of a linear differential transformer to record the relative motion between floor slabs in a building in the fault zone at a magnification of 30. A recording paper drive with a speed of 610 mm/day was used. A third instrument involved a mechanical linkage with a magnification of 14 recorded at a speed of 40 mm/day. With these devices, floor slab displacements totaling some 25 mm were recorded over a period of $1\frac{1}{2}$ years.

A recent investgation of fault motion along the Hayward fault involves the measurements across a crack in a culvert under the Memorial Stadium at the University of California, Berkeley (Bolt and Marion, 1966). A linear differential transformer is used in a system having an overall magnification of 30, recording on a strip chart at a speed of $\frac{1}{4}$ in./hr. In this way a cumulative displaceemnt of 0.053 in. was recorded in 13 weeks.

As more regions are located in which slow fault slip motions may be occurring, measurements of the above type undoubtedly will become more prevalent, and more detailed studies of the optimum types of instrumentation for such investigations may be expected.

6.12 CRUSTAL STRAIN MEASUREMENTS

Mention also should be made of improvements in crustal strain measurements that may be of direct interest to the engineer because of the light they may ultimately throw on local seismicity and the earthquake mechanism problem.

Direct strain measurements of the Earth over gage lengths of the order of 100 ft have been made for some years by seismologists using various forms of the Benioff (1935) strain seismograph (Benioff, 1935). A recent version of the Benioff device has a gage length of 24.08 m, a fused quartz tube as the reference length, and a variable capacity-type displacement transducer (Benioff, 1959). With this system 1 mm on the recording paper equals a strain of 5.18×10^{-12} in the range of 0.01 to 10 cps (Shopland, 1966). It is usually not possible to operate at such high sensitivities because of background noise levels, and problems of calibration may be troublesome. One recent instrument includes a built-in Fabry–Perot interferometer calibration unit with an automatic daily recording of a calibration signal (Blayney and Gilman, 1965).

In connection with fault slip studies, shorter gage length strain instruments of a lower sensitivity are being developed. One such device now being tested at the Seismological Laboratory of the California Institute of Technology for studies on the San Andreas fault system has a 10 ft gage length and a strain resolution of 10^{-8}.

Also to be considered as a type of strain gage is the proposed laser strain meter suggested as a part of the overall instrumentation system outlined in the earthquake prediction program (Office of Science and Technology, 1965). These devices are likely to involve optical paths of approximately 1 km in buried pipes that are evacuated or filled with dry nitrogen, with the possibility of measuring strain changes over that distance of the order of 10^{-9}

6.13 PRECISION GEODETIC MEASUREMENTS

The standard techniques of first-order triangulation and leveling have yielded much interesting information as to fault motions accumulating over periods of time of several years (Richter, 1958; Whitten and Claire, 1960; Pope, Stearn and Whitten, 1966). First-order triangulation, however, is reliable only to one part in 10^5, whereas it would be very desirable to extend this accuracy to one part in 10^7 so that shorter term displacements could be detected.

Measurements based on electromagnetic travel times between fixed points on the Earth can theoretically attain the desired accuracy, but in practice such accuracy cannot be reached under field conditions with optical paths in the atmosphere because of variations in the index of refraction with changes in atmospheric temperature, pressure, and humidity. One approach to the problem that is now being investigated at the Seismological Laboratory of the California Institute of Technology involves a correction for meteorological variations by monitoring air temperatures along the light paths during measurement. Another suggested approach is to confine the optical path to a buried pipeline with a controlled atmosphere. This technique would become very expensive for path lengths exceeding a mile or so.

The methods of precision geodetic surveying are also applicable to many local problems closely related to earthquake ground motions. Notable studies have been made, e.g., of ground motions associated with subsidence (Gilluly and Grant, 1949). In at least one such case, subsidence caused by oil withdrawal has been suggested as being directly related to local seismic activity (Richter, 1958). The use of precision surveying to monitor landslide conditions and local earth movements around dams and other large structures also may be expected to become increasingly important.

REFERENCES

Amin, M., and A. H. S. Ang (April 1966). *A Nonstationary Stochastic Model for Strong-Motion Earthquakes*, Urbana: Department of Civil Engineering, University of Illinois.

Anderson, J. A., and H. O. Wood (March 1925). "Description and Theory of the Torsion Seismometer," *Bull. Seism. Soc. Am.*, **15** (1).

Beckwith, T. G., and N. L. Buck (1961). *Mechanical Measurements*, Reading: Addison-Wesley.

Benioff, H. (1935). "A Linear Strain Seismograph," *Bull. Seism. Soc. Am.*, **25**.

Benioff, H. (1955). "Earthquake Seismographs and Associated Instruments," in H. E. Landsberg, ed., *Advances in Geophysics*, New York: Academic Press.

Benioff, H. (1959). "Fused Quartz Extensometer for Secular, Tidal, and Seismic Strains," *Bull. Geol. Soc. Am.*, **70**.

Berg, G. V. (1963). *A Study of Error in Response Spectrum Analyses*, Santiago: First Chilean Conference on Seismology and Earthquake Engineering.

Berg, G. V., and G. W. Housner (April 1961). "Integrated Velocity and Displacement of Strong Earthquake Ground Motion," *Bull. Seism. Soc. Am.*, **51** (2).

Blayney, J. L., and R. Gilman (December 1965). "A Portable Strain Meter with Continuous Interferometric Calibration," *Bull. Seism. Soc. Am.*, **55** (6).

Bolt, B. A., and W. C. Marion (April 1966). "Instrumental Measurement of Slippage on the Hayward Fault," *Bull. Seism. Soc. Am.*, **56** (2).

Borisevich, E. S., M. L. Goldfarb, and M. S. Mosyagina (1961). "Recorder with Luminescent Memory," in D. P. Kirnos and E. S. Borisevich, eds., *Seismic Instruments*, Ann Arbor: University of Michigan.

Bradley, W., and E. E. Eller (1961). "Introduction to Shock and Vibration Measurements," in C. M. Harris and C. E. Crede, eds., *Shock and Vibration Handbook*, New York: McGraw-Hill.

Brady, A. G. (1966). *Studies of Response to Earthquake*

Ground Motion, Pasadena: Earthquake Engineering Research Laboratory, California Institute of Technology.

Caughey, T. K., D. E. Hudson, and R. V. Powell (1960). *The C.I.T. Mark II Response Spectrum Analyzer for Earthquake Engineering Studies, Proceedings of the Second World Conference on Earthquake Engineering*, Tokyo and Kyoto.

Cloud, W. K. (1963). *Maximum Accelerations During Earthquakes, Proceedings of Primeras Jornadas Chilenas de Sismologia e Ingenieria Antisismica*, Vol. 1, Asociacion Chilena de Sismologia e Ingenieria Antisismica.

Cloud, W. K., and D. E. Hudson (April 1961). "A Simplified Device for Recording Strong Motion Earthquakes," *Bull. Seism. Soc. Am.*, **51** (2).

Cloud, W. K., and V. Perez (December 1967). "Accelerograms—Parkfield Earthquake," *Bull. Seism. Soc. Am.*, **57** (6).

Cluff, L. S., and K. V. Steinbrugge, (April 1966). "Hayward Fault Slippage in the Irvington–Niles Districts of Fremont, California," *Bull. Seism. Soc. Am.*, **56** (2).

Duflou, P. C. J., and R. I. Skinner (1965). *New Strong-Motion Accelerographs, Proceedings of the Third World Conference on Earthquake Engineering*, New Zealand.

Earthquake Research Institute (February 1966). *Strong-Motion Earthquake Records in Japan*, Vol. 4, Tokyo: University of Tokyo.

Edwards, A. T., and T. D. Northwood (1959). *Experimental Blasting Studies on Structures*, Ottawa: Hydro-Electric Power Commission of Ontario, and the National Research Council.

Fremd, V. M. (1963). "Installation with Memory for the Recording of Strong Earthquakes," *AN SSSR, Trudy Institute Fiziki Zemii*, No. 26 (193), Moscow: Academy of Sciences, USSR. (English Trans.).

Galitzin, B. (June 1913). "Über Eine Skala zur Schätzung von Makroseismischen Bewegungen," St. Petersburg, 1911, review and translated abstract by H. O. Wood, *Bull. Seism. Soc. Am.*, **3** (2).

Gilluly, J., and U. S. Grant (March 1949). "Subsidence in the Long Beach Harbor Area, California," *Bull. Geol. Soc. Am.*, **60**.

Halverson, H. T. (1965). *The Strong Motion Accelerograph, Proceedings of the Third World Conference on Earthquake Engineering*, New Zealand.

Hansen, W. R. (1965). *Effects of the Earthquake of March 27, 1964 at Anchorage, Alaska*, Washington, D.C.: U.S. Geological Survey Professional Paper 542-A.

Heck, N. H., H. E. McComb, and F. P. Ulrich (1936). "Strong-Motion Program and Tiltmeters," *Earthquake Investigations in California, 1934–35*, Special Pub. No. 201, Washington, D.C.: U.S. Department of Commerce, Coast and Geodetic Survey.

Housner, G. W. (1965). *Intensity of Earthquake Ground Shaking Near the Causative Fault, Proceedings of the Third World Conference on Earthquake Engineering*, New Zealand.

Housner, G. W., and M. D. Trifunac (December 1967). "Analysis of Accelerograms—Parkfield Earthquake," *Bull. Seism. Soc. Am.*, **57** (6).

Hudson, D. E. (1956). *Response Spectrum Techniques in Engineering Seismology, Proceedings of the World Conference on Earthquake Engineering*, Berkeley.

Hudson, D. E. (1959). *Ground Motion Measurements in Earthquake Engineering, Proceedings of the Symposium on Earthquake Engineering*, Roorkee, U. P., India: University of Roorkee.

Hudson, D. E. (1960). *A Comparison of Theoretical and Experimental Determinations of Building Response to Earthquakes, Proceedings of the Second World Conference on Earthquake Engineering*, Tokyo and Kyoto.

Hudson, D. E. (April 1962). "Some Problems in the Application of Spectrum Techniques to Strong-Motion Earthquake Analysis," *Bull. Seism. Soc. Am.*, **52** (2).

Hudson, D. E. (February 1963). "The Measurement of the Ground Motion of Destructive Earthquakes," *Bull. Seism. Soc. Am.*, **53** (2).

Hudson, D. E. (1965). *Ground Motion Measurements in Earthquake Engineering, Proceedings of the Symposium on Earthquake Engineering*, Vancouver, B.C.: The University of British Columbia.

Hudson, D. E., and W. K. Cloud (December 1967). "An Analysis of Seismoscope Data from the Parkfield Earthquake of June 27, 1966," *Bull. Seism. Soc. Am.*, **57** (6).

Hudson, D. E., and R. F. Scott (February 1965). "Fault Motions at the Baldwin Hills Reservoir Site," *Bull. Seism. Soc. Am.*, **55** (1).

International Institute of Seismology and Earthquake Engineering (February 1965). *The Niigata Earthquake 16 June, 1964 and Resulting Damage to Reinforced Concrete Buildings*, Report No. 1, Tokyo.

Krishna, J. and A. R. Chandrasekaran (1965). *Structural Response Recorders, Proceedings of the Third World Conference on Earthquake Engineering*, New Zealand.

Medvedev, S. V. (1962). *Engineering Seismology*, Moscow; English trans., Israel Program for Scientific Translations, Jerusalem, 1965 (Available from U.S. Dept. of Commerce Clearinghouse for Scientific and Technical Information).

Muto, K., R. Takahasi, I. Aida, N. Ando, T. Hisada, K. Nakagawa, H. Umemura, and Y. Osawa (1960). *Non-Linear Response Analyzers and Application to Earthquake Resistant Design, Proceedings of the Second World Conference on Earthquake Engineering*, Tokyo and Kyoto.

Nazarov, A. G. (1959). *The Method of Engineering Analysis of Seismic Forces*, Armenian Academy of Sciences (in Russian).

Neumann, F. (1942). *United States Earthquakes, 1940*, Serial No. 647, Washington, D. C.: U. S. Department of Commerce, Coast and Geodetic Survey.

Neumann, F. (1958). "Damaging Earthquake and Blast Vibrations," *The Trend in Engineering*, Seattle: University of Washington.

Neumann, F. (1959). *Seismological Aspects of the Earthquake*

Engineering Problem, *Proceedings of the Third Northwest Conference of Structural Engineering*, State College of Washington.

Office of Science and Technology (1965). "Earthquake Prediction—A Proposal for a Ten Year Program of Research," Washington, D. C.

Osawa, Y., and M. Murakami (March 1966). "Response Analysis of Tall Buildings to Strong Earthquake Motions. Part 2, Comparison with Strong Motion Accelerograms (1)," *Bull. Earthquake Res. Inst.*, **44** (1).

Pope, A. J., J. L. Stearn, and C. A. Whitten (April 1966). "Surveys for Crustal Movement Along the Hayward Fault," *Bull. Seism. Soc. Am.*, **56** (2).

Richter, C. F. (1958). *Elementary Seismology*, San Francisco: W. H. Freeman.

Savarensky, E. F., and D. P. Kirnos (1955). *Elements of Seismology and Seismometry*, Moscow (English trans., U.S. Dept. of Commerce, Office of Technical Services).

Schiff, A., and J. L. Bogdanoff (October 1967). "Analysis of Current Methods of Interpreting Strong-Motion Accelerograms," *Bull. Seism. Soc. Am.*, **57** (5).

Shi-Yuan, E., D. P. Kirnos, and V. N. Solovyev (1961). "A Simplified Recording Unit for Instrumental Observations in Epicentral Zones of Strong Earthquakes," in D. P. Kirnos and E. S. Borisevich, eds., *Seismic Instruments*, Ann Arbor: University of Michigan.

Shopland, R. C. (April 1966). "Shallow Strain Seismograph Installations at the Wichita Mountains Seismological Observatory," *Bull. Seism. Soc. Am.*, **56** (2).

Steinbrugge, K. V., and E. G. Zacher (July 1960). "Creep on the San Andreas Fault—Fault Creep and Property Damage," *Bull. Seism. Soc. Am.*, **50** (3).

Suyehiro, K. (1926). "A Seismic Vibration Analyzer and the Records Obtained Therewith," *Bull. Earthquake Res. Inst.*, Tokyo, **1**.

Suyehiro, K. (1932). "Engineering Seismology—Notes on American Lectures," *Proc. Am. Soc. Civil Eng.*, **58** (4).

Takahasi, R. (1956). *The SMAC Strong Motion Accelerograph and Other Latest Instruments for Measuring Earthquakes and Building Vibrations*, *Proceedings of the World Conference on Earthquake Engineering*, Berkeley.

Tocher, D. (July 1960). "Creep on the San Andreas Fault—Creep Rate and Related Measurements at Vineyard, California," *Bull. Seism. Soc. Am.*, **50** (3).

Vanmarcke, E. H., and C. A. Cornell (December 1967). "Optimum Location of a Network of Strong-Motion Accelerometers," *Research Report R67–63*, Department of Civil Engineering, Massachusetts Institute of Technology, Cambridge.

Whitten, C. A., and C. N. Claire (July 1960). "Creep on the San Andreas Fault—Analysis of Geodetic Measurements Along the San Andreas Fault," *Bull. Seism. Soc. Am.*, **50** (3).

Wilt, J. W. (April 1958). "Measured Movement Along the Surface Trace of an Active Thrust Fault in the Buena Vista Hills, Kern County, California," *Bull. Seism. Soc. Am.*, **48** (2).

Chapter 7

Dynamic Tests of Full-Scale Structures

DONALD E. HUDSON

Professor of Mechanical Engineering and Applied Mechanics
California Institute of Technology
Pasadena, California

7.1 INTRODUCTION

The problem of the determination of the response of structures to prescribed exciting forces in theory can be formulated and solved in very general terms, even for situations involving plastic deformations. To make any practical use of the analysis, however, requires that quantitative information be available on such basic structural dynamic properties as natural periods of vibration, mode shapes, energy dissipation, and yield limits. Such dynamic properties depend in turn on many details of material behavior and structural configuration that are not amenable to a fully analytical treatment. Direct experimental determination of such dynamic characteristics is thus a necessity at the present stage of development of the subject.

The main concern of the present review study is the dynamic testing of full-scale structures from the special point of view of the knowledge needed for earthquake-resistant design. There have been important applications in which model studies have been made to explore dynamic behavior, and models may be a useful adjunct to full-scale testing. In general, however, structural dynamic responses are so dependent upon such details as the stiffness and energy losses in individual joints and connections that it is not possible to answer all of the important questions by model tests alone. In fact, suitable models themselves may be large structures and may present the same problems of testing as the full-scale prototypes.

In view of the important advantages that may be possessed by models for many investigations, one of the objectives of full-scale testing at the present time should be to establish the correct basis for such model studies. For this purpose it is essential to carry along model and full-scale testing in close coordination, so that a true assessment of the applicability and accuracy of various size models can be made (Hudson, 1961a). Such investigations unfortunately are very rare. If the study begins with a model, the model tests usually point to design changes in the full-scale structure that thus turns out to be different from the model. Once the full-scale structure is completed, of course, the interest in restudying a modified model has usually disappeared. There thus is a need for specific research investigations of these modeling problems by a group whose main interest would be the basic problems of modeling rather than the design of specific structures.

7.2 DYNAMIC TESTS

7.2.1 Objectives of Dynamic Tests

In one general category of dynamic tests, the linear dynamic properties of the structure are to be determined. This implies relatively small displacement, and hence no consideration of structural failure is usually involved. The major parameters desired are the natural periods of vibration of all significant modes, the corresponding mode shapes, and the amount of energy dissipation or damping associated with each mode.

In a second category of tests come studies of nonlinear behavior, such as investigations of yield conditions and the determination of energy dissipation under such yielding. Included in this category are studies of criteria of failure and of failure details involving excessive yielding, fracture, impact, and fatigue.

7.2.2 Theoretical Background of Dynamic Tests

It is essential that structural dynamic testing be guided and interpreted by the basic theory. The tests themselves cannot be designed or carried out without a complete understanding of the underlying dynamic principles. As an example, consider the problem of exciting a pure normal mode in a linear system. This can be done only by a certain distribution of generalized exciting force throughout the structure. The proper distribution, however, will not be known until the mode shapes have been determined. An exact solution of the problem thus would involve an iterative experimental technique requiring successive changes in force distribution to approximate the resulting mode shapes. Such experimental techniques in fact have been used in aircraft structural investigations (Lewis and Wrisley, 1950).

Another complication in testing that requires some theoretical consideration is the fact that for certain types of damping and damping distribution, simple normal modes in the classical sense do not exist (Foss, 1958; Caughey, 1960; Berg, 1958; O'Kelly, 1961; Caughey and O'Kelly, 1955). It seems likely that for the magnitude of damping involved in most structural tests such special considerations will not be critical, but it is necessary to be aware of the possibilities.

The experimental determination of the damping in structures presents other difficulties. If the natural frequencies of the various modes of vibration are close together, it may be difficult to get an accurate determination of the damping from the shape of the steady-state resonance curve (Kennedy and Pancu, 1947). Under these same conditions the measurement of damping from the decay of free vibrations also may be rendered difficult by the appearance of beats in the record (Scruton and Harding, 1957). For all such problems the theory always must be fully employed to extract the maximum amount of information from the experimental results (Berg, 1962; de Veubeke, 1956). A recent investigation has examined in some detail the theoretical problems involved in deriving system properties from experimentally measured quantities (Nielsen, 1966a). In this connection the chapter in the present volume by Bouwkamp and Rea, "Dynamic Testing and the Formulation of Mathematical Models," should be consulted for additional details and examples.

7.2.3 Types of Dynamic Tests

The main types of dynamic tests suitable for full-scale structures may be summarized in the following outline:

- I. Free vibration tests.
  - A. Initial displacement tests. Pull-back and quick-release tests.
  - B. Initial velocity tests. Impact, rockets, etc.
- II. Forced vibration tests.
  - A. Resonance tests—steady-state sinusoidal excitation.
    - 1. Rotating eccentric weight exciters.

2. Man-excited vibrations.

B. Variable frequency sinusoidal excitation.

C. Transient excitations.

1. Natural earthquakes.
2. Blasts and explosions.
3. Microtremor excited.
4. Wind excited.

III. Vibration table tests.

Each of these tests will now be discussed, giving first the general principles and then specific examples.

7.2.3.1 Free vibration tests

7.2.3.1.1 *Pull-back tests.* The simplest type of dynamic test consists of deforming a structure by pulling on it with a cable that is then suddenly released, thus causing the structure to perform free vibrations about its static equilibrium position. By recording the vibration curve vs time, the natural period of the structure can be directly determined. Because of energy dissipation in the structure, the vibration amplitudes will decay, and from the ratios of amplitudes of successive cycles of motion the damping can be calculated. Figure 7.1 shows displacement-time curves obtained from such pull-back tests on elevated water tanks made by the U.S. Coast and Geodetic Survey (Carder, 1965). A difficulty in such tests is to apply the pull and release in such a way that the structure will vibrate in one plane only. What often happens is that two different modes of vibration may be simultaneoulsy excited. If the periods of the two modes are close together, as will be the case with structures such as water tanks, which usually have at least an approximate structural symmetry, the resulting motion will show beat phenomena, making it difficult to obtain any meaningful measure of damping. This is clearly indicated by the first two curves of Fig. 7.1 and would evidently interfere with damping measurements on any of the curves shown there. A recent application of the pull-back technique to some steel stacks in Chile after the 1960 earthquake illustrates the same procedures and the same difficulties (Cloud, 1963).

7.2.3.1.2 *Initial velocity tests.* Free vibrations also can be set up in a structure by initial velocities rather than by the initial displacements described above in the pull-back tests. This can be done by impact forces, caused by falling weights or by a pendulum that can strike a horizontal blow. Impulsive loads for structural tests also have been generated by explosive cartridges (Mazet, 1956) or by small rockets (Scruton and Harding, 1957). In such impulsive tests the total time duration of the applied force is preferably short compared with the natural periods of significant structural modes, so that the resulting motions are functions of the total impulse or the initial velocity rather than of the force magnitudes.

Fig. 7.1. Pull-back tests of elevated water tanks. (From Carder, 1965.)

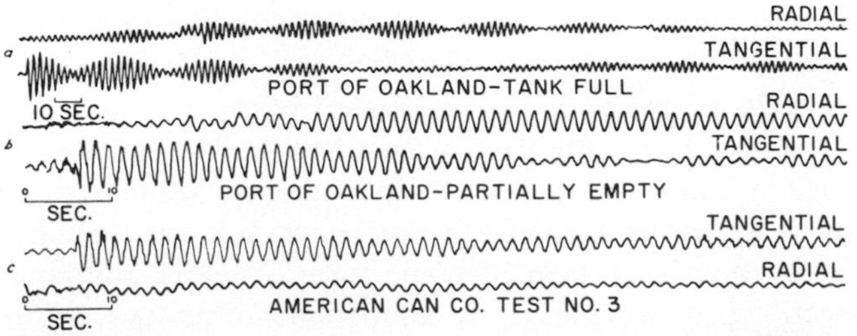

An interesting example of an impulsive or initial velocity-type test is afforded by an investigation made in England of a 450-ft high 36-ft diameter concrete chimney (Scruton and Harding, 1957). The chimney was excited by firing groups of six, twelve, and eighteen 1000-lb thrust rockets transversely from the top of the chimney. The impulse provided by each rocket was about 385 lb-sec, and the burning time duration was about $\frac{1}{2}$ sec, which was sufficiently short compared to the fundamental period of the chimney of approximately $1\frac{1}{2}$ sec so that the load acted essentially as a single impulse to excite an initial velocity. Care was taken to orient the rockets so that motions would be excited in one plane only, and the test was made at a time when there were no wind-excited motions in evidence. The chimney motions were measured by piezoelectric accelerometers, and the double integrated displacement record was recorded vs time. A smooth free-vibration decay curve was obtained from which the damping was determined to be somewhat less than 1% of critical damping.

7.2.3.2 Forced vibration tests. Although the free-vibration tests described above may appear to have the virtue of a relative simplicity, forced vibration tests will usually produce more complete and more accurate information, and the increased complexity and costs of such forced vibration tests are usually justified. The most generally useful type of forced vibration test involves a steady-state sinusoidal excitation, and it is this test that will be discussed in the most detail in the following sections.

7.2.3.2.1 *Steady-state sinusoidal excitation.* A steady-state resonance test involves the application to the structure of a sinusoidally varying, uni-directional force whose period can be held accurately constant at one value while measurements are made of the resulting motion of the structure. The period is then adjusted to a new value, the measurements are repeated, and so on to describe the whole period-response curve. By thus measuring the amplitude of motion of the structure at various periods covering the whole range of natural periods of the structure, the resonance curves can be plotted. From these resonance curves, accurate values of

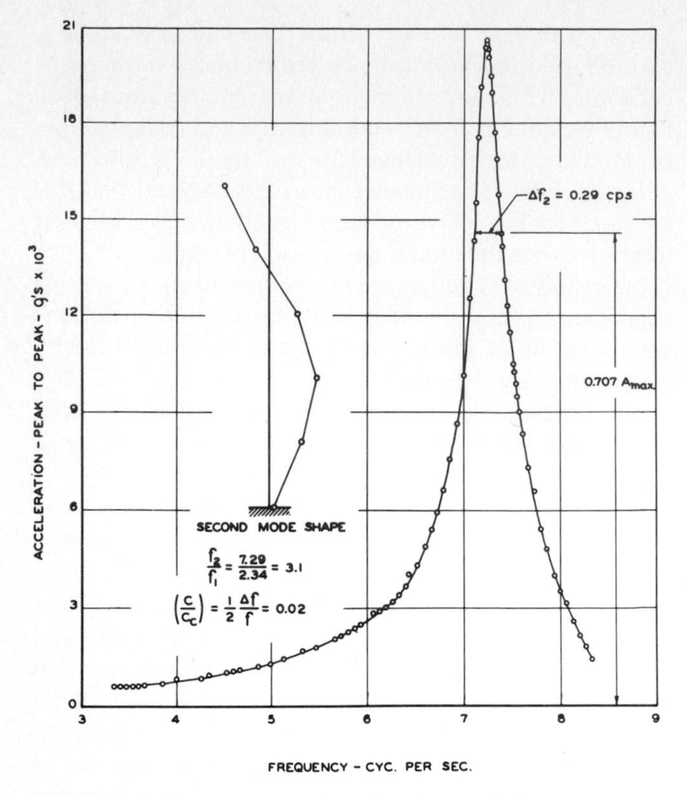

Fig. 7.2. Resonance test of a five-story reinforced concrete building. (From Hudson, 1964.)

natural period and damping can be obtained. By varying the magnitude of the exciting force as well as its period, various nonlinear characteristics of the structure can be studied.

Figure 7.2 shows a typical resonance curve obtained from a five-story reinforced concrete building under small forces for which the system remains linear. The approximate symmetry of the curve about the resonant

Fig. 7.3. Hysteresis-type resonance curves. (From Jennings, 1964.)

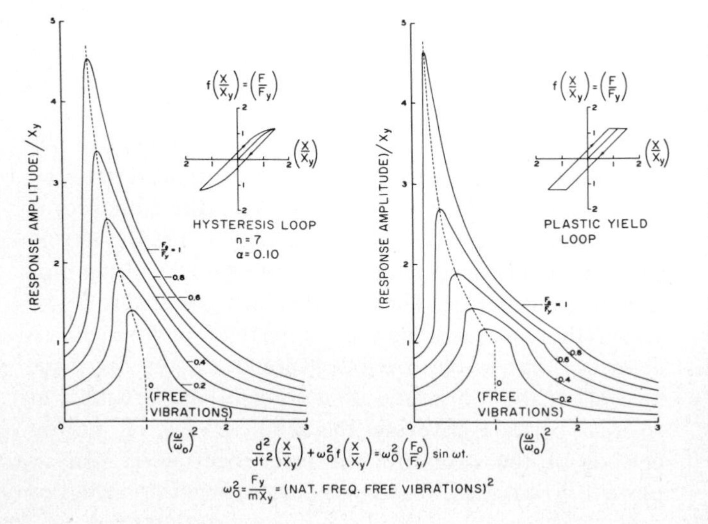

peak is typical of linear systems. The width of the resonance curves at some prescribed fraction of the peak amplitude is a measure of the energy dissipation in the system, and for small damping this width can be directly related, through the concept of an equivalent viscous damping, to the fraction of critical damping in the system (Jacobsen, 1930; 1960; Hudson, 1965; Jennings, 1968).

The two resonance curves of Fig. 7.3 illustrate the distinctive shape of resonance curves for nonlinear systems with hysteretic damping (Caughey, 1960b; Iwan, 1961; Jennings, 1964; Iwan, 1965). Characteristic of these hysteretic-type resonance curves are the vertical slopes below the natural frequency. These curves may be compared with those shown in Fig. 7.4, which are the experimentally obtained resonance curves for a five-story reinforced concrete building. It is interesting to note that the concrete building exhibits distinctly nonlinear behavior even at the relatively low force levels of the test.

A comparison of Figs. 7.2, 7.3, and 7.4 indicates how the shape of the resonance curve can be used to give additional information about the structure. On the other hand, a comparison of the two sets of curves of Fig. 7.3 makes it evident that it would not be easy to say much about the shape of the hysteresis loop from the resonance curves alone, since the curved hysteresis loop and the elasto-plastic hysteresis loop have very similar resonance curves. It thus is clear that a good understanding of the basic theory of such structural behavior is necessary in order to extract the maximum amount of information from the resonance tests.

7.2.3.2.2 *Sinusoidal force generation.* Most test structures are so large that it is not feasible to apply suitable forces or displacements by interposing some mechanism between the structure and a fixed point. A simple crank and connecting rod attached to a fixed point, e.g., might be an appropriate way to excite motion in a small model in the laboratory, whereas for large structures it ordinarily would be quite impractical. The generation of a force independent of any fixed point is accomplished by giving an acceleration to a mass, using the corresponding inertia force as the exciting force for the structure.

For ease of interpretation of the results, it is desired that simple motions be excited in the test structure, and hence the applied inertia force ideally should be a unidirectional force. Similarly, there should be only one frequency component in the force, i.e., the force should vary according to a simple single frequency sine wave.

Perhaps the simplest mechanical way of achieving such a unidirectional sinusoidal motion would be to move a mass back and forth along a straight track. Such a mass could be driven by some form of crank and connecting rod mechanism, by compressed air pistons, by an electromagnetic system, or by a hydraulic cylinder. Considering the fact, however, that the most convenient source of power is a rotating electric motor, the mechan-

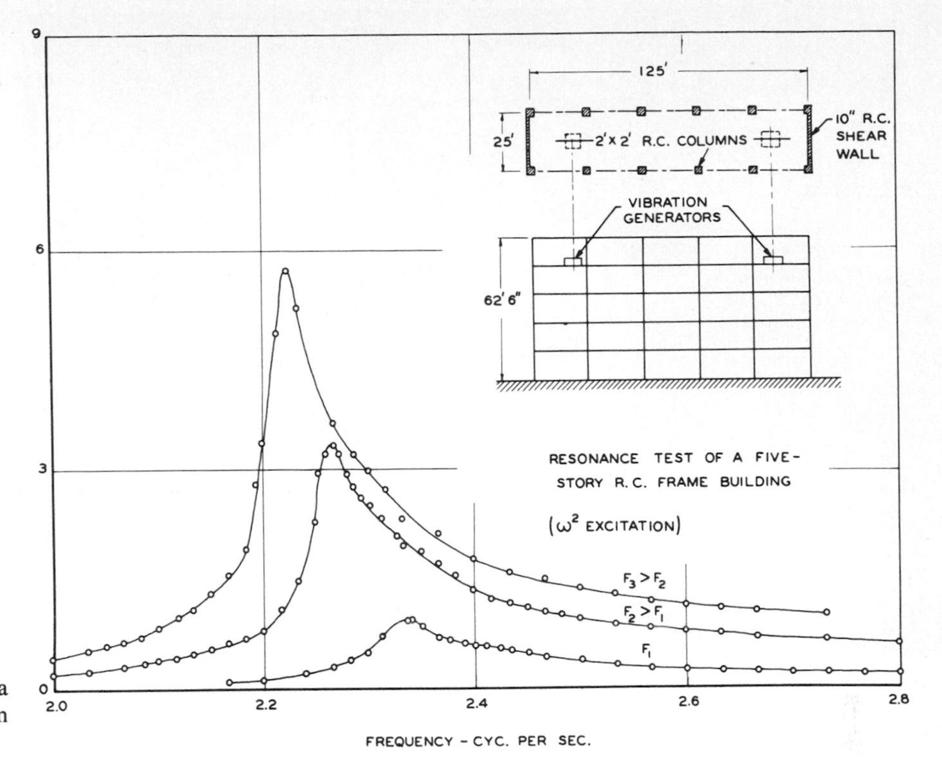

Fig. 7.4. Nonlinear resonance in a reinforced concrete building. (From Hudson, 1964.)

ical design details of any such reciprocating inertia generator never have been worked out in a form suitable for economic construction and field convenience. Additional difficulties present themselves in the reciprocating type when questions of precision speed control and synchronization of individual units are raised.

The common solution of the problem has been to adopt an eccentric rotating weight arrangement, which avoids many of the mechanical design problems of a reciprocating system. A single weight of mass m rotating at a radius r at a frequency ω rad/sec would produce a rotating inertia force of magnitude ($mr\omega^2$). This would not satisfy the requirement of a unidirectional force, however, since the radial inertia force would be continually changing direction as the unbalanced weight rotates. The component of the rotating inertia force along a fixed direction is, however, a sinusoidally varying force ($mr\omega^2 \sin \omega t$). The final solution of the problem, therefore, is simply to arrange two equal eccentric weights so that they rotate in opposite directions. In this way the force components cancel in one direction and add in the perpendicular direction, the net result being a sinusoidally varying force in one direction.

This principle has been used for many years in a wide variety of vibration generators (Späth, 1938). Early work done in the United States with rotating weight vibration generators exciting building structures is summarized in a reissued publication of the U.S. Coast and Geodetic Survey (Carder, 1965; Blume, 1935; Patterson, 1940).

The rotating weight exciter even has been used in its simplest form as a single arm device. Figure 7.5 shows a view of the large single arm machine of the Japanese Building Research Institute (Hisada and Nakagawa, 1956). This machine will produce a force of 3000 lb at 1 cycle/sec, and has a total weight of approximately 3300 lb. This machine has been used to excite high levels of dynamic load in small full-scale structures. Although the force is not unidirectional, certain types of tests can be successfully carried out.

7.2.3.2.3 *A rotating weight vibration generator.* As an example of a modern version of the eccentric weight exciter, a description will be given of the vibration generator system developed for the California State Division of Architecture. This system was developed under specifications formulated by a special committee of the Earthquake Engineering Research Institute. The design, construction, and field testing were carried out at the California Institute of Technology (Hudson, 1961; 1964).

Figure 7.6 shows a general view of one of the vibration generator units. Eccentric weights having the form of flat baskets rotate about a vertical axis. Counterrotation of the two baskets is produced by driving pulleys off the two sides of a roller chain. By placing various numbers of lead weights in the baskets the magnitude of the unbalanced weights can be altered. The single unit shown can produce an alternating force of the order of 1000 lb at 1 cycle/sec, with a total force capability of 5000 lb. The vertical shaft system was used so that there would be no large rotating gravity force couples to impose varying loads on the constant speed drive. This of course limits the generators to horizontal forces. By keeping

Fig. 7.5. Rotating arm vibration exciter of the Japanese Building Research Institute. (From Hisada and Nakagawa, 1956.)

the vertical dimensions small, rotating couples are reduced to negligible values.

The electric drive system is made detachable for ease of transportation and installation. The whole machine can be disassembled into easily manageable parts for handling under adverse field conditions.

7.2.3.2.4 *Electric drive and speed control system.* The requirements of the electric drive for the vibration generator system are particularly stringent because of the variable torques imposed on the drive and the necessity to ensure stability when operating near resonance of lightly damped structures. The ability to hold an accurate speed control at and near resonance peaks requires that the speed-torque curve of the drive system be unusually flat, with essentially a constant speed maintained under relatively large torque variations. It is also required that the speed be variable over a very wide range.

To meet these requirements a $1\frac{1}{2}$ hp DC drive motor is used along with a servo-controlled electronic amplidyne system. The general features of the system are shown in Fig. 7.7. A tachometer driven directly from the drive motor supplies a speed signal that is compared with a standard set voltage. The difference between these voltages provides an input signal to the amplidyne amplifier, which then acts upon the drive motor to adjust its speed to correspond to the set speed. By this means the speed can be controlled to 0.1% and stable operation can be attained in systems with damping as low as 0.5% of critical.

Fig. 7.6. Counterrotating eccentric weight vibration generator. (From Hudson, 1964.)

7.2.3.2.5 *Synchronized operation of vibration generators.* One of the main advantages of the electronic speed control described above is the ease with which it can be adopted to synchronized control of multiple machines (Herwald, Gemmell and Lazan, 1946; Hudson, 1962b). Accurate synchronization makes it possible to increase the total forces applied to a structure by conveniently sized units by adding the effects of several machines. This is a more practical way to generate large forces than to design a larger machine in one unit, since the larger machine becomes difficult to transport and to install.

Even more important, synchronized machines can be distributed throughout a structure so as to excite in the

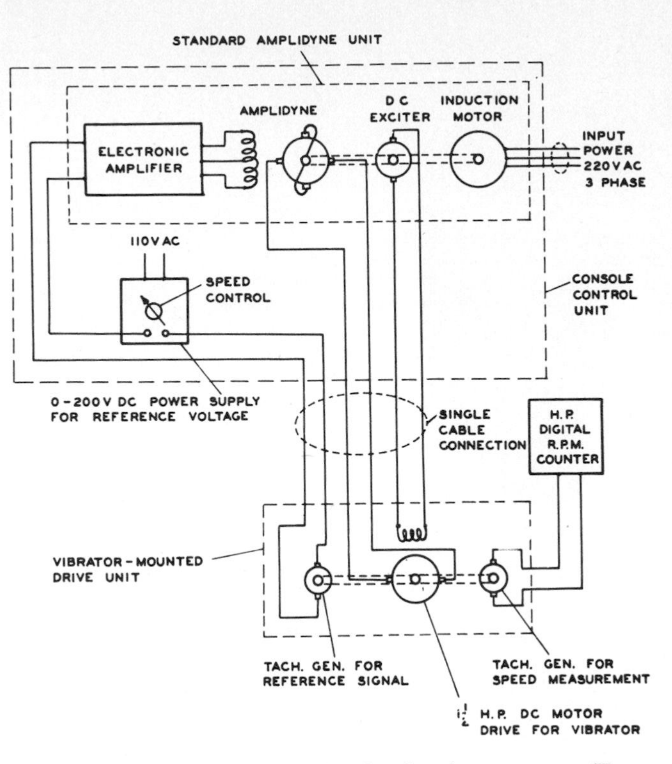

Fig. 7.7. Speed control system for the vibration generator. (From Hudson, 1962.)

Fig. 7.8. Synchronizing circuits for the vibration generator. (From Hudson, 1964.)

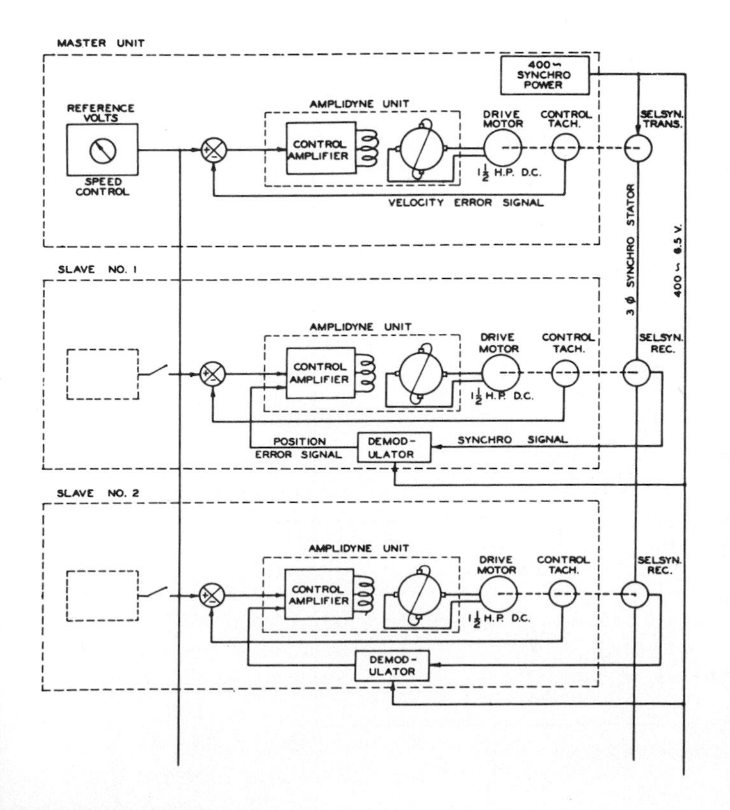

most efficient way various modes of vibration. For example, two machines at the ends of a long building, if in phase, will excite lateral motions. By arranging the two machines 180° out of phase, torsion would be excited.

Figure 7.8 shows the basic scheme for arranging suitable synchronization. A master control unit drives a selsyn transmitter from the drive motor. A selsyn receiver is attached to the drive motor of each separate vibration generator to be synchronized with the master unit. Any difference in angular position between the control unit and the slave unit goes to the control amplifier of the speed control amplidyne as an error signal and suitably adjusts the speed. The system thus has both velocity and position control, and accurate synchronization can be maintained under a wide range of conditions.

Figure 7.9 shows four synchronized vibration generators with their control consoles arranged for tests of an earth-filled dam. The vibration generators have been bolted down to a reinforced concrete slab cast in the compacted earth at the crest of the dam. Figure 7.10 gives typical resonance curves from the test of the dam. From the close spacing of the points and the sharpness of the resonance peaks, an idea of the capabilities of the test system can be gained.

The system in its present form can be used either as four independent units, as a set of four synchronized units, or as two independent sets of two synchronized units each.

7.2.3.2.6 *Vibration generator tests.* The sinusoidal vibration generators have been used in a wide variety of structural tests and have made it possible to learn a great deal about the structural dynamic behavior of typical structures. These investigations include: tests of two concrete intake towers for a dam, with associated bridge structures (Keightley, Housner and Hudson, 1961; Keightley, 1963; 1966 a). A nine-story steel-frame building (Nielsen, 1966 b), a five-story reinforced concrete building (Nielsen, 1964 a), a fifteen-story steel-frame building (Rea, Bouwkamp and Clough, 1966), a two-story steel-frame building (Bouwkamp and Blohm, 1966), an eight-story reinforced concrete building (Englekirk and Matthiesen, 1957), two earth-filled dams (Keightley, 1963; 1966), a concrete arched dam (Rouse and Bouwkamp, 1967), a steel space-frame structure with a stressed skin covering (Hanson, 1965), a nuclear reactor structure (Matthiesen and Smith, 1966), and a multistory building and foundation soil interaction study (Kuroiwa, 1967). Related tests using similar methods also have been carried out recently in Japan (Umemura, 1967). Figures 7.11 and 7.12 show some of these typical tests to indicate in a general way typical field situations involved in these studies.

An example of another type of dynamic test is afforded by a recent investigation of the postyield dynamic behavior of a steel structure (Hanson, 1965 a;

Fig. 7.9. Four synchronized vibration generators exciting an earth-filled dam. (From Keightley, 1966.)

Jennings, 1967). The small structure shown in Fig. 7.13 was designed with properties such that the forces available from two of the synchronized vibration generators would excite the structure well into the plastic range. Maximum dynamic deflections several times as great as initial yielding displacements were attained. By means of a jack and a strain-gage load cell, reversed static loads were applied, giving an accurately defined hysteresis loop. From simultaneous acceleration and force measurements made during steady-state sinusoidal tests, the corresponding dynamic hysteresis loop was derived. Figure 7.14 shows a typical comparison between these static and dynamic hysteresis curves, from which it will be seen that the common assumption that the area of the hysteresis loop is independent of frequency is approximately true at the test frequency (Hanson, 1966).

7.2.3.2.7 *Electromagnetic vibration exciters.* A number of electromagnetic vibration exciters of a large size have been produced for vibration testing in the aircraft and

Fig. 7.10. Resonance curves for an earth-filled dam. (From Keightley, 1966.)

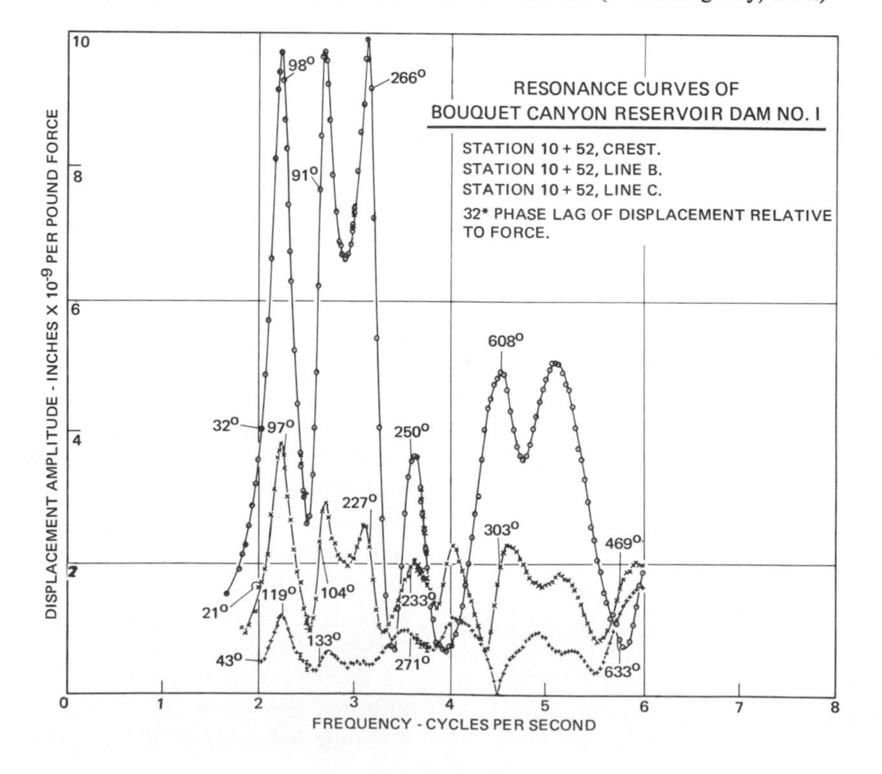

Fig. 7.11. Vibration generator installation on concrete intake tower. (From Keightley, Housner and Hudson, 1961.)

missile industries. These shakers are usually used as fixed point devices for testing missile components of a moderate size at frequencies that are relatively high compared to earthquake work. Experiments have been made using an electromagnetic shaker driving a concentrated mass to produce a sinusoidally varying inertia force for structural testing. Considering the force magnitudes desired at low frequencies, the size of equipment involved poses many practical problems of transportation, handling, and installation.

A notable example of the use of electromagnetic vibration generators for the full-scale test of a large structure is the test work on the 160-ft high Saturn space vehicle (Watson, 1962). For this test the elaborate test stand shown in Fig. 7.15 was constructed, so that fixed point vibration generators of the type shown in Fig. 7.16 could be used. The tests were supported by experiments on the one-fifth scale model shown in Fig. 7.17 and by theoretical calculations of frequencies and mode shapes (Mixson and Catherine, 1962; Bullock, 1962).

7.2.3.2.8 *Hydraulic-type vibration actuators.* A considerable amount of work has been done in the development of large hydraulic piston actuators that can be driven sinusoidally under electronically controlled conditions. For missile tests, such hydraulic cylinders are connected between various points of the missile and fixed points, and programmed forces or displacements having sinusoidal or random characteristics are imposed. Such devices are now commercially available in a wide range of sizes and have been used in such applications as fatigue testing of aircraft structures and road simulation tests of automotive vehicles. Such hydraulic systems could

Fig. 7.12. Static test of a space-frame structure. Vibration generator mounted in roof structure. (From Hanson, 1965a.)

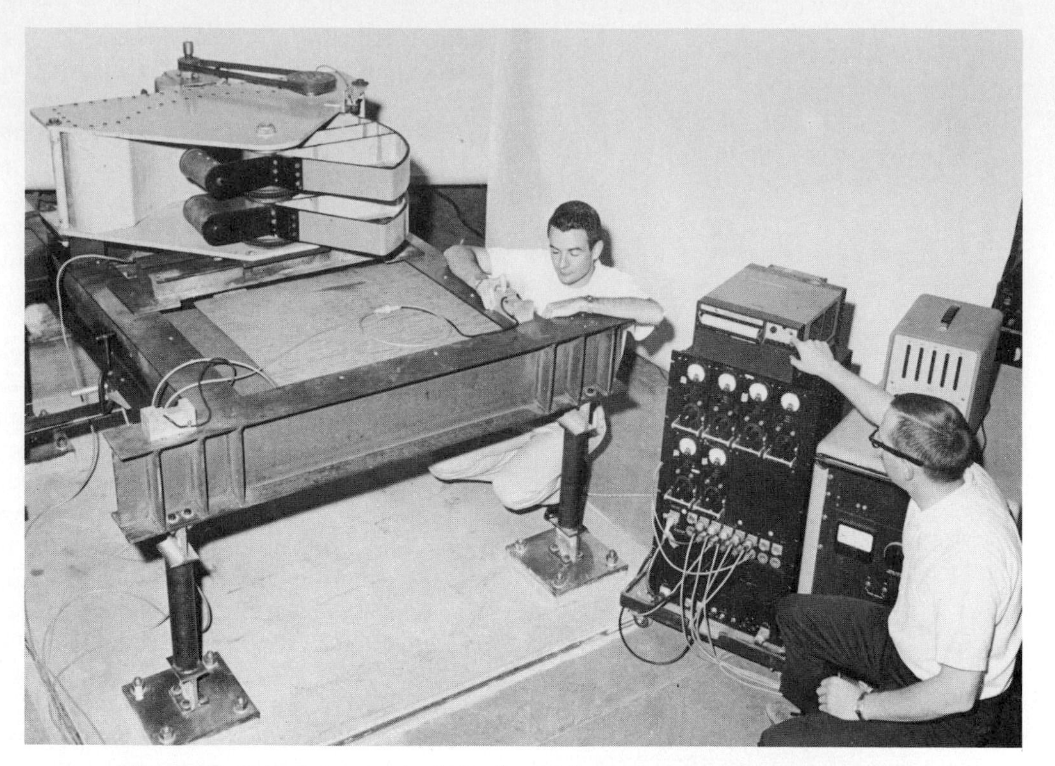

Fig. 7.13. Vibration generator tests of a yielding structure. (From Hanson, 1965b.)

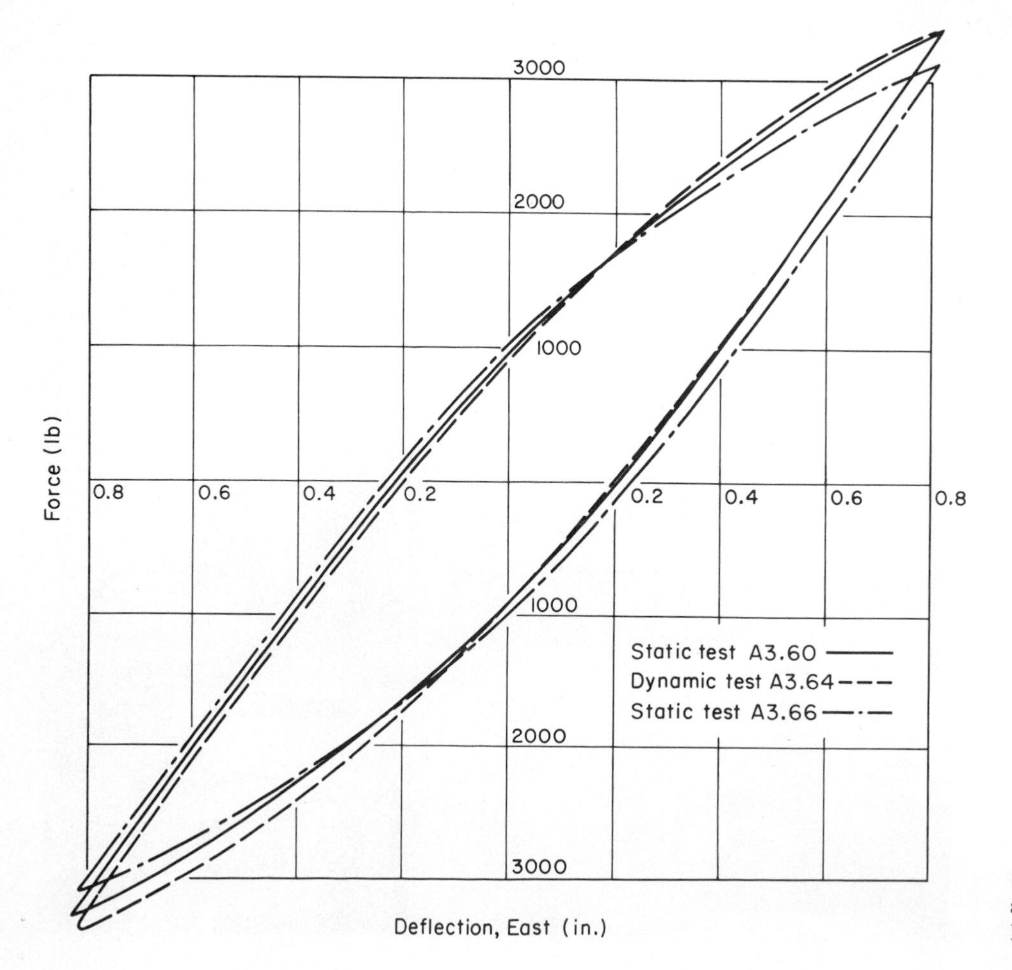

Fig. 7.14. Comparison of static and dynamic hysteresis curves. (From Hanson, 1966.)

Fig. 7.15. Test stand for a full-scale Saturn space vehicle. (From Watson, 1962.)

be used to drive rotating or reciprocating vibration generators of the inertia type, but so far no decisive advantages over the previously described methods have appeared.

7.2.3.2.9 *Man-excited vibrations.* All of the above methods of resonant vibration testing require relatively complex equipment and hence are limited in various ways to very special field conditions. There is thus a need for simpler tests that would give some information on building characteristics without the elaborate preparations that often make more complete dynamic tests impractical.

In the course of measuring wind-excited vibrations at the top of a concrete tower, it was observed that the motions of the instrument operator himself produced a measurable motion (Keightley, 1963). This pointed naturally to the idea that if the operator could move his body back and forth in synchronism with the natural period of the tower, it might be possible to build up vibration amplitudes suitable for resonance testing. It was found that this could in fact easily be done, particularly if it is possible to visually monitor the vibrations on a pen-writing oscillograph so that the course of the resonant buildup could be followed.

At first thought it may seem highly unlikely that a sufficient inertia force could be generated in this way so that large structures such as multistory buildings could be measurably excited. It turns out, however, that it is just for such large structures with their relatively low natural frequencies that the method is most useful. For example, if a 150-lb man moves his center of mass back and forth with a double amplitude of 6 in. at a frequency of 1 cps, an inertia force magnitude of some 46 lb is generated. Since at resonance with 1% of critical damping there is a dynamic amplification factor of 50, this means that an effective force of some 2300 lb is being applied to the building.

Figure 7.18 shows two vibration records obtained in this way in a five-story reinforced concrete building. During the first few cycles the motion builds up to a maximum, at which point the operator stops his motion, leaving the structure to perform damped free oscillations. In this way, both the natural period and the damping can be determined. Figure 7.19 shows a sketch of a nine-story steel-frame building in which similar measurements were made. Figures 7.20, 7.21, and 7.22 show man-excited vibration curves from which eight different modes of vibration could be studied (Hudson, Keightley, and Nielsen, 1964).

Fig. 7.16. Electromagnetic-type vibration generators in space vehicle test. (From Watson, 1962.)

Fig. 7.17. Model test of a Saturn space vehicle. (From Mixson and Catherine, 1962.)

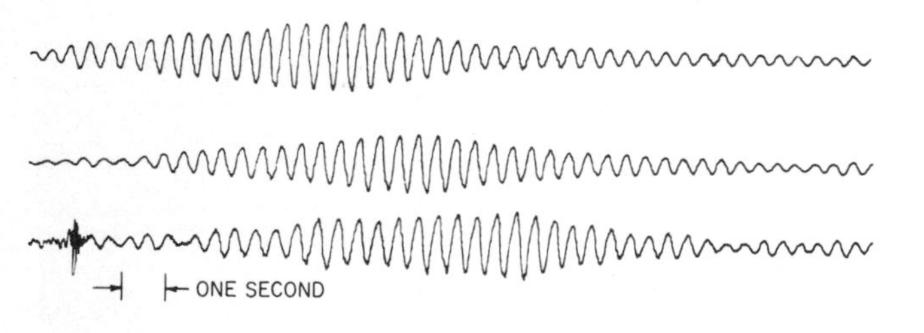

Fig. 7.18. Man-excited vibrations of a reinforced concrete building. (From Hudson, Keightley, and Nielsen, 1964.)

Fig. 7.19. Nine-story steel-frame structure showing points for man-excited excitation. (From Hudson, Keightley, and Nielsen, 1964.)

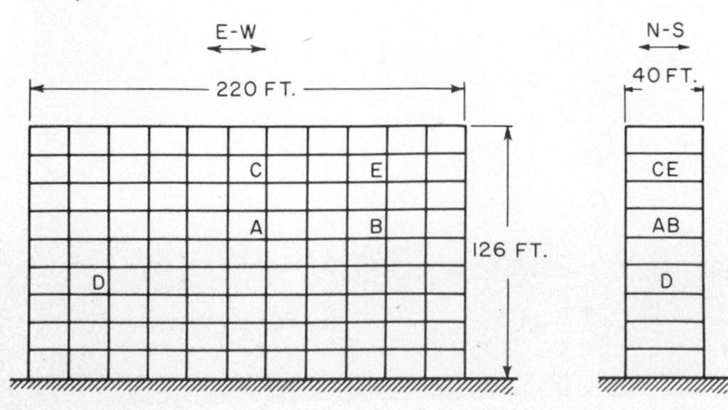

ONE SECOND

(a)

(b)

FUNDAMENTAL E-W TRANSLATION, 1.04 CPS

SECOND MODE E-W TRANSLATION, 3.0 CPS

Fig. 7.20. E–W lateral translation modes. Man-excited vibrations. (From Hudson, Keightley, and Nielsen, 1964.)

Such man-excited tests with their very low vibration amplitudes of course are no substitute for more complete dynamic investigations. The method may make it possible, however, to secure some useful information in situations for which no tests otherwise could be performed.

7.2.3.2.10 *Variable frequency sinusoidal excitations.* In steady-state resonance tests of the type described above, the object is to hold the exciting frequency fixed for a time sufficient so that all transient motions have died out and a uniform steady-state motion has been established. Because of the practical difficulties of holding fixed frequencies, which result in the need for the relatively complex control apparatus described above,

Fig. 7.21. N–S lateral translation modes. Man-excited vibrations. (From Hudson, Keightley, and Nielsen, 1964.)

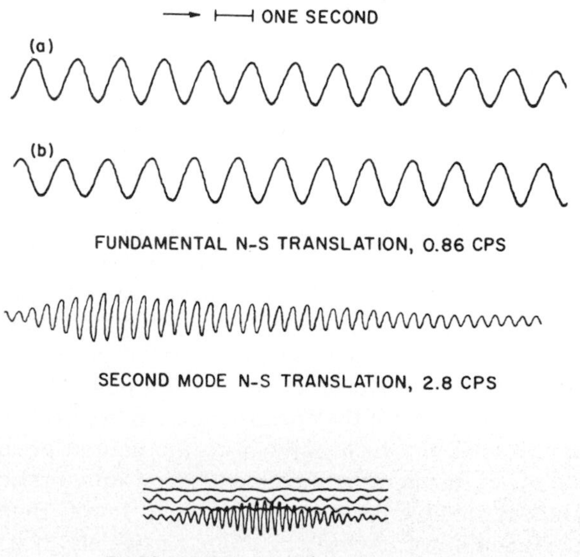

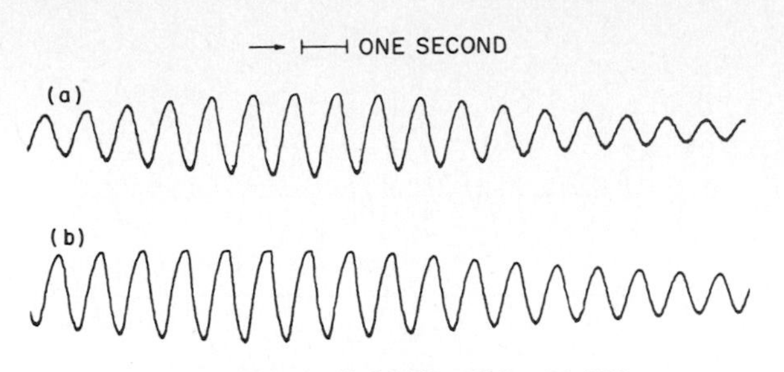

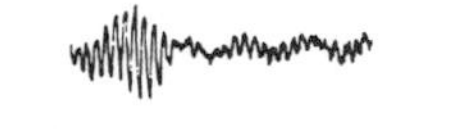

Fig. 7.22. Torsional modes. Man-excited vibrations. (From Hudson, Keightley, and Nielsen, 1964.)

there has been considerable study of the use of continuously varying frequency excitations.

For example, if a rotating eccentric weight is brought up to a speed higher than any natural periods of the system and the power is then cut off, the machine will coast down through the range of periods. If the friction in the machine is low, this coasting period will be long and an appreciable amplitude of vibration will be built up at each natural period of vibration. The advantages of such a test are first, the absence of a need for an elaborate speed control and second, the rapidity with which the whole test can be carried out. The disadvantage is the difficulty of analyzing the data to find energy dissipation in the system. Unless the machine coasts down very slowly, the peaks will not build up to as large a value as in a steady-state resonance curve, and the whole method of calculating damping from the data becomes much more difficult. Fortunately, this problem has been studied from a theoretical point of view in connection with the problem of vibration in such rotating machinery as steam turbines as they are started and stopped (Lewis, 1932; Barber and Ursell, 1948; Cronin, 1965).

Another difficulty with such "run-down" tests is that in lightly damped systems a type of beat phenomena can appear, which may lead one to suppose that resonant peaks are present that are in fact not there (Cronin, 1965). Figure 7.23 shows a continuous varying frequency test of a single-degree-of-freedom system. The biggest peak in the center of the figure is the true resonance peak. The later peaks are beats between the frequency of the free vibration of the system, which has been excited once the forcing frequency has passed the natural frequency, and the continuously varying exciting frequency.

Run-down tests are useful in preliminary investigations because of the speed with which they can be carried out. As a guide to the frequency regions in which more detailed steady-state studies should be made, such run-down tests can constitute an important first stage of vibration investigations. It thus is useful to design vibration generation equipment so that the power can be cut off and any high friction transmission links removed from the system so that the run-down times can be made as long as possible (Nielsen, 1964b).

7.2.3.2.11 *Transient excitations.* In contrast to the above steady-state forced vibration experiments are tests

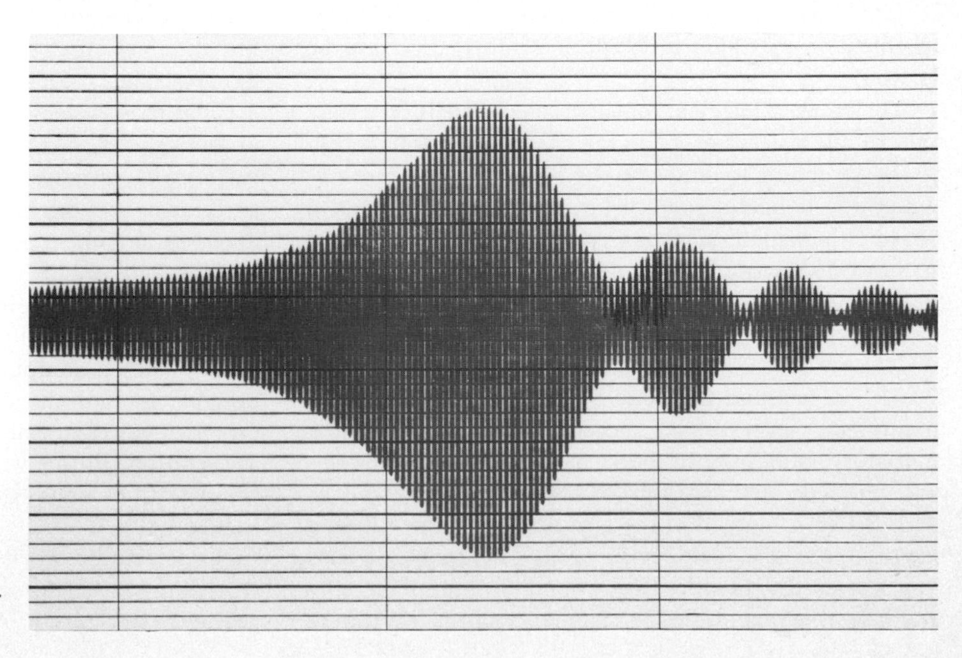

Fig. 7.23. Beat phenomenon in a run-down test. (From Cronin, 1965).

involving some kind of transient load such as an impact, blast, explosion, or an actual earthquake. Most important in this category is the natural earthquake itself, and the methods by which earthquakes can be used to make structural dynamic tests therefore will be considered in some detail.

7.2.3.2.12 *Earthquakes considered as dynamic tests.* A destructive earthquake is a dynamic test of full-scale structures on a vast scale, and it is of the utmost importance that preparations before the earthquake and investigations after the earthquake be carried out in such a way that the maximum amount of information is obtained.

No structural test can be interpreted in any very meaningful way unless the input forces are accurately known. In order to make use of natural earthquakes as structural tests it thus is necessary that the true ground acceleration vs time be accurately known at the site of the test structure. It is for this reason that so much importance attaches to the expansion of the networks of strong-motion accelerographs throughout the world (Hudson, 1963). Even for the most recent destructive earthquakes in the world there has been a lamentable lack of measured ground acceleration data. For example, not a single ground acceleration measurement was obtained in Mexico (1957), Chile (1960), Agadir (1960), Iran (1962), Skopje (1963), or Alaska (1964). Among recent major earthquakes, it is only for Niigata (1964) that important ground acceleration records were obtained. In this connection, the chapter "Ground Motion Measurements" in this volume should be consulted.

In the absence of ground acceleration data, the nature of the input excitation can be inferred only from structural behavior. As far as using the earthquake as a dynamic test is concerned, this just results in a circular argument from which little specific information can be derived.

The best arrangement for earthquake structural tests is to have one accelerograph in the basement or foundation of the structure to measure the input ground motions and additional accelerographs in upper level positions to measure the structural responses. In addition, as discussed in more detail later, it would be highly advantageous to have instrumentation to directly measure structural strains and relative displacements.

As a part of the strong-motion accelerograph network, the U.S. Coast and Geodetic Survey for some years has maintained several such accelerograph instrumented multistory buildings in both San Francisco and Los Angeles. The first earthquake for which simultaneous records were obtained of the ground and of the building response was the October 2, 1933, earthquake in Los Angeles, at the 14-story Hollywood Storage Company Building (Neumann, 1935). Similar records also were

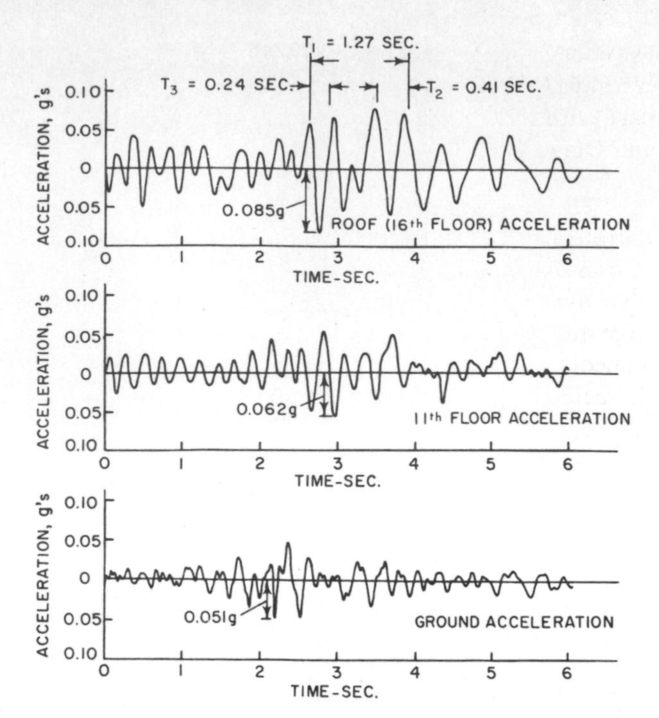

Fig. 7.24. Measured earthquake accelerations in Alexander Building, San Francisco earthquake of March 1957. (From Hudson, 1960.)

obtained in the same building during the 1952 Kern County earthquake (Murphy and Cloud, 1954; Housner, 1961). The most complete records, however, were obtained during the 1957 San Francisco earthquake in the 17-story Alexander Building, as shown in Fig. 7.24 (Brazee and Cloud, 1959; Hudson, 1960). A detailed analysis was made of this building, in which the dynamic properties of the building as previously determined by forced vibration tests were used to calculate the earthquake response (Hudson, 1960). The agreement between the calculated earthquake response and the actual measured earthquake behavior gave a very direct confirmation of the applicability of modern structural dynamic analysis to such problems.

The most recent results of this type were obtained at the city of Akita, Japan, during the 1964 Niigata earthquake. Analysis has shown again an excellent agreement between calculation and experiment (Osawa and Murakami, 1966). At Akita, good accelerograph records were obtained in the basement and at the ninth floor of a modern reinforced concrete frame structure. By using the basement accelerogram as the input for an electric analog computer model of the structure, it was possible to find the damping and periods that would give the best agreement with the measured upper-story acceleration. In this way a direct calculation of building characteristics under the actual earthquake motions was carried out.

Considerable work has been done in Japan on the recording of earthquakes on dams and on adjacent geologic structures. Although no large earthquakes as yet have been studied in this way, a number of smaller earthquakes have supplied data on dam response to compare with theoretical studies (Okamoto, Tamura, Kato, and Otawa, 1966).

7.2.3.2.13 *Explosive generated ground motions.* Large earthquakes are fortunately rare, so other sources of transient ground motion also must be sought as possible dynamic test excitations. It has been found that large quarry blasts generate ground accelerations in their immediate vicinity that are comparable in size and other characteristics to earthquake ground motions (Hudson, Alford, and Iwan, 1961; Hudson, 1961b).

In one instance, response measurements were made in a steel-frame mill building located some 1000 ft from a 370,000-1b explosive charge (Hudson, Alford and Housner, 1954; Hudson and Housner, 1957). The general results are indicated in Fig. 7.25. The maximum ground acceleration was about 0.12 g, and the wave shape is similar to that of the initial portion of a strong earthquake. Calculations of building response agreed well with accelerograph measurements made during the blast in an upper story location.

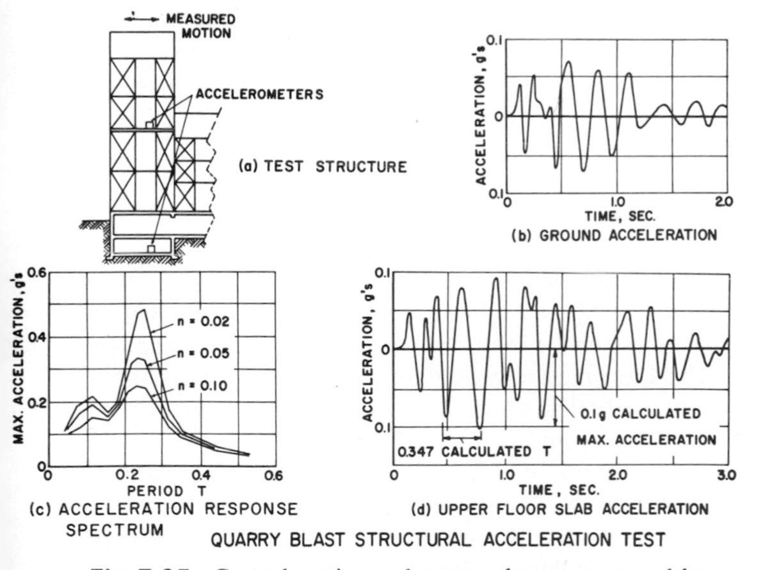

Fig. 7.25. Ground motion and structural response caused by a quarry blast. (From Hudson, Alford, and Housner, 1954.)

Another example of explosive generated ground shock is afforded by investigations of building damage caused by quarry blasting carried out by the National Research Council of Canada involving acceleration measurements of both ground and structural response (Edwards and Northwood, 1960; Northwood, Crawford, and Edwards, 1963). A recent study has summarized a number of investigations of damage caused by quarry blasting (Duvall and Fogelson, 1962). Explosion-excited motions also have been used as exciting forces for the testing of dams (Keightley, 1968).

It would be supposed that a number of opportunities for similar explosive-generated dynamic tests would appear in connection with the underground nuclear explosion program. The pattern of ground accelerations from such underground nuclear explosions has been well defined (Carder and Cloud, 1959; Sauer, Clark, and Anderson, 1964; Power, 1966). However, it appears that little work has been done on the structural response aspect of the problem in previous tests. It is hoped that some of the future tests can be used in this way to advance knowledge in the general structural dynamics field. Such investigations would require the construction of special test structures at various distances from the detonation point and would no doubt represent relatively costly programs compared with past studies in the field.

7.2.3.2.14 *Microtremor excited vibrations.* Although large earthquakes are infrequent, there are many small earthquakes and there is a continuous motion of the ground caused by microseismic activity and by various man-made disturbances such as machinery, vehicle traffic, etc. (Hudson, 1961b). Although these motions are usually of a very low magnitude, they can be used in certain circumstances to throw some light on structural properties.

The basic instrumentation and techniques for such microtremor studies have been developed at the Earthquake Research Institute of the University of Tokyo (Kanai, Tanaka, and Osada, 1954). Although the method has been used most extensively for studies of ground period, it also has been used to determine structural periods of buildings (Kanai, Suzuki, and Yoshizawa, 1953). Because of the very low levels of vibration involved, methods of data reduction based on autocorrelation techniques have been developed to separate information from background noise (Takahasi and Husimi, n.d.; Hatano and Takahashi, 1957; Takahashi, Tsutsumi, and Mashuko, 1959; Shima, Tanaka, and Den, 1960; Cherry and Brady, 1965).

7.2.3.2.15 *Wind-excited vibrations.* A useful source of excitation for natural period determinations of full-scale structures is the wind. Even on a relatively quiet day, there are usually enough gusts of wind to excite a tall building into measurable vibrations at its fundamental period. Figure 7.26 shows a typical record of a wind-excited vibration of a concrete intake tower of a dam and indicates how clearly the fundamental period appears (Keightley, Housner, and Hudson, 1961). In the same figure is shown a second mode of vibration excited by the wind. This is a rarer occurrence, and ordinarily it is not expected that anything but the lowest modes will be evident without a detailed analysis of the data.

The Seismological Field Survey of the U.S. Coast and Geodetic Survey for some time has been carrying out a program of period determination of multistory buildings using wind-excited vibrations (Carder, 1965). The suggestion often has been made that hidden earth-

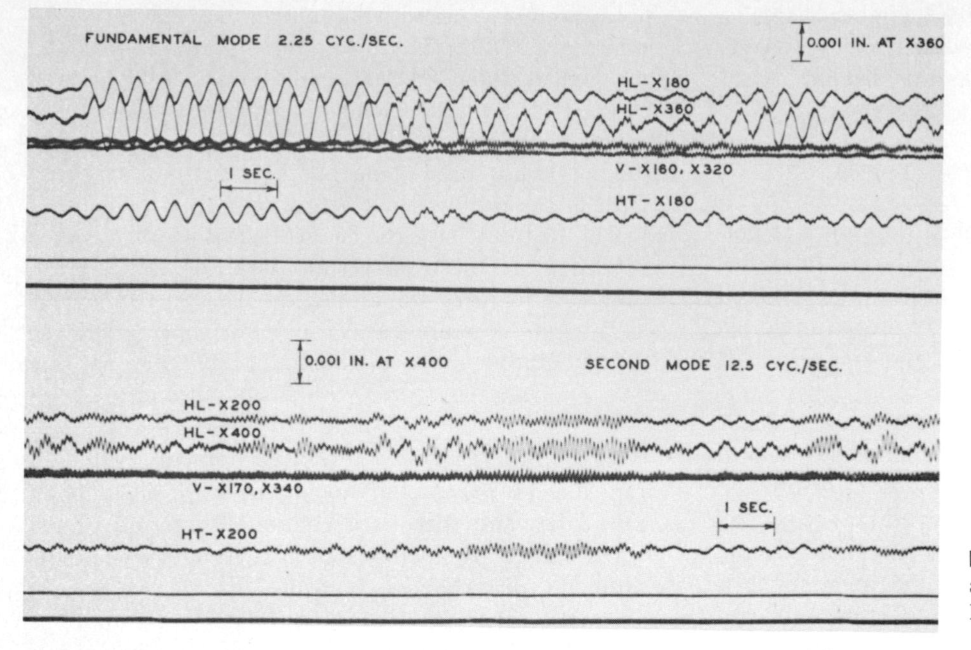

Fig. 7.26. Wind-excited vibrations of a concrete intake tower. (From Keightley, Housner, and Hudson, 1961.)

quake damage in structures might reveal itself through a measurable alteration in natural period. The U.S. Coast and Geodetic Survey has now accumulated many such pre-earthquake measurements of buildings in Los Angeles, San Francisco, Alaska, Mexico, Chile, Peru, etc., so that in the event of a destructive earthquake in those areas a study of such period changes will be possible. Perhaps the clearest indication so far of significant period changes during earthquakes was that noted during the Peruvian earthquake of October 17, 1966 (Kuroiwa, 1964; Esteva and Nieto, 1967). There are, of course, many other ways in which a knowledge of the fundamental period will throw light on the dynamic characteristics of a structure (Housner, 1962).

Because of the very low level of excitation and the complex nature of the exciting force, the same type of data analysis based on autocorrelation techniques that was mentioned above for microtremor vibrations also has been applied to wind-excited motions (Cherry and Brady, 1965; Crawford and Ward, 1964; Ward and Crawford, 1966).

Many structures for which wind-excited vibration tests are feasible also would seem to be suitable for the type of man-excited forced vibration tests discussed above. In fact, wind excitation may interfere with such man-excited tests or with free-vibration decay tests. In such cases the operators will need to wait for a sufficiently quiet air condition to obtain undisturbed data.

Another type of wind-excited structural vibration problem that has been described in the literature is the vibration of tall stacks and chimneys (Dockstader, Swiger, and Ireland, 1956; Dickey and Woodruff, 1956; Ozker and Smith, 1956). Instances have been observed of large vibratory motions set up by a vortex shedding mechanism.

Mention also should be made of the problems of wind-excited oscillations of suspension bridges and of the notable test of a full-scale structure that occurred in 1940 at the Tacoma Narrows Bridge (American Society of Civil Engineers, 1961; Ammann *et al.*, 1941; Farquharson *et al.*, 1954; Bleich *et al.*, 1950; Nishkian, 1947).

Another type of dynamic structural loading that has received a considerable amount of attention of late is that associated with sonic booms. A number of measurements of structural response under such excitation have been made (Mayes and Edge, 1964; Newberry, 1964).

7.2.3.3 Vibration table tests. A number of large vibrating tables have been constructed on which various structures can be mounted for dynamic tests. Although such tables are usually used for model tests, some of them are of a size that small full-scale structures or structural components could be handled.

An example of one of the larger and more elaborate installations is the seismic test table of the Japanese Building Research Institute, shown in Fig. 7.27 (the Japan Atomic Power Company, 1959; Building Research Institute, 1960). Although this table was specifically designed to test a one-third scale model of the core assembly of a graphite-pile nuclear power reactor, it is evidently of a size to produce large forces in many full-scale structural components. The whole table tilts upward through 45° by hydraulic jacks so that static component gravity forces can be applied to the test structure. Dynamic inertia forces are set up by displacing the table horizontally against 20 large coil springs and then suddenly releasing a tripping mechanism. In this way horizontal accelerations up to 2 g can be obtained. By changing the number and characteristics of the springs, the natural period of vibration of the table system can be varied between 0.14 and 0.96 sec and amplitudes up to 18 cm can be obtained. Test structure weights up to 17 tons can be accommodated.

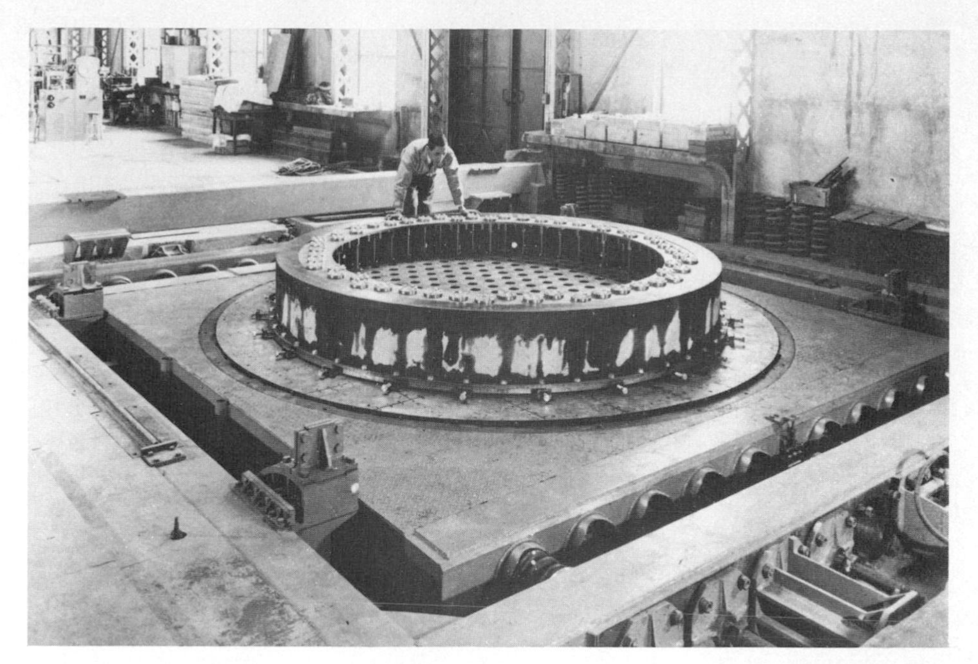

Fig. 7.27. Vibration test table of the Japanese Building Research Institute. (From The Japan Atomic Power Company, 1959.)

Another large shaking table has been constructed in Japan primarily for the study of soil dynamic properties (Kubo and Sato, 1967). This machine will shake a 170-ton weight sandbox of $2 \times 4 \times 10$ m size.

A large hydraulically operated vibration test table is at present being designed by the Japanese National Research Center for Disaster Prevention (National Research Center for Disaster Prevention, 1966). This is planned to be a square horizontal table about 50 ft on a side, weighing some 150 tons. With a 300-ton structural test load, horizontal accelerations of 0.8 g are to be attained. Vertical accelerations of 0.3 g are to be imposed on a total load of 270 tons, but horizontal and vertical motions are not to be simultaneous. A frequency range of 0.1 to 50 cycles/sec will be covered, with maximum displacement amplitudes of 15 cm. The basic drive will be a hydraulic actuator system, electronically controlled so that sine waves or programmed transients of various forms can be imposed.

The problems of designing such large vibration test tables with accurately programmed acceleration inputs are formidable ones. Such undertakings indicate that structural dynamic testing for earthquake engineering applications is entering a new era, in which outlays of manpower and money commensurate with the social and economic importance of the problem may be expected.

7.3 INSTRUMENTATION

7.3.1 Instrumentation for Dynamic Tests

The main attention in the above summary has been on the means for producing suitable dynamic loads for testing of full-scale structures. This is not the only problem. It is also required that accurate measurements of structural response be simultaneously made at various points and recorded vs time. Because of this need for simultaneous recording to establish mode shapes and other structural relationships, electrical transducers recording on multichannel oscillographs are necessary for most structural dynamic tests.

A large variety of modern instrumentation is available for structural dynamic testing. Unfortunately, no single transducer or instrumentation system is ideal for all applications, and hence the experimenter must expect to become familiar with a number of different devices and systems.

7.3.2 Accelerometers for Structural Response Measurements

Since relatively low frequencies are encountered in most building tests it is convenient to use systems with DC response so that there will be no question about low frequency distortion. This DC response carries an additional advantage for accelerometer systems because of the ease with which a 1-g steady acceleration signal can be imposed in the Earth's gravitational field for calibration purposes. Accelerometers have the following advantages for structural dynamic testing: (1) Because of their relatively high natural frequencies, accelerometers are inherently rugged, reliable devices, of a small size and weight, that are easily handled in the field. (2) Field installation is simplified because of low tilt sensitivity and the lack of a requirement for accurate orientation. (3) An accurate absolute static calibration can be quickly made by rotating the accelerometer in the Earth's gravitational field. Thus an overall system calibration can be made frequently under field conditions that will give a comprehensive check of all parts of an instrumentation system. (4) Maximum information is derived from acceleration measurements as compared with velocity or displacement measurements. Velocities and displacements can be obtained from acceleration readings by integration, but the inverse differentiations are difficult to carry out with the required accuracy.

The main disadvantage of the accelerometer is the fact that certain extraneous high frequency vibrations that are not structurally significant may have associated with them sufficiently high accelerations to complicate the record. If tests are to be made in a building that has machinery running in it, or if the vibration exciter itself generates large high frequency vibrations as in a gear drive, it may be difficult to get good accelerograms of the desired low frequency building vibrations. The unwanted high frequency vibrations could be removed from the record by filtering, but this would increase the complexity of the instrumentation system, since a suitable filter would be about as complicated as the amplifying equipment itself.

A second disadvantage of accelerometers is their relatively low output signal, which requires a considerable amount of electronic amplification. Such amplifying equipment is commercially available in many forms, but it is inherently complicated in nature and is subject to difficulties in rough field use.

For most structural vibration work, the instrumentation response should be essentially constant from DC to about 50 cps, and the measurable acceleration magnitudes should extend from 0.005 to 1 g. Since phase measurements are of importance, the overall phase shift characteristics of the whole system must be accurately known over the whole frequency range.

As an example of an accelerometer-type system, a description will be given of the instrumentation assembled for the structural dynamic tests described above using the synchronized rotating eccentric weight vibration generators (Keightley, 1963; 1966b). The system consisted of an unbonded wire strain gage-type accelerometer transducer, a carrier-type amplifier, and a moving coil-type multichannel oscillograph. For building structure work in which relatively high levels of acceleration could be excited, a transducer having a maximum range of 2 g and a natural frequency of about 100 cps was used. For tests of dams that involved low level accelerations, a transducer having a 0.25 g range and a natural frequency of about 15 cps was used. The carrier amplifier system involved a 3000-cps carrier frequency and a 10-volt maximum bridge voltage. The recording galvanometers of the moving coil type had natural frequencies of 20 cps and were electromagnetically damped to about 0.60 of critical damping. The direct writing oscillograph used ultraviolet light to produce records that would become visible in ordinary room light without processing. With this system an overall sensitivity of 0.1 in. peak-to-peak amplitude per 10^{-6} g over the frequency range 0 to 10 cps was attained. The noise level at a signal level of 0.1 in. peak to peak varied from 0.03 to 0.07 in. peak to peak. Thus it was possible to accurately measure accelerations as low as 2×10^{-6} g.

With the above arrangement, cable lengths as long as 900 ft were successfully used between the transducers and the amplifiers, provided that special symmetrical cables were used.

A point of special importance is the need for a direct writing recording system so that the records are available for inspection in the field during the course of the experiments. It often happens in structural dynamic tests that the results of one test will have an important bearing on the planning of the remaining field program. An immediate inspection of the test results also may reveal malfunctions or improper adjustments that then can be corrected.

There are many advantages from the data processing point of view in having the basic data recorded on magnetic tape. The data then can be entered directly into analog-digital converters and computers for processing. Such operations as filtering, integrating, compensating, and calibrating all can be programmed into the computer. If data are recorded on tape, it is still important, however, to have at least one monitoring direct write-out channel to give the above advantages of instant inspection.

7.3.3 Instruments for Period Measurement

In connection with experimental measurements of structural periods by wind, microtremors, or by the man-excited techniques described above, there is a need for a highly sensitive system that is at the same time compact and easily portable. The simplicity of such tests would be defeated if an instrumentation system as complicated as the accelerometer–recorder system described above would be required.

For the type of period and damping tests associated with these low level excitations, an exact calibration of the equipment is not essential, since relative amplitude values alone are sufficient. It has been customary to use transducer elements of about a 1 sec period since these will give considerably greater sensitivity than accelerograph elements, and for the essentially steady-state sinusoidal motions involved in such tests the accuracy will be adequate. The output of this velocity-type transducer then operates a recorder through an AC amplifier. As mentioned before, it is an important advantage if the recorder can be a direct write-out type like a pen and ink, heated stylus, or ultraviolet printout device. For man-excited tests it is in fact essential that the structural vibrations be visually monitored during the test so that the process of resonant buildup can be directly observed.

It also is important that the instrument contain its own source of power and that it can be easily carried by one man under the sometimes difficult conditions of construction jobs.

Unfortunately there is at present no commercially available package that meets all of the above requirements. All of the separate components exist in a satisfactory state of development, but the particular combination needed for the above tests has not yet been assembled in a standard form. One instrument that comes close to meeting the requirements is a version of the lunar seismograph developed for the Ranger moon shots (Hudson, Keightley, and Nielsen, 1964). This transducer, used along with a small drum recorder developed as a demonstration unit for the Ranger seismometer, has been successfully used for a number of structural tests.

The microtremor equipment developed in Japan also has been useful for such investigations, although it is not as portable as would be desired (Kanai, Tanaka, and Osada, 1954). This equipment combines a very high overall sensitivity with a convenient visible record made with a stylus on smoked paper.

Another instrument commonly used for wind-excited structural period measurements is a commercially available, portable, three-component seismograph developed primarily for explosion studies. This instrument has the advantage of a self-contained power supply but has the disadvantage of recording on photographic paper so that the success of the test is not known until the paper has been developed.

7.3.4 Instruments for Earthquake-Excited Tests

As discussed above, one of the most important structural dynamic tests is that imposed by a strong natural earthquake, and it is of the utmost importance that a number of characteristic structures in seismic zones be instrumented to yield the maximum information in such an event.

The value of the strong-motion accelerograph for such tests already has been mentioned, and since the properties of these accelerographs have been covered in detail in another chapter in this volume, no further mention will be made of them here. However, there are other types of instruments that also are of value in such studies, and since they have not been as frequently used as would be desirable, some details will be presented.

It would be of the utmost value to record directly strains in structural members during an actual earthquake. A building in which a relatively complete instrumentation system has been installed is the Engineering Building of the University of California at Los Angeles (Duke and Brisbane, 1955). This four-story reinforced concrete building was equipped during construction with 14 Carlson strain meters to measure concrete strain, 40 SR-4 bonded wire strain gages to measure strain in the reinforcing steel, and two Carlson joint meters for measuring the relative displacement between the building and an adjacent older unit. The various gages are arranged in bridge circuits and record on a multichannel oscillograph. The system is triggered by the inertia mechanism of a standard U.S. Coast and Geodetic Survey Strong-Motion Accelerograph that gives the ground acceleration to which the structure was subjected. Calibration of the system was accomplished with all gages in place by applying measured static loads to the second-floor girders. Since the installation of the equipment there have been several small earthquakes that have shown that the overall operation of the system is satisfactory and that have yielded strain measurements of considerable structural interest. There is no doubt but that it ultimately would be very much worthwhile to similarly instrument other buildings in the various seismic regions of the world.

Another building in which a complete instrumentation system consisting of strain gages, extensometers, earth pressure gages, and seismic recorders has been installed is the new building of the Earthquake Research Institute at the University of Tokyo (Osawa, Tanaka, Murakami, and Hosoda, 1966).

A notable installation for the direct measurement of structural response to earthquakes are the instruments for measuring relative story displacement in the Latino-Americana Tower in Mexico City (Zeevaert and Newmark, 1956; Cuevas Barajas, 1962; Zeevaert, 1957). The building was completed at the time of the destructive earthquake of July 28, 1957, and excellent records were obtained. The maximum relative diagonal displacements between the first and second floor during the earthquake were measured to be 0.637 cm, which corresponds to a computed shear force of 500 metric tons. It is of interest to note the comparison of this 500-metric ton figure with the dynamic design figure of 600 metric tons that had been based on a study of the seismicity of the region.

The displacement instrumentation in the Latino-Americana Tower represents a very important approach to the structural test problem that should be much more widely adopted. There is a need for the continued development of simple devices of this type to supplement the accelerograph installations that are now increasing in number.

7.4 CONCLUSION

The above review, which has been prepared with the problems of earthquake excitation mainly in mind, has omitted two important fields in which structural dynamic tests are of great importance—ships

(McGoldrick *et al.*, 1951; Voigt, 1958) and aircraft (Bisplinghoff *et al.*, 1955; Fung, 1955; Scanlan and Rosenbaum, 1951). There are, of course, many common problems of excitation and instrumentation that cut across all technical fields, and it is important for workers in any paritcular specialty to keep aware of developments in related fields.

It is evident that the field of dynamic testing of full-scale structures has made a very rapid progress in the past few years. This means, however, that in one sense the easy problems have now been solved, and what remains will involve an increasing level of effort and expenditure. Because large test programs of the kind described above are relatively costly ventures, it will be increasingly necessary to plan such programs with a clear economic justification in mind. There can be no doubt, however, that in our present state of knowledge or of ignorance, a greatly increased effort on well-conceived experimental investigations of the above type would be a very desirable development.

REFERENCES

American Society of Civil Engineers, Task Committee on Wind Forces (1961). "Wind Forces on Structures," *Trans. ASCE*, **126**, Pt. II.

Ammann, O. H., T. von Kármán, and G. B. Woodruff (March 1941). *The Failure of the Tacoma Narrows Bridge*, Washington, D.C.: Federal Works Agency.

Barber, N. F. and F. Ursell (1948). "The Response of a Resonant System to a Gliding Tone," *Phil. Mag.*, Ser. 7, **39**, 345–361.

Berg, G. V. (May 1958). *The Analysis of Structural Response to Earthquake Forces*, Ann Arbor: The University of Michigan, Industry Program of the College of Engineering, Report IP-291.

Berg, C. V. (1962). "Finding System Properties from Experimentally Observed Modes of Vibration," *Primeras Jornadas Argentinas de Ingenieria Antisismica*, San Juan-Mendoza, Argentina.

Bisplinghoff, R. L., H. Ashley, and R. L. Halfman (1955). *Aeroelasticity*, Cambridge: Addison-Wesley.

Bleich, F., C. B. McCullough, R. Rosecrans, and G. S. Vincent (1950). "The Mathematical Theory of Vibration in Suspension Bridges," Washington, D.C.: Bureau of Public Roads, Department of Commerce.

Blume, J. A. (October 1935). "A Machine for Setting Structures and Ground into Forced Vibration," *Bull. Seism. Soc. Am.*, **25**(4).

Bouwkamp, J. G., and J. K. Blohm (December 1966). "Dynamic Response of a Two-Story Steel Frame Structure," *Bull. Seism. Soc. Am.*, **56**(6).

Brazee, R. J., and W. K. Cloud (1959). *United States Earthquakes, 1957*, Washington, D.C.: U.S. Department of Commerce, Coast and Geodetic Survey.

Building Research Institute (July 1960). *Facilities for Earthquake Engineering Research in Building Research Institute*, Japan: Ministry of Construction.

Bullock, T. (April 1962). "A Comparison of Theoretical Bending and Torsional Vibrations with Test Results of the Full-Scale Saturn and the One-Fifth Scale Test Vehicle," *Shock and Vibration Bulletin No. 30*, Washington, D.C.: Department of Defense.

Carder, D. S. (ed.) (1965). *Earthquake Investigations in the Western United States, 1931–1964*, Pub. 41-2, Washington, D.C.: U.S. Department of Commerce, Coast and Geodetic Survey.

Carder, D. S. and W. K. Cloud (October 1959). "Surface Motion from Large Underground Explosions," *J. Geophys. Res.*, **64**(10).

Caughey, T. K. (June 1960a). "Classical Normal Modes in Damped Linear Dynamic Systems," *J. Appl. Mech.*, **27**(2), 269–271.

Caughey, T. K. (December 1960b). "Sinusoidal Excitation of a System with Bilinear Hysteresis," *J. Appl. Mech.*, **27**(4), 640–643.

Caughey, T. K., and M. E. J. O'Kelly (September 1955). "Classical Normal Modes in Damped Linear Dynamic Systems," *J. Appl. Mech.*, **32**(3).

Cherry, S., and A. G. Brady (1965). *Determination of Structural Dynamic Properties by Statistical Analysis of Random Vibrations, Proceedings of the Third World Conference on Earthquake Engineering*, New Zealand.

Cloud, W. K. (February 1963). "Period Measurements of Structures in Chile," *Bull. Seism. Soc. Am.*, **53**(2).

Crawford, R., and H. S. Ward (December 1964). "Determinations of Natural Periods of Buildings," *Bull. Seism. Soc. Am.*, **54**(6).

Cronin, D. L. (August 1965). *Response of Linear, Viscous Damped Systems to Excitations Having Time-Varying Frequency*, Pasadena: Dynamics Laboratory, California Institute of Technology.

Cuevas Barajas, L. (1962). "Comportimento de la estructura de la Torre Latinoamericana," *Ingenieria*, **32**(1).

Dickey, W. L., and G. B. Woodruff (1956). "The Vibration of Steel Stacks," *Trans. ASCE*, **121**, 1054–1087.

Dockstader, E. A., W. F. Swiger, and E. Ireland (1956). "Resonant Vibration of Steel Stacks," *Trans. ASCE*, **121**, 1088–1112.

Duke, C. M., and R. A. Brisbane (April 1955). "Earthquake Strain Measurements in a Reinforced Concrete Building," *Bull. Seism. Soc. Am.*, **45**(2).

Duvall, W. I., and D. E. Fogelson (1962). "Review of Criteria for Estimating Damage to Residences from Blasting Vibrations," RI-5968, Washington, D.C.: U.S. Department of the Interior, Bureau of Mines.

Edwards, A. T., and T. D. Northwood (September 1960).

"Experimental Studies of the Effects of Blasting on Structures," *The Engineer*, **210** (Research Paper No. 105, Ottawa: Division of Building Research, National Research Council).

Englekirk, R. E., and R. B. Matthiesen (June 1957). "Forced Vibration of an Eight-Story Reinforced Concrete Building," *Bull. Seism. Soc. Am.*, **57**(3).

Esteva, L., and J. A. Nieto (1967). "El Temblor de Lima, Perú, October 17, 1966," *Rev. Ingen.*, Mexico, **XXXVII**.

Farquharson, F. B., et al. (1949–1954). "Aerodynamic Stability of Suspension Bridges," *Bulletin No. 116*, Seattle: University of Washington Engineering Experiment Station, Pts. I–V.

Foss, K. A. (September 1958). "Coordinates which Uncouple the Equations of Motion of Damped Linear Systems," *J. Appl. Mech.*, **25**(3), 361–364.

Fung, Y. C. (1955). *An Introduction to the Theory of Aeroelasticity*, New York: John Wiley.

Hanson, R. D. (June 1965a). *Post-Elastic Dynamic Response of Mild Steel Structures*, Pasadena: Earthquake Engineering Research Laboratory, California Institute of Technology.

Hanson, R. D. (August 1965b). *Static and Dynamic Tests of a Full-Scale Steel-Frame Structure*, Pasadena: Earthquake Engineering Research Laboratory, California Institute of Technology.

Hanson, R. D. (October 1966). "Comparison of Static and Dynamic Hysteresis Curves," *Engr. Mech., ASCE.*

Hatano, T., and T. Takahashi (February 1957). "The Stability of an Arch Dam Against Earthquakes," Tech. Rept. C-5607, *Central Res. Inst. Elect. Power.*

Herwald, S. W., R. W. Gemmell, and B. J. Lazan (October 1946). "Mechanical Oscillators and Their Electrical Synchronization," *Trans. ASME.*

Hisada, T., and K. Nakagawa (1956). *Vibration Tests on Various Types of Building Structures up to Failure, Proceedings of the World Conference on Earthquake Engineering*, Berkeley: Earthquake Engineering Research Institute and University of California.

Housner, G. W. (1962). "The Significance of the Natural Periods of Vibration of Structures," *Primeras Jornadas Argentinas de Ingenieria Antisismica*, Argentina: San Juan-Mendoza.

Housner, G. W. (1961). "Vibration of Structures Induced by Seismic Waves, Part I: Earthquakes," in C. M. Harris and C. E. Crede (eds.), *Shock and Vibration Handbook*, New York: McGraw-Hill.

Hudson, D. E. (1960). *A Comparison of Theoretical and Experimental Determinations of Building Response to Earthquakes, Proceedings of the Second World Conference on Earthquake Engineering*, Tokyo and Kyoto.

Hudson, D. E. (1961a). "Scale-Model Principles," in C. M. Harris and C. E. Crede (eds.), *Shock and Vibration Handbook*, New York: McGraw-Hill.

Hudson, D. E. (1961b). "Vibrations of Structures Induced by Seismic Waves, Part II: Man-Made Ground Motions," in C. M. Harris and C. E. Crede (eds.), *Shock and Vibration Handbook*, New York: McGraw-Hill.

Hudson, D. E. (1962a). *A New Vibration Exciter for Dynamic Tests of Full-Scale Structures*, Earthquake Engineering Research Laboratory, California Institute of Technology, 1961; *Primeras Jornadas Argentinas de Ingenieria Antisismica*, San Juan-Mendoza, Argentina.

Hudson, D. E. (1962b). *Synchronized Vibration Generators for Dynamic Tests of Full-Scale Structures*, Pasadena: Earthquake Engineering Research Laboratory, California Institute of Technology.

Hudson, D. E. (February 1963). "The Measurement of the Ground Motion of Destructive Earthquakes," *Bull. Seism. Soc. Am.*, **53**(2).

Hudson, D. E. (June 1964). "Resonance Testing of Full-Scale Structures," *Proc. ASCE, J. Engr. Mech.*

Hudson, D. E. (1965). *Equivalent Viscous Friction for Hysteretic Systems with Earthquake-Like Excitations, Proceedings of the Third World Conference on Earthquake Engineering*, New Zealand.

Hudson, D. E., J. L. Alford, and G. W. Housner (July 1954). "Measured Response of a Structure to an Explosive-Generated Ground Shock," *Bull. Seism. Soc. Am.*, **44**(3).

Hudson, D. E., J. L. Alford, and W. D. Iwan (April 1961). "Ground Accelerations Caused by Large Quarry Blasts," *Bull. Seism. Soc. Am.*, **51**(2).

Hudson, D. E. and G. W. Housner "Structural Vibrations Produced by Ground Motion," *Trans. ASCE*, **122**, 705–721.

Hudson, D. E., W. O. Keightley, and N. N. Nielsen (February 1964). "A New Method for the Measurement of the Natural Periods of Buildings," *Bull. Seism. Soc. Am.*, **54**(1).

Iwan, W. D. (1961). *The Dynamic Response of Bilinear Hysteretic Systems*, Pasadena: Earthquake Engineering Research Laboratory, California Institute of Technology.

Iwan, W. D. (March 1965). "The Steady-State Response of a Two-Degree-of-Freedom Bilinear Hysteretic System," *J. Appl. Mech.*, **32**(1).

Iwan, W. D. (December 1966). "A Distributed-Element Model for Hysteresis and Its Steady-State Dynamic Response," *J. Appl. Mech.*, **33**(4).

Jacobsen, L. S. (1930). "Steady Forced Vibration as Influenced by Damping," *Trans. ASME*, **APM-52-15**, 169–181.

Jacobsen, L. S. (1960). *Damping in Composite Structures, Proceedings of the Second World Conference on Earthquake Engineering*, Tokyo and Kyoto.

The Japan Atomic Power Company (January 1959). *Seismic Tests of Graphite Pile of Calder Hall Type Power Reactor.*

Jennings, P. C. (April 1964). "Periodic Response of a General Yielding Structure," *Proc. ASCE, J. Engr. Mech.*

Jennings, P. C. (April 1967). "Force-Deflection Relations from Dynamic Tests," *J. Engr. Mech., ASCE.*

Jennings, P. C. (February 1968). "Equivalent Viscous Damping for Yielding Structures," *J. Engr. Mech.*, ASCE.

Kanai, K., T. Suzuki, and S. Yoshizawa (December 1953).

"Relation Between the Property of Building Vibration and the Nature of the Ground (Observation of Earthquake Motion at Actual Buildings) I," *Bull. Earthquake Res. Inst.*, **31**, Pt. 4.

Kanai, K., T. Tanaka, and K. Osada (July 1954). "Measurement of the Microtremor I," *Bull. Earthquake Res. Inst.*, **32**, Pt. 2.

Keightley, W. O. (1963). *Vibration Tests of Structures*, Pasadena: Earthquake Engineering Research Laboratory, California Institute of Technology.

Keightley, W. O. (September 1966a). *Vibrational Characteristics of Dams and Bridges, Proceedings of the Symposium for Design for Earthquake Loadings*, Montreal: McGill University.

Keightley, W. O. (December 1966b). "Vibrational Characteristics of an Earth Dam," *Bull. Seism. Soc. Am.*, **56**(6).

Keightley, W. O. (February 1968). "Response of Earth Dams to Seismic Disturbances. Part I: Motion Measurements on Fort Peck Dam," Lawrence Radiation Laboratory.

Keightley, W. O., G. W. Housner, and D. E. Hudson (1961). *Vibration Tests of the Encino Dam Intake Tower*, Pasadena: Earthquake Engineering Research Laboratory, California Institute of Technology.

Kennedy, C. C., and C. D. P. Pancu (November 1947). "Use of Vectors in Vibration Measurement and Analysis," *J. Aerospace Sci.*, **14**(11), 603–625.

Kubo, K., and N. Sato (December 1967). "Vibration Test of a Structure Supported by Pile Foundation," *Bull. Earthquake Resis. Struct. Res. Ctr.*, No. 1 (The Institute of Industrial Science, University of Tokyo).

Kuroiwa, J. (1964). "Periodo de Vibracion y Caracteristicas Estructurales de los Edeficios Ubicados en Lima y sus Aldrededores," *Bol. No. 9*, Lima; Instituto de Estructuras, Universidad Nacional de Ingenieria.

Kuroiwa, J. H. (June 1967). *Vibration Test of a Multistory Building*, Pasadena: Earthquake Engineering Research Laboratory, California Institute of Technology.

Lewis, F. M. (1932). "Vibration During Acceleration Through a Critical Speed," *Trans. ASME*, **APM-54-24**, 253–261.

Lewis, R. C., and D. L. Wrisley (November 1950). "A System for the Excitation of Pure Natural Modes of Complex Structure," *J. Aerospace Sci.*, **17**(11), 705–722.

Matthiesen, R. B., and C. B. Smith (October 1966). "A Simulation of Earthquake Effects on the UCLA Reactor Using Structural Vibrators," UCLA NEL-105 (Rep. No. 66-56), Los Angeles: Nuclear Energy Laboratory—Earthquake Laboratory, University of California.

Mayes, W. H., and P. M. Edge (November 1964). "Effects of Sonic Boom and Other Shock Waves on Buildings," *Mater. Res. Std.*, **4**(11). American Society for Testing and Materials.

Mazet, R. (1956). "Quelques Aspects des Essais de Vibration au Sol et en Vol," *Note Technique No. 34*, Office National D'Etudes et de Recherches Aéronautiques.

McGoldrick, R. T., A. N. Gleyzal, R. L. Hess, and G. K. Hess, Jr. (February 1951). "Recent Developments in the Theory of Ship Vibration," *Report 739*, The David W. Taylor Model Basin, Navy Department.

Mixson, J. S., and J. J. Catherine (April 1962). "Investigation of Vibration Characteristics of a One-Fifth Scale Model of Saturn SA-1," *Shock and Vibration Bulletin No. 30*, Washington, D. C.: Department of Defense.

Murphy, K. M., and W. K. Cloud (1954). *United States Earthquakes, 1952*, Washington, D.C.: U.S. Department of Commerce, Coast and Geodetic Survey, Serial No. 773.

National Research Center for Disaster Prevention (March 1966). "On the Plan of Establishing a Large-Sized Earthquake Engineering Experimental Apparatus," Tokyo: Science and Technology Agency.

Neumann, F. (1935). *United States Earthquakes, 1933*, Washington, D.C.: U.S. Department of Commerce, Coast and Geodetic Survey, Serial No. 579.

Newberry, C. W. (November 1964). "Measuring the Sonic Boom and Its Effect on Buildings," *Mater. Res. Std.*, **4**(11).

Nishkian, L. H. (April 1947). "Vertical Vibration Recorders for the Golden Gate Bridge," *Bull. Seism. Soc. Am.*, **37**(2).

Northwood, T. D., R. Crawford, and A. T. Edwards (May 1963). "Blasting Vibrations and Building Damage," *The Engineer*, **215** (Research Paper No. 186, Ottawa: Division of Building Research, National Research Council).

Nielsen, N. N. (1964a). *Dynamic Response of Multistory Buildings*, Pasadena: Earthquake Engineering Research Laboratory, California Institute of Technology.

Nielsen, N. N. (December 1964b). "Steady-State Versus Run-Down Tests of Structures," *Proc. ASCE, J. Struct. Div.*

Nielsen, N. N. (January 1966a). "Theory of Dynamic Tests of Structures," *Shock and Vibration Bulletin, No. 35*, Washington, D.C.: Department of Defense.

Nielsen, N. N. (February 1966b). "Vibration Tests of a Nine-Story Steel Frame Building," *Proc. ASCE, J. Engr. Mech.*

Okamoto, S., C. Tamura, K. Kato, and M. Otawa (October 1966). "Dynamic Behavior of Earth Dam During Earthquakes," *Rept. Inst. Indust. Sci.*, Tokyo, **16**(4).

O'Kelly, M. E. J. (1961). *Normal Modes in Damped Systems*, Pasadena: Dynamics Laboratory, California Institute of Technology.

Osawa, Y., and M. Murakami (March 1966). "Response Analysis of Tall Buildings to Strong Earthquake Motions. Part 2. Comparison with Strong Motion Accelerograms (1)," *Bull. Earthquake Res. Inst.*, **44**(1).

Osawa, Y., T. Tanaka, M. Murakami, and Y. Hosoda (June 1966). "Earthquake Strain Measurements in the ERI Main Building," *Bull. Earthquake Res. Inst.*, **44** (2).

Ozker, M. S., and J. O. Smith (1956). "Factors Influencing the Dynamic Behavior of Tall Stacks Under the Action of Wind," *Trans. ASME*, **78**(6).

Patterson, W. D. (April 1940). "Determination of Ground Periods," *Bull. Seism. Soc. Am.*, **30**(2).

Power, D. V. (December 1966). "A Survey of Complaints of Seismic-Related Damage to Surface Structures Following the Salmon Underground Nuclear Detonation," *Bull. Seism. Soc. Am.*, **56**(6).

Rea, D., J. G. Bouwkamp, and R. W. Clough (September 1966). *The Dynamic Behavior of Steel Frame and Truss Buildings*, Report No. 66-24, Berkeley: Structural Engineering Laboratory, College of Engineering, University of California.

Rouse, G. C., and J. G. Bouwkamp (1967). *Vibration Studies of Monticello Dam*, Res. Rept. No. 9, Denver: Bureau of Reclamation, U.S. Department of the Interior.

Sauer, F. M., G. B. Clark, and P. C. Anderson (May 1964). *Nuclear Geoplosics*, Washington, D.C.: Part Four, Defense Atomic Support Agency. (Stanford Research Institute.)

Scanlan, R. H., and R. Rosenbaum (1951). *Introduction to the Study of Aircraft Vibration and Flutter*, New York: Macmillan.

Scruton, C., and D. A. Harding (1957). "Measurement of the Structural Damping of a Reinforced Concrete Chimney Stack at Ferrybridge 'B' Power Station," Report NPL/Aero/323, National Physical Laboratory.

Shima, E., T. Tanaka, and N. Den (1960). *Some New Instruments Used in Earthquake Engineering in Japan, Proceedings of the Second World Conference on Earthquake Engineering*, Tokyo and Kyoto.

Späth, W. (1938). *Theorie und Praxis der Schwingungsprüfmaschinen*, Springer, 1934, English trans. issued as David Taylor Model Basin Translation No. 51.

Takahashi, T., H. Tsutsumi, and Y. Mashuko (December 1959). "Behaviors of Vibration of Arch Dam," Tech. Rept. C-5905, *Central Res. Inst. of Elect. Power.*

Takahasi, K., and K. Husimi (n.d.) "Vibrating Systems Exposed to Irregular Forces," *J. Inst. Phys. Chem. Res. (Japan)*, **14**(4).

Umemura, H., et al. (December 1967). *Vibration and Loading Tests of Old Tokyo Kaijo Building*, Report of Special Research Committee, Tokyo: Department of Architecture, Faculty of Engineering, University of Tokyo.

de Veubeke, B. F. (1956). "A Variational Approach to Pure Mode Excitation Based on Characteristic Phase Lag Theory," Report 39, NATO Advisory Group for Aeronautical Research and Development.

Voigt, H. (February 1958). "Recent Findings and Empirical Data Obtained in the Field of Ship Vibrations," *Translation 268*, The David Taylor Model Basin, Navy Department.

Ward, H. S., and R. Crawford (August 1966). "Wind-Induced Vibrations and Building Modes," *Bull. Seism. Soc. Am.*, **56**(4).

Watson, C. E. (April 1962). "Experimental Vibration Program on a Full-Scale Saturn Space Vehicle," *Shock and Vibration Bulletin No. 30*, Washington, D.C.: Department of Defense.

Zeevaert, L., and N. M. Newmark (1956). *Aseismic Design of Latino Americana Tower in Mexico City, Proceedings of the World Conference on Earthquake Engineering*, Berkeley.

Zeevaert, A. (1957). *Latino American Building, Proceedings of the Twenty-Sixth Annual Convention, Structural Engineering Association of California.*

Chapter 8

Dynamic Testing and the Formulation of Mathematical Models

J. G. BOUWKAMP

Professor of Civil Engineering
University of California
Berkeley, California

DIXON REA

Assistant Research Engineer
University of California
Berkeley, California

8.1 INTRODUCTION

Conventional earthquake-resistant design involves ensuring that a structure under design can carry a set of static lateral loads. The magnitude and distribution of lateral loads are specified by codes (Uniform Building Code, 1967) and, insofar as possible, simulate dynamic forces that the structure would experience in medium-sized earthquakes. The increasing availability of high-speed digital computers has initiated a trend whereby earthquake loads are represented by dynamic loads. Such methods involve dynamic analysis requiring, first, the idealization of the structure so that a mathematical model can be formulated and, second, the determination of the response of the mathematical model to suitable ground motion. For the purposes of this discussion it will be assumed that suitable ground motion is available (Housner and Jennings, 1964), and attention will be concentrated upon the idealization of structures and the formulation of mathematical models.

The computation involved in performing the dynamic analysis associated with even a simple mathematical model is extensive. Hence, it is important that the mathematical model should be as simple as possible without omitting any features of the prototype that affect its dynamic behavior appreciably. Therefore, a considerable number of dynamic tests of real structures (Bouwkamp and Blohm, 1966; Englekirk and Matthiesen, 1967; Keightley, 1961; 1964; Keightley, Housner, and Hudson, 1961; Nielsen, 1964; 1966; Rea, Bouwkamp, and Clough, 1966; Rouse and Bouwkamp, 1967) has been conducted in order to determine dynamic properties and establish mathematical models that can represent the dynamic behavior of the prototype structures.

A second purpose of these dynamic tests has been to accumulate a body of experimental results on the damping capacity of structures. Unlike the stiffness and mass properties of a structure, damping capacity cannot be calculated. Therefore, in formulating a mathematical model it is important that some experimental results on the damping capacity of structures similar to the one being modeled are available.

Since tests on real structures of necessity are conducted at low amplitudes, little information is found from such tests on the nonlinear behavior of structures. Thus dynamic tests have been conducted on small model structures vibrating at large amplitudes to study nonlinear behavior and energy absorption characteristics (Hanson, 1965). Hopefully, data from such tests then can be extended to an analytical evaluation of the nonlinear behavior of full-scale models.

The experimental procedures required to obtain the dynamic properties of structures using eccentric mass-type vibration generators (Hudson, 1964) are described below. The reduction of data and the accuracy of results are discussed. Results from dynamic tests (Rouse and Bouwkamp, 1967) on the Monticello Dam, California, and on a model steel frame are presented for illustrative purposes.

The formulation of mathematical models is then discussed and, as an example, a mathematical model of a building at the University of California Medical Center in San Francisco is formulated. Parameters in the model are varied in order to match the dynamic behavior of the model and prototype. Thus some of the problems of formulating a mathematical model in a particular case are revealed. Subsequently, the mathematical model is subjected to ground motion to predict the prototype behavior under strong-motion earthquakes.

8.2 DYNAMIC TESTING

The quantities normally determined by a dynamic test of a structure are resonant frequencies, mode shapes, and damping capacities. At present the equipment most commonly employed in dynamic tests are eccentric mass-type generators. These machines were developed by the Earthquake Engineering Laboratory at the California Institute of Technology, under the supervision of the Earthquake Research Institute, through a grant from the California State Division of Architecture. The vibration generators are described in Chapter 7, and the experimental procedures used in the determination of dynamic properties of structures are discussed below.

8.2.1 Resonant Frequencies

Resonant frequencies are determined by sweeping the frequency range of the vibration generators from 0.5 to 10 cps. The exciting frequency is increased slowly until acceleration traces on the recording chart are large enough for measurement. Above this level the frequency is increased in steps until the upper speed limit of the machine is reached. Near resonance, where the slope of the frequency-response curve is changing rapidly, the frequency-interval steps are as small as the speed control permits. These steps are relatively large in regions away from resonance. Each time the frequency is set to a particular value, the vibration response is given sufficient time to become steady state, before the acceleration traces are recorded. At the same time, the frequency of vibration as recorded on a digital counter is observed and written on the chart with its corresponding trace. Plotting the vibration response at each frequency step results in a frequency-response curve.

Frequency-response curves, in the form of acceleration amplitude vs exciting frequency, may be plotted directly from the data on the recording chart. However, the curves are for a force that increases with the square of the exciting frequency, and each acceleration amplitude should be divided by the corresponding square of its exciting frequency to obtain so-called normalized curves equivalent to those for a constant force (assuming linear stiffness and damping for the structural system). If the original acceleration amplitudes are divided by the frequency to the fourth power, displacement frequency-response curves for constant exciting force are obtained. In cases of fairly low damping (under 5%), there is little difference between results obtained for resonant frequencies and damping capacities measured from the different curves.

In general, resonant frequencies may be obtained with sufficient accuracy by reading the frequencies corresponding to peak amplitudes from resonant curves. Factors that limit the precision with which a resonant frequency may be specified are (1) nonlinear behavior of the structure, and (2) lack of suitable exciting forces.

Most structures behave even at low amplitudes as a spring-softening system with a weak nonlinearity. Resonant frequencies thus decrease slowly with amplitude so that resonant frequencies are a function of vibration amplitude unlike the case of classical linear systems. Lack of suitable excitation for multidegree-of-freedom systems causes different parts of the structure to attain their peak amplitudes at slightly different frequencies. The problem of inadequate excitation for multidegree-of-freedom systems is well known in resonance testing and has been discussed in detail (Gauzy, 1959; Bishop and Gladwell, 1963). Only a brief description will be given here.

If normal modes are not excited in a structure, the maximum amplitude of different points occurs at slightly different frequencies. The excitation of normal modes requires (1) the damping matrix of the structure to be such that it does not couple modes and (2) the excitation to be such that it balances the damping forces at all times at all points of the structure. Nothing can be done about the first requirement except to hope that the damping does not couple the modes or, if it does, that it will be too weak to do so effectively.

The second requirement implies that the same number of exciters should be used as the number of degrees of freedom of the structure. Gauzy (1959) points out that this requirement is not essential in practice if the locations of the exciter stations are chosen judiciously. He suggests that if it is desired to excite a particular mode, the exciters should be deployed to feed a maximum of energy to this mode and a minimum of energy to its neighboring modes, since only these will appreciably affect the response of that particular mode. The last observation can be of considerable benefit when trying to excite higher modes using only one exciter station.

8.2.2 Mode Shapes

Once the resonant frequencies of a structure have been found, the mode shapes at each of these frequencies may be determined. Generally, there are insufficient accelerometers, or insufficient recorder channels, to measure the vibration amplitude of all the required points simultaneously. Thus, it is necessary, after recording the amplitudes of a number of points, to stop the vibration, shift the accelerometers to new positions, and then vibrate the structure at resonance once more. This procedure is repeated until the vibration amplitude of all required points has been recorded.

The structure may not vibrate at exactly the same amplitude in each test run because it is impossible to vibrate the structure at precisely the same frequency each run. Therefore, it is necessary to maintain one reference accelerometer (preferably at a point of maximum displacement) through all the mode shape measurements for a particular mode. Subsequently, all vibration amplitudes can be adjusted to a constant modal amplitude.

In addition it is necessary to make corrections to the recorded amplitudes to compensate for differences between calibration factors. Absolute calibration is not required for mode shapes, and cross calibration is sufficient. The accelerometers and all equipment associated with them in their respective recording channels may be cross calibrated simply by placing them all together so that they measure the same vibration. Cross calibration should be carried out before and after measuring each mode shape. The average calibration factors as derived from the pre- and post-test cross calibration runs should be used to adjust the recorded amplitudes.

The number of points required to define a mode shape accurately depends on the mode and the number of degrees of freedom in the system. For example, in a dynamic test on a 15-story building (Rea, Bouwkamp, and Clough, 1966), four points were sufficient to define the first mode, whereas it required measurements of the vibration of all 14 floors and the roof to define the 5th mode shape accurately.

8.2.3 Damping Capacities

Damping capacities may be found from resonance curves in the normalized frequency-response curves by the formula:

$$\zeta = \frac{\Delta f}{2f} \tag{8.1}$$

where ζ = damping factor, f = resonant frequency, and Δf = difference in frequency of the two points on the resonance curve with amplitudes of $1/\sqrt{2}$ times the resonant amplitude. Strictly, the expression for ζ is only applicable to the displacement-resonance curve of a linear, single-degree-of-freedom system with a small amount of viscous damping. However, it has been used widely for systems differing appreciably from that for which the formula was derived and it has become accepted as a reasonable measure of damping. In this respect it should be remembered that in the case of full-size civil engineering structures it is not necessary to measure damping accurately in a percentage sense. It is sufficient if the range in which an equivalent viscous damping coefficient lies is known. Meaningful ranges might be defined as under 1%, 1–2%, 2–5%, 5–10%, and over 10% of critical damping.

The bandwidth method just described is extremely useful when damping lies in the range 1–10% of critical. However, if the damping lies below 1%, difficulties may be encountered in observing sufficient points on the resonance curve. Also, the small frequency difference between two relatively large frequencies becomes difficult to measure accurately. Above 10% of critical damping, resonance curves often become poorly defined due to interference between modes, and the results from the bandwidth method have little meaning.

If the frequency-response curves are plotted in the form of Kennedy and Pancu (1947) plots, in which the phase angle between excitation and response is plotted as argument and the amplitude of vibration as modulus in an Argand diagram, a more accurate measure of damping sometimes may be obtained. This method has advantages if two modes have resonant frequencies sufficiently close that each contributes substantially to the total response in their neighborhood. In civil engineering structures the resonant frequencies of different modes are usually well separated, and the method is not used as frequently as in aircraft engineering where it originated.

Time responses also have been used to determine the damping capacities of structures. A time response may be obtained by vibrating the structure at a resonant frequency and recording the resulting transient vibration when the machine is stopped. The method is particularly successful for the fundamental mode when the damping is small, under 2% of critical. At high values of damping the vibration ceases too quickly, often before the excitation stops having an effect on the response. In trying to obtain time responses for higher modes it is difficult not to excite, at the same time, lower modes that interfere with the desired time response.

8.2.4 Dynamic Tests of Monticello Dam

In order to illustrate the type of results obtainable from dynamic tests of multidegree-of-freedom systems,

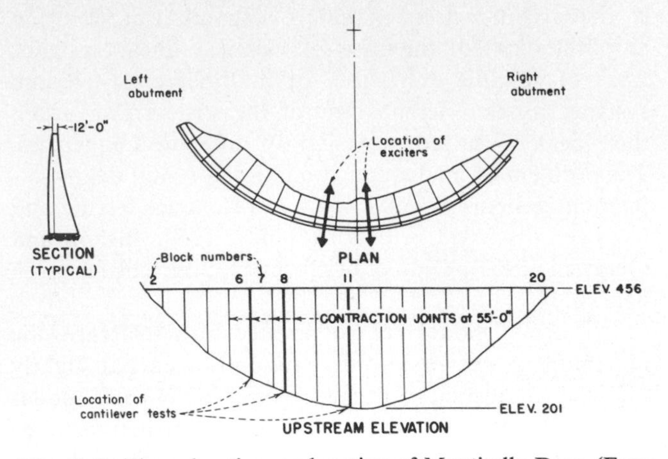

Fig. 8.2. Plan, elevation, and section of Monticello Dam. (From Rouse and Bouwkamp, 1967.)

some results from a test on the Monticello Dam, California, are described (Rouse and Bouwkamp, 1967).

The Monticello Dam is a 300-ft high mass concrete arch structure, shown in Fig. 8.1.* A plan, elevation, and section of the dam are shown in Fig. 8.2, which also indicates the location of the vibration generators on the crest of the dam.

Frequency responses for the dam are shown in Fig. 8.3. Symmetrical modes were excited by operating the machines in phase and asymmetrical modes, by operating the machines 180° out of phase. The forcing axis in all instances was radial.

The crest mode shapes of the dam for the first four

*Figures 8.1 through 8.5 are reproduced from Rouse and Bouwkamp (1967).

Fig. 8.1. Monticello Dam. (From Rouse and Bouwkamp, 1967.)

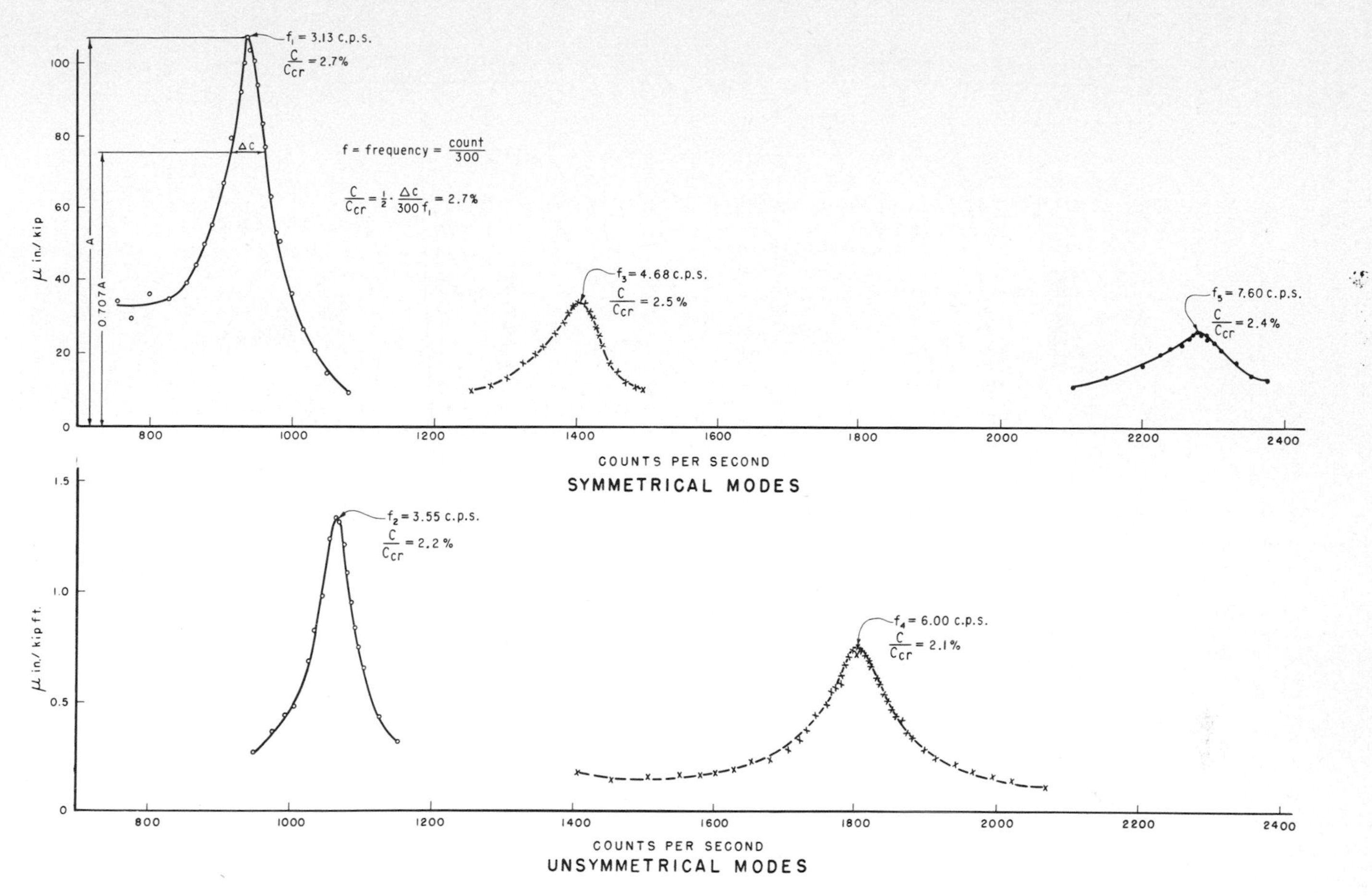

Fig. 8.3. Frequency response curves, Monticello Dam. (From Rouse and Bouwkamp, 1967.)

Fig. 8.4. Crest mode shapes, Monticello Dam. (From Rouse and Bouwkamp, 1967.)

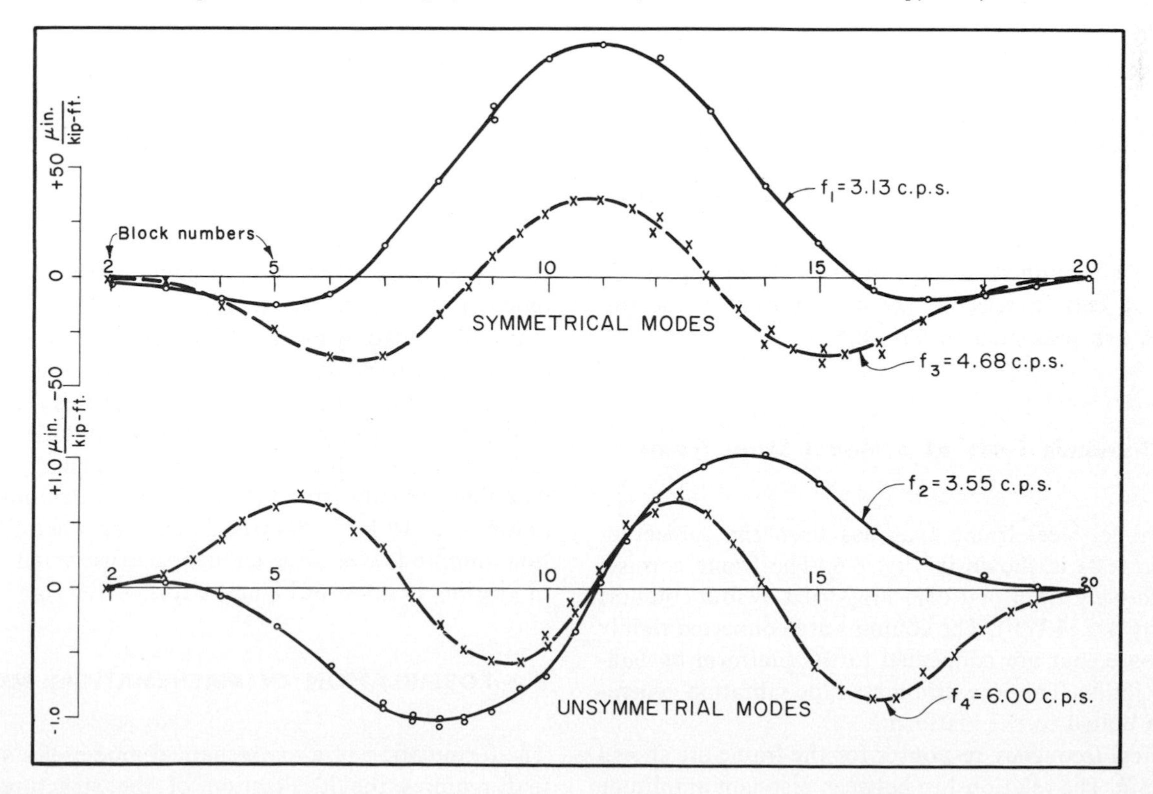

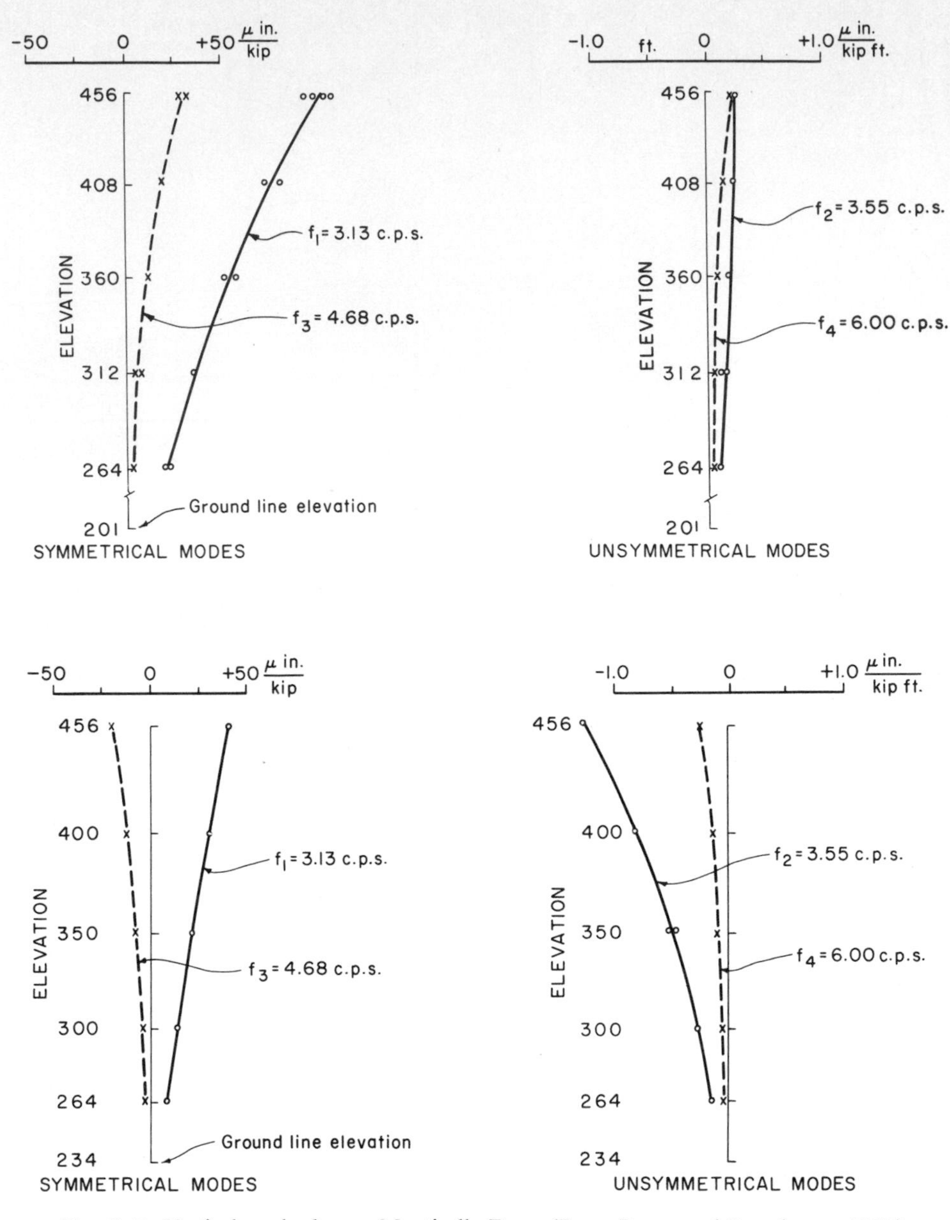

Fig. 8.5. Vertical mode shapes, Monticello Dam. (From Rouse and Bouwkamp, 1967.)

modes of vibration are shown in Fig. 8.4. Vertical mode shapes at certain selected sections and resonant frequencies are presented in Fig. 8.5.

8.2.5 Dynamic Tests of a Model Steel Frame

A model steel frame that has been the subject of dynamic tests is shown in Fig. 8.6. The frame consists of a relatively rigid platform supported by four columns (typically 4 × 4 WF). The columns are connected rigidly to the base, but are connected to the platform by ball-bearing joints. The eccentric mass type vibration generators are bolted to the platform.

Typical frequency responses for the frame are shown in Fig. 8.7. The relationship between resonant amplitude and exciting force amplitude is shown in Fig. 8.8, the change in slope of this curve being attributable to the onset of yielding in the columns. The effect of yielding is also illustrated by means of an energy-input vs resonant amplitude curve on a log–log scale in Fig. 8.9. The increase in slope of this curve also indicates the increase of damping capacity with vibration amplitude. Material yielding has been found to be capable of increasing the damping capacity from 0.1% of critical damping in the linear range to 1.2% of critical damping when the vibration amplitude was large enough to cause small amounts of yielding in areas of highest stress.

8.3 FORMULATION OF MATHEMATICAL MODELS

The formulation of a mathematical model of a structure first requires the idealization of the structure into a mechanical model, usually the lumped-mass type. Next

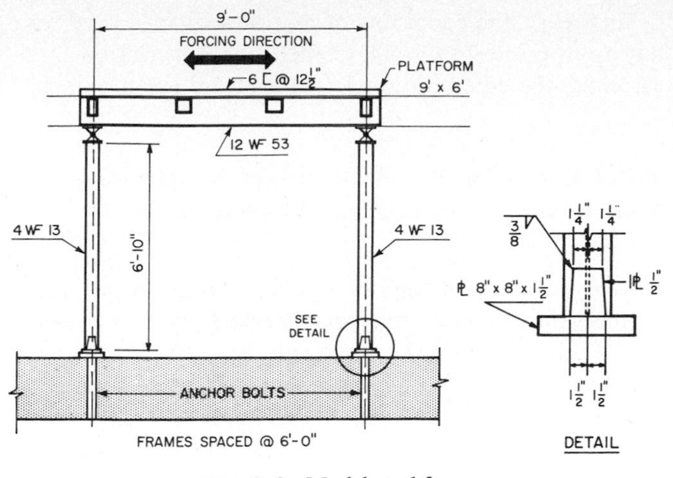

Fig. 8.6. Model steel frame.

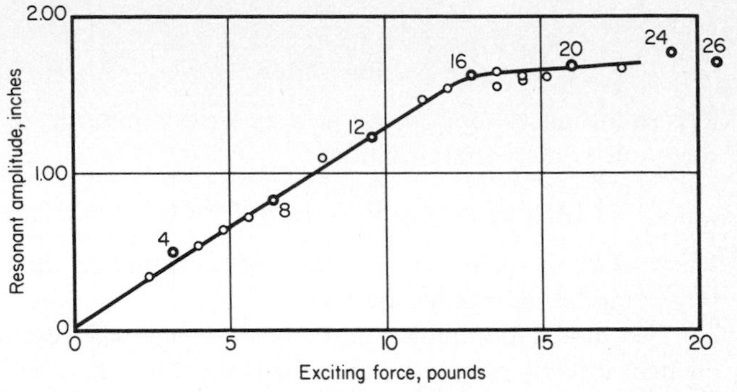

Fig. 8.8. Resonant amplitude vs excitation load, model steel frame.

the differential equations of motion governing the dynamic behavior of the model are written. These have the following form:

$$[m]\{\ddot{x}\} + [c]\{\dot{x}\} + [k]\{x\} = 0 \tag{8.2}$$

where $[m]$ = a mass matrix, $[c]$ = a damping matrix, $[k]$ = a stiffness matrix, and $\{x\}$ = a vector of displacements. The matrices $[m]$, $[c]$, and $[k]$ must be specified, the difficulty of which depends on whether the analysis to be performed is linear or nonlinear.

Fig. 8.7. Frequency-response curves, model steel frame.

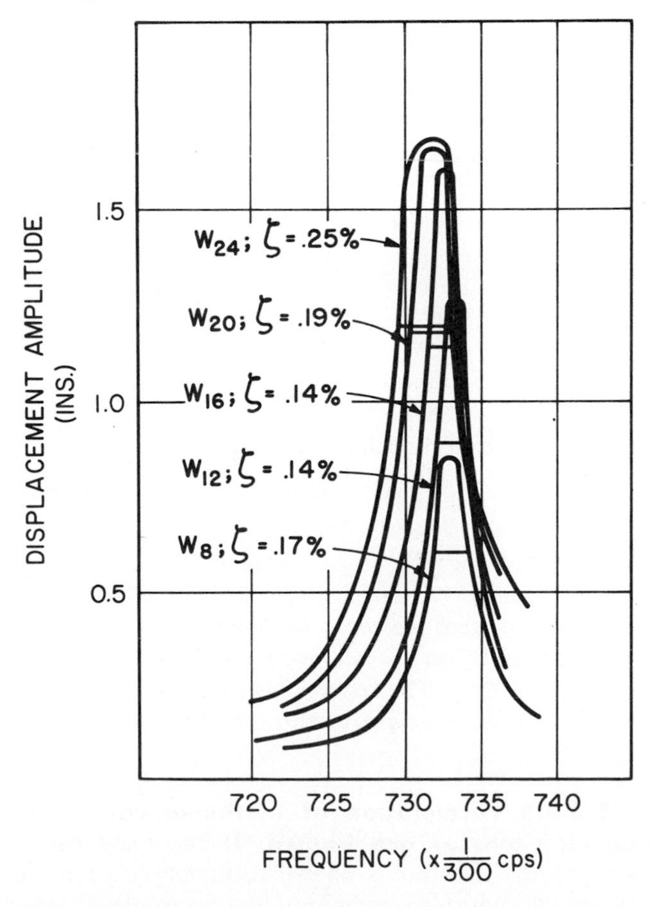

8.3.1 Linear Analysis

In linear analysis the matrices $[m]$ and $[k]$ may be derived from the detailed plans of the structure. The resulting eigenvalue problem

$$[[k] - \omega^2[m]]\{x\} = 0 \tag{8.3}$$

may be solved by standard techniques to yield the mode shapes (Φ_i) and associated natural frequencies (ω_i). The generalized mass $[M]$ and generalized stiffness $[K]$ may be formed from

$$[M] = [\Phi]^T[m][\Phi] \tag{8.4}$$

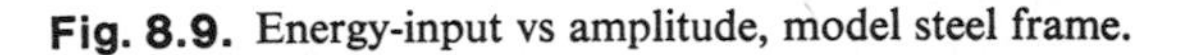

Fig. 8.9. Energy-input vs amplitude, model steel frame.

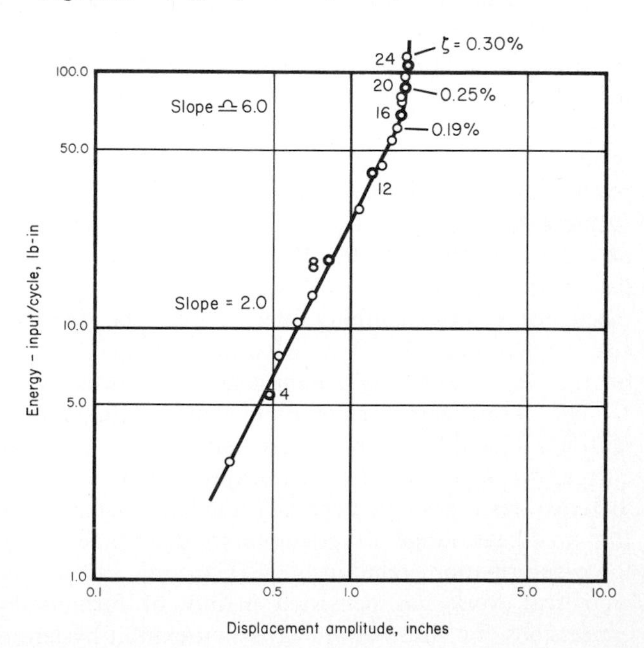

and

$$[K] = [\Phi]^T[k][\Phi] \tag{8.5}$$

The equations of motion can now be written in terms of uncoupled differential equations:

$$[M]\{\ddot{Y}\} + [C]\{\dot{Y}\} + [K]\{Y\} = 0 \tag{8.6}$$

where $\{Y\}$ = vector of generalized coordinates and $[C]$ = modal damping matrix.

The modal damping matrix is diagonal with each element directly proportional to the percentage of critical damping in respective modes. Thus the values of $[C]$ may be specified at least intuitively or more accurately specified from the results of dynamic tests conducted on similar structures in the past.

The uncoupled equations of motion then may be solved individually for any generalized forcing function, and the total response may be found by mode superposition.

8.3.2 Nonlinear Analysis

Nonlinear analysis is inherently more difficult than linear analysis because the original set of coupled differential equations must be solved directly by a step-by-step integration procedure, since mode superposition is no longer valid.

Again the mass matrix $[m]$ may be calculated from the design specifications, but neither the stiffness nor damping matrix may be defined as simply as in the linear case.

Ideally, the stiffness and damping matrices should be combined to form a nonlinear force-deformation relationship. There have been a number of suggestions for such force-deformation relationships (e.g., Jennings, 1964), none of which is entirely satisfactory.

If the damping capacity is not incorporated in a force-deformation relationship, then the damping matrix $[c]$ must be specified. It is difficult to specify because all measurements of damping yield the modal damping matrix $[C]$ and there is no satisfactory method of transforming this matrix to a $[c]$ matrix for use in the nonlinear analysis (Caughey, 1960). Even in the case of large deformations, when energy is being dissipated by yielding mechanisms, this damping matrix must be specified since it has been found to have an appreciable effect on the response (Clough, Benuska, and Wilson, 1965).

The unsatisfactory methods of defining the stiffness matrices $[c]$ and $[k]$ for nonlinear analyses have led to the current work in this area. The experimental work has included testing single-degree-of-freedom systems into the nonlinear range in attempts to determine actual force-deformation relationships (Hanson, 1966). The analytical work has consisted mainly of formulating expressions for stress–strain laws that exhibit hysteresis loops of appropriate shape.

In theory, the equations of motion may be solved for any appropriate form of force-deformation relationship, assuming the computational problems are not too great.

8.3.3 Checking the Accuracy of Mathematical Models with Results from Dynamic Tests

The accuracy of the $[m]$ and $[k]$ matrices of a linear mathematical model may be checked by a dynamic test of the structure. The check may be (1) a direct determination of resonant frequencies and generalized masses, (2) an independent mathematical model assuming the mass matrix is known, and (3) the adjustment of the idealization parameters in the model to match the experimental mode shapes and resonant frequencies.

8.3.3.1 Direct determination of dynamic properties. The resonant frequencies of a structure may be determined from the frequency-response curves as described in Section 8.1.1. The generalized mass of a particular mode can be determined experimentally by adding a known quantity of mass to the structure and measuring the consequent reduction in resonant frequency. The mass must be sufficient to lower the resonant frequency an amount that can be measured accurately. Also, it should not alter the mode shape of the structure, so that the added masses at each point should be proportional to the original mass at that point. If the masses are not added in the correct form to maintain the mode shape, a method to correct for this error is described by Simpson (1966). Since the stiffness of the structure remains unaltered with the addition of the extra mass,

$$M_i\omega_i^2 = K_i = (M_i + \Delta M_i)(\omega_i - \Delta\omega_i)^2 \tag{8.7}$$

and, hence,

$$M_i = -\frac{1}{2}\Delta M_i \cdot \frac{\omega_i}{\Delta\omega_i} \tag{8.8}$$

The generalized stiffness could be determined by altering the stiffness instead of the mass, but this is generally more difficult experimentally and hardly necessary since

$$K_i = M_i\omega_i^2 \tag{8.9}$$

The method is not feasible for most full-scale structures because of the large amount of added mass (although small relative to the total weight of the structure) required. However, the method has been used quite successfully on a small model frame having a single degree of freedom. The reduction in resonant frequency caused by adding weight to this structure is shown in Fig. 8.10.

8.3.3.2 Formulation of mathematical model assuming masses are known. If the mass distribution of the structure is known accurately (e.g., in the case of a multistory building), the generalized mass may be calculated from the measured mode shapes,

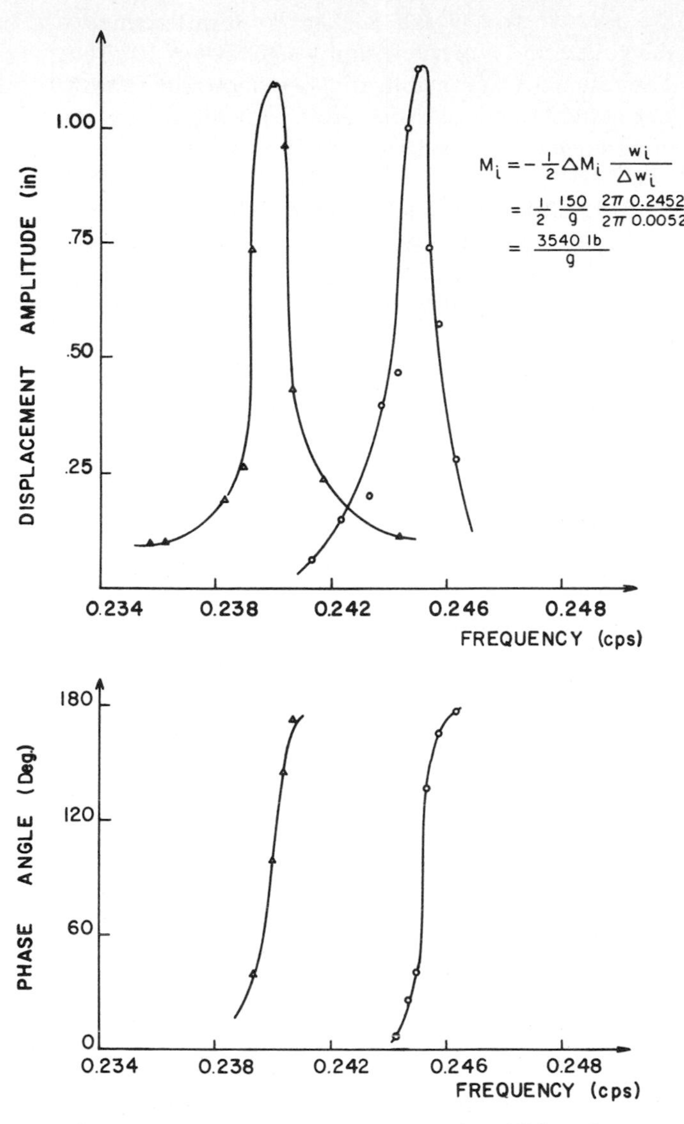

Fig. 8.10. Change of frequency caused by addition of mass.

$$M_i = \Sigma m_i \Phi_i^2 \tag{8.10}$$

and the generalized stiffness may be obtained from

$$K_i = M_i \omega_i^2 \tag{8.11}$$

A general treatment of this case is given by Nielsen (1966).

8.3.3.3 Adjusting parameters of model to match experimental mode shapes and frequencies. The ultimate objective of dynamic testing is to enable adequate mathematical models of a structure to be formulated confidently in the design stages. Therefore, it is considered preferable to adjust the parameters in the idealization of the structure so that the dynamic properties of the model match those of the prototype. This may be done by adjusting various parameters whose values are not known exactly until the resonant frequencies and mode shapes of the model match the experimental results. The end result, of course, is equivalent to matching the stiffness and mass matrices of the model and prototype. The virtue of the method is that it keeps the ultimate objective of dynamic testing in sight, and the relative importance of the various parameters becomes clear. An example of matching the dynamic properties of a mathematical model to the experimental results of the prototype is given below.

8.4 MATHEMATICAL MODEL OF A BUILDING AT THE UNIVERSITY OF CALIFORNIA MEDICAL CENTER, SAN FRANCISCO

A mathematical model of the East Building of a new building complex (Fig. 8.11) at the University of California Medical Center was formulated using results from a dynamic test (Rea, Bouwkamp, and Clough, 1966). The East Building is 195 ft (15 stories) high with outside plan dimensions of 107 × 107 ft. The steel frame is of the moment resistant type with columns placed only on the perimeter of the building, as shown in Fig. 8.11. The formulation consisted of adjusting stiffness, mass, and damping characteristics until the model's response to harmonic excitation matched that of the real structure. Since resonant frequencies depend primarily on mass and stiffness properties, and resonant amplitudes primarily on damping properties, it was possible to develop the model in two stages. First, the estimated mass and stiffness were adjusted in order to match the resonant frequencies of the model and prototype, and then the damping capacities of the first five modes were varied until the model's resonant amplitudes equaled those of the real structure for equivalent excitation.

All computations in the formulation of the analytical model were performed by an IBM 7090/7094 Direct Coupled System using programs developed in the Division of Structural Engineering and Structural Mechanics, Department of Civil Engineering, at Berkeley. The programs were used for the following purposes:

Fig. 8.11. General plan view of U.C. Medical Center. (From Rea, Bouwkamp, and Clough, 1966.)

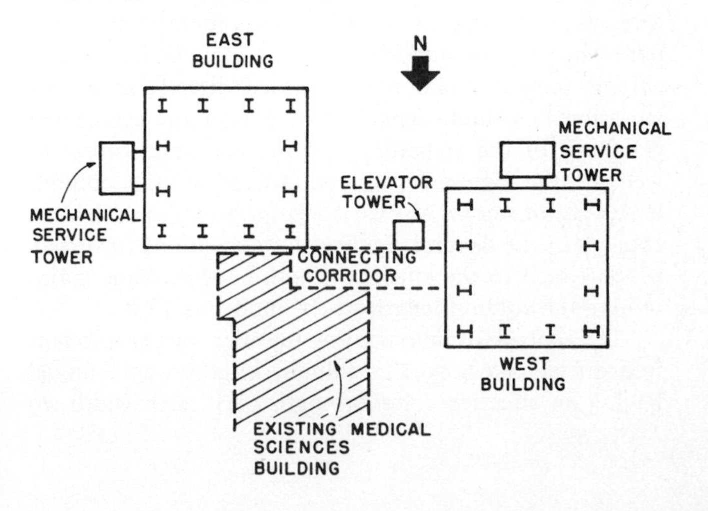

1. To compute the translational vibration mode shapes, natural frequencies, and steady-state response of the model in the various stages of its formulation. Also, to compute the linear response of the model to ground acceleration records.
2. To compute the nonlinear translational response of the model after yielding had occurred in some of the members.

These programs are restricted to structures composed of a number of parallel rectangular frames with horizontal girders and vertical columns. They assume (1) the geometry of the structure is defined by center-line to center-line dimensions and (2) the mass of the structure and lateral forces are concentrated at the floor levels.

8.4.1 Matching Resonant Frequencies

Resonant frequencies of a structure are determined primarily by its stiffness and mass characteristics, and most of the data required to assemble the stiffness and mass matrices of the building were known to a high degree of accuracy, e.g., the geometry of the structure, flexural rigidity, and mass of structural members. However, there were some quantities that were not known to sufficient accuracy, and errors in their assigned values might have a significant effect on the resonant frequencies. These were (1) effective flexural length of columns, (2) width of floor slab acting compositely with the girders, and (3) unit weight of the lightweight concrete in floor slabs. The first two quantities affect the stiffness matrix, and the third affects the mass matrix. The changes in resonant frequencies resulting from variations in these quantities were computed, and these changes are discussed below.

8.4.1.1 Effective length of columns. The 42-in. deep girders spanning 97 ft between exterior columns increased the column stiffness over the girder depth, reducing the effective length of the columns to some length less than the story length of 13 ft. It therefore was necessary to consider the column length as a parameter in the model's formulation. However, the column lengths could not be varied directly since any variation in column lengths would alter the center-line geometry of the structure and the dynamic forces, as well as the column stiffness. Therefore, the column length parameter (h) was varied by introducing equivalent changes in the flexural rigidity of the columns (inversely proportional to the effective length cubed), while maintaining the column length of the model as 13 ft.

The control for investigating the effect on the resonant frequencies of changes in column lengths was a model having an effective column length 11 ft, slab width (b) 8 ft ($16t$ where t = slab thickness = 6 in.) acting with the girder, and a concrete unit weight (w) of 107 lb/ft^3. The results of this investigation are shown in Table 8.1. The natural frequencies decreased by 5–8% for a change in effective column length from 11 to 13 ft.

8.4.1.2 Effective slab width. The flexural rigidity of the steel girders was increased by composite action with the floor slab, and the effect of this composite action on the resonant frequencies was investgated by assuming various widths of floor slab to act with the girders. The control model was the same as that used to investigate the effective column lengths. Table 8.2 shows that a reduction in effective slab width from 8 to 4 ft resulted in a decrease of natural frequencies of 4 to 5%; an increase in effective slab width from 8 to 16 ft increased the natural frequencies from 3 to 4%.

The elastic modulus of the lightweight concrete in the floor slabs was not known accurately but was assumed to be 2.9×10^6 lb/in.^{2} (modular ratio of 10). Any variations in the modulus, however, are equivalent to changes in effective slab width and thus may be considered incorporated in the slab width parameter.

8.4.1.3 Unit weight of concrete. The unit weight of the lightweight concrete in the floor slabs was not known exactly, and it also was treated as a parameter. Using the same control model as previously, a decrease in unit weight from 107 to 100 lb/ft^3 increased the natural frequencies by about 6%, and an increase in unit weight from 107 to 130 lb/ft^3 decreased the natural frequencies by about 6% (see Table 8.3).

8.4.1.4 Final choice of model. Clearly there is no unique combination of values for effective length of columns, effective slab width, and concrete unit weight for which the resonant frequencies of the model will match those of the real structure. The unique combination only could be found by matching all the resonant frequencies and mode shapes of the model and prototype. In this case, the experimentally determined mode shapes were not considered sufficiently accurate for matching with the model mode shapes, and the stiffness and mass properties of the model were finalized by matching the first five resonant frequencies of the model and prototype. In matching the model's resonant frequencies to the experimental resonant frequencies, particular attention was paid to the second and third resonant frequencies, since these were considered the most accurate. The final choice of values for the parameters is a matter for engineering judgment, and the combination chosen was: effective column length for all stories = 11 ft, effective slab width = 8 ft, and density of lightweight concrete = 120 lb/ft^3. Table 8.4 compares the resonant frequencies of the model and prototype.

Table 8.1. EFFECT ON RESONANT FREQUENCIES OF CHANGES IN COLUMN LENGTHS

Description	Natural frequency, cps, and (% difference)					Remarks
	f_1	f_2	f_3	f_4	f_5	
$h = 11$ ft $b = 8$ ft $w = 107$ lb/ft^3	0.811	2.34	4.07	5.88	7.90	Control model
$h = 9$ ft at first floor only	0.830 (2.2)	2.38 (1.7)	4.15 (2.0)	6.00 (2.0)	8.11 (1.9)	Slab unchanged Base fixity influence
$h = 13$ ft	0.765 (−5.7)	2.20 (−5.1)	3.78 (−7.1)	5.41 (−8.0)	7.25 (−8.2)	$h = 13$ ft = story height

Table 8.2. INFLUENCE OF COMPOSITE ACTION BETWEEN FLOOR SLABS AND GIRDERS

Description	Natural frequency, cps, and (% difference)					Remarks
	f_1	f_2	f_3	f_4	f_5	
$h = 11$ ft $b = 8$ ft $w = 107$ lb/ft^3	0.811	2.34	4.07	5.88	7.90	Control model
$b = 4$ ft	0.776 (−4.3)	2.24 (−4.5)	3.87 (−4.8)	5.64 (−4.1)	7.62 (−3.5)	$h = 11$ ft $w = 107$ lb/ft^3
$b = 16$ ft	0.845 (4.1)	2.43 (3.9)	4.21 (3.4)	6.05 (2.9)	8.11 (2.8)	$h = 11$ ft $w = 107$ lb/ft^3

Table 8.3. INFLUENCE OF CONCRETE UNIT WEIGHT

Description	Natural frequency, cps, and (% difference)					Remarks
	f_1	f_2	f_3	f_4	f_5	
$h = 11$ ft $b = 8$ ft $w = 107$ lb/ft^3	0.811	2.34	4.07	5.88	7.90	Control model
$w = 100$ lb/ft^3	0.858 (5.8)	2.48 (6.0)	4.33 (6.4)	6.25 (6.3)	8.40 (6.3)	$h = 11$ ft $b = 8$ ft
$w = 130$ lb/ft^3	0.760 (−6.3)	2.19 (−6.3)	3.82 (−6.2)	5.55 (−5.6)	7.46 (−4.3)	$h = 11$ ft $b = 8$ ft

Table 8.4. COMPARISON OF RESONANT FREQUENCIES OF MODEL AND PROTOTYPE

	Resonant frequencies, cps				
	f_1	f_2	f_3	f_4	f_5
Prototype (N–S)	0.85	2.25	3.90	5.55	7.20
Model (N–S)	0.78	2.25	3.92	5.68	7.62
Prototype (E–W)	0.85	2.25	3.90	5.55	7.20
Model (E–W)	0.79	2.25	3.92	5.68	7.64

8.4.2 Matching Resonant Amplitudes

The second phase in the formulation of the analytical model consisted of adjusting the values of damping capacity in the first five modes to give equal resonant amplitudes in the model and prototype for equivalent sinusoidal excitation. The exciter force was simulated in the model by applying a horizontal force $\omega^2 P \sin \omega t$ to the roof. In the first trial, the damping capacities of the first five modes of the model were made equal to the corresponding experimental values. The circular frequency ω was increased from zero to a frequency above the fifth resonant frequency; then the model's resonant amplitudes were compared with those found experimentally. The damping values were adjusted; the resonant amplitudes were recomputed and compared with the experimental results. This process was continued until the resonant amplitudes of the model matched the experimental resonant amplitudes of the building to the required degree of accuracy. Table 8.5 compares the values of the damping capacity derived for the model with those found in the building in the summer of 1964 tests.

Table 8.5. COMPARISON OF MODEL'S DAMPING RATIOS WITH THOSE FOUND FOR THE EAST BUILDING IN THE SUMMER 1964 TEST

	N–S		E–W	
Mode	Model	Experiment	Model	Experiment
1	1.70	1.8	1.05	2.0
2	0.35	0.7	0.40	0.9
3	0.38	0.4	0.40	0.7
4	0.50	0.5	0.45	—
5	0.75	0.7	—	—

8.4.3 Response of Analytical Model to Ground Accelerations

The analytical model for the N–S direction was subjected to ground acceleration records in order to assess the behavior of the real structure in an earthquake. A linear analysis was employed at first, but since it was found for certain cases of damping that the yield moments of some members were being exceeded, a nonlinear analysis also was conducted.

8.4.3.1 Linear analysis. The computer program used for the linear analysis employs the mode superposition procedure. In this case the first six modes were considered. The response of the model was computed for two earthquakes: El Centro of May 18, 1940 (N–S direction), and Taft of 1952 (N 69°W). Figures 8.12 and 8.13 show, for the first 15 sec of El Centro and Taft, respectively, the variation with time of roof displacement, base shear, and base overturning moment.

The stresses generated in the model by the ground acceleration are compared with those predicted by the Uniform Building Code (UBC) in Table 8.6. As has been found in many previous analyses, the UBC stresses are much smaller than those calculated theoretically. It should be remembered, however, that the values of damping employed in the computer calculations are experimental values for very small amplitudes. It is reasonable to expect, therefore, that in the larger stress amplitudes that would occur in an earthquake, the values

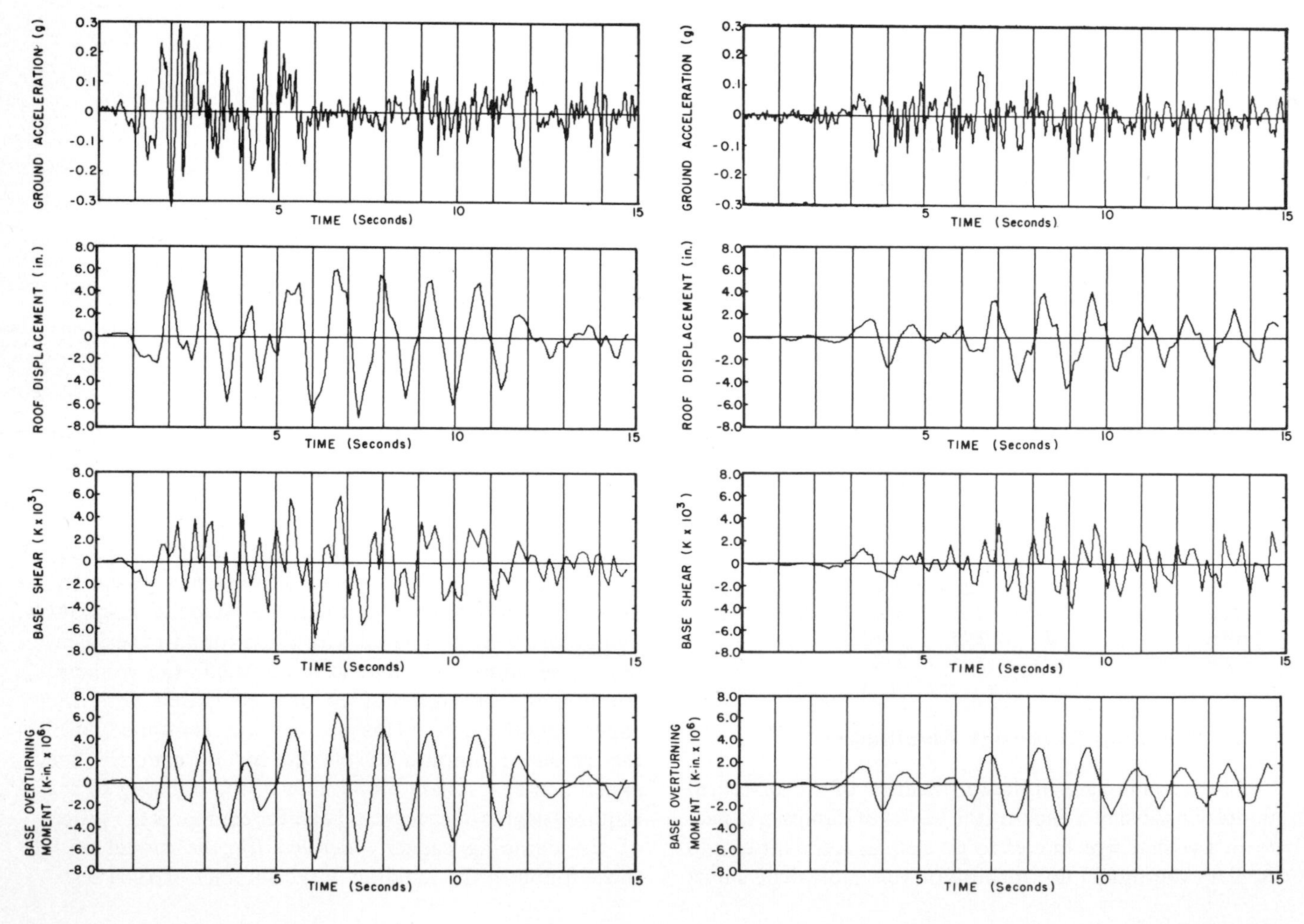

Fig. 8.12. Dynamic response of model to El Centro earthquake. (From Rea, Bouwkamp, and Clough, 1966.)

Fig. 8.13. Dynamic response of model to Taft earthquake. (From Rea, Bouwkamp, and Clough, 1966.)

Table 8.6. MAXIMUM RESPONSE VALUES FOR VERTICAL AND HORIZONTAL LOADS, EAST BUILDING

	Vertical	Horizontal, N–S			
			Dynamic		
		UBC,	El Centro		Taft
Description	Static	earthquake	Case I*	Case II†	Case I
Maximum roof displacement, in.	—	1.45	7.09	5.53	4.44
Maximum overturning moment, k-in.	—	1.22×10^6	6.80×10^6	5.52×10^6	4.05×10^6
Maximum base shear, k	—	934	6760	3760	4595
Bending stresses, ksi					
Corner column (strong direction)	1.93	5.23	31.8	20.3	
Intermediate column (strong direction)	2.91	3.39	23.4	13.5	
Intermediate column (weak direction)	3.24	6.39	46.2	25.8	
Shear stress, ksi					
Corner column (strong direction)	0.01	0.68	18.2	10.0	
Intermediate column (strong direction)	1.00	2.16	15.9	8.8	
Intermediate column (weak direction)	0.06	2.30	6.0	3.3	

*Case I: 1.70, 0.35, 0.38, 0.58, and 0.75% critical damping in the first five modes.
†Case II: 5% critical damping in first five modes.

of damping capacity would be substantially larger than the ones used in this calculation.

Table 8.6 also shows that for an earthquake of El Centro intensity, yielding should be expected to occur if the damping values used were of the same magnitude as measured in the tests. Yielding would occur first in the flanges of the intermediate columns around their weak axis. On increasing the values of damping capacity to 5% of critical in all modes, yield stresses were not reached in any part of the structure. The yield mechanism of the structure under earthquake loading was studied by means of a special program.

8.4.3.2 Nonlinear analysis. The response of the model to ground acceleration when yielding of the members occurred was computed by means of a program that can accommodate nonlinear structural behavior. This program employs a step-by-step numerical integration procedure to solve the differential equations governing the motion of the structure.

It was necessary to consider only one quarter of the model structure in the analysis since the structure has two axes of symmetry. The quarter structure analyzed is shown in Fig. 8.14a, and for purposes of this analysis, the two frames *AB* and *CD* were linked to form an equivalent plane frame shown in Fig 8.14b.

The equivalent plane frame of the East Building with zero damping was subjected to the El Centro ground acceleration. The resulting yield mechanism is shown in Fig 8.15 .The ductility ratio shown beside each yielding

Fig. 8.14. Idealization of structure for nonlinear analysis. (From Rea, Bouwkamp, and Clough, 1966.)

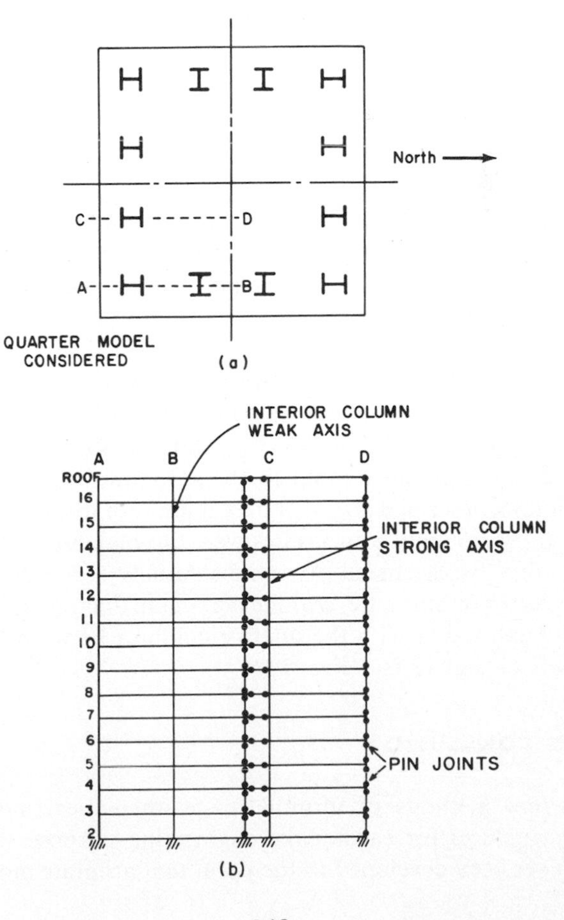

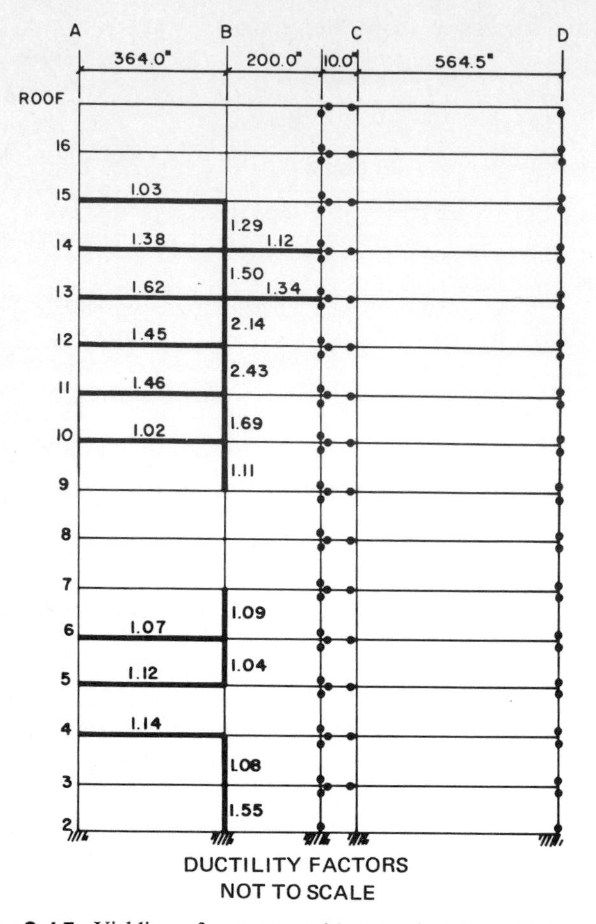

Fig. 8.15. Yielding of structure with zero damping, El Centro earthquake. (From Rea, Bouwkamp, and Clough, 1966.)

member is defined as the maximum ratio of plastic rotation to elastic rotation capacity of that element. Yielding was severest between floor levels 11 and 13, around the weak axis of the interior columns.

When the damping was increased to 5% of critical in the first mode, the yield moment of no member was exceeded.

In order to cause yielding with 5% of critical damping in the first mode, the acceleration ordinates of the El Centro record were all increased by a factor of 1.3. Under these conditions, the maximum lateral story displacements were approximately equal to those for the case of zero damping described above (a maximum roof displacement of approximately 8 in.). However, the maximum story-to-story displacements were substantially reduced from the zero damping case, and the maximum ductility factors were associated with the interior columns between floor levels 11 and 13 (see Fig. 8.16).

8.5 CONCLUSION

Current methods of formulating mathematical models of structures for earthquake engineering purposes have not yet been developed to the point that accurate models can be formulated for most types of structures. However, with additional experimental and analytical work, progress can be made in this area.

The present experimental equipment and techniques are adequate to determine the significant dynamic properties of a majority of structures vibrating at small amplitudes. Further tests are required to aid (and to stimulate) the development of mathematical models for more complex types of structures. In addition, such tests will add to the body of experimental data on damping capacities which, as has been revealed by tests, are frequently smaller than had previously been assumed.

Suitable mathematical models are already available to simulate the behavior of simple types of structures. Even in these cases the models are complex, requiring a substantial amount of computation in any dynamic analysis. Efforts should be made to simplify these models while still maintaining sufficient accuracy for engineering design purposes. There are no suitable mathematical models for more complex types of structures—e.g., arch dams, buildings with shear walls, cores and setbacks, and nonlinear structures, etc.—and further work is required in the development of models for these types of structures.

Fig. 8.16. Yielding of structure with 5% damping, 1.30 × El Centro earthquake. (From Rea, Bouwkamp, and Clough, 1966.)

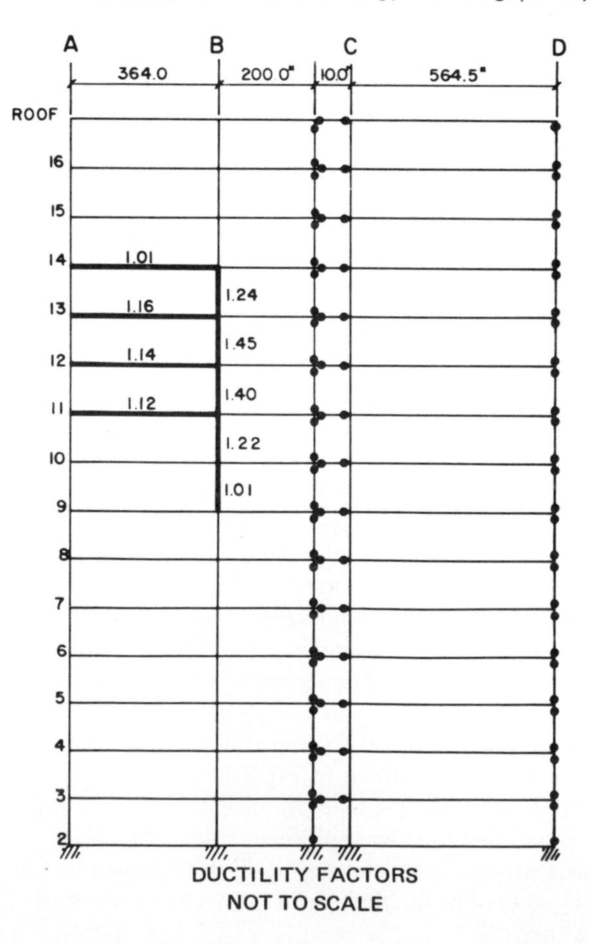

REFERENCES

Bishop, R. E. D., and G. M. L. Gladwell (1963). "An Investigation into the Theory of Resonance Testing," *Phil. Trans. Roy. Soc.*, Ser. A, *255*: 1055.

Bouwkamp, J. G., and J. K. Blohm (December 1966). "Dynamic Response of a Two-Story Steel Frame Structure," *Bull. Seism. Soc. Am.*, **56**(6), 1289–1303.

Caughey, T. K. (June 1960). "Classical Normal Modes in Damped Linear Dynamic Systems," *J. Appl. Mech.*, **27**(2).

Clough, R. W., K. L. Benuska, and E. L. Wilson (1965). *Inelastic Earthquake Response of Tall Buildings*, *Proceedings of the Third World Conference on Earthquake Engineering*, New Zealand.

Englekirk, R. E., and R. B. Matthiesen (June 1967). "Forced Vibration of an Eight-Story Reinforced Concrete Building," *Bull. Seism. Soc. Am.*, **57**(3).

Gauzy, H. (1959). "Measurement of Inertia and Structural Damping," *Manual on Aeroelasticity*, **IV**.

Hanson, R. D. (1965). *Static and Dynamic Tests of a Full-Scale Steel Frame Structure*, Pasadena: Earthquake Engineering Research Laboratory, California Institute of Technology.

Hanson, R. D. (1966). "Comparison of Static and Dynamic Hysteresis Curves," *Proc. ASCE, J. Engr. Mech. Div.*, **92**(EM 5).

Housner, G. W., and P. C. Jennings (1964). "Generation of Artificial Earthquakes," *Proc. ASCE, J. Engr. Mech. Div.*, **90**(EM 1).

Hudson, D. E. (1964). "Resonance Testing of Full-Scale Structures," *Proc. ASCE, J. Engr. Mech. Div.*, **90**(EM 3).

Jennings, P. C. (1964). "Periodic Response of a General Yielding Structure," *Proc. ASCE, J. Engr. Mech. Div.*, **90**(EM 2).

Keightley, W. O. (1961). *Vibration Tests of Structures*, Pasadena: Earthquake Engineering Research Laboratory, California Institute of Technology.

Keightley, W. O. (1964). *A Dynamic Investigation of Bouquet Canyon Dam*, Pasadena: Earthquake Engineering Research Laboratory, California Institute of Technology.

Keightley, W. O., G. W. Housner, and D. E. Hudson (1961). *Vibration Tests of the Encino Dam Intake Tower*, Pasadena: Earthquake Engineering Research Laboratory, California Institute of Technology.

Kennedy, C. C., and C. D. P. Pancu (November 1947). "Use of Vectors in Vibration Measurement and Analysis," *J. Aeronautical Sci.*, **14**.

Nielsen, N. N. (1964). *Dynamic Response of Multistory Buildings*, Pasadena: Earthquake Engineering Research Laboratory, California Institute of Technology.

Nielsen, N. N. (February 1966). "Vibration Tests of a Nine-Story Steel Frame Building," *Proc. ASCE, J. Engr. Mech.*

Rea, D., J. G. Bouwkamp, and R. W. Clough (1966). *The Dynamic Behavior of Steel Frame and Truss Buildings*, Structures and Materials Research Report No. 66-24, Berkeley, California: University of California.

Rouse, G. C., and J. G. Bouwkamp (1967). "Vibration Studies of Monticello Dam," *Research Report No. 9*, Washington, D.C.: Bureau of Reclamation, United States Department of the Interior.

Simpson, A. (June 1966). "An Improved Displaced Frequency Method for Estimation of Dynamical Characteristics of Mechanical Systems," *J. Roy. Aeronautical Soc.*, **70**(666).

Uniform Building Code (1967). International Conference of Building Officials, Pasadena, California.

Chapter 9

Earthquake Damage and Structural Performance in the United States

KARL V. STEINBRUGGE

Chief Engineer, Earthquake Department
Pacific Fire Rating Bureau,
San Francisco, California

9.1 INTRODUCTION

The rapid developments being made in the mathematical theory of structural dynamics as they apply to earthquake engineering make it very important to critically evaluate the validity of these theories by actual experience in large earthquakes. Furthermore, earthquake records from strong-motion seismic instruments must be reconcilable with observed earthquake damage.

Strong earthquakes provide an excellent test of the state of the art of earthquake resistive construction. Building codes' earthquake provisions, which reflect consensus judgment in some design areas having inadequately developed theory or theory that is unconfirmed by records from seismic instruments, must be updated on the basis of new experience. No present building code can cover all of the possible problems and difficulties that arise in earthquake resistive design, and experience is particularly vital for new material assemblies and techniques previously untested by a major earthquake.

As in any profession, relevant experience is a vital component in making judgment decisions.

It is quite impossible in the limited space of one chapter to discuss detailed earthquake engineering problems such as modes of failure of structures in particular earthquakes. The analysis of the building damage in any one earthquake such as the 1964 Alaskan earthquake (*The Prince William Sound, Alaska, Earthquake of 1964 and Aftershocks*, U.S. Environmental Science Services Administration, vol. II, Part A) fills a volume many times exceeding the size of this chapter. The purposes of this chapter, then, are to develop a brief overview of earthquake damage patterns in the United States, to give summary damage statistics, to point to a few critical areas in building design, and to show some of the bases for judgment required in competent earthquake resistive design. The reader must turn to the references in the bibliography for any discussion in depth for the topics or buildings discussed in this chapter.

From the standpoint of damage evaluation, only the strongest earthquakes need to be considered. However, for seismicity studies, persons wishing to examine the historical records for all known earthquakes in a particular region can consult catalogs which, unfortunately, are scattered among many publications. Stronger earthquakes in the United States are listed in the following U.S. Coast and Geodetic Survey publication: *Earthquake History of the United States*, Part I (1965) and Part II (1966). More detailed information is available in the U.S. Coast and Geodetic Survey's *United States Earthquakes*, which is issued annually and contains all United States earthquakes for the year under study. The various volumes of the *Bulletin of the Seismological Society of America* are the best sources for very detailed regional studies. Foreign earthquake catalogs have been listed by Hollis in his *Bibliography of Engineering Seismology*.

9.2 DEFINITION AND DISCUSSION OF TERMS

It is appropriate to define certain terms from an engineering standpoint and to develop briefly their meanings in the context of this chapter. It is a commonly held theory that when the strains become too great within the Earth's crust, then rupture (faulting) will take place. Faulting can generate seismic waves that define an earthquake. This faulting will occur in the weakest zone in the Earth's crust. This weakest zone is normally one that has had previous movements (earthquakes) on it, and this zone is called a fault zone. Any single event normally takes place on a plane or in a narrow belt within what may be a much wider fault zone, with the zone possibly having been formed by thousands of earthquakes over long geologic time. The most probable place for the next rupture in the fault zone is where the faulting occurred in the last event; however, by its very nature of formation, any place within the fault zone could have breakage. Geologists often can determine if future ruptures are to be expected on a fault, and if so, the fault is often termed an active fault. The fault displacement may be vertical, horizontal, or any combination of these. The amount of the displacement in a single event can vary from about zero upward, with 20 or more feet of displacement not being uncommon in large shocks. However, it must be pointed out that large earthquakes that have caused severe damage have not always been accompanied by reported surface faulting.

The design engineer, then, can evaluate three degrees of seismic risk with respect to a site on or in an active fault mapped by a geologist:

1. *Not in the fault zone:* Man-made structures on these sites will be subjected only to vibrational forces.
2. *Within the fault zone, but not adjacent or across the last rupture:* While it is improbable that the next rupture will go through the particular site, some degree of increased risk exists. Competent geologic advice is mandatory for any engineering decision.
3. *Across or adjacent to the last rupture:* It is quite imprudent to build on these sites when life safety is involved, unless unusual and often very costly design features are included.

The focus of an earthquake is an instrumentally located point below the Earth's surface where rupture first occurs, presumably on a fault. The depth of the focus below the Earth's surface is called the focal depth. In California, the focal depth is commonly taken to be about 10 mi, with the range usually being between 5 and 20 mi. Recent studies of small earthquakes (such as aftershocks) show that focal depths can be considerably less than 10 mi in parts of California. In some regions of the world, focal depths may be in terms of hundreds of miles. Effects on the ground surface may be far different in these regions where earthquakes have deep focal depths than where they have shallow focal depths since the seismic energy must travel greater distances for the deeper shocks. This must be borne in mind when extrapolating strong motion record data from one region to another, and it also applies to damage observations.

The instrumental epicenter is defined as the point on the Earth's surface beneath which initial rupture (focus) first started. The distance of any location on the Earth's surface from the epicenter, or the epicentral distance, is a convenient figure to state in reports, and it may be related to damage in some cases. However, the epicenter may, or may not, be above the center of the rock's strain energy that was released by the fault movement. If the epicenter is at one end of faulting that extends for many miles, then the center of the energy release may be at or near the midpoint of the fault rupture rather than at the epicenter. It is the released energy, reduced by damping and dissipation as it travels by wave

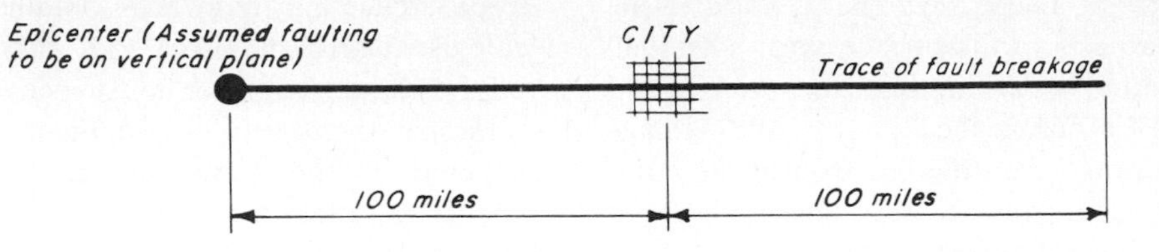

Fig. 9.1. Epicentral distance may be misleading from a damage standpoint. In this diagram, the epicenter is 100 mi from the city, but the faulting through the city could be destructive.

motion from its source, that causes vibrational damage to a building. With respect to Fig. 9.1, suppose that an earthquake has an epicenter 100 mi from a city and that the fault rupture extended from the epicenter to, through, and beyond the city for 100 mi. The epicentral distance of 100 mi would certainly not be a measure of the damage in the city, which might be located at the center of the energy release and also suffer damage from faulting.

The field epicenter, as contrasted to the instrumental epicenter described above, is usually defined as the location of the greatest earthquake intensity. In many cases the field epicenter may be located at some point on the surface faulting. Alternately, the field epicenter may be some miles away from the instrumental epicenter, and it often is. When no surface faulting occurs, the field epicenter may be some distance away from the causative fault. The field epicenter is relatively little used today.

Earthquake intensity is an arbitrary measure of earthquake effects. The Modified Mercalli Intensity scale is in use in the United States; a condensed version of this scale may be found in Table 4.9 in Chapter 4. The earthquake intensities from a particular shock observed at various locations may be plotted on a map. Lines, called isoseismal lines, can be drawn on the map so as to separate one group of equal intensities from another. Examples of isoseismal maps will be found throughout this chapter. With reference to Table 4.9, the lowest intensity values rely heavily on human reactions, the middle range intensity values principally relate to building damage, and the highest intensity values are strongly influenced by geologic effects. Human reactions, building damage, and geologic effects are not truly compatible. For example, items have not fallen from shelves in buildings adjacent to major fault scarps. New building materials, new construction techniques and new design methods have complicated the application of the Modified Mercalli scale. For example, long period ground motions can selectively damage taller multistory buildings, leaving the small one- and two-story "collapse hazard" buildings undamaged; this information is not reflected in the present intensity scale. For another example, the phrase "good construction" used in the scale has different meanings in different areas: In some areas, brick walls must be heavily reinforced with steel to be classified as "good construction," while in other areas the walls require no reinforcement to be classified as "good construction." Despite these criticisms and despite several proposed instrumental intensity scales, the Modified Mercalli Intensity scale is the best current vehicle for defining intensity in the United States. Isoseismal maps are still by far the best method for summarizing the geographical distribution of damage, and they are quite important in earthquake insurance research.

The magnitude scale was invented and originally developed by Charles F. Richter. The magnitude of an earthquake can be quickly obtained by seismographic methods, and it is expressed in ordinary numbers and decimals. The magnitude of an earthquake is defined in terms that include the logarithm of the amplitude of the motion recorded on a seismogram. Because the scale is logarithmic, a magnitude increase from 5 to 6, e.g., means the multiplying of the recorded amplitude on the seismogram by 10. Relationships have been developed that roughly correlate magnitude with the earthquake's energy. According to Richter, one unit increase in magnitude corresponds to about a thirtyfold increase in energy.

Since magnitude is related to seismic energy, it is therefore also related to damage and earthquake intensity. Richter has correlated Modified Mercalli Intensity with the earthquake's Richter magnitude as follows:

Magnitude	2	3	4	5	6	7	8
Maximum intensity	I–II	III	V	VI–VII	VII–VIII	IX–X	XI

The foregoing is a rough correlation "for ordinary ground conditions in metropolitan centers in California," and it must be used with caution (Richter, 1958, p. 353). Examples of some of the limitations are as follows:

1. Effect of soils: The 1957 Mexican Earthquake caused extensive damage to the multistory buildings in Mexico City but not to the small buildings. The earthquake was located from 170 to 220 mi from Mexico City. The reasons for the pocket of damage in Mexico City are attributable to the poor ground and also to the long-period motions that did not damage the weak but short-period structures (Steinbrugge and Bush, 1960). Another example is the damage distribution from the 1906 San Francisco shock (Fig. 9.13).

2. Focal depth: The deeper the energy release, the less energy is available per unit surface area. The 1949 and 1965 Puget Sound, Washington, earthquakes had deeper than normal California focal depths, and damage patterns in Washington were different from those commonly observed in California (Steinbrugge and Cloud in U.S. Coast and Geodetic Survey, 1965c).

3. Duration of shaking: The 1964 Alaskan earthquake, with its 3-min duration of damaging intensity, caused progressive and cumulative damage. This "duration" may be the result of one shock or the cumulative effect of several shocks. For example, the 1952 Bakersfield shock of August 22 found many structures that had suffered slight cracking or loosening in the principal shock of a month before. The apparent intensity of the aftershock was increased as the result of cumulative effects (Steinbrugge and Moran, 1954). An example of the other extreme is the March 18, 1957, earthquake at Port Hueneme, California, which was essentially a single pulse with all its energy concentrated in the pulse (Housner and Hudson, 1958).

Relationships between intensity and acceleration are found in some textbooks, but great care should be used before accepting these correlations for engineering purposes. One should not overlook the study by Hershberger (1956) who compared the maximum accelerations recorded on 108 strong-motion records obtained in 60 earthquakes with intensity ratings; he concluded that there was no definite quantitative relationship between acceleration and intensity.

A large earthquake may be viewed from the differing standpoints of engineering, geology, and seismology. A geologically large earthquake, having large fault displacements or substantial changes in land levels, does not necessarily lead to extensive building damage, as will be discussed in later paragraphs on the 1959 Hebgen Lake earthquake. A large magnitude earthquake may be considered to be noteworthy from the seismological standpoint, but building damage may be light if the seismic energy is released at a great depth. On the other hand, a low magnitude earthquake with no surface faulting that releases its energy close to the ground surface can cause extensive damage to buildings immediately above the energy release. The Agadir, Morocco, earthquake of February 29, 1960, may be cited as such an example. In this earthquake, the poor unit masonry construction and the exceptionally poor reinforced concrete construction materially contributed to the estimated minimum of 12,000 killed and 12,000 injured of the more than 33,000 inhabitants in the city of Agadir.

9.3 SEISMIC RISK

The earthquake provisions of a building code, plus the design engineer's judgment, normally determine the seismic risk for any particular building or structure. Expert advice may have been obtained from engineering geologists, seismologists, soil engineers, and others, but the design engineer must evaluate all reports and synthesize them into a judgment decision sometimes tempered by the minimum standards of the building code. Too often all efforts are directed toward just meeting the minimum earthquake standards of a building code; just meeting these code provisions is, in reality, placing a building on the verge of being legally unsafe.

The seismic risk may be quite different for life than it is for property. For example, in the 1964 Alaskan earthquake, major multistory buildings up to 14 stories in height suffered damage ranging up to 40% of the buildings' replacement values without accompanying life loss (Table 9.9). Conversely, the failure of light fixtures and shelving, or the loss of lighting in an auditorium that results in panic, are examples of life hazards in what otherwise might be undamaged structures. The basic philosophy behind the seismic provisions of most American building codes is stated in the "Recommended Lateral Force Requirements and Commentary" by the Seismology Committee of the Structural Engineers Association of California (1967). This publication states that "in most structures it is expected that structural damage, even in a major earthquake, could be limited to repairable damage"; in certain occupancies such as hotels and hospitals it is quite possible to receive 50% property loss without serious structural damage using certain types of flexible but "safe" framing systems. More than once it has been a difficult position for a structural engineer who, when viewing his heavily damaged building (but safe structure) with his client after an earthquake, is asked to explain why his design permitted so much damage. Design for damage control automatically includes life safety, but design for life safety (i.e., minimum code standards) does not necessarily include damage control.

Another aspect of seismic risk is the evaluation of the potential frequency and intensity of destructive shocks. Figure 9.2 shows the location of the larger earthquakes in historic times in the United States. Major building codes in the United States often include a map, which usually shows the country divided into four zones of seismic probability or risk. The means for implementing the information on this map into design calculations are adequately covered in building codes.

The interpretation of this map (Fig. 9.2) is not quite as simple as the zones would imply. The length of the historical record is not at all uniform throughout the United States. The east coast of the United States has a continuous historical earthquake record from 1638, while western Nevada history has little more than a 100-year time base. Indeed, the Wonder, Nevada, earthquake of 1903 with its surface faulting was not recorded in the literature until 1959. Undoubtedly other shocks of similar strength and of similar age remain unrecorded in uninhabited regions. Coastal California earthquake history begins with an entry dated July 28, 1769, but other

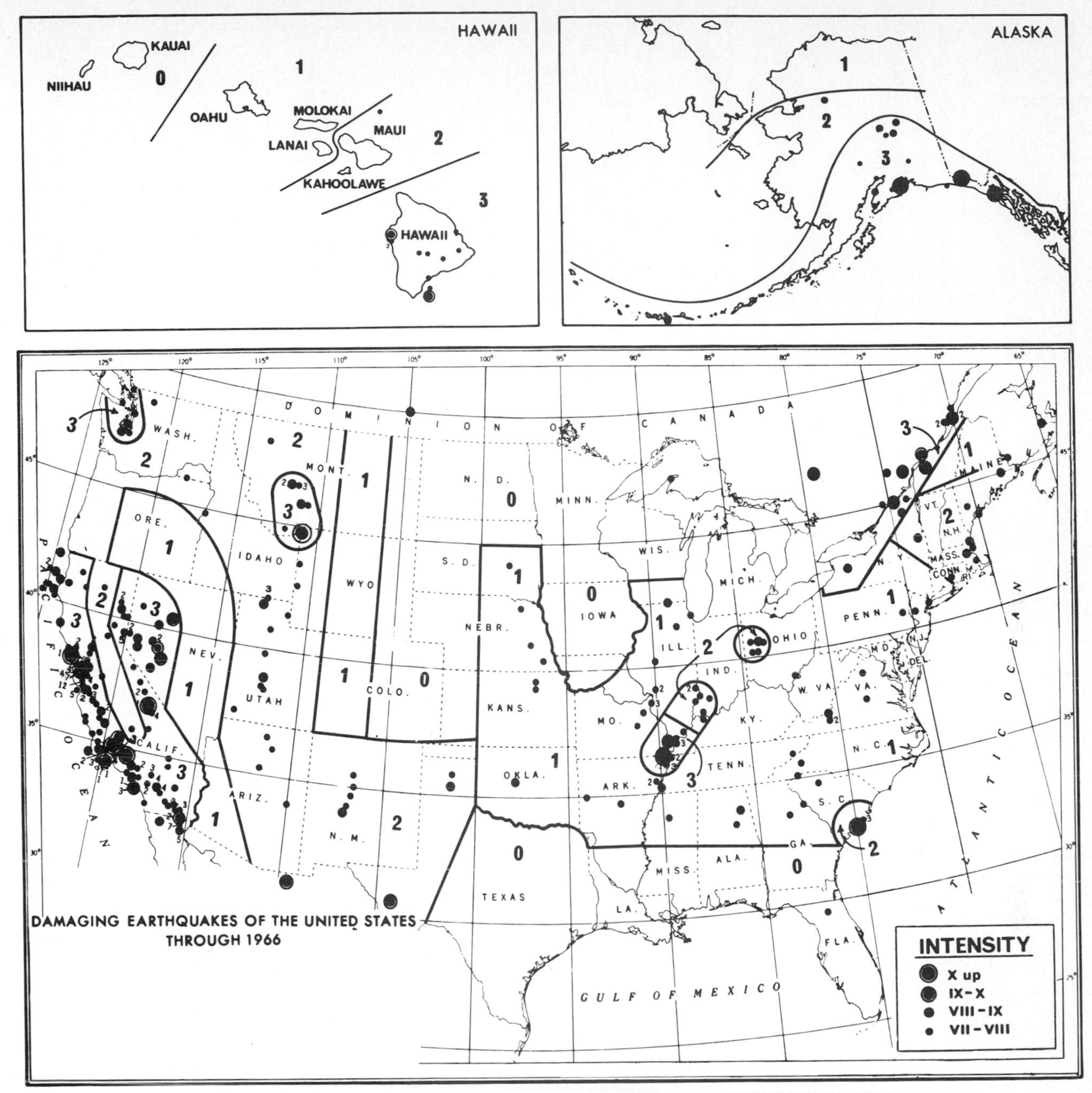

Fig. 9.2. Location of damaging historic earthquakes through 1966. Zone 0 represents minimum risk, while Zone 3 represents maximum risk. Similar maps exist in building codes.

regions within the state have little more than a 100-year history. The absence of historic destructive earthquakes in large portions of the western United States is no reason for immunity since the destructive earthquakes may occur 100 or more years apart. The 1811–1812 New Madrid earthquakes discussed in the following paragraphs emphasize this point. While not politically expedient, it would be reasonable from a cost and safety standpoint to design buildings in almost all areas west of the Rocky Mountains for earthquake Zone 3 forces.

The earthquakes discussed in the following sections were chosen for their significance to American earthquake engineering practice and are case histories of various aspects of seismic risk. Emphasis therefore is principally on recent earthquakes in which the performance of earthquake resistive construction can be evaluated. A comparative summary of these earthquakes may be found in Table 9.1.

9.4 CASE HISTORIES

9.4.1 New Madrid, Missouri, Earthquakes

Three great earthquakes of approximately equal strength were centered near New Madrid, Missouri

Table 9.1. U.S. EARTHQUAKES SELECTED FOR ENGINEERING INTEREST

Name of earthquake	Date and (local) time	Epicenter location*	Maximum Modified Mercalli Intensity†	Richter magnitude*	Approximate length of surface faulting, mi	Lives lost‡	Dollar loss§	Remarks
New Madrid, Missouri	December 16, 1811 (about 2:15 AM) January 23, 1812 (about 8:50 AM) February 7, 1812 (about 10:10 AM)	36 N, 90 W	XII (for each shock)	Over 8	See remarks	1 death		Richter assigned a magnitude of greater than 8 based on observed effects. Surface faulting possibly occurred. U.S. Geological Survey (1912).
Charleston, South Carolina	August 31, 1886 (9:51 PM)	32.9 N, 80.0 W	X		None	27 killed outright, plus 83 or more from related causes	$5,000,000–6,000,000	
San Francisco, California	April 18, 1906 (5:12 AM, PST)	38 N, 123 W	XI	8.3	190 minimum 270 possible	700 to 800 deaths	$400,000,000 including fire; $80,000,000 earthquake only	Portions of the San Andreas fault are under the Pacific Ocean.
Santa Barbara, California	June 29, 1925 (6:42 AM)	34.3 N, 119.8 W	VIII–IX	6.3	None	12 to 14 deaths	$6,500,000	The dollar loss is for the City of Santa Barbara; losses elsewhere were slight.
Long Beach, California	March 10, 1933 (5:54 PM, PST)	33.6 N 118.0 W	IX	6.3		Coroner's report: 86. 102 killed is more Probable	$40,000,000–50,000,000	Epicenter in ocean. Associated with Inglewood fault.
Helena, Montana	October 12, 1935 (12:51 AM, MST)	46.6 N, 112.0 W	VII	—	None		$50,000	First of three destructive shocks: October 12, 18, and 31.
Helena, Montana	October 18, 1935 (9:48 PM, MST)	46.6 N, 112.0 W	VIII	6.25	None	2 killed, "score" injured	$3,000,000–4,000,000 (October 18 and 31 combined)	
Helena, Montana	October 31, 1935 (11:38 AM, MST)	46.6 N, 112.0 W	VIII	6.0	None	2 killed, "score" injured		
Imperial Valley, California	May 18, 1940 (8:37 PM, PST)	32.7 N, 115.5 W	X	7.1	40 minimum	8 killed outright, 1 died later of injuries	$5,000,000–6,000,000	M. M. IX for building damage and M. M. X for faulting.
Santa Barbara, California	June 30, 1941 (11:51 PM, PST)	34.4 N, 119.6 W	VIII	5.9		None killed, 1 hospitalized	$250,000	Epicenter in ocean.
Olympia, Washington	April 13, 1949 (11:56 PM, PST)	47.1 N, 122.7 W	VIII	7.1	None	8 deaths	$15,000,000–25,000,000	
Kern County, California	July 21, 1952 (4:52 AM, PDT)	35.0 N, 119.0 W	XI	7.7	14	10 of 12 deaths in Tehachapi	$37,650,000 to buildings $48,650,000 total (including August 22 aftershock)	M. M. XI assigned to tunnel damage from faulting; vibration intensity to structures generally VIII, rarely IX. Faulting probably longer, but covered by deep alluvium.

Table 9.1. U.S. EARTHQUAKES SELECTED FOR ENGINEERING INTEREST (CON'T)

Bakersfield, California	August 22, 1952 (3:41 PM, PDT)	35.3 N, 118.9 W	VIII	5.8	None	2 killed and 35 injured in Bakersfield	See above	Aftershock of July 21, 1952.
Fallon–Stillwater, Nevada	July 6, 1954 (4:13 AM, PDT)	39.4 N, 118.5 W	IX	6.6	11	No deaths, several injuries	$500,000–700,000, including $300,000 to irrigation system	M. M. IX assigned along fault trace; vibration intensity VIII. First of two shocks on same fault.
Fallon-Stillwater, Nevada	August 23, 1954 (10:52 PM, PDT)	39.6 N, 118.5 W	IX	6.8	19	No deaths		M. M. IX assigned along fault trace; vibration intensity VIII. Second of two shocks on same fault.
Fairview Peak, Nevada	December 16, 1954 (3:07 AM, PST)	39.3 N, 118.1 W	X	7.1	35	No deaths		M. M. X assigned along fault trace; vibration intensity VII. Two shocks considered as a single event from the engineering standpoint.
Dixie Valley, Nevada	December 16, 1954 (3:11 AM, PST)	39.8 N, 118.1 W	X	6.8	30	No deaths		
Eureka, California	December 21, 1954 (11:56 AM, PST)	40.8 N, 124.1 W	VII	6.6	None	1 killed	$1,000,000	
Port Hueneme, California	March 18, 1957 (10:56 AM, PST)	34.1 N, 119.2 W	VI	4.7	None	No deaths		Epicenter in ocean.
San Francisco, California	March 22, 1957 (11:44 AM, PST)	37.7 N, 122.5 W	VII	5.3	None	No deaths, about 40 minor injuries.	$1,000,000	
Hebgen Lake, Montana	August 17, 1959 (11:37 PM, MST)	44.8 N, 111.1 W	X	7.1	14	19 presumed buried by landslide, plus probably 9 others killed, mostly by landslide	$2,334,000 (roads and bridges) $150,000 (Hebgen Dam) $1,715,000 (landslide correction)	M. M. X assigned along fault trace. Vibrational intensity was VII maximum. Faulting complex, and regional warping occurred. Dollar loss to buildings relatively small.
Prince William Sound, Alaska	March 27, 1964 (5:36 PM, AST)	61.1 N 147.5 W		8.4	400 to 500	110 killed by tsunami; 15 killed from all other causes	$311,192,000 (incl. tsunami)	Also known as the "Good Friday Earthquake." Fault length derived from seismic data.
Puget Sound, Washington	April 29, 1965 (8:29 AM, PDT)	47.4 N, 122.3 W	VIII	6.5	None	3 killed outright, 3 died from heart attacks	$12,500,000	M. M. VII general, M. M. VIII rare.
Parkfield, California	June 27, 1966 (9:26 PM, PDT)	35.54 N, 120.54 W	VII	5.5	$23\frac{1}{2}$ and $5\frac{1}{2}$	No deaths	Less than $50,000	Damaging earthquakes in same area in 1901, 1922, and 1934. The 1966 shock had peak acceleration of 50% g.

Abbreviations: M.M. = Modified Mercalli intensity; PST = Pacific Standard Time; PDT = Pacific Daylight Time (subtract 1 hour for Pacific Standard Time); MST = Mountain Standard Time; and AST = Alaska Standard Time.

*Slight variations will be found in various publications.

†Modified Mercalli Intensities are those assigned by the U.S. Coast and Geodetic Survey when available.

‡Original sources do not always clearly indicate if deaths include those attributable to exposure, unattended injury, heart attacks, and other nonimmediate deaths.

§Value of dollar at time of earthquake. Use of these figures requires a critical examination of reference materials since the basis for the estimates varies.

(Fig. 9.2) and occurred on December 16, 1811, and January 23 and February 7, 1812. Many aftershocks were also reported for fully a year after the last large shock [U.S. Geological Survey (1912)].

About 30,000–50,000 mi^2 of land mostly west of and usually adjacent to the Mississippi River had spectacular geologic effects such as sunken lands, uplifted areas, fissures, sand blows, and landslides. Waves on the Mississippi swamped boats and washed others upon the shore. The earthquakes were felt from Canada on the north to New Orleans on the south. They were also felt as far away as Washington, D.C., which is 700 mi away, and possibly at Boston, which is 1100 mi away. Only a single life was reported lost, and property damage was very slight because the epicentral area was lightly inhabited. These three major shocks rank with the greatest to have occurred in the United States in historic times.

Of engineering significance is the seismic risk problem that arises from the fact that there have been no historic great earthquakes in the New Madrid area before or since the 1811–1812 sequence. The moderate earthquakes in the area have not been particularly damaging. New construction in Missouri and in neighboring states along the Mississippi River, which were affected by the 1811–1812 shocks, generally is not specifically designed to resist major earthquake forces. Does this long time interval since 1811–1812 negate the need for earthquake resistive design on the basis of great earthquakes possibly being a remote risk?

The seismic risk problem is not unique to the New Madrid area. For example, Charleston, South Carolina, has had only one historic destructive earthquake—that of August 31, 1886. Improved techniques for assessing the seismic risk for most of the United States are necessary.

Meanwhile, in potentially earthquake active regions having no seismic building code provisions, a prudent engineer should avoid the consequences of ignoring the peril. One-, two-, and three-story buildings, which constitute the bulk of the nonresidential construction, can be designed to resist major earthquake forces with essentially no increase in cost, provided the design engineer is qualified in earthquake engineering and, of equal importance, is allowed to select the structural materials and the layout of earthquake resisting elements. The architectural trend of "no means of visible support" for roofs and supported floors is too often poor earthquake design.

9.4.2 San Francisco, California, Earthquake

The great earthquake of April 18, 1906, is one for which considerable geologic, seismologic, and engineering data are available, and much of these data remains of considerable importance today. An isoseismal map indicating how widespread the damage was is shown in Fig. 9.3.

The earthquake's large magnitude of 8.3 ranks it among the highest ones to have been recorded by seismographs. Therefore, on a seismological basis, this earthquake can be considered as approaching the maximum credible earthquake to be anticipated for building design purposes.

The San Andreas fault rupture extended 190 mi from San Juan in San Benito County to Point Arena in Mendocino County; then it may have continued under the Pacific Ocean to enter land at Shelter Cove in Humboldt County. The faulting certainly extended for 190 mi, and possibly as far as 270 mi (Lawson et al., 1908). The fault displacement was right lateral, meaning that a fence crossing the fault would have that portion on the other side of the fault trace displaced horizontally to an observer's right as he faced the fault. This horizontal fault displacement was not less than 10 ft for most of its length; in places it measured more than 15 ft; in one marshy ground area it measured as much as 21 ft (Lawson et al., 1908). Figure 9.4 shows a right lateral offset of 8.5 ft in the formerly straight fence; this offset was at Woodville in Marin County and occurred in the 1906 shock.

The closest San Andreas fault breakage to San Francisco was $1\frac{1}{2}$ mi from the city limits. The financial and commercial center of the city, which was 9–10 mi from the fault rupture, contained a number of multistory buildings, many of which are still in existence. From the geological and seismological standpoints, the 1906 shock ranks as a great earthquake and one that can be expected to approach the maximum credible ground motions for which buildings should be designed in California and in many other states in the United States. The duration of the severe shaking in San Francisco has been estimated to be 40 sec to possibly 1 min; the duration of the 1964 Alaskan earthquake may have been 3 min for forces at 2% g or longer. Longer duration of heavy shaking obviously will accentuate damage and life hazard, and in this regard the 1906 San Francisco shock is not a maximum design criteria.

Statistics regarding life loss vary widely, and many contemporary publications quote figures that are unsubstantiated. The statistics given in Table 9.2 are believed to be the most accurate of those known to the author.

Table 9.2. Life Loss in San Francisco*

Killed outright and accounted for at the Coroner's office	315
Shot for crime	6
Shot by mistake	1
Reported missing and not accounted for	352
Total	674

*From *Report of the Sub-Committee on Statistics*, Marsden Manson, Chairman (1907?).

In addition to Table 9.2 another source which should be authoritative is the report of the army relief operations (Greely, 1906), which makes its count on a somewhat different basis from that given in Table 9.2. The following paragraph (p. 176) is quoted from the Greely report:

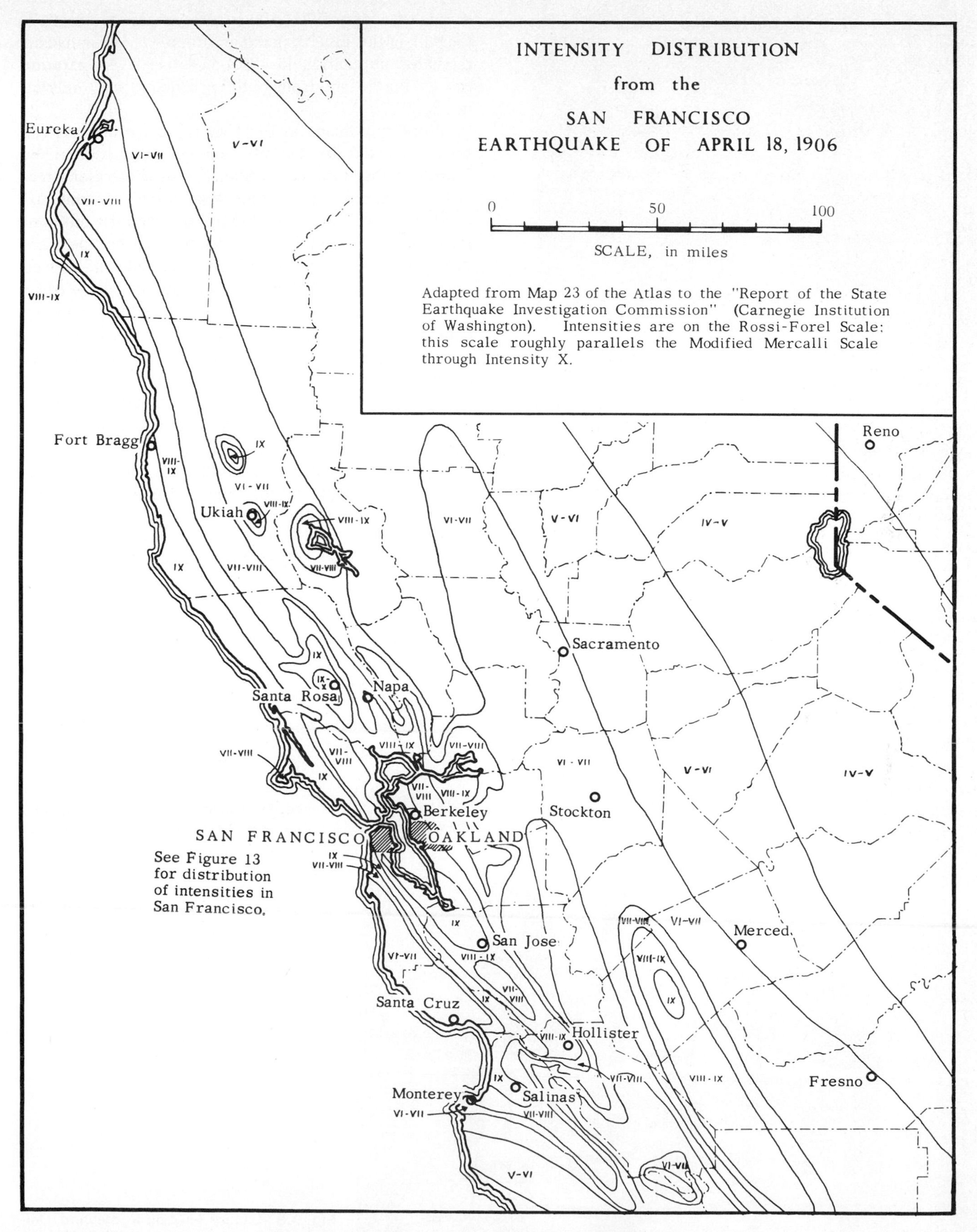

Fig. 9.3. Isoseismal map of the 1906 San Francisco Earthquake.

Fig. 9.4. Fence offset shows 8.5-ft right lateral movement that occurred on the San Andreas fault near Woodville in Marin County north of San Francisco. Note wood frame structures in background are not destroyed. San Francisco Earthquake of 1906. (G. K. Gilbert photo.)

Of deaths and injuries from earthquake and fire, which were enormously exaggerated in current dispatches, the roll, including all bodies discovered and those who have since died of injuries, is as follows: San Francisco, 304 known; 194 unknown (largely bodies recovered from the ruins in the burned district); in addition 415 were seriously injured. In Santa Rosa there were 64 deaths and 51 seriously injured; in San Jose, 21 deaths and 10 seriously injured; and at Agnew's Asylum, near San Jose, 81 deaths.

A total life loss of 700–800 is a reasonable figure, with the bulk of this loss being in San Francisco, which had an estimated population in 1905 of 400,000. An extreme case of life hazard from building collapse may be seen in Fig. 9.5.

Property damage in the City of San Francisco has been estimated by various reliable authorities.* The Manson Subcommittee on Statistics used assessor's records and placed the building loss (excluding contents) at $105,008,480 (Manson, 1907). The Chamber of Commerce in their report (1906) approached the problem differently, using extrapolated insurance data, and derived a loss of about $350 million for buildings and their contents for San Fransisco. The probable loss, including consequential damages of all kinds, was estimated by the Committee of Five (1906) to the "Thirty Five (Insurance) Companies" at $1 billion. It is reasonable to use a figure of $400 million for direct earthquake and fire loss to buildings and to their contents for San Francisco and the outlying areas.

The 3-day conflagration following the earthquake caused substantially more damage than did the earthquake. The area of the burned district covered 4.7 mi^2, comprising 521 blocks of which 13 were saved and 508 burned. One count of burned buildings was as follows:

Wooden framed buildings	24,671
Brick—Classes C and B	3,168
Brick and wood (unclassified)	259
Fireproof Class A	42
Stone	15
Corrugated iron (wooden frame)	33
Total	28,188

*All dollar figures used throughout this chapter are based on the value of the dollar at the time of the earthquake under discussion.

Fig. 9.5. Library building at Stanford University, Palo Alto. San Francisco Earthquake of 1906. (Walter Huber photo.)

Conflagration following earthquake is a distinct hazard for all cities in earthquake-prone areas. However, fire does not automatically follow a major earthquake; if it does, the reasons should have been apparent before the event. Therefore, it is of value to briefly review the background for the San Francisco fire. The National Board of Fire Underwriters (1905) published a report before the earthquake and summarized their findings as follows:

> In view of the exceptionally large areas, great heights, numerous unprotected openings, general absence of fire-breaks or stops, highly combustible nature of the buildings, many of which have sheathed walls and ceilings, frequency of light wells and the presence of interspersed frame buildings, the potential hazard is very severe.
>
> The above features combined with the almost total lack of sprinklers and absence of modern protective devices generally, numerous and mutually aggravating conflagration breeders, high winds, and comparatively narrow streets, make the probability feature alarmingly severe.
>
> In fact, San Francisco has violated all underwriting traditions and precedent by not burning up. That it has not done so is largely due to the vigilance of the fire department, which cannot be relied upon indefinitely to stave off the inevitable.

Actuality was worse than the prediction since portions of the water system were severely damaged by the earthquake. Two of the three main storage reservoirs for San Francisco were located on the San Andreas fault to the south of the city on the San Francisco Peninsula. The third was nearby but not on the fault. The reservoirs, the earth-fill dams, and one concrete dam survived excellently. For example, the Crystal Springs Dam was built in stages (1887, 1888, and 1890) to a height of 146 ft using irregular and interlocking concrete blocks, formed in place, and intended to be substantially monolithic; no earthquake damage was found although the fault rupture was within a quarter of a mile of it (Schussler, 1906). However, all of the three conduits from the main storage reservoirs to San Francisco were damaged or destroyed where they crossed the San Andreas fault (Fig. 9.6) and where they crossed marshy areas. Only the Lake Honda Reservoir of the distributing reservoirs (of a total of 3) was damaged by the earthquake; however, when the fire in San Francisco was under control, this reservoir still contained more than one-sixth of its capacity. One supply conduit from the main storage reservoirs was repaired in 3 days, and at no time during the conflagration were all of the distribution reservoirs empty.

Hundreds of pipe breaks occurred in the city distributing system, principally where the lines crossed filled ground and former swamps. Equally serious was the fact that probably thousands of service pipes were broken by earthquake motions and by the collapse of burning buildings. Water in vital portions of the distribution system therefore was not available to fight the fire, although it was available in the Western Addition residential section of San Francisco during the entire conflagration (Schussler, 1906).

Fig. 9.6. Rupture of the 30-in. water supply conduit to San Francisco where it crossed the San Andreas fault. San Francisco Earthquake of 1906.

While today's cities may not be the tinderbox that San Francisco was at the time of the 1906 shock, the problems of supply lines crossing potentially active faults and distribution systems located in poor ground areas still exist in many communities.

In the years that followed the 1906 disaster, it became "proper" to call it the 1906 fire and to omit any reference to the earthquake since eastern United States investment capital had a greater fear of earthquake than fire. As a result, the estimates of the ratio of earthquake to fire damage tended to decrease after the earthquake. Recently, a realistic restudy of all available records indicated that the earthquake losses amounted to perhaps as much as 20% of the total earthquake and fire loss in San Francisco, based on today's methods of estimating losses. Earthquake damage often is not spectacular up to the point of building collapse, but usually smoke damage is spectacular even when not costly. Figures 9.7, 9.8, and 9.9 are typical street scenes in San Francisco after the earthquake but before the fire; obviously the city was far from being leveled, even though many buildings had been significantly damaged. Figures 9.10 and 9.11 show the spectacular ruins after the fire.

Serious differences of opinion have existed, particularly among geologists and engineers, as to the earthquake intensity at locations on structurally poor ground such as swamps and uncontrolled fills (Fig. 9.12). American building codes presently ignore any increase

Fig. 9.7. Mission Street in San Francisco. Note earthquake debris. Fire advancing toward photographer. San Francisco Earthquake of 1906. (E. A. Rogers photo.)

Fig. 9.8. Some, but not all, unreinforced brick buildings were destroyed. Fire burning downtown. San Francisco Earthquake of 1906. (Arnold Genthe photo.)

Fig. 9.9. Note earthquake damage that occurred before fire reached the buildings. San Francisco Earthquake of 1906. (Bear Photo Collection.)

Fig. 9.10. The fire swept away the wood frame construction. San Francisco Earthquake of 1906. (Bear Photo Collection.)

in seismic risk on structurally poor ground, despite substantial criticism of them that has been partially based on intensity studies made after the 1906 shock.

Figure 9.13 is a map of a portion of San Francisco showing 1906 apparent earthquake intensities. The San Andreas fault is located about 10 mi west of the Ferry Building. Note the large variations in intensity for areas approximately the same distance from the fault. "Violent" areas are associated with the structurally poorest ground. In many cases of earthquake damage to brick buildings in the poor ground areas, differential settlements both before and during the earthquake greatly accentuated the damage. This is not to say that the ground motions, as apart from foundation failures, did not have substantially greater amplitudes and possibly longer periods than those on nearby rock outcrops. However, buildings known to have foundations of a type adequate to prevent significant differential settlements, such as those under the tall buildings, had no foundation damage; friction piling was the normal support for these taller structures when located on bay muds. The differential settlement damage to pipelines and to buildings in poor ground

Fig. 9.11. The two tall buildings are located at California and Montgomery Streets in San Francisco, and both are still used as office buildings. The Kohl Building (with flag) is a steel frame fireproof structure that had only a few floors burned out. Referring to the unburned stories, "The earthquake did but little damage, breaking a few panes of glass and loosening the marble wainscoting." San Francisco Earthquake of 1906. (Bear Photo Collection.)

Fig. 9.12. Crack due to structurally poor ground in the Mission District of San Francisco. San Francisco Earthquake of 1906. (Bear Photo Collection.)

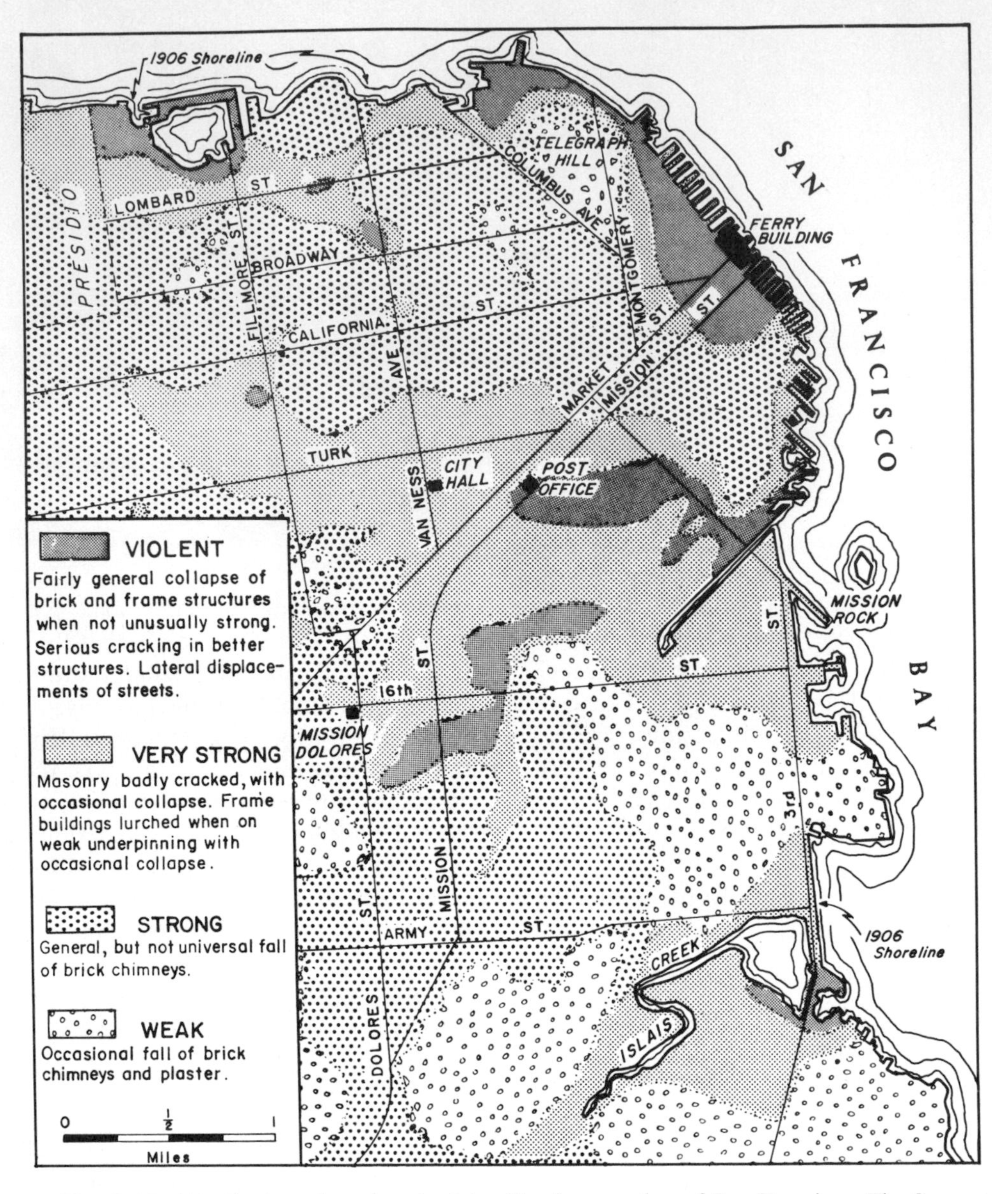

Fig. 9.13. Distribution of earthquake intensities in a portion of San Francisco. The San Andreas fault is located about 10 mi west of the Ferry Building. Note the large variations in intensity for areas approximately the same distance from the fault. "Violent" areas are associated with poorest ground areas.

areas can be better understood by reading Hittell's account of *A History of the City of San Francisco* (1878), in which he states:

> The peat in the marshes that had their heads near the site of the new city hall was strong enough to sustain a small house or a loaded wagon, though a man . . . by jumping on it could give it a perceptible shiver. . . . When the streets were first made the weight of the sand pressed the peat down, so that the water stood where the surface was dry before. . . . More than once a contractor had put on enough sand to raise the street to official grade, and gave notice to the city engineer to inspect the work, but in the lapse of a day between notice and inspection, the sand had sunk down six or eight feet. . . .

The 1906 earthquake marked the first test of multistory steel frame buildings and the largest test to date in the United States of this construction type near to a great earthquake. A total of 17 structures ranging in height from 8 to 16 stories, with one at 19 stories, experienced the earthquake. Four additional structures were under construction. Extensive nonstructural earthquake damage was common, and a few had known structural damage in the form of sheared bolts, bent I-beams, torn gusset plates, and the like. The actual extent of the earthquake damage (as opposed to the well-documented fire damage) is inadequately known and has been the subject of some dispute. Obviously the 3 days between the earthquake and the end of the conflagration did not allow for adequate inspections. None of these multistory buildings was so heavily damaged as to be unsafe (American Society of Civil Engineers, 1907). The earthquake clearly showed that total destruction to multistory steel frame structures of the type then in existence was not to be expected. (Their counterparts in reinforced concrete did not exist at that time.)

Adequate statistics are not available for other construction types. Wood frame construction performed excellently in the earthquake but provided the "tinder" for the fire. Unreinforced sand-lime mortar brick-bearing walls performed poorly, but this was somewhat obscured by the ensuing fire.

9.4.3 Santa Barbara, California, Earthquake

The time of occurrence of the June 29, 1925, Santa Barbara earthquake (6:42 AM) minimized the life loss to 12 or 14 persons and to about 50 injured since few people were on the streets during the earthquake. The earthquake was not a large one, but it was centered close to Santa Barbara and damage was considerable in this city of about 30,000 population. The Richter magnitude has been established as 6.3, while the maximum Modified Mercalli Intensity was VIII to IX.

A comparatively large number of reports were published for this earthquake. The well-printed and liberally distributed reports published by some materials-oriented organizations tended to minimize the damage to their products and implied that competing products performed poorly. Additionally, some sweeping statements were made in some published reports; M. M. O'Shaughnessy (City Engineer, San Francisco) stated in his unpublished report "the almost total failure of brick and hollow tile construction."

Probably the best published objective report of damage to specific buildings was written by Dewell and Willis (1925). Their summary evaluation of the construction is worth quoting:

> The larger buildings were, however, designed and constructed by men experienced in the methods of building according to current practice and familiar with ordinances as they exist in other cities. Thus, while poor design and poor construction, especially with reference to resistance to earthquakes, were evident in Santa Barbara, it is probable that the more modern buildings were neither better nor worse than similar buildings in other cities.

Except for plumbing and electrical regulations, there had been no building code or inspection until 30 days before the earthquake.

The insurance industry examined all important buildings, most of the wood frame mercantile buildings, and a few of the damaged dwellings. Table 9.3 recaps their damage surveys (Board of Fire Underwriters of the Pacific, 1925). The building classifications given in Table 9.3 are based on insurance definitions and will vary slightly from those used by structural engineers. However, Table 9.3 clearly shows that C class construction, which was mostly unreinforced brick with sand–lime mortar, was unsuited as a structural material for earthquake resistive construction (see also Fig. 9.14). This fact was not fully accepted until the 1933 Long Beach earthquake repeated the results. Of current engineering significance is the fact that many similarly constructed buildings exist in many seismic areas in the United States, and the hazard will remain as long as the buildings exist.

Table 9.3. DAMAGE TO BUILDINGS IN SANTA BARBARA

	†Building classification						
*Damage class	Incomplete A	B	C	D	Adobe	All steel	Total
Undamaged	1	4	6	85	0	2	98
Slightly damaged	1	9	88	98	3	0	199
Moderately damaged	0	4	110	22	4	0	140
Seriously damaged	0	7	98	19	7	0	131
Demolished	0	0	38	6	6	0	50
Total	2	24	340	230	20	2	618

*Damage definitions. (1) Undamaged: No damage. (2) Slightly damaged: Loss of chimneys, glass fronts, cornice, interior embellishments, excessive amount of cracked or fallen plaster, small or hairline cracks—not serious. (3) Moderately damaged: Parapets fallen, cracked or shattered walls capable of repair, no serious damage to frame. (4) Seriously damaged: Frame cracked or shattered, walls shattered, floors cracked. (5) Demolished: Building fallen or necessary to raze. Repairs not economically possible.

†Building classification. Incomplete A: Incomplete steel frame. B: Reinforced concrete construction. Panel walls may be unit masonry or concrete. C: Unit masonry or reinforced concrete walls with wood interiors. D: Wood frame. Mixed construction and other variants in construction were apparently placed in the most nearly similar category.

The major amount of earthquake damage was suffered by the mercantile district, which was centered on State Street and occupied approximately 36 city blocks. On the lower portion of State Street and on the blocks nearer the ocean there had been a ravine that had opened into a marsh. This area had been filled, and this undoubtedly contributed to the earthquake damage.

In the residential districts, more than 90% of the buildings were wood frame. The principal loss to residences was caused by the fall of chimneys and by plaster cracking. An occasional wood frame dwelling failed as a result of rotten underpinnings or a lack of bracing between the ground and the supported floor.

An engineering committee convened at the request of the Board of Safety and Reconstruction of Santa Barbara inspected 411 structures (Inard, 1925). They estimated the damage to buildings (excluding residences) as follows:

Business and semipublic structures	\$5,000,000
Schools	700,000
County buildings	530,000
Total	\$6,230,000

Fig. 9.14. Californian Hotel, a newly built hotel in Santa Barbara. Santa Barbara Earthquake of 1925. (Walter Huber photo.)

Fig. 9.15. San Marcos Building, Santa Barbara. It was an L-shaped structure of reinforced concrete. Santa Barbara Earthquake of 1925. (Walter Huber photo.)

Therefore, total building losses may be estimated at about \$6.5 million. This is a substantial figure for a city having property values then estimated at \$46 million and in which the numerous wood frame dwellings survived with little loss.

Dewell and Willis (1925) described the damage to specific buildings in detail, using the unpublished Matthews and Stocklmeir report to the Home Insurance Company. The following is condensed from their paper and describes two of the most seriously damaged major structures (see also Figs. 9.15 and 9.16):

> *San Marcos Building* was a four story reinforced concrete, beam-girder-column, floor slab type of construction utilized for stores and office purposes. The walls were of the concrete panel type, having fifty to sixty per cent openings on the street sides. The structure was L-shaped, extending with wings of equal height. It had been built in several sections, at different times, and there was a gap of a few inches between the older structure and the newer part on the corner. The corner section was totally destroyed by collapse of the columns and the fall of the heavy concrete walls and floors. It is reported that the concrete, which was designed for a stress of 2,000 pounds to the square inch, failed in tests made after the earthquake under 760 pounds pressure. Better work had apparently been done in the older section. The conspicuous site and pretentious character of the San Marcos Building, the spectacular collapse of the corner, and the loss of several lives in the ruins, have combined to attract a great deal of attention to it. It is probably not an exceptionally striking example of poor design or construction.
>
> *The Arlington Hotel* was a large building, apparently of brick, but in fact of several different types of construction. The southern wing was a wood frame structure, veneered with brick, which was nailed on. The central building and north wing contained a reinforced concrete frame of beams and columns. The latter were concealed within false walls of brick veneer built up around them, but set out two or three inches from them and not bonded to them. Between the columns were light, hollow panel walls of double brick. The effect of this construction was to give an impression of massiveness at a minimum cost. A large water tank was placed under the roof over the southeast corner, near the junction of the central section with the north wing.
>
> The central section was severely shattered and the southeastern corner under the water tank entirely collapsed. Elsewhere throughout the central building the concrete frame was severely strained and cracked. Portions of the walls and floor fell. The false walls of the columns were split open, and the brick veneer fell off of much of the surface of the south wing.

Fig. 9.16. Arlington Hotel, Santa Barbara. Santa Barbara Earthquake of 1925. (Walter Huber photo.)

The Sheffield Dam was an earth fill structure for strong 40 million gallons of water in a distribution reservoir for the Santa Barbara Municipal Water Department. It was built in Sycamore Canyon north of Santa Barbara in the winter of 1917. The dam was 25 feet high and had a 720 foot crest. The upstream slope had a 4 foot clay blanket which extended into the ground 10 feet; this blanket was overlain with a 3 inch concrete facing. Seepage through the dam had been noted before the earthquake. The earthquake caused the center third of the dam to move downstream as a unit, pivoting on one end. Some piping was broken in the distribution system, but despite all of the foregoing, there was no interruption of water supply for more than a few minutes.

9.4.4 Long Beach, California, Earthquake

The Long Beach earthquake of March 10, 1933, marks a major turning point in the field of earthquake resistive design and construction for much of California. Earthquake bracing provisions up to that time were not contained in any of the metropolitan Los Angeles building ordinances, including that of Long Beach.

Indeed, wind bracing also was not generally required. A strong impetus toward earthquake engineering research had been given earlier when, on January 31, 1925, by Act of Congress, the U.S. Coast and Geodetic Survey was authorized to make investigations and reports in seismology. The 1925 Santa Barbara earthquake underscored the need for pure and applied research. The 1927 edition of the Uniform Building Code, which was used in the far west, included earthquake design provisions in its appendix for optional use. The California Chamber of Commerce proposed code of 1928 was "dedicated to the safeguarding of buildings against earthquake disaster." These points of view, however, had not been generally accepted in building ordinances. For example, there had been controversy regarding the potential earthquake hazard to the Los Angeles area. One book, authored by a prominent geologist and published in 1928, stated: "The accumulative weight of data substantiates beyond a doubt my deduction that Los Angeles is in no danger of a great earthquake disaster." The 1933 Long Beach disaster brought debate to a close.

The earthquake had a Richter magnitude of 6.3. Its instrumental epicenter was about 15 mi from downtown Long Beach and $3\frac{1}{2}$ mi offshore in the Pacific Ocean from Newport Beach. The focal depth was about 6 mi. The location of the main epicenter and its aftershocks, which ranged northwesterly from the main epicenter into the city limits of Long Beach, suggests movement at depth along the Inglewood fault and also that the center of the energy release was much closer to Long Beach than the epicenter implies (see the general discussion on this topic in previous paragraphs). Property damage, apparently including oil tanks and the like as well as buildings, was estimated to be from $40 to $50 million, with perhaps the most carefully made estimate stating $41 million.

This shock was not of major proportions from the seismological standpoint. However, the Modified Mercalli Intensity reached IX, and the occurrence of the earthquake in a highly populated area makes it of engineering significance. The shock ranks third to the 1906 San Francisco earthquake and to the 1964 Alaskan shock as the most destructive earthquake in United States history. Estimates of loss of life have ranged up to 120. The coroner's report placed the loss of life at 86, with 48 of these in Long Beach and 12 in Compton. The most authoritative estimate was by R.W. Binder (1952), who placed the life loss at 102. Most of the loss of life was occasioned by persons being struck by falling cornices, parapets, and ornamentation as they tried to leave shaking buildings. The duration of severe shaking has been estimated from 10 to 20 sec.

9.4.4.1 Building damage. The ratio of damage to value on business, industrial, and residential property was as follows (National Board of Fire Underwriters, 1933): Long Beach, 9.4%; Compton, 29.0%; and Huntington Park, 3.0%. The ratio of damage to value on schools and other public buildings was estimated to be considerably larger.

Building damage by class of construction followed age-old patterns. Structures with walls of brick masonry having sand–lime mortar and with wood roofs and floors suffered severely. Table 9.4 and Figs. 9.16 through 9.24 show the severe damage that occurred to these sand–lime mortar brick buildings. The 1933 Long Beach earthquake brought to an end the practice of laying unit masonry without reinforcing steel. In his study on 1264 brick-bearing walled structures, R. R. Martel (U.S. Coast and Geodetic Survey, 1965b) has concluded that damage to buildings of this type was somewhat less on soft, water-logged soil than those on more firmly consolidated soil. This conclusion is curious and interesting in view of contrary observations in other United States earthquakes and one that was not shared in other less detailed studies. Martel also found that the percentage of damage decreased with an increase in the number of stories: 23% for one story, 21% for two stories, 16% for three stories, and 12% for four stories.

Table 9.4. DAMAGE TO MASONRY WALLED BUILDINGS IN COMPTON*

Damage, %	Commercial: Number of buildings	Commercial: Fraction of total number, %	Residential: Number of buildings	Residential: Fraction of total number, %
0–4	2	2	13	47
5–24	5	4	3	16
25–49	26	21	4	27
50–75	25	20	1	10
100 (demolished)	64	53	0	0
Total	122	100	21	100

*For all practical purposes, the masonry-walled construction consisted of unreinforced brick-bearing walls. Interiors were wood, although occasionally some steel or brick might have existed. Data from *Earthquake Investigations in the Western United States, 1931–64*, in the chapter by R. R. Martel, Table 5.

Public schools with unit masonry construction deserve special mention. Exterior walls often were brick, or in some cases hollow clay tile. Roofs and supported floors were wood. The destruction to this type of school construction was most spectacular (Figs. 9.21 through 9.24). Fortunately, the earthquake occurred after school hours and a potentially catastrophic situation was averted. However, the destruction was so extensive that the legislature of the State of California passed a bill that became law on April 10, 1933. This law, known as the Field Act, required all new public school construction to be highly earthquake resistive. Particularly important in the code that implements this law is the requirement for superior field supervision of construction. Structures built under this law have performed excellently in subsequent shocks, and their records will be discussed on the following pages.

A larger than normal number of wood frame dwellings had serious damage due to failures at or near the foundation level (Figs. 9.25 and 9.26). This was a result of a lack of lateral force bracing or was due to the deterioration of such bracing as might have existed. However, performance compared to other construction materials was excellent. Table 9.5 shows that of 4575 wood frame residences studied in Compton, about 95% of them had less than 5% damage. These statistics should be cautiously used and in the proper frame of reference—if, e.g., repainting and patching of a plastered room is paid for by insurance or by government, then losses will be substantially higher than if repairs were made by the homeowner as is commonly the case.

Fig. 9.17. Apartment house in Long Beach. Long Beach Earthquake of 1933. (W. N. Ball Collection.)

Fig. 9.18. Compton, California. Long Beach Earthquake of 1933. (W. N. Ball Collection.)

Fig. 9.19. Damage from the Long Beach Earthquake of 1933. (W. N. Ball Collection.)

Fig. 9.20. Compton City Hall. Long Beach Earthquake of 1933. (H. M. Engle photo.)

Fig. 9.21. Jefferson Junior High School, Long Beach. Long Beach Earthquake of 1933. (*Los Angeles Times.*)

Fig. 9.22. Alexander Hamilton Junior High School, Long Beach. Long Beach Earthquake of 1933. (*Los Angeles Times.*)

Fig. 9.23. Roosevelt School, Long Beach. Long Beach Earthquake of 1933. (*Los Angeles Times.*)

Fig. 9.24. Polytechnic High School, Long Beach. Long Beach Earthquake of 1933. (H. M. Engle photo.)

Fig. 9.25. Wood frame dwelling damage in Compton. One of several hundred wood frame dwellings thrown off their foundations. Long Beach Earthquake of 1933. (H. M. Engle photo.)

Fig. 9.26. Wood frame dwelling failure at the foundation line. Long Beach Earthquake of 1933. (R. W. Binder Collection.)

Table 9.5. DAMAGE TO WOOD FRAME DWELLINGS IN COMPTON, CALIFORNIA*

Damage, %	Number of buildings	Fraction of total number, %
0–4	4,334	94.7
5–24	131	2.9
25–49	63	1.4
50 or more	36	0.8
Demolished	11	0.2
Total	4,575	100.0

*Data from *Earthquake Investigations in the Western United States, 1931–64*, in chapter by R. R. Martel, p. 221.

Multistory building damage was common in Long Beach and in Los Angeles. It usually consisted of cracked walls and partitions as well as damage to other nonstructural elements. The majority of the multistory steel and concrete frame buildings had brick or tile exterior panel walls and often interior tile partitions. Panels in exterior walls were generally loosened, but the walls were not shaken out except possibly in a few instances. Damage was always greater in the lower stories. Some observers held that in Long Beach the nonstructural damage was greater in tall, reinforced concrete frame buildings than steel frame buildings, with the reverse being true in Los Angeles. Pounding damage between multistory buildings was frequently noted. A comparison of the damage to selected multistory buildings experiencing the 1933 Long Beach and the 1952 Southern California (Kern County) earthquakes may be found in the report by Steinbrugge and Moran (1954).

It should be pointed out that the few examples of buildings known to have been specifically designed to resist strong seismic forces performed excellently. For example, the then-new Southern California Edison Building had no damage.

In summary, the damage patterns for nonresistive buildings were those to be expected and noted from previous earthquakes. Perhaps the most important result gained from the destruction was the requirement that all new buildings be earthquake resistive in southern California. Building ordinances established since that time have very materially reduced the earthquake damage potential for buildings constructed since 1933.

9.4.4.2 Water supply and fire hazard. Fire disaster did not follow the earthquake disaster, but the margin of safety was greatly reduced. The Long Beach Fire Department suffered severe damage to its fire stations and fire-alarm systems. The water mains had numerous breaks, with the great majority of the breaks occurring in sand, silt, or filled ground. The Long Beach Water Department reported 127 breaks in the distributing mains, not counting the places at which pipes were pulled apart. The water in the reservoirs fell dangerously low but never gave out. An interesting instance of tank failure is shown in Fig. 9.27. Disrupted gas services and broken

Fig. 9.27. Steel water tank located in Los Angeles. Note how top rings were torn off and thrown into the field. Long Beach Earthquake of 1933. (H. M. Engle photo.)

devices caused 7 of the 19 fires reported in Long Beach during the night of March 10. It is somewhat surprising to note that in Long Beach, having an estimated 46,000 gas consumers at the time of the shock, more fires did not occur since evening meals were being prepared at the time of the shock. None of the fires that occurred resulted in major losses. A detailed description of the performance of fire departments, damage to gas and water systems, and damage to the telephone system may be found in the *Report on the Southern California Earthquake of March 10, 1933* by the National Board of Fire Underwriters (1933).

Twenty-five elevated steel tanks on steel towers in use for municipal water supply were in the heavily shaken area. Two collapsed, two were out of service because of broken risers, one was on the verge of collapse, and practically all the rest had elongated or broken rods. Elevated steel tanks for private use, such as those for sprinkler supply, also performed poorly. On the other hand, the only known elevated steel tank designed to resist earthquake forces was a 75,000-gal structure near the Long Beach waterfront; it survived with no damage. The excellent performance of the earthquake resistive tank as compared to those not so designed was repeated in the 1940 Imperial Valley earthquake and in the 1952 Southern California (Kern County) shocks.

Of about 25 wooden tanks on wood towers, about one-third were destroyed.

9.4.5 Helena, Montana, Earthquakes

The Helena, Montana, area had not been considered to be a seismically active zone by Montana residents, and the shocks came as a distinct surprise. The series of earthquakes of October 12, 18 and 31, 1935, emphasized again that most of the western United States is seismically active. Typical damage is shown in Figs. 9.28, 9.29 and 9.30.

Four lives were lost, two on October 18 and two on October 31. Property damage estimates ranged from $3 to $4 million. Since there was no specifically designed earthquake resistive construction in Helena, the damage to unreinforced brick masonry with sand–lime mortar was readily understood (Engle, 1936). Extensive damage occurred to the new buildings, such as the new Helena High School, as well as to older buildings. Damage distribution was uneven, with the worst wreckage occurring on the alluvial soils toward the valley.

The cumulative damage during this series of earthquakes was quite apparent. This cumulative effect is not always well understood by the public who, in time, may consider a slightly loosened brick-bearing wall structure to be "safe" since it survived without collapsing.

9.4.6 Imperial Valley, California, Earthquake

The 1940 Imperial Valley earthquake of May 18, 1940, with its 40 mi of surface faulting, had a Richter magnitude of 7.1. The right lateral displacement on the Imperial fault reached almost 15 ft at the United States–Mexico border, and Richter reports one location with a 19 ft displacement. The maximum Modified Mercalli Intensity was X along the fault trace and IX in the most heavily shaken cities of Imperial and Brawley. A total of seven persons were killed by building collapse, one was burned to death, and one died several days later of injuries. At least 20 were seriously injured.

Of major engineering importance was the record obtained from the strong-motion instrument located in the city of El Centro in Imperial Valley. This accelerograph record has been of exceptional influence since it was a good recording of the ground motion in the epicentral region of a damaging earthquake. It is the most useful record obtained to date in the United States and is discussed in detail in other chapters. The intensity in El Centro was only VII to VIII.

Fig. 9.28. Failure of brick veneer on wood frame structure. Helena Earthquakes of 1935. (H. M. Engle photo.)

Fig. 9.29. New Helena High School. The portion shown collapsed in the shock of October 31. This part of the school had a reinforced concrete frame, roof, and floors, with tile walls faced with brick. Helena earthquakes of 1935. (H. M. Engle photo.)

Fig. 9.30. New Helena High School. Note lack of adequate ties. Helena earthquakes of 1935. (H. M. Engle photo.)

It is to be regretted that the engineering aspects of this earthquake have never been studied in detail. It would be of great value to be able to have some quantitative evaluation of damage patterns in El Centro that might be related to the spectrum of the earthquake record. One summary report, published by Ulrich (1941), describes the damage to nonresistive buildings as follows:

> *Damage at Imperial* (M. M. Intensity of IX). This community of 2,000 population was hit hard. It is estimated that 80 percent of the buildings were damaged to some degree. In the business district almost all buildings were damaged, many so seriously that they were condemned. Older residences suffered appreciable damage; new ones, very little. The city water tank collapsed. Apparently there was no great damage to the water mains and the sewage system in Imperial.
>
> *Damage at Brawley* (M. M. Intensity of IX). This city, with a population of 12,000 to 15,000, suffered more damage altogether, but probably not so much relatively as Imperial. In the business district all buildings were damaged, and it was estimated that 50 percent of them had to be condemned. This estimate may be a little high. . . . Estimates of damage to residences range from 25 to 75 percent. It is believed that the lower figure is more nearly the correct one. . . .
>
> There were a large number of breaks in the city water mains, and water was shut off in hundreds of homes because of broken plumbing. . . .
>
> *Damage at El Centro* (M. M. Intensity of VII to VIII). In El Centro a number of old brick buildings were so much damaged that they were condemned. Damage was chiefly to old brick construction, to walls that were not reinforced or tied into the structure, and to balconies projecting over sidewalks. There were a few breaks in the water mains, but practically no damage to the water plant or sewer lines.

Damage in Imperial and Brawley was greater than in El Centro, although the epicenter was closer to El Centro. The aftershock of 9:53 PM on May 18 caused more damage at Brawley than did the main shock; the reverse was true for El Centro. Figs. 9.31 through 9.34 show examples of damage.

The 1940 Imperial Valley earthquake was the first strong test of public schools designed to resist earthquake forces under California's Field Act, which was enacted after the 1933 Long Beach earthquake. A total of 15 "Field Act" public schools in the area had no apparent damage, including the several new reinforced concrete buildings at Brawley's Union High School. The few privately owned buildings known to be earthquake resistive also performed well.

Fig. 9.31. Hotel Dunlack in Brawley, side and rear elevations. Details of damage may be seen in next figure. El Centro Earthquake of 1940. (U.S. Coast and Geodetic Survey photo.)

Fig. 9.32. Detail of damage to north elevation of Dunlack Hotel. Note diagonal tension failures in shear walls and tendency for moment failures in more flexible elements. El Centro Earthquake of 1940. (H. M. Engle photo.)

At least eight elevated steel tanks on steel towers were in the strongly shaken area. Two of these tanks that were designed to resist heavy lateral forces were undamaged. Of the other six tanks that only had wind bracing, two collapsed and four were seriously damaged. Property damage estimates ranged between $5 and $6 million, plus indirect loss to crops due to damage to the irrigation systems.

9.4.7 Santa Barbara, California, Earthquake

The Santa Barbara earthquake of June 30, 1941, had its epicenter offshore in the Santa Barbara Channel. The maximum Modified Mercalli Intensity on shore was VIII. Unpublished information indicates that the monetary loss was well in excess of the $100,000 stated in published sources and probably approached $250,000 (Ulrich, 1941b). The shock was not large from a seismological standpoint or from the amount of damage, but the type of damage was of importance.

Two significant lessons were repeated in this earthquake. Unrepaired earthquake damage was cumulative from one earthquake to another, and a time interval of 16 years between damaging shocks did not alter this cumulative process. The second lesson was that "paint and plaster" structural repairs may allay public fears, but this type of repair does not make a building safe again. Summary records of damage have been published on three multistory buildings (Steinbrugge and Moran, 1954).

One of these three buildings was a six-story nonresistive, reinforced concrete frame building having tile panel walls; its damage summary is similar to those of the other two multistory buildings:

> *June 29, 1925, earthquake.* Exterior panel walls and interior partitions seriously shattered and cracked. Columns and girders fractured or cracked in 1st floor. Damage estimated at 50% of value of

Fig. 9.33. Typical damage to brick buildings in El Centro. El Centro Earthquake of 1940. (H. M. Engle photo.)

Fig. 9.34. Former two-story hotel in Brawley. El Centro Earthquake of 1940. (H. M. Engle photo.)

building. Exterior masonry veneer fell off street corner column to a height of over 2 stories. Shear cracks at main entrance and at window sills. Exterior panel cracks, as high as 4 stories on the north street elevation. Similar cracks on west alley wall. Estimated repair cost was $100,000.

February 18, 1926, earthquake. About $10,000 damage, location not specified.

June 29, 1926, earthquake. Diagonal crack in two corners. Possibly the columns were affected. All window corners cracked diagonally. Arch on east building entrance badly cracked. Partition cracks on all floors.

June 30, 1941, earthquake. Shear cracks at entrance and at window sill panels. Repairs estimated at over $50,000.

July 21, 1952, earthquake. Cold jointed masonry broke loose in many locations of the 1st story. Many shear cracks developed in the upper exterior walls. Repairs estimated at 22% of the value of the structure.

The above history has included the damage from the 1952 Southern California (Kern County) shock. This 1952 shock caused more damage in percentage terms to multistory buildings in Santa Barbara than it did in Bakersfield. Bakersfield was much closer to the epicenter of the 1952 shock than was Santa Barbara. The importance of cumulative earthquake damage is difficult to overemphasize.

In summary, the damage patterns were those of a moderate earthquake rather than those of a great earthquake. Therefore, strong motion instrumental results from this earthquake require an upward estimate in order to establish a reasonable design standard for a great earthquake.

9.4.8 Olympia, Washington, Earthquake

The April 13, 1949, earthquake had its instrumental epicenter between Tacoma and Olympia, Washington. The Richter magnitude has been given as 7.1. The maximum Modified Mercalli Intensity of VIII was mapped up to the unusual distance of about 85 mi from the epicenter. Loss of life has been placed at 8. Property damage has been estimated to be from $15 to $25 million in an area containing about 1 million population. The duration of the strongest shaking has been estimated at 20–30 sec. Figures 9.35 through 9.37 show interesting examples of damage.

There was very little construction that had been specifically designed to resist strong seismic forces (Edwards, 1951). The damage patterns were familiar. The buildings with brick-bearing walls with sand–lime mortar gave their usual poor performances. Public buildings, particularly public schools and the state capital buildings, suffered extensively. Pipeline damage was common. Fire was not a significant problem. Transformers shifted, transmission lines tangled, and telephone circuits became overloaded with calls. About 10,000 brick chimneys in the northwestern section of the state required repair. It was unusual to note that one free-standing, steel radio tower buckled.

Fig. 9.35. Damage to Pier 42 in Seattle. Olympia Earthquake of 1949. (*Seattle Times* photo.)

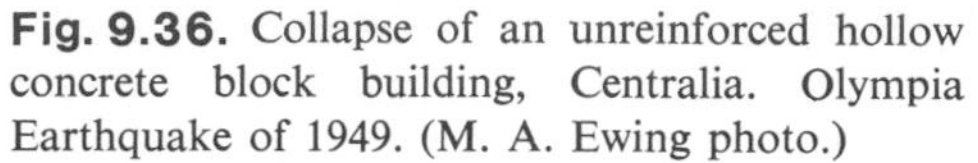

Fig. 9.36. Collapse of an unreinforced hollow concrete block building, Centralia. Olympia Earthquake of 1949. (M. A. Ewing photo.)

The 1949 Olympia earthquake provided an interesting example of a roof tank failure. A 54,000-gal wood stave tank on a 20 ft high reinforced concrete column and beam-type tower was located on top of a seven-story reinforced concrete refrigeration warehouse. The warehouse was located on the filled tideflat area near the mouth of the Duwamish River. In the 1946 earthquake, the top and the base of the reinforced concrete tank columns were cracked; repairs were made by cleaning out the breaks and filling them with gunite. The April 13, 1949, shock collapsed the tank structure, with the falling tank structure crashing through the concrete roof slab to the floor below. This resulted in the breaking of the refrigeration lines and the destroying of two elevators, which put the refrigeration warehouse out of service for weeks.

Fig. 9.37. Damage to radio station KJR, Seattle. Olympia Earthquake of 1949. (Associated Press photo.)

Probably the principal lesson learned from the 1949 earthquake was that the earthquake hazard truly exists in the Pacific Northwest. Building ordinances were strengthened as a result.

9.4.9 Kern County, Southern California, Earthquakes

The July 21, 1952, Kern County (Southern California) earthquake and its aftershocks are of particular engineering interest since they constituted the first major test in the United States of structures of earthquake resistive design now in common use. The engineering aspects have been studied and reported on by Steinbrugge and Moran; their data are the basis for this section.

The July 21 shock had a Richter magnitude of 7.7 and developed at least 14 mi of surface faulting. The time of its occurrence, 4:52 AM, PDT, undoubtedly was a factor in keeping the life loss to the relatively low figure of 12. An isoseismal map of this earthquake is shown in Fig. 9.38. The August 22 aftershock, with a Richter magnitude of 5.8, was not the largest earthquake on the magnitude scale in the aftershock sequence. However, its epicenter was located near Bakersfield, and the shock therefore caused extensive damage to many already weakened structures. Other aftershocks were quite intense in sparsely populated regions. For example, the January 12, 1954, aftershock caused slight additional damage to one earthquake resistive tank. Property damage was extensive for the series of earthquakes.

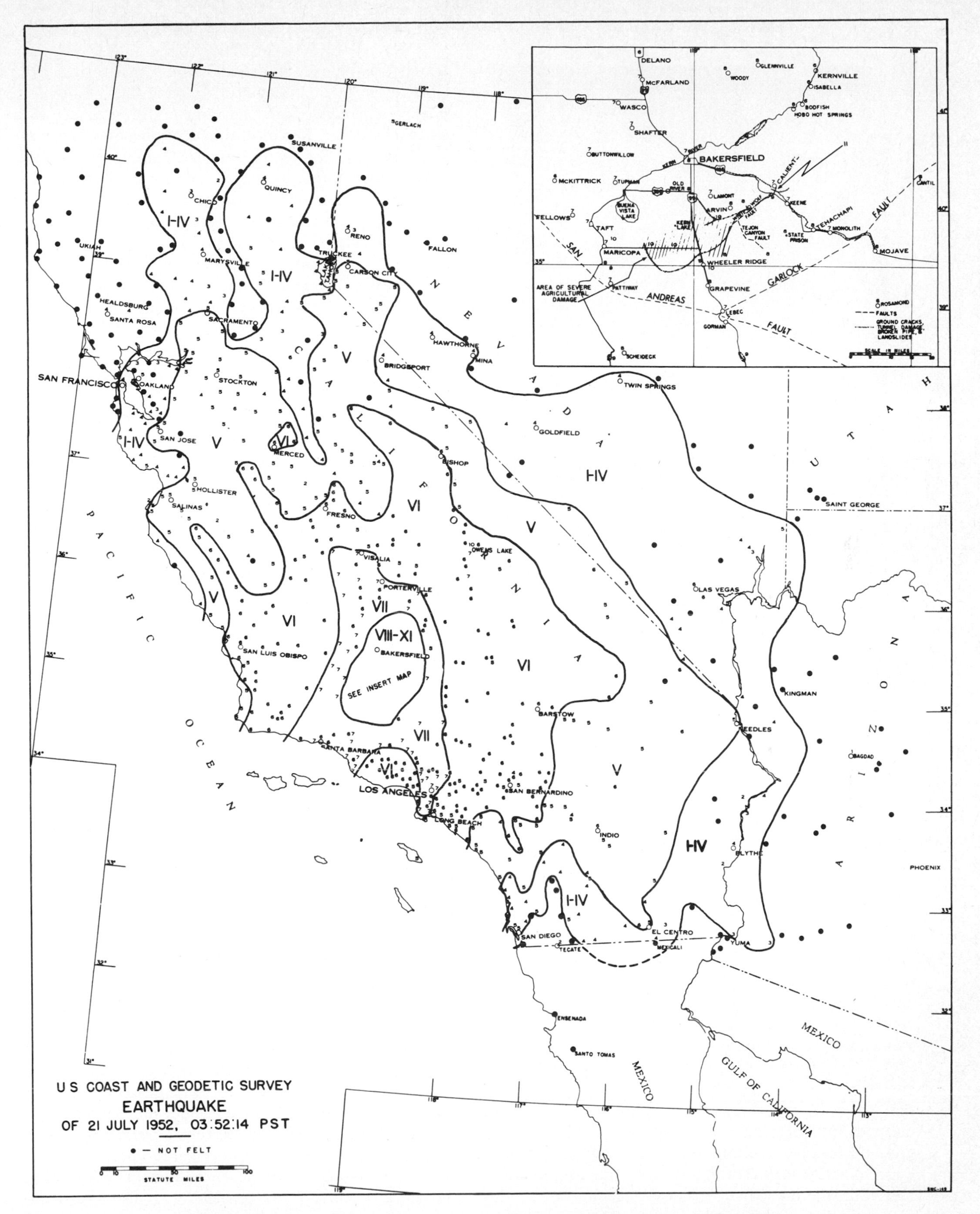

Fig. 9.38. Isoseismal map of the July 21, 1952, Kern County Earthquake. Prepared by the U.S. Coast and Geodetic Survey.

Fig. 9.39. Railroad tunnel walls have crushed together due to faulting. Scale may be determined from hard hat in photo. Kern County Earthquake of 1952. (Southern Pacific Railroad photo.)

Building damage was estimated to be in excess of $37 million, while other losses amounted to an estimated additional $11 million.

Damage in the White Wolf fault zone as a result of the surface ruptures was primarily confined to the Southern Pacific Railroad tunnels, since very few other man-made structures were in the fault zone. The tunnels were originally timber lined and later concrete lined without removing the timbers. Of the then total of 15 tunnels between Bakersfield and Tehachapi, the four seriously damaged ones were in the fault zone. Other tunnels had slight damage; nearby bridges suffered no damage. At the time of the shock, passenger and freight trains were in the vicinity, but none were in the damaged tunnel area. The ground motion was felt in the trains, and the engineers brought all safely to a halt. It is to be expected that generally very heavy loss, if not total loss, must occur to buildings, canals, tunnels, etc. that are astride ground breakage resulting from fault movement, as may be seen in Fig. 9.39. This need not be true for structures located in the immediate vicinity of the fault trace; this subject is discussed in more detail for the 1959 Hebgen Lake shock.

Damage effects were cumulative, and therefore the main shock plus all aftershocks usually are treated as a single event from an engineering standpoint. When it is possible and convenient, the damage will be related to a particular shock. For most practical purposes, the main shock of July 21 plus the August 22 aftershock are the only ones requiring attention. The towns of Tehachapi and Arvin were severely hit by the July 21 earthquake. The August 22 aftershock was particularly damaging in Bakersfield.

It is convenient to consider the damage by class of construction material, pointing out the comparative performance of earthquake resistive structures vs those not so designed. "Kern County" in the topic headings below relates to that portion of Kern County roughly bounded by Tehachapi, Bakersfield, and Grapevine.

9.4.9.1 Wood frame structures, Kern County. Wood frame construction was extremely common for dwellings. In some contrast to the 1933 Long Beach shock, rarely did dwellings of wood frame construction suffer more than plaster cracks and destruction of unreinforced brick chimneys. The rare wood frame dwelling that suffered significant structural damage invariably had obvious deficiencies such as no anchorage to its foundations, decayed wooden studs between the foundation and the first floor, or no lateral force bracing whatever. Whenever current acceptable construction practices were followed, damage was, at most, 5% of the structure's value.

9.4.9.2 Steel frame structures, Kern County. All steel structures, such as the conventional gasoline service stations, had negligible damage at most. These structures were of course highly resistant to seismic damage due to their small mass and the bracing required by wind forces (wind force design requirements almost always greatly exceed the earthquake design requirements).

Multistory steel frame buildings were not too numerous. The five-story Haberfelde Building in Bakersfield had considerable nonstructural damage, principally to the partitions and to the exterior facing. The damage amounted to about 7% of the estimated value of the building. About half of the monetary loss was related to the pounding that occurred between the original and the 1929 addition. While the structure had no specific lateral force bracing, it probably had no greater percentage of loss than the more heavily damaged multistory buildings in downtown Los Angeles located 70–80 mi away from the July 21 epicenter and about 100 mi from Bakersfield. This interesting observation will be discussed in more detail in later paragraphs.

The earthquake experience of the three-story Kern General Hospital in Bakersfield was also instructive. The 1938 addition was specifically designed to be highly earthquake resistant; it was primarily of reinforced concrete construction with only a minor amount of structural steel framing. The 1938 addition had negligible damage. The other units, built in 1924 and 1929, were so badly damaged that much of them had to be torn down. These nonresistive units had concrete floors on an interior steel frame; exterior walls were unreinforced brick with poor quality mortar. Structural damage was principally in the form of a separation between the unreinforced brick load-bearing walls and the concrete floors; in other words, the walls moved outwards. The wings had particularly heavy damage at their ends due in part to torsional effects. The difference in damage between earthquake resistive construction and nonresistive construction was extreme in this case.

9.4.9.3 Reinforced concrete structures, Kern County. Several multistory structures, such as the eight-story Hotel Padre in Bakersfield, had only minor damage (Fig. 9.40). Total damage to this hotel represented about 1% of the value of the building. The damage was in the form of cracking of nonstructural partitions, plus some slight movement along a few construction joints. The Hotel Padre was relatively rigid, and appreciable torsional forces must have existed in the first story due to an unbalanced shear wall arrangement. Concrete cores taken after the earthquakes indicated ultimate compressive stresses as low as 1100 lb/in.^{2}. It is interesting to note that there was evidence that the entire building rocked as a unit, with a measured single amplitude displacement of $\frac{1}{4}$ to $\frac{1}{2}$ in. found at the sidewalk level.

Fig. 9.40. Padre Hotel, Bakersfield. Earthquake damage to this reinforced concrete building was about 1% of the building value. Kern County Earthquake of 1952.

Fig. 9.41. Brock's Department Store, Bakersfield. Serious structural damage occurred to wall in alley—see next figure. Kern County Earthquake of 1952. (Walter Dickey photo.)

The extent that tall rigid buildings may rock on their foundations due to compressible soils is not yet fully understood.

Brock's Department Store, three stories and a basement in height, is of interest since it was the only monolithic, reinforced concrete structure in Bakersfield to experience serious structural damage (Fig. 9.41). It was constructed in about 1920 and in a period when earthquake resistive construction was not practiced. The building survived the July 21 shock with apparently negligible damage except for the roof tank supports. However, the August 22 aftershock caused serious structural damage to the south wall of the second story. The damage to this reinforced concrete wall was in the form of 45-deg diagonal tension cracks (Fig. 9.42). Examination showed improperly designed wall reinforcement, and core tests found the concrete quality to be as low as 1300 lb/in.^{2}. There was also some structural damage elsewhere, including one fractured interior reinforced concrete column. The cost of rehabilitation plus betterments approached 40% of the value of the structure. This building is a good example of the rather common condition that structural damage may be very serious but not spectacular up to the point of collapse; this condition has led untrained observers to erroneously report buildings as being but little damaged when they actually were on the verge of collapse.

The Monolith Portland Cement plant located 4 mi east of the badly damaged town of Tehachapi had numerous buildings and structures of reinforced concrete. Much of the plant was not specifically designed to resist large earthquake forces. A 180-ft reinforced concrete stack, already in a damaged condition prior to the July 21 shock, received additional cracking in the lower 15–20 ft of its height; but it did not fail and it remained plumb.

Fig. 9.42. Typical and characteristic diagonal tension failures in south wall of Brock's Department Store, Bakersfield. Kern County Earthquake of 1952.

The most severely damaged building was a 3-story structure of exceptionally poor reinforced concrete construction; damage consisted of wall slippages along construction joints that in one case amounted to $\frac{3}{4}$ in. Tall cement silos of reinforced concrete survived with minor damage at construction joints. Foundation conditions were not the best, as evidenced by piling under some of the buildings.

There were no cases of collapse or even near collapse of multistory buildings of poured-in-place reinforced concrete construction in Kern County, even when poorly designed and built.

The case history of one pair of precast reinforced concrete buildings is worth summarizing. The Lockheed plant in Bakersfield survived the July 21 shock, but the August 22 aftershock caused significant structural damage to one of the two buildings at this site. Both structures had precast reinforced concrete roof slabs on precast reinforced concrete beams and girders, in turn supported by precast reinforced concrete columns (Fig. 9.43). Exterior walls were also precast reinforced concrete. The design engineer considered the roof to be a rigid diaphragm taking lateral forces to the exterior walls, which in turn were to be used as shear walls. However, the joints between the precast concrete roof panels were filled with mastic instead of grout, and other significant deviations from the plans were noted. As a result of the inadequate roof diaphragm, the building swayed violently, breaking interior columns at the point of high bending moment at the floor line (Fig. 9.44) and leaving the structure out of plumb. Precast concrete requires careful attention to the joinery between precast elements,

Fig. 9.43. Lockheed Plant in Bakersfield is shown under construction. Shown are the precast concrete roof slabs resting on precast concrete beams, girders, and columns. Building was completed and occupied prior to earthquake. Kern County Earthquake of 1952.

Fig. 9.44. Lockheed Plant in Bakersfield. Bending stress damage to interior column at floor line from the August 22, 1952, aftershock. Damage due to excessive roof deflection caused by faulty diaphragm construction. Kern County Earthquake of 1952. (Lockheed Aircraft photo.)

and good field supervision by the design engineer is necessary to ensure the proper execution of a good design.

9.4.9.4 Reinforced concrete walls with various roof materials, Kern County. Numerous one-story structures of this type were found throughout Kern County. At most, minor or negligible damage was noted in buildings having specifically designed lateral force bracing systems. Roof bracing systems included metal X-bracing and diaphragms of plywood or diagonal sheathing. The performance of those buildings not having specific earthquake bracing was reasonably good on the average; admittedly, most of them were inherently strong due to small wall openings and small roof areas. A classic exception was the partial collapse of the poorly constructed Cummings Valley School built in about 1910. Elsewhere, some structures lost part of their roof systems when the roof had no strong wall ties. Damage was occasionally found along construction joints in the concrete walls, and evidence of laitance and foreign material in these damaged joints was very noticeable.

A number of one-story structures having precast reinforced concrete walls with poured-in-place reinforced concrete pilasters performed well. This was true even when the roof system was a poorly designed and constructed diaphragm. Certainly there was no evidence that precast concrete was not an acceptable material in the hands of competent designers and contractors using the best accepted methods of earthquake resistive construction.

9.4.9.5 Unreinforced brick walls with various roof materials, Kern County. This class of construction, with its sand-lime mortar, was common in the older sections of the cities of Tehachapi, Bakersfield, and Arvin (Figs. 9.45 through 9.47). Damage to this class of construction was severe, just as it has been in all previous major earthquakes. Destruction in Tehachapi was particularly severe. See Table 9.6 for pertinent damage statistics for Bakersfield. The performance of non-resistive brick masonry is of interest since it may be compared with the successful performance of masonry materials assembled in a manner to resist seismic forces (see the following paragraphs).

9.4.9.6 Reinforced brick walls with various roof materials, Kern County. Brick masonry walls built with a technique known as reinforced grouted brick masonry performed excellently. The building technique involves two wythes of brick laid in cement mortar. The wythes are separated by several inches, and the space is filled with small aggregate reinforced concrete.

The best-known example of reinforced grouted brick masonry construction is the Arvin High School in the city of Arvin. This school consisted of about 15 buildings constructed in the period 1949–1951. The three construction contracts, totaling $2.8 million, indicate the extent of the one- and two-story buildings at this site. Reinforced grouted brick masonry was used as the principal structural wall material on most major buildings. The design and construction were carried out under the requirements of California's Field Act. No more than minor or negligible damage was found in

Fig. 9.45. Downtown Tehachapi. Note the structural value of "nonstructural" partitions after collapse of bearing walls. Numerous wood stud partitions tend to reduce life hazard, although property damage may remain high. Kern County Earthquake of 1952.

Fig. 9.46. Two-story lodge hall in Tehachapi. Ceiling of second-story lodge hall now supported by "piano and chairs." Absence of partitions increases life hazard. Kern County Earthquake of 1952.

Fig. 9.47. Note cracks in the reentrant corners of the window, indicating in this case that the building has become slightly wider. Merely repointing the masonry joints would conceal damage until next shock. Kern County Earthquake of 1952.

Fig. 9.48. One of the structures at Arvin High School, Arvin. Damage occurred to the reinforced grouted brick wall over the roof of the one-story section—see arrow. See next three figures for details. Kern County Earthquake of 1952. (H. J. Degenkolb photo.)

Fig. 9.49. With respect to Fig. 9.48, photo taken from roof of one-story building looking at damage to second-story wall. Note that shear failures (i.e., X-cracks) go through brick as well as mortar. Kern County Earthquake of 1952.

most buildings, and none of it constituted a significant life hazard. However, the two-story administration building had significant damage to one $8\frac{1}{2}$ in. thick, reinforced grouted brick wall as a result of the July 21 earthquake. The damage to this second-story wall consisted of X-cracks from diagonal tension forces, plus separations at the building corners due to diaphragm deflections causing torsional stresses in the damaged wall. Subsequent aftershocks increased the damage. The damage and the construction details are shown in Figs. 9.48 through 9.51. While seriously damaged, collapse was not imminent. The principal workmanship error in the damaged wall was extensive voids in the grout, clearly suggesting that the grout was placed too dry.

Fig. 9.50. Cores cut from the damaged wall of the Arvin High School shown in the last two figures. Note that grout has voids and probably was placed much too dry. Kern County Earthquake of 1952. (Ralph Taylor photo.)

The overall damage to all buildings was less than 1% of their value. The performance of these buildings was a milestone in the development of a material previously associated with building collapse and large loss of life.

The few other known examples of buildings of earthquake resistive design using reinforced grouted brick masonry had no structural damage.

9.4.9.7 Hollow concrete block with various roof materials, Kern County. Hollow concrete block was a relatively new material insofar as its general use in Kern County. As a result, most of it contained reinforcing steel in selected grout-filled cells, and the buildings constructed of hollow concrete block normally had some degree of earthquake resistance. It was therefore not surprising to find that this material was rarely associated with serious damage. This is clearly evident in Table 9.6.

9.4.9.8 Public schools, Kern County. Public school performance in California is always of interest due to the Field Act. The performance of several schools has already been discussed. Table 9.7 is a summary, and it clearly shows the value of earthquake bracing.

9.4.9.9 Elevated steel water tanks, Kern County. Elevated steel water tanks with no lateral force design other than wind suffered badly, as they always have in severe shocks. Of the 12 wind-designed tanks in the strongly shaken area, 2 collapsed, 7 had broken

Fig. 9.51. Construction details of damaged wall shown in Figs. 9.48 through 9.50. Kern County Earthquake of 1952.

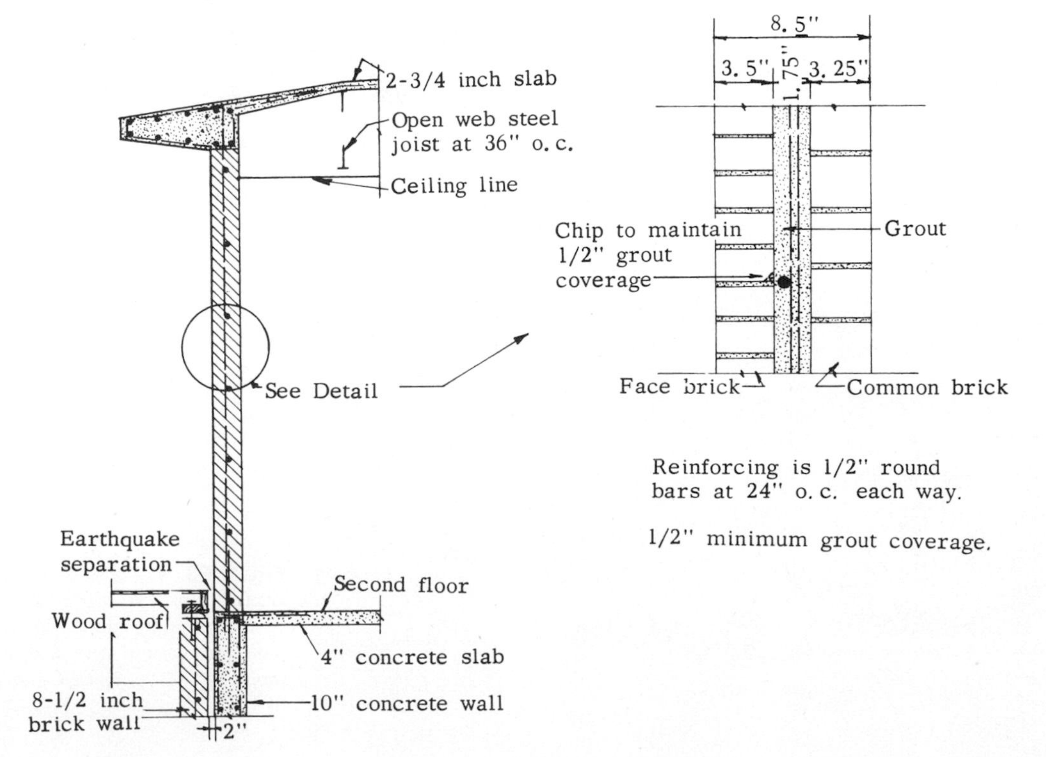

Table 9.6. EARTHQUAKE DAMAGE IN BAKERSFIELD; FLOOR AREAS OF STRUCTURES WITH MASONRY WALLS, WOOD FLOORS, AND WOOD ROOFS

Wall*	Torn down, %	Repaired, %	Repair or demolition undecided, %	Undamaged, %	Total, %
Brick	16	42	20	22	100 (2,717,410 ft²)
Concrete brick	20	40	36	4	100 (230,950 ft²)
Concrete	6	12	6	76	100 (1,186,680 ft²)
Hollow concrete block	2	6	Negligible	92	100 (488,525 ft²)

*Brick and concrete brick usually were old structures not specifically designed to be earthquake resistive. Hollow concrete block were usually new structures specifically designed to be earthquake resistive. Tabulation made July 1953.

Fig. 9.52. Grout had never been placed in the cell of this fallen hollow concrete block parapet. Kern County Earthquake of 1952. (George Simonds photo.)

Table 9.7. COMPARISON OF DAMAGE ON JULY 21 TO EARTHQUAKE RESISTIVE AND NONRESISTIVE MASONRY PUBLIC SCHOOLS OF KERN COUNTY WEST OF MOJAVE

	Number of schools damaged	
Extent of damage	Earthquake resistive	Earthquake nonresistive
None	21	1
Slight	6	9
Moderate	1*	9
Severe	0	13
Collapse	0	1

*Administration Building, Arvin High School.

or stretched rods, and 3 were not damaged in the July 21, 1952 shock. No additional damage was found after the August 22 shock.

The characteristic position of a collapsed tank is approximately inverted and usually within the area bounded by the foundations. Collapse usually begins with the failure of a rod X-brace, creating severe torsional

Fig. 9.53. Collapse of a 100,000-gal steel riveted tank on 100 ft steel tower. Note that tank is upside down and within the tank foundation area. See next figure for mode of failure. Kern County Earthquake of 1952.

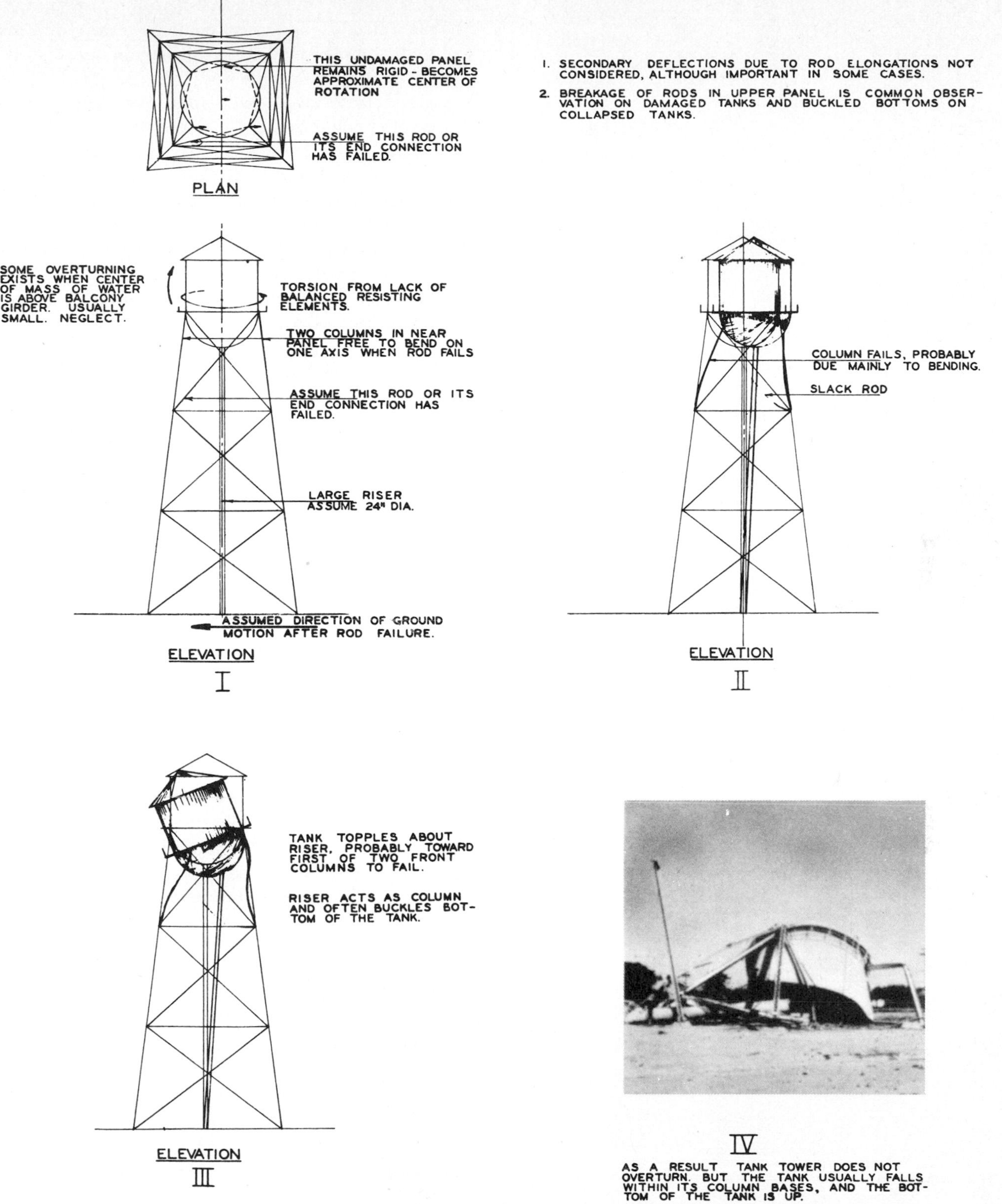

Fig. 9.54. Failure mode of many elevated tanks.

forces and thereby twisting the tank until eccentricity of the column load causes column failure. The tank then collapses vertically upon its standpipe, which acts as the equivalent of a single column support which, of course, is unstable. Inversion of the tank results from the unstable condition of a single support; see Figs. 9.53 and 9.54.

Performance of earthquake resistive tanks was superior to that of wind-designed tanks. Some cases of damage

Fig. 9.55. Damage to these tall processing towers at the Paloma L.P.G. plant occurred to their anchor bolts as shown in next figure. Kern County Earthquake of 1952. (D. F. Moran photo.)

were noted, and important lessons can be learned from these. Probably the most interesting of the earthquake resistive tanks was the one at Maricopa Seed Farms. There was evidence that this structure was in one of the highest intensity areas for the main shock and that the area was also severely shaken by several aftershocks. This 100,000-gal tank on a 100-ft tower was supported by four welded tubular steel columns resting on piling. The tank was designed for a horizontal force of 10% gravity with an allowable rod stress of 18,000 lb/in.^{2} with no increase for lateral forces. The X-bracing rods between the columns stretched, the reinforced concrete bases were damaged, and anchor bolts became elongated. Other signs of high stress were not as spectacular. Calculations indicated that the actual lateral forces were in terms of 20% of gravity or more in order to cause the observed rod stretching and anchor bolt elongations.

9.4.9.10 Oil refineries and related plants, Kern County. Extensive oil fields exist around Bakersfield. Plants for pumping oil from the fields to the refineries were largely undamaged. Underground pipelines were broken in a few instances in poor ground locations, and this damage was usually noted in corroded or otherwise weakened pipe. Ground level, steel, cylindrical oil-storage tanks at pumping plants, refineries, and elsewhere were occasionally damaged at their tops due to sloshing oil. Understandably, floating roofs suffered more damage than fixed roofs. An important factor in the design of piping entering these large diameter ground level tanks is that the tank shell will deflect vertically as well as horizontally due to the earthquake induced oscillations of the contents. Tank shells were found to have deflected vertically as much as 1 in. Obviously, any piping entering the sides of a cylindrical tank is subject to breakage if it is not sufficiently flexible for these horizontal and vertical motions.

Overall earthquake damage at the Paloma Cycling Plant was slight, but an ensuing explosion and fire caused extensive damage (Figs. 9.55, 9.56, and 9.57). The July 21 earthquake broke some high pressure process piping and an incoming field gas line. Apparently the supports of two out of five butane spheres also collapsed,

Fig. 9.56. Anchor bolt damage to processing towers shown in last figure. Kern County Earthquake of 1952. (D. F. Moran photo.)

Fig. 9.57. Fire following earthquake at Paloma L.P.G. plant. Kern County Earthquake of 1952. (R. L. Johnston photo.)

and the heavier-than-air gas from the two formerly full (2500-barrel) spheres spread throughout the plant. In about 3 min the gases ignited, and the resulting explosion set fire to the plant. This plant had been designed to be earthquake resistive, with the apparent exception of the butane spheres and several other minor items. Of considerable engineering interest was the stretch of anchor bolts of the processing vessels that were up to $112\frac{1}{2}$ ft high. Anchor bolts stretched up to $1\frac{9}{16}$ in. on one vessel that was 52 ft high by 5 ft in diameter. These vessels have been the subject of considerable study, and references may be found in Steinbrugge and Moran (1954).

9.4.9.11 Damage in Los Angeles and Long Beach. Damage in Los Angeles as a result of the July 21, 1952, shock was generally confined to steel and concrete frame fire-resistive structures over five or six stories high. A few isolated instances of minor damage to one- and two-story buildings were noted, but they were not significant. This pattern of damage was opposite to that experienced in Kern County on July 21 and in Bakersfield on August 22, 1952, in that there the one- and two-story nonreinforced brick-bearing wall buildings were much more affected than the multistory steel and reinforced concrete frame buildings.

The explanation for this difference is that the short-period ground motions die out more rapidly with distance than do long-period motions. Additionally, long-period ground motion tends to adversely affect taller buildings that have longer natural periods than those of low rigid buildings. Therefore, the ground motions in Los Angeles (70–80 mi from the epicenter) were such as to excite vibrations of crack-producing magnitudes in tall structures while not affecting the lower, more rigid buildings. A contributing factor was the previous damage to these tall buildings in past shocks, particularly the Long Beach shock of 1933, since effective repairs had generally never been made (Steinbrugge and Moran, 1954). No cases of structural damage were noted, and principal damage was to partitions, masonry filler walls, ceilings, marble trim, veneer, and exterior facing. It should be added that the buildings under discussion are the older ones without adequate earthquake bracing. The newer earthquake resistive structures behaved well with the exception of one relatively flexible design that suffered damages of $150,000 to interior partitions and trim.

Behavior of tall buildings in Long Beach was similar to that in Los Angeles. However, it is disquieting to note rather extensive nonstructural damage to major structures in some cases, when one considers that the structures were located about 100 mi south of the epicenter. In the 1933 Long Beach shock these buildings, in general, suffered more extensive damage than those in Los Angeles, and the methods of repair were often equally ineffective.

The occurrence of long-period ground motion at long distances from an earthquake's epicenter has been well known, but building damage therefrom was not well recognized until the 1952 earthquakes. Since then, other examples of quasi-resonance have been clearly identified. The full significance of quasi-resonance was vividly displayed by the serious structural damage to multistory buildings (with a few collapses) in the Mexican earthquake of July 28, 1957. The shock's epicenter has been reported as 170–220 mi from Mexico City by various authorities. The period of the ground motion in Mexico City was about 1.5–2.0 sec, and this adversely affected long-period structures such as tall buildings (Steinbrugge and Bush, 1960). Another factor in the intensified damage at Mexico City was the poor foundation conditions. The fact that the collapsed as well as the seriously damaged multistory structures were usually quite weak or of defective design should not obscure the observation that the ground motion did not damage weak and poorly designed, low, rigid buildings in the same proportion as tall structures.

9.4.10 Churchill County, Nevada, Earthquakes

These 1954 earthquakes in Churchill County, Nevada, form an interesting sequence from the structural damage standpoint (Steinbrugge and Moran, 1956 and 1957). Table 9.8 compares some of the data regarding these shocks.

Table 9.8. CHURCHILL COUNTY, NEVADA, EARTHQUAKES OF 1954

Date	Magnitude	Surface faulting, approximate mi	Distance east of Fallon, approximate mi
July 6, 1954	6.6	11	15
August 23, 1954	6.8	19	15
December 16, 1954:			
3:07 AM	7.1	35	30
3:11 AM	6.8	30	30

The July 6 shock caused damage in the city of Fallon. However, it is interesting to note that the August 23 earthquake was the same distance from Fallon, had a slightly larger magnitude, and had longer surface faulting; surprisingly, it caused somewhat less damage in Fallon despite the existence of buildings loosened by the previous earthquake. This anomaly has not been satisfactorily explained. The damage pattern in Fallon from these two earthquakes was that which might be expected from unit masonry buildings not specifically designed to resist strong earthquakes. Extensive ground breakage in irrigated fields and damage to the irrigation system around

Fig. 9.58. No damage occurred to this wood frame structure located in the fault zone. Dixie Valley (Churchill County) Earthquake of December 1954.

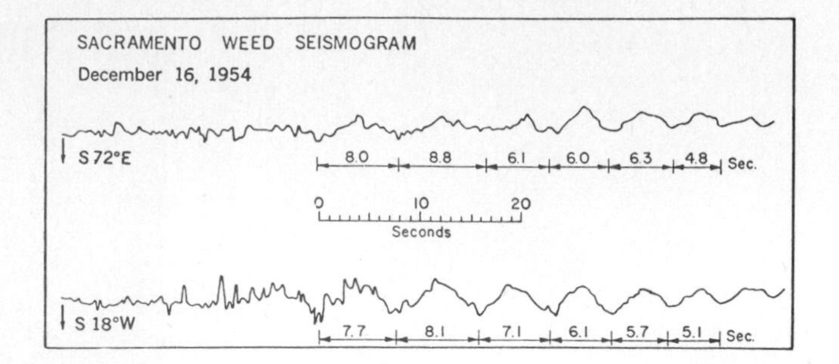

Fig. 9.59. Reproduction of Weed seismogram from instrument located at Sacramento, California, of the Dixie Valley (Churchill County) Earthquake of December 1954. Note predominance of long-period motion.

Fallon and Stillwater were due to the poor structural qualities of the soil, which had been swampland or nearly so prior to development.

As compared with the two earlier shocks, the first of the two December 16, 1954, Dixie Valley–Fairview Peak earthquakes was of somewhat greater magnitude but was twice as far from Fallon. The second of the Dixie Valley–Fairview Peak earthquakes was of the same magnitude as the August 23 shock and about twice as far from Fallon as the July 6 and August 23 shocks. Spectacular surface faulting was found along Fairview Peak and in Dixie Valley. However, Fallon was far enough away so that damage was confined to only a further loosening of the masonry in the brick buildings. The very few stone and hollow concrete block buildings located near the surface faulting also survived without damage, although they were lacking in all forms of customary earthquake bracing. One wood frame structure in the fault zone survived quite well (Fig. 9.58).

It is of engineering significance that structural damage from the December 16 earthquakes occurred in Sacramento, California, some 185 air-miles away. Thus, the seismic waves, which originated with the earthquakes, passed beneath Fallon, Reno, and Carson City without damaging them only to damage tanks and reservoirs in Sacramento (Steinbrugge and Moran, 1957b). Fortunately, a strong-motion seismograph in Sacramento recorded the ground motion (Fig. 9.59). The record clearly shows a number of cycles of ground motion ranging from about 5 to 8 sec. This caused quasi-resonance to the liquids in tanks at three different locations in Sacramento and in one case destroyed reinforced concrete walls (Fig. 9.60). The computed natural period of the fluids in the tanks and in the reservoir fell in the same range as the period of the ground motion, and the quasi-resonance that occurred caused large-amplitude water waves. The 5-to-8-sec ground-motion period was far too long to damage the nonresistive structures in Sacramento. As in the case of the selective damage to multistory buildings in Los Angeles and Long Beach in the 1952 shocks and in Mexico City in 1957, the dynamic prop-

Fig. 9.60. Damage to concrete walls and gunite panels in reservoir in Sacramento due to sloshing water. Dixie Valley (Churchill County) Earthquake of December 1954.

erties of a structure (or its contents) responding to long-period motions may create damage patterns reversed from those found in epicentral regions.

9.4.11 Eureka, California, Earthquake

The Eureka earthquake of December 21, 1954, had a magnitude comparable to that of the July and August 1954 earthquakes in Fallon, Nevada. Several nonresistive brick buildings in Eureka were seriously damaged, notably the Humboldt County Court House, which had been damaged in a previous shock.

The number of pipeline breaks, both gas and water, is of interest since the earthquake was of relatively low intensity. Many of these breaks could be associated with former swamplands, former stream channels, and filled-in areas. This experience, of course, has been common in many other larger earthquakes, notably that of 1906 in San Francisco.

Referring to water systems in general, there is a need for adequate valving of the principal distribution mains in order to bypass potentially hazardous poor ground areas in the event of a major shock or for taking other suitable precautions. Unfortunately, many cities in earthquake-prone areas have not planned their underground systems with earthquake damage in mind.

9.4.12 Port Hueneme, California, Earthquake

The Port Hueneme earthquake of March 18, 1957, with a small Richter magnitude of 4.7, caused exceptional damage for a shock of such low magnitude (Housner and Hudson, 1958). The maximum ground motion as recorded by a Coast and Geodetic Survey accelerometer was 18% of gravity. An analysis of the strong-motion record showed the earthquake to be essentially a single pulse of energy. This type of single-pulse energy release appears to be unusual from a damage standpoint in the United States and is quite the opposite to the more commonly known large-magnitude earthquakes having a destructive duration of perhaps 1–3 min or more. The Agadir, Morocco, earthquake of February 29, 1960, which caused severe destruction to a poorly built city, also appeared to have its energy release in a single primary pluse.

The full engineering implications of an eathquake releasing its energy in a single pulse are not well understood, and the subject obviously needs more study.

9.4.13 San Francisco, California, Earthquake

The instrumental epicenter of this 5.3 Richter magnitude earthquake on March 22, 1957, was in the San Andreas rift zone near Mussel Rock. Mussel Rock is on the coast of the Pacific Ocean about 3 mi south of the city limit boundary of San Francisco and Daly City. This earthquake received considerable newspaper publicity due to its proximity to San Francisco. Since there were large property values within a radius of 10 mi from the epicenter, the generally slight damage to buildings and structures of all types amounted to about $1 million. Although this shock was a relatively weak earthquake from both the magnitude and intensity standpoints, there was interesting damage that has been reported on by Steinbrugge, Bush, and Zacher (1959).

Building damage, including damage to numerous wood frame dwellings, was usually considerably less than 5% of the building value. Highest earthquake intensities occurred in areas having new construction, and this construction was largely earthquake resistant.

It was of interest that the greatest damage to dwellings was concentrated in one housing tract in Daly City. At the time of the earthquake, this housing tract contained more than 550 wood frame dwellings, about 90% of which were two stories high. Almost all had some damage, with perhaps 50% having pronounced plaster cracking in the front wall. This damage generally did not exceed $200 per dwelling. An analysis of the structural design did not account for the increased damage to these wood frame dwellings over those located in nearby tracts of almost identical construction. Damage to this housing tract could not be correlated with the deep cuts and fills made at the time the tract was started or with the soil types that were reasonably uniform over both damaged and undamaged housing tracts.

The ground surface in the damage area indicated compressional and tensional effects. It has been suggested that the ground movements were a form of lurch movements since the tract was located on a bluff overlooking the Pacific Ocean, the highest point being some 550 ft above sea level. This damage indicates the need for increased engineering geology studies with particular reference to seismic problems. This is becoming more important with the increased use of topographically difficult areas.

Minor damage in the form of partition cracking occurred in several multistory buildings in the financial center of San Francisco some 10 mi away. Interestingly, none of the nearby rigid one- and two-story brick buildings were reported damaged, although they were not earthquake resistive and many had pre-earthquake differential settlement cracks due to poor foundation material.

9.4.14 Hebgen Lake, Montana, Earthquake

An extensive and complex fault scarp system was formed during the Hebgen Lake earthquake of August 17, 1959, which had a Richter magnitude of 7.1. A landslide, containing about 43 million cu yd of material, dammed Madison Canyon and caused a lake with a peak storage capacity of about 80,000 acre-ft. An estimated 19 persons were buried by the Madison Canyon slide. The engineering aspects of this earthquake have been reported on by Steinbrugge and Cloud (1962).

Bedrock beneath Hebgen Lake warped and rotated, causing a seiche in the lake. The surface of Hebgen Lake, which contained 324,000 acre-ft at the time of the earthquake, dropped slightly more than 10 ft due to the changes in bedrock levels. Precise relevel measurements found a maximum drop of a bench mark near the north shore to be 18.8 ft. Figures 9.61 and 9.62 show some of the geologic effects.

Hebgen Dam, an earth-fill structure with an unreinforced concrete corewall located on the Madison River at its entrance to Madison Canyon, suffered significant earthquake damage. The dam was constructed in the period 1909–1914. The concrete corewall had a maximum height of about 113 ft; it generally extended from the bedrock to the top of the dam, except that 20% of its length rested on earth. A major vertical fault scarp formed during the earthquake was less than 1000 ft from the spillway. A resurvey indicates that the dam dropped about 9 ft, and this drop was related to tectonic movements. Horizontal compressional effects in the dam were exemplified by an S-curve in the corewall. Vertical compressional effects in the dam were indicated by the slumping of the earth along both sides of the corewall (Fig. 9.63). The major damage to the dam occurred where man-made fill was over old slide material and where the corewall did not extend to bedrock. The seiche in the lake, which overtopped the dam four times, caused little damage. All factors considered, the dam performed quite well.

Fig. 9.61. Entire north shore of Hebgen Lake dropped due to tectonic movements. Therefore, all buildings and the dam on the north shore also dropped during the earthquake in addition to being shaken. These vertical movements amounted to 18 ft in some places. The fault scarp was located several hundred yards left of this road. Hebgen Lake Earthquake of 1959.

Fig. 9.62. Earthquake scarp formed during Hebgen Lake Earthquake of 1959. A scarp of varying heights was near the dam and buildings shown in Figs. 9.63 through 9.66.

Building damage in the epicentral region (which included the town of West Yellowstone) was singularly unspectacular insofar as vibratory forces were concerned. The wooden buildings can be considered as earthquake resistant, except for their unreinforced masonry chimneys and for the apparent fact that these structures were rarely anchored to their foundations. Chimney damage was general but far from universal, and few wooden structures left their foundations. Hollow concrete block buildings did not perform as well as did wooden structures. Again damage was remarkably small considering the usually complete lack of construction features held mandatory in earthquake resistant design. However, the instances of masonry veneer failure on wood walls would substantiate present conservative design methods with respect to veneer anchorage.

Vibratory intensity of the earthquake along the fault scarp was not severe with respect to buildings. Based only on building damage from vibratory forces, and thereby excluding the subjective (or human) factors and also excluding geologic effects, the Modified Mercalli Intensity was VIII in some instances and VII rather generally. This was despite the fact that some structures dropped vertically as much as 19 ft, in addition to being subjected to horizontal vibratory forces. Figures 9.64, 9.65, and 9.66 are examples of damage near or on the fault.

The intensity in the town of West Yellowstone, as measured only by building damage, was not significantly different from that near the trace of the fault, considering comparable soils and comparable construction. West Yellowstone is about 8 mi from the nearest major surface faulting. This is not to state that the ground motions were the same throughout the epicentral region. For example, the large changes in elevation along the north shore of Hebgen Lake were not experienced in West Yellowstone. Figure 9.67 shows veneer damage at West Yellowstone.

It is not reasonable to conclude that all earthquakes will show similar damage patterns for similar earthquake magnitudes and similar geologic phenomena. The relation to damage of factors such as focal depth, epicenter location, center of energy release, duration of motion, predominant motion (i.e., horizontal or vertical), predominant period of ground motion, area over which the energy is released, and modifying effects of local geology is not yet fully understood. For example, pure vertical motion would allow many unit masonry buildings to escape major damage, although horizontal motion would damage them severely; this is one possible explanation for the apparent lack of high intensities with respect to nonresistive unit masonry buildings around Hebgen Lake.

Fig. 9.63. Crest of Hebgen Dam. Concrete corewall originally below crest, but emerged as shown when earth settled during the earthquake. Arrows indicate location of fault scarp. Hebgen Lake Earthquake of 1959.

Fig. 9.64. Hebgen Lake Lodge. Fault scarp in center left at arrow. Buildings were undamaged, including the chimneys. Nearby bench mark showed a drop of 18.45 ft. Hebgen Lake Earthquake of 1959.

Fig. 9.65. Fault scarp at right continues through this building at Culligan's Blarneystone Ranch. Building at top of scarp in total collapse. Hebgen Lake Earthquake of 1959.

Fig. 9.66. Failure of unanchored brick veneer. Fault near treeline in rear. Hebgen Lake Earthquake of 1959.

Fig. 9.67. Failure of brick veneer from walls of a flexible wood frame gymnasium in West Yellowstone. Hebgen Lake Earthquake of 1959. (B. J. Morrill photo.)

Fig. 9.68. Air view of downtown Anchorage after the March 27, 1964, Alaskan Earthquake. A: Mt. McKinley Building; B: Anchorage–Westward Hotel; C: Hill Building; D: Cordova Building; and E: Penney store.

9.4.15 Alaskan Earthquake

The Alaskan earthquake of March 27, 1964, at 5:36 PM, with its large Richter magnitude of 8.4, is of major importance in the field of earthquake engineering. The structural engineering aspects have been discussed in detail in *The Prince William Sound, Alaska, Earthquake of 1964 and Aftershocks* (vol. II, Part A), and it forms the basis for this section.

Of particular interest is the damage that occurred to some of the many relatively new buildings in Anchorage that were designed to be earthquake resistive. Also of considerable importance is the extensive landsliding that caused building damage in Anchorage due to liquefaction of the supporting soils. Figure 9.68 is an air view of Anchorage after the earthquake that shows that the city was not destroyed.

The epicenter of the earthquake was about 75 mi in an easterly direction from the city of Anchorage and in the Prince William Sound area. Length of faulting, derived from seismic data, has been given as being from 400 to 500 mi. The faulting extended from about the epicenter in Prince William Sound to south of Kodiak Island. Figure 9.69 shows the epicenter of the main shock, plus stronger aftershocks; the fault trace would be under water for most of its length and its position would be roughly centered among the aftershocks as plotted in Fig. 9.69.

Property damage in Alaska has been officially estimated at $311,192,000; this figure includes tsunami damage. Of this sum, $77 million was private property, with highways, railroads, docks, public buildings, and military facilities constituting the greater share. For the city of Anchorage, private property loss amounted to $67 million, while municipal property loss was $19 million.

Life loss has been placed at 125 known or presumed dead. Considering the Richter magnitude of the earthquake and the large amount of property damage, life loss was low. Excluding the lives lost from the seismic sea wave, the death toll of 15 persons from landslide and from vibration damage to buildings was remarkably low and significant. This figure was low despite the fact that many persons were in masonry mercantile stores at the time of the earthquake. Except for schools being closed, the time of day tended to be an unfavorable factor for life safety. There are a number of reasons for this low death figure, and not the least of these was the earthquake bracing found to some degree in practically all of the substantial buildings. One exception to the general rule that buildings were intended to be earthquake resistive may be seen in Fig. 9.70.

The earthquake has been extensively studied from the standpoints of geology, seismology, and soils engineering, as well as structural engineering, and the available material is profuse. The bibliography lists the principal published reports. In order to restrict this chapter to reasonable limits, the balance of the discussion on this earthquake will be limited to the city of Anchorage and

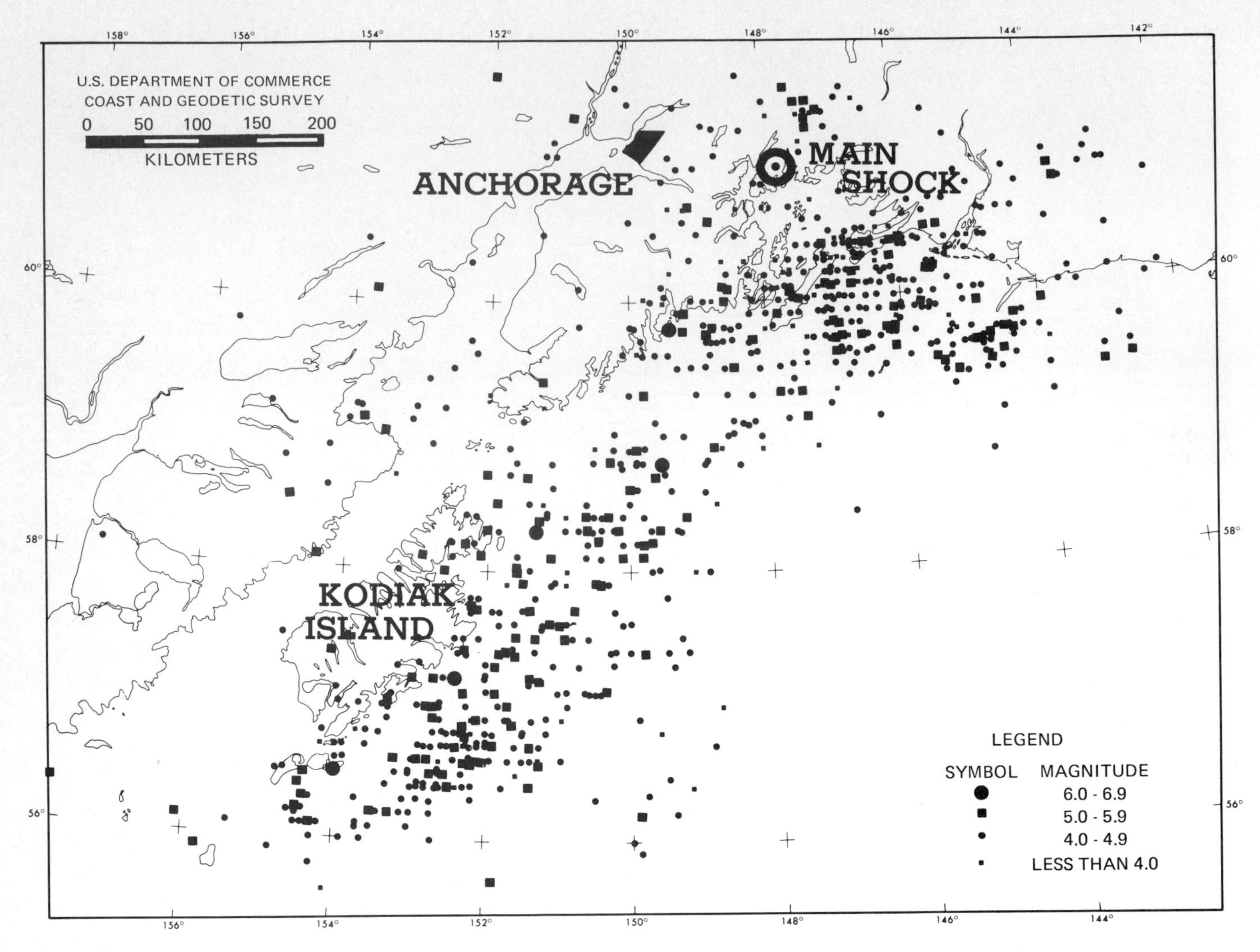

Fig. 9.69. Epicenters of main shock and aftershocks through December 31, 1964. Alaskan Earthquake of 1964. (U. S. Coast and Geodetic Survey diagram.)

Fig. 9.70. Hillside Manor Apartment House in Anchorage was not designed to be earthquake resistive. Walls were nonbearing and nonreinforced hollow concrete block. Alaskan Earthquake of 1964. (Frank McClure photo.)

Fig. 9.71. Interior of a music store across the street from a major landslide in Anchorage. Stock on the shelves was little disturbed, and only several items fell. This represents one instance of the lower limits of vibrational intensity in Anchorage. Alaskan Earthquake of 1964.

its suburbs, with the principal emphasis being placed on vibration damage. This is not to say that the landslide investigations were not important (they were, indeed—see Chapter 10), but that vibration damage occurs in all large earthquakes while landslides need not be of engineering concern in many earthquakes.

It has already been pointed out that the duration of the damaging intensity is important. Unfortunately, there were no strong-motion seismic instruments in Anchorage to give this information or data on the spectrum. Based on the reactions of persons, including a tape recording of the event, the duration of forces of 2 % g or more was perhaps as long as 3 min. Persons timed the duration of the felt motions, and their reports ranged from $4\frac{1}{2}$ to 7 min. Seismic data, including the length of faulting and its rupture rate, allow calculations that are consistent with a 3-min duration of damaging intensity. The 1906 San Francisco earthquake, which is considered by many to approach a design earthquake, had perhaps a time duration of one-third that of Anchorage. The longer the earthquake's duration of damaging intensity, the greater the damage. Many repeated excursions into the yield range will eventually bring destruction to steel. Hairline shear cracks in reinforced concrete become larger with an extended duration. Observers' accounts of the performance of buildings often stated that the collapse or partial collapse occurred in the latter stages of this long duration earthquake.

The horizontal accelerations have been estimated as being not over, and probably less than, 16 % g. Vertical accelerations probably were substantially smaller.

The ground motion in Anchorage did not contain the significant short-period motion that has been commonly observed in epicentral regions of destructive shocks. The evidence for this is in the numerous instances of small, rigid masonry buildings that were undamaged, although such buildings lacked recognizable earthquake bracing in some cases. Merchandise on shelves, in some instances, was practically undisturbed. China often did not fall from shelves in dwellings (Fig. 9.71). On the other hand, tall buildings in Anchorage were significantly damaged in the majority of cases. The foregoing effects are consistent with observation made in other earthquakes where the epicenters were located at some distance from buildings. The predominant period range of ground motions in Anchorage was in the order of $\frac{1}{2}$ sec and longer. Instrumental records of strong aftershocks with epicenters near that of the main shock give similar results.

Practically all of the major buildings, and most of the smaller buildings, were built since 1950 when the first building code was adopted. The building code has always included seismic provisions, with Zone 2 being in effect until about 1956 and Zone 3 thereafter. This means that earthquake resistance was legally required for practically all construction in Anchorage at the time of the earthquake; Fig. 9.70 shows one notable exception.

Most of the spectacular damage was clearly related to land movements that caused foundation failures. Land movements and landslides encompassing many city blocks caused severe to total damage to large numbers of structures. Another cause of damage associated with earth movements was the numerous ground cracks and fissures often parallel to the slide areas and grabens. In the downtown area, e.g., cracks were found in streets and sidewalks, buildings were separated from sidewalks, and some buildings had cracks best explained by tensional forces in the ground. Thus, vibratory effects on many of the buildings were intermixed with ground displacements.

A qualification to any broad classification of earthquake damage is workmanship. As compared to best-known practice, workmanship was almost invariably poor in unit masonry structures wherever damage was noted. Similar statements can be made for poured-in-

Fig. 9.72. These wood frame dwellings and their chimneys remained standing in Anchorage. The snowman also survived the earthquake. Alaskan Earthquake of 1964.

place reinforced concrete, precast concrete, and mixed construction (such as connections of wood roofs to masonry walls).

Following is a description of building damage by construction materials—a method that is convenient rather than being an effort to indicate superiority:

Wood frame dwellings and similar small wooden structures performed excellently as a class of construction when located in an area not subjected to land movement (Fig. 9.72). Also, in many cases, wood frame dwellings behaved remarkably well in areas that were subject to land movement. Unreinforced masonry chimneys, usually of hollow concrete block, were often intact. Interior wood stud partitions, with gypsumboard as well as other materials, usually suffered no more than minor cracking. The dwellings had various types of foundations: concrete (apparently not always reinforced), hollow concrete block (apparently not always reinforced), and wood sleepers. Observers generally found no significant positive anchorage between the wood framing of a structure and its masonry foundation. Foundations, including basement walls, were usually intact where no land movements occurred. Dishes, books, and similar objects to a large degree remained on the shelves although they shifted. Glass in windows was broken. The foregoing is a general description, and numerous obvious exceptions were found. However, it would be difficult to assign a high earthquake intensity to dwellings when land movement effects were not present.

Small all-metal structures, such as the light mass conventional gasoline service stations, also survived excellently. Larger structures of principally metal construction did have structural damage. Of particular interest is the Knik Arm plant of the Chugach Electric Association located in the northern section of Anchorage. This plant is near Ship Creek, and evidences of differential settlements were noted. Heavy equipment was located on mezzanines and upper floors as well as on the ground floor. The earthquake, with its accompanying differential settlements, caused failures to some of the angle X-bracing in the walls, twisted steel columns, shifted and broke some piping, moved equipment, and broke and buckled some of the metal skin on an exterior wall. An elevated silo on steel supports collapsed (Fig. 9.73). On the other hand, major equipment such as the turbines was undamaged.

Another power plant, owned by the city of Anchorage and located near the previously mentioned plant, was also of all metal construction. It had no supported floors. Damage to the building and to the equipment was slight. Steel X-bracing in the walls was the lateral force-resisting system.

Hollow concrete block was a common wall construc-

Fig. 9.73. Collapsed ash silo at the Knik Arm Plant of the Chugach Electric Association, Anchorage. Alaskan Earthquake of 1964.

tion material for many mercantile buildings as well as for the buildings for other types of occupancies. Roof and supported floors were usually of wood. When small in floor area, when one-story high, and when not located in land movement areas, damage was usually no more than slight to moderate. Some parapets fell, and unanchored roof diaphragms punched out wall sections, but collapse was uncommon. The hollow concrete block usually had some form of reinforcing steel, often in the form of wire webbing laid horizontally in the mortar joints, although heavier reinforcing bars were placed in some bond beams when bond beams were used. The use of vertical reinforcement was erratic and often

Fig. 9.75. Note that the cell containing the reinforcing bar was never grout-filled in this collapsed building. The importance of competent job supervision cannot be overemphasized. Alaskan Earthquake of 1964.

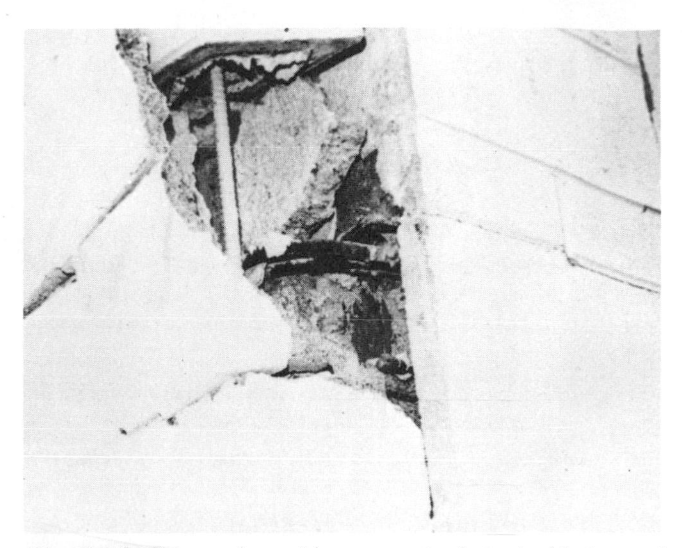

Fig. 9.74. Poor workmanship was a major factor in this damaged hollow concrete block wall. Note that cells containing steel were not grout-filled. Alaskan Earthquake of 1964.

lacking. Design details and workmanship were definitely inferior in the great majority of damage cases. Bars were found in hollow concrete block cells that had not been grouted (Figs. 9.74 and 9.75); alternately, there were grouted cells that did not have steel. Mortar, although hard, too often did not adequately bond to the block. Some buildings had essentially no reinforcing steel. Many of the buildings were older, and they may have been built at a time when no consideration was given to lateral force design.

Poured-in-place reinforced concrete wall construction performed well for small buildings. Usually the roof and supported floor materials were wood for the small buildings, although not always so. Instances of metal deck and metal open web joist roof and floor systems were found. The performance of these structures was generally good, consistent with structures of similar size and of different materials when not located in the land movement areas.

Structures in Anchorage containing precast, prestressed reinforced concrete T-beam floor and roof elements were given their first test in this earthquake, and the performance of 20 of them in Anchorage has been published. A review of their performance shows that the largest completely undamaged building had a diaphragm area of about 6500 ft^2, which certainly is not large. The only instance of *internal* diaphragm damage not related to building collapse was to a 13,000 ft^2 roof; metal connections between precast T-beams failed. Diaphragm boundary connections were often troublesome, either at the actual connection or adjacent and within the supporting system. The prestressed T-beams performed excellently as individual structural components, but buildings containing these members did not perform as well as similar area buildings having similar wall materials but different roof or floor systems. In summary and conclusion, the standard interconnection methods for precast T-beams commonly used in Anchorage and elsewhere in the United States did not perform satisfactorily.

The multistory buildings of poured-in-place reinforced concrete more often than not had structural damage, some of which was quite serious and significant. This also was true of structural steel frame buildings, although complete steel frames were rare. Several abbreviated case histories of building damage may be found in later paragraphs. These and many others are described in much greater detail in *The Prince William Sound, Alaska, Earthquake of 1964 and Aftershocks.* A tabular summary of principal multistory building damage is given in Table 9.9.

The Mt. McKinley Apartment House and the 1200 "L" Street Apartment House are almost identical 14-story buildings that are interesting examples of consistent damage (Figs. 9.76 through 9.81). Exterior walls are of reinforced concrete and are bearing. The walls around the stairs and the interior columns also are of rein-

Table 9.9. Damage to multistory buildings in Anchorage, Alaska*

Building name and (occupancy)	Year built	Stories	U.B.C. seismic zone	Structural system: Frame	Structural system: Floors	Structural system: Exterior walls	Structural system: Core	Principal lateral force bracing system	Percent damage (of replacement value)	Remarks
Airport Control Tower	1952	6 and bsmt.	?	R/C	5 and 6 in. R/C	Insulated metal	None	R/C frame	100	Also damaged in 1954 shock.
Anchorage–Westward (hotel)	1960, 1964	14 and bsmt.	3	Steel with some R/C columns	$5\frac{1}{2}$–$6\frac{1}{2}$ in. R/C on MD on steel beams	Insulated metal and R/C	See remarks	R/C shear walls	12	Landslide shifted building about 1 ft. R/C around elevators not a major core.
Cordova (office)	1960	6 and bsmt.	2 (?)	Steel	$2\frac{1}{2}$ in. R/C on MD on steel joist and beams	Insulated metal and 4 in. R/C	R/C	Steel moment connection also shear walls in R/C core	20	
Elmendorf Hospital	1955	7 and bsmt.	3	R/C	6 in. R/C	Nonstructural hollow concrete block	R/C	R/C shear walls	1 (remarks)	Lower height buildings not listed. Structural damage 1%; nonstructural larger.
Four Seasons (apartments)	1964	6	3	None	8 in. prestressed post-tensioned R/C, tendons not grouted	Plastered studs	R/C	Shear walls in R/C central core	100	Lift slab using steel cols.
Hill (office)	1962	8	3	Steel (remarks)	5 in. R/C on steel beams	Insulated metal	R/C	Shear walls in R/C central core	20–25	Contral core was R/C bearing.
Knik Arms (apartments)	1950	6 and bsmt.	2	Incomplete R/C	$5\frac{1}{2}$ in. R/C	R/C	R/C	R/C shear walls	Negligible	Building moved 10 to 11 ft due to landslide.
Mt. McKinley (apartments)	1951	14 and bsmt.	2	R/C (remarks)	$5\frac{1}{2}$ in. R/C on R/C beams	R/C bearing	R/C	R/C shear walls	40	R/C interior beams and columns. Walls bearing. Almost identical to 1200"L" building.
1200 "L" (apartments)	1951	14 and bsmt.	2	R/C (remarks)	$5\frac{1}{2}$ in. R/C on R/C beams	R/C bearing	R/C	R/C shear walls	30	R/C interior beams and columns. Walls bearing. Almost identical to Mt. McKinley building.
Penney (department store)	1962	5	3	None	10 in. R/C slabs on R/C columns	Precast R/C on 2 sides, R/C on 4 sides	Essentially none	R/C exterior walls	100	Some hollow concrete block exterior walls.
Providence Hospital	1961	5 and bsmt.	3	Steel	$5\frac{1}{4}$ in. R/C on MD on steel beams	Insulated metal	R/C	Shear walls in R/C central core	$2\frac{1}{2}$	Stair and elevator tower, and lower height buildings not listed.

**Abbreviations:* U.B.C. = Uniform Building Code; R/C = Reinforced concrete. Poured in place unless otherwise specified. MD = Metal deck, usually having tradename "Corruform" or "Corfar."

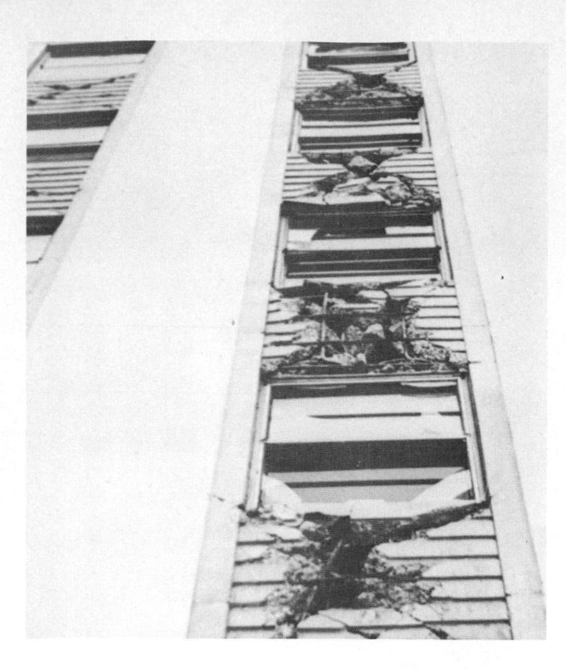

Fig. 9.78. Vertical alignment of shear damage in rusticated spandrels. 1200 L Apartment Building in Anchorage. See Fig. 9.79 for interpretation. Alaskan Earthquake of 1964.

Fig. 9.79. Vertical shear diagrams. While the problem may be self-evident, the actual structural design is quite complex for walls with irregular openings.

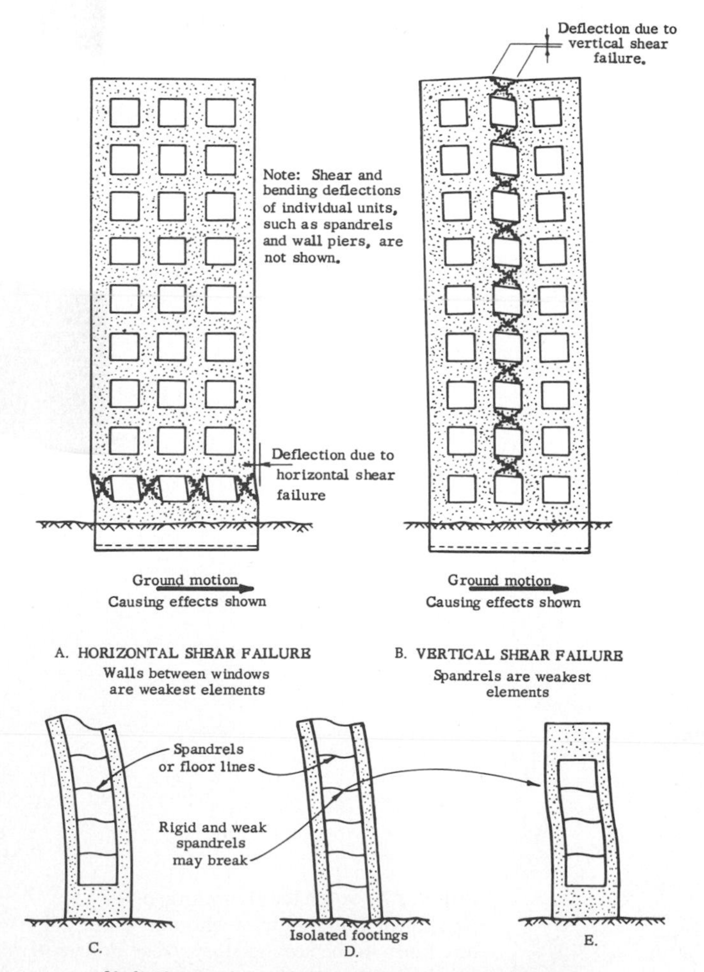

Fig. 9.80. Broken pier in end wall of Mt. McKinley Building; note damage similarity with Fig. 9.77. Air gap extends full length of this load carrying pier. Alaskan Earthquake of 1964.

Fig. 9.81. Top floor of Mt. McKinley Building. The intensity of the shock as defined by overturned objects in this figure was greater in top stories of tall buildings than at grade level. Alaskan Earthquake of 1964.

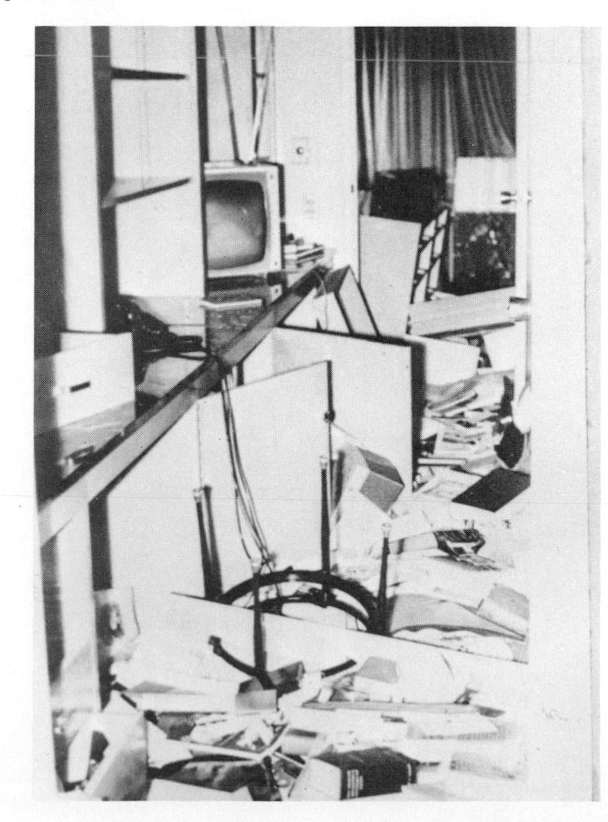

Fig. 9.76. East (front) elevation of the Mt. McKinley Building in Anchorage. Alaskan Earthquake of 1964.

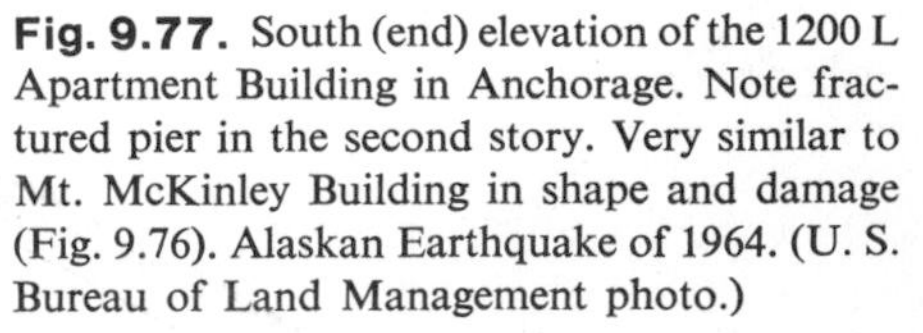

Fig. 9.77. South (end) elevation of the 1200 L Apartment Building in Anchorage. Note fractured pier in the second story. Very similar to Mt. McKinley Building in shape and damage (Fig. 9.76). Alaskan Earthquake of 1964. (U. S. Bureau of Land Management photo.)

forced concrete. The earthquake resistance was provided by the exterior bearing walls and by the interior reinforced concrete walls, all acting as shear walls. Damage was usually in the form of X-cracking in the reinforced concrete spandrels (typical reinforced concrete shear failures). The vertical alignment of shear failures, for convenience termed "vertical shear," was particularly noticeable in the stories of the middle third of the face of the exterior walls. One explanation for this damage pattern is given in Fig. 9.79. One major bearing wall in each building completely failed in a lower story (Figs. 9.77 and 9.80). Both buildings had a similar orientation and had very similar damage patterns, although located about 1 mi apart. Shaking machine tests have been run on one structure, and detailed theoretic/computer analysis has been made by various organizations and persons; the literature on this structure is important. Loss to the Mt. McKinley Building was placed at 40% of its replacement value.

Fig. 9.82. West Anchorage High School. Shear walls had diagonal tension failures (or slippage along a construction joint) while columns had moment failures, consistent with theory. Alaskan Earthquake of 1964.

Fig. 9.83. Slippage along a construction joint of a shear wall. Unremoved laitenance facilitated failure. Alaskan Earthquake of 1964.

Other multistory, poured-in-place reinforced concrete structures were of lower height, but usually they were not as extensively damaged. Both shear cracking and construction joint movement (or working) were commonly observed.

An example of a low, reinforced concrete building that partially collapsed is the West Anchorage High School. The failure of shear walls probably was followed by column bending failure. Extensive pounding between structurally separated units occurred. Eyewitness accounts tell of major oscillations before the partial collapse occurred. Figures 9.82 and 9.83 show typical damage.

The Four Seasons Apartment Building, which was a six-story reinforced concrete lift slab building, totally collapsed (Figs. 9.87 and 9.88). The building was structurally complete prior to the earthquake, but it had not yet been occupied. It has not been determined how much, if any, of the soil beneath the building was affected by the nearby land movements (Fig. 9.85). This rectangular

Fig. 9.84. Four Seasons Apartment House in Anchorage before the earthquake. Two interior core towers still to be built. Alaskan Earthquake of 1964.

Fig. 9.85. Aerial view of collapsed Four Seasons Apartment House. The L Street landslide graben cuts diagonally through this photograph. Alaskan Earthquake of 1964. (U.S. Air Force photo.)

Fig. 9.86. Four Seasons Apartment House. Typical upper-story floor plan. Alaskan Earthquake of 1964.

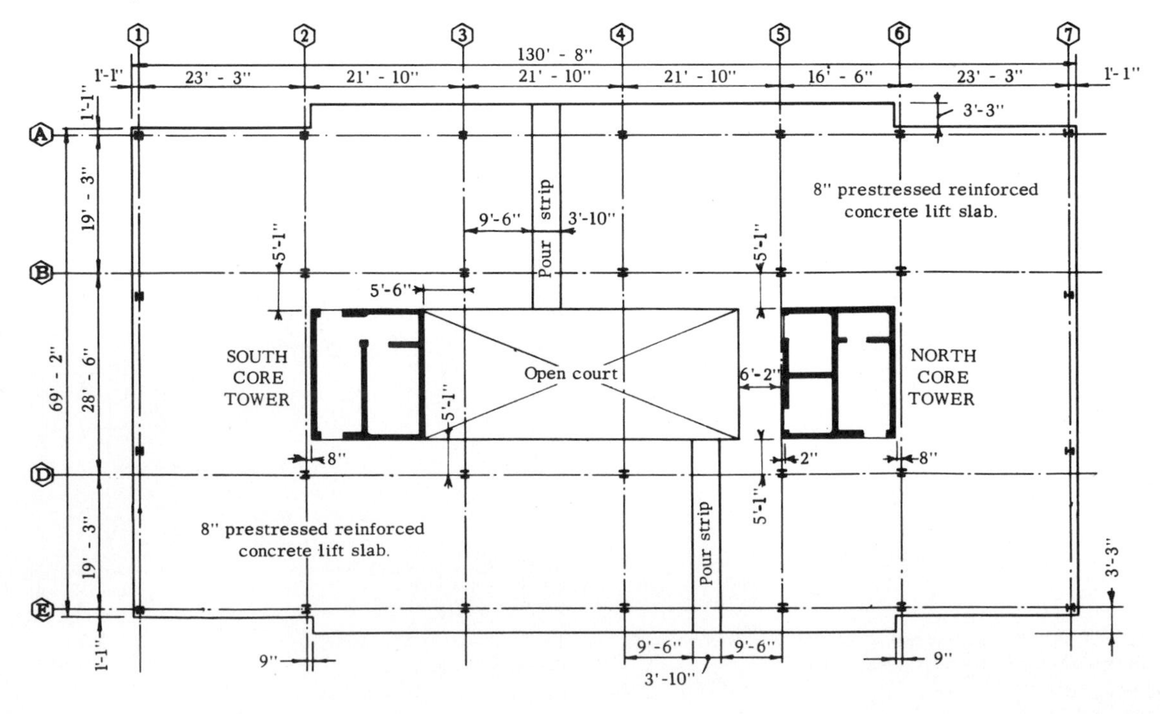

Fig. 9.87. North core tower, Four Seasons Apartment House. Note the apparently undamaged wood frame dwelling. Arrow points to a prestressing strand, part of which is on the roof of a dwelling. Alaskan Earthquake of 1964.

Fig. 9.88. Floor slabs stacked like pancakes. Four Seasons Apartment House. Note that many of the prestressing rods have failed. Alaskan Earthquake of 1964.

Fig. 9.89. Undamaged dowels from foundations to failed core tower. Four Seasons Apartment House. Debris removed by the author to take this photo. Alaskan Earthquake of 1964.

Fig. 9.90. Alaska Sales and Service Building, Anchorage. Alaskan Earthquake of 1964. (T. F. Moore photo.)

Fig. 9.91. Anchorage–Westward Hotel complex in Anchorage. The 14-story building suffered the principal damage. Alaskan Earthquake of 1964.

Fig. 9.92. Typical floor plan, Anchorage–Westward Hotel. Alaskan Earthquake of 1964.

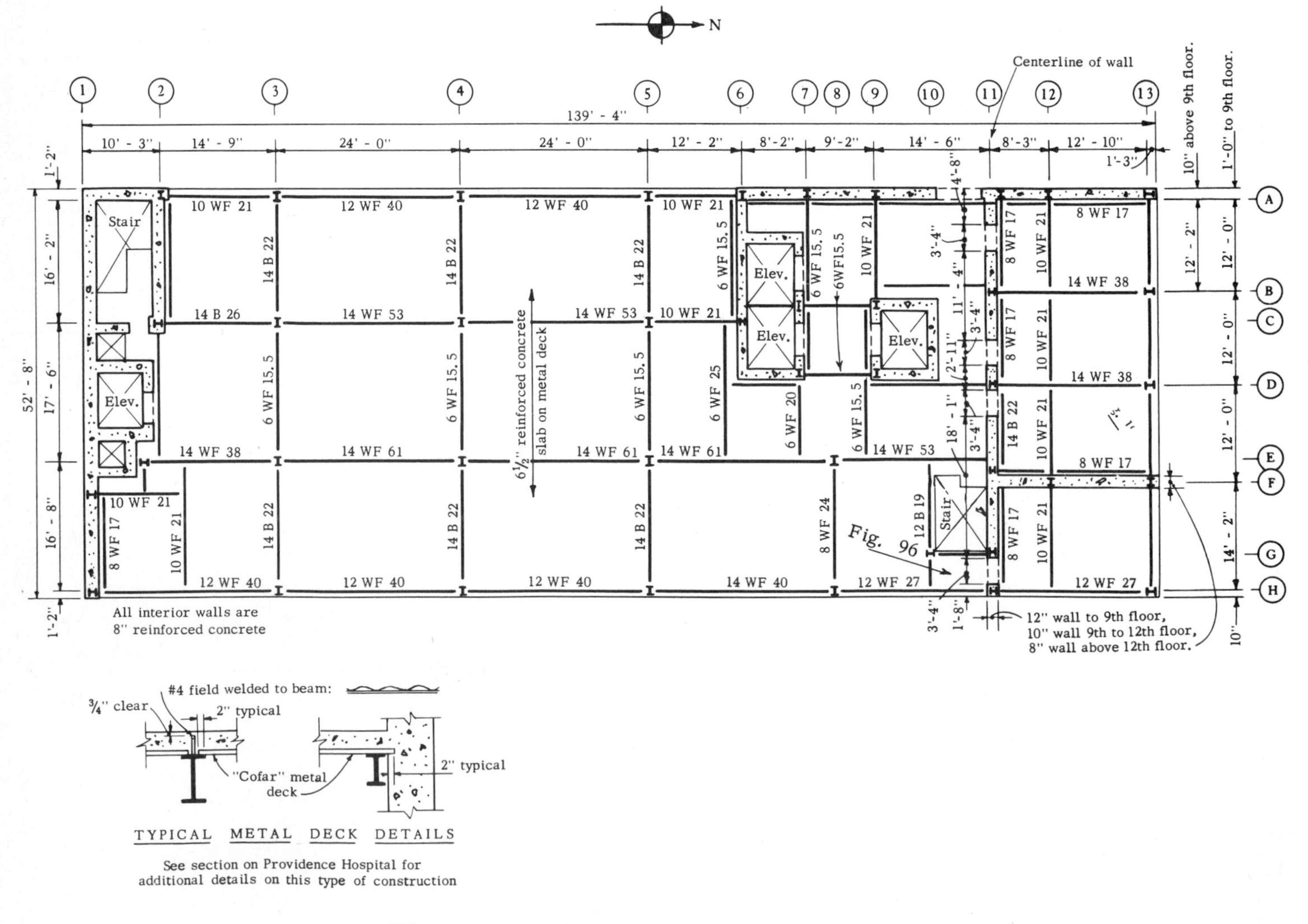

building was stabilized against lateral forces by two poured-in-place reinforced concrete stairwell and elevator cores used as shear walls. Floors were of prestressed and post-tensioned reinforced concrete, while the columns were of structural steel. The prestressing steel had not been grouted. The collapse is often attributed to overturning forces causing bond failure in the reinforcing steel splices at the first-floor level (Fig. 9.89), followed by tower overturning. An alternate point of view attributes initial failure to the failure of end anchorages of unbonded prestressing tendons at a corner column. This structure, too, has been extensively studied, and results may be found in the previously cited publication.

Precast reinforced concrete construction had more than its share of failures. An interesting one-story failure in this category, the Alaska Sales and Service Building, had a precast T-beam roof system on precast reinforced concrete bents (Fig. 9.90). Walls were generally of precast concrete. Interconnections between precast elements were usually made by means of welded connections. The building was in the final stages of the structural construction, and all of the bracing elements were not in. Large sections of the building collapsed, while much of the remainder was out of plumb and on the verge of collapse. Generally, connections failed rather than the members.

The Anchorage–Westward Hotel, a 14-story structure built in two stages, had a partial steel frame. Steel columns buckled, reinforced concrete spandrels failed in shear due to vertical shear, and the metal deck floor separated from the steel frame on at least one upper floor. Damage occurred at the roof level of two adjoining buildings due to pounding. Other significant structural damage was noted throughout, as may be seen in Figs. 9.91 through 9.96. The loss to this building has been placed at 12% of its replacement value.

Fig. 9.93. Pounding damage between 14-story hotel and ballroom of Anchorage–Westward Hotel. Alaskan Earthquake of 1964. (J. D. Simpkins photo.)

9.4.16 Puget Sound, Washington, Earthquake

The Puget Sound earthquake of April 29, 1965, had its epicenter between Seattle and Tacoma and had a magnitude of 6.5. Damage occurred in these two cities as well as in Olympia to the south (Steinbrugge and Cloud in U.S. Coast and Geodetic Survey, 1965c). The 1949 shock, with a magnitude of 7.1, had its epicenter between Tacoma and Olympia and was located somewhat southwest of the 1965 earthquake. The damage pattern from the 1965 earthquake resembled that of the 1949 shock, with damage surveys indicating that the 1949 shock was the more destructive of the two. Figure 9.97 is an intensity map for the 1965 shock.

Fig. 9.94. Lintel damage at a doorway, Anchorage–Westward Hotel, along column line 11 in Fig. 9.92. See also Fig. 9.95. Alaskan Earthquake of 1964.

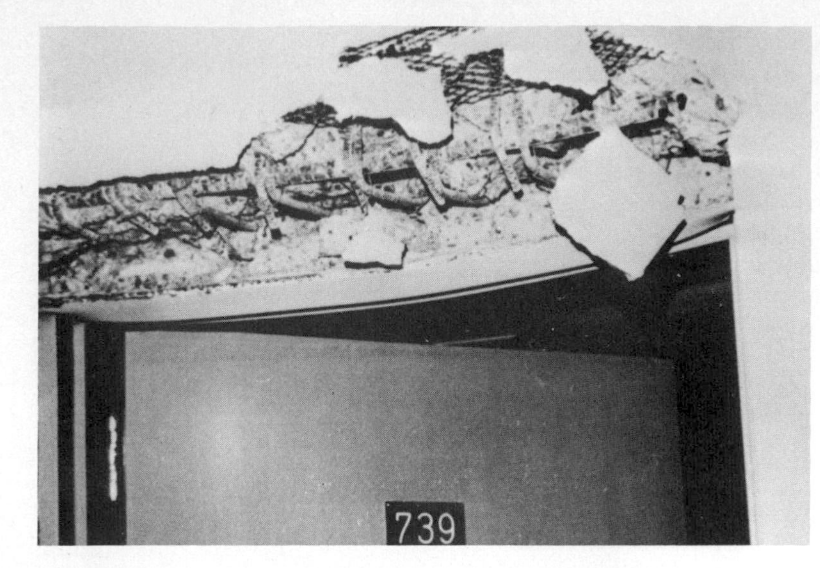

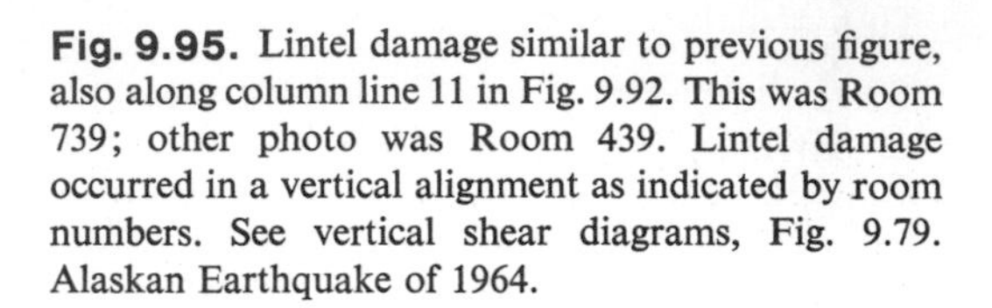

Fig. 9.95. Lintel damage similar to previous figure, also along column line 11 in Fig. 9.92. This was Room 739; other photo was Room 439. Lintel damage occurred in a vertical alignment as indicated by room numbers. See vertical shear diagrams, Fig. 9.79. Alaskan Earthquake of 1964.

The focal depth of the 1965 earthquake has been placed at about 36 mi, and the 1949 shock had a focal depth "slightly greater than normal." In view of the 10 mi or less focal depth commonly given for the majority of California earthquakes, the deeper focal depths in these two Washington shocks suggest a more moderate surface intensity over a wider area than comparable magnitude but shallower California earthquakes. Observed damage confirmed this view.

Building damage was generally light and followed the damage patterns described in this chapter for other shocks. Damage, as usual, was intensified in poor ground regions such as the Duwamish River Basin, and particularly at its mouth in Puget Sound. Figure 9.98 shows

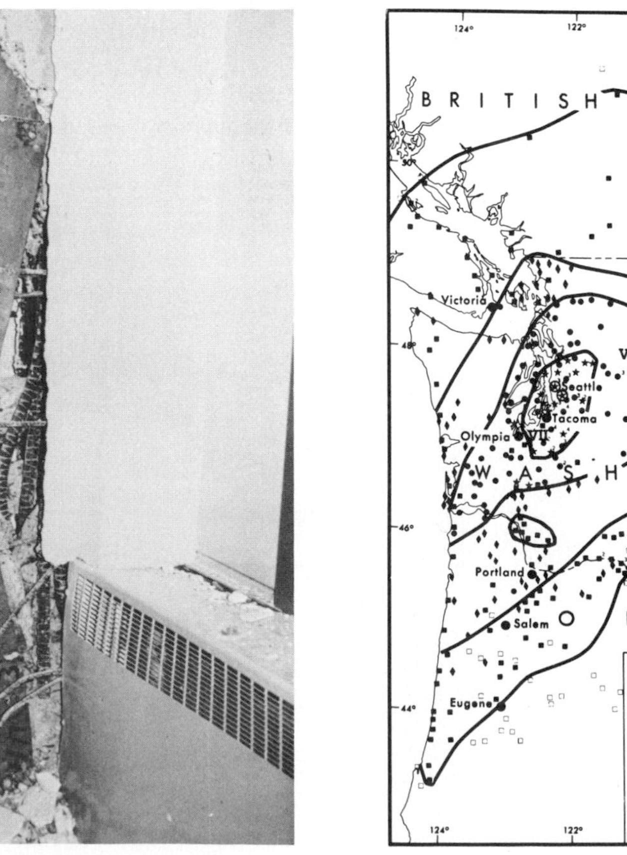

Fig. 9.96. Buckled column in second story of Anchorage–Westward Hotel due to overturning forces. Alaskan Earthquake of 1964.

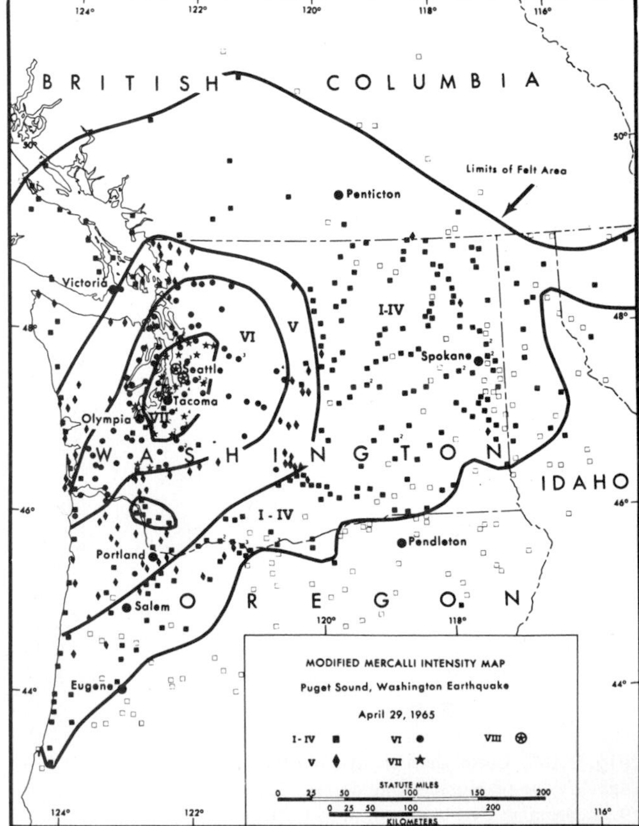

Fig. 9.97. Intensity map of the Puget Sound Earthquake of 1965. (By U.S. Coast and Geodetic Survey.)

Fig. 9.98. Chimney damage in western Seattle. Puget Sound Earthquake of 1965.

typical dwelling damage. Figure 9.99 shows damage on structurally poor ground.

9.4.17 Parkfield, California, Earthquake

The Parkfield earthquake of June 27, 1966, was centered on a known active segment of the San Andreas fault. The preliminary magnitude determination of the main shock has been given as about 5.5. Strong earthquakes in the same area have occurred in 1901, 1922, and 1934, and these shocks possibly were accompanied with some surface faulting. The 1966 event had surface faulting that consisted of a main fracture zone $23\frac{1}{2}$ mi long and a subsidiary zone $5\frac{1}{2}$ mi long, both in the San Andreas fault zone.

An array of 16 strong-motion seismic recording stations were in the area, including one station almost on the trace of the 1966 rupture. Maximum short duration transient accelerations were recorded at 50% of gravity, and this may be compared with 30% during the 1940 El Centro earthquake. Maximum response of the seismic equipment was approximately at right angles to the strike of the fault.

Little building damage occurred, in part due to the wood frame construction. Items often fell from shelves, but some locations within a mile of the fault trace had very little damage of this type. Some headstones overturned in two cemeteries at Parkfield. Bridges in the epicentral area had some structural damage such as cracks in concrete abutments and buckled steel X-braces. However, none of the bridges were damaged to the extent that they had to be closed. Certainly, the high accelerations from this very short duration earthquake were not particularly destructive.

9.5 SURFACE FAULTING

The study of active earthquake faults properly falls into the field of geology and its related sciences. However, engineers in planning utilities, highways, canals, etc. often are required to lay out facilities which cross faults.

Fig. 9.99. Ground dropped due to the pier at the left shifting toward the water. Location is the Harbor Island in Seattle. Puget Sound Earthquake of 1965.

Table 9.10. Surface faulting accompanying earthquakes in California and Nevada, *1900–1968**

Date, local time	Fault	Location	Principal references
Fall 1903	Gold King	Churchill County, Nevada	*BSSA* **49**, 251–265 (1959)
April 18, 1906	San Andreas	†San Francisco, California	State EQ Invest. Comm. (1908).
October 2, 1915	Pearce, Tobin	Pleasant Valley, Nevada	*BSSA* **5**, 190–205; *J. Geol.* **43**, 690–707 (1935)
December 20, 1932		Cedar Mountain, Nevada	*BSSA* **24**, 345–384 (1934)
January 30, 1934		Excelsior Mountains, Nevada	*BSSA* **25**, 161–168 (1935)
§June 7, 1934	San Andreas	Parkfield, California	*BSSA* **25**, 223–246 (1935)
May 18, 1940	Imperial	Imperial Valley, California	Richter (1958), pp. 487–495; *BSSA* **31**, 13–31 (1941)
April 10, 1947	Manix	Mojave Desert, California	Richter (1958), pp. 516–518
December 14, 1950		Fort Sage Mountains, California	*BSSA* **47**, 173–177 (1957)
January 23, 1951	Superstition Hills	Imperial Valley, California	*BSSA* **55**, 768 (1965)
July 21, 1952	White Wolf	Kern County, California	Oakeshott (1955); *BSSA* **44**, 283–299 (1954)
July 6, 1954	Rainbow Mountain	Churchill County, Nevada	*BSSA* **46**, 10–14 (1956)
August 23, 1954	Rainbow Mountain	Churchill County, Nevada	*BSSA* **46**, 10–14 (1956)
‡December 16, 1954	Fairview, Dixie Valley	Churchill County, Nevada	*BSSA* **47**, 353–375 (1957)
March 4, 1966	Imperial	Imperial Valley, California	*BSSA* **53**, 501–514 (1967)
§June 27, 1966	San Andreas	Parkfield, California	*BSSA* **56**, 961–971 (1966), *USGS Prof. Paper 579*
April 9, 1968	Coyote Creek	Borrego Mt., California	*BSSA* **58**, 1183–1186 (1968)

*BSSA = *Bulletin of the Seismological Society of America*. (Some instances of faulting are borderline. Consult references for any critical study.)
†Includes coastal California to the north and south of San Francisco.
‡Two earthquakes, each with surface faulting, occurred about 4 min apart. From an engineering standpoint, they are considered as a single event.
§The June 7, 1934, and June 27, 1966, shocks were in the same general location. Faulting also may have occurred here on March 2, 1901, and March 10, 1922

The problem is by no means a remote one; e.g., the metropolitan areas of Los Angeles, San Francisco, and Salt Lake City have active faults in their environs.

Structures in or across a fault zone that undergoes surface faulting can expect to have severe damage, as did the railroad tunnels in the 1952 Kern County earthquake and the hollow concrete block structures across the fault scarp in the 1959 Hebgen Lake earthquake.

The design engineer may have the option to choose from several sites. Speaking broadly, and assuming that all other factors are equal, the farther that a structure is from a known active fault, the better. However, in many populated regions of coastal California and elsewhere the farther that a building is placed from one known fault, the closer it comes to another. Except for avoiding fault zones, the designer normally does not let the distance from the fault to his structure become a major factor in site location.

Earthquakes accompanied by surface faulting appear to be more common in the United States than previously supposed. Table 9.10 lists California and Nevada earthquakes with verified surface faulting since 1900. The larger number of instances of surface faulting since 1940 probably is due to more active earthquake investigations than to increased seismic activity.

Evidence exists that faults may have slippage, or creep, and progressively destroy buildings and other man-made works on the fault. This creep may occur without recorded earthquakes. Periodic measurements made over many years at the Almaden Winery near Hollister, California, indicates that one segment of the San Andreas fault is creeping right laterally relative to the other segment at a rate of about $\frac{1}{2}$ in./year. There was some evidence that the creep may have been going on at the same rate for perhaps 50 years.

It is also known that the Hayward fault, which goes through populated sections of Berkeley, Oakland, Hayward, and Fremont, has been moving, apparently without earthquakes. San Francisco and Oakland have water supply conduits that have bent and in some cases broken. The stadium at Berkeley has deformed. Fault creep damage can be viewed as an expensive maintenance problem, but it is also an ill omen for the future. The two principal sources of information of fault creep may be found in the collections of papers published in the *Bulletin of the Seismological Society of America* (**50**, 389–415, 1960; **56**, 257–323, 1966).

ACKNOWLEDGMENTS

Any abridged study, such as this one, must rely heavily on the published works of many persons. Liberal use has been made of the various issues of the *Bulletin of the Seismological Society of America*, of the files of the Pacific Fire Rating Bureau, and of the publications of the U.S. Coast and Geodetic Survey, and their permission to use this material is gratefully acknowledged.

REFERENCES

The bibliography below has been selected on the basis of importance to engineering interest. It is further restricted to the earthquakes discussed in this chapter. Obviously important geological and seismological references usually have been omitted unless they significantly supplement the engineering studies. The bibliography by E. P. Hollis, listed below, will be of particular importance to the student wishing to explore the subject further. The extensive references in Richter's *Elementary Seismology* also will be of particular value. The various issues of the *Bulletin of the Seismological Society of America* are the best general source for American earthquakes as well as for some major foreign shocks.

Allen, C.R., et al. (1968). "The Borrego Mountain, California, Earthquake of 9 April 1968: A Preliminary Report," *Bull. Seism. Soc. Am.*, **58**, 1183–1186.

American Society of Civil Engineers (1907). "The Effects of the San Francisco Earthquake of April 18, 1906 on Engineering Constructions," *Trans. Am. Soc. Civil Engr.*, **59**, 208–329.

Anonymous (June 1940). "Imperial Valley Earthquake," *West. Constr. News.*

Binder, R. W. (1952). "Engineering Aspects of the 1933 Long Beach Earthquake," *Proc. Symp. on Earthquake and Blast Effects on Structures*, 186–211.

Board of Fire Underwriters of the Pacific (1925). *Notes of Interest to Underwriters on Earthquake Damage to City of Santa Barbara, California.*

Chamber of Commerce of San Francisco (1906). *Report of the Special Committee of the Board of Trustees of the Chamber of Commerce of San Francisco.*

Committee of Five (1906). *Report of the Committee of Five to the Thirty-Five Companies on the San Francisco Conflagration.*

Degenkolb, H. J. (1955). "Structural Observations of the Kern County Earthquake," *Trans. Am. Soc. Civil Engr.*, **120**, 1280–1294.

Dewell, H. D., and B. Willis (1925). "Earthquake Damage to Buildings," *Bull. Seism. Soc. Am.*, **15**, 282–301.

Duke, C. M., and D. J. Leeds (1959). "Soil Conditions and Damage in the Mexico Earthquake of July 28, 1957," *Bull. Seism. Soc. Am.*, **49**, 179–191.

Dutton, C. E. (1887–1888). "The Charleston Earthquake of August 31, 1886," *Ninth Annual Report*, U.S. Geological Survey.

Edwards, H.H. (February, March, and April 1951). "Lessons in Structural Safety Learned from the 1949 North-west Earthquake," *West. Construction.*

Engle, H. M. (1936). "The Montana Earthquakes of October, 1935: Structural Lessons," *Bull. Seism. Soc. Am.*, **26**, 99–109.

Greely, A. W. (1906). *Special Report of Maj. Gen. Adolphus W. Greely, U.S.A., Commanding the Pacific Division, on the Relief Operations Conducted by the Military Authorities of the United States at San Francisco and other Points.*

Hershberger, J. (1956). "A Comparison of Earthquake Accelerations with Intensity Ratings," *Bull. Seism. Soc. Am.*, **46**, 317–320.

Hiriart, M., and E. Rosenblueth (1958). *Los Efectos del Terremoto del 28 de Julio y la Consiguiente Revisión de los Criterios para el Diseño Sísmico de Estructuras*, Universidad Nacional Autonoma de Mexico, Escuela Nacional de Ingenieria.

Hollis, E. P. (1958). *Bibliography of Engineering Seismology*, 2nd ed., San Francisco: Earthquake Engineering Research Institute.

Housner, G. W., and D. E. Hudson (1958). "The Port Hueneme Earthquake of March 18, 1957," *Bull. Seism. Soc. Am.*, **48**, 163–168.

Inard, C. D. (1925). "Report of Engineering Committee on the Santa Barbara Earthquake," *Bull. Seism. Soc. Am.*, **15**, 302–304.

Lawson, A. C. et al. (1908). *The California Earthquake of April 18, 1906. Report of the State Earthquake Commission*, 2 vols. and atlas, Washington, D.C.: Carnegie Institution of Washington.

Manson, M. (1907). *Report of the Sub-Committee on Statistics to the Chairman and Committee on Reconstruction.*

Miller, A. L. (1952). "Earthquake Lessons from the Pacific Northwest," *Proc. Symp. on Earthquake and Blast Effects on Structures*, 212–223.

National Board of Fire Underwriters (May 1906). *The San Francisco Conflagration of April 1906, Special Report to the National Board of Fire Underwriters (by a) Committee of Twenty.*

National Board of Fire Underwriters (1933). *Report on the Southern California Earthquake of March 10, 1933.*

National Board of Fire Underwriters and Pacific Fire Rating Bureau (1964). *The Alaska Earthquake.*

Oakeshott, G. B., ed. (1955). "Earthquakes in Kern County, California, During 1952," California Division of Mines, *Bull. 171.* (A collection of 34 papers.)

Oakeshott, G. B., ed. (1959). "San Francisco Earthquake of March 1957," California Division of Mines, *Special Report 57.*

Richter, C. F. (1958). *Elementary Seismology*, San Francisco: W. H. Freeman.

Richter, C. F. (1959). "Seismic Regionalization," *Bull. Seism. Soc. Am.*, **49**, 123–162.

Schussler, H. (1906). *The Water Supply of San Francisco, California, Before, During and After the Earthquake of April 18, 1906 and the Subsequent Conflagration.*

Shannon & Wilson, Inc. (1964). *Report on Anchorage Area Soil Studies, Alaska to U.S. Army Engineer District, Anchorage, Alaska.*

Steinbrugge, K. V., and V. R. Bush (1960). *Earthquake Experience in North America, 1950–1959, Proceedings of the Second World Conference on Earthquake Engineering*, **I**, 381–396.

Steinbrugge, K. V., and V. R. Bush (1965). "Review of Earthquake Damage in Western United States, 1933–1964," in *Earthquake Investigations in the Western United States, 1931–1964*, pp. 223–256.

Steinbrugge, K. V., V. R. Bush, and E. G. Zacher (1959). "Damage to Buildings and Other Structures During the Earthquake of March 22, 1957," in California Division of Mines *Special Report 57*, pp. 73–106.

Steinbrugge, K. V., and W. K. Cloud (1962). "Epicentral Intensities and Damage in the Hebgen Lake, Montana, Earthquake of August 17, 1959," *Bull. Seism. Soc. Am.*, **52**, 181–234, map.

Steinbrugge, K. V., and D. F. Moran (1954). "An Engineering Study of the Southern California Earthquake of July 21, 1952, and Its Aftershocks," *Bull. Seism. Soc. Am.*, **44**, 199–462.

Steinbrugge, K. V., and D. F. Moran (1956). "The Fallon–Stillwater Earthquakes of July 6, 1954, and August 23, 1954," *Bull. Seism. Soc. Am.*, **46**, 15–33.

Steinbrugge, K. V., and D. F. Moran (1957a). "An Engineering Study of the Eureka, California, Earthquake of December 21, 1954," *Bull. Seism. Soc. Am.*, **47**, 129–153, maps.

Steinbrugge, K. V., and D. F. Moran (1957b). "Engineering Aspects of the Dixie Valley–Fairview Peak Earthquakes," *Bull. Seism. Soc. Am.*, **47**, 335–348.

Structural Engineers Association of California (1967). *Recommended Lateral Force Requirements and Commentary.*

Tocher, D. (1960). *Movement on Faults, Proceedings of the Second World Conference on Earthquake Engineering*, **I**, 551–564.

Ulrich, F. P. (1936). "Helena Earthquakes," *Bull. Seism. Soc. Am.*, **26**, 323–339, maps.

Ulrich, F. P. (1941a). "The Imperial Valley Earthquakes of 1940," *Bull. Seism. Soc. Am.*, **31**, 13–31, map.

Ulrich, F. P. (October 1941b) "The Santa Barbara Earthquake," *Building Standards Monthly.*

U.S. Coast and Geodetic Survey (1964). *Preliminary Report, Prince William Sound, Alaskan Earthquakes, March–April 1964.*

U.S. Coast and Geodetic Survey (1965a). *Earthquake History of the United States, Part I*, Washington, D.C.: U.S. Government Printing Office.

U.S. Coast and Geodetic Survey (1965b). *Earthquake Investigations in the Western United States, 1934–1964*, D. S. Carder, ed.

U.S. Coast and Geodetic Survey (1965c). *The Puget Sound, Washington, Earthquake of April 29, 1965*, Washington, D.C.: U.S. Government Printing Office.

U.S. Coast and Geodetic Survey (1966a). *Earthquake History of the United States, Part II*, Washington, D.C.: U.S. Government Printing Office.

U.S. Coast and Geodetic Survey (1966b). *The Parkfield, California, Earthquake of June 27, 1966*, Washington, D.C.: U.S. Government Printing Office.

U.S. Coast and Geodetic Survey (1967). *The Prince William Sound, Alaska, Earthquake of 1964 and Aftershocks*, Washington, D.C.: Environmental Sciences Services Administration, U.S. Government Printing Office.

Vol. 1: *Operational Phases of the USC & GS, Including Seismicity*, 1966.

Vol. 2, Part A: *Engineering Seismology*, 1967. (This is the volume of greatest engineering interest.)

Vol. 2, Part B: *Seismology*, in press.

Vol. 3: *Geodesy and Photogrammetry Research Studies*, in press.

U.S. Coast and Geodetic Survey (annual). *United States Earthquakes*, Washington, D.C.: U.S. Government Printing Office.

U.S. Geological Survey (1907). "The San Francisco Earthquake and Fire of April 18, 1906 and their Effects on Structures and Structural Materials, by G. K. Gilbert, R. L. Humphry, J. S. Sewell and F. Soulé," *Bull. 324.*

U.S. Geological Survey (1912). "The New Madrid Earthquake," by Myron L. Fuller, *Bull. 494.*

U.S. Geological Survey (1964–1968). "The Alaska Earthquake, March 27, 1964," *Professional Papers 541, 542, 543, 544, and 545.*

U.S. Geological Survey (1964). Circular 491: *Alaska's Good Friday Earthquake, March 27, 1964.*

U.S. Geological Survey (1967). "The Parkfield–Cholame, California, Earthquakes of June–August 1966," *Professional Paper 579.*

Wood, H. O. (1933). "Preliminary Report on the Long Beach Earthquake," *Bull. Seism. Soc. Am.*, **23**, 43–56.

Chapter 10

Soil Problems and Soil Behavior

H. BOLTON SEED

Professor of Civil Engineering
University of California
Berkeley, California

10.1 INTRODUCTION

The development of an analytical solution to a civil engineering problem usually involves a sequence of five steps:

1. A clear recognition of the problem so that it may be defined as precisely as possible.
2. The development of a qualitative picture of the mechanics controlling the problem.
3. The development, for a somewhat idealized form of the problem, of a method of analysis by means of which the problem may be expressed and analyzed in quantitative terms.
4. The development of suitable test procedures to determine the material properties involved in the analytical procedure.
5. Observations or investigations of the performance of prototype or large-scale structures to determine whether the somewhat idealized analysis is capable of predicting the behavior of actual structures under field conditions with sufficient accuracy for design purposes.

In connection with the analysis of soil behavior during earthquakes, important progress has been and is being made in all the five phases of problem solution listed above. However, while much remains to be accomplished in connection with items 2 to 5, it is possible at the present time to delineate reasonably well the soil stability problems that occur during earthquakes and to illustrate them by means of examples from recent as well as older earthquakes. In fact many of the types of problems might well be illustrated by examples occurring during the 1964 Alaskan earthquake alone.

The delineation of problems in this way offers the advantage that once they are clearly recognized, adequate design measures for dealing with them can invariably be developed even though no quantitative methods of analysis are as yet available. A survey of common problem types also provides a background against which to judge the singularity of the effects produced by any given earthquake. Accordingly, an outline of the types of problems confronting the soil engineer in seismically active areas of the world is presented in the following pages.

10.2 SETTLEMENT OF COHESIONLESS SOILS

It has long been recognized that vibration is an effective means of compacting cohesionless soils. Thus it is not surprising that the ground vibrations caused by earthquakes often lead to compaction of cohesionless soil deposits and associated settlement of the ground surface.

A quantitative measure of ground settlement of this type was provided by the behavior of a well casing at Homer (Grantz, Plafker, and Kachadoorian, 1964) during the 1964 Alaska Earthquake. The casing had been installed to firm rock before the earthquake and projected about 1 ft above the ground surface. Following the earthquake the casing projected some $3\frac{1}{2}$ ft above the ground surface, indicating a decrease in thickness of the soil layer of about 2.5 ft. As shown in Fig. 10.1, tectonic movements caused the rock surface to be lowered by 2 ft and this, together with the 2.5 ft of settlement caused by soil compaction, resulted in a total settlement of the ground surface of 4.5 ft.

Fig. 10.1. Ground settlement around well casing at Homer during Alaska earthquake (1964). (After Grantz et al.)

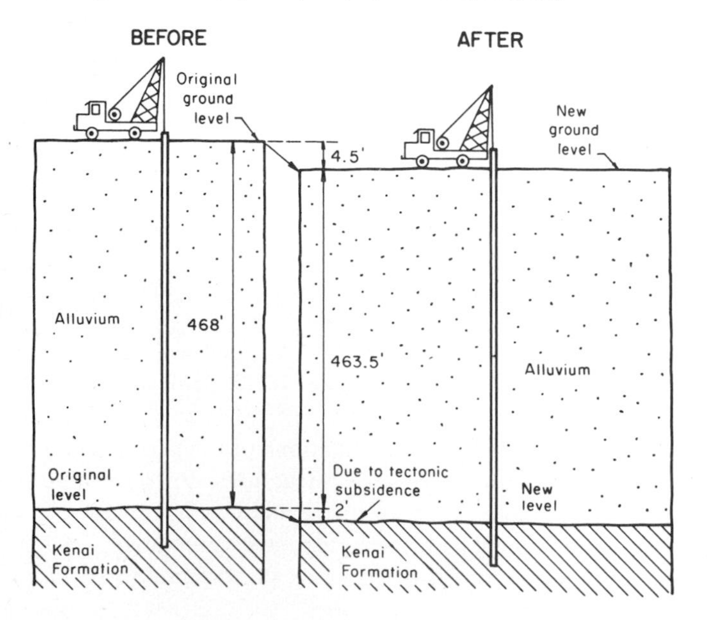

Fig. 10.2. Anchorage–Portage highway after earthquake (1964).

A similar combination of effects in the Portage area of Alaska (1964) led to the town being inaccessible during periods of high tide. Here the combination of about 4 ft settlement of the rock due to tectonic movements together with about 4 ft of settlement due to compaction of the overlying soil led to a ground surface settlement of about 8 ft. As a result the highway from Anchorage to Portage, which was constructed on an embankment fill along the coastline, was submerged during periods of high tide (see water on both sides of embankment in Fig. 10.2). As a result of the general flooding in this area during high tide periods, the township had to move to a new location.

Similar problems of flooding and inundation of land due to settlement of soil by compaction or a combination of compaction and tectonic movements also occurred in the Chilean Earthquake (1961) and the Niigata, Japan, Earthquake (1964). Figure 10.3 shows an island near Valdivia, Chile, which was partially submerged as a result of these effects.

Ground settlements due to compaction often lead to differential settlements of engineering structures—a phenomenon that is particularly well illustrated by the performance of bridge abutments. Often the abutment is supported on firm materials or on a pile foundation and undergoes relatively small settlements compared with the backfill material for the abutment, which rests

Fig. 10.3. Submerged island near Valdivia (1960).

Fig. 10.4. Differential settlement between bridge abutment and backfill, Niigata (1964).

directly on the ground surface and settles due to compaction of the soil on which it rests. Figure 10.4 shows a differential movement of several feet between an abutment and its backfill for a railroad bridge in Niigata, while Fig. 10.5 shows differential settlement among the abutment, piers, and backfill for a bridge near Portage, Alaska. In the latter case the piers settled more than the abutment, and the bridge deck was left spanning across the stream bed, unsupported by the piers on which it had previously rested.

10.3 SOIL LIQUEFACTION—LEVEL GROUND

If saturated cohesionless materials are subjected to earthquake ground vibrations, the resulting tendency to compact must be accompanied by an increase in pore water pressure in the soil and a resulting movement of water from the voids. Water is thus caused to flow upward to the ground surface, where it emerges in the form of mud spouts or sand boils. The development of high pore water pressures due to ground vibration and the resulting upward flow of water may turn a sand into a "quick" or liquefied condition (it may be recalled that quicksand is a condition induced by an upward flow of water through a sand under a hydraulic gradient in excess of about unity).

Liquefaction of saturated sands has been reported in a number of earthquakes, but nowhere has it been more dramatically illustrated than in the town of Niigata, Japan (Japan Society of Soil Mechanics and Foundation Engineering, 1966; International Institute of Seismology and Earthquake Engineering 1965), during the earthquake of June 16, 1964. The epicenter of the earthquake (magnitude about 7.3) was located some 35 mi from Niigata, but nevertheless the earthquake induced extensive liquefaction of the sand deposits in the low-lying areas of the town. The proximity of the town to the coast and a soil profile through the general area are shown in Fig. 10.6.

Fig. 10.5. Differential settlement of bridge pier, abutment, and backfill, Portage (1964).

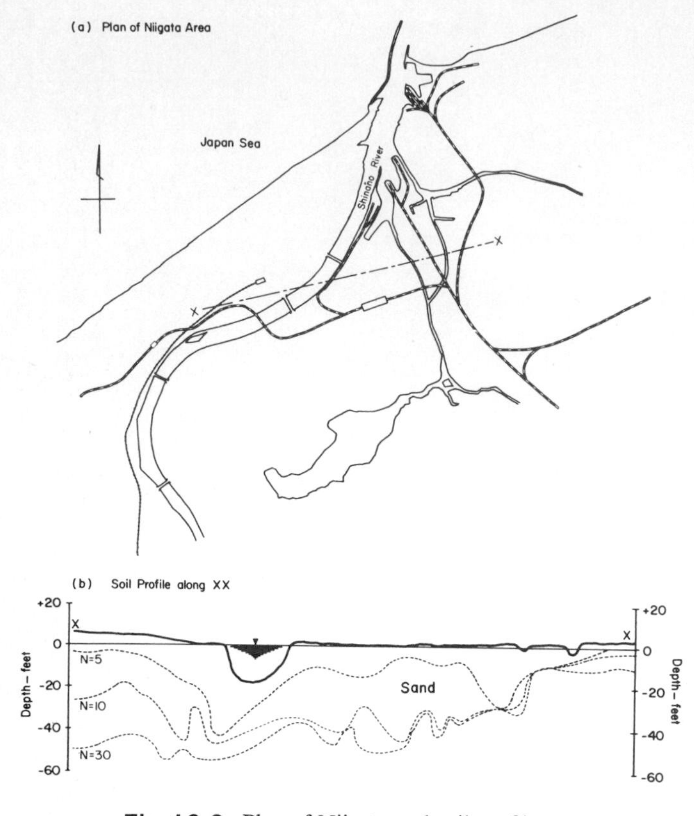

Fig. 10.6. Plan of Niigata and soil profile.

Fig. 10.7. Initial stages of water flow from ground, Niigata (1964).

Fig. 10.8. Later stages of water flow from ground, Niigata (1964).

Cracking of the ground surface associated with the initial emergence of water from the ground during the earthquake is shown in Fig. 10.7, and the considerable depth of water accumulated at the same location some little time later is shown in Fig. 10.8. In some areas numerous sand vents could be observed after the earthquake (Fig. 10.9), sometimes surrounded by rings of sand carried to the surface by the upward-flowing water (Fig. 10.10).

As liquefaction developed over extensive areas, automobiles, structures, and other objects gradually settled into the resulting quicksand. In several cases, lightweight buried structures floated to the surface. The sewage treatment tank shown in Fig. 10.11 was originally buried with its base some 15 ft below the ground surface, but it projected about 10 ft above the surface when the ground motions subsided.

Figure 10.12 shows a truck that sank into the liquefied soil. Similar settlement of a heavy concrete structure is shown in Fig. 10.13, while Fig. 10.14 shows the cracking of a relatively small building induced when its heavier neighbor settled into the ground, tending to pull one side of the lighter structure down with it. Many structures settled more than 3 ft, and the settlement was often accompanied by severe tilting, as shown in Fig. 10.15. Undoubtedly the most dramatic case of settling and tilting was that experienced by the apartment buildings at Kawagishi-cho (see Fig. 10.16). One of

Fig. 10.9. Area of sand vents, Niigata (1964).

Fig. 10.10. Area of sand vents, Niigata (1964).

Fig. 10.11. Sewage treatment tank which floated to surface during Niigata earthquake (1964).

Fig. 10.12. Settlement of truck into liquefied sand, Niigata (1964).

Fig. 10.13. Settlement of building due to soil liquefaction, Niigata (1964).

Fig. 10.14. Building damage induced by settlement of neighboring structure, Niigata (1964).

these buildings tilted through an angle of 80° and the occupants were able to evacuate, after the motion stopped, by walking down the face of the building. Neighboring buildings suffered varying lesser degrees of tilting.

Settlement and accumulation of ground water around the airport building at Niigata are shown in Fig. 10.17. The airport building settled 3 ft, and extensive liquefaction occurred in the surrounding area causing severe damage to runways (Fig. 10.18) and other facilities.

Fig. 10.15. Tilting of building during Niigata earthquake (1964).

Fig. 10.16. Tilting of apartment buildings, Niigata (1964).

Similar effects to those described above occurred at Puerto Montt (Duke and Leeds, 1963) during the Chilean Earthquake of May 1960. The southerly part of the city is located on low flat land adjacent to the waterfront. Along the waterfront itself structures are constructed on a granular fill, placed in some areas by dumping and in others by hydraulic methods. Both types of fill liquefied extensively during the earthquake. A view of an area of liquefaction, showing building tilting and uneven ground displacements, at the Puerto Montt Naval Base is shown in Fig. 10.19. Similar liquefaction of sands and silts underlying level ground is reported to have occurred at Valdivia and other locations.

Fig. 10.17. Settlement of Niigata airport building (1964).

Fig. 10.18. Damage to airport runways, Niigata (1964).

Fig. 10.19. Building settlements in liquefied sand area, Puerto Montt (1960).

10.4 SOIL LIQUEFACTION IN SLOPING GROUND—FLOW SLIDES

If liquefaction occurs in or under a sloping soil mass, the entire mass will flow or translate laterally to the unsupported side in a phenomena termed a flow slide. Such slides also develop in loose, saturated, cohesionless materials during earthquakes and are reported at Chile (1960), Alaska (1964), and Niigata (1964).

Flow slides developed in loose fill along the waterfront at Puerto Montt (described above), as may be seen from the lateral separation of about 9 ft that occurred between two halves of a building, one part of which rested on liquefied soil (Fig. 10.20). A similar phenomena occurred at a school building constructed near the river bank at Niigata.

More extensive flow slides occurred during the Alaskan Earthquake at Valdez (Coulter and Migliaccio, 1966), Seward (Shannon, 1966), Kenai Lake (McCulloch, 1966), and various other locations. At Valdez and Seward, although large parts of the slide masses were under water, extensive sections of the waterfront areas were carried away by the flow slides. An artist's concept of the flow slide at Valdez is shown in Fig. 10.21. The extent of damage by a similar slide involving about 4000 ft of coastline at Seward is illustrated by Fig. 10.22, which shows the dock area before the earthquake, and by Fig. 10.23, which shows the same area after the earthquake. The complete disappearance of the boat harbors, docks, and portions of the coastline is readily apparent. In Fig. 10.23 railroad tracks are seen to be running directly into the water where dock facilities were formerly located. The slide also carried away warehouses and fuel storage tanks. A view of damage to a warehouse at the edge of the slide is shown in Fig. 10.24.

Probably the largest flow slides in recorded history occurred in Kansu Province, China (Close and McCormick, 1922), during an earthquake in 1920. Again the soil involved was a loose, loess-type granular deposit. Large flow slides in this material occurred over an area of 100 mi^2. Cities were buried, and farmsteads were carried away to be set down in new locations once the movements stopped. In one area a piece of a road-

Fig. 10.20. Lateral separation of building at Puerto Montt (1960).

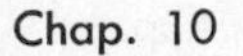

Fig. 10.21. Artist's concept of flow slide at Valdez (1964) (After Coulter and Migliaccio.)

way, with its border of poplar trees, was transported and set down 1 mi from its original location. It is reported that some 200,000 people were killed during these enormous land movements.

However, flow slides also may be induced on a smaller scale by relatively small earthquakes, as evidenced by the slides that occurred along the highway bordering Lake Merced in San Francisco during an earthquake of magnitude 5.5 in 1957. The soil conditions in the area are illustrated in Fig. 10.25; the highway was constructed on several feet of fill resting on a saturated loose sand deposit. As may be seen from the damage shown in Figs. 10.26 and 10.27, sections of the highway were carried some distance into the lake, illustrating the large lateral translations characteristic of flow slides.

10.5 WATERFRONT BULKHEAD FAILURES DUE TO BACKFILL LIQUEFACTION

Since liquefaction develops most frequently in loose, saturated, cohesionless soils, it is not surprising that many instances of waterfront bulkhead failures during earthquakes have been reported due to liquefaction of the backfill. Such bulkheads are often backfilled with loose sand, since it is difficult to compact the backfill

Fig. 10.22. Coastline at Seward before earthquake.

Fig. 10.23. Coastline at Seward after Alaska earthquake (1964).

Fig. 10.24. Damage to port facilities, Seward (1964).

below the water level and the soil is naturally saturated for the most part due to its proximity to the adjacent water. Thus quay walls and bulkheads in dock areas often suffer major damage during earthquakes due to the fact that liquefied backfills exert higher pressures than those for which the walls are designed.

Extensive failures of quay walls and bulkheads as a result of backfill liquefaction occurred at Puerto Montt and Niigata. At Puerto Montt (Duke and Leeds, 1963), during the Chilean Earthquake, failure of retaining structures occurred over a length of 2000 feet. Quay walls, consisting typically of 16-ft reinforced concrete sections constructed on about 35-ft high soil-filled concrete caissons, overturned completely over a length of about 900 ft; over another section 700 ft long, the reinforced concrete upper section overturned completely while the lower caisson section tilted outward. Views of the damaged zones are shown in Figs. 10.28 and 10.29.

Fig. 10.25. Soil conditions adjacent to Lake Merced, San Francisco.

Fig. 10.26. Slope failures around Lake Merced, San Francisco (1957).

Fig. 10.27. Slope failures around Lake Merced, San Francisco (1957).

Fig. 10.28. Quay wall failures at Puerto Montt (1960).

Fig. 10.29. Quay failures at Puerto Montt (1960).

Fig. 10.30. Building damage due to outward movement of sheet pile bulkhead, Niigata (1964).

Anchored sheet pile bulkheads along the waterfront were pushed outward up to 3 ft over a length of about 1250 ft, with resulting settlement and distortion of the railway tracks behind them.

Similar damage occurred to the sheet pile bulkheads with loose sand backfills in the harbor area of Niigata. Extensive sections moved outward under the increased backfill pressures developed by the backfills during the earthquake, with resulting damage to structures behind them due to settlement and lateral translation of the foundations. A view of such wall movement and the collapsed building behind it is shown in Fig. 10.30. Damage to the facilities in the harbor area of Niigata was estimated at many millions of dollars.

Extensive failures of walls involved in flow slides at Valdez and Seward in the Alaskan Earthquake already have been described.

10.6 SLIDES CAUSED BY LIQUEFACTION OF THIN SAND LAYERS

Loose saturated silts and sands often occur as essentially horizontal thin seams or layers underlying firmer materials. In such cases liquefaction of the sand induced by earthquake ground motions may cause an overlying sloping soil mass to slide laterally along the liquefied layer at its base. When this happens, a zone of soil at the back end of the sliding mass sinks into the vacant space formed as the mass translates (see Fig. 10.31) resulting in a depressed zone known as a graben. During motions of this type structures on the main slide mass are translated laterally, often without significant damage, but buildings in the graben area are subjected to such large differential settlements that they are often completely destroyed. Furthermore, buildings near the toe of the slide area, where the soil pushes

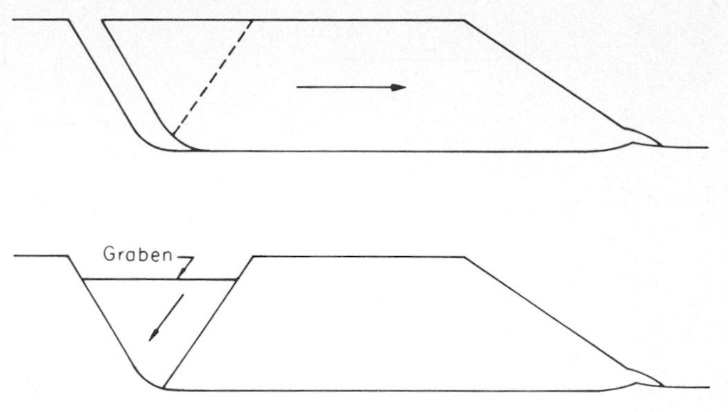

Fig. 10.31. Mechanism of graben formation due to sliding on horizontal layer.

outward, are often heaved upward or pushed over by the lateral thrust.

Several slides of this type occurred in Anchorage during the Alaskan Earthquake of 1964 (Shannon and Wilson, Inc., 1964; Hansen, 1965). A view of a graben, about 10 ft deep and 100 ft wide, where it crossed a

Fig. 10.32. Graben behind L-Street slide area, Anchorage (1964).

Fig. 10.33. Aerial view of graben behind L-Street slide area, Anchorage (1964).

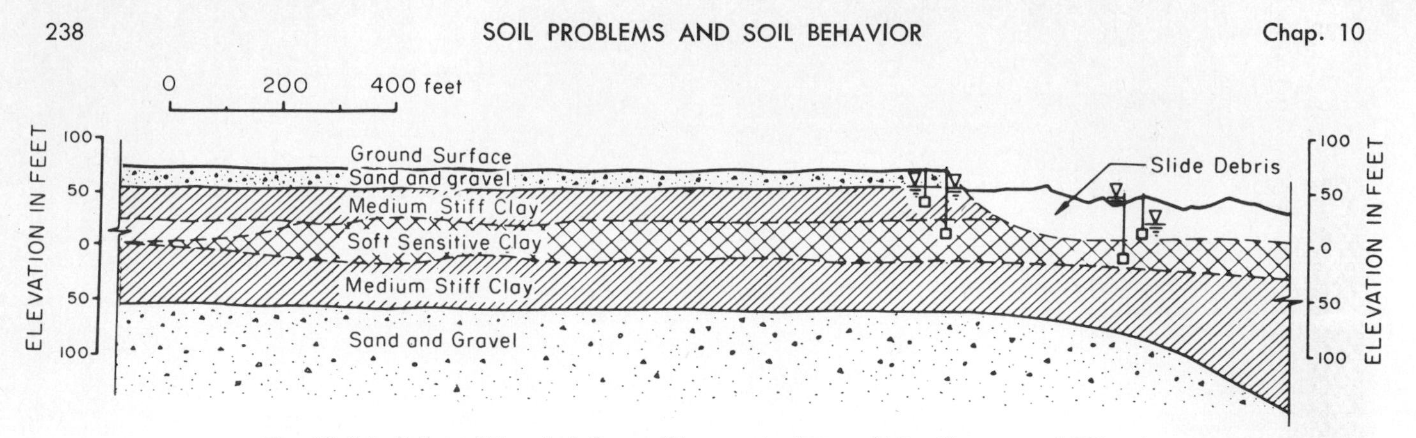

Fig. 10.34. Soil conditions in L-Street slide area, Anchorage. (After Shannon and Wilson.)

Fig. 10.35. Undamaged houses alongside graben, L-Street slide area, Anchorage (1964).

field behind the L-Street slide area in Anchorage, is shown in Fig. 10.32. An aerial view of part of the same slide area is shown in Fig. 10.33. The graben can be readily discerned, winding its way across the landscape. In this slide the land mass, about 4000 ft long and 1200 ft wide, moved laterally about 14 ft. A cross section through the slide area is shown in Fig. 10.34.

Undamaged houses along the edge of the graben are shown in Fig. 10.35, while Fig. 10.36 shows a five-story apartment building that moved 14 ft on the main slide mass with no discernible damage except a severing of underground utilities. The relatively minor distortion of a building that settled some 10 ft but entirely within the graben is shown in Fig. 10.38, but the disastrous effect of a similar slide along Fourth Avenue in Anchorage on a building located on the edge of the resulting graben may be seen in Fig. 10.38. Many buildings suffered in this way.

A similar type of slide with rather more extensive land movement producing two grabens also occurred in the Government Hill area. An aerial view of the Government Hill slide is shown in Fig. 10.39, and the damage to a school building located in the slide area is shown in Fig. 10.40.

The upward heaving of the center portion of a building located near the toe of the L-Street slide is shown in Fig. 10.41. At the toe of the Fourth Avenue slide area, the movement of the soil caused the complete collapse of a number of warehouse buildings.

Fig. 10.36. Apartment building that translated 14 ft during L-Street slide, Anchorage (1964).

Fig. 10.37. Settlement of building into graben, Anchorage (1964).

Fig. 10.38. Differential settlement of building at edge of graben, Anchorage (1964).

Fig. 10.40. Damage to school building in Government Hill, slide area, Anchorage (1964).

Fig. 10.39. Aerial view of Government Hill slide, Anchorage (1964).

Fig. 10.41. Heaving of building at toe of L-Street slide area Anchorage (1964).

10.7 LANDSLIDES IN CLAY SOILS

Major slide movements during earthquakes also can result from failure in clay deposits. However, soft clay deposits often contain sand lenses, and liquefaction of these lenses may well contribute significantly to the slide development even in such cases.

One of the largest slides of this type was that which occurred along the coastline of the Turnagain Heights area of Anchorage (Seed and Wilson, 1967) during the Alaskan Earthquake of 1964. An aerial view of the slide area is shown in Fig. 10.42. The coastline in this area was marked by bluffs some 70 ft high and sloping at about $1\frac{1}{2}$: 1 down to the bay. The slide extended about 8500 ft from east to west along the bluff line and retrogressed inland an average distance of about 900 ft. The total area within the slide zone was thus about 130 acres.

The soil conditions at the east end of the slide area are shown in Fig. 10.43. Within the slide area the original ground surface was completely devastated by displacements that broke up the ground into a complex system of ridges and depressions. In the depressed areas between the ridges the ground dropped an average of 35 ft during the sliding.

The east end of the slide area had been developed for residential purposes, and about 75 houses in this section of the area were destroyed. A general view of the central part of the slide is shown in Fig. 10.44, and a similar view at the east end is shown in Fig. 10.45. Destruction of the houses is readily apparent in these figures. A study conducted by the Engineering Geology Evaluation Group in Anchorage revealed that some of the houses moved laterally as much as 500–600 ft during the sliding. Two houses that were originally separated by several hundred feet but came together during the sliding are shown in Fig. 10.46.

Three somewhat similar slides occurred on the San Pedro River near Lake Rinihue during the Chilean

Fig. 10.42. Turnagain Heights slide area, Anchorage (1964).

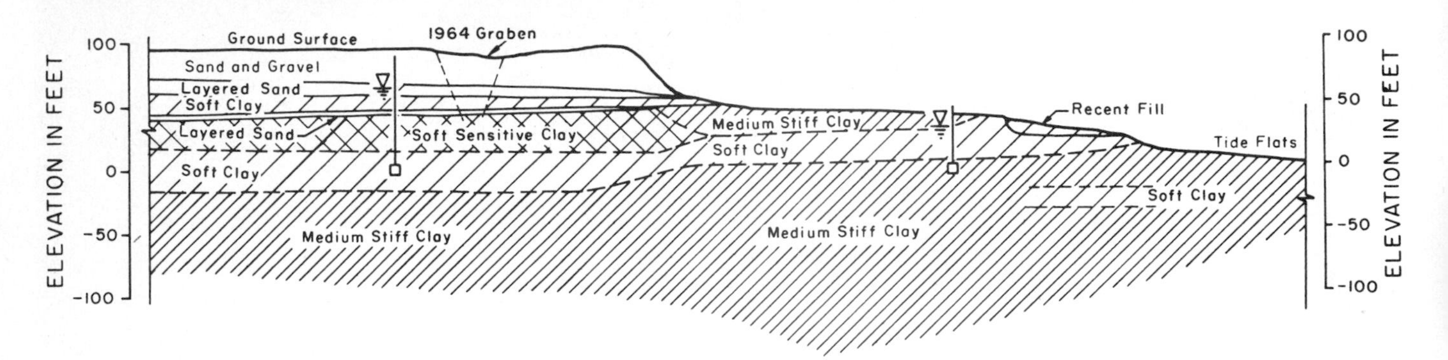

Fig. 10.43. Section through east end of Turnagain Heights slide area, Anchorage (1964).

Fig. 10.44. View in central part of Turnagain Heights slide area, Anchorage (1964).

Fig. 10.45. View in east end of Turnagain Heights slide area, Anchorage (1964).

Fig. 10.46. Damaged houses in Turnagain Heights slide area, Anchorage (1964).

Fig. 10.47. Slide near Lake Rinihue, Chile (1960).

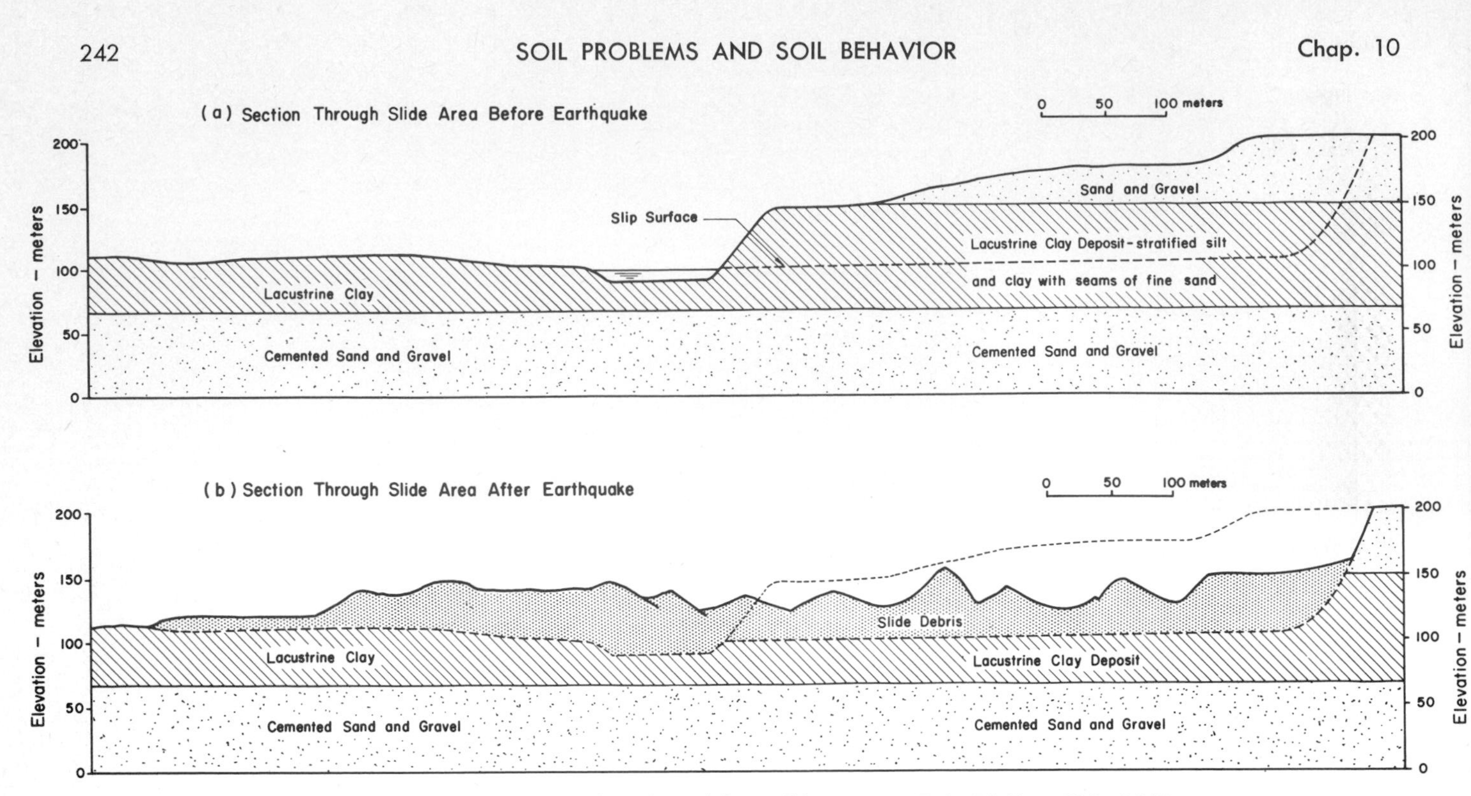

Fig. 10.48. Approximate sections through large slide area near Lake Rinihue, Chile (1960). (After Davis and Karzulovic, 1961)

Earthquake of 1960 (Davis and Karzulovic, 1961). The largest of these involved about 30 million yd^3 of material that moved vertically about 60 ft and laterally about 1000 ft during the sliding. The other two slides involved about 6 and 2 million yd^3 of soil respectively. A view of the smallest slide is shown in Fig. 10.47, and a cross section through the largest slide is shown in Fig. 10.48.

The slides blocked the course of the river and raised the water level in the lake by 80 ft, thereby endangering the City of Valdivia some 35 mi to the west. It is of interest to note that a similar effect occurred when an adjacent area slid into the river during an earthquake in 1575.

Detailed studies (Davis and Karzulovic, 1961) of the largest slide have indicated that the main sliding surface was located in a clay deposit containing numerous silt and sand seams and that liquefaction of these seams was probably the main cause of the slide movement.

Six other similar landslide areas are reported (Duke and Leeds, 1963) to have developed during the earthquake in the provinces of Osorno and Valdivia, while

Fig. 10.49. Slumping of highway fill in Kern County, California (1952). (After Perry 1955)

Fig. 10.50. Slumping of fill in Chilean earthquake (1960).

Fig. 10.51. Slumping of Hebgen Dam relative to concrete core wall (1957).

a number of slides in cohesive soils occurred in the Anchorage bluffs during the Alaskan Earthquake of 1964.

10.8 SLOPING FILLS ON FIRM FOUNDATIONS

The effect of earthquakes on banks of well-compacted fill constructed on firm foundations in which no significant increases in pore-water pressure develop during the earthquake is characteristically a slumping of the fill varying from a fraction of an inch to several feet. Slumping of six to eight inches of a highway fill (Perry, 1955), which occurred during a severe aftershock of the Kern County, California, Earthquake in 1952, is shown in Fig. 10.49, while similar slumping of a fill in the Chilean Earthquake is shown in Fig. 10.50. Slumping of about 3 ft of the compacted fill shell of the Hebgen Dam (Sherard, 1959) relative to the concrete core wall that occurred during the Hebgen Lake Earthquake of 1957 is shown in Fig. 10.51.

10.9 SLOPING FILLS ON WEAK FOUNDATIONS

In contrast to the behavior of fills on firm foundations, many cases have been reported in which fills on weak foundations have either failed completely or slumped severely with associated longitudinal cracking.

An excellent example is provided by the behavior of a fill section of the north–south highway between Puerto Varas and Puerto Montt, Chile. A 4-ft high fill, about 1500 ft long, was constructed where the highway passed over a swampy area of ground and surfaced with an 8-in. thick concrete pavement. During the earthquake of 1960 the fill collapsed completely, as shown in Figs. 10.52 and 10.53, with the concrete pavement slabs being deposited at the level of the swamp. The undamaged highway, where it enters and leaves the swamp area, is readily apparent in the figures.

Similar behavior of a dike, built to a height of 7 ft along the bank of a canal in northern Mexico, which sank completely into the underlying soft foundation soil during the El Centro Earthquake of 1940, is shown in Fig. 10.54, and the complete collapse of the side fill of a highway bridge approach ramp is shown in Fig. 10.55.

Such failures seem to be characterized by lateral spreading of the base of the fill, and where movement is less severe the sliding leads to severe longitudinal cracking of the fill. This type of behavior developed along many sections of highway fills in the Alaskan and Niigata Earthquakes of 1964. Extensive cracking of fills constructed on a silt deposit on the highway from Anchorage to Portage, Alaska, is shown in Figs. 10.56 and 10.57.

One of the most dramatic examples of the failure of an entire fill resulting from loss of strength of the foundation soil during an earthquake is the failure of the Sheffield Dam (Ambraseys, 1960) during an earthquake near Santa Barbara in 1926. A view of the completed dam after construction but before filling the

Fig. 10.52. Collapse of highway fill in Chilean earthquake (1960).

Fig. 10.53. Collapse of highway fill in Chilean earthquake (1960).

Fig. 10.54. Collapse of dike along bank of canal during El Centro earthquake (1940).

Fig. 10.55. Collapse of highway bridge approach fill during El Centro earthquake (1940).

Fig. 10.56. Cracking of embankment on Anchorage–Portage highway (1964).

Fig. 10.57. Cracking of embankment on Anchorage–Portage highway (1964).

Fig. 10.58. Sheffield Dam, Santa Barbara, before earthquake.

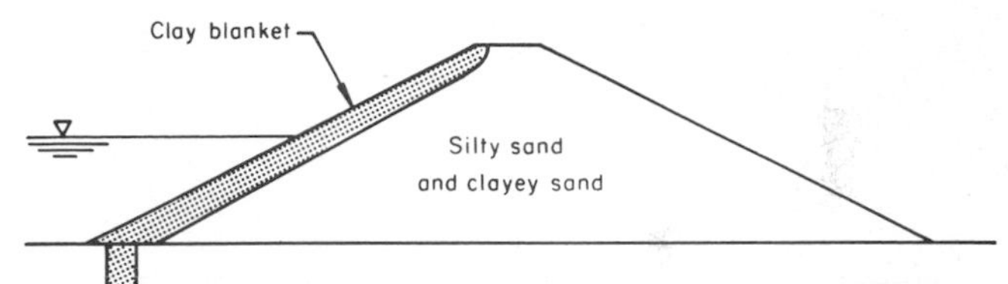

Fig. 10.59. Section through Sheffield Dam.

Fig. 10.60. Failure of Sheffield Dam during Santa Barbara earthquake (1925).

reservoir is shown in Fig. 10.58, and a cross section of the dam after filling the reservoir is shown in Fig. 10.59. During the earthquake observers reported that a section of the dam, about 300 ft long, slid downstream with a complete loss of water from the reservoir; a view of the embankment after failure is shown in Fig. 10.60.

10.10 CANAL BANKS, RESERVOIRS, AND EARTH DAMS

Some of the most critical types of soil instability problems induced by earthquakes are those that affect water storage or conveyance structures such as canal banks, reservoir banks, or earth dams. Canal banks in particular have a long history of slope failures during earthquakes. Failure of the banks of the All America Canal during the El Centro Earthquake of 1940 is shown in Fig. 10.61, and the bank disruption and associated flooding along a length of the Solfatara Canal is shown in Fig. 10.62.

The ground motions in the Alaska Earthquake caused severe cracking of the concrete lining and fill for a small reservoir in Anchorage. The reservoir was about 300 by 600 ft in plan and was formed by construction of a fill

Fig. 10.61. Failure of bank of All-America Canal, El Centro earthquake (1940). (U.S. Bureau of Reclamation Photograph.)

Fig. 10.62. Failure of banks of Solfatara Canal, El Centro earthquake (1940).

varying up to about 15 ft in height. During the earthquake transverse cracks developed extending through the crest of the fill (Fig. 10.63) and the concrete lining. The cracks were about $\frac{1}{2}$ in. wide across the crest of the embankment and could be readily traced down into the water (Fig. 10.64). Such cracking in some types of soils could lead to complete failure of a reservoir bank.

The failure of a small earth dam, about 15–20 ft high, in the vicinity of Anchorage during the 1964 Alaskan Earthquake is shown in Fig. 10.65. Nothing is known about the method of construction of this small dam, but it indicates the potential dangers when such structures are not adequately designed and constructed. Extensive longitudinal and some traverse cracking of dams has been reported in a number of major earthquakes.

10.11 RETAINING WALLS AND BRIDGE ABUTMENTS

The tendency for earth fills to slide downhill during earthquakes necessarily results in increased pressures

Fig. 10.63. Transverse cracks across bank of reservoir, Anchorage (1964).

Fig. 10.64. Extension of crack through concrete reservoir lining into water, Anchorage (1964).

on retaining walls. Such pressures often result in wall displacements. The displacement of a retaining wall during the Chilean Earthquake of 1960 is shown in Fig. 10.66, and similar, though smaller, movement of the wing walls for a highway bridge during the Niigata Earthquake of 1964 is shown in Fig. 10.67. In spite of the small wall movements shown in Fig. 10.67, they resulted in extensive cracking of the backfill, as may be seen from Fig. 10.68.

Similar displacements occur in bridge abutments, which serve the dual purpose of supporting the bridge deck and retaining the approach fill. Inward movements of bridge abutments during earthquakes causing a distortion of the bridge structure often have been reported. Figure 10.69 shows the slumping of soil adjacent to the wing wall of the Isla–Teja Bridge at Valdivia, Chile (Duke and Leeds, 1963), in the 1960 earthquake. The resulting pressures on the abutment caused the structural distortion shown in Fig. 10.70.

Similar abutment movements caused distortion and collapse of numerous bridges in the Alaskan Earthquake. In many cases these were timber structures that were sufficiently flexible to buckle under the imposed axial loadings (see Fig. 10.71). However, in many cases the loading caused a complete collapse of the deck, as shown in Figs. 10.72 and 10.73. The Alaska Highway Department reported that 22 bridges were destroyed in a 30-mi length of highway between Potter and Portage alone (Grantz, Plafker, and Kachadoorian, 1964).

10.12 CONCLUSION

In the preceding pages an attempt has been made to classify various types of soil instability problems that may develop during earthquakes and to illustrate them by examples from recent major earthquakes. The similarity in types of problems encountered in such widely separated locations as Alaska, Niigata (Japan), and Chile is illustrated by the summary of some of these problems presented in Table 10.1 (p. 250). The list of problems discussed is not exhaustive; e.g., no discussion is presented of stability problems involving surface sliding of dry sands or mass movement of tailings piles from mining operations. However, it is hoped that the examples cited will serve to illustrate the variety of soil problems confronting the civil engineer working in seismically active regions of the world and also to demonstrate the need for careful consideration of the soil conditions in

Fig. 10.65. Failure of small earth dam, Anchorage (1964).

Fig. 10.66. Displacement of retaining wall, Chile (1960).

Fig. 10.68. Cracking of fill due to displacement of wing walls, Niigata (1964)

Fig. 10.67. Displacement of wing walls of highway bridge, Niigata (1964).

Fig. 10.69. Slumping of soil adjacent to Isla-Teja Bridge, Valdivia, Chile (1960).

Fig. 10.70. Distortion of Isla-Teja Bridge due to soil pressure on abutment, Chile (1960).

Fig. 10.71. Buckling of bridge deck due to inward movement of abutments, Alaska (1964).

earthquake-resistant design and construction of engineering structures.

ACKNOWLEDGMENTS

Grateful acknowledgment is made to the following colleagues and organizations who kindly provided the author with photographs of earthquake damage included in this report: C. R. Allen and G. W. Housner (California Institute of Technology); C. K. Chan, C. M. Duke, and J. Penzien (University of California); P. St. Amand (U.S. Naval Weapons Center); the Japanese Society of Soil Mechanics and Foundation Engineering; U.S. Army Corps of Engineers; U.S. Bureau of Reclamation; and the Water Department, City of Santa Barbara.

Fig. 10.72. Collapse of bridge during Alaska earthquake (1964).

Fig. 10.73. Collapse of bridge during Alaska earthquake (1964).

Table 10.1. EXAMPLES OF SOIL INSTABILITY IN RECENT EARTHQUAKES

Type of instability	Location	Earthquake		References
Settlement of cohesionless soils	Homer and Portage	Alaskan Earthquake	1964	Grantz *et al.* (1964)
	Niigata	Niigata Earthquake	1964	Japanese Society of Soil Mechanics and Foundation Engineering (1966); International Institute of Seismology and Earthquake Engineering (1965)
	Valdivia	Chilean Earthquake	1960	
Liquefaction of saturated sands	Niigata	Niigata Earthquake	1964	Japanese Society of Soil Mechanics and Foundation Engineering (1966); International Institute of Seismology and Earthquake Engineering (1965)
	Puerto Montt and Valdivia	Chilean Earthquake	1960	Duke and Leeds (1963)
Flow slides due to liquefaction of cohesionless soils	Valdez, Seward, and Kenai Lake	Alaskan Earthquake	1964	Coulter and Migliaccio (1966); Shannon (1966); McCulloch (1966)
	Niigata	Niigata Earthquake	1964	Japanese Society of Soil Mechanics and Foundation Engineering (1966); International Institute of Seismology and Earthquake Engineering (1965)
	Puerto Montt	Chilean Earthquake	1960	Duke and Leeds (1963)
	San Francisco	San Francisco Earthquake	1957	
Bulkhead failures due to backfill liquefaction	Niigata	Niigata Earthquake	1964	Japanese Society of Soil Mechanics and Foundation Engineering (1966); International Institute of Seismology and Earthquake Engineering (1965)
	Puerto Montt	Chilean Earthquake	1960	Duke and Leeds (1963)
Slides caused by liquefaction of thin sand layers	Anchorage	Alaskan Earthquake	1964	Shannon and Wilson, Inc. (1964); Hansen (1965); Seed and Wilson (1966)
	Rinihue	Chilean Earthquake	1960	Davis and Karzulovic (1961)
Failures of fills on weak foundations	Anchorage–Portage	Alaskan Earthquake	1964	Grantz *et al.* (1964)
	Niigata	Niigata Earthquake	1964	Japanese Society of Soil Mechanics and Foundation Engineering (1966); International Institute of Seismology and Earthquake Engineering (1965)
	Puerto Montt–Puerto Varas	Chilean Earthquake	1960	Duke and Leeds (1963)
Lateral movement of bridge abutments	Anchorage–Seward Highway	Alaskan Earthquake	1964	Grantz *et al.* (1964)
	Niigata	Niigata Earthquake	1964	Japanese Society of Soil Mechanics and Foundation Engineering (1966); International Institute of Seismology and Earthquake Engineering (1965)
	Valdivia	Chilean Earthquake	1960	Duke and Leeds (1963)

REFERENCES

Ambraseys, N. N. (July 1960). *On the Seismic Behaviour of Earth Dams, Proceedings of the Second World Conference on Earthquake Engineering*, Tokyo, Japan.

Close, U., and E. McCormick (May 1922). "Where the Mountains Walked," *Natl. Geographic* Mag., **XLI** (5).

Coulter, H. W., and R. R. Migliaccio (1966). *Effects of the Earthquake of March 27, 1964 at Valdez, Alaska*, Washington, D.C.: Geological Survey Professional Paper 542-C, U.S. Department of the Interior.

Davis, S., and J. K. Karzulovic (1961). *Deslizamientos en el valle del rio San Pedro Provincia de Valdivia Chile*, Anales de la Faculted de Ciencieas Fisical y Matematicos, Santiago, Chile: University of Chile, Institute of Geology, Publication No. 20.

Duke, C. M., and D. J. Leeds (February 1963). "Response of Soils, Foundations, and Earth Structures to the Chilean Earthquakes of 1960," *Bull. Seism. Soc. Am.*, **53** (2).

Grantz, A., G. Plafker, and R. Kachadoorian (1964). *Alaska's Good Friday Earthquake, March 27, 1964*, Washington, D.C.: Department of the Interior, Geological Survey Circular 491.

Hansen, W. R. (1965). *Effects of the Earthquake of March 27, 1964 at Anchorage, Alaska*, Washington, D.C.: Geological Survey Professional Paper 542-A, U.S. Department of the Interior.

Henderson, J. (July 1923). "The Geological Aspects of the Hawke's Bay Earthquakes," *New Zealand J. Sci. Technol.*, No. XV.

International Institute of Seismology and Earthquake Engineering (February 1965). *The Niigata Earthquake of 16 June, 1964 and Resulting Damage to Reinforced Concrete Building*, Tokyo, Japan: OIISEE Report No. 1.

Japanese Society of Soil Mechanics and Foundation Engineering (January 1966), "Soil and Foundation," Vol. VI, No. 1.

McCulloch, D. S. (1966). *Slide Induced Waves, Seiching and Ground Fracturing Caused by the Earthquake of March 27, 1964, at Kenai Lake, Alaska*, Washington, D.C.: Geological Survey Professional Paper 543-A, U.S. Department of the Interior.

Perry, O. W. (November 1955). *Highway Damage Resulting from the Kern County Earthquakes*, Bulletin 171, Division of Mines, State of California, Department of Natural Resources.

Seed, H. B. and S. D. Wilson (July, 1967). *The Turnagain Heights Landslide in Anchorage, Alaska*, Journal of the Soil Mechanics and Foundations Division, ASCE.

Shannon, W. L. (August 1966). "Slope Failures at Seward, Alaska," Paper prepared for ASCE Conference on Stability and Performance of Slopes and Embankments, Berkeley, California.

Shannon and Wilson, Inc. (1964). *Report on Anchorage Area Soil Studies, Alaska, to U.S. Army Engineer District, Anchorage, Alaska*, Seattle, Washington.

Sherard, J. L. (August 17, 1959). "A Report on the Damage to Hebgen Dam in the West Yellowstone Earthquake," *Eng. News Rec.*

Chapter 11

Tsunamis

ROBERT L. WIEGEL

Professor of Civil Engineering
University of California
Berkeley, California

11.1 CAUSES AND NATURE OF TSUNAMIS

Tsunamis are the long water waves (with wave "periods" in the 5 to 60 min, or longer, range) generated impulsively by mechanisms such as exploding islands (Krakatoa in 1883, see Wharton and Evans, 1888), submerged landslides (Sagami Bay, Japan, in 1933, see Shepard, 1933, and Gutenberg, 1939), rockfalls into bays or the ocean (Lituya Bay, Alaska, in 1958, see Miller, 1960), tectonic displacements associated with earthquakes (Alaskan tsunami of 1964, see U.S. Coast and Geodetic Survey, 1964), and underwater explosions of nuclear devices. The word tsunami is now generally used to describe these waves, in preference to either of the terms tidal wave or seismic sea wave. It is likely that the major cause of catastrophic tsunamis is a rapidly occurring tectonic displacement of the ocean bottom. It appears that the displacements should have a substantial vertical component (dip-slip) if they are to be able to generate tsunamis of large magnitudes.

Before After

Fig. 11.1. Damage at Hilo, Hawaii, due to tsunami of May 23, 1960.

One would expect that strike-slips would have to occur through a seamount, submarine cliff, etc. to generate a tsunami, and owing to the rapid decrease of the ground displacement with distance from the fault it is unlikely that major tsunamis would be generated by this mechanism.

Tsunamis are important because of the loss of life and vast property damage that have resulted from the larger ones. One of the worst tsunamis in historical times occurred in Japan during the evening of June 15, 1896. The biggest wave rushed onto nearby land to an elevation of from 75 to 100 ft above the tide level, engulfing entire villages. More than 27,000 persons were killed and 10,000 houses destroyed (Leet, 1948). Many other tsunamis have caused appreciable, but less, loss of life. For example, the April 1, 1946, wave killed more than 150 persons, badly injured another 163 persons, and caused about $25 million in property damage in the Hawaiian Islands, largely in Hilo (Shepard, MacDonald, and Cox, 1950).

On May 23, 1960, a disastrous tsunami struck Chile, Hawaii, and Japan. Some of the damage suffered in Hilo, Hawaii, can be seen in Fig. 11.1. Frame buildings were largely destroyed, but reinforced concrete buildings were only damaged. The force of the wave can be estimated from the bent parking meters shown in the photograph. The turf in the area between the parking lot and the concrete building in the background remained in fair condition.

On March 27, 1964, a major tsunami originated in the Gulf of Alaska and caused loss of life and extensive damage (about $11 million) in Crescent City, California (Griffin *et al.*, 1964; Magoon, 1966). Major damage occurred in a number of coastal towns in Alaska; much of the damage was due to a combination of the earthquake, soil failures, and the tsunami (Grantz, Plafker, and Kachadoorian, 1964).

Tsunamis have occurred in the Bay of Bengal, Caribbean Sea, Celebes Sea, Java Sea, Mediterranean Sea, Sea of Japan, and the South China Sea (Cueller, 1953; Leet, 1948). They have also occurred in the Atlantic Ocean; in fact, one of the most devastating tsunamis in recorded history was caused by the Lisbon earthquake on November 1, 1755 (Pereira, 1888). However, the Pacific Ocean has been subject to more large tsunamis than any other region. A glance at a map on which the locations of large shallow earthquakes are plotted shows why (see, e.g., Gutenberg and Richter, 1949). The Pacific Ocean is ringed with active seismic zones, with many of

Fig. 11.2. Relationship between magnitude M and focal depth h. Submarine earthquakes during the period from 1900 to 1960. (From Iida, 1963a.)

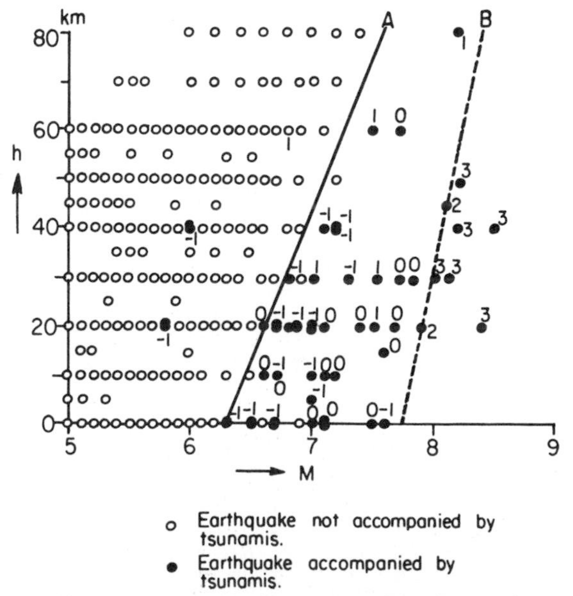

The numeral outside of the circle is the tsunami magnitude.

the most severe earthquakes occurring offshore in relatively shallow water. There are numerous papers on tsunamis, many of which are referred to in several bibliographical publications (Heck, 1947; Cueller, 1953; Ambraseys, 1962; Berringhausen, 1962; Japanese Organization for Tsunami Investigations, 1962; Spaeth, 1964), and the reader is referred to these publications for access to detailed papers on specific tsunamis or specific locations.

Most earthquakes that occur offshore are of small enough magnitude, and/or are of such deep focal depth that no noticeable tsunamis accompany them. The probable reason for this is that this type of earthquake is unlikely to be associated with an appreciable tectonic displacement. Iida's (1963a) results for tsunamis generated near Japan are shown in Fig. 11.2. To the left of line *A*, no tsunamis of any appreciable height have been observed. The data between lines *A* and *B* are important to areas in which the elevations are in the range of 2 to about 10 ft above the normal sea level at the time of the tsunami. The tsunamis designated by the magnitudes *m* of 2 and 3, which are to the right of line *B*, are major ones, with the water running up on land to an elevation of 20 ft, or more, above the normal sea level at the time of the tsunami (in Japan for locally generated tsunamis). In Table 11.1 the relationship between the tsunami magnitude *m* and the maximum tsunami run-up elevation is given. The relationship between tsunami magnitude *m* and the Richter scale earthquake magnitude *M* is given in Fig. 11.3. This figure is derived from data of the type shown in Table 11.2, which relates tsunami energy to earthquake energy.

As an oversimplification, the displacement of the Earth's surface can be visualized as acting as a huge paddle, shoving a vast amount of water. This causes a series of tsunami waves, with the greater the magnitude of the earthquake (hence, the tsunami), the longer the period of the waves (see Fig. 11.4), as well as the higher the waves.

The magnitude of an earthquake appears to be related to the size of the area in which the aftershocks

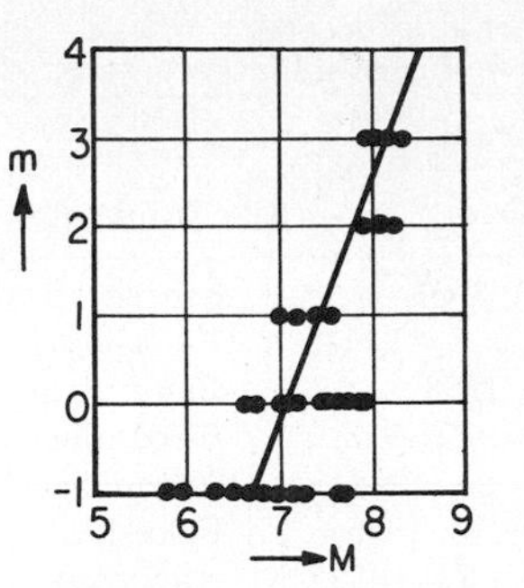

Fig. 11.3. Relationship between earthquake magnitude *M* and tsunami magnitude *m*. (From Iida, 1963a.)

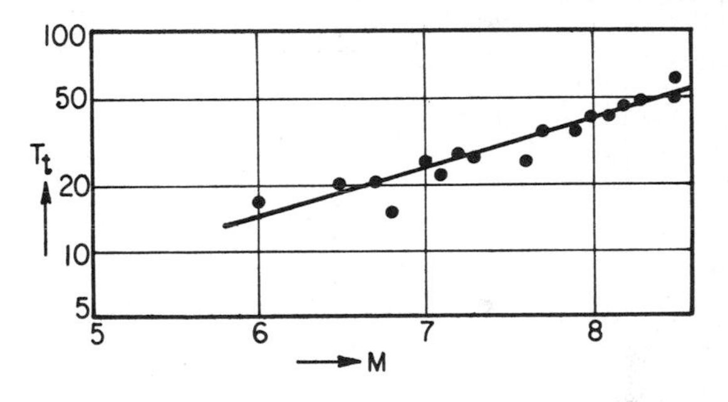

Fig. 11.4. Relationship between the maximum period of a tsunami and earthquake magnitude. (From Iida, 1963b.)

Table 11.1. MAGNITUDE, ENERGY, AND RUN-UP ELEVATION OF TSUNAMIS IN JAPAN*

Tsunami magnitude classification	Tsunami energy		Maximum run-up elevation	
m	ergs	ft-lb	m	ft
5	25.6×10^{23}	18.9×10^{16}	> 32	> 105
4.5	12.8	9.4	24–32	79–105
4	6.4	4.7	16–24	52.5–79
3.5	3.2	2.4	12–16	39.2–52.5
3	1.6	1.2	8–12	26.2–39.2
2.5	0.8	0.59	6–8	19.7–26.2
2	0.4	0.29	4–6	13.1–19.7
1.5	0.2	0.15	3–4	9.9–13.1
1	0.1	0.074	2–3	6.6–9.9
0.5	0.05	0.037	1.5–2	4.9–6.6
0	0.025	0.018	1–1.5	3.2–4.9
−0.5	0.0125	0.0092	0.75–1	2.5–3.2
−1	0.006	0.0044	0.50–0.75	1.6–2.5
−1.5	0.003	0.0022	0.30–0.50	1.0–1.6
−2	0.0015	0.0011	<0.30	<1.0

*After Iida (1963a).

Table 11.2. Tsunami energy*

	Date	Tsunami	Earthquake energy ergs	Earthquake energy ft-lb	Tsunami energy ergs	Tsunami energy ft-lb	Tsunami energy / Earthquake energy
1	March 3, 1933	Sanriku	20.0×10^{23}	14.7×10^{16}	17×10^{22}	12.5×10^{15}	0.085
2	November 3, 1936	Fukushima	2.2	1.6	0.2	0.15	0.0091
3	May 23, 1938	Fukushima	0.28	0.21	0.04	0.03	0.011
4	November 5, 1938	Fukushima	2.2	1.6	0.2	0.15	0.0091
5	December 7, 1944	Tonankai	6.4	4.7	7.9	5.8	0.12
6	February 10, 1945	Fukushima	0.56	0.41	0.04	0.03	0.007
7	December 21, 1946	Nankaido	9.0	6.6	8.0	5.9	0.098
8	March 2, 1952	Tokachi	9.0	6.6	8.0	5.9	0.098
9	November 4, 1952	Kamchatka	13.0	9.6	15	11	0.11
10	November 25, 1953	Boso	1.0	0.7	0.7	0.5	0.07

*After Iida (1963a).

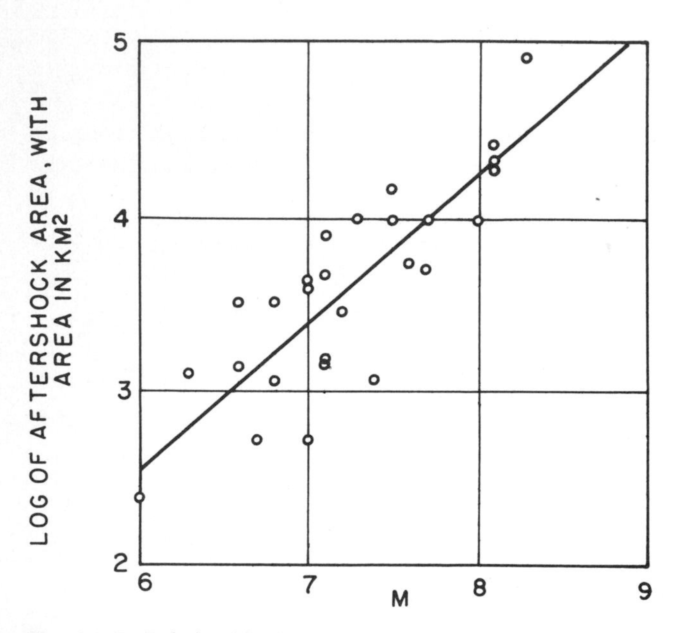

Fig. 11.5. Relationship between aftershock area and magnitude M of tsunamigenic earthquakes. (From Iida, 1963a.)

occur (Fig. 11.5) and to the maximum resultant ground displacement (Fig. 11.6). As the magnitude of a tsunami is related to the magnitude of the earthquake with which it is associated, the tsunami must be related to the aftershock area, the resultant ground displacement, and the water depth in this region. The larger the aftershock area, and the larger the resultant ground displacement, the greater the magnitude of the earthquake and the greater the magnitude of the associated tsunami. In addition, the size of a tsunami at a particular location depends upon the shape and orientation of the displaced ground. Except for certain remarkable circumstances, it would appear that relatively substantial vertical displacements of the Earth's surface must occur to generate large tsunamis.

The ground displacements for the recent large earthquakes in Chile (May 22, 1960) and Alaska (March 27, 1964) were roughly rectangular in shape, so that they acted more like an elliptical source for the tsunami waves rather than a circular source. Owing to this, these waves moved out as a spreading ellipse, rather than as a spreading circle as would be the case for waves generated by a rock thrown into a pool of water. If the ocean were of uniform depth the maximum wave heights would lie nearly along a great circle route drawn normal to the center of source line, but modified slightly by the Coriolis force (Wiegel, 1964a). Refraction modifies this, but it is still an approximation to the real

Fig. 11.6. Trend of dependence of maximum resultant ground displacement on earthquake magnitude. (From Wilson, Webb, and Hendrickson, 1962.)

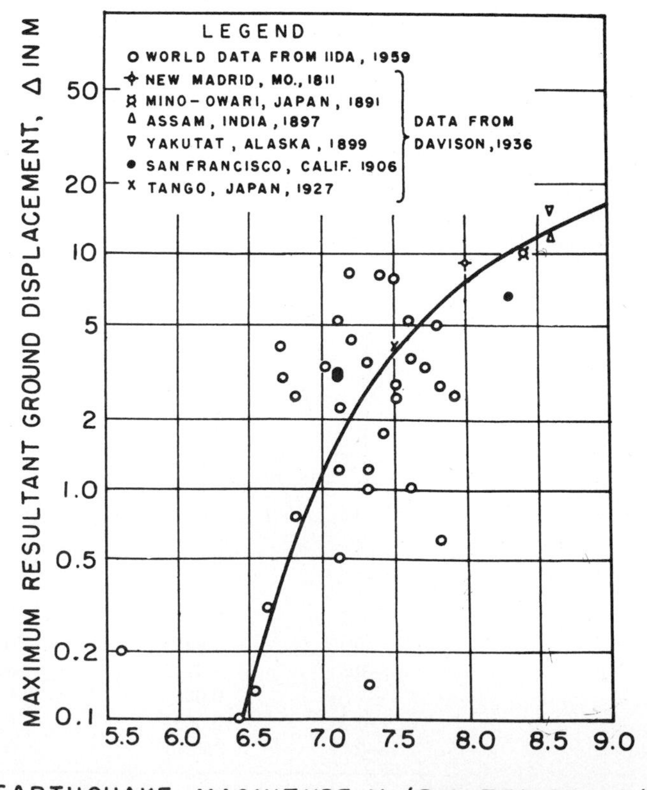

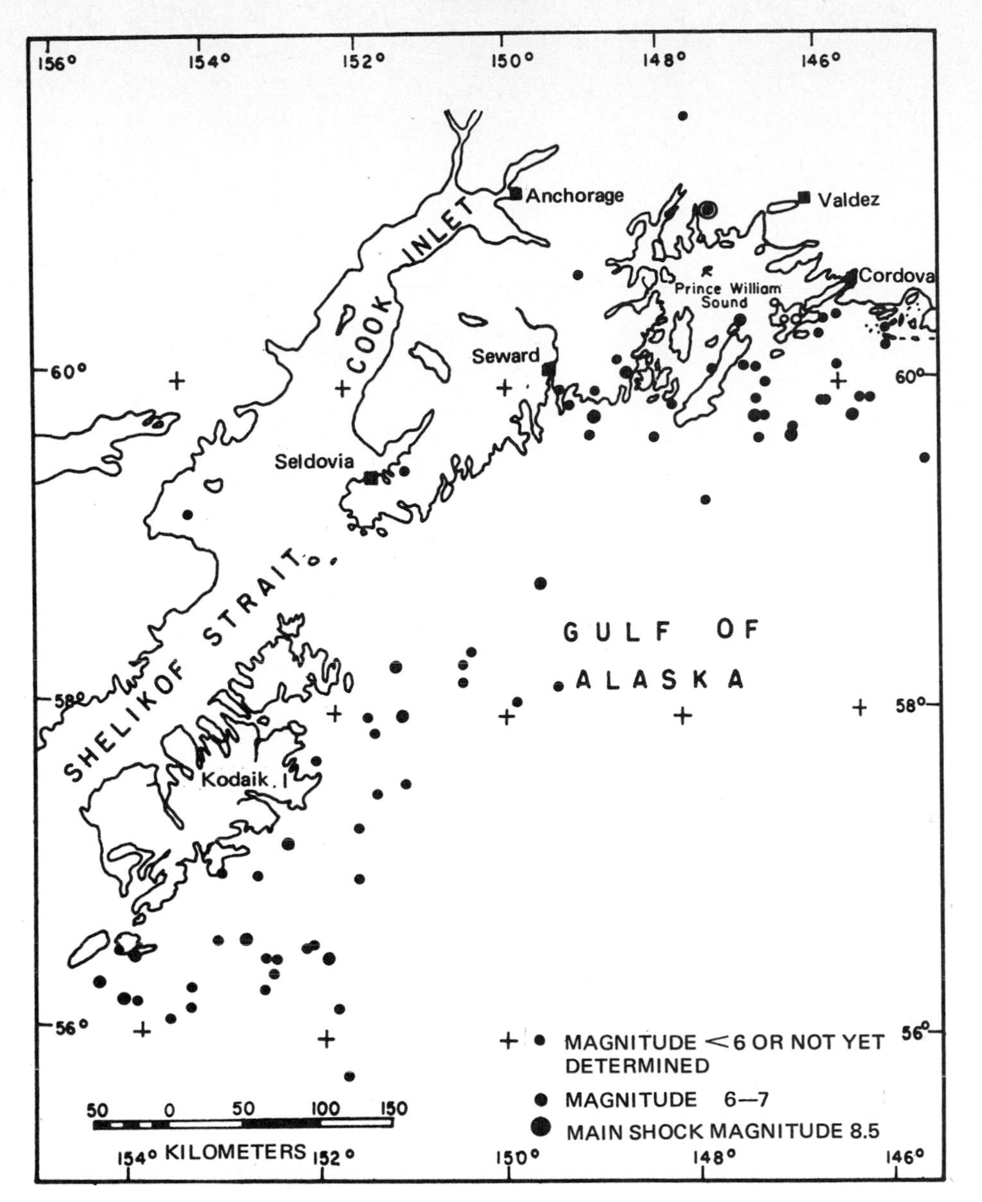

Fig. 11.7. Epicenter map. Prince William Sound Earthquake of March 28, 1964, and after-shocks.(From U.S. Coast and Geodetic Survey, 1964.)

solution. Using this concept, one can easily explain the high tsunami waves in the general vicinity of Crescent City due to the March 27, 1964, Alaskan Earthquake. If one draws a line through the epicenters of the after-shocks (Fig. 11.7), plotted on a terrestrial globe, and then draws a great circle normal to the line, starting from the center of the source line, it can be seen that it heads to the area comprised of northern California and southern Oregon (Fig. 11.8). This would explain the relatively low tsunami waves in Hawaii and Japan and the relatively high waves in the vicinity of Crescent City, even though the great circle approaches the area at glancing incidence.

It is generally agreed that in the open ocean tsunami waves are very long and low; e.g., of the order of 50 mi in length and less than about 2 ft in height. They travel at a speed of about $\sqrt{gd}$ ft/sec, where g is the acceleration of gravity and d is the water depth. (Thus, in the mid-Pacific area they travel at speeds of from 350 to 500 mph.) Owing to the effect of the water depth on the speed of the wave, the wave bends in water of varying depth. This process is known as refraction. In some areas the offshore topography causes the waves to refract in such a manner that the wave energy converges, so that the waves are higher than the average while in other areas the hydrography is such that the wave energy diverges so that the waves are lower than the average.

When tsunami waves reach a coast, the water runs up onto the land. The elevation above the tide level (at the time of the tsunami) reached by this water is called the run-up elevation. The run-up elevation is

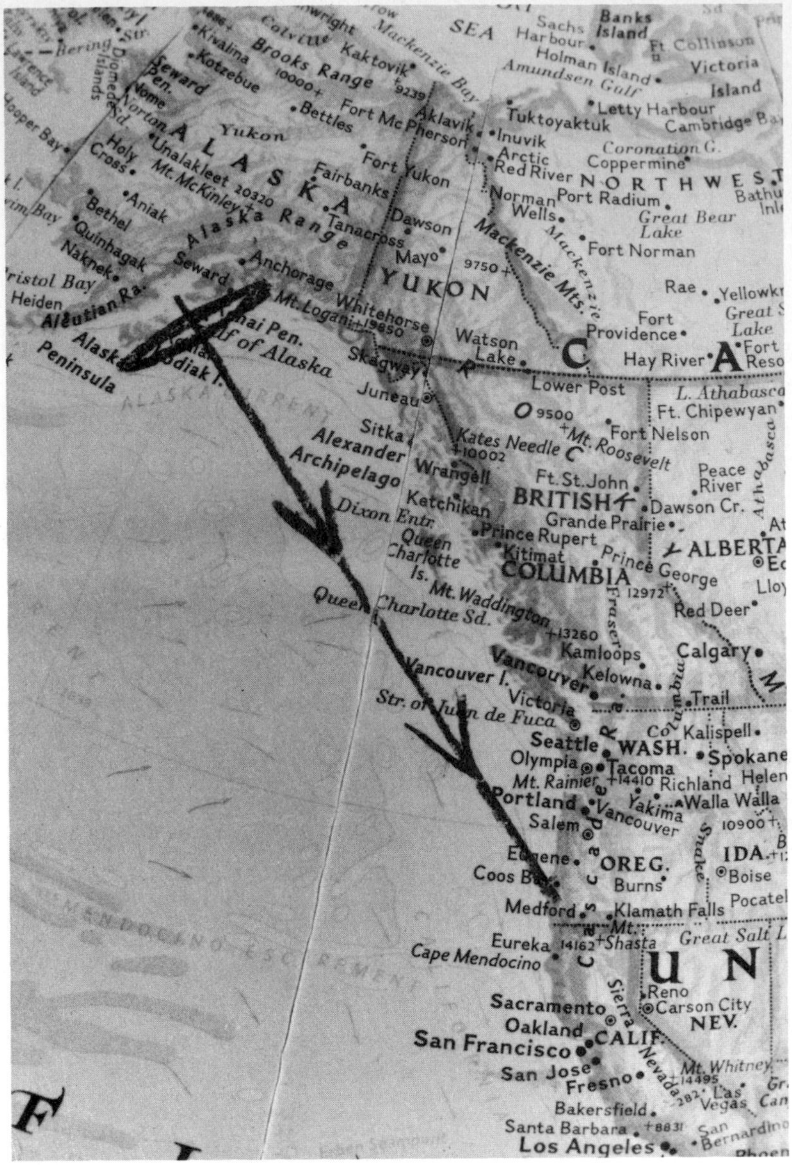

Fig. 11.8. Photograph of globe showing great circle path normal to center of March 28, 1964, Alaskan Earthquake epicenter locations.

Fig. 11.9. Sketch showing definitions of tsunami wave height and tsunami wave run-up.

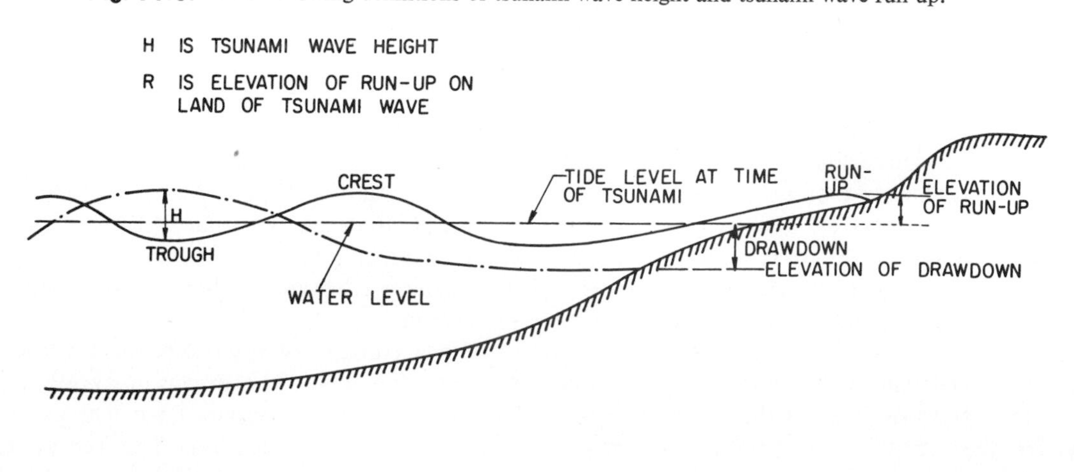

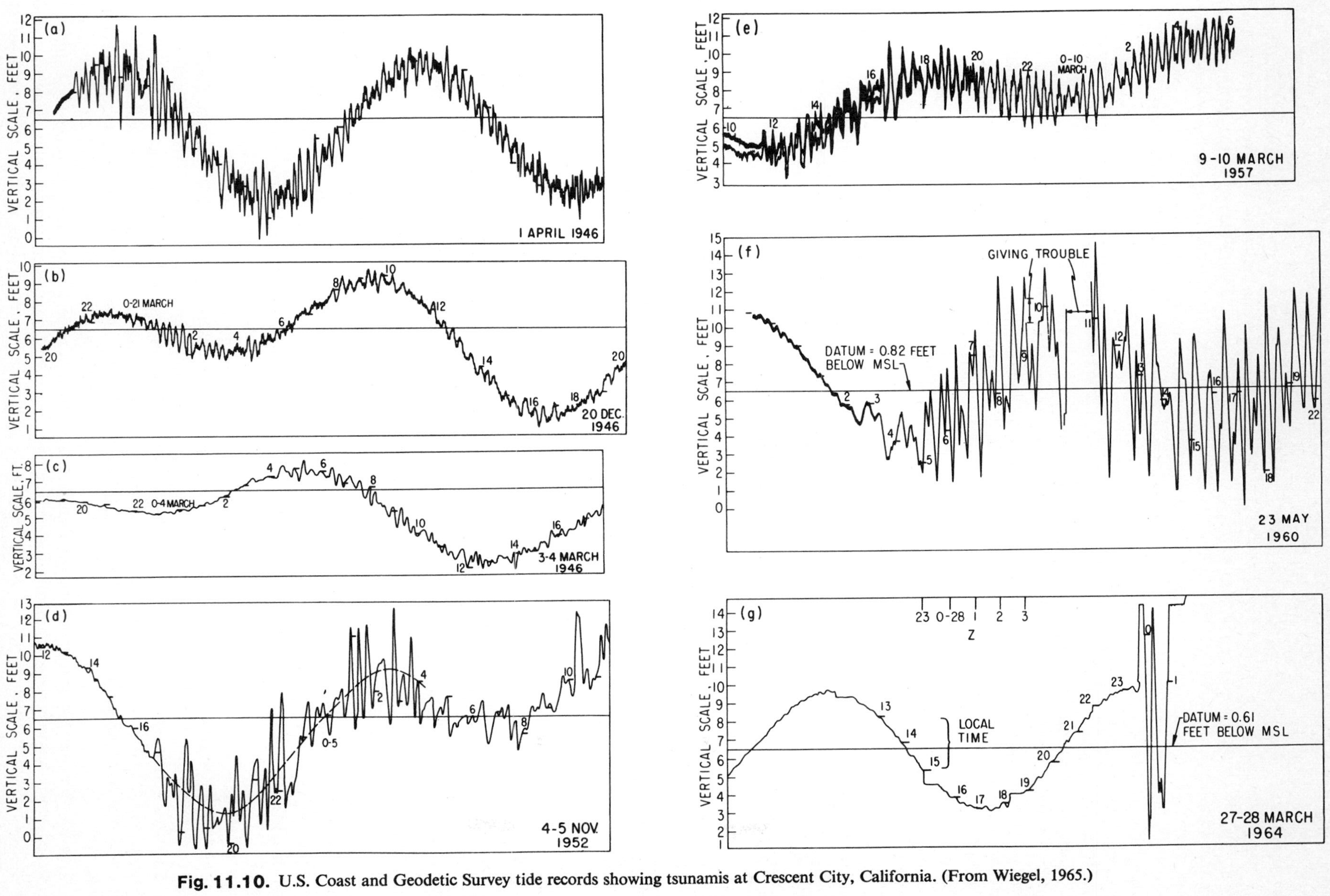

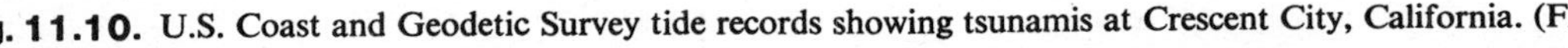

Fig. 11.10. U.S. Coast and Geodetic Survey tide records showing tsunamis at Crescent City, California. (From Wiegel, 1965.)

not the same as the wave height, which is the vertical distance between the crest and trough of a tsunami wave. The two terms have been illustrated in Fig. 11.9. As tsunamis are not waves of translation, the water level draws down along the coast as well as runs up onto the land. The vertical distance of the drawdown is probably as great, on the average, as the vertical distance of the run-up, although no observations are available with which to substantiate this statement.

There is almost no accurate information on drawdown elevations. However, most tide gages are located close to shore—usually within a fraction of one tsunami wave length of the shore. Because of this, an examination of tide gage records should be indicative of the relative importance of wave run-up and drawdown. Examples of tide gage records taken during a tsunami at Crescent City, California, are shown in Fig. 11.10.

An examination of the tide gage records for a number of tsunamis at many locations (Green, 1946; Zorbe, 1953; Committee for Field Investigations of the Chilean Tsunami of 1960, 1961; Spaeth and Berkman, 1965) shows that the relationship between the elevation of the crest above the tide level and the height of the particular wave is quite complicated. It appears to depend upon local response to the tsunami waves and to the presence, or absence, of other long-period oscillations. It appears that a conservative estimate is to double the elevation of the crest above the tide level at the time of the wave and to call this value the wave height at the site. This is a reasonable assumption, because, as stated previously, the tide gages are usually located a small fraction of a tsunami wave length from shore.

Both the run-up and drawdown elevations vary considerably from point to point along a coast. Van Dorn (1959) appears to feel that it is essentially a random phenomena, except for certain circumstances. The author does not agree with this completely; it is believed that the gross features can be predicted by the use of refraction drawings. The main difficulties are that refraction is complicated by diffraction, the trapping of waves, and by certain little-understood nonlinear phenomena that occur in very shallow water near shore where the wave height is no longer small compared with the water depth.

In most areas the water rises like a rapidly rising tide, while in a few areas a bore has formed during some large tsunamis (Hilo, Hawaii, is one of these), provided the area is a relatively great distance from the source of the tsunami. In the immediate vicinity of the source it is likely that bores are quite common (see, e.g., Tudor, 1964).

A substantial part of the literature on tsunamis is concerned with the very large tsunamis that affect an entire ocean. Smaller tsunamis are of great importance to local areas. For example, the tsunami of June 15, 1896, that was so disastrous in Japan was barely detectable at San Francisco. As another example, on a much smaller scale, many of the lives lost as a result of the eruption of Taal Volcano in the Philippines in September 1965 were due to drowning when small tsunamis swamped boats overloaded with people fleeing the island (De Roos and Spiegel, 1966).

11.2 DAMAGE BY TSUNAMIS

Damage caused by tsunamis can be very great, as the photograph in Fig. 11.1 shows. In addition to loss of life and damage to buildings, extensive damage often occurs to boats that break or drag their moorings and pound against other boats and docks or against buildings as they are carried ashore by the tsunami waves (see, e.g., Magoon, 1966).

A number of reports have been written on several tsunamis that include information on the damage and a partial evaluation of protective systems and measures. The reader is referred to them for details (Wharton and Evans, 1888; Tokyo Imperial University, 1934; Shepard, MacDonald, and Cox, 1950; Savarensky, Tischchenko, and Sviatlovsky, 1958; U.S. Army Corps of Engineers, 1960; Matlock, Reese, and Matlock, 1961; Solov'ev and Ferchev, 1961; Committee for Field Investigations of the Chilean Tsunami of 1960, 1961). Essentially three types of damage are caused by tsunamis. These types have been described by John D. Isaacs (1946) in a report on the damage that occurred on Oahu, Hawaii, during the April 1, 1946 tsunami, as follows: (1) Damage or effect of tsunami did not exceed that which would be expected from an equal tidal inundation without surf (houses either floated off their foundations or merely flooded, and the vegetation was disturbed to a small extent). (2) Damage or effect of tsunami was intermediate between conditions 1 and 3 (houses were moved some distance and damaged, and the ground was eroded to a certain extent). (3) Damage or effect of the tsunami seemed disproportionately great compared with that which would be expected from a tidal inundation of similar height (evidences of high velocity everywhere, with buildings destroyed, considerable erosion, automobiles rolled about, and in level regions the water moved great distances inland).

In describing the type of damage that has been done to buildings by tsunamis, it is desirable to retain the flavor of the words of the original investigators. Owing to this, a series of quotations will be used. A study of these observations will help the engineer to design structures that will be resistant to tsunami waves and to minimize loss of life and property damage.

The damage caused in Hilo, Hawaii, by the 1960 Chilean tsunami was studied by a team of structural engineers (Matlock, Reese, and Matlock, 1962). They made detailed analyses of several relatively clear-cut cases of structural damage in order to get some numerical estimations of the forces exerted by the tsunamis. Their general conclusions were as follows:

> All of the evidence gathered in the survey of structural damage indicates that the third wave of the tsunami approached the shoreline of Hilo Harbor as a bore approximately 15 ft above normal sea level, and swept inland at high velocity as a sheet of water. It was this mass of high-velocity water moving laterally which inflicted the heavy damage to structures on shore. Evidence of structural damage indicates that the height of this high-velocity water ranged from 8 to 12 ft above ground level, and then it traveled at 25 to 40 ft/sec.
>
> Analysis of specific structural elements indicates that a lateral force greater than 400 lb/ft^2 and less than 1800 lb/ft^2, with a mean average of approximately 700 lb/ft^2, was exerted by the water.
>
> The force of the water completely demolished all light frame buildings and most heavy timber structures and inflicted varying degrees of damage to structural steel and reinforced concrete structures. Properly designed reinforced concrete construction seemed to withstand the force of the water with less serious damage to the structural frame. However, damage was particularly severe where the fronts of buildings were open or glassed, but where the rear walls were relatively continuous.
>
> There was definite evidence that a strong structure will tend to reduce the damage to weaker structures which might be downstream from the strong structure. Shielding in this manner was observed in regard to lightly constructed buildings which lay behind the Mooheau Pavilion area, the Cow Palace, the Hilo Theater, and the Power Plant. On the other hand, the reservation of unobstructed open areas, such as much of that between Shore Drive and Kamehameha Street, offers little or no protection to structures further inland.
>
> From the practical standpoint the ground floors of even the strong structures whose framework suffered no significant damage can be said to have been greatly harmed by the passage of the wave. Substantial pieces of machinery and equipment were swept away, furnishings were ruined, and water damage to floors and walls was severe. It would seem to be impractical to attempt to design a structure so that the fixtures and furnishings on the ground floor would resist a tsunami similar to this one; however, on the basis of the findings in this report, it does appear possible to design a structural framework which would resist such waves. Certainly, adequate design foresight would greatly reduce the possibility of damage.
>
> In designing a structure for resistance to tsunamis, consideration should be given to several points. First, if possible, the structure should be oriented with its long dimension parallel with the anticipated direction of the wave. This will offer the greatest strength in resisting the water pressure, especially if special features can be incorporated. For example, "shear walls" could be included parallel to the anticipated wave direction. These would afford maximum strength without increasing the load on the structure. Of course, the possibility of building on stilts should be considered, thus reducing the total load to be withstood. Many of our most modern structures have been built in this fashion with the first floor area being utilized for parking, or for attractive gardens. If it is impractical to abandon the ground level area, then it would be possible to enclose the first floor with light exterior walls which would be ripped away by the water. In this event, the first floor would be considered expendable in order to safeguard upper floors.

Many of the conclusions of Matlock, Reese, and Matlock were in agreement with some of the conclusions reached by a team of geologists (Shepard, MacDonald, and Cox, 1950), who investigated another tsunami that hit the Hawaiian Islands on April 1, 1946, while some conclusions were at variance. Their findings are as follows:

> Structural damage—A considerable part of the damage to buildings was due to the light types of construction customary in the island. No reinforced concrete buildings seen were seriously damaged, even at places where the attack was severe. Of scores of buildings on the water front in Hilo, for example, only two stood in place and intact, and both of these were of reinforced concrete.
>
> Even steel-framework and sheet-iron-siding buildings generally stood up well except where the attack was severe and where there was a great deal of floating wreckage. However, in Hilo many buildings of this type were damaged. The wharf sheds on Piers 2 and 3 were badly battered by some large bridge-type pontoons that were moored in the harbor. The greatest single loss through building damage was the destruction of the boilerhouse and powerhouse sections of the Hakalau sugar mill on Hawaii. The steel frame collapsed; corrugated iron siding was ripped off; and machinery, pipes, and tanks were swept out of the building. A large clarifier tank was carried 800 feet inland. The loss at Hakalau was estimated at $375,000.
>
> Most island houses are of wood-frame construction. They are generally supported on stilts which raise them a foot to several feet off the ground, but some, especially the newer ones, are anchored to a concrete floor platform, which is usually low and sometimes is flush with the ground. Unless they were struck by floating debris, or unless the water rose sufficiently to float them off their foundations, the houses on stilts were not seriously damaged. Even some which were floated consider-

able distances were little damaged. Some houses of poor construction apparently broke up as they were lifted or while they were floating. Most of the houses that were destroyed were pushed against trees or rocks. Many were set down again without great damage, except as they were strained because the ground on which they came to rest was uneven. The gentleness of the floating was sometimes amazing. Thus, a house at Kawela Bay on Oahu was carried 200 feet inland and deposited gently in a cane field, leaving breakfast still cooking on the stove and dishes intact on shelves. Another house, at Kainalu, Molokai, was moved 50 feet without shaking any of the dishes from the shelves. It was possible to move many of the houses back to their original foundations without much repairing.

Those houses on concrete floor platforms that were in areas of severe attack or that had weak frameworks were ripped off their foundations and broken up in the process. Those with strong frameworks and relatively light siding commonly had part of the siding pushed in or out, but stood with their frames and roofs intact. Floating debris and rolling or saltatory rocks broke in many walls and windows that might have been spared by water alone.

In a large proportion of the two-story houses—not a common type on the beaches—the second story was wholly saved from damage by the destruction of siding on the ground floor, since the skeleton structure remaining offered little resistance to the passage of the water. Some two-story houses were reduced to one story by the destruction of the first floor and collapse of the second to the ground, a type of damage common during earthquakes.

There was, of course, great destruction of furnishings, personal property, and merchandise in buildings from breakage or wetting. Probably the largest item was 8,824 tons of sugar awaiting shipment on the docks in Hilo. Unlike most of the other losses, however, this one was covered by insurance. Many articles were lost simply by floating away or by burial.

Damage to roads and railroads was mostly of a nature common in any flood. In many places railroad and highway fills were cut partly or completely across. Blocks of undermined road pavement were rolled around like rocks. Railroad tracks usually still attached to ties, were pushed out of line and in places were wrapped around trees and other obstacles.

An unusual type of damage was noted in pavement on the abandoned Kakuku airfield on Oahu, which was completely submerged. In roughly circular areas 3 to 5 feet across, blocks of pavement were tilted so as to make conical hills a foot or so high. The pavement rests on sand, and the raising of the blocks was doubtless a result of hydraulic pressure, but how the water penetrated the pavement, and why the pressure was greater under the pavement than over it, is not known.

The foundations of some bridges were undermined, and the bridges collapsed. Others were pushed or floated off their foundations. One span of the steel railroad bridge across the Wailuku River in Hilo was pushed off its piers and carried about 750 feet upstream. Two of the four concrete piers under the center pier of the high steel railroad trestle over the Kolekole Valley on Hawaii were undermined. The trestle collapsed and was carried 500 feet upstream, leaving the bridge deck hanging by the rails. High flumes at Kakaiau and Papikou, Hawaii, were demolished, probably in a similar manner.

At several places, railroad cars were swept off the tracks or rolled over. Many automobiles by the shores of the islands were rolled or pushed around, thrown against obstacles, and battered by floating debris.

Damage to piers, other than to the sheds on them, was slight, even in Kahului and Hilo harbors, where the water was high and the currents swift. In those harbors and in Honolulu Harbor most of the damage was done by ships battering the docks at which they were moored. At Waianae, Oahu, a pier was pushed shoreward several inches by the waves. The inner end of the pier and the pavement which abutted it were buckled. The movement was made possible by bending of the long offshore piles and by shearing of the pier deck over the inshore piles. Marine railways at Kahului and Hilo were smashed.

Most breakwaters were not damaged, or only very slightly damaged. Only the Hilo breakwater which has often been damaged by storm waves, suffered greatly during the tsunami. Of the part of the breakwater above sea level, 6,040 feet, or about 61 percent, was destroyed. The cap and outside face were composed of rocks weighing 8 tons or more, and the inside face of rocks weighing 3 tons or more. The rocks were thrown both shoreward and seaward by the waves. The average depth of scour in the gaps in the breakwater was 3 feet. The breakwaters at Ahukini and Nawiliwili on Kauai, at Kahului on Maui, and even at Hilo, in spite of the damage, probably reduced greatly the severity of the attack of the waves inside the harbors. The effect is easily shown by the drop in wave heights from 17 and 22 feet outside the breakwaters at Kahului to 7 and 11 feet inside the breakwaters, and from 29 feet outside at Hilo to an average of about 21 feet inside. The effect was similar to that of a coral reef with fairly deep water behind it.

Sea walls, where well constructed, were not materially damaged. Many small sea walls of loose or poorly cemented rocks were smashed by the waves. There was some erosion behind sea walls at Pier 1 in Kahului Harbor, where the water came from the sea side and cascaded over the walls into the harbor. Much more erosion occurred behind short discontinuous sea walls at Hilo which had been designed merely to protect sewer outfalls. At the Puu Maile Hospital, east of Hilo, the water

> poured in great volume over the sea wall, causing considerable erosion of the highway pavement just inside the wall. The wall undoubtedly protected the main buildings of the hospital from the full force of the waves, and probably saved them from severe damage.

Most of the findings on structural damage by tsunamis that has occurred in Japan is in Japanese. A few of the results have been given in English, however. The general conclusions in regard to the 1933 tsunami, by Nasu (1934), are as follows: (1) The maximum water surface does not always increase in height as the wave moves inland; in some cases it was found that the water decreased in height as it proceeded landward. (2) The inclination of the ground is an important factor that determines the velocity of the water. (3) The ordinary Japanese wooden type of house (both one- and two-storied houses) was partially damaged when the water reached a height of 3–4½ ft above the ground. When the water reached heights of about 4½ ft the houses that were not firmly connected to their foundation began to float. When the height of the water exceeded 6 ft, the ground floor collapsed and the upper floor came down.

This conclusion applies only when the velocity of the water was not much greater than 30 ft/sec, which was the case for the tsunami studied by Nasu. If the velocity is less than this, the houses might be able to resist the tsunami waves. It is interesting to note that at Crescent City, California, where most of the buildings near the waterfront are of wood, that damage was severe in the 1964 Alaskan tsunami in areas where the water depth exceeded 4–6 ft above the ground level (Magoon, 1966). Furthermore, the damage was similar in type to that described by Nasu.

The conclusions by the Architectural Institute of Japan (1961) in regard to the 1960 Chilean tsunami are as follows:

> The damages caused by tsunami are quite different at each location mainly due to its topographical features. Generally speaking the tsunami wave increases in height intensively inside a bay owing to the fact that the depth is shallower and the width is narrower at the recess of the bay comparing with that at the entrance of the bay. Afterwards the tsunami runs up the shore and inundates the background. The height of tsunami at the recess of a bay is greatly influenced by the shape of bay and reaches more than twice the height at the bay entrance in the case of V shaped or U shaped bay. [Hence, regions which have frequently suffered tsunami damage in the past are likely to suffer damage during future tsunamis.] For example, Sanriku district and Kii peninsula, the coast of which is jagged very much, have been frequently hit by tsunamis. In the case of Chilean tsunami the damages in these districts were tremendous.
>
> The amount of damages caused by the Chilean tsunami in each prefecture is listed in [the original publication]. Among them, the building damages can be classified into two types: one is the destruction of buildings due to water pressure induced by tsunamis, and the other is the secondary destruction of buildings caused by the collision of floating materials such as timber pieces and boats. In any district and in any town both types of destruction could be found at the same location.
>
> As for the floating materials, the existence of destroyed houses and fishing boats should be kept in mind. Most of the villages suffered from the tsunami damages are fishery ports, therefore many fishing boats were pushed toward land by tsunami and collided with buildings. Most of the buildings standing at water front of the port sector in Sanriku district [were severely damaged by] floating material. In Owase City the situation was almost the same.
>
> The damages caused by floating timber pieces also cannot be ignored as it was observed at Shizukawa, Miyagi prefecture, and Susaki, Kochi prefecture.
>
> As a result of the investigations, it is concluded that the buildings, which are rather superannuated or improperly constructed, shall have a great damage on the structures due to tsunami.
>
> In addition to the building damages caused by floating materials, the severe scouring at the foundations of buildings owing to the strong current induced by tsunami can be a cause of the building destruction as shown in many cases at the Chilean tsunami. Generally speaking the continuous footing had enough resistance against the scouring action.

Other conclusions in regard to the effects of the 1960 Chilean tsunami on Japan are as follows (Horikawa, 1961):

> Scouring by Currents—During the latest tsunami in 1960, a fairly strong current developed at several places such as inside a bay and at a river mouth. It will be rather helpful understanding such phenomena to cite here some examples of observation at the scenes.
>
> (1) According to H. Kanno, Port Development Office at Hachinoe, a dredger was torn from a lightly moored position and carried out of the harbor basin at a velocity of approximately 10 knots by a stream which took place inside the harbor during the reflux of the tsunami.
>
> (2) A current which took place at the breakwater gap of the Miyako port during the flux of the tsunami, was so swift that a fisherman barely succeeded in traversing the current in a boat equipped with a 20 HP diesel engine at the top speed of 8 knots. He estimates that the current velocity was approximately 5 knots. It was also

observed at Miyako and Shizukawa that the drifting timbers floated at an amazing speed toward the lengthwise direction, and that the current was so violent that it looked as if a [turbulent] river had suddenly emerged in the midst of the still basin. A similar story was reported at the Bay of Corral in Chile.

(3) The tsunami took on a bore-shaped pattern at the river mouth port of Ishinomaki, as it advanced upstream pushing drifting boats along the wave front.

Though variable with topographical and many other factors, a current caused by a tsunami is often reported to attain a velocity of [3–10 ft/sec]. There is a frequent story that the reflux was even more violent than the flux. Failure of various coastal structures occurred in many places due primarily to scouring of the foundation by the receding water during the reflux and secondary to subsequent increase of earth pressure against a retaining structure owing to loosening of the bond between the backfilling concrete and soils. The patterns of failure varied in minor details according to the level and period of inundation, height of structure, drainage conditions of the inundated area, direction and pattern of the receding water.

A reflux current seems to play an intricate procedure of failure at some places and against some types of structure. [For example] a quay-wall [failed] which has the capacity of berthing a 10,000-ton vessel at Ofunato where the backfilling was scoured. This structure had the crown level 1.65 m T.P.* and was inundated for approximately 20 minutes by the tsunami which had the maximum level of 3.85 m T.P., i.e., 2.20 m above the crown level. The next wave flooded the quay-wall again about 0.70 m above the crown level, but the subsequent waves remained below the crown. The sheet piles, 15 m long and resting at the level −14.35 m T.P., had the embedment depth of approximately 11 m below the sea bed of −3.35 m T.P. The failure is attributed to the plunging water during the reflux which caused a deep scour at the front bed to start suction of the back-filling soils through gaps of the sheet piles. The water level in front of the quay-wall remained less than −2.35 m T.P. or 1 m deep during the reflux.

On the other hand, the sheet piles of the Fuji Steel quay-wall at Kamaishi were 11 m long with only 3m of embedment depth. The construction of this wall dates back to 1937 and supposedly the age had reduced it vulnerable to an unusual condition. It overturned toward the sea. At the Konakano fishery wharf at Hachinoe the front bed was severely scoured due to raging water which rushed to and fro inside the anchorage basin in accordance with the flux and the reflux of the tsunami waves. The wharf collapsed toward the sea.

*T.P. is the tidal datum in Japan.

11.3 THEORY AND LABORATORY STUDIES OF TSUNAMI GENERATION

No attempt will be made here to present a theory for the generation of a real tsunami, owing to the complicated motions of the ocean bottom during a tectonic displacement (or other mechanism), the variable hydrography, and the irregular coastal features that exist in the ocean. Rather, results of theoretical and laboratory studies are presented to illustrate some of the physics of the generation and movement of tsunami waves. The theory for the generation of waves by complicated motions of the ocean bottom, however, has been developed by Kajiura (1963).

The time histories of tsunamis in the open ocean have not been measured. The nearest thing to such a measurement has been made with a specially designed long-period wave recorder on Wake Island in the Pacific Ocean (Van Dorn, 1959). The size and hydrography of this island are such that it is similar to a large pile rising from the floor of the ocean. The diameter of the island is about 5 mi, whereas the lengths of the early waves in the tsunami were from 20 to 30 mi. It is believed that such an island may give reliable records for long-period waves, from the standpoint of measuring their open ocean characteristics (Munk, Snodgrass, and Tucker, 1959). The record of the March 9, 1957 tsunami is shown in Fig. 11.11. It will be seen that this record is not too different in appearance from the wave groups predicted by the following theories. Records obtained at a number of other locations are also presented in the figure to show how different the records appear at different sites, owing to trapped waves, local resonance conditions, etc.

A simple way to induce gravity waves on the surface of a body of water is either suddenly to add some water or to remove some water. Prins (1958a, 1958b) has done this in a consistent manner. He installed an airtight transparent plastic box at one end of a long water channel. The front of the box consisted of a sliding gate. The box was the same width as the channel, but was very short compared with the length of the channel. The box was partially evacuated to cause an initial surface elevation relative to the water level in the channel, and compressed air was added to obtain an initial surface depression. When the sliding gate was raised suddenly, the vertical wall of water collapsed, generating waves in the process.

When the length (λ) and height (h) of the initial elevation, or depression, were not large compared with the water depth, waves were generated, which in general had the characteristics of a linear wave system (see Fig. 11.12). The main features of the waves system were fairly well described by the theory of Unoki and Nakano (1953), although the measured wave amplitudes were

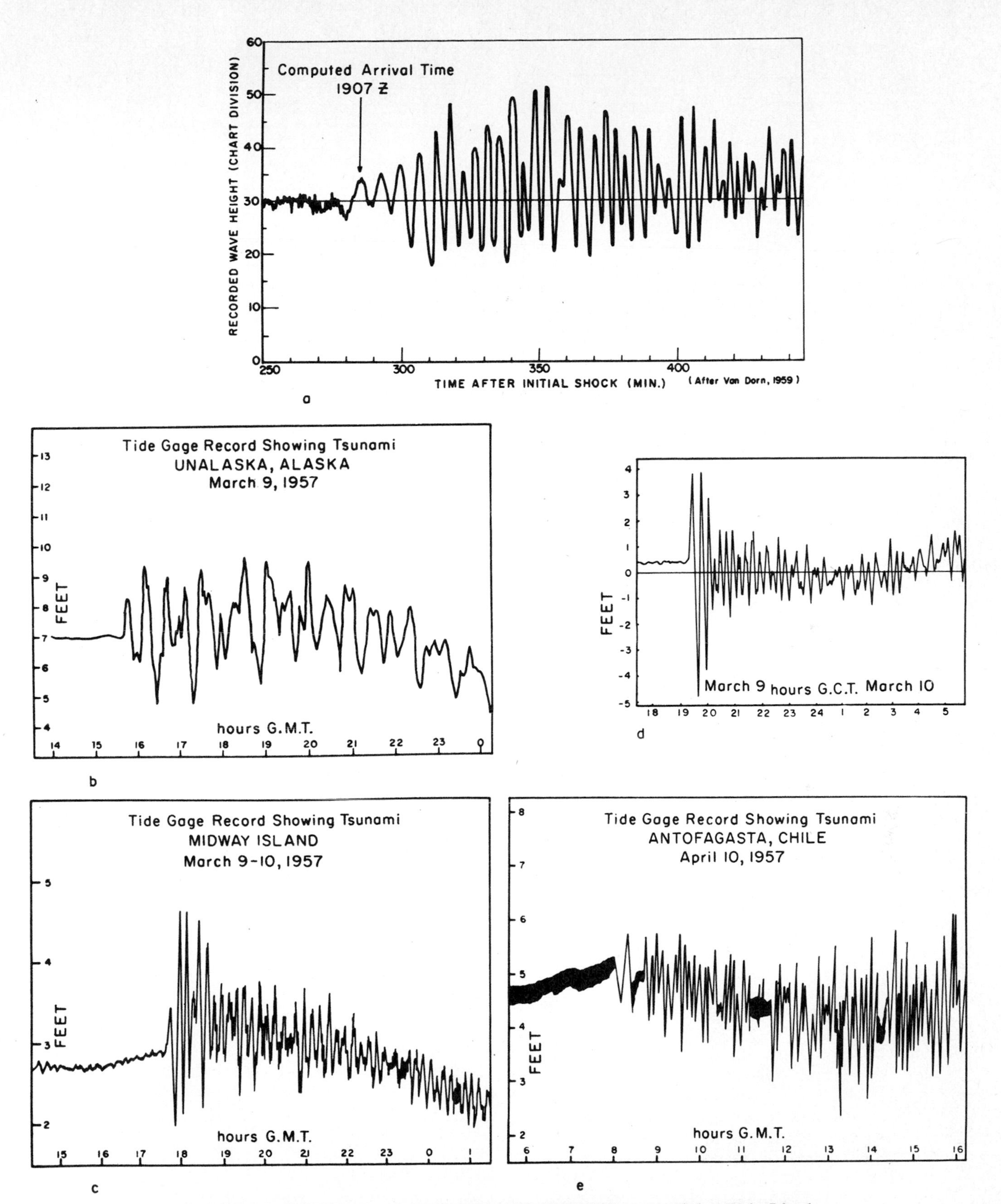

Fig. 11.11. Tsunami records: (a) Tsunami of March 9, 1957, as recorded at Wake Island. (b) Period first to second crest; 27 min; third wave highest. (After Salsman, 1959.) (c) Period first to second crest; 12 min; first wave highest. (After Salsman, 1959.) (d) Record of the tsunami of March 9, 1957, from the Hilo tide gage. Time is 10 hr ahead of Hawaiian time. Period first to second crest; 19 min; first wave highest. (Courtesy, U.S. Coast and Geodetic Survey.) (e) Period first to second crest; 14 min; seventh wave highest. (After Salsman, 1959.)

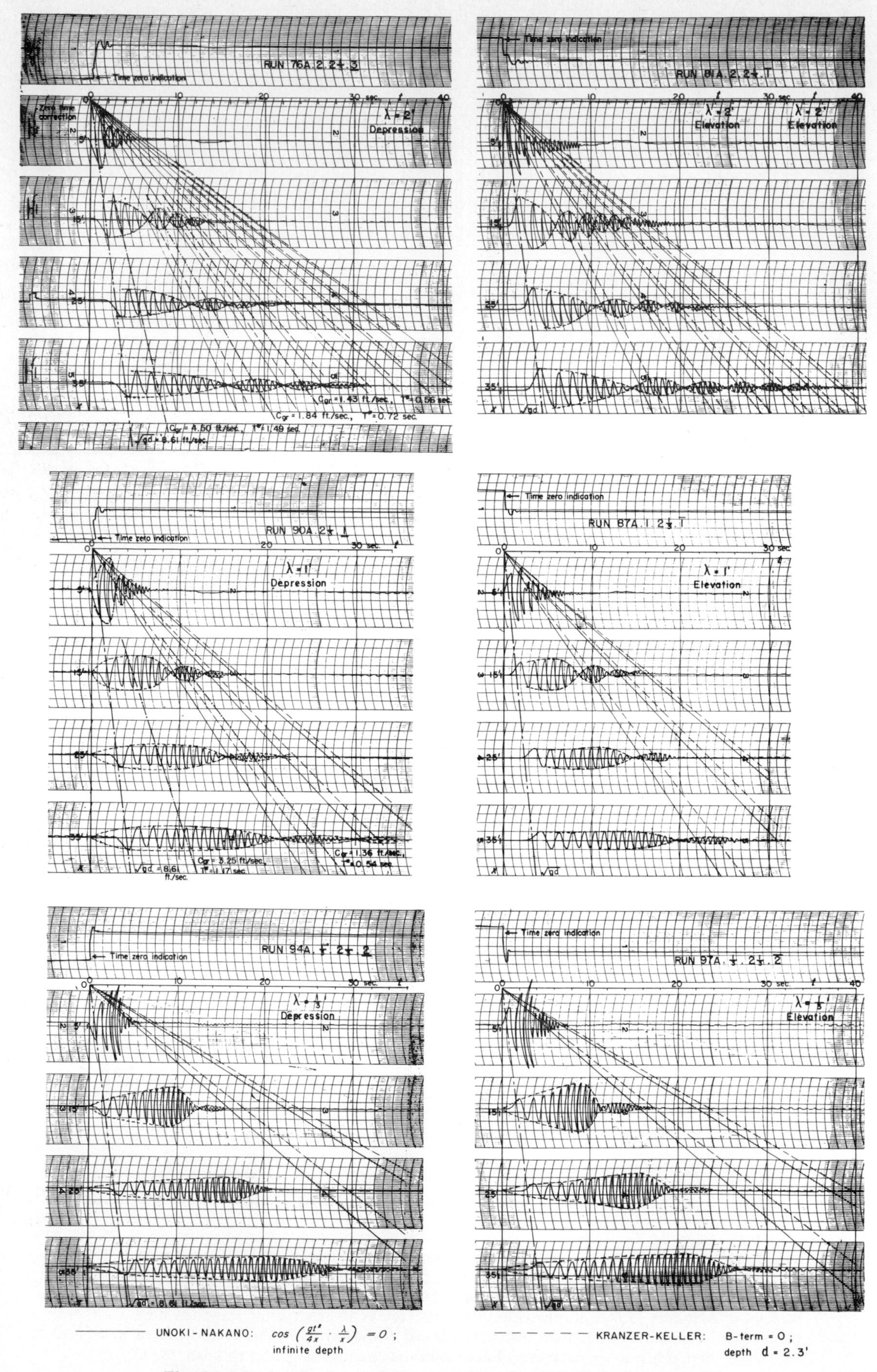

Fig. 11.12. Amplitude envelope zero-points, $d = 2.3'$. (From Prins, 1958b.)

lower than the amplitudes calculated from the equations. The theory is for an initial elevation of the water surface of $y_s = h$, extending over the region $-\lambda \leq x \leq \lambda$, and $y_s = 0$ for $x > \lambda$ for the two-dimensional case, where x is the horizontal coordinate in the direction of wave advance. The solution for the water surface elevation y_s is

$$y_s = \frac{4h}{t}\sqrt{\frac{x}{\pi g}} \sin\left(\frac{gt}{4x} \cdot \frac{\lambda}{x}\right) \cos\left(\frac{gt^2}{4x} - \frac{\pi}{4}\right) \quad (11.1)$$

for $x \gg \lambda$. This describes individual waves with characteristics

$$T = \frac{4\pi x}{gt} \qquad L = \frac{8\pi x^2}{gt^2} \qquad C = \frac{2x}{t} = \sqrt{\frac{gL}{2\pi}} \quad (11.2)$$

and groups, or amplitude enevelopes, with characteristics

$$T_G = \frac{4\pi x}{gt} \cdot \frac{x}{\lambda} \qquad L_G = \frac{8\pi x^2}{gt^2} \cdot \frac{x}{2\lambda}$$

$$C_G = \frac{x}{t} = \text{constant} \quad (11.3)$$

In Eqs. 11.2 and 11.3, T, L, and C are the period, length, and phase speed of the individual waves and T_G, L_G, and C_G are the group period, group length, and group speed. The group speed is equal to one-half the phase speed at any particular x,t, as is the case for uniform periodic waves in deep water linear wave theory. The leading part of the disturbance is a "long wave," so that portion travels with a group speed approximately equal to the phase speed; however, the first wave "stretches" as it moves and is represented by a more complicated equation. (For a complete discussion, see Kajiura, 1963.)

Equation 11.1 can be written as

$$y_s = \frac{h}{\pi}\sqrt{\frac{2L}{x}} \sin\frac{2\pi\lambda}{L} \cos\left(\frac{2\pi x}{L} - \frac{\pi}{4}\right) \quad (11.4)$$

For an initial depression, rather than an elevation, only the sign of Eq. 11.4 changes.

These equations show that the phase wave period, length, and speed depend only upon x and t; they are independent of the size of the source of the initial disturbance. The period and length of the groups are dependent upon the length of the initial disturbance, with the greater the λ, the shorter the group period and length. The amplitude of the waves depends upon the height of the initial disturbance, being proportional to h. The greater λ relative to d, the closer the maximum wave height in the first group is to the leading wave. The height of the maximum wave in this group depends upon λ in a complicated way, but in general the greater the λ, the higher is the maximum wave, all other conditions being equal. The waves travel away from the region of the initial disturbance with a finite speed in water that is not infinitely deep. The laboratory study of Prins showed that the first detectable variation from the undisturbed water surface of the leading wave traveled at a speed a little in excess of $\sqrt{gd}$, while the first crest (or trough, if the initial disturbance were a depression) traveled at a speed of about $\sqrt{gd}$. This is in agreement with the theoretical predictions of Ursell (1958) and Kajiura (1963), provided the parameter $\lambda/(x^{1/3}d^{2/3})$ is relatively small. Thus, there is no disturbance at a distance a little greater than $t\sqrt{gd}$. It is interesting to note that a study of the April 1, 1946, Aleutian tsunami by Green (1946) showed that the initial disturbance arrived at most locations a little in advance of the time predicted using a speed based upon actual ocean depths and assuming $C = \sqrt{gd}$.

Kranzer and Keller (1959) obtained a theoretical solution for waves generated by an initial distribution of impulse applied on the water surface and by an initial elevation or depression of the water surface. Solutions were obtained for both the two-dimensional and three-dimensional cases. The solution is based upon linear wave theory, so that the wave heights must be small compared with both the wavelengths and water depth. Furthermore, the solution is for $x \gg \lambda$. The solution for the case of an initial elevation or depression of the water surface is

$$y_s \approx \frac{1}{d\sqrt{kx}}\left[\frac{kd\Phi(kd)}{-\Phi'(kd)}\right]^{1/2} \bar{E}(k) \sin 2\pi \left(\frac{t}{T} - \frac{x}{L} + \frac{1}{8}\right) \quad \text{for } x < t\sqrt{gd} \quad (11.5a)$$

$$\approx \frac{6^{5/6}\Gamma(\frac{4}{3})\bar{E}(0)}{4\sqrt{\pi}\, d^{2/3}x^{1/3}} \quad \text{for } x = t\sqrt{gd} \quad (11.5b)$$

$$= 0 \quad \text{for } x > t\sqrt{gd} \quad (11.5c)$$

where $\Phi'(kd)$ denotes differentiation of Φ with respect to kd and Γ is the gamma function. $\bar{E}(k)$ is the Fourier cosine transform of the initial elevation $E(x)$, as defined by

$$\bar{E}(k) = \sqrt{\frac{2}{\pi}} \int_0^\infty E(x) \cos(kx)\, dx \quad (11.6)$$

The auxiliary variable kd is a function of $x/(t\sqrt{gd})$ and is defined as the unique nonnegative root of the equation

$$\Phi(kd) \equiv \frac{1}{2}\left(\frac{\tanh kd}{kd}\right)^{1/2} + \frac{1}{2(\cosh kd)^{3/2}}\left(\frac{kd}{\sinh kd}\right)^{1/2} = \frac{x}{t\sqrt{gd}} \quad (11.7)$$

This equation is obtained by use of the method of stationary phase, which requires that

$$0 = \frac{d}{dk}(kx - kCt) = x - t\frac{d(kC)}{dk} = x - tC_G \quad (11.8)$$

where C_G is the speed at which energy is transmitted by a wave of wave number k ($k = 2\pi/L$, where L is the wavelength). The phase speed of the wave is given by C.

The usual form for expressing C_G (Wiegel, 1964d) is

$$C_G = \frac{1}{2}C\left(1 + \frac{2kd}{\sinh 2kd}\right) = \sqrt{gd}\left(\frac{1}{2}\sqrt{\frac{\tanh kd}{kd}} + \frac{1}{2\cosh^{3/2} kd}\sqrt{\frac{kd}{\sinh kd}}\right) \quad (11.9)$$

Using this, together with the relationship

$$C = \sqrt{gd}\left(\frac{\tanh kd}{kd}\right)^{1/2} \quad (11.10)$$

results in being able to express Eq. 11.7 as

$$\Phi(kd) = \frac{C_G}{\sqrt{gd}} = \frac{x}{t\sqrt{gd}} \quad (11.11)$$

Equation 11.11 can be described physically as follows: At some distance x from the source there is no wave disturbance until the time $t = x/\sqrt{gd}$. After this time $(x/\sqrt{gd})$, component waves arrive with continually increasing wave number k. Equation 11.11 permits the calculation of C_G as a function of time. For a particular water depth d there is only one wave component of wave number k, whose energy can be transmitted with this particular speed C_G. Equation 11.7 permits the calculation of the wave number k for the C_G obtained from Eq. 11.11.

Using the relationship

$$\frac{\Phi(kd)}{\Phi'(kd)} = \frac{dC_G(k)}{C'_G(k)} \quad (11.12)$$

permits Eq. 11.5a to be expressed as

$$y_s(x, t) = \frac{1}{\sqrt{x}}\left(\frac{-C_G(k)}{C'_G(k)}\right)^{1/2} \bar{E}(k) \sin 2\pi \times \left(\frac{t}{T} - \frac{x}{L} - \frac{1}{8}\right) \quad (11.13)$$

The Fourier cosine transform of $E(x)$ is $\bar{E}(k)$ and is given by Eq. 11.6 (Kranzer and Keller, 1959); for the case of $E(x) = h$ for $0 \leq x \leq \lambda$ and $E(x) = 0$ for $x > \lambda$, $\bar{E}(k)$ is (Sneddon, 1951)

$$\bar{E}(k) = \sqrt{\frac{2}{\pi}}\frac{h \sin k\lambda}{k} \quad (11.14)$$

For this particular case, the wider the λ, the narrower is the Fourier cosine transform, i.e., the greater the amount of "energy" concentrated in the low wave numbers (long waves). Furthermore, as waves that are long compared with the water depth are not very dispersive, the leading wave will not decrease in amplitude as rapidly with increasing x as will the succeeding waves.

Now, for the particular case wherein $\bar{E}(k)$ is given by Eq. 11.14, Eq. 11.13 becomes

$$y_s(x, t) = \left[\frac{-C_G(k)/C'_G(k)}{x}\right]^{1/2}\frac{h\sqrt{2/\pi}}{k} \times \sin k\lambda \sin 2\pi\left(\frac{t}{T} - \frac{x}{L} + \frac{1}{8}\right) \quad (11.15)$$

The last term, $\sin 2\pi(t/T - x/L + 1/8)$, represents the individual waves, which fluctuate rather rapidly compared with the envelope. The middle term, which is a part of the envelope, expresses the distribution of energy with respect to component wave number. The first term, $[-C_G(k)/C'_G(k)]^{1/2}/\sqrt{x}$, represents the dispersion of wave energy as it is transported away from the source. The product of the first two terms gives the envelope of the waves as a function of t for any given value of x and as a function of x for any given value of t.

Examples of the envelope and individual waves, as observed in the laboratory, are given in Fig. 11.12. It can be seen that the maximum occurs near the leading part of the envelope when λ/d is large and occurs near the trailing end of the envelope when λ/d is small. This is because for large values of λ/d most of the energy is in the long waves and for small values of λ/d most of the energy is in the short waves; as the long waves travel faster than the short waves, the energy of the long waves is in the leading part of the group while the energy of the short waves is in the rear part of the group.

In a paper by Keller (1963; also see Kajiura, 1963), it was shown that waves caused by a sudden displacement of the bottom could be described by Eq. 11.15, provided the right-hand side of the equation was multiplied by $1/\cosh kd$.

Equation 11.13 has been programmed for a high-speed digital computer (Wiegel *et al.*, 1968) for the initial condition of $E(x) = h$ for $0 > x < \lambda$ and $E(x) = 0$ for $x > \lambda$. The Fourier cosine transform of this initial condition is given by Eq. 11.14. The results for a portion of the envelope described by the equation are given in Fig. 11.13 for four values of λ/d for each of two values of x/d. The abscissa is given in terms of the dimensionless time scale $t\sqrt{g/d}$. Comparison of Fig. 11.13 with Fig. 11.12 shows that, except for $t\sqrt{g/d} \approx x/d$, the general trend of the shape of the envelope, with respect to λ/d, is as predicted by the theory. Comparison of calculated results for the individual waves with one set of measurements by Prins (1958b) are shown in Fig. 11.14a for $\lambda/d = 0.145$ and $x/d = 15.2$. Again, theory and measurements are remarkably good except in the vicinity of $t\sqrt{g/d}$ slightly greater than x/d. Comparisons of theory and measurements of elevation of the maximum point of the envelope of the first group have been made for a number of values of x/d (Wiegel *et al.*, 1968). It was found that the comparison was excellent even for a value of x/d as small as 2 for small values of λ/d and h/d. For relatively large values of λ/d and h/d the comparison was not nearly so favorable, and there was a definite effect of h/d, with the greater the h/d, the greater was the discrepancy between measured values and those calculated by linear theory. For values of $\lambda/d > 1$, the nonlinear effect was even more apparent than for $\lambda/d = 0.857$.

It is evident that the solution is not valid for values of $t\sqrt{g/d}$ slightly greater than x/d, nor for $t\sqrt{g/d} < x/d$.

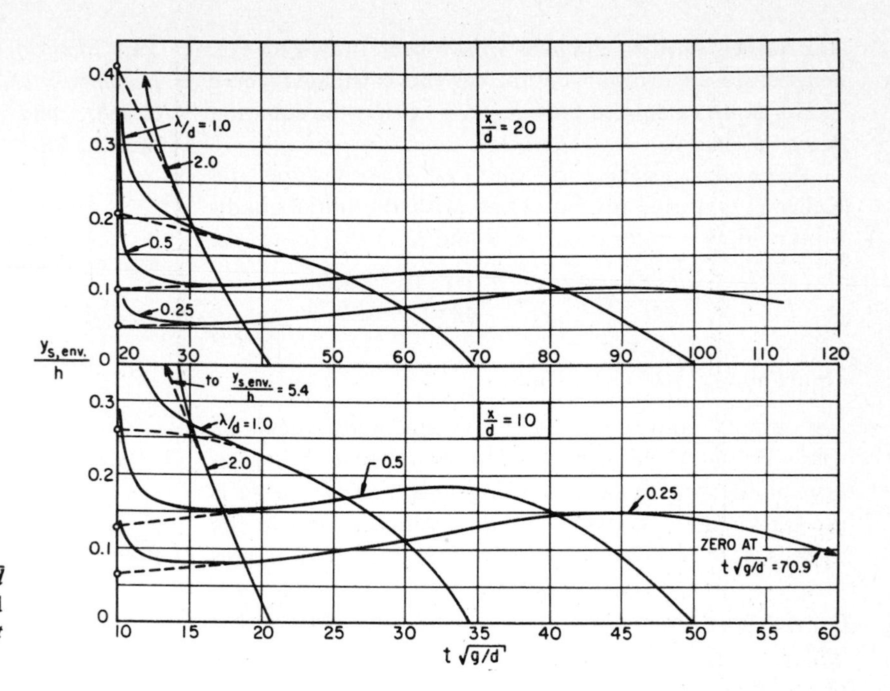

Fig. 11.13. Envelope as a function of $t\sqrt{g/d}$ calculated with Eq. 11.9 for x/d of 10 and 20 and for λ/d of 2.0, 1.0, 0.5, and 0.25. (From Wiegel *et al.*, 1968.)

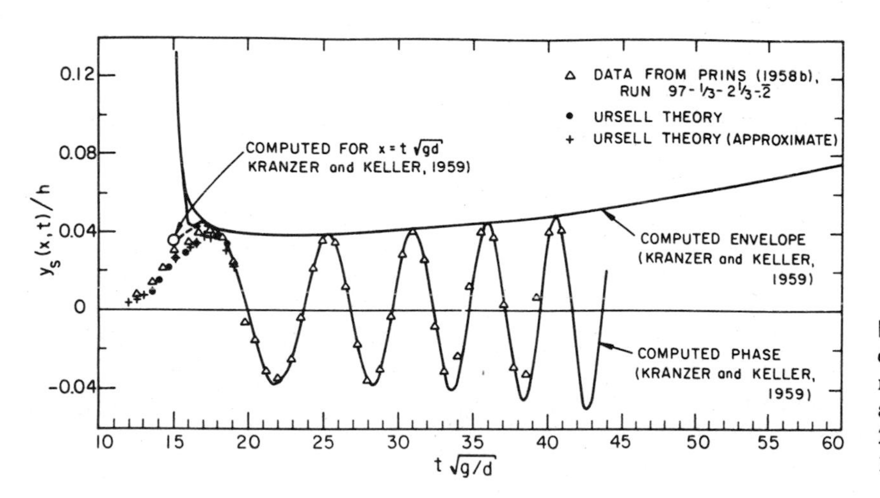

Fig. 11.14a. Comparison of calculated values of $y_s(x, t)/h$ vs dimensionless time, $t\sqrt{g/d}$ with measured values of Prins (1958b) for $\lambda/d = 0.145$ and $x/d = 15.2$ ($h = 0.2$ ft, $\lambda = 0.333$ ft, $d = 2.333$ ft, and $x = 35.0$ ft). (From Wiegel *et al.*, 1968.)

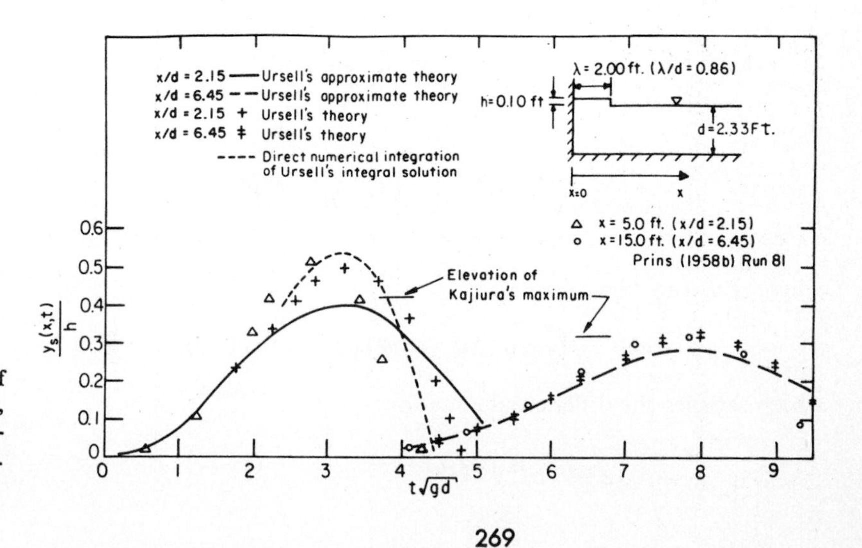

Fig. 11.14b. Comparison of measured values of first wave for $x/d = 2.15$ and 6.45 (with $h = 0.10$ ft, $\lambda = 2.00$ ft, and $d = 2.33$ ft) with theoretical calculations for Ursell's theory and Kajiura's theory. (From Wiegel *et al.*, 1968.)

The dashed lines on Fig. 11.13 have been drawn by eye, connecting an arbitrary point on the calculated curve with a point calculated for $t\sqrt{g/d} \approx x/d$ by the equation described below.

At $x = t\sqrt{gd}$ (i.e., for $x/d = t\sqrt{g/d}$), Kranzer and Keller (1959) give for Eq. 11.5b with the initial conditions of $E(x) = h$ for $0 < x < \lambda$ and $E(x) = 0$ for $x = \lambda$

$$y_s(x, t) \approx \frac{6^{5/6}\Gamma(\frac{4}{3})\bar{E}(0)}{4\pi^{1/2}d^{2/3}x^{1/3}} = \frac{6^{5/6}\Gamma(\frac{4}{3})\sqrt{2/\pi}\,h\lambda}{4\pi^{1/2}d^{2/3}x^{1/3}} \approx \frac{0.45h\lambda}{d^{2/3}x^{1/3}} \tag{11.16}$$

The points shown for $t\sqrt{g/d} = x/d$ were calculated using Eq. 11.16. Equation 11.16 shows that the elevation of the wave phase traveling at a speed of $\sqrt{gd}$ is proportional to both the initial elevation h and to the length of the initial elevation λ and that it decreases slowly with x, being proportional to $1/x^{1/3}$. When λ/d is relatively large, the first wave in the group is the highest, and Eq. 11.16 is of considerable practical importance. However, as pointed out by Kranzer and Keller (1959), it is unlikely that their solution is valid in the vicinity of $x = t\sqrt{gd}$. Rather, something analogous to the Airy Integral form will exist (Eckart, 1948; Ursell, 1958; Kajiura, 1963).

Ursell (1958) obtained the following solution for the first wave for the initial condition described prior to Eq. 11.14:

$$y_s(x, t) = \frac{2h}{\pi}\int_0^\infty \cos kx \cos\left(t\sqrt{gk \tanh kd}\right) \times \frac{\sin k\lambda}{k}\,dk \tag{11.17}$$

Except for the "wave front," Kelvin's principle of stationary phase (see Lamb, 1945) was used by Ursell to obtain a solution, which was the same as the one obtained by Kranzer and Keller (1959) and is given by Eq. 11.16. It should be pointed out that the inequality for Eq. 11.5a, $x < \sqrt{gd}$, must be determined by matching numerical results from Eq. 11.15 with those obtained from an equation to be described below. Near the "wave front" the Method of Steepest Descents was used by Ursell to evaluate the integral (Chester, Friedman, and Ursell, 1957). Ursell found for $x/t\sqrt{gd} < 1$ and $0 < \beta < \infty$ that

$$y_s(x, t) \approx \frac{h}{(t\sqrt{g/d})^{1/3}}p_0(\beta)A_i\left[-\left(t\sqrt{\frac{g}{d}}\right)^{2/3}\alpha(\beta)\right] \tag{11.18}$$

where A_i is the Airy Integral.

$$A_i(X) = \int_0^\infty \cos\left(Xu + \tfrac{1}{3}u^3\right)du \tag{11.19}$$

which satisfies the differential equation

$$\left(\frac{d^2}{dX^2} - X\right)A_i(X) = 0 \tag{11.20}$$

This integral has been described and tabulated in the *Handbook of Mathematical Functions with Formulas, Graphs, and Mathematical Tables* (Abramowitz and Stegun, 1964). β is defined by

$$\frac{x}{t\sqrt{gd}} = \frac{1}{2}\left(1 + \frac{2\beta}{\sinh 2\beta}\right)\sqrt{\frac{\tanh\beta}{\beta}} \tag{11.21}$$

with $\alpha(\beta)$ and $p_0(\beta)$ given by

$$\alpha(\beta) = \left(\frac{\beta}{\sinh 2\beta}\right)^{2/3}\left(\frac{\tanh\beta}{\beta}\right)^{1/3} \times \left(\frac{-2\beta + \sinh 2\beta}{\frac{4}{3}\beta^3}\right)^{2/3}\beta^2 \tag{11.22}$$

$$p_0(\beta) = \frac{\sin(\beta\lambda/d)}{\beta}\left[\frac{2\sqrt{\alpha(\beta)}}{-\frac{d^2}{d\beta^2}(\sqrt{\beta\tanh\beta})}\right]^{1/2} \tag{11.23}$$

where

$$\frac{d^2}{d\beta^2}(\sqrt{\beta\tanh\beta}) = -(\beta\tanh\beta)^{1/2} \times \left[\frac{1}{\cosh^2\beta} + \frac{1}{4\beta^2}\left(1 - \frac{2\beta}{\sinh 2\beta}\right)^2\right] \tag{11.24}$$

For $x/t\sqrt{gd} > 1$, substitution of $\gamma = i\beta$ (for $0 < \gamma < \frac{1}{2}\pi$) into Eq. 11.18 results in

$$y_s(x, t) = \frac{h}{(t\sqrt{g/d})^{1/3}}p_0(\gamma)A_i\left[-\left(t\sqrt{\frac{g}{d}}\right)^{2/3}\alpha(\gamma)\right] \tag{11.25}$$

Here, γ is defined by

$$\frac{x}{t\sqrt{gd}} = \frac{1}{2}\left(1 + \frac{2\gamma}{\sin 2\gamma}\right)\sqrt{\frac{\tan\gamma}{\gamma}} \tag{11.26}$$

and $\alpha(\gamma)$ and $p_0(\gamma)$ are given by

$$\alpha(\gamma) = -\left(\frac{\gamma}{\sin 2\gamma}\right)^{2/3}\left(\frac{\tan\gamma}{\gamma}\right)^{1/3}\left(\frac{2\gamma - \sin 2\gamma}{\frac{4}{3}\gamma^3}\right)^{2/3}\gamma^2 \tag{11.27}$$

$$p_0(\gamma) = \frac{\sinh(\gamma\lambda/d)}{\gamma}\left[\frac{2\sqrt{\alpha(\gamma)}}{-\frac{d^2}{d\gamma^2}(\sqrt{-\gamma\tan\gamma})}\right]^{1/2} \tag{11.28}$$

where

$$\left[\frac{2\sqrt{\alpha(\gamma)}}{-\frac{d^2}{d\beta^2}(\sqrt{-\gamma\tan\gamma})}\right]^{1/2} = \left[\frac{2\left(\frac{\gamma}{\sin 2\gamma}\right)^{1/3}\left(\frac{\tan\gamma}{\gamma}\right)^{1/6}\left(\frac{2\gamma - \sin 2\gamma}{4\gamma^3/3}\right)^{1/3}\gamma}{(\gamma\tan\gamma)^{1/2}\left[\frac{1}{\cos^2\gamma} - \frac{1}{4\gamma^2}\left(1 - \frac{2\gamma}{\sin 2\gamma}\right)^2\right]}\right]^{1/2} \tag{11.29}$$

For small values of β, Ursell found that:

$$\beta^2 \longrightarrow 2\left(1 - \frac{x}{t\sqrt{gd}}\right) \tag{11.30a}$$

$$\alpha(\beta) \longrightarrow \beta^2/2^{2/3} \tag{11.30b}$$

$$-\frac{d^2}{d\beta^2}(\sqrt{\beta\tanh\beta}) \longrightarrow \beta \tag{11.30c}$$

with similar approximations occurring for small values of γ, so that Eqs. 11.18 and 11.25 become

$$y_s(x, t) \approx \frac{2^{1/3}\lambda h}{(t\sqrt{g/d})^{1/3}\, d} A_i\left[\frac{2^{1/3}}{(t\sqrt{g/d})^{1/3}\, d}(x - t\sqrt{gd})\right] \tag{11.31a}$$

$$\approx \frac{1.26\lambda h}{(t\sqrt{g/d})^{1/3}\, d} A_i\left[\frac{1.26}{(t\sqrt{g/d})^{1/3}}\left(\frac{x}{d} - t\sqrt{g/d}\right)\right] \tag{11.31b}$$

For the point on the "wave front" where $x = t\sqrt{gd}$, $A_i(0) = 0.355$, and Eq. 11.31a becomes

$$y_s(x, t) \approx \frac{(0.355)2^{1/3}\lambda h}{[(x/\sqrt{gd})(\sqrt{g/d})]^{1/3}\, d} = \frac{0.45h\lambda}{d^{2/3}x^{1/3}} \tag{11.31c}$$

which is the same as Eq. 11.16. A numerical example, calculated from Eq. 11.31b, showed that the wave crest travels at a speed of less than $\sqrt{gd}$ and that the wave front elevation at the point where $t = x/\sqrt{gd}$ is considerably lower than the crest elevation. However, numerical calculations of the leading wave for a number of values of $(\lambda/d)(d/x)^{1/3}$ reveal that for relatively large values of this parameter the velocity of the first crest actually exceeds $\sqrt{gd}$. The theory developed by Kajiura also shows this.

Equations 11.18 and 11.31a were used to calculate the time history of a wave front, which was then compared with one set of Prins' (1958b) measurements. The comparison, for conditions which meet the criteria of small λ/d, small h/d, and large x/d, are shown in Fig. 11.14a. It can be seen that the theory is good for these conditions. Comparison of measurements and theory for relatively large values of λ/d and small values of x/d, but still for small values of h/d, are shown in Fig. 11.14b. The results are not as good as for the conditions shown in Fig. 11.14a. Equation 11.18 predicted the crest height correctly for $x/d = 2.15$ but not the value of $t\sqrt{g/d}$ at which the crest occurred. Equation 11.31a was adequate for prediction purposes for $x/d = 6.45$. A numerical solution of the integral equation, Eq. 11.17, shows closer agreement, as would be expected.

Except for the first wave, theory shows that the amplitudes of the waves decrease in proportion to $1/\sqrt{x}$. Numerical solutions of Eqs. 11.17, 11.25, and 11.31 show there is no simple relationship between the amplitude of the first wave and x.

It appears that as long as h/d is small, the theories are quite good, provided λ/d is not too large. It can be seen from Prins' diagram (Fig. 11.15) that the larger the λ/d, the smaller must be h/d for the waves to have "oscillatory characteristics." For a given value of λ/d, as h/d is increased the waves gradually change from oscillatory (0) to a "solitary with an oscillatory tail" (ST), to a solitary wave (SS) to a "complex solitary" (CS) wave and finally to a bore. The characteristics

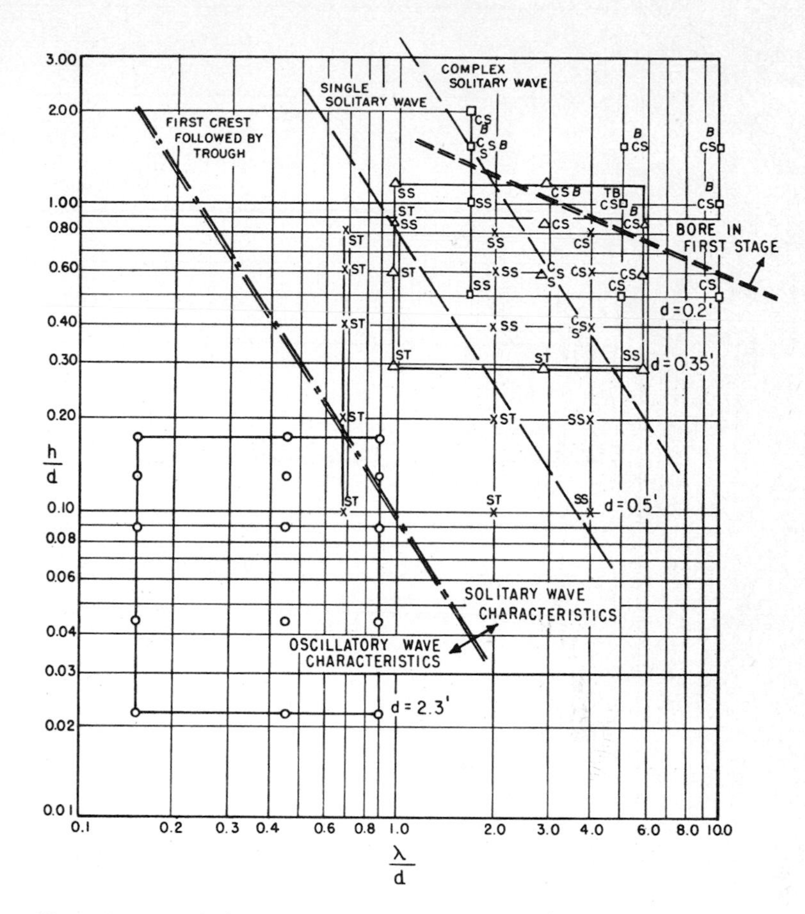

Fig. 11.15. Relation between λ/d, h/d, and the characteristics of the leading wave in the case of elevation. (After Prins, 1958.)

described in Fig. 11.15 are for positive values of h. Prins found that as long as the waves were in the oscillatory region it made no difference whether h was positive or negative, but for large values of h/d the wave characteristics were quite different for negative values of h than for positive values (Fig. 11.16); for negative values of h the waves were always dispersive. In this figure, negative values of h refer to an initial depression of the water surface and positive values of h refer to an initial elevation of the water surface. One practical conclusion that might be reached from this is that if, during an earthquake, there were both a substantial area of uplift and depression of the ocean bottom in relatively shallow water (as apparently occurred in the March 27, 1964 Alaskan Earthquake), both nondispersive and dispersive wave systems would form.

The most important parameter to describe the regime in which a uniform periodic wave system belongs was shown by Stokes (1880) and Ursell (1953) to be $y_{s_0}L^2/d^3$, where y_{s_0} is the value of y_s at the wave crest. This parameter has been discussed in detail by Wilson, Webb, and Hendrickson (1962), and the result of their analysis of the numerical ranges of the regimes, presented in graphical form, is shown in Fig. 11.17.

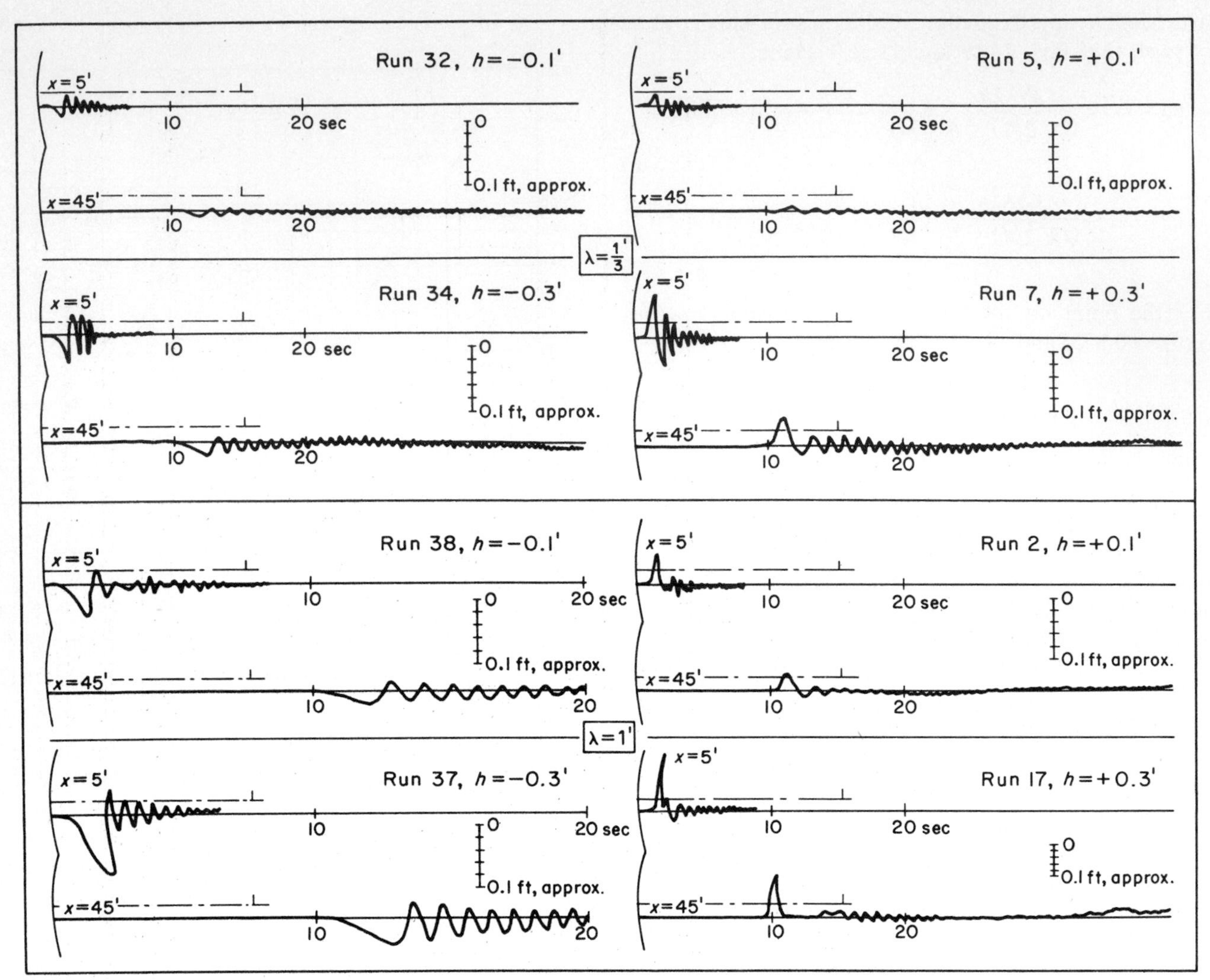

(a) Leading parts of the wave patterns at depth $d = 0.5'$

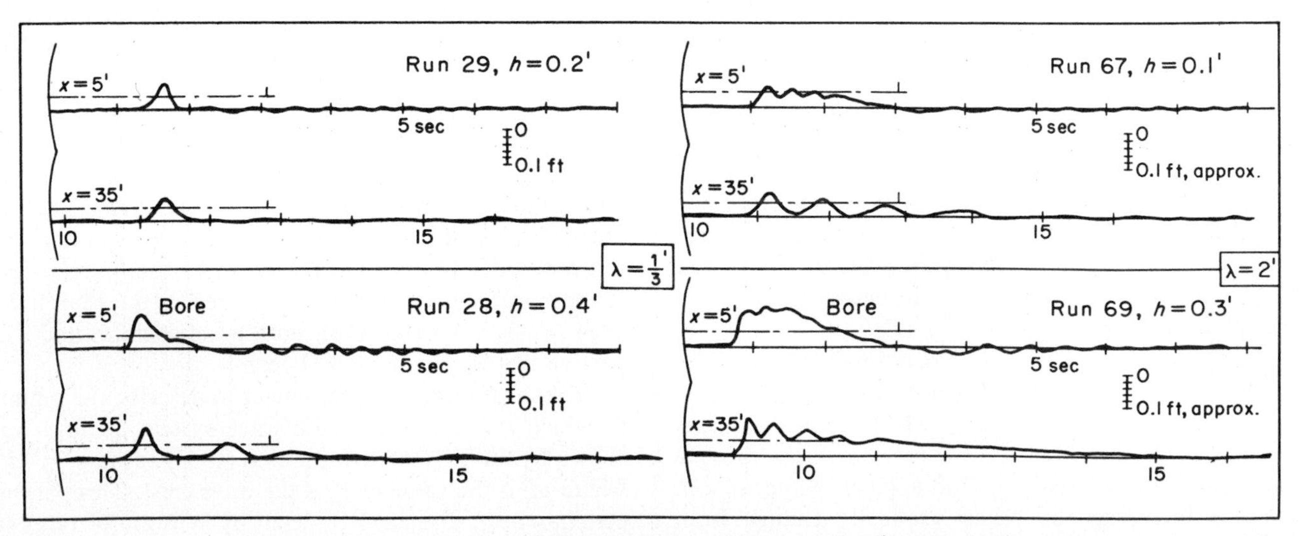

(b) Leading parts of the wave patterns at depth $d = 0.2'$ (after Prins, 1958)

Fig. 11.16. Leading parts of the wave patterns. (After Prins, 1958.)

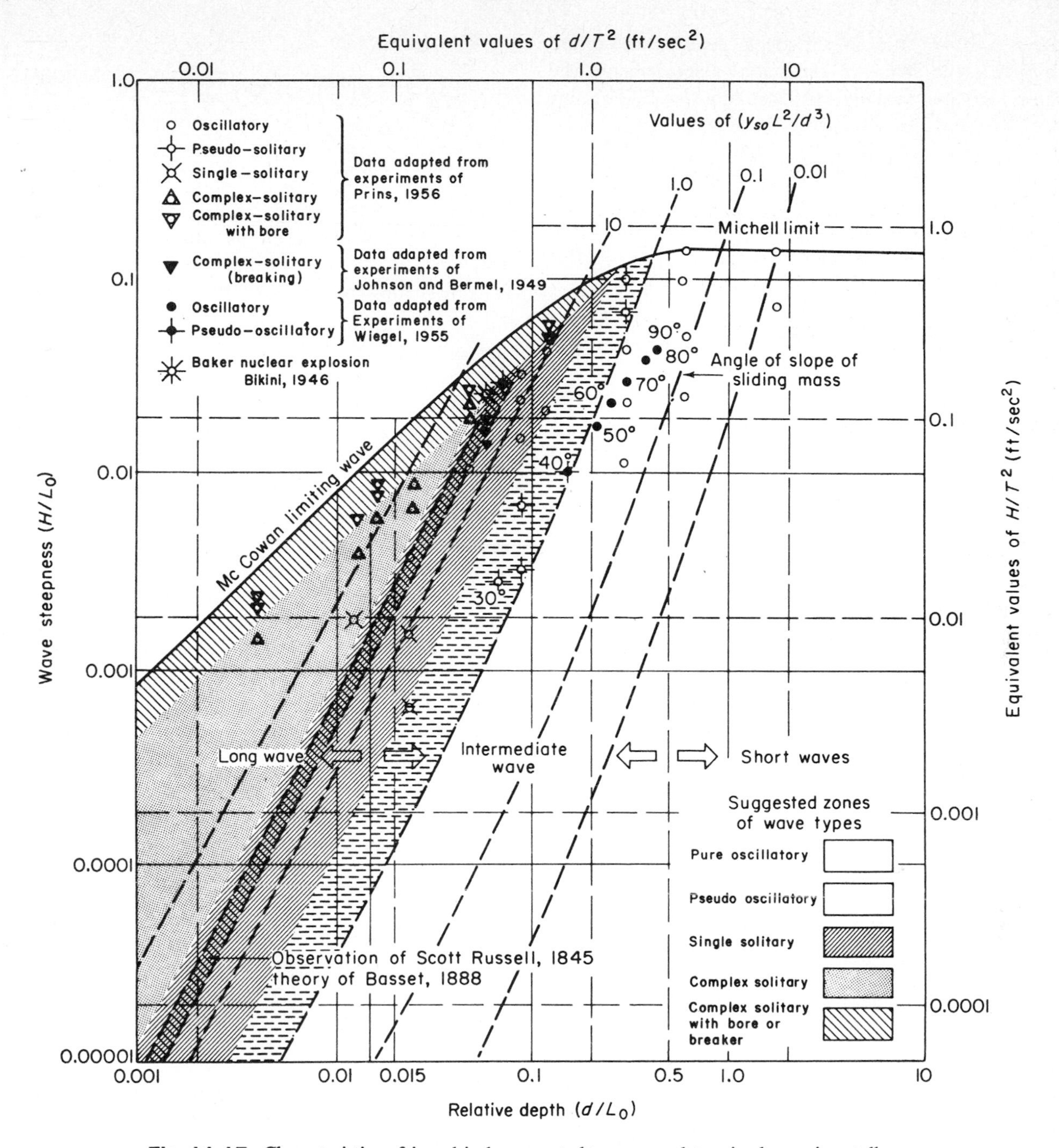

Fig. 11.17. Characteristics of impulsively generated waves, as determined experimentally, in relation to their theoretical prediction. (From Wilson, Webb, and Hendrickson, 1962.)

For the case of a circular initial disturbance of radius R and elevation h, the waves spread radially from the source; Unoki and Nakano found that

$$y_s = \frac{hR\sqrt{2}}{r}\cos\left(\frac{gt^2}{4r}\right)J_1\left(\frac{gt^2}{4r}\frac{R}{r}\right) \tag{11.32a}$$

$$= \frac{hR\sqrt{2}}{r}\cos(2\pi r/L)J_1(2\pi R/L) \tag{11.32b}$$

for $r \gg \lambda$. Here r is the radial distance from the source and J_1 is the Bessel function of the first kind. This equation shows that the amplitude of the waves decreases in proportion to $1/r$. Thus, one-half of the decrease is due to dispersion and the other half to radial spreading.

For tsunamis of any appreciable size the water is shallow in comparison with the wavelength (Kranzer and Keller, 1955, 1959). The effective radius of the initial displacement R is defined by

$$\frac{1}{2}|h|R^2 = \int_0^\infty |E(r)|\, r\, dr \tag{11.33}$$

where h is the elevation of the displacement at the origin and $E(r)$ is the shape of the initial displacement. The volume of water initially displaced is $\pi\,|h|\,R^2$. The surface elevation, for $r \gg R$, is

$$y_s = \frac{hR}{r} B \cos 2\pi\left(\frac{t}{T} - \frac{r}{R}\right) \tag{11.34}$$

where

$$B = 0, \text{ for } r > t\sqrt{gd} \tag{11.35}$$

and

$$B = \frac{1}{hRd}\bar{E}\left(\frac{2\pi}{L}\right)\sqrt{\frac{(2\pi d/L)\Phi(2\pi d/L)}{-\Phi'(2\pi d/L)}}, \quad \text{for } r < t\sqrt{gd} \tag{11.36}$$

Here, d is the water depth and $\bar{E}(2\pi/L)$ is the zero order Hankel transformation of the initial elevation $E(r)$ and is given by

$$\bar{E}\left(\frac{2\pi}{L}\right) = \int_0^\infty E(r) J_0\left(\frac{2\pi r}{L}\right) r\,dr \tag{11.37}$$

The group velocity C_G is given by Eq. 11.9.

The variable $2\,d/L$ (i.e., kd) is a function of $r/t\sqrt{gd}$, and is defined by Eq. 11.7, with r being substituted for x.

For the case of $E(r) = h$ for $r < R$ and zero for $r > R$,

$$\bar{E}\left(\frac{2\pi}{L}\right) = \frac{hR}{(2\pi/L)} J_1\left(\frac{2\pi R}{L}\right) \tag{11.38}$$

and

$$y_s = \frac{hR}{r(2\pi d/L)}\sqrt{\frac{(2\pi d/L)\Phi(2\pi d/L)}{-\Phi'(2\pi d/L)}} \cos 2\pi \times \left(\frac{t}{T} - \frac{r}{R}\right) J_1(2\pi R/L) \tag{11.39}$$

It appears that the shallower the water the higher will be the waves in relatively shallow water. Equation 11.39 often appears in a slightly different form, with the transformation being made by substituting $d[kC_G/-C_G']^{1/2}$ for $[kd\Phi(kd)/-\Phi'(kd)]^{1/2}$, where $k = 2\pi/L$, and C_G' denotes differentiation of C_G with respect to k.

A numerical example of this equation has been calculated by Kranzer and Keller. For the conditions of the calculation ($h = 28$ m, $d = 5$ km, $R = 2$ km, and $r = 20$ km) the highest wave was the third rather than the first wave in the group. As pointed out by Kranzer and Keller there really is no bore at $r = t\sqrt{gd}$, as is shown by his calculation. Rather, something analogous to the Airy Integral form of the two-dimensional case will exist. (See also Kajiura, 1963.) In a similar manner, it has been found theoretically that the amplitude of the first wave decreases as $1/r^{5/6}$, rather than as $1/r$ (Takahasi, 1961). Although this has apparently been verified by both prototype and laboratory tests (Van Dorn, 1961; Wilson, Webb, and Hendrickson, 1962), it is in disagreement with the theoretical findings of Kajiura (1963a).

For deep water,

$$\sqrt{\left(\frac{2\pi d}{L}\right)\Phi\left(\frac{2\pi d}{L}\right)\Big/-\Phi'\left(\frac{2\pi d}{L}\right)} \longrightarrow \sqrt{2}\left(\frac{2\pi d}{L}\right)$$

and

$$y_s \approx \frac{\sqrt{2}\,hR}{r}\cos 2\pi\left(\frac{t}{T} - \frac{r}{L}\right) J_1\left(\frac{2\pi R}{L}\right) \tag{11.40}$$

which is the same result as obtained by Unoki and Nakano.

For shallow water,

$$\sqrt{\left(\frac{2\pi d}{L}\right)\Phi\left(\frac{2\pi d}{L}\right)\Big/-\Phi'\left(\frac{2\pi d}{L}\right)} \longrightarrow 1$$

and

$$y_s \approx \frac{hR}{r(2\pi d/L)}\cos 2\pi\left(\frac{t}{T} - \frac{r}{L}\right) J_1\left(\frac{2\pi R}{L}\right) \tag{11.41}$$

The theories and experimental results described in the previous portion of this section have been given to indicate the general features of impulsively generated waves. Generation by surface displacements was considered as there were sufficient experimental results to indicate the regions of applicability of the theories. A more realistic model is the vertical displacement of a portion of the bottom of the ocean. Takahasi (1963) and Keller (1963) considered the sudden vertical movement of a portion of the ocean bottom of amplitude h and radius R.

For the main body of waves, they found that

$$\begin{aligned} y_s &= \frac{hR}{rk\cosh(2\pi d/L)}\sqrt{\frac{kC_G}{-C_G'}}\cos 2\pi\left(\frac{t}{T} - \frac{r}{R}\right) \times J_1\left(\frac{2\pi R}{L}\right) \\ &= \frac{hR}{r\cosh(2\pi d/L)}\sqrt{\frac{C_G}{-kC_G'}}\cos 2\pi\left(\frac{t}{T} - \frac{r}{R}\right) \times J_1\left(\frac{2\pi R}{L}\right) \end{aligned} \tag{11.42}$$

where C_G' refers to the differentiation of C_G with respect to the wave number k. Equation 11.42 is identical to the alternate form of Eq. 11.39, with the exception of the term $1/\cosh(2\pi d/L)$.

Takahasi found that in the neighborhood of the front of the disturbance as it moved, the wave amplitude decreased as $1/r^{5/6}$, while for the body of waves the amplitudes decreased as $1/r$ [this is a little in disagreement with the theoretical findings of Kajiura (1963a)]. He also found theoretically that no bore formed in the front (in his linear theory).

Takahasi performed some laboratory experiments to check his theory. A circular cylinder was placed in the bottom of a large basin; the cylinder could be either raised or depressed very rapidly. It was found that for relatively deep water it made no difference whether the cylinder was raised or depressed, except for the change in sign of y_s. For shallow water, results similar to those of Prins for a surface disturbance occurred.

A number of tests were made with two circular cylinders, oriented as shown in Fig. 11.18a. The results are given in Fig. 11.18b. It is evident that the "beaming" of the energy was strongly directional. The relatively high waves in the S direction was expected, but the difference in wave period with direction was unexpected. When cylinder A was raised and cylinder B was depressed simultaneously, the direction of greatest energy was SE.

In order to check the directional effect further, a series of six cylinders were installed on an E–W line and operated simultaneously. When all were raised, it was found that the wave amplitude decreased as $1/r^{1/3}$ to $1/r^{1/4}$ toward S and $1/r^{3/2}$ toward E. This is extremely important in understanding the relative effects of tsunamis generated by nearly rectangular ocean bottom displacements (May 22, 1960, Chile, and March 27, 1964, Alaska tsunamis) and those generated by nearly circular ocean bottom displacements (June 19, 1933, and October 13, 1935, Japan tsunamis).

Theoretical studies of more general bottom displacements have been made by Kajiura (1963a), and studies of time-dependent bottom motions have been made by Wilson, Webb, and Hendrickson (1962). The form of the equations are such, however, that one cannot see simple physical relationships in them.

Wiegel (1955, 1963) performed some laboratory experiments representative of a two-dimensional submarine landslide. The first experiments were made by permitting a submerged body to fall vertically. The potential energy of the body was changed by varying the elevation of the fall and by placing additional lead

Fig. 11.18. Distribution of wave amplitude and period as a function of direction from source. (From Takahasi, 1963.)

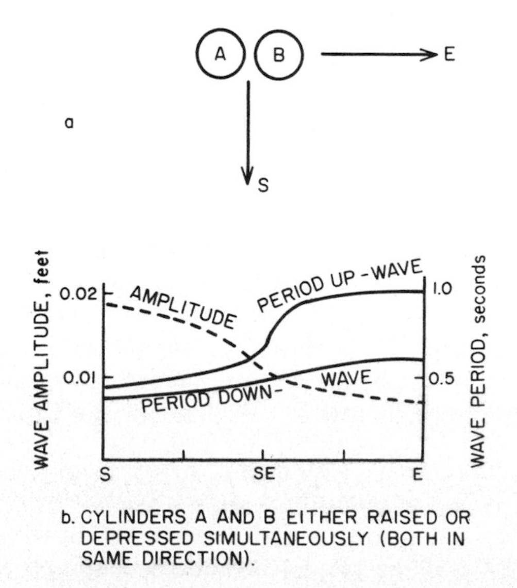

b. CYLINDERS A AND B EITHER RAISED OR DEPRESSED SIMULTANEOUSLY (BOTH IN SAME DIRECTION).

weights in the body. The effect of the length of such an initial disturbance is shown in Fig. 11.19, together with the type of waves generated in relatively deep water. The amount of energy in the wave disturbance was found to be of the order of 1–2% of the initial potential energy (net, submerged) of the body.

A more realistic model of a landslide is a section of material moving down a slope, rather than vertically. The results of an experiment with this type of mechanism are shown in Fig. 11.20. The period of the main wave increases with increasing slope, with the rate of increasing period becoming large for slopes less steep than about one in three ($18\frac{1}{2}$ deg).

It is interesting to note that the relationship between the length of the vertically falling body and the initial period of the waves, when extrapolated, indicates that an underwater disturbance of the order of a few thousand feet generates waves with an initial period of the order of 10–15 min. If the body were sliding down a one in three slope the dimensions needed to generate long waves would be much less. The waves would have the general appearance of the tsunamis observed in nature. Slides of the dimensions necessary to generate small tsunamis have occurred at the boundary to the ocean (Miller, 1966), and there is little reason to expect that they would not occur in such submarine regions as the Aleutian Trench.

11.4 TRAVEL OF TSUNAMIS IN THE OCEAN

11.4.1 Introduction

The transformation of wave groups described in the previous section was for the special case of water of uniform depth. Certain of these observations apply to tsunamis in the ocean. For example, a fairly good approximation of the travel time of the initial wave disturbance can be obtained using $C = \sqrt{gd}$. For the April 1, 1946, Aleutian tsunami it was found that the average wave speeds ranged from 375 to 490 mph between the origin and the station, depending upon the ocean depths. The arrival times were computed to within a few minutes for the stations relatively near the epicenter (1100 statute miles with a travel time of just under 3 hr) and within 30 min for a distant station (8000 mi with a travel time just over 18 hr; Green, 1946), with the travel time usually being a little less than that predicted from the use of $C = \sqrt{gd}$, for the reason discussed previously. Travel times to a number of places have been computed for tsunamis originating in several areas. Figure 11.21 is an example of a tsunami travel time chart.

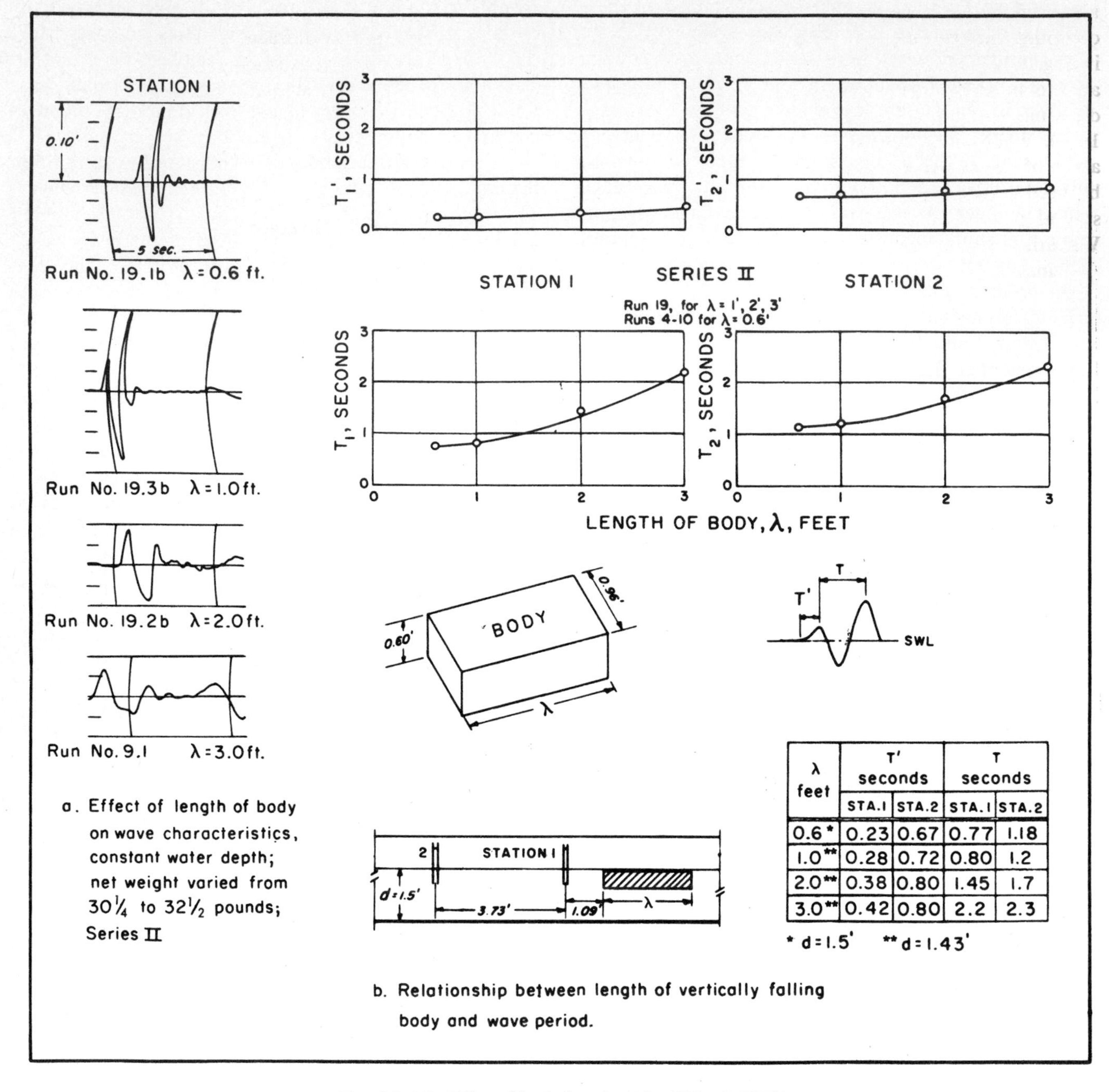

λ feet	T' seconds STA.1	T' seconds STA.2	T seconds STA.1	T seconds STA.2
0.6*	0.23	0.67	0.77	1.18
1.0**	0.28	0.72	0.80	1.2
2.0**	0.38	0.80	1.45	1.7
3.0**	0.42	0.80	2.2	2.3

* d = 1.5' ** d = 1.43'

Fig. 11.19. Effect of body length. (After Wiegel, 1955.)

In addition to the dispersion and radial spreading effects, described briefly in the previous section, waves are subject to refraction and diffraction. Detailed discussions of these two phenomena are available in *Oceanographical Engineering* (Wiegel, 1964d).

11.4.1.1 Refraction. In the linear theory of progressive water gravity waves, the phase velocity is given by Eq. 11.10. This phase velocity depends upon both the water depth and the wavelength. When d/L is less than about 1/25, tanh $2\pi d/L \longrightarrow 2\pi d/L$ (within 2%), and Eq. 11.10 reduces to $C \approx \sqrt{gd}$. Each part of the wave travels with a phase velocity that is dependent upon the water depth under it; if the water depth is not constant, the wave must bend. This bending of the wave is known as wave refraction. In treating the phenomenon of wave refraction, the assumption has been made that what is known in optics as Snell's Law can be applied:

$$\frac{\sin \alpha_2}{\sin \alpha_1} = \frac{C_2}{C_1} \tag{11.43}$$

where α_1 and α_2 are angles between adjacent wave front positions and the respective bottom contours. Refraction drawings can be constructed from a knowledge of the original shape of the wave front and the bottom contours. An example of a refraction drawing is shown in Fig. 11.22. Methods have been developed to calculate and to plot the orthogonals (rays) of wave refraction drawings using high-speed digital computers (Wilson, 1966). When constructing refraction drawings for large areas of the ocean, regular hydrographic charts cannot be used owing to the curvature of the ocean surface; special projection charts are used instead (see, e.g., Wadati, Hirono, and Hisamoto, 1963).

Lines constructed normal to the wave fronts are known as orthogonals. Providing that no energy is dissipated, reflected, or diffracted, the wave power transmitted between a pair of orthogonals remains constant. If the distance between a pair of orthogonals changes from b_1 to b_2, the wave height also changes, as

$$\frac{H_2}{H_1} \alpha \sqrt{\frac{b_1}{b_2}} = K_d \qquad (11.44)$$

so that a convergence of orthogonals results in an increase in wave height and a divergence of orthogonals results in an increase in wave height. The term $K_d = \sqrt{b_1/b_2}$ is called the refraction coefficient.

Fig. 11.20. Effect of incline angle. (After Wiegel, 1955.) (a) Effect of slope on wave characteristics, constant water depth, and constant body dimensions; Series II. (b) Relationships among slope of sliding body, period, and amplitude of waves.

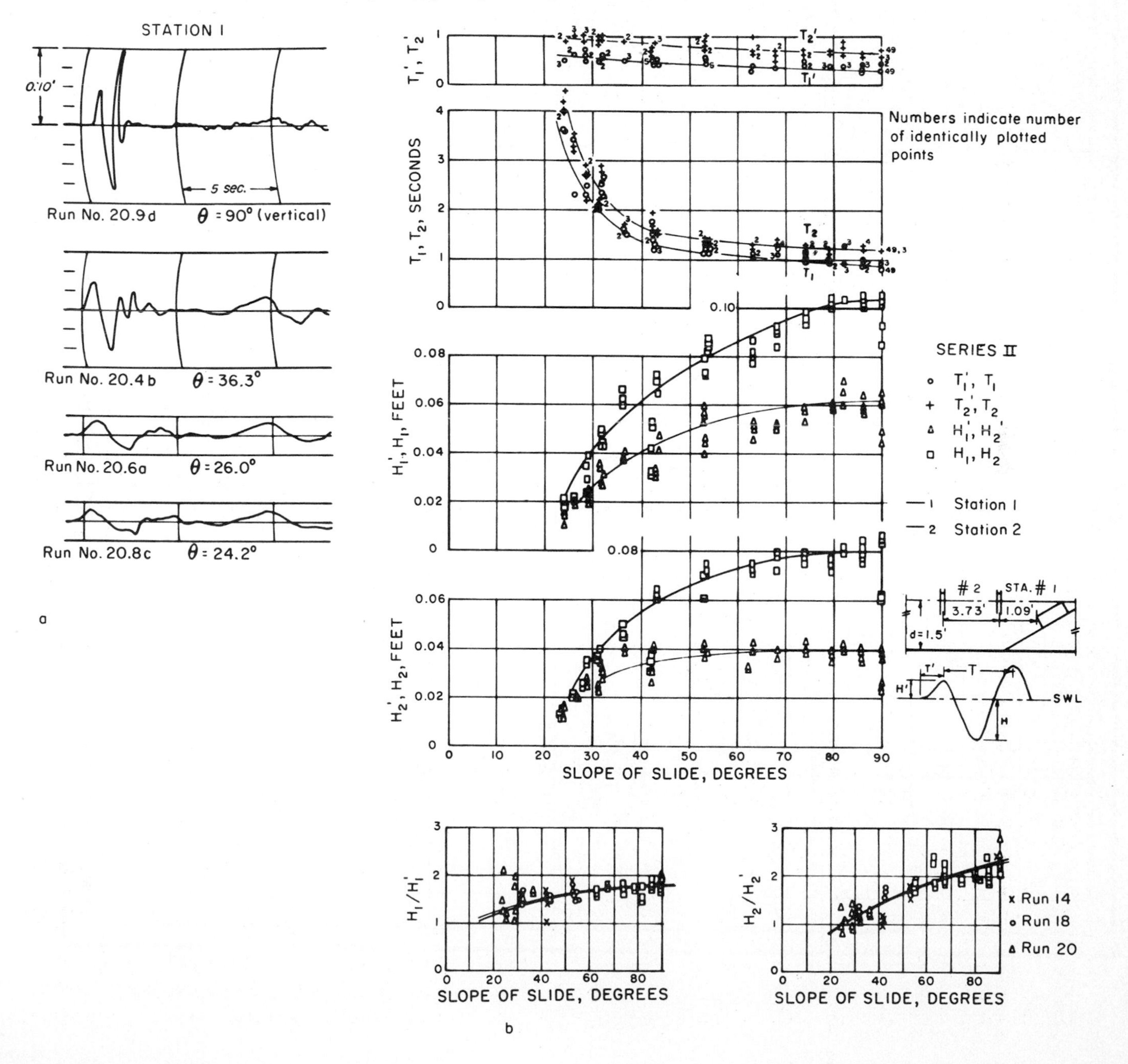

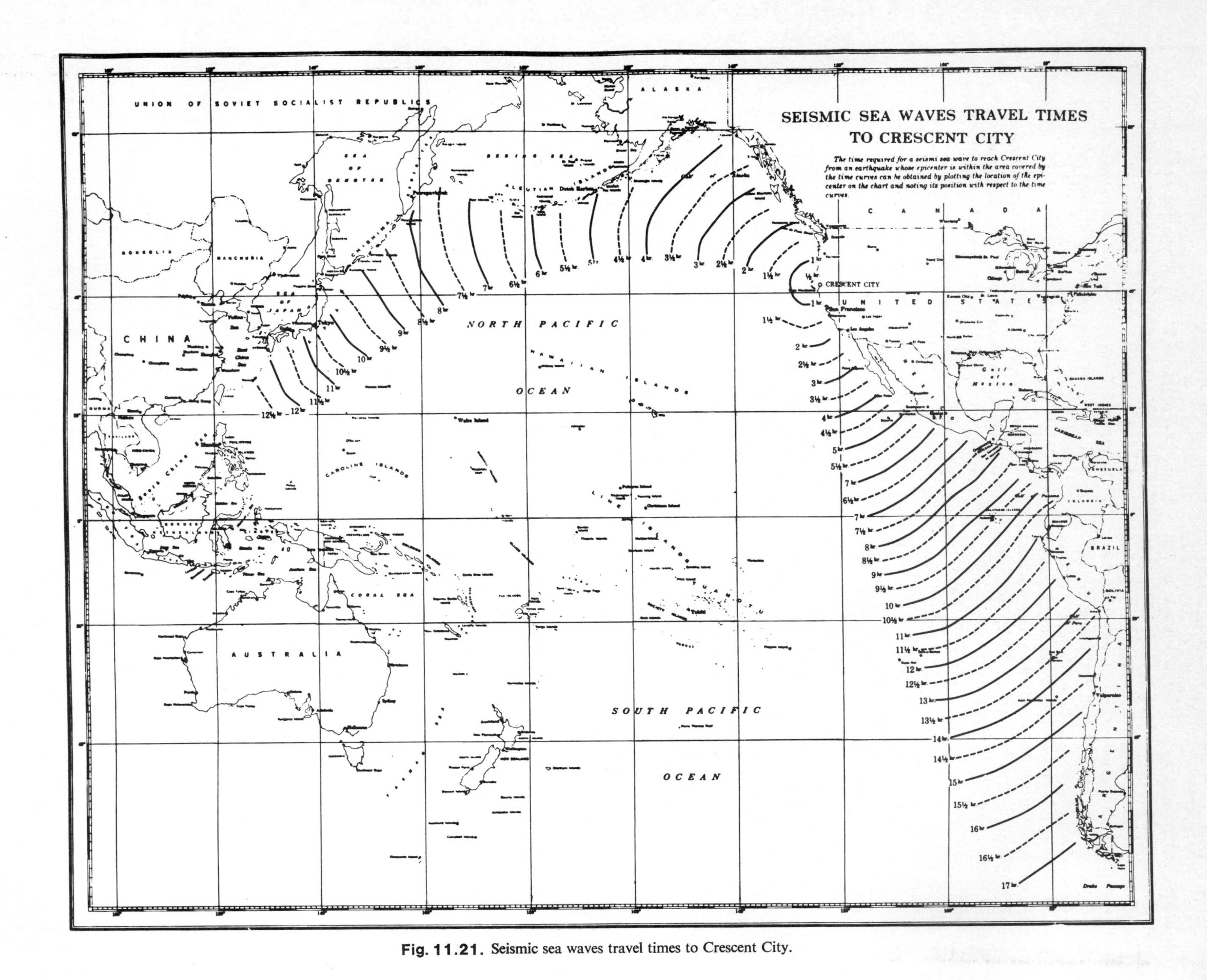

Fig. 11.21. Seismic sea waves travel times to Crescent City.

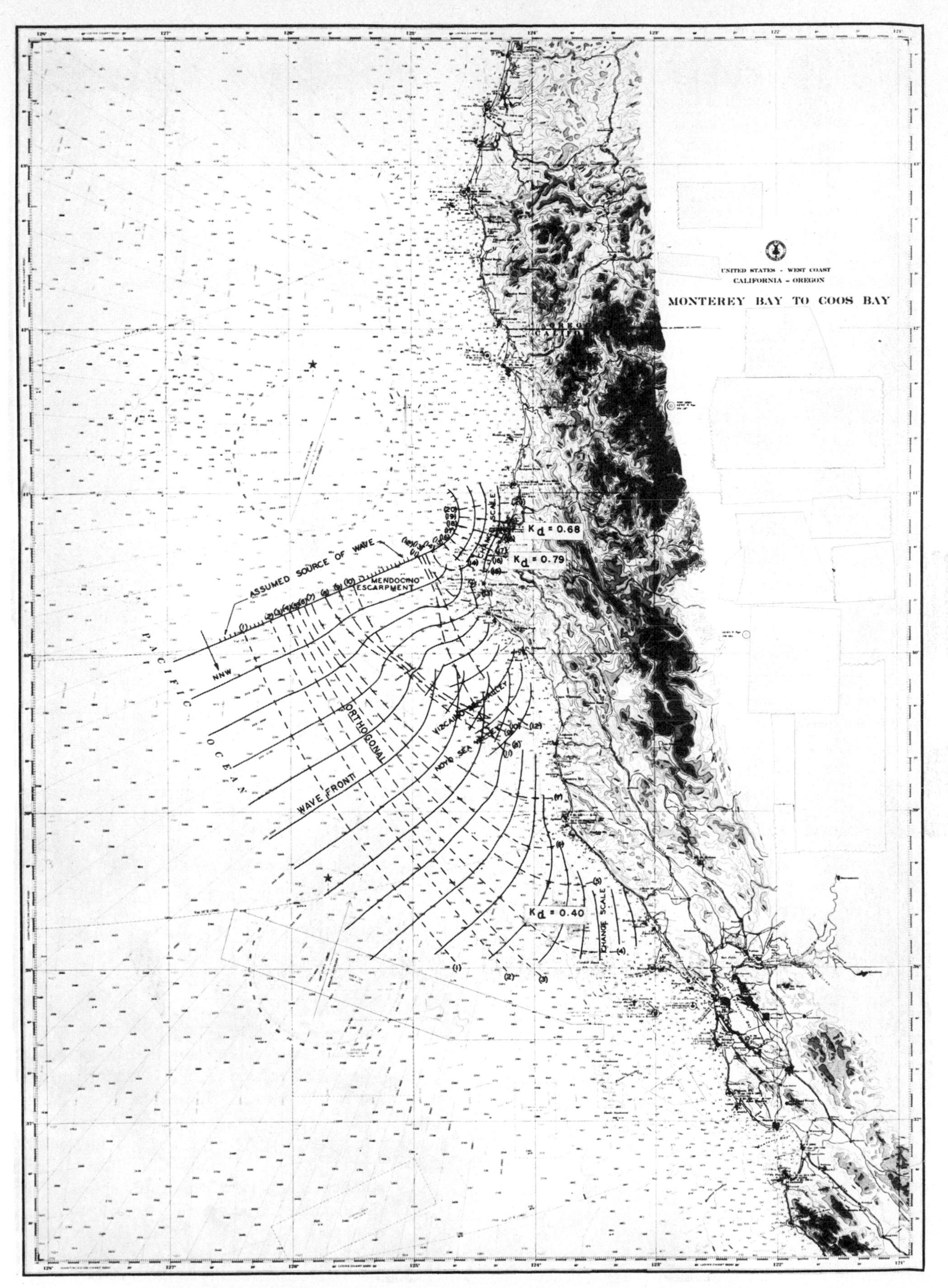

Fig. 11.22. Refraction drawing. Mendocino Escarpment and Bodega Head, California. (From Wiegel, 1964b.)

Owing to refraction, regions of relatively high waves and regions of relatively low waves will develop. One would expect a transfer of energy because of this difference in potential. The author is not aware of any studies that have been made of this problem.

Phase speed decreases continuously with decreasing water depth (Eq. 11.10), but the group speed C_G at first increases and then decreases (see Eq. 11.9). As C_G is the speed at which energy is propagated, one would expect the group of waves to refract at a different rate than the individual waves (Williams and Isaacs, 1952). Practically no studies have been made of this phenomenon. As tsunamis are long waves, $\sinh 4\pi d/L \longrightarrow 4\pi d/L$ and $C_G \longrightarrow C$ so that this phenomenon is not as important for tsunamis as it is for wind-generated waves. For example, for a tsunami wave with a period of 20 min and a water depth of 15,000 ft, $C_G/C = 0.995$.

11.4.2 Reflections from Submerged Reefs and Shelves

Shallow water waves moving over submerged reefs, the continental shelf, and other "shoal" regions are partially reflected. Kajiura (1963b) has developed the theory for such reflections for a number of specific conditions. The simplest case is the one of a long wave moving in water of constant depth d_2 and traveling over a vertical cliff into water of constant depth d_1. In this case the reflection coefficient (the ratio of the reflected to incident wave height, H_R/H_I) is

$$\left|\frac{H_R}{H_I}\right| = \left|\frac{\sqrt{d_2} - \sqrt{d_1}}{\sqrt{d_2} + \sqrt{d_1}}\right| \tag{11.45}$$

The transmitted wave height H_T is given by

$$\left|\frac{H_T}{H_I}\right| = 1 - \left|\frac{H_R}{H_I}\right| \tag{11.46}$$

A more general case is a gradual change in the bottom $d(x)$ given by the equation

$$\frac{1}{d(x)} = \frac{1}{2}\left(\frac{1}{d_1} + \frac{1}{d_2}\right) - \frac{1}{2}\left(\frac{1}{d_1} - \frac{1}{d_2}\right)\tanh\left(\frac{nx}{2}\right) \tag{11.47}$$

where $n = 2\pi/\lambda$, with λ being the representative width. A sketch of the bottom, giving definitions, is shown in Fig. 11.23a. The reflection coefficient is given by

$$\left|\frac{H_R}{H_T}\right| = \left|\frac{\sinh \pi(1/L_1 - 1/L_2)}{\sinh \pi(1/L_1 + 1/L_2)}\right| \tag{11.48}$$

where L_1 and L_2 are the wavelengths in water depths d_1 and d_2, respectively. The results are plotted in Fig. 11.23b for a number of values of λ/L_2 and $\sqrt{d_2/d_1}$. The transmitted wave height is given by Eq. 11.46. It is evident from Fig. 11.23b that when the length of the transition λ is of the order of half a wavelength that reflection is relatively small. When $\lambda/L_2 \longrightarrow 0$, Eq. 11.48 can be approximated by Eq. 11.45.

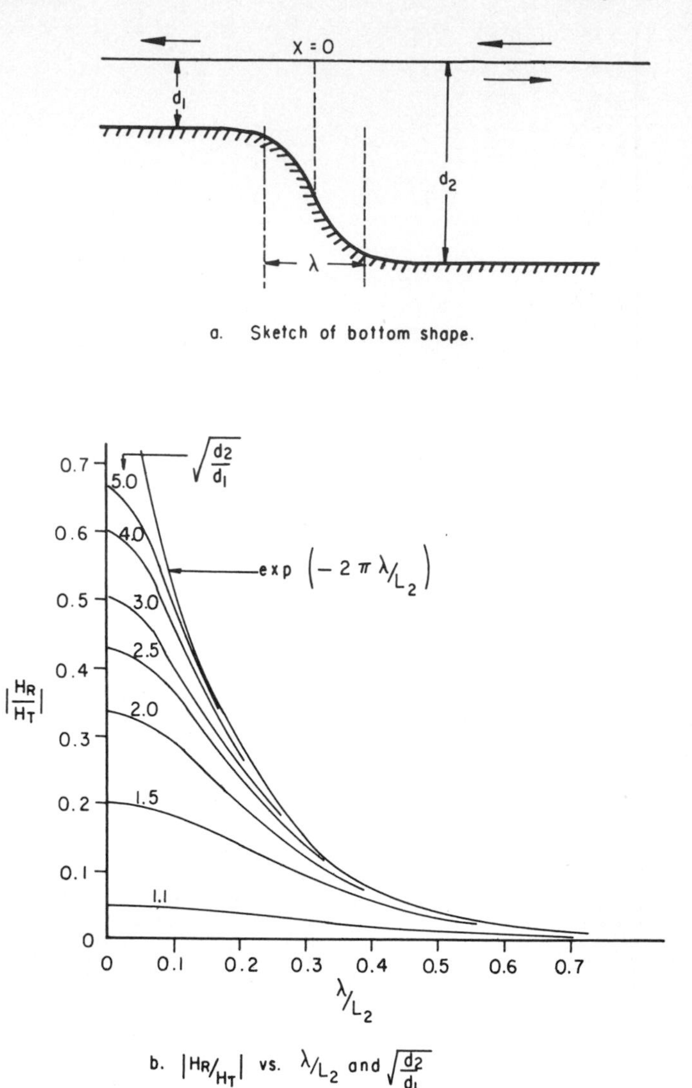

Fig. 11.23. The reflection coefficient of long waves for a continuous transition. (From Kajiura, 1963b.)

11.5 TSUNAMI WAVES ALONG THE SHORE

11.5.1 Wave Trapping

It has been shown theoretically that waves reflected from shore can be refracted in such a manner that they become trapped in the near shore zone or on the continental shelf (Isaacs, Williams, and Eckart, 1951; Williams and Isaacs, 1952); the term "total reflection by deep water" also is used in describing this phenomenon. A practical example of this can be seen in Fig. 11.24. This phenomenon has been observed in a model study of tsunami waves in Hilo Harbor, Hawaii (Palmer, Mulvihill, and Funasaki, 1966). A fundamental difficulty is encountered in constructing refraction drawings to check for wave trapping. The assumption is made that the angle of reflection at the coast is equal to the angle

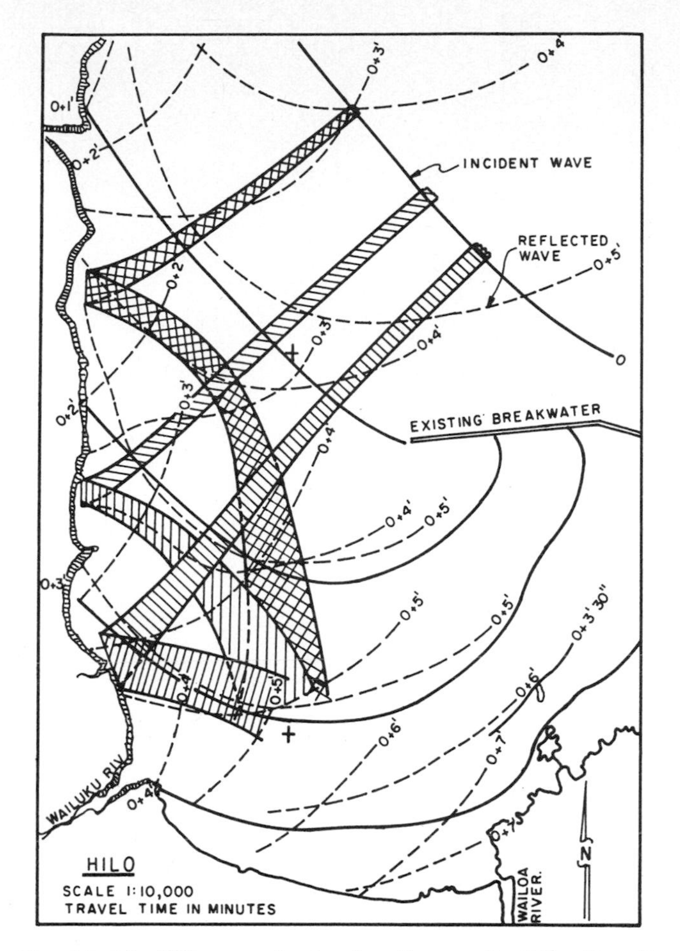

Fig. 11.24. 1960 tsunami refraction diagram for Hilo, Hawaii. (From Palmer, Mulvihill, and Funasaki, 1966.)

of incidence. The incident wave fronts cannot be constructed to the zero depth contour as the phenomena is no longer linear in this region; in fact, a wave may transform into a bore. All other things being equal, the angle of incidence at the point of reflection depends upon the depth of water at that point. At the present time a subjective decision must be made by the person constructing the drawing. It is generally agreed that this problem needs further study.

11.5.2 Mach-Reflection

Along certain types of shores a phenomenon has been found to occur, which is analogous to the Mach-reflection of blast waves. The possibility of this was first suggested by John D. Isaacs. A number of aspects of this phenomenon have been investigated experimentally by several graduate students under the author's direction (Perroud, 1957; Chen, 1961; Sigurdsson and Wiegel, 1962; Nielsen, 1962). A summary of these studies, together with a practical example concerned with tsunamis, has been presented by Wiegel (1964c).

It has been found that periodic waves (shallow water and transitional water) and solitary water gravity waves do not reflect from a wall (or steep slope), in the manner commonly supposed, when the angle between the direction of wave advance and the wall is less than about 35–45 deg. The wave front bends near the wall, becoming normal to the wall, with a small reflected wave forming. For angles less than about 20 deg the reflected wave becomes almost negligible. The portion of the wave near the wall (called the Mach-stem in air blast waves) increases in height as the wave continues to move along the wall. Once the Mach-stem is formed, it continues to grow even when the wall is curved around through almost 90 deg; for periodic waves a Mach-reflected wave also develops. The Mach-stem is insensitive to undulations of the wall.

Hilo, Hawaii, has been subject to severe damage from tsunamis, which originate in the vicinity of Alaska and Chile. The orientation, topography, and hydrography of the region (Fig. 11.25) are such that it appeared likely that a Mach-stem might have been associated with the April 1, 1946 tsunami that originated in the Aleutian Islands of Alaska. The height of the tsunami should increase by a factor of 4–5 as it moves onto the shallow portion of the reef off Hilo (Fig. 11.24), owing to shoaling effects alone, and it was believed that this, together with the Mach-stem effect, could account for the characteristics of several of the waves of the tsunami as they were observed. It was observed (M. L. Child of Hilo, Hawaii) that the two waves that did the most damage moved as a bore in a southerly direction along the cliff that forms the west border of the bay, swinging easterly and running up through the streets of the town and into the lee area of the breakwater. At the same time waves overtopped the breakwater. A photograph taken at the time shows a wave that looks remarkably like a Mach-stem (Fig. 11.26). Another photograph, showing the wave rolling into the town, is shown in Fig. 11.27.

In order to study the gross characteristics of tsunamis at Hilo, a 1: 15,000 ($1:\sqrt{15,000} = 1:122$ time and velocity scale) undistorted model was constructed of fiberglass. The model was approximately 8 ft on a side, so that the entire bay could be included as well as the reef. A portion of the ocean was included, to a depth of 6000–7000 ft, prototype. The model was placed at one end of an 8-ft wide by 6-ft deep by 200-ft long tank so that a number of waves could be measured before reflections would be of importance. A series of runs were made with periods ranging from 8 to 24 min, prototype, and with waves from the N, E, and SE directions.

It was found, for waves from the north, that a wave which had the appearance of a Mach-stem was generated along the west cliff and rolled into the town of Hilo in a manner that was similar to observations. It was also

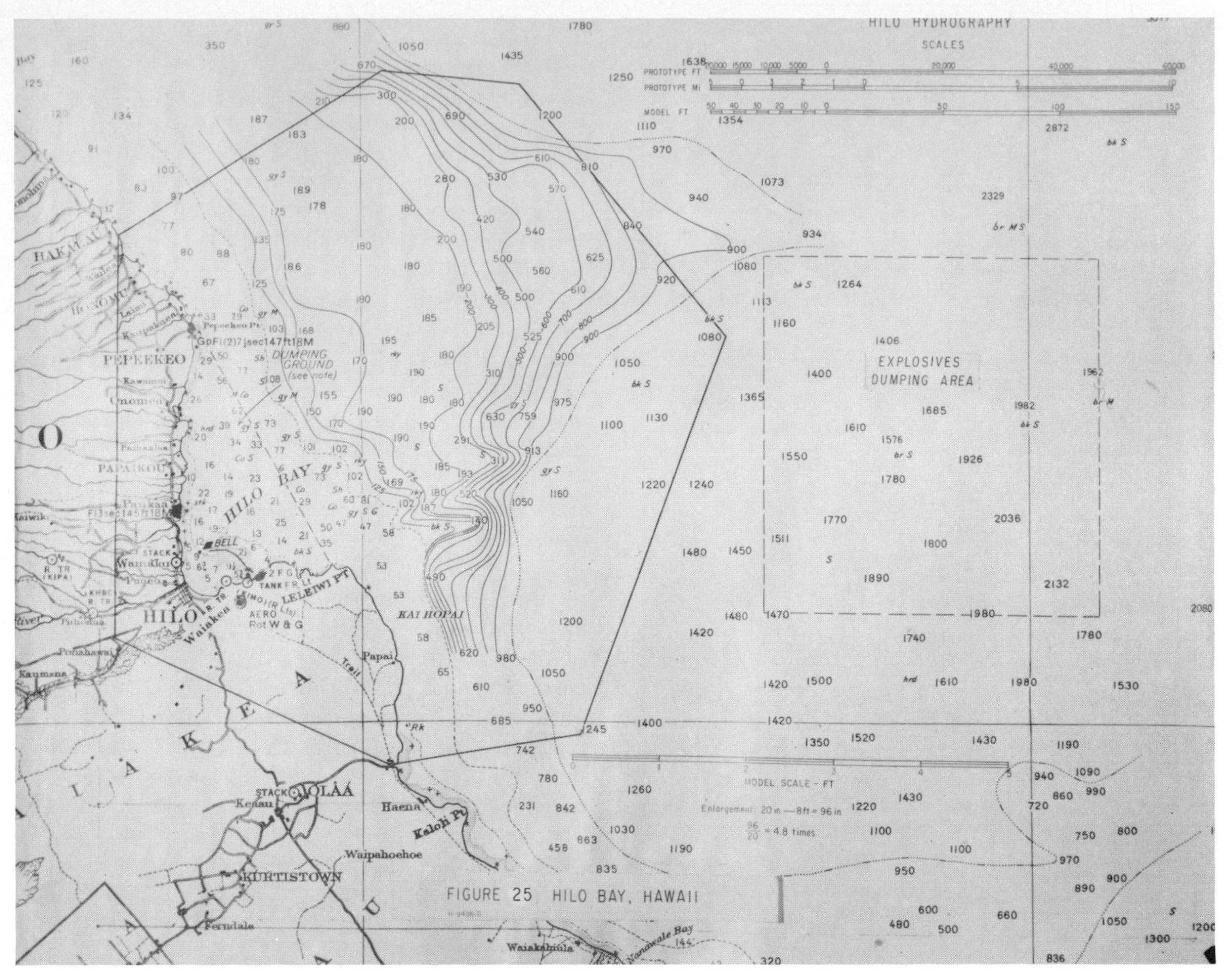

Fig. 11.25. Hilo Bay, Hawaii.

Fig. 11.26. 1946 tsunami at Hilo. Wave in the form of a bore approaching mouth of Wailuku River. Note: Mach-stem effect increased wave height along coastal bluff. Water in foreground has receded several feet. (Courtesy, Modern Camera Center, Hilo.)

Fig. 11.27. 1946 tsunami at Hilo. View down Waianuenue Street. This was second wave according to Yoshio Shigenaga, employee of American Trading Company, Ltd., which was then located in ground floor of two-story building on right at end of street. Bus was at intersection of Kamehameha Street. Breakwater in distance was breached by third or fourth wave. (Courtesy, Modern Camera Center, Hilo.)

found that the shoaling effect was about as theory predicted; i.e., the wave height increased by a factor of about 4 over the reef, with respect to the wave height in the water portion of the tank.

After the tests had been run, it was brought to the investigator's attention that due to refraction in the ocean, the tsunamis generated off Chile would most likely approach Hilo Bay from an easterly direction, rather than from a southeasterly direction as was originally supposed. Owing to this, the results of the model tests for the waves from the east will be described herein. A remarkable phenomenon was observed in the 12–20-min (prototype) period range. Referring to Fig. 11.28a, the initial wave refracted to about the position shown as (1)-(1). The northerly portion started to reflect from the coast while the southerly portion continued to move toward shore. This resulted in the pattern (2A)-(2A) as the reflected portion and (2)-(2) as the continuing portion. As the reflected portion (2A)-(2A) moves down the coast, it became independent of (2)-(2). At the same time the southerly tip of (2)-(2) diffracted into the harbor, raising the water level. About the same time (2A)-(2A) progressed to position (3)-(3) with the portion near the coast being considerably higher than the portion offshore. The portion near the coast ran right along the coast, reaching positions (4)-(4) and (5)-(5) as a high wave running on top of the water that had diffracted into the harbor from (2)-(2). It then ran into the town of Hilo. The author believes that something similar to this must have happened during the actual tsunami.

The transformation of (2A)-(2A) to (3)-(3) was probably caused by a combination of refraction and reflection, together with some nonlinear effects because of the relatively large wave height to water depth ratio along the coast, and the height of the tsunami at Hilo was probably due to this combined with the diffracted wave. At some place between (2A)-(2A) and (3)-(3) a Mach-stem type of phenomenon evolved and because of its strength became independent of the normally reflected portion of the wave. In Fig. 11.28b are shown the successive positions of this Mach-stem-type of wave as it moves along the coast. These positions were traced from enlargements of a 16-mm motion picture taken

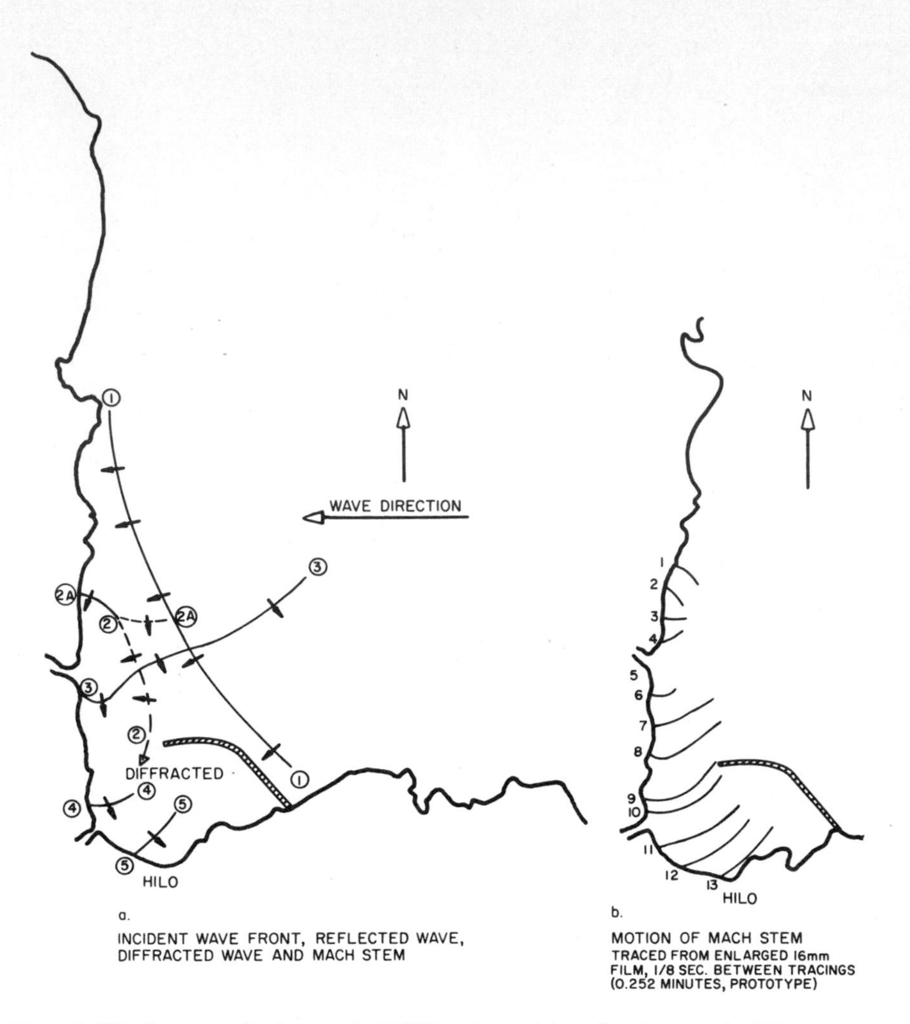

Fig. 11.28. Some results from a 1: 15,000 scale model study of tsunamis (16 min, prototype) at Hilo, Hawaii. Run *X*, February 8, 1963, wave from the east (in 7000 ft of water, prototype). a. Incident wave front, reflected wave, diffracted wave, and Mach-stem. b. Motion of Mach-stem traced from enlarged 16-mm film, 1/8 sec between tracings (0.252 min, prototype).

during the model study, for a wave of 8-sec period (16-min period in the prototype).

11.5.3 Diffraction

Consider a wave system interrupted by an impervious structure such as a breakwater or headland. The portion of the waves incident to the structure will reflect or break or both (and possibly overtop the structure), whereas the portion moving past the tip of the structure will be the source of the flow of energy in the direction essentially along the crest and into the region in the lee of the structure. The "end" of the wave will act somewhat as a potential source, and the wave in the lee of the breakwater will spread out in approximately a circular arc. This physical phenomenon is known as diffraction.

Only two aspects of diffraction will be considered here. One is the application of the diffraction of sound waves incident to a circular cylinder (Wiener, 1947) to water waves. An approximate distribution of wave relative wave heights can be made using this theory. Omer and Hall (1949) found that the pattern of heights for the April 1, 1946, tsunami at Kauai Island, Hawaii, was very similar to the pattern predicted by diffraction theory (see also Van Dorn, 1959). The theoretical and measured pressure distribution of acoustic waves around a circular cylinder are shown in Fig. 11.29 as a function of $\pi D/L$. Here D is the diameter of the cylinder and L is the wavelength. Some laboratory measurements by Laird (1955), using water waves, have shown that Wiener's theory can be applied to water waves with fair results to predict the wave elevation distribution around a cylinder. It can be seen in the figure that when a wave is long compared with the diameter of a cylinder, there is essentially no "lee."

The second aspect of diffraction that will be considered is the amount of energy that is transmitted through a breakwater gap. Use of geometric optics is beneficial to show one of the difficulties inherent in using a breakwater to protect a region from long waves. Consider a bay with an entrance width of 4000 ft. Suppose that a breakwater extends to a distance of 2000 ft from shore. Considering only geometric optics, the wave height transmitted past the breakwater can be expressed as

$$\frac{H_T}{H_I} = \sqrt{\frac{2000}{4000}} = 0.71$$

If another breakwater were to be built from the other shore, leaving an opening of 600 ft,

$$\frac{H_T}{H_I} = \sqrt{\frac{600}{4000}} = 0.39$$

Thus, even if the entrance were decreased as much as possible while still allowing for the safe navigation of cargo ships through the entrance, the wave height would still be 39% of the height that would exist if there were no breakwater and 55% as high as the waves that would have existed in the lee if the additional breakwater had been constructed.

Application of diffraction theory, rather than geometric optics, would show even less effect. A transmission factor has been computed (Morse and Rubenstein, 1938) that is the ratio of the energy transmitted by a wave of angle of incidence to the energy that would be predicted from geometric optics (the ratio of wave heights is proportional to the square root of the energy ratios) for a wave of 90 deg angle of incidence. The limited case for an opening that is small compared with the wavelength has been solved by Lamb (1945), in which transmission factor is given by:

$$\frac{\pi^2/4}{(\pi B/L)\{[\ln(\pi B/4L) + \gamma]^2 + \frac{1}{4}\pi^2\}} \tag{11.49}$$

where γ is Euler's number (0.577) and the incident waves are normal to the breakwater with an opening of width B. For L/B of 10, 100, and 1000, the transmission factor is 1.24, 3.80, and 17.2. In practice the transmission factor would not continue to increase with decreasing ratio of opening width to wavelength, as the energy lost by eddies formed at the two edges of the opening would predominate for very small openings.

Model studies made by the U.S. Army Corps of Engineers of Hilo Harbor, Hawaii, have shown that the phenomenon is at least qualitatively as predicted by the theory. It should be pointed out that an important by-product of making a narrow opening is a greatly increased tsunami-induced current through the opening.

11.5.4 Shoaling and Run-up

Most shore slopes are relatively steep when compared with waves as long as tsunamis, and the waves act like

Fig. 11.29. Acoustic wave pressure distribution around a circular cylinder, $|p/p_0|$ in decibels as a function of $\pi d/L$. Circles and triangles denote the data for the median plane while squares show the pressures near the edge of the cylinder. (After Weiner, 1947.)

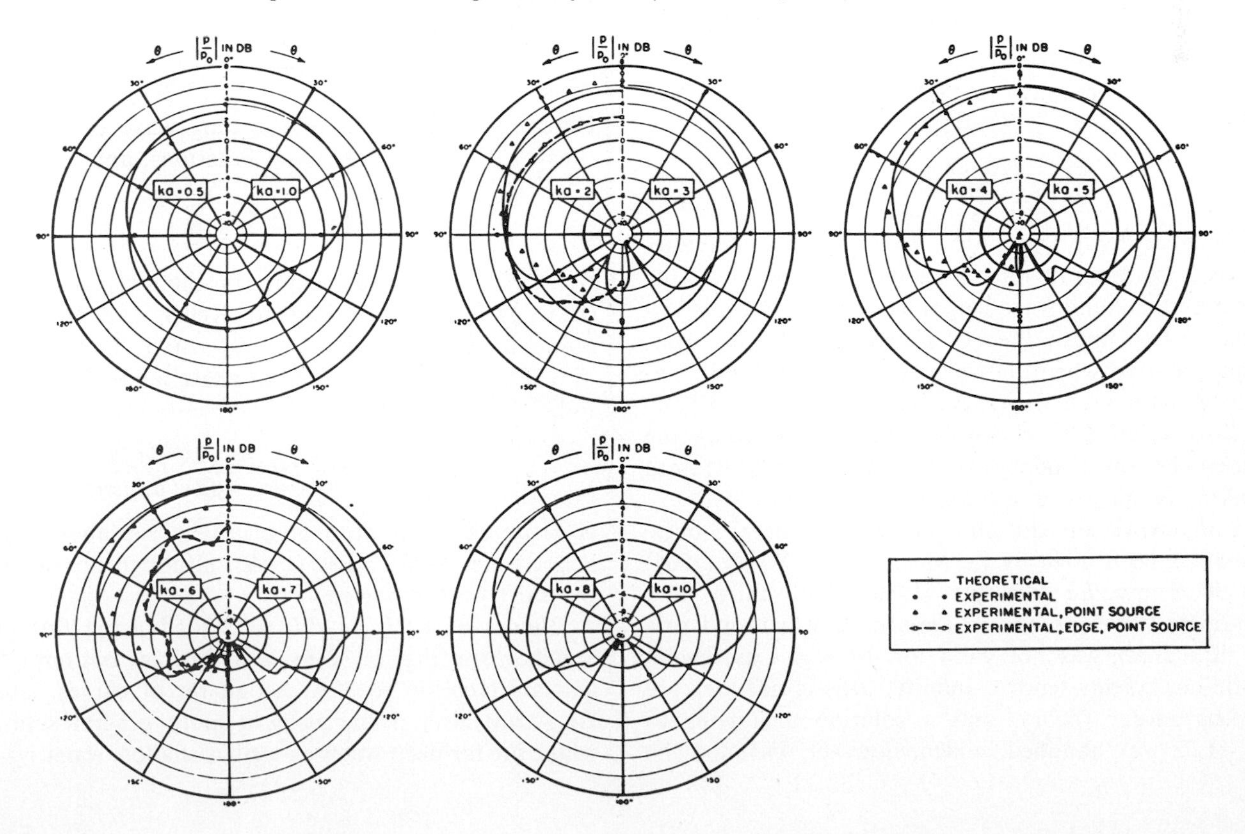

rapidly rising tides. It is only under certain rather rare conditions that bores are formed in areas that are far removed from the source of the tsunami. Although the run-up on shore can be many times the height of the waves in oceanic depths, the run-up is not much greater than the wave height as measured on a tide gage that is usually located a very short distance offshore.

In the linear theory of waves the power transmitted per unit wave crest length is

$$P = \frac{1}{8}\rho g H^2 C \frac{1}{2}\left(1 + \frac{4\pi d/L}{\sinh 4\pi d/L}\right) \quad (11.50a)$$

$$= \frac{1}{8}\rho g H^2 C_G \quad (11.50b)$$

where ρ is the mass density of the water. It can be seen that the power transmitted is the product of the energy per unit of water surface and the group velocity of the waves.

The change in height of waves traveling in water of variable depth can be obtained for the case of no energy loss and no reflection (the conditions for which reflection is negligible have been discussed in a previous section) by assuming the power to be constant as the wave moves from one region to another. The resulting equation is

$$\frac{H_2}{H_1} = \sqrt{\frac{C_{G_1}}{C_{G_2}}\frac{b_1}{b_2}} \quad (11.51)$$

where $\sqrt{b_1/b_2}$ is the refraction coefficient and $\sqrt{C_{G_1}/C_{G_2}}$ is the shoaling coefficient, which can be calculated using Eq. 11.9. In shallow water, neglecting refraction,

$$\frac{H_2}{H_1} = \sqrt{\frac{C_{G_1}}{C_{G_2}}} \approx \sqrt{\frac{C_1}{C_2}} = \left(\frac{d_1}{d_2}\right)^{1/4} \quad (11.52)$$

which is known as Green's Law. Suppose a tsunami wave train travels from water 15,000 ft deep into water 30 ft deep; then

$$\frac{H_2}{H_1} = \left(\frac{15{,}000}{30}\right)^{1/4} = 4.7$$

If the shore depth were 10 ft, $H_2/H_1 = 6.3$. Therefore, it is easy to account for an appreciable portion of the increase of tsunami wave height as it moves into shallow water, using linear theory. Linear theory, however, cannot be used when the wave height is not small compared with the water depth. As the wave height eventually increases relative to the water depths near shore the process becomes nonlinear. It may reflect without breaking or may form a bore.

Some work on the shoaling of nonlinear Stokes' waves has been done by Le Mehaute and Webb (1964) and Koh and Le Mehaute (1966). Both third- and fifth-order solutions were examined. It was found that the fifth order was not valid for $d/L < 0.1$ as several of the coefficients tend to infinity. This is not the case for third-order theory, and a solution analogous to Eq. 11.52 was obtained, which, however, included the convective inertia term. Koh and Le Mehaute (1966) concluded that for practical purposes the linear theory (Eqs. 11.50 and 11.51) were as satisfactory for predicting H_2/H_1 as was third-order theory. They caution that third-order theory is necessary to predict the ratio of crest height above SWL to trough depth beneath SWL, however.

Keller and Keller (1964) developed a linear theory for the run-up of waves over a uniformly sloping bottom and also over a slowly varying bottom, assuming both the wave front and the bottom contours to be parallel to a straight shoreline. An explicit solution was obtained for the case of a uniform periodic wave system moving from infinitely deep water over a uniform sloping bottom, making an angle α with the horizontal, assuming no energy losses. Implicit in the assumption of no energy loss is the assumption that no bore forms. Their solution was a modification of an earlier result by Isaacson (1950):

$$\bar{R} = \frac{1}{2}H_0\sqrt{\frac{2\pi}{\alpha}} \quad (11.53)$$

where H_0 is the wave height in infinitely deep water and $\bar{R}$ is the run-up onto land, measured vertically from the still water level (SWL) (see Fig. 11.9). For the case of a vertical wall, $\alpha = \pi/2$ and

$$\bar{R} = H_0 \quad (11.54)$$

which is the classic linear solution for a standing wave at a vertical wall.

For the case of linear shallow water waves moving from an "open ocean" of constant depth d_u over a bottom of uniform slope α, Keller and Keller found that

$$\bar{R} = H_u\left[J_0^2\frac{2\sqrt{\beta d_u}}{\alpha} + J_1^2\frac{2\sqrt{\beta d_u}}{\alpha}\right]^{-1/2} \quad (11.55)$$

Here H_u is the height of the long wave in the "open ocean," J_0 and J_1 are Bessel functions, and

$$\beta d_u = \frac{4\pi^2 d_u}{gT^2} = \left(2\pi/T\sqrt{\frac{g}{d_u}}\right)^2 = \frac{2\pi d_u}{L_0} \quad (11.56)$$

where L_0 is the wavelength in infinitely deep water ($L_0 = gT^2/2\pi$). According to Keller and Keller this approximation is valid for $\beta d_u < 0.5$ (i.e., for $d_u/L_0 < 0.08$). The equation for the phase angle θ was found to be

$$\theta = \frac{\sqrt{\beta d_u}}{\alpha} - \tan^{-1}\frac{J_1(2\sqrt{\beta d_u}/\alpha)}{J_0(2\sqrt{\beta d_u}/\alpha)} \quad (11.57)$$

A phase shift has been observed in the laboratory by Schoemaker and Thijsse (1949), although no values of the shift were reported.

Values of $\bar{R}/H_u$ vs βd_u calculated using Eq. 11.55 are shown in Fig. 11.30 for a number of bottom slopes and are labeled "A_2." A second set of curves, labeled "A_1," are also given, using a more general solution, which are for deep water as well as shallow water waves;

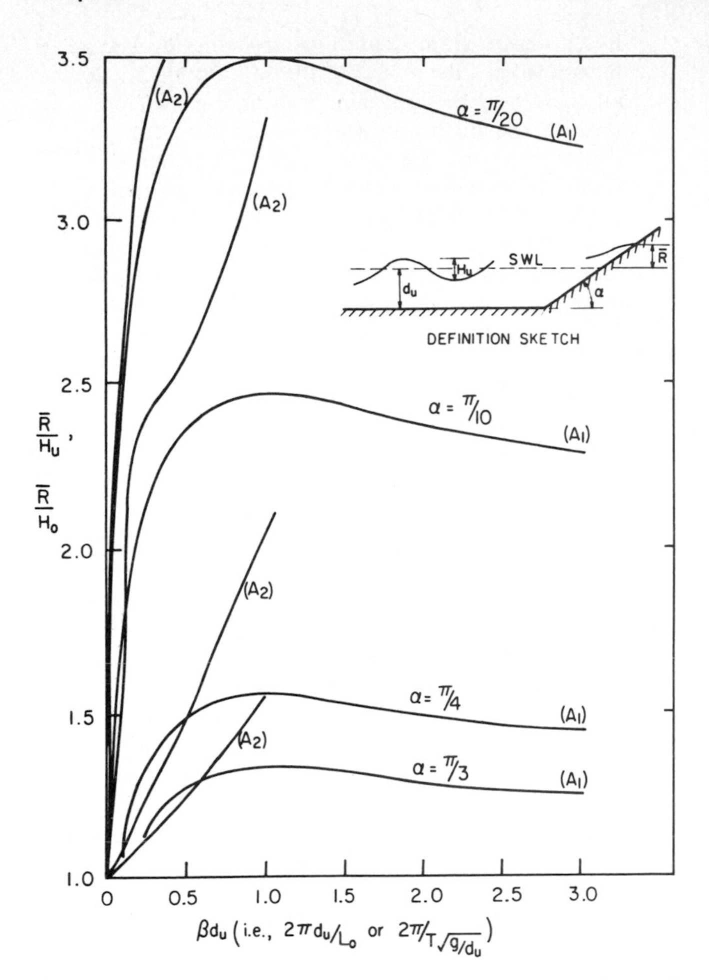

Fig. 11.30. Run-up amplification factor. (From Keller and Keller, 1964.)

these curves are for $\bar{R}/H_0$ vs βd_u, where H_0 is the wave height in infinitely deep water.

Keller and Keller gave solutions for more general types of bottom conditions and, in addition, they made a numerical study of the run-up of a nonlinear shallow water wave train (see also Keller and Keller, 1965). The results will not be presented herein as Eq. 11.53 is adequate to show the approximate effect of a bottom slope on the run-up.

Van Dorn (1966) tested Eq. 11.55, using the laboratory measurements of Savage (1959) for bottom slopes between $\alpha = 0.10$ rad and $\alpha = \pi/2$ rad ($S = \tan\alpha$, from $\frac{1}{10}$ to ∞) and found the equation to be satisfactory.

Van Dorn found, in addition, that the equation could be used for very steep (hence, nonlinear) waves, providing a simple substitution was made for H_u, as described below.

Finite amplitude Stokes' waves are not symmetrical about the still water level. Rather, the crest is a greater distance above the still water level than the trough is below it by an amount (Wiegel, 1964d)

$$\Delta H_u = \frac{\pi H_u^2}{4L}\left(1 + \frac{3}{2\sinh^2 2\pi d_u/L}\right)\coth\frac{2\pi d_u}{L} \qquad (11.58)$$

where L is the length of the wave of period T in water of depth d_u. Thus, the crest lies a distance $\frac{1}{2}H_u + \Delta H_u$ above the still water level. Van Dorn found that Eq. 11.55 could be used to predict the run-up, provided twice this value was used rather than twice the wave amplitude (twice the wave amplitude is the wave height in linear theory). Thus,

$$\bar{R} = 2\left(\frac{1}{2}H_u + \Delta H_u\right)\left(J_0^2\frac{2\sqrt{\beta d_u}}{\alpha} + J_1^2\frac{2\sqrt{\beta d_u}}{\alpha}\right)^{-1/2} \qquad (11.59)$$

for nonlinear Stokes' waves. A similar correction could be made for cnoidal waves.

As was pointed out by Van Dorn, Eqs. 11.55 and 11.56 can be used to calculate wave run-up providing the waves do not break in the process. Criteria for breaking were discussed by Van Dorn, and he prepared a nomograph chart to be used in determining whether breaking would occur and in obtaining values of run-up for the case of breaking waves. Van Dorn found that the empirical equation proposed by Hunt (1959) in a slightly modified form could be used to predict the run-up of breaking waves. Hunt had stated that

$$\begin{aligned}\frac{\bar{R}}{H} &= K_1\left(\frac{H}{L}\right)^{-1/2}(\tan\alpha)\left(\tanh\frac{2\pi d}{L}\right)^{-1/2} \\ &= K_1\tan\alpha\frac{T^2}{H}\sqrt{\frac{g}{2\pi}} \approx \tan\alpha\frac{T^2}{H}\sqrt{\frac{g}{2\pi}} \\ &\approx 2.3\tan\alpha\frac{T^2}{H}\end{aligned} \qquad (11.60a)$$

Van Dorn suggests that this should be modified by substituting α for $\tan\alpha$

$$\frac{\bar{R}}{H} = \alpha\frac{T^2}{H}\sqrt{\frac{g}{2\pi}} = \frac{\alpha T\sqrt{g/d}}{2\pi}\left(\frac{2\pi d}{H}\right)^{1/2} \qquad (11.60b)$$

where H and L are the wave height and length in water of depth d. It is not clear from the paper just where these are measured. However, as the tests to check the validity of Eq. 11.60a were made in a tank filled with water to a depth d_u, with a slope of angle α at one end of the tank, it would appear that $d = d_u$ and $H = H_u$.

Hunt examined the theoretical studies of Miche (1944, 1951), the theoretical and laboratory studies of Iribarren and Nogales (1947), and the laboratory studies of Granthem (1953) and Caldwell (1954) and concluded that the criterion for the breaking of a wave on a slope is

$$\sqrt{\frac{H}{T^2}} > \tan\alpha \qquad (11.61)$$

That is, if $\sqrt{H/T^2} > \tan\alpha$, the wave will break. If $\sqrt{H/T^2} < \tan\alpha$, the wave will reflect without breaking. There is, of course, no sharp boundary between the two conditions. It is believed that the value of H to be used in the above inequality is the "deep-water" wave height H_0. It must be emphasized that the laboratory studies were for slopes and waves that were relatively steep compared with tsunamis and many natural offshore slopes.

For relatively flat slopes, frictional losses must be important. For example, Eq. 11.55 and Fig. 11.30 show that the flatter the slope the greater will be the run-up. On the other hand, the flatter the slope, the greater will be the distance that the waves must move over a very shallow bottom. As an example of this effect, Matuo (1934) found for the 1933 tsunami in Japan that an offshore slope of 1: 50 was associated with a run-up seven times as high as for an offshore slope of 1: 500. It also should be pointed out that unpublished results of a laboratory study by Alavi (1964) of surges moving along a dry bottom and then up a uniform slope showed that the steeper the slope the greater the run-up for bottom slopes between $\frac{1}{5}$ and $\frac{1}{20}$.

11.5.5 Bores

We know little of the circumstances under which a bore forms. They are not the same as just a gigantic wave breaking in the manner of ordinary sea waves, fortunately. According to Ishimoto and Hagiwara (1934) the entire wave does not break; rather, in the manner of the tidal bores of Whangchow in China, only a front part of the wave surface breaks. Further, in Japan it was found in the 1933 tsunami that these occurred only where the water wave was very high or the ground had very little slope, such as along rivers. In a later report, the one on the 1960 Chilean tsunami, Kato *et al.* (1961) state that during the Chilean tsunami the rise in water level along the coast of Japan occurred very gradually, unlike the borelike invasion and highly turbulent motion of the 1933 tsunami. During the Yakutat Bay, Alaska, Earthquake of September 10, 1899, several prospectors were camped near the epicenter; J. P. Fults reported a "wall of water 20 feet high," and Dr. L. A. Cox reported, "We heard a terrible roar in the direction of the bay, and on looking that way we saw a tidal wave coming toward us which appeared to be about 20 feet high and was preceded by some great geysers shooting into the air, some of which were several feet across and 30 or 40 feet high" (Tarr and Martin, 1912).

The following quotation from Eaton, Richter, and Ault (1961) is enlightening in regard to a bore formed at the front of the 1960 Chilean tsunami at Hilo, Hawaii:

> Water continued to pour out of the estuary, and its level beneath the bridge continued to drop until 1: 00 a.m. when measurement with the tape showed it to be nearly 7 feet below the pre-wave water level. For a short while a strange calm prevailed as ground water cascaded from among rocks that are rarely exposed to view along the shore.
>
> At first there was only the sound, a dull rumble like a distant train, that came from the darkness far out toward the mouth of the bay. By 1: 02 a.m. all could hear the loudening roar as it came closer through the night. As our eyes searched for the source of the ominous noise, a pale wall of tumbling water, the broken crest of the third wave, was caught in the dim light thrown across the water by the lights of Hilo. It advanced southward nearly parallel to the coast north of Hilo and seemed to grow in height as it moved steadily toward the bayshore heart of this city. [Note: See Fig. 11.25 for a plan view of Hilo.]
>
> At 1: 04 a.m. the 20-foot-high nearly vertical front of the in-rushing bore churned past our lookout, and we ran a few hundred feet toward safer ground. Turning around, we saw a flood of water pouring up the estuary. The top of the incoming current caught in the steel-grid roadway of the south half of the bridge and sent a spray of water high into the air. Seconds later, brilliant blue-white electrical flashes from the north end of Kamehameha Avenue a few hundred yards south of where we waited signalled that the wave had crossed the sea wall and buffer zone and was washing into the town with crushing force. Flashes from electrical short circuits marked the impact of this wave as it moved swiftly southeastward along Kamehameha Avenue. Dull grating sounds from buildings ground together by the waves and sharp reports from snapped-off power poles emerged from the flooded city now left in darkness behind the destroying wave front. At 1: 05 a.m. the wave reached the power plant at the south end of the bay, and after a brief greenish electrical arc that lit up the sky above the plant, Hilo and most of the Island of Hawaii was plunged into darkness.

Some mathematical studies recently have been made on the formation of bores, and the reader is referred to them for details (Keller, Levine, and Whitham, 1960; Le Mehaute, 1963; Amein, 1964).

11.6 DISTRIBUTION ON RUN-UP ELEVATIONS AND WAVE HEIGHTS ALONG A COAST

It is known that the elevation reached by a tsunami and the characteristics of rising water (gentle, fast current, bore, etc.) depend upon the offshore hydrography, on the orientation, slope, and configuration of the shore, and on resonance.

A general feature of resonance is that the greater the build-up of an oscillation in a harbor, the greater will be the required number of waves necessary to cause the peak response of a bay. Thus, if the highest waves occur shortly after the arrival of the first wave, then damping is of considerable importance. At Hilo, e.g., the highest wave was the third (or perhaps it was the

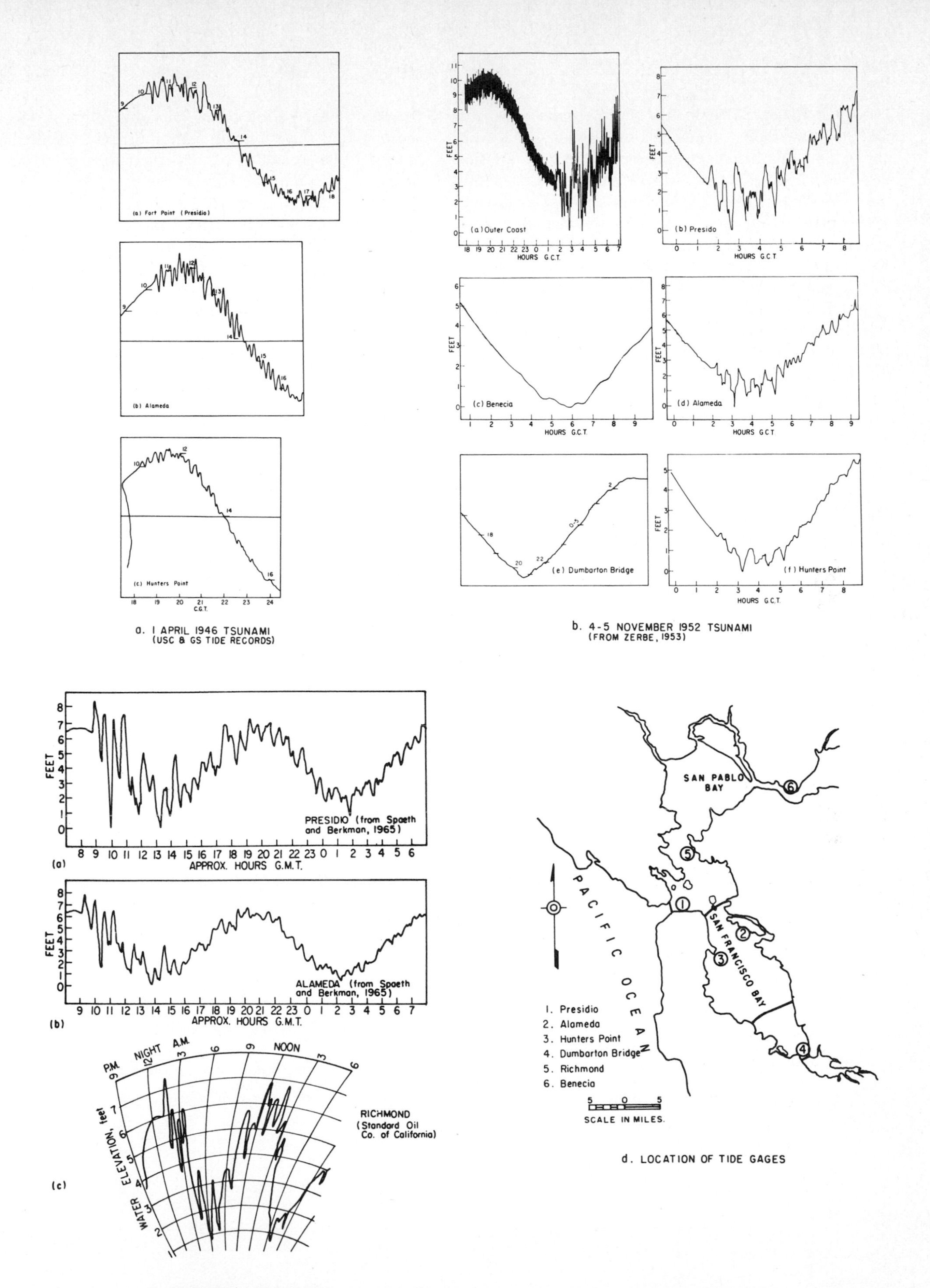

Fig. 11.31. Tide gage records of three tsunamis in San Francisco Bay, California.

fifth or sixth, depending on how one interprets the record) for the May 1960 tsunami (Eaton, Richter, and Ault, 1961), and this also was apparently the case for that of April 1, 1946 (Shepard, MacDonald, and Cox, 1950), whereas in both the 1952 tsunamis (MacDonald and Wentworth, 1959) the first wave was the highest. It would appear that, for this bay, damping is of considerable importance. An example of a bay in which damping is of great importance is San Francisco Bay. It can be seen from the tide gage records of the tsunamis of April 1, 1946, November 4–5, 1952, and March 28–29, 1964, that the individual waves were propagated into the bay with essentially no change of form, except for damping (Fig. 11.31).

It was found in Japan (Takahasi, 1963a) that the heights of the tsunamis at the heads of triangular bays are considerably higher than at the mouths of the bays. This relationship, however, appears to be a function of the ratio of the period of the tsunamis to the period of the bay, and for tsunamis with a period several times as long as the periods of the bays there was little increase in the amplitude at the head of the bay.

Some peculiar observations of tsunami heights in open and hook bays have been made. Along the California coast, the bays that had the most protection from the direction of approach of the April 1, 1946, tsunami had the largest rise in water level (Bascom, 1946). For example, at Monterey Bay there was practically no wave at the south side of the bay, but the water level rose 10 ft at Santa Cruz on the north side of the bay.

Much more detailed information is available for the Hawaiian Islands and Japan on the distribution of tsunami heights along the coast than for other regions. The run-up for three tsunamis for the island of Hawaii is shown in Fig. 11.32. Examples from Japan for two of their largest tsunamis, the ones occurring in 1896

Fig. 11.32. Map of the island of Hawaii showing water heights, in feet above mean lower low water for the tsunamis of May 23, 1960; March 9, 1957; and April 1, 1946. Data for 1957 from Fraser *et al.* (1959) and for 1946 from MacDonald *et al.* (1947). (From Eaton, Richter, and Ault, 1961.)

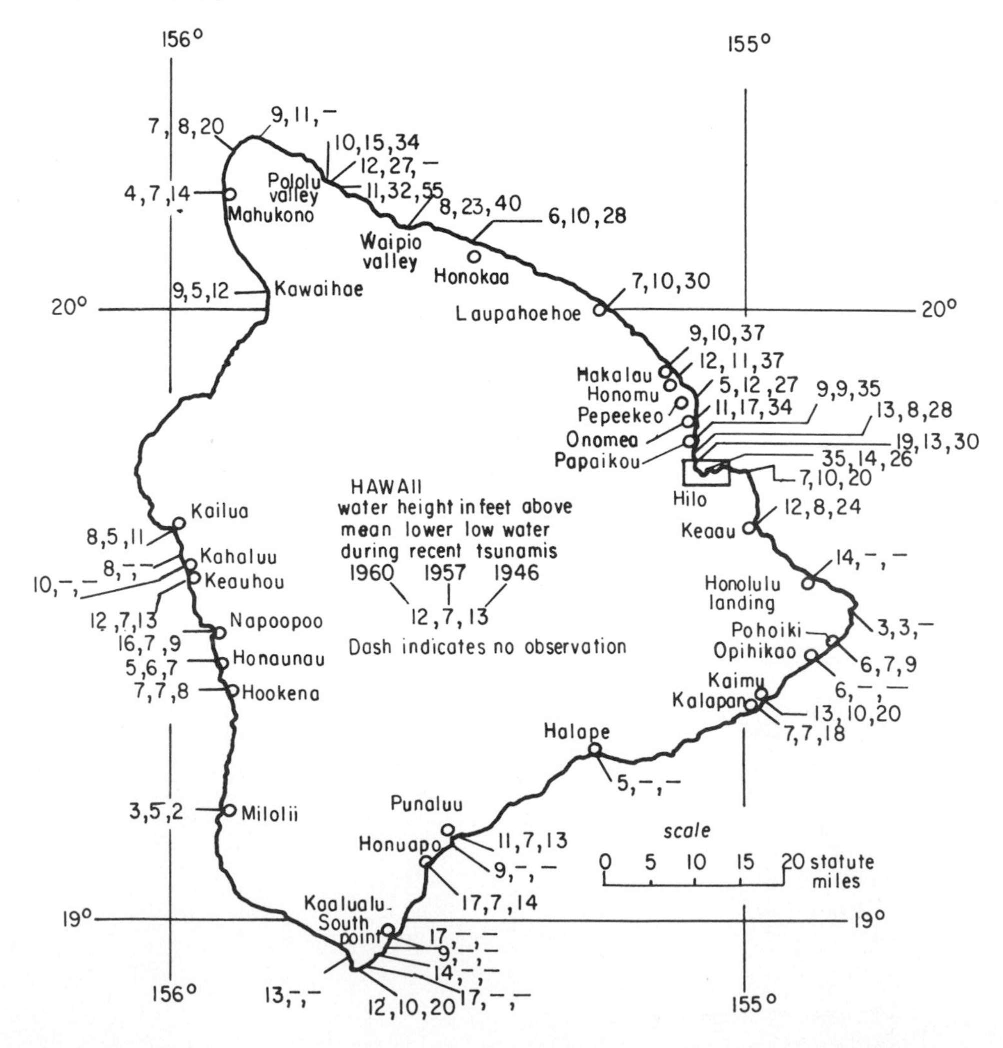

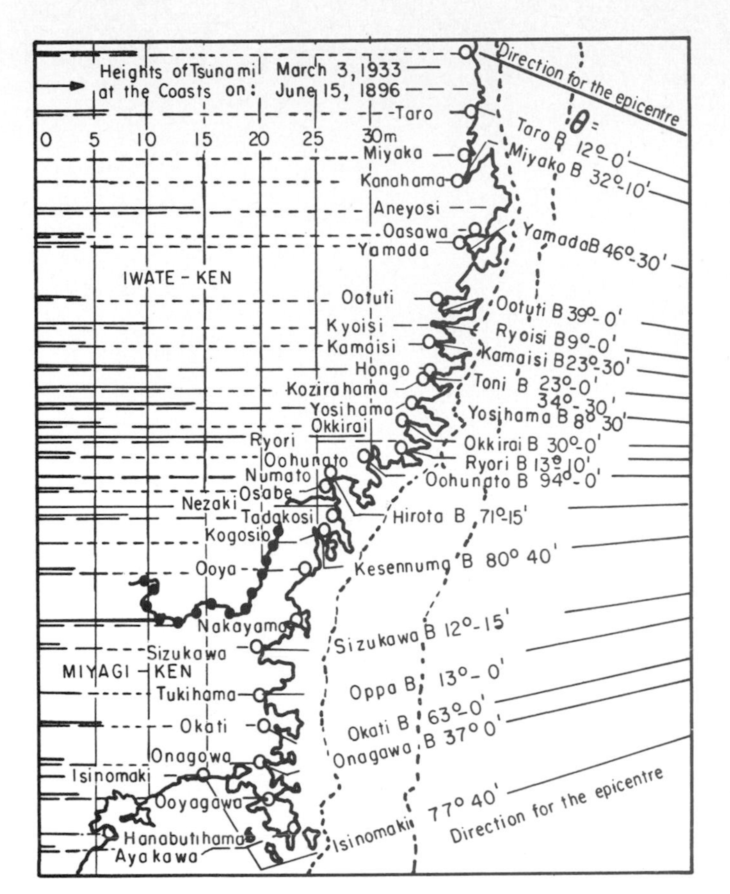

Fig. 11.33. Height of the tsunamis at Sanriku coasts. (From Matuo, 1934.)

and in 1933, are shown in Fig. 11.33. The epicenters of the two earthquakes were at nearly the same place (about 150 mi offshore). The wave height distributions along the coast were very similar.

There are few generalities that can be drawn with respect to the elevation on shore reached by a tsunami. In general, coastal regions facing the area in which a tsunami originated usually suffered high run-up of the water; this refers to large-scale phenomenon, not the "lee" side of a relatively small promontory. Regions in Hawaii with large coral reefs had lower wave heights, and in Japan the ends of V-shaped bays had higher waves.

When tsunamis are generated locally, the wave heights along the coast are usually higher along the portion of the coast nearest to the epicenter if the tsunami originated from a roughly circular source, and near the intersection of the coast and the line drawn perpendicular from a line through the epicenters of aftershocks for the case of an elongated elliptical source. Several examples, taken from a paper of Watanabe (1964), are shown in Fig. 11.34. In the examples of the locally generated tsunami (Fig. 11.34a and b), wave run-up heights drop off fairly fast some 300 km (180 mi) along the coast measured from the nearest position of the coast to the epicenter. Run-ups resulting from the Chilean tsunami of May 23, 1960, a tsunami originating a great distance from the coast of Japan, were more nearly uniform (Fig. 11.34c).

Detailed observations of three major tsunamis that occurred offshore in another part of Japan also show that the run-up is greatest, on the average, along the shore opposite the epicenter, and it decreases to about one-quarter of this value at a distance of about 100 mi from the point on shore opposite the epicenter (Iida, 1965). This can be seen in Fig. 11.35 for the Tonankai and Nankaido tsunamis off Japan (some local effects of the 1960 Chilean tsunami also are shown in Fig. 11.35 for comparision purposes; Iida, 1963b). A more recent example has been given by Iida (1968) for the 1964 Niigata, Japan, Earthquake.

As has been shown by at least two studies (Shepard, MacDonald, and Cox, 1950; Watanabe, 1964), wave refraction diagrams can be used to determine some of the overall characteristics of wave height distributions along a coast, although they do not explain all of the observations. If more detailed analyses had been made, including the total reflection of waves that sometimes occur (at Hilo, Hawaii, for the 1960 Chilean tsunami), some of the anomalies probably could have been explained. In general, offshore submarine ridges tend to cause an increase in wave height.

As an example of how refraction drawings can be used for studying the effect of a "locally generated" tsunami, consider the seismically active area in the vicinity of the Mendocino Escarpment. Refraction drawings were made for two types of tsunamis that possibly might be generated in the vicinity of the Mendocino Escarpment, an elongated quasi-elliptical initial wave (see Fig. 11.22) and a circular initial wave. The wave height offshore from Bodega Head (at about 124°W longitude) is from one-half to one-third as high as it is just north of Cape Mendocino. It is interesting to note that the waves would be highest along the coast to the north of Point Cabrillo. In order to examine the details of a particular site (in this case, the Bodega Head area) a second set of refraction drawings were made. It could be seen from these drawings that Bodega Canyon, offshore from Bodega Head, offers considerable protection to Bodega Head owing to its causing the wave orthogonals to diverge. This would result in a refraction coefficient at Bodega Head of from 0.08 to 0.13; i.e., the waves would be about one-sixth as high as in the area just to the north of Cape Mendocino. However, because of the severe convergence upcoast from Bodega Head, wave energy would flow along the wave crest and a higher wave would result, say an overall refraction coefficient of about 0.25.

Van Dorn (1959) studied 254 run-up observations in the Hawaiian Islands obtained for the April 1, 1946, and the March 9, 1957, tsunamis and plotted the data

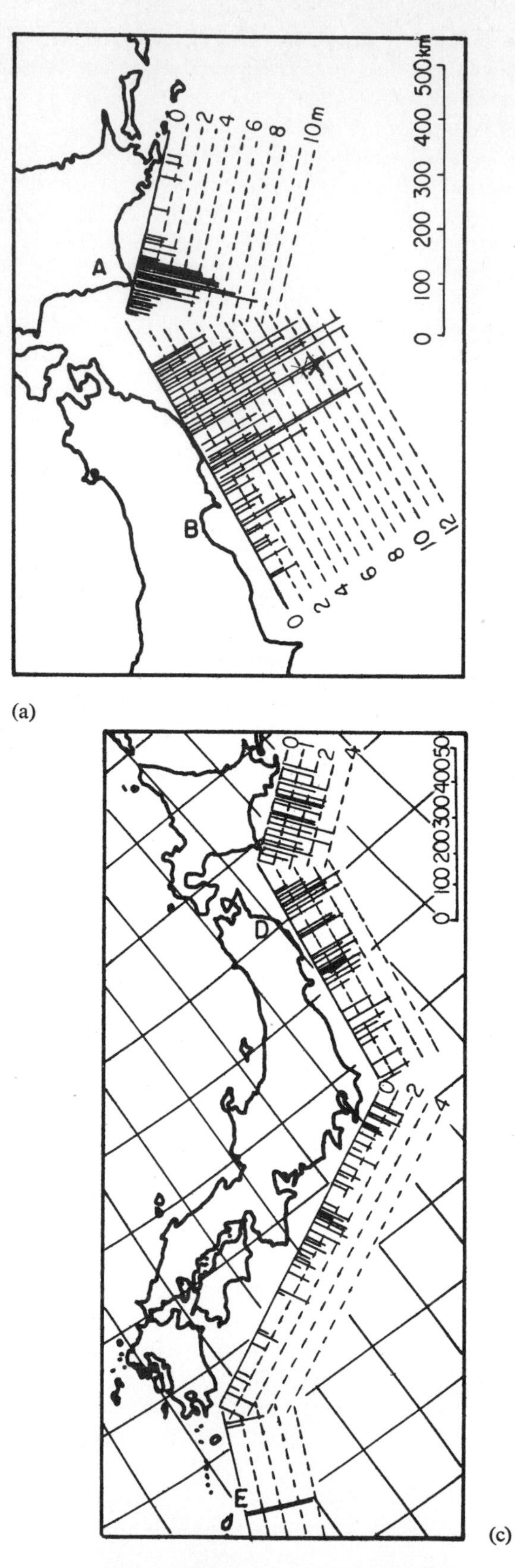

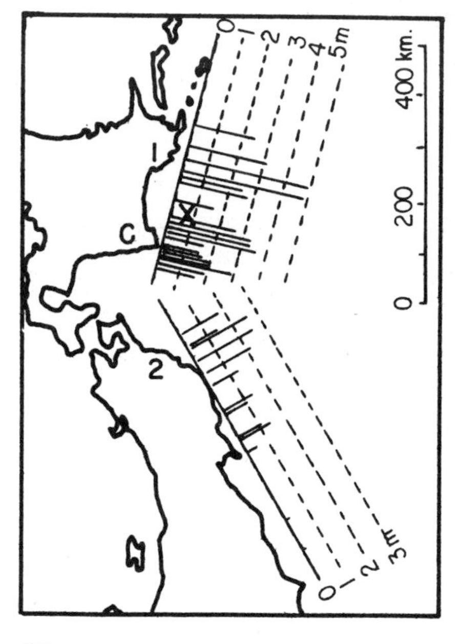

Fig. 11.34. Tsunami height distribution along coast of Japan. (From Watanabe, 1964.) (a) The distribution of maximum height along the open coast for the Sanriku-Oki tsunami of March 2, 1933. *X*: origin of tsunami. (b) The distribution of maximum height along the open coast for the Tokachi-Oki tsunami of March 4, 1952. *X*: the origin of tsunami. (c) The distribution of maximum height along the open coast for the Chilean tsunami of May 23, 1960.

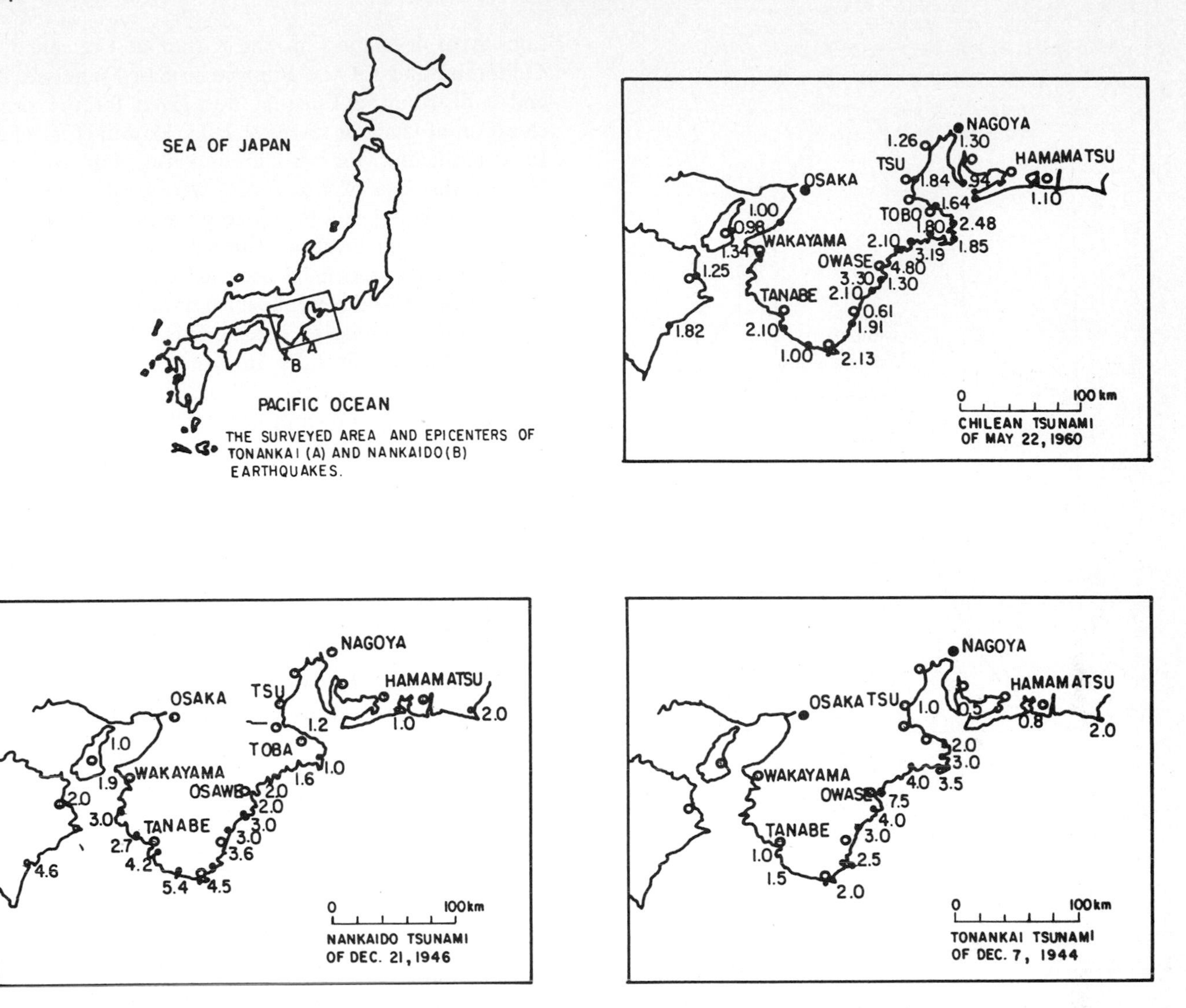

Fig. 11.35. Geographical distribution of tsunami heights on the Pacific Coast of Japan. Tsunami height in meters. (From Iida, 1963b.)

as shown in Fig. 11.36. He attempted to correlate the run-up height with such factors as local offshore slope, coastal indentations and prominences, and the presence or absence of coral reefs; good correlations were obtained only with respect to reefs. Van Dorn plotted the ratio of deviations from the mean curve shown in Fig. 11.36 to the mean values as represented by the cosine curve in the figure and arrived at Fig. 11.37. The data for both tsunamis apparently were described by a Gaussian distribution. A similar curve resulted from a plot of data measured along the Japanese coast for the 1933 tsunami. In interpreting this type of a distribution it must be remembered that for similar tsunamis similar run-ups were experienced at each place in Japan—the 1896 and 1933 tsunamis as shown in Fig. 11.33.

A study was made by Yamaguti (1934) of the relationship between the orientation of a bay and the azimuth of the direction of tsunami movement for the

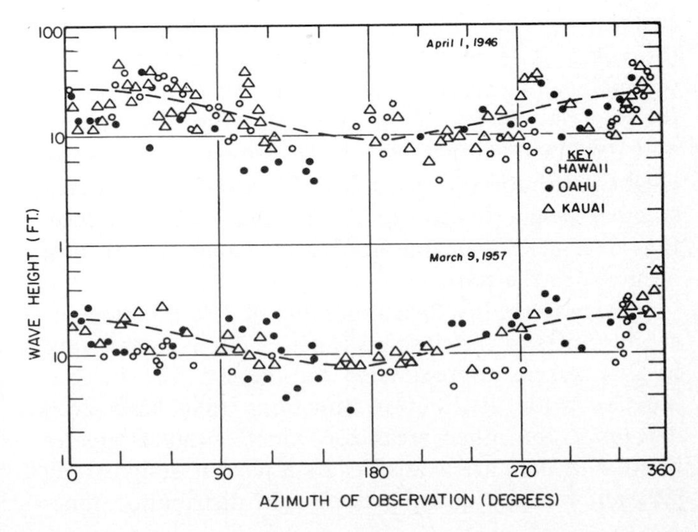

Fig. 11.36. Azimuthal distribution of observed run-up heights in Hawaiian Islands for two tsunamis. (From Van Dorn, 1959.)

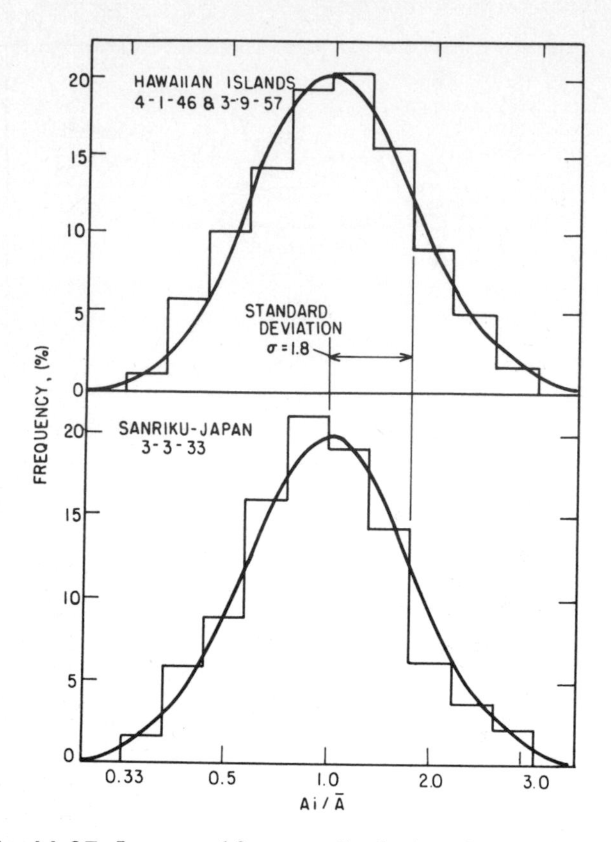

Fig. 11.37. Log-normal frequency distribution of run-up heights for Hawaiian Islands and coastal Japan. (From Van Dorn, 1959.)

1933 tsunami in Japan. If the few bays directly opposite the epicenter of the earthquake causing the tsunami are exempted, the scatter of data is such that it would be difficult to predict the effect of the orientation of a small promontory on the height of a tsunami along its "lee" coast.

11.7 TSUNAMI DISTRIBUTION FUNCTIONS

In locating and designing a structure to be located on the coast, it is of vital importance to determine the probability of occurence of tsunamis in the future and the possible heights of the tsunamis. The most important item in this regard is what is known as a tsunami height distribution function. This function is the expected tsunami height related to its frequency of occurence. It is developed from the heights of tsunamis that have occurred in the past.

Owing to the insufficient number of data for tsunamis at most sites, a distribution function for tsunami wave heights can be approximated for the site only in conjunction with distribution functions that have been developed for other areas for which a much greater number of data are available. As a part of design study (Wiegel, 1965a) tsunami wave height distribution functions were developed for the harbor at Crescent City, California, and for the Presidio in San Francisco Bay, and a distribution function was given for the run-up elevation of tsunami waves at Hilo, Hawaii (Fig. 11.38). In constructing the distribution function for Hilo, Hawaii, the data of Cox (1964) were used. These data, given in Table 11.3, are probably the most detailed that exist for a specific location. These data are also important in that the dates can be used as a guide to searching through files of old newspapers and the like to find information on tsunamis at specific sites. The distribution function for the Presidio in San Francisco Bay was developed by using the U.S. Coast and Geodetic Survey tide gage records shown in Fig. 11.39.

In addition, a distribution function has been drawn through the data on run-up in Japan (no locations specified) using the data given by Iida to the Hilo Technical Tsunami Advisory Council (1962). The data for Japan were in the form of the probabilities of occurrence as a function of tsunami magnitude (m), so that only ranges of run-up elevations were given. The ranges are plotted in Fig. 11.38, and a curve was drawn through the midpoints of the ranges.

In a previous section the difficulty of estimating the

Fig. 11.38. Distribution function for maximum tsunami waves. (From Wiegel, 1965a.)

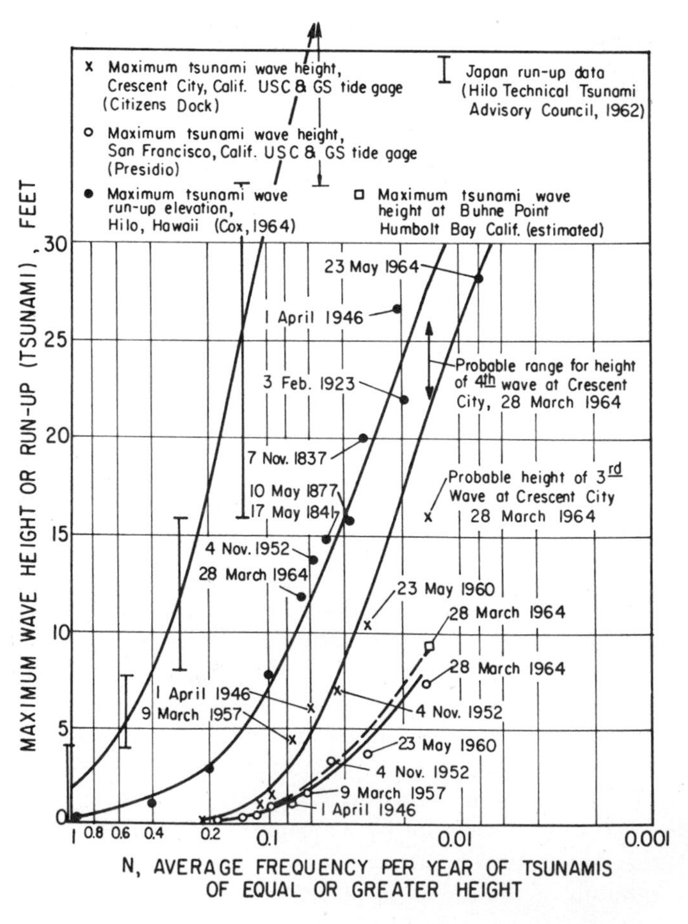

Table 11.3. TSUNAMI HEIGHTS AT HILO 1837–1964*

Year	Month	Day	Source area	Observer	Place	Visual inundation height, ft	Datum	Tide amplitude, ft	h_R, ft	Method of computation and note
1837	November	7	Chile	Coan	Hilo	20	hwm		20	V
1841	May	17	Kamchatka	Lyman	Hilo	15	hwm		15	V
1868	April	2	Hawaiian Islands	Coan	Hilo	6	hwm			
				PCA	Hilo	10	hwm		11	V
1868	August	13	Peru-Chile	estimated					12	V
1872	August	23	?	Coan	Hilo	4	sl		4	V
1877	May	10	Chile	Severance	Hilo	12	lw			
					Waiakea	16	?		16	V
1896	June	15	Japan Islands	HT	Hilo	8	ht		9	V
1906	January	31	Columbia	HT	Hilo	$2\frac{1}{2}$	ht		$3\frac{1}{2}$	V
1906	August	17	Chile	PCA	Hilo	(5)	R		$2\frac{1}{2}$	Vr
1918	September	7	Kuril Islands	HT	Waiakea	4 or 5	sl?		5	Va
1919	April	30	Tonga Islands	HT	Hilo	(4 or 5)	R		$2\frac{1}{2}$	Vr
1922	November	11	Chile	HT	Kuhio	6	sl?		6	V
1923	February	3	Kamchatka	HT	?	25?	?			
				Jaggar	Waiakea	20+	?		22	V
1923	April	13	Kamchatka	HTH	Waiakea	1	sl?		1	V
1927	November	4	California	mg	Kuhio	(0.8)	R			
				Jaggar	Bkwtr	> 1	sl?	0.4	1	V
1927	December	28	Kamchatka					0.2	0.5	M
1928	June	17	Mexico					0.7	1.8	M
1929	March	6	Aleutian Islands	mg	Kuhio	(0.7)	R			
				Jaggar	Waiakea	(1.3)	R	0.3	$\frac{1}{2}$	V
1931	October	3	Solomon Islands					0.3	0.8	M
1932	June	3	Mexico					1.2	3.0	M
1932	June	18	Mexico					0.1	0.1	M
1932	June	22	Mexico					0.1?	0.25	M
1933	March	2	Japan	Jaggar		(2 or 3)	R		$1\frac{1}{2}$	V
1938	November	10	Alaska	HTH		1	sl?		1	V
1946	April	1	Aleutian Islands	Shepard	Reeds	9				
				and others	Wainaku	30	mllw		27	D
1951	August	21	Hawaiian Islands					0.1	0.25	M
1952	March	3	Japan					0.4	1.0	M
1952	November	4	Kamchatka	Macdonald	Wainaku	6				
				and Wentworth	Hilo	12	mllw	4.0	12	D
1956	March	30	Kamchatka					0.5	1.2	M
1957	March	9	Aleutian Islands	Fraser	Waiakea	7				
				and others	Hilo	14	mllw	4.5	14	D
1958	July	10	Alaska					0.1	0.25	M
1958	November	6	Kuril Islands					0.5	1.2	M
1959	May	4	Kamchatka					0.2	0.5	M
1960	May	21	Chile					0.2	0.5	M
1960	May	23	Chile	Eaton	Reeds	10				
				and others	Waiakea	35	mllw		28	D
1960	November	21	Peru					0.4	1.0	M
1963	October	13	Kuril Islands					1.0	2.5	M
1963	October	20	Kuril Islands					0.2	0.5	M
1964	March	28	Alaska	Loomis	Waiakea	$2\frac{1}{2}$				
				Wickland	Radio	10	mllw	6.9	8	D

Abbreviations and notes

Observer
HT = *Hilo Tribune*
HTH = *Hilo Tribune Herald*
PCA = *Pacific Comm. Adv.*

Datum
hwm = high water mark
ht = high tide
sl = sea level
msl = mean sea level
lw = low water
mllw = mean lower low water
R = range from high to low
H_R = "normal" maximum height above sea level

Methods of computation and notes
E = estimated
V = maximum of a few visual observations
D = upper decile value of many visual observations
M = tide record amplitude times 2.5
a = seismographs installed at HVO September1918; interest in tsunamis developed in HVO (Hawaiian Volcano Observatory)
b = Hilo tide gage operated February 1927 to late 1932; December 1946 to date

*From Cox (1964).

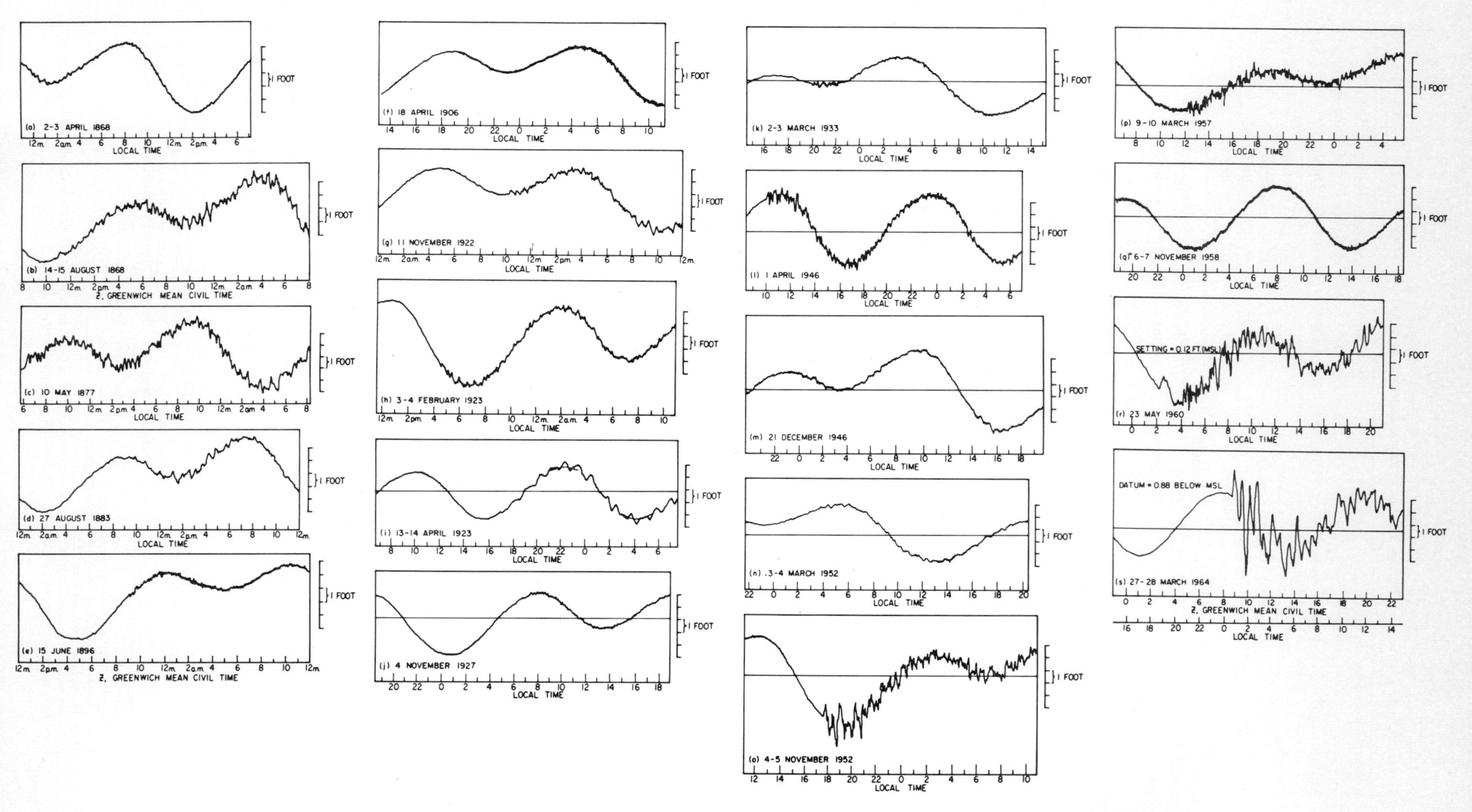

Fig. 11.39. Traces of U.S. Coast and Geodetic Survey tide gage records for a number of tsunamis, Presidio, San Francisco Bay, California.

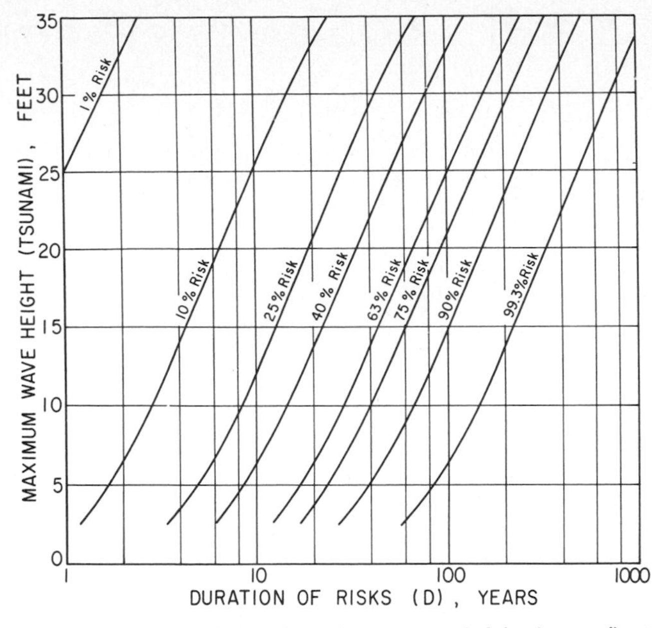

Fig. 11.40. Probability of maximum wave height (tsunami) at Crescent City, California, exceeding a given value in a given duration.

height of a tsunami wave from an observation of the run-up was mentioned. However, most tide gages are located within a harbor, usually a fraction of one tsunami wavelength from shore. Owing to this, the elevation of the wave crest at a "close-to-shore" tide gage site is about the same as the elevation of the maximum run-up, especially for the case of a tsunami wave that does not transform into a bore. The trough of the wave would be the "drawdown" within the harbor.

The distribution functions given in Fig. 11.38 can be put in a more useful form, following the work of Wemelsfelder (1961) on storm surges along the coast of the Netherlands. These curves, shown in Fig. 11.40, were constructed using Poisson's Law (Gumbel, 1958; Wemelsfelder, 1961; Borgman, 1963):

$$R = 1 - e^{-ND} \tag{11.62}$$

where R is the chance of a given height being exceeded in D years and N is as shown in Fig. 11.40. Although the curves are extrapolations, these are the best estimates that the author knows of at this time.

11.8 OVERTOPPING OF SEAWALL

One method that is used for the protection of an area from tsunami waves is a seawall between the site and the ocean or bay. If a seawall is to be constructed it is necessary to know the quantity of water that would flow over the top of the seawall if the crest of a tsunami wave were to be higher than the crown elevation of the seawall. An approximation of the flow can be made for a nonbore-type tsunami by considering the tsunami to be slowly varying in elevation and the water particle velocities in the tsunami near the seawall to be negligible; i.e., to consider the problem as a simple flow over a broad-crested weir (Wiegel, 1965b). Furthermore, this approximation assumes that the wall has a negligible effect on the tsunami waves.

The flow rate per foot of seawall q would then be

$$q = C_w \frac{2}{3}\sqrt{2g}E^{3/2}$$
$$\approx 0.577 \times \frac{2}{3}\sqrt{64.4}E^{3/2} = 3.09E^{3/2}\ \text{ft}^3/\text{sec/ft} \tag{11.63}$$

where C_w is the weir coefficient and E is the head above the top of the weir as shown in Fig. 11.41.

$$E = y_s - (S - \Delta S) \tag{11.64}$$

where S is the elevation of the crown of the seawall above mean lower low water, ΔS is the tide level at the time of the tsunami wave above mean lower low water, and y_s is the elevation of the tsunami wave surface above the tide level. As a simple approximation,

$$y_s = \frac{1}{2}H\cos\frac{2\pi t}{T} \tag{11.65}$$

where H is the height of the tsunami wave, T is the wave period, and t is time.

$$q \approx 3.09\left[\left(\frac{1}{2}H\cos\frac{2\pi t}{T}\right) - (S - \Delta S)\right]^{3/2}\ \text{cfs/ft} \tag{11.66}$$

provided $y_s > S - \Delta S$.

The total flow over the top of the seawall during one wave cycle can be approximated, assuming the variation in tide level during the time of overtopping is small, by the equation

$$\bar{q} \approx 3.09\int_{t_1}^{t_2}\left[\left(\frac{1}{2}H\cos\frac{2\pi t}{T}\right) - (S - \Delta S)\right]^{3/2} dt \quad \text{cfs/ft per wave cycle} \tag{11.67}$$

Here t_1 is the time at which $y_s = S - \Delta S$ as the wave surface rises and t_2 is the time at which $y_s = S - \Delta S$ as the wave surface falls. This equation can be solved easily using graphical means.

Consider a periodic wave 30 ft high with a 40 min

Fig. 11.41. Sketch showing definitions. (From Wiegel, 1965.)

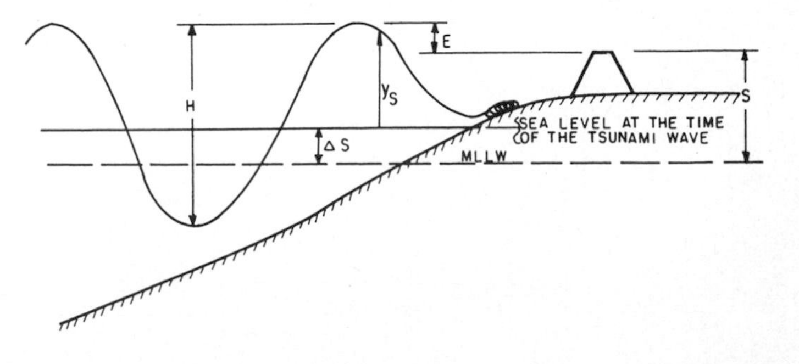

Table 11.4. $E^{3/2}$ AS A FUNCTION OF t

t, min	0	2	4	6	8	10
t, sec	0	120	240	360	480	600
t/T	0	0.05	0.10	0.15	0.20	0.25
$\cos 2\pi t/T$	1.00	0.93	0.81	0.59	0.31	0.00
$\frac{1}{2} \times 30 \cos 2\pi t/T$	15.0	13.9	12.1	8.9	4.7	0
$E = \frac{1}{2} \times 30 \cos 2\pi t/T - 10$	5.0	3.9	2.1	-1.1		
$E^{3/2} = (\frac{1}{2} \times 30 \cos 2\pi t/T - 10)^{3/2}$	11.2	7.9	3.04			

(2400 sec) period. Assume $S - \Delta S = 10$ ft. The easiest way to obtain t_1 and t_2 is to plot the wave profile and then measure the values on the plot. Only half of the wave has to be plotted, using the sea level at the time of the tsunami as the datum.

The next step is to calculate values of $E^{3/2}$ as shown in Table 11.4. $E^{3/2}$ is then plotted as a function of t, for several values of $S - \Delta S$. The area under one of the curves is

$$\int_{t_1}^{0} E^{3/2}\, dt \tag{11.68}$$

Fig. 11.42. Relationship between overtopping flow and overtopping height of tsunami waves. (After Wiegel, 1965.)

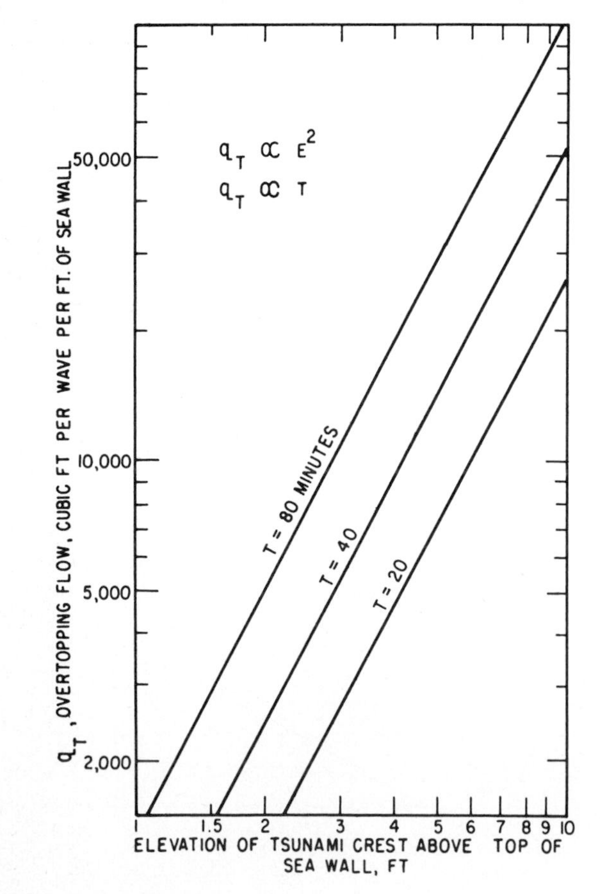

so that the total flow over the top of the seawall during one wave cycle is

$$\bar{q} = 3.09 \times 2 \int_{t_1}^{0} E^{3/2}\, dt \text{ ft}^3\text{/ft per wave cycle} \tag{11.69}$$

That is, the total flow over the top of the seawall, per foot of wall, during one wave cycle is 3.09 times twice the area under the $S - \Delta S$ curve. The flow for $S - \Delta S = 10$ ft is approximately $3.09 \times 2 \times 1900 =$ 11,700 ft³/ft in one wave cycle of 40 min.

Flows for a number of values of $S - \Delta S$ are shown in Fig. 11.42. Rather than using $S - \Delta S$ for the abscissa, the height of the wave crest above the top of the seawall, $\frac{1}{2}H - (S - \Delta S)$ is used. It is evident that the flow increases very rapidly with $\frac{1}{2}H - (S - \Delta S)$; in fact, it increases as the square of $\frac{1}{2}H - (S - \Delta S)$. This is because both $t_1 - t_2$ and E increase with $\frac{1}{2}H - (S - \Delta S)$.

A few calculations also were made for $H = 30$ ft and $T = 20$ min and $H = 30$ ft and $T = 80$ min. As long as the wave height is the same, the flow over the top of the seawall is proportional to the wave period.

The effect of wave height is not quite as simple as the effect of wave period. It is necessary to plot the new wave profile. This was done for the case of $H = 20$ ft and $T = 40$ min. New graphs must be made and the area under the curves measured. The results are also shown in Fig. 11.43.

How can these data be used for a practical problem? Consider a land area 1000 ft in width protected by a seawall, as shown in Fig. 11.43. Suppose the elevation of the seawall is 10 ft above the sea level at the time of a tsunami, with a wave height of 30 ft and a period of 40 min. The total flow of water over the top of the seawall in one wave cycle would be about 11,700 ft³/ft of wall. If all of this water were trapped in an area 1000 ft wide, the total depth landward of the seawall would be $11{,}700/1000 = 11.7$ ft, which would be $(11.7 + 12) - (6 + 15) = 2.7$ ft above the top of the wave crest. This, of course, could not be the case. It means that the wave would act almost as if the seawall were not there and that the control would be the cliff at the landward edge of the site.

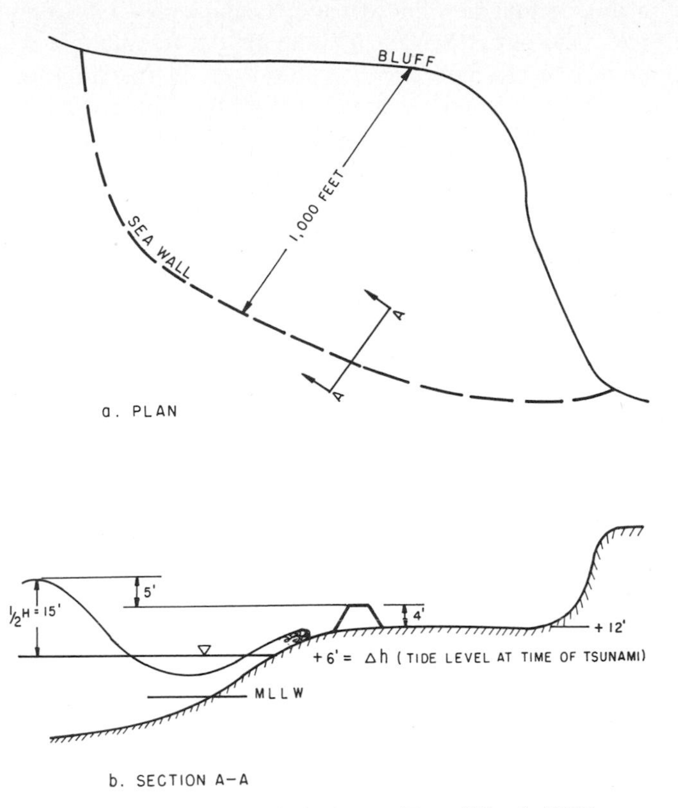

Fig. 11.43. Hypothetical case. (From Wiegel, 1965.)

The action of the March 27–28, 1964, tsunami at Crescent City, California, is an example of an event somewhat of the type described above. There was a small seawall built within the harbor, and the area landward of it was filled with sand. The elevation of the top of the seawall was from about +8 ft above mean lower low water in the most protected region to +13 ft in the least protected region. The fill landward of this wall was from about +10 to +14 ft above mean lower low water. The highest two tsunami waves apparently went over the top and flooded the towns as if the wall were not present. It is evident from Fig. 11.42 that a seawall, to be effective, must be designed to permit only a few feet of overtopping for the "improbable" tsunami and for no overtopping for the "design" tsunami. This is especially true in light of some laboratory experiments by Iwasaki (1965) that showed that the height would be increased at a seawall, at least for waves that are relatively short compared with tsunami waves.

11.9 TSUNAMI WAVE FORCES

Few studies have been made of the forces exerted by tsunami waves. In the ocean, and in bays, the waves probably can be treated in the manner of the shorter progressive and standing waves such as swell and seiches. Forces exerted by these types of waves will not be treated here as there are a number of papers on the subject (see, e.g., Wiegel, 1964d). The forces exerted on structures by tsunami waves running onto the land present a much more difficult problem. No actual measurements of these forces have been made, and only a few estimates of the forces are available.

Matlock, Reese, and Matlock (1962) made a study of the damage to structures in Hilo, Hawaii, caused by the May 23, 1960, Chilean tsunami. They treated the problem as if a bore moved over a dry bed in a manner similar to a surge running downstream from a dam that had failed. They used the equation given by Keulegan (1949) for the approximate speed V_s of such a surge for the case in which bottom friction is of major importance:

$$V_s = 2\sqrt{gd_s} \tag{11.70}$$

where g is the acceleration of gravity and d_s is the height of the surge. A further assumption was made that the water particles, from top to bottom, all moved with the speed of the surge. They examined the numerous observations made of the maximum elevation reached by the highest wave over the land submerged by the highest wave and decided that the crest was at about 15 ft above mean lower low water datum. Finally, they took the vertical distance from the ground to the plus 15-ft level for each particular point of interest in the region from the line of maximum inland inundation ($d_s = 0$) to the line of maximum withdrawal (−7 ft below mean lower low water, $d_s = 7 + 15 = 22$ ft) and used this as d_s in Eq. 11.70 to calculate the maximum velocity at that point. Thus, the speed of the bore, and the water within the bore, was assumed to move at speeds between 0 and 53 ft/sec. The observations made by Eaton, Richter, and Ault (1961) indicated that the bore traveled from the breakwater to shore, a distance of about 7000 ft, in from $2\frac{1}{2}$ to 3 min, at a speed of from 40 to 45 ft/sec. This would fix the upper limit of the surge speed, which is not too different from the estimate made by Matlock, Reese, and Matlock. It must be cautioned, however, that practically nothing is known of the velocity distribution within a tsunami as it moves over land.

Matlock, Reese, and Matlock examined in detail 14 cases of structural failure, or near failure, for which they were reasonably certain that secondary causes, such as a drifting log or automobile hitting the structure, were not involved. In all cases but one, they neglected hydrostatic forces and assumed that the horizontal fluid force intensity (pressure) exerted by the flowing waters on the structure was given by the equation

$$p = \tfrac{1}{2} C_D \rho V_s^2 \tag{11.71}$$

The values of C_D used in their calculation were the ones normally used in steady flow problems in which the object was completely submerged in a fluid. For one case, a reinforced concrete wall of a building, they included the hydrostatic force. Their approach was to calculate the forces necessary to cause structural failure and then to use Eq. 11.72 to calculate the velocity necessary to obtain this force. They then compared this velocity with the velocities calculated from Eq. 11.70 and found reasonable correlation.

A theoretical and laboratory study was made by Fukui, Nakamura, Shiraishi, and Sasaki (1963), and they found that the tip of a bore advancing over a dry bed, or in a channel with an initial water depth $\bar{d}$, traveled at a speed of

$$V_s = \left(\frac{q\bar{D}(\bar{D} + \bar{d})}{2(\bar{D} - \eta H)}\right)^{1/2} \tag{11.72}$$

where $\bar{D}$ is the total depth of water in the bore, H is the bore height $(\bar{D} - \bar{d})$, and η is a resistance term. It was found experimentally that η was equal to about 0.85 (equivalent to a Manning's n of 0.13) for a dry bed and increased with increasing $\bar{d}/\bar{D}$ to a value of about 1.03 at $\bar{d}/\bar{D} = 0.5$ and then remained constant for greater values of $\bar{d}/H$.

For an initially dry bed, $\bar{d} = 0$ and $\bar{D} = H$ (i.e., the equivalent of d_s); then Eq. 11.72 becomes

$$V_s \approx 1.8\sqrt{g\bar{D}} = 1.8\sqrt{gH} = 1.8\sqrt{gd_s} \tag{11.73}$$

which agrees rather well with the approximation given by Keulegan (Eq. 11.70). It is interesting to note that all of the bore tip speeds, for the case in which there was an initial depth of water in the channel, were in the region between $V_s = 2\sqrt{gd_s}$ and $V_s = \sqrt{gH}$. They found for the case in which there was an initial water depth in the channel that the bore had a relatively steep front and that the top of the bore was nearly horizontal. It should be pointed out in regard to the "nearly horizontal" top of the bore that the reservoir in the channel was of a fairly limited extent. They found the maximum pressure developed on a vertical wall, which extended the entire width of the laboratory channel, to be

$$p_{\max} = \frac{K_0 \rho g V_s^4}{g^2 d_s} = K_0 V_s^2\left(\frac{V_s^2}{gd_s}\right) \tag{11.74}$$

This can be expressed as

$$p_{\max} = \tfrac{1}{2} C_D \rho V_s^2 N_F^2 \tag{11.75}$$

where $C_D/2 = K_0$ and the Froude number N_F is given by $V/\sqrt{gd_s}$. They found for a vertical wall that $K_0 \approx 0.5$, which would be the equivalent of $C_D \approx 1$.

It is not clear how Eqs. 11.74 and 11.75 can be used in practice. Three pressure cells were used in one set of tests and six pressure cells were used in another set of tests. The maximum measured pressure was used to determine the exponential of V_s in Eqs. 11.74 and 11.75, and this maximum might have occurred at a different cell for each bore height.

Dressler (1952, 1954) and Whitham (1955) developed theories for the speed and shape of a surge moving over a dry bottom in which bottom friction plays a major role. The initial height of water in the reservoir is d_0 above the channel, and d is the depth of the surge in the channel at some distance x from the dam at time t. In the portion of the flow substantially removed from the tip of the surge, the flow can be considered to be the same as for the frictionless case, and the set of parametric equations for d and the speed V at this x and t can be solved to give the water speed as

$$V = 2\sqrt{gd_0} - 2\sqrt{gd} \tag{11.76}$$

If the flow were frictionless, the speed of the surge tip would be

$$V_s = 2\sqrt{gd_0} \tag{11.77}$$

This, however, is not the case. In fact, the shape and speed of the tip are controlled largely by the friction of the bottom. The results for the tip speed from the theories of Whitham and Dressler are very similar. An average of the results of the two theories is given in Table 11.5. In this table the ratio of the tip speed to $2\sqrt{gd_0}$ is given as a function of $Rt\sqrt{g/d_0}$, where R is a dimensionless resistance coefficient ($R = g/C^2$, in which C is the Chezy roughness coefficient in the square root of feet per second).

Table 11.5. TIP SPEED FOR SURGE IN A DRY BED

$Rt\sqrt{g/d_0}$	0	0.1	0.2	0.3	0.4	0.5	0.6
$V_s/2\sqrt{gd_0}$	1.0	0.48	0.40	0.35	0.32	0.29	0.26

Cross (1967) made a laboratory and theoretical study of a surge running over a dry bottom (and also over a bottom with a film of water on it) and of the forces exerted by the surge on a structure placed in a channel. He found that the tip speeds were generally a little faster than those shown in Table 11.5, using the appropriate value of the Chezy roughness coefficient. He used both a smooth and a rough bottom in his tests and in his calculations.

In his studies of the shape of the surge tip, Cross found that the theory predicted the shape reasonably well in the immediate region of the tip of the smooth bottom after the surge had run several feet down the channel. However, after the surge had traveled 15 ft or so, the depth of the surge became nearly constant a few feet back from the tip, while the theory showed a continually increasing depth. It should be pointed out that the reservoir used in the experiment was of limited extent, while the theory is for the case of a reservoir of unlimited extent.

Cross also found that when the channel bed had a thin film of water over it (0.015 ft deep) the tip became steeper and the speed of the tip was less, compared with the dry bottom case.

The theory of Cumberbatch (1960) for the force exerted by a fluid wedge impinging on a wall was

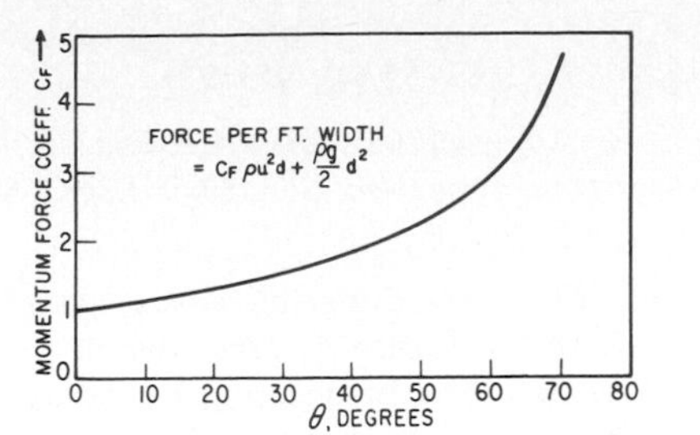

Fig. 11.44. Plot of force coefficient C_f vs θ. (From Cross, 1967.)

modified by Cross to include the hydrostatic force and was given as

$$F = \tfrac{1}{2}\rho g d_s^2 + C_F \rho V^2 d_s \qquad (11.78)$$

where F is the force per unit width of wall and d_s is taken as the height of the surge at the structure, if the structure were not present. Equation 11.78 is for the case of a surge striking a vertical wall extending the entire width of a channel. Cross calculated values of C_F as a function of the slope of the water surface relative to the horizontal ϕ. The results are shown in Fig. 11.44. The value of ϕ for any time t at which the surge would be moving past the obstruction were the obstruction not present can be obtained from measurements, or approximately by using the theory of a surge given by Cross.

As an example, one set of measurements is shown in Fig. 11.45, together with the profile calculated from theory using the measured surge tip speed. In this figure d_0 is the original water level in the reservoir prior to the opening of the gate to cause the surge and V_0 is the theoretical velocity of the surge front at the structure were the structure not present. The term "predicted from measured surge" refers to the measurement of a surge that was developed in the channel in a prior test, under identical conditions, but without the vertical wall being installed.

Fig. 11.45. Surge profiles (5/26 data) with predicted and measured force profiles (6/2 data), $x = 16.33$ ft. (From Cross, 1967.)

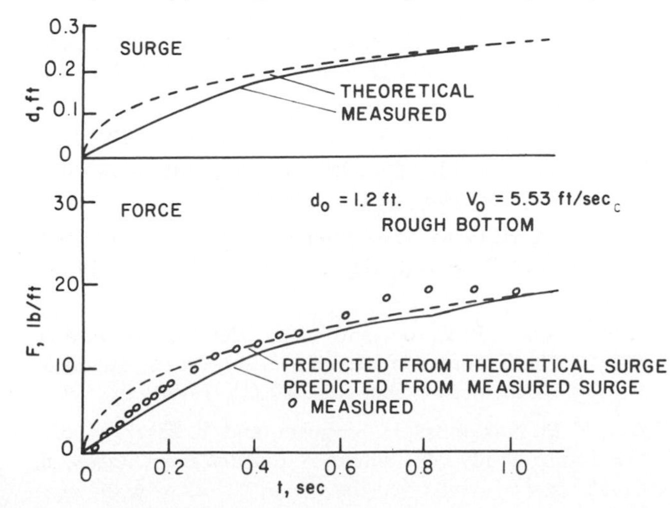

Several force records are shown in Fig. 11.46. It can be seen that a force peak occurs at about 0.75 sec after the initial force rise and that then the force remains at about a constant magnitude. The reason for this force peak was studied in detail. The surge, upon striking the vertical wall, ran up the wall a distance approximately equal to V_0^2/g for the wet bottom case and from $V_0^2/2g$ to V_0^2/g for the dry bottom case. This run-up curled back to some extent and then fell, hitting the surface of the reflecting surge. The peak force occurred at the moment this mass of water hit the surface of the reflecting surge.

If the peak force, described above, is neglected, it was found that the "steady" maximum force that occurred (and predicted with a good degree of accuracy by Eq. 11.78 together with Fig. 11.44) is also equal to the hydrostatic force computed using the depth of the reflected surge. Figure 11.44 shows, except near the immediate tip of the surge, that ϕ is small enough for

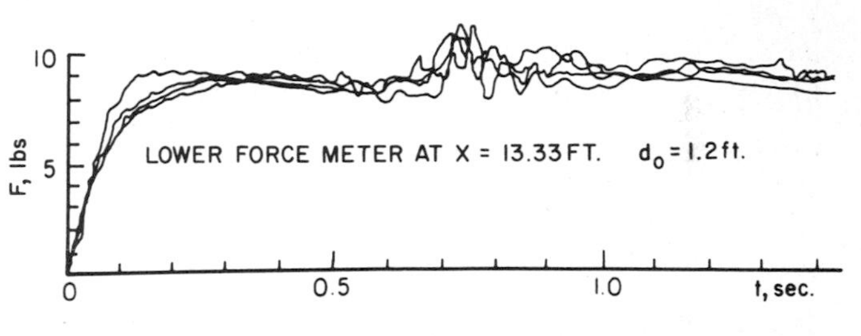

Fig. 11.46. Typical force profiles; smooth bottom. (From Cross, 1967.)

$C_F \approx 1$ to be a valid approximation. Using the further approximation that $V \approx 2\sqrt{gd_s}$ and substituting this together with $C_F \approx 1$ into Eq. 11.78 results in

$$F = 4.5\rho g d_s^2 \qquad (11.79)$$

Now, consider a surge of depth d_s and velocity $2\sqrt{gd_s}$ being reflected by a vertical wall. If no energy is lost, the top streamline will be displaced vertically by $V^2/2g$ (i.e., by $2d_s$), so that the depth of the reflected surge should be $d_r = 3d_s$. Some previous work by the author showed that this was approximately correct, and the studies by Fukui *et al.* (1963) also showed this to be approximately correct for a surge moving in a dry channel. Then,

$$F = \tfrac{1}{2}\rho g d_r^2 = \tfrac{1}{2}\rho g(3d_s)^2 = 4.5\rho g d_s^2 \qquad (11.80)$$

which is the same as Eq. 11.77. It must be emphasized that the surge depth d_s referred to here is taken as the depth of the nearly horizontal portion of the surge a few feet to the rear of the tip.

Cross also made a limited study of the forces exerted by a surge on vertical wall that extended only part way across the channel and found that when the width of the wall was less than about twice the height of the surge, the force started to decrease rather rapidly, with $C_F \approx \frac{1}{2}$ for a section about one-half the surge height in width.

Laboratory studies similar to Cross's were made by Alavi (1964) for the author, in which the characteristics of surges in a dry bed were studied, together with the forces exerted by the surge on square and circular plies 0.145 ft in diameter. Owing to the restricted width of the channel (0.5 ft) compared with the size of the piles, and owing to the fact that the surge flows are supercritical, the results are only indicative of the forces. The equation

$$F = \tfrac{1}{2} C_D \rho D d_s V^2 \tag{11.81}$$

was used to express the force, where D is the pile diameter and C_D is the coefficient of drag. It was found that C_D averaged 1.1 for the circular pile and 1.8 for the square pile for Reynolds numbers (DV/ν, where ν is the kinematic viscosity) between about 10^4 and 10^5.

In these studies, Alavi found that there was a linear relationship between d_s and d_0, with $d_s = 0.26d_0$, at the point where he measured d_s, about 20 ft downstream from the dam. It was found that $V_s \approx 2\sqrt{gd_s}$ for values of $d_s < 0.2$ ft and $V_s \approx 2.2\sqrt{gd_s}$ for $0.2 < d_s < 0.35$ ft. It was also found that although the maximum run-up on a vertical wall was a function of d_s (hence, V_s), $d_r/d_s = 3$.

REFERENCES

Abramowitz, M., and I. A. Stegun, eds. (June 1964). *Handbook of Mathematical Functions with Formulas, Graphs, and Mathematical Tables*, National Bureau of Standards, Applied Mathematics Series No. 55.

Alavi, M. (1964). *Surge Run-up on Sloping Beach, and Surge Forces on Piles*, University of California, Berkeley, College of Engineering, memo to R. L. Wiegel (unpublished).

Ambraseys, N. N. (October 1962). "Data for the Investigation of the Seismic Sea-Waves in the Eastern Mediterranean," *Bull. Seism. Soc Am.*, **52** (4), 895–913.

Amein, M. (1964). *Bore Inception and Propagation by the Nonlinear Wave Theory, Proceedings of the Ninth Conference on Coastal Engineering, ASCE*, pp. 70–81.

Architectural Institute of Japan (December 1961). *General State of Damage of Building, Report on the Chilean Tsunami of Japan, May 24, 1960, as Observed Along the Coast of Japan*, Committee for Field Investigation of the Chilean Tsunami of 1960, pp. 151–164.

Barber, N. F. (1961). *Experimental Correlograms and Fourier Transforms*, New York: Pergamon Press.

Bascom, W. (April 16, 1946). *Effect of Seismic Sea Wave on California Coast*, University of California IER, Tech. Rept. 3-204 (unpublished).

Basset, A. B. (1961). *A Treatise on Hydrodynamics*, 2 vols., New York: Dover Publications.

Berringhausen, W. H. (October 1962). "Tsunami Reported from the West Coast of South America, 1562–1960," *Bull. Seism. Soc. Am.*, **52** (4), 915–921.

Borgman, L. E. (August 1963). "Risk Criteria," *J. Waterways and Harbors Div., Proc. Am. Soc. Civil Engr.*, **89** (WW3), 1–36.

Caldwell, J. M. (1955). *The Design of Wave Channels, Proceedings of the First Conference on Ships and Waves*, Council on Wave Research, The Engineering Foundation, and the Society of Naval Architects and Marine Engineers, pp. 271–287.

Chen, T. C. (March 1961). *Experimental Study on the Solitary Wave Reflection Along a Straight Sloped Wall at Oblique Angle of Incidence*, U.S. Army Corps of Engineers, Beach Erosion Board, Tech. Memo. No. 124.

Chester, C., B. Friedman, and F. Ursell (July 1957). "An Extension of the Method of Steepest Descents," *Proc. Cambridge Phil. Soc.*, **53** (3), 599–611.

Committee for Field Investigation of the Chilean Tsunami of 1960 (December 1961). *Report on the Chilean Tsunami of May 24, 1960, as Observed Along the Coast of Japan*, Tokyo, Japan.

Cox, D. C. (November 1964). *Tsunami Height–Frequency Relationship at Hilo*, University of Hawaii, Hawaii Institute of Geophysics.

Cross, R. H. (November 1967). "Tsunami Surge Forces on Coastal Structures," *J. Waterways and Harbors Div., Proc. ASCE*, **93** (WW4), 201–231.

Cueller, M. P. (February 1953). *Annotated Bibliography on Tsunamis*, U.S. Army Corps of Engineers, Beach Erosion Board, Tech. Memo No. 30.

Cumberbatch, E. (March 1960). "The Impact of a Water Wedge on a Wall," *J. Fluid Mech.*, **7** (3), 353–373.

Davison, C. (1936). *Great Earthquakes*, London: T. Murby and Co.

De Roos, R. and T. Spiegel (September 1966). "The Philippines, Freedom's Pacific Frontier," *Natl. Geograph.*, **130** (3), 301–351.

Dressler, R. F. (September 1952). "Hydraulic Resistance Effects Upon the Dam-Break Functions," *J. Res. Natl. Bur. Stds.*, **49** (3), 217–225.

Dressler, R. F. (1954). "Comparisons of Theories and Experiments for the Hydraulic Dam-Break Wave," *IUGG, Rome*, **3**, 319–327.

Eaton, J. P., D. H. Richter, and W. U. Ault (April 1961). "The Tsunami of May 23, 1960, on the Island of Hawaii," *Bull. Seism. Soc. Am.*, **51** (2), 135–157.

Eckart, C. (April 1948). "The Approximate Solution of One-Dimensional Wave-Equations," *Rev. Mod. Phys.*, **20** (1), 399–417.

Fraser, G. D., J. P. Eaton, and C. K. Wentworth (January 1959). "The Tsunami of March 9, 1957, on the Island of Hawaii," *Bull. Seism. Soc. Am.*, **49** (1), 79–90.

Fukui, Y. M. Nakamura, H. Shiraishi, and Y. Sasaki (1963). "Hydraulic Study on Tsunami," *Coastal Engr. Japan*, **6**, 67–82.

Granthem, K. N. (October 1953). "Wave Run-up on Sloping Structures," *Trans. Am. Geophys. Union*, **34** (5), 720–724.

Grantz, A., G. Plafker, and R. Kachadoorian (1964). *Alaska's Good Friday Earthquake, March 27, 1964; A Preliminary Evaluation*, U.S. Geological Survey Circular 491.

Green, C. K. (1946). "Seismic Sea Wave of April 1, 1946, as Recorded on Tide Gages," *Trans. Am. Geophys. Union*, **27** (4), 490–500.

Griffin, W. *et al.* (1964). *Crescent City's Dark Disaster*, Crescent City, California.

Gumbel, E. J. (1958). *Statistics of Extremes*, New York: Columbia University Press.

Gutenberg, B. (October 1939). "Tsunamis and Earthquakes," *Bull. Seism. Soc. Am.*, **29** (4), 517–526.

Gutenberg, B., and C. F. Richter (1935). "On Seismic Waves (Second Paper)," *Gerlands Beitr. Geophys.*, **45**, 280–360.

Gutenberg, B., and C. F. Richter (1949). "Seismicity of the Earth and Associated Phenomena," Princeton, New Jersey: Princeton University Press.

Heck, N. H. (1947). *List of Seismic Sea-Waves*, Doc. Scient. Comm. Raz de Maree, IUGG.

Hilo Technical Tsunami Advisory Council (April 8, 1962). *Protection of Hilo from Tsunamis*, reprinted in the *Sunday Tribune-Herald*, Hilo, Hawaii.

Horikawa, Kiyoshi (December 1961). *Tsunami Phenomena in the Light of Engineering View-point, Report on the Chilean Tsunami of May 24, 1960, as Observed Along the Coast of Japan*, the Committee for the Field Investigation of the Chilean Tsunami of 1960, pp. 136–150.

Housner, G. W. (June 1952). *Spectrum of Intensities of Strong-Motion Earthquakes, Proceedings of the Symposium on Earthquake and Blast Effects on Structures*, Earthquake Research Institute, pp. 20–36.

Hunt, I. A. (September 1959). "Design of Seawalls and Breakwaters," *J. Waterways and Harbors Div., ASCE*, **85** (WW3), 123–152.

Iida, K. (1959). "Earthquake Energy and Earthquake Fault," *J. Earth Sci., Nagoya Univ.*, **7** (2), 98–107.

Iida, K. (July 1963a). *Magnitude, Energy and Generation Mechanisms of Tsunamis and a Catalog of Earthquakes Associated with Tsunamis, Proceedings of the Tsunami Meetings Associated with the Tenth Pacific Science Congress, IUGG*, Monograph No. 24, pp. 7–18.

Iida, K. (July 1963b). *On the Heights of Tsunamis Associated with Distant and Near Earthquakes, Proceedings of the Tsunami Meetings Associated with the Tenth Pacific Science Congress, IUGG*, Monograph No. 24, pp. 105–123.

Iida, K. (April 1965). *Behavior of Tsunami in the Nearshore, Recent Studies on Tsunami Run-up II, U.S.–Japan Cooperative Scientific Research Seminars on Tsunamis Run-up*, Sapporo, Japan, pp. 18–24.

Iida, K. (1968). The Niigata Tsunami of June 16, 1964, *General Report on the Niigata Earthquake of 1964*, Hirosi Kawasumi, Editor-in-Chief, Tokyo Electrical Engineering College Press, pp. 97–127.

Iribarren, R. C., and C. Nogales y Olano (1947). *Protection of Ports*, Sec. II, Comm. 4, Ocean Navigation, Seventeenth Congress, International Association of Navigation Congresses, Lisbon.

Isaacs, J. D. (May 3, 1946). *Field Report on the Tsunami of April 1, 1946*, University of California, Berkeley, Department of Engineering (unpublished).

Isaacs, J .D., E. A. Williams, and C. Eckart (February 1951). "Reflection of Surface Waves by Deep Water," *Trans. Am. Geophys. Union*, **32** (1), 37–40.

Isaacson, E. (March 1950). "Water Waves Over a Sloping Bottom," *Comm. Pure Appl. Math.*, **3** (1), 11–32.

Ishimoto, M., and T. Hagiwara (March 1934). "The Tsunami Considered as a Phenomenon of Sea Water Overflowing the Land, Papers and Reports on the Tsunami of 1933 on the Sanriku Coast," *Japan, Bull. Earth. Res. Inst., Tokyo Univ.*, Suppl., **I**, 17–24.

Iwasaki, T. (April 1965). *On the Run-up of Tsunamis Against the Vertical Sea-Walls, Recent Studies on Tsunami Run-up II, U.S.–Japan Cooperative Scientific Research Seminars on Tsunami Run-up*, Sapparo, Japan, pp. 18–24.

Jaggar, T. A. (February 19, 1931). *Hawaiian Damage from Tidal Waves*, The Volcano Letter, Hawaiian Volcano Observatory, No. 321, pp. 1–3.

Japanese Organization for Tsunami Investigations (1962). *The Annotated Bibliography of Tsunamis, 1889–1962*, Tokyo: Japanese Organization for Tsunami Investigations.

Johnson, J. W., and K. J. Bermel (April 1949). "Impulsive Waves in Shallow Water as Generated by Falling Weights," *Trans. Am. Geophys. Union*, **30** (2), 223–230.

Kajiura, K. (September 1963a). "The Leading Wave of Tsunami," *Bull. Earth. Res. Inst.*, Tokyo University, **41** (3), 535–571.

Kajiura, K. (1963b). *On the Partial Reflection of Water Waves Passing Over a Bottom of Variable Depth, Proceedings of the Tsunami Meetings Associated with the Tenth Pacific Science Congress*, IUGG, Monograph No. 24, pp. 206–230.

Kaplan, K. (1955). *Generalized Laboratory Study of Tsunami Run-up*, U.S. Army Corps of Engineers, Beach Erosion Board, Tech. Memo No. 60.

Kato, Y., Z. Suzuki, K. Nakamura, A. Takagi, K. Emura, M. Ito, and H. Ishida (December 1961). *The Chile Tsunami of May 24, 1960, Observed Along the Sanriku Coast, Japan, Report on the Chilean Tsunami of May 24, 1960, as Observed Along the Coast of Japan, the Committee for Field Investigation of the Chilean Tsunami of 1960*, pp. 67–76.

Keller, J. B. (July 1963). *Tsunamis-Water Waves Produced by Earthquakes, Proceedings of the Tsunami Meetings Associated with the Tenth Pacific Science Congress, August–September 1961*, International Union of Geodesy and Geophysics, Monograph No. 24, pp. 154–166.

Keller, H. B., D. A. Levine, and G. B. Whitham (February 1960). "Motion of a Bore on a Sloping Beach," *J. Fluid Mech.*, **7** (2), 302–316.

Keller, J. B., and H. B. Keller (June 1964). *Water Wave*

Run-up on a Beach, I, Service Bureau Corporation, New York, Contract NONR-3828(00) (unpublished).

Keller, J. B. and H. B. Keller (1965). *Water Wave Run-up on a Beach*, II, Service Bureau Corporation, New York, Contract NONR-3828(00) (unpublished).

Keulegan, C. H. (1949). Wave Motion, Chapter 11 of Engineering Hydraulics, Hunter Rouse, ed., New York: John Wiley, pp. 711–768.

Koh, R. C. Y., and B. LeMehaute (January 1966). *Wave Shoaling*, National Engineering Science Company, Report No. SN 134-9, Contract NONR-4177(00) (unpublished).

Kranzer, H. C., and J. B. Keller (1955). *Water Waves Produced by Explosions*, New York University, New York Institute of Mathematical Sciences, IMM-NYU 222.

Kranzer, H. C., and J. B. Keller (March 1959). "Water Waves Produced by Explosions," *J. Appl. Phys.*, **30** (3), 398–407.

Laird A. D. K. (April 1955). "A Model Study of Wave Action on a Cylindrical Island," *Trans. Am. Geophys. Union*, **36** (2), 279–285.

Lamb, H. (1945). *Hydrodynamics*, 6th ed., New York: Dover Publications.

Leet, D. L. (1948). *Causes of Catastrophe: Earthquakes, Volcanos, Tidal Waves, Hurricanes*, New York: McGraw-Hill.

LeMehaute, B. (1963). *On Non-saturated Breakers and the Wave Run-up, Proceedings of the Eighth Conference on Coastal Engineering*, Berkeley, California: Council on Wave Research, The Engineering Foundation, 77–92.

LeMehaute, B., and L. M. Webb (1964). *Periodic Gravity Waves Over a Gentle Slope at a Third Order of Approximation, Proceedings of the Ninth Conference on Coastal Engineering*, ASCE, pp. 23–40.

MacDonald, G. A., and C. K. Wentworth (July 1954). "The Tsunami of November 4, 1952, on the Island of Hawaii," *Bull. Seism. Soc. Am.*, **44** (3), 463–469.

Magoon, O. T. (1966). *Structural Damage by Tsunamis, Coastal Engineering: Santa Barbara Specialty Conference, October 1965*, American Society of Civil Engineers, 35–68.

Matlock, H., L. C. Reese, and R. B. Matlock (March 1962). *Analysis of Structural Damage from the 1960 Tsunami at Hilo, Hawaii*, prepared for the Defense Atomic Support Agency, Contract DA-49-146-XZ-028, University of Texas (unpublished).

Matuo, H. (1934). "Estimation of Energy of Tsunami and Protection of Damage of Coasts, Papers and Reports on the Tsunami of 1933 on the Sanriku Coast, Japan," *Bull. Earth. Res. Inst., Tokyo Univ., Suppl.*, **I**, 55–64.

Miche, R. (1944). "Mouvements Ondulatoires des Mers en Profondeur Constante ou Décroissante," *Annal. Ponts Chaussées*, pp. 25–78, 131–164, 270–292, 369–406.

Miche, R. (April 1953). "The Reflecting Power of Maritime Works Exposed to Action of the Waves," *Bull. Beach Erosion Bd., Corps Engrs, U.S. Army*, **7** (2), 1–7. (Note: This is a translation of the original article that appeared in the *Ann. Ponts Chaussées* in 1951.)

Michell, J. H. (1893). "The Highest Waves in Water," *Phil. Mag.*, 5th Ser., **36** (5), 430–437.

Miller, D. J. (1960). *Giant Waves in Lituya Bay, Alaska*, U.S. Geological Survey Professional Paper 354-C.

Morse, P. M., and P. J. Rubenstein (December 1938). "The Diffraction of Waves by Ribbons and Slits," *Phys. Rev.*, **54**, 895–898.

Nasu, N. (March 1934). "Heights of Tsunamis and Damage to Structures," *Bull. Earth. Res. Inst., Tokyo Univ.*, Suppl., **I**, 218–227.

Niazi, M. (April 1964). "Seismicity of Northern California and Western Nevada," *Bull. Seism. Soc. Am.*, **54**(2), 845–850.

Nielsen, A. H. (December 1962). *Diffraction of Periodic Waves Along a Vertical Breakwater for Small Angles of Incidence*, University of California, Berkeley, IER Tech. Rept. HEL-1-2.

Omer, G. C., Jr., and H. H. Hall (October 1949). "The Scattering of a Tsunami by a Cylindrical Island," *Bull. Seism. Soc. Am.*, **39** (4), 257–260.

Palmer, R. Q., M. E. Mulvihill, and G. T. Funasaki (1966). *Hilo Harbor Tsunami Model-Reflected Waves Superimposed, Coastal Engineering: Santa Barbara Specialty Conference, October 1965*, American Society of Civil Engineers, pp. 24–31.

Pereira, E. J. (1888). "The Great Earthquake of Lisbon," *Trans. Seism. Soc. Japan*, **12**, 5–19.

Perroud, P. H. (September 1957). *The Solitary Wave Reflection Along a Straight Vertical Wall at Oblique Incidence*, University of California, Berkeley, IER Tech. Rept. 99-3 (unpublished).

Prins, J. E. (October 1958a). "Characteristics of Waves Generated by a Local Disturbance," *Trans. Am. Geophys. Union*, **39** (5), 865–874.

Prins, J. E. (1958b). *Water Waves Due to a Local Disturbance, Proceedings of the Sixth Conference on Coastal Engineering*, Berkeley, California, The Engineering Foundation Council on Wave Research, pp. 147–162.

Russell, J. S. (1844). *Report on Waves*, Fourteenth Meeting of the British Association on the Advancement of Science, pp. 311–390.

Salsman, G. G. (July 1959). *The Tsunami of March 9, 1957, as Recorded at Tide Stations*, U.S. Department Commerce, Coast and Geodetic Survey, Tech. Bull. No. 6.

Savage, R. P. (March 1959). *Laboratory Data on Wave Run-up on Roughened and Permeable Slopes*, U.S. Army Corps of Engineers, Beach Erosion Board, Tech. Memo No. 109.

Savarensky, E. F., U. G. Tischchenko, and A. E. Sviatlovsky (1958). "The Tsunami of 4–5 November, 1952," *Bull. Council Seism., Acad. Sci. USSR*, (4), 1–60, W. G. Van Campen, trans., University of Hawaii, East–West Center, Hawaii Institute of Geophysics, Trans. Series No. 10.

Schoemaker, H. J., and J. Th. Thijsse (September 1949). *Investigations of the Reflection of Waves*, Third Meeting of the International Association of Hydraulic Structures Research, I-2.

Shepard, F. P. (July–August 1933). "Depth Changes in the Sagami Bay During the Great Japanese Earthquake," *Jour. Geol.*, **41** (5), 527–536.

Shepard, F. P., G. A. MacDonald, and D. C. Cox (1950). "The Tsunami of April 1, 1946," *Bull. Scripps Inst. Oceanog., Univ. Calif.*, **5** (6), 391–528.

Sigurdsson, C., and R. L. Wiegel (October 1962). *Solitary Wave Behavior at Concave Barriers*, The Port Engineer (Calcutta), pp. 4–8.

Sneddon, I. N. (1951). *Fourier Transforms*, New York: McGraw-Hill.

Solov'ev, S. L., and M. D. Ferchev (1961). "Summary of Data on Tsunamis in the USSR," *Bull. Council Seism., Acad. Sci. USSR*, (9), 23–55, W. G. Van Campen, trans., University of Hawaii, East–West Center, Hawaii Institute of Geophysics, Trans. Series No. 9.

Spaeth, M. G. (July 1964). *Annotated Bibliography on Tsunamis*, IUGG Monograph No. 27.

Spaeth, M. G., and S. C. Berkman (April 1965). *The Tsunami of March 28, 1964, as Recorded at Tide Stations*, U.S. Department of Commerce, Coast and Geodetic Survey.

Stokes, G. G. (1880). *On the Theory Oscillatory Waves, Mathematical and Physical Papers, I*, Cambridge: Cambridge University Press.

Takahasi, R. (1963a). *A Summary Report on the Chilean Tsunami of May 24, 1960, as Observed Along the Coast of Japan, Proceedings of the Tsunami Meetings Associated with the Tenth Pacific Science Congress*, IUGG, Monograph No. 24, pp. 77–86.

Takahasi, R. (July 1963b). *On Some Model Experiments on Tsunami Generation, Proceedings of the Tsunami Meetings Associated with the Tenth Pacific Science Congress*, IUGG, Monograph No. 24, pp. 235–248.

Tarr, R. S. and L. Martin (1912). *The Earthquakes at Yakutat Bay, Alaska in September, 1899*, U.S. Geological Survey Professional Paper No. 69.

Tokyo Imperial University (March 1934). "Papers and Reports on the Tsunamis of 1933 on the Sanriku Coast, Japan," *Bull. Earth. Inst. Japan*, Suppl., **I**.

Tsuboi, C. (1957). "Energy Accounts of Earthquakes in and Near Japan," *Geophys. Notes, Geophys. Inst., Tokyo Univ.*, **10** (1).

Tudor, W. J. (November 1964). *Tsunami Damage at Kodiak, Alaska, and Crescent City, California, from Alaskan Earthquake of 27 March 1964*, U.S. Naval Civil Engineering Laboratory, Tech. Note N-622.

U.S. Army, Corps of Engineers, Honolulu District (November 15, 1960). *Hilo Harbor, Hawaii, Report on Survey for Tidal Wave Protection and Navigation.*

U.S. Coast and Geodetic Survey (April 17, 1964a). *Preliminary Report: Prince William Sound, Alaska Earthquakes, March–April 1964*, U.S. Department of Commerce, Coast and Geodetic Survey.

U.S. Coast and Geodetic Survey (July 1964b). *Annotated Bibliography on Tsunamis*, International Union of Geodesy and Geophysics, Monograph No. 27.

Unoki, S., and M. Nakano (1953). *On the Cauchy–Poisson Waves Caused by the Eruption of a Submarine Volcano (1st paper), Oceanograph. Mag.*, **4** (4), 119–141.

Ursell, F. (1953). "The Long-Wave Paradox in the Theory of Gravity Waves," *Proc. Cambridge Phil. Soc.*, **49** (4), 685–694.

Ursell, F. (July 1958). "On the Waves Generated by a Local Surface Disturbance" (personal communication).

Van Dorn, W. G. (August 12, 1959). *Local Effects of Impulsively Generated Waves*, Report No. II, Scripps Institute of Oceanography, University of California, Contract NONR 233 (35) (unpublished).

Van Dorn, W. G. (November 1961). "Some Characteristics of Surface Gravity Waves in the Sea Produced by Nuclear Explosions," *J. Geophys. Res.*, **66** (11), 3845–3862.

Van Dorn, W. G. (1965). *Tsunamis, Advances in Hydroscience*, vol. 2, New York: Academic Press, pp. 1–48.

Van Dorn, W. G. (May 31, 1966). *Theoretical and Experimental Study of Wave Enhancement and Run-up on Uniformly Sloping Impermeable Beaches*, University of California, Scripps Institute of Oceanography, Report No. SIO 66-11, Contract NONR. 2216 (16) (unpublished).

Wadati, K., T. Hirono, and S. Hisamoto (July 1963). *On the Tsunami Warning Service in Japan, Proceedings of the Tsunami Meetings Associated with the Tenth Pacific Science Congress*, IUGG, Monograph No. 24, pp. 138–145.

Watanabe, H. (March 1964). "Studies on the Tsunamis on the Sanriku Coast of the Northeastern Honshu in Japan," *Geophys. Mag., Tokyo*, **22** (1), 1–64.

Wemelsfelder, P. J. (1961). *On the Use of Frequency Curves of Storm Floods, Proceedings of the Seventh Conference on Coastal Engineering*, Berkeley, California: Council on Wave Research, The Engineering Foundation, pp. 617–632.

Wharton, W. J. L., and F. J. Evans (1888). *The Eruption of Krakatoa and Subsequent Phenomena. Part III, on the Seismic Sea Waves Caused by the Eruptions of Krakatca, August 26 and 27*, 1883, Royal Society of London, pp. 89–151.

Whitham, G. B. (January 1955). "The Effects of Hydraulic Resistance on the Dam-Break Problem," *Proc. Roy. Soc. (London)*, Ser. A, **227** (1170), 399–407.

Wiegel, R. L. (October 1955). "Laboratory Studies of Gravity Waves Generated by the Movement of a Submerged Body," *Trans. Am. Geophys. Union*, **36** (5), 759–74.

Wiegel, R. L. (July 1963). *Research Related to Tsunamis Performed at the Hydraulic Laboratory, University of California, Berkeley, Proceedings of the Tsunami Meetings Associated with the Tenth Pacific Science Congress, August–September 1961*, International Union of Geodesy and Geophysics, Monograph No. 24, pp. 174–197.

Wiegel, R. L. (May 6, 1964a). *Tsunami Information in Regard to Proposed Nuclear Power Plant Site, Pacific Gas and Electric Company at Bodega Head, California*, Consultant Report to Pacific Gas and Electric Company, San Francisco, California.

Wiegel, R. L. (September 1964b). *Possibility of Tsunamis at Bodega Head, and Forces Exerted by Such Tsunamis*, Consultant Report to the Pacific Gas and Electric Company, San Francisco, California.

Wiegel, R. L. (1964c). *Water Wave Equivalent of Mach-Reflection, Proceedings of the Ninth Conference on Coastal Engineering, Lisbon, Portugal, June 1964*, American Society of Civil Engineers, pp. 82–102.

Wiegel, R. L. (1964d). *Oceanographical Engineering*, Englewood Cliffs, New Jersey: Prentice-Hall.

Wiegel, R. L. (March 5, 1965a). *Protection of Crescent City, California, from Tsunami Waves*, Report for the Redevelopment Agency of the City of Crescent City.

Wiegel, R. L. (November 19, 1965b). *Flow over Seawall: Tsunami Waves*, Consulting Report to Pacific Gas and Electric Company, San Francisco, California.

Wiegel, R. L., E. K. Noda, E. M. Kuba, D. M. Gee, and G. F. Tornberg (April 1968). *Water Waves Generated by Landslides in Reservoirs*, University of California, Berkeley, Hyd. Eng. Lab., Tech. Rept. 19-1 (unpublished).

Wiener, F. M. (May 1947). "Sound Diffraction by Rigid Spheres and Cylinders," *J. Acoust. Soc. Am.*, **19** (3), 444–451.

Williams, E. A., and J. D. Isaacs (August 1952). "The Refraction of Groups and of the Waves They Generate in Shallow Water," *Trans. Am. Geophys. Union*, **33** (4), 523–530.

Wilson, B. W., L. M. Webb, and J. A. Hendrickson (August 1962). *The Nature of Tsunamis, Their Generation and Dispersion in Water of Finite Depth*, National Engineering Science Company, Tech. Rept. No. SN 57-2.

Wilson, W. S. (February 1966). *A Method for Calculating and Plotting Surface Wave Rays*, U.S. Army Corps of Engineers, Coastal Engineering Research Center, Tech. Memo. 17.

Yamaguti, S. (March 1934). "Abnormally High Waves, or 'Tsunami,' on the Coast of Sanriku in Japan on March 3, 1933, Papers and Reports on the Tsunami of 1933 on the Sanriku Coast, Japan," *Bull. Earth. Res. Inst., Tokyo Univ.*, Suppl. **I**, 36–54.

Zerbe, W. B. (1953). *The Tsunami of November 4, 1952, as Recorded at Tide Stations*, U.S. Department of Commerce, Coast and Geodetic Survey, Special Publ. No. 300.

Chapter 12

Earthquake Response of Structures

BY RAY W. CLOUGH

Professor of Civil Engineering
University of California
Berkeley, California

12.1 INTRODUCTION

Although structural damage may result from several basically different effects of an earthquake—wave forces during tsunamis, foundation failure due to loss of soil strength by liquefaction, foundation displacements associated with fault break or landslide movements, etc.—the principal loading mechanism recognized by seismic design requirements in building codes is the response to the earthquake ground motions applied at the base of the structure. The purpose of this chapter is to summarize the theory of structural dynamics as it applies to the calculation of such vibratory response. The specific objective of this theory is to predict the stresses and deflections that will be developed in any given structural system as a result of any specified ground motion history applied at the base of the structure.

It should be noted that the problem is completely defined by the physical properties of the structural system, i.e., by its mass, stiffness, and damping characteristics,

and by the time-varying displacements introduced at its foundation support points. Thus the evaluation of these structural properties and the selection of an appropriate earthquake input are the most critical factors in the earthquake response analysis. However, for the purpose of this discussion, it will be assumed that the structural and earthquake characteristics are specified, and attention will be focused exclusively on the techniques of dynamic response analysis. On this basis, an earthquake response analysis may be considered a classic problem of the theory of structural dynamics.

In the formulation of any dynamic response analysis, it must be recognized that the structure generally will be subjected to static loadings (e.g., gravitational forces) in addition to the dynamic excitation which is the subject of immediate interest. If the structure is linearly elastic, so that the principle of superposition is applicable, it is convenient to consider separately the static and dynamic loadings; then the total structural response is obtained by adding the static stresses and deflections to the results of the dynamic analysis. However, if the structure yields or is subject to some other nonlinear behavior during the dynamic loading, superposition is not valid and the static loads must be considered in the analysis concurrently with the dynamic effects.

12.1.1 Basic Characteristic of a Dynamic Problem

In the present context, the term "dynamic" means merely "time-varying." Thus, in a structural dynamics problem, the loading and all aspects of the structural response (deflections, internal forces, stresses, etc.) all vary with time. Accordingly, it is clear that one important distinction between a static and a dynamic structural analysis is that the dynamic analysis does not have a single solution—instead, a separate solution must be obtained for each instant of time during the entire history of interest. Thus, a dynamic analysis obviously must entail a greater expenditure of effort than a static analysis of the same structure.

However, this is not the most significant distinction between a static and a dynamic problem. The basic feature of a dynamic problem may be recognized with reference to the simple beam structure shown in Fig. 12.1. When it is subjected to a static load, as shown in Fig. 12.1a, the internal forces resisting the load may be evaluated by simple statics, and from these the resulting stresses and deflections may be determined. If the same load were applied dynamically, however, the time-varying deflections would involve accelerations, and by d'Alembert's principle these would engender inertia forces resisting the motion as shown in Fig. 12.1b. Thus the beam may be considered to be subjected to two loadings: the external loading $P(t)$ that causes the motion and the inertia forces $f_I(t)$ that resist its acceleration. The internal forces in the beam are called upon to equilibrate this combined load system; consequently, it is necessary to determine the inertia forces before the internal stresses may be determined.

From this discussion it may be concluded that inertia forces are the essential characteristic of a structural dynamics problem. The magnitude of these forces depends upon the rate of loading of the structure and on its flexibility and mass characteristics. If the loading is applied slowly enough, the inertia forces will be small in relation to the applied load and may be neglected; in this case, the problem may be treated as static. If the loading is rapid, however, the inertia forces will be significant, relative to the external load, and will have an important effect on the resulting stresses; thus, dynamic analysis procedures must be applied to evaluate the response. The basic difficulty of a dynamic analysis results from the fact that the deflections which lead to the development of inertia forces are themselves influenced by these inertia forces. To break this closed cycle of cause and effect, the problem must be formulated in terms of differential equations—expressing the inertia forces in terms of the time derivatives of the structural deflections.

Fig. 12.1. Essential difference between a static and a dynamic loading.

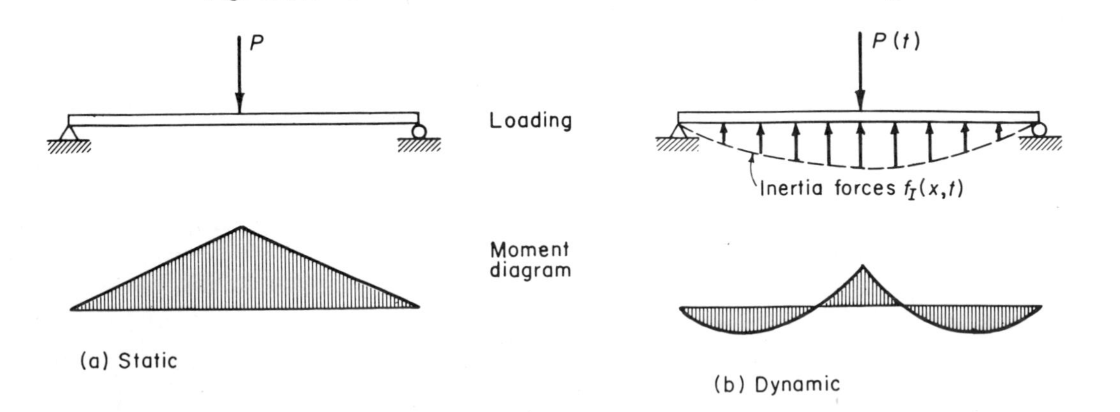

12.1.2 Degrees of Freedom

The complete system of inertia forces acting in a structure can be determined only by evaluating the accelerations (and therefore the displacements) of every mass particle. In any real structure, this means that the displacements must be calculated for every point in the structure, which is a large computational task even in a static analysis. The analysis can be simplified greatly if the deflections of the structure can be specified adequately by a limited number of displacement components or coordinates. Two different assumptions concerning the manner of specifying the deflected shape of the structure frequently are made in dynamic structural analyses. These will be referred to here as the lumped mass approach and the generalized coordinate approach. In either case, the number of displacement components or coordinates required to specify the position of all significant mass particles in the structure is called the number of degrees of freedom of the structure.

In the lumped mass idealization (as its name implies) it is assumed that the entire mass of the structure is concentrated in a number of discrete points, located appropriately. A three-degree-of-freedom lumped mass idealization of a simple beam is shown in Fig. 12.2. If axial distortions of the beam are neglected, the displacements of the three masses are given by the three coordinates v_a, v_b, and v_c, and it is evident that it is necessary to evaluate the accelerations at only these three points in order to define the inertial forces developed in this system.

An alternative means of approximating the deflections of the beam by a limited set of displacement coordinates is provided by the Fourier series representation. For example, the deflections of a simple beam can be expressed as a sine series,

$$v(x) = \sum_{n=1}^{\infty} Y_n \sin \frac{n\pi x}{L} \tag{12.1}$$

where Y_n represents the amplitude of the nth component of the series. To represent any arbitrary deflected shape exactly would require an infinite number of terms in the series, but a reasonable approximation of the deflection in a typical simple beam often can be obtained with only the first few terms. For example, if only the first three terms of the series were considered, the deflection would be given by

$$v(x) = Y_1 \sin \frac{\pi x}{L} + Y_2 \sin \frac{2\pi x}{L} + Y_3 \sin \frac{3\pi x}{L} \tag{12.2}$$

This type of approximation is sketched in Fig. 12.3; it should be noted that the three amplitude coefficients (Y_1, Y_2, and Y_3) completely define the shape of the beam in this case. Thus, they are the generalized coordinates of this three-degree-of-freedom system.

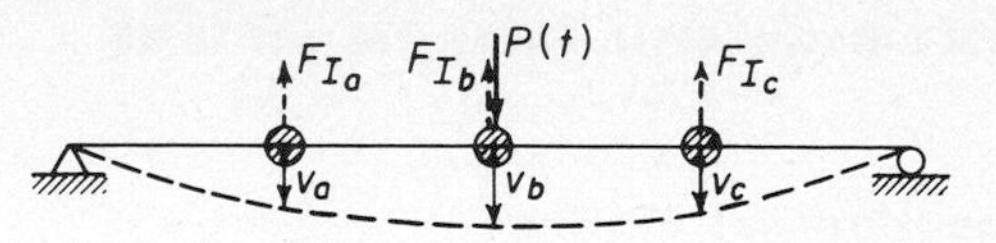

Fig. 12.2. A 3-degree-of-freedom lumped mass system.

The sine wave shapes used in defining the displacements in Eq. 12.1 are merely a specific example of a general concept. In principle, the shapes could have been any arbitrary independent shapes. Thus, if the nth term shape function of the displacement series is designated $\psi_n(x)$, the general form of an N degree of freedom generalized coordinate representation of the structural displacements may be written

$$v(x) = \sum_{n=1}^{N} \psi_n(x) Y_n \tag{12.3}$$

where Y_n is the amplitude of the nth shape function.

In either of these methods of limiting the number of displacements to be considered in a dynamic analysis, the accuracy of the idealization will increase with the number of degrees of freedom considered and will approach the exact solution as the number approaches infinity. The number of terms required to obtain an adequate solution depends on the complexity of the structure and of its loading, but in many cases only two or three degrees of freedom are required and in some cases a single coordinate will provide useful results. For a given number of degrees of freedom, the generalized coordinate approach generally gives greater accuracy, but at a somewhat larger expenditure of computational effort.

In the following discussion, both techniques will be considered. Single-degree-of-freedom systems will be discussed first to introduce the concepts of structural dynamics in their simplest form; the development will then be extended to systems having many degrees of freedom.

Fig. 12.3. Generalized coordinate representation of beam deflections.

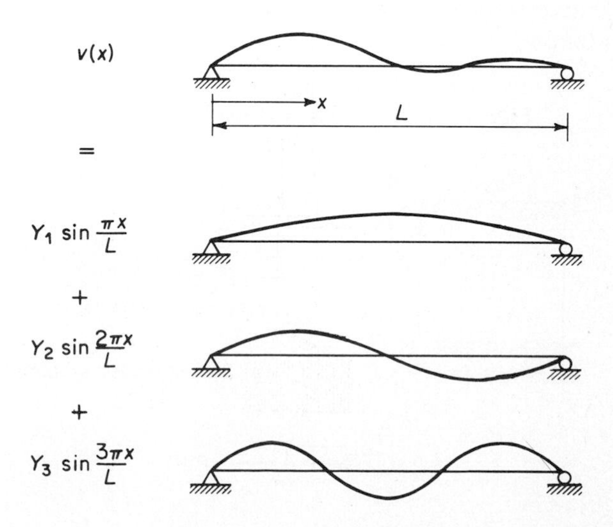

12.2 SINGLE-DEGREE-OF-FREEDOM SYSTEMS

12.2.1 Equations of Motion

The dynamic response of a structure is defined primarily by its displacement history, i.e., by the time variation of the coordinates which represent its degrees of freedom. The equations from which these displacements are determined are called the *equations of motion* of the system. They may be looked upon as expressions of the dynamic equilibrium of all forces acting on the structure. One such equation may be written corresponding to each degree of freedom; thus, only a single equation of motion need be formulated for a single-degree-of-freedom system. In this discussion it will be convenient to consider first the equation for a lumped mass system and then to consider the generalized coordinate formulation.

12.2.1.1 Lumped mass system. A general single mass system may be pictured either as a one-story frame (Fig. 12.4a) or as an academic spring-mass system (Fig. 12.4b). In either case it consists of a single rigid mass M so constrained that it can move only with one component of simple translation. Motion in that component is resisted by weightless elastic elements having a total spring constant K and also by a damping device which absorbs energy from the system. In the present discussion it will be assumed that the damping force is proportional to the velocity of the mass, the damping coefficient being denoted by C.

In a typical dynamic problem, the motion is excited by an external load $P(t)$, and the complete set of forces acting on the mass may be represented as shown in Fig. 12.4c. The equation of dynamic equilibrium thus is given by

$$F_I + F_D + F_S = P(t) \tag{12.4}$$

in which the forces resisting motion are proportional to the acceleration, velocity, and displacement of the mass as follows:

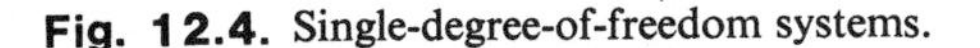
Fig. 12.4. Single-degree-of-freedom systems.

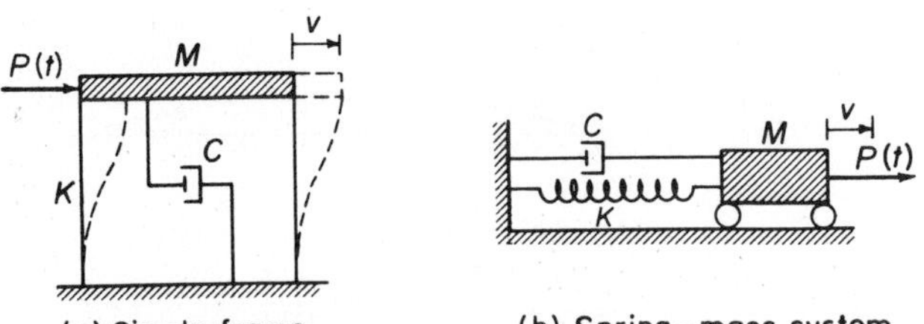

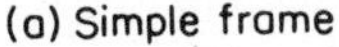

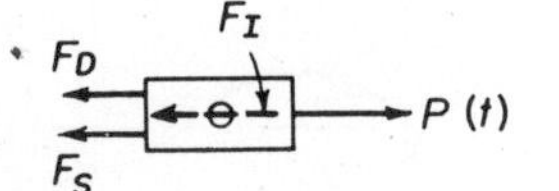

(c) Forces acting on mass

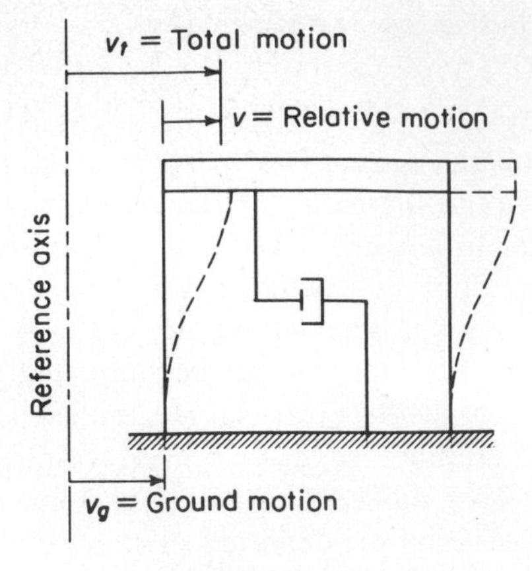

Fig. 12.5. Earthquake excitation of building frame.

$$\begin{aligned} \text{Inertia force:} \quad & F_I = M\ddot{v} \\ \text{Damping force:} \quad & F_D = C\dot{v} \\ \text{Elastic force:} \quad & F_S = Kv \end{aligned} \tag{12.5}$$

where the dots denote differentiation with respect to time. Introducing Eq. 12.5 into Eq. 12.4, the equation of motion for the system may be written

$$M\ddot{v} + C\dot{v} + Kv = P(t) \tag{12.6}$$

The dynamic problem of Fig. 12.4 differs from the earthquake excitation problem in that a dynamic load is shown applied directly to the structure. In the earthquake problem, the excitation is provided by the motion introduced at the supports of the structure, $v_g(t)$, as shown in Fig. 12.5, and there is no external loading (i.e., $P = 0$). Thus, Eq. 12.4 becomes in this case

$$F_I + F_D + F_S = 0 \tag{12.7}$$

However, it must be recognized that the inertia force depends on the total acceleration of the mass $\ddot{v}_t$, which includes a component relative to the base $\ddot{v}$ plus the acceleration of the base $\ddot{v}_g$:

$$\ddot{v}_t = \ddot{v} + \ddot{v}_g \tag{12.8}$$

(Clearly an equivalent statement can be made with regard to the displacement as well.) On this basis, the inertia force may be expressed

$$F_I = M\ddot{v}_t = M\ddot{v} + M\ddot{v}_g \tag{12.9}$$

Using this expression in Eq. 12.7 (together with the unchanged expressions for damping and elastic force from Eq. 12.5) leads to the earthquake equation of motion:

$$M\ddot{v} + M\ddot{v}_g + C\dot{v} + Kv = 0$$

or transferring the term containing the specified ground acceleration input to the right hand side,

$$M\ddot{v} + C\dot{v} + Kv = P_{\text{eff}}(t) \tag{12.10}$$

where

$$P_{\text{eff}}(t) = -M\ddot{v}_g \tag{12.11}$$

is an *effective* load resulting from the ground motion. Comparing Eqs. 12.6 and 12.10 it is apparent that the earthquake input is exactly equivalent to a dynamic load equal to the product of the ground acceleration and mass of the structure. The negative sign in Eq. 12.11 merely indicates that the effective load opposes the direction of ground acceleration. It generally is of little interest in an earthquake response analysis.

12.2.1.2 Generalized coordinate system. The formulation of the equation of motion for a system having its displacements defined by a single generalized coordinate will be illustrated first by the column on a hinge support shown in Fig. 12.6. In this analysis it will be assumed that the column is rigid and merely rotates about the pivot during its dynamic response. Its displacement thus can be expressed

$$v(x, t) = \psi(x)\, Y(t) = \frac{x}{L}\, Y(t) \tag{12.12}$$

in which the generalized coordinate $Y(t)$ represents the motion of the top of the column and the shape function $\psi(x)$ is the linear displacement expression x/L. The mass per unit of length of this column is assumed to vary arbitrarily along its length and to be defined by the function $\mu(x)$. Elastic and damping constraints are provided by discrete elements positioned along the column, as shown in the sketch, and it is loaded by the arbitrary force $P(t)$ as indicated.

Forces generated by the motion of this system depend on the total displacement, velocity, and acceleration at the positions where the constraining elements are located. Thus, the elastic and damping forces may be represented

$$F_S = Kv_a = k\frac{a}{L}\,Y \tag{12.13a}$$

$$F_D = C\dot{v}_b = C\frac{b}{L}\,\dot{Y} \tag{12.13b}$$

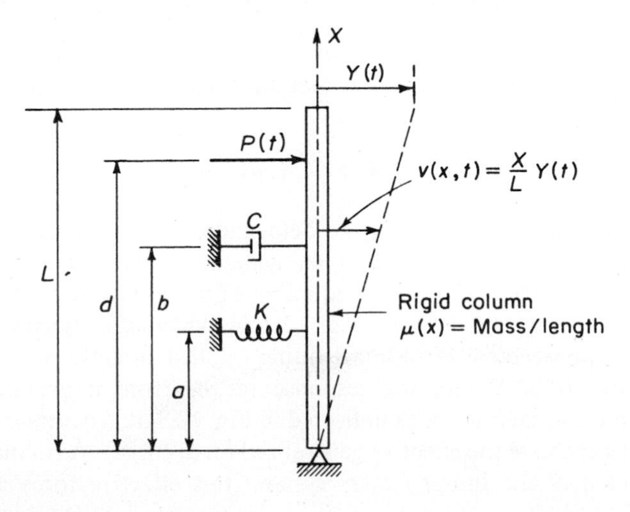

Fig. 12.6. Generalized coordinate system with single degree of freedom.

where a and b are dimensions indicated in the sketch. Because the mass is distributed along the length of the column, the inertia force is also a distributed quantity, which is expressed as follows:

$$f_I(x) = \mu(x)\ddot{v}(x) = \mu(x)\frac{x}{L}\ddot{Y} \tag{12.14}$$

The equation of motion of the system may now be formulated by expressing the dynamic equilibrium of the moments $\mathcal{M}$ produced by these forces about the hinge support:

$$\mathcal{M}_I + \mathcal{M}_D + \mathcal{M}_S = \mathcal{M}_P(t) \tag{12.15}$$

where

$$\mathcal{M}_S = F_a \cdot a \tag{12.16a}$$

$$\mathcal{M}_D = F_D \cdot b \tag{12.16b}$$

$$\mathcal{M}_P(t) = P(t) \cdot d \tag{12.16c}$$

and the inertia moment is given by the integral

$$\mathcal{M}_I = \int_0^L f_I(x)x\,dx \tag{12.16d}$$

Introducing Eqs. 12.13 and 12.14 into Eq. 12.16, substituting into Eq. 12.15, and dividing by the length L leads to the equation of motion,

$$\ddot{Y}\int_0^L \mu(x)\left(\frac{x}{L}\right)^2 dx + \dot{Y}C\left(\frac{b}{L}\right)^2 + YK\left(\frac{a}{L}\right)^2 = P(t)\frac{d}{L} \tag{12.17}$$

Now it may be noted that Eq. 12.17 can be put in the form of Eq. 12.16 by substituting new symbols as follows:

$$M^*\ddot{Y} + C^*\dot{Y} + K^*Y = P(t) \tag{12.18}$$

in which

$$M^* = \int_0^L \mu(x)\left(\frac{x}{L}\right)^2 dx$$

$$C^* = C\left(\frac{b}{L}\right)^2 \tag{12.19}$$

$$K^* = K\left(\frac{a}{L}\right)^2$$

$$P^*(t) = P(t)\frac{d}{L}$$

These are the generalized mass, damping, stiffness, and load properties of the system corresponding to the specified generalized coordinate Y.

The formulation of the equations of motion for the preceding example was carried out easily by direct application of equilibrium relationships because of the simple form of the deflected shape. In a more general case, wherein the deflected shape function $\psi(x)$ is more complicated, it usually is more convenient to apply the principle of virtual displacements to establish the equilibrium of the forces associated with the generalized coordinates. According to this principle, if the work done by the external loads during a virtual displacement is equal to the work done on the resisting forces during this motion, the forces corresponding to the virtual displacement are in equilibrium. The example structure shown in Fig. 12.7 will be used to demonstrate the virtual work formulation of the equations of motion. This structure is assumed to be a general cantilever column having arbitrary mass per unit length $\mu(x)$, arbitrary flexural rigidity $EI(x)$, and subjected to a general distributed loading $P(x, t)$. Damping could also be included similarly but will be omitted from this discussion because it is more easily represented by an assumed damping ratio, as will be explained later.

To make a single-degree-of-freedom analysis of this system, it is first necessary to approximate its deflections by means of a generalized coordinate and shape function,

$$v(x, t) = \psi(x) Y(t) \tag{12.20}$$

where the generalized coordinate $Y(t)$ may be taken conveniently as the deflection of the tip of the column and $\psi(x)$ is the function describing the assumed shape as shown in Fig. 12.7. The resisting forces developed in the system during this motion may be expressed as follows:

Inertia force:

$$f_I(x) = \mu(x)\ddot{v}(x) = \mu(x)\ddot{Y}\psi(x)$$

Elastic moments: (12.21)

$$m(x) = EI(x)\frac{d^2 v}{dx^2} = EI(x)\, Y \frac{\partial^2 \psi(x)}{\partial x^2}$$

It is assumed herein that only flexural deformations occur, and by the principles of mechanics elastic moments develop that are proportional to the distortion curvature $\partial^2 v/\partial x^2$.

The equation of motion for the system can now be obtained by applying a virtual displacement corresponding to the generalized coordinate and equating the work done externally and internally. The virtual displacement may be expressed

Fig. 12.7. Generalized coordinate idealization of cantilever column.

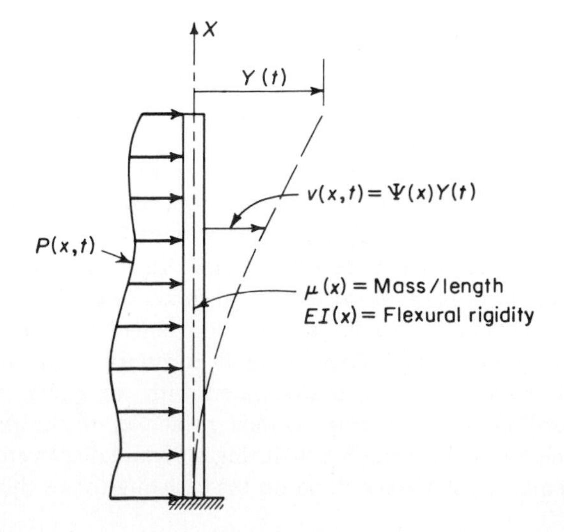

$$\delta v(x) = \psi(x)\delta Y \tag{12.22}$$

The external work done during this displacement then is given by

$$W_E = \int_0^L P(x, t)\delta v(x)\, dx = \delta Y \int_0^L P(x, t)\psi(x)\, dx \tag{12.23}$$

The internal work done on the system (inertia and elastic forces) is

$$\delta W_I = \int_0^L f_I(x)\delta v(x)\, dx + \int_0^L m(x)\delta\frac{\partial^2 v}{\partial x^2}\, dx$$

or, substituting from Eqs. 12.21 and 12.22,

$$\delta W_I = \ddot{Y}\delta Y \int_0^L \mu(x)[\psi(x)]^2\, dx + Y\,\delta Y \int_0^L EI(x)\left(\frac{\partial \psi}{\partial x^2}\right)^2 dx \tag{12.24}$$

Equating external and internal work and dividing by the arbitrary virtual displacement δY leads to the following:

$$\ddot{Y}\int_0^L \mu(x)[\psi(x)]^2\, dx + Y\int_0^L EI(x)\left(\frac{\partial^2 \psi}{\partial x^2}\right)^2 dx = \int_0^L P(x, t)\psi(x)\, dx \tag{12.25}$$

Finally, by introducing the following definitions of the generalized properties,

Generalized mass:

$$M^* = \int_0^L \mu(x)[\psi(x)]^2\, dx \tag{12.26a}$$

Generalized stiffness:

$$K^* = \int_0^L EI(x)\left[\frac{\partial^2 \psi(x)}{\partial x^2}\right]^2 dx \tag{12.26b}$$

Generalized force:

$$P^*(t) = \int_0^L P(x, t)\psi(x)\, dx \tag{12.26c}$$

the generalized coordinate equation of motion is obtained:

$$M^*\ddot{Y} + K^* Y = P^*(t) \tag{12.27}$$

It should be noted that Eq. 12.26b accounts for only the flexural strain energy of the system. If spring elements K_i also were present the complete generalized stiffness expression would become

$$K^* = \int_0^L EI(t)\left(\frac{\partial \psi}{\partial x^2}\right)^2 dx + \sum_i K_i \psi_i^2 \tag{12.26d}$$

in which ψ_i represents the deformation of spring i due to a unit displacement of the generalized coordinate Y.

Equation 12.17 is the equation of motion for a system of the type shown in Fig. 12.7 in which the dynamic response results from a loading applied directly to the structure. Where the response results from a ground motion excitation, as indicated in Fig. 12.8, it is necessary to evaluate the effective generalized force $P^*_{\text{eff}}(t)$. As in the case of the lumped mass system, this effective force is

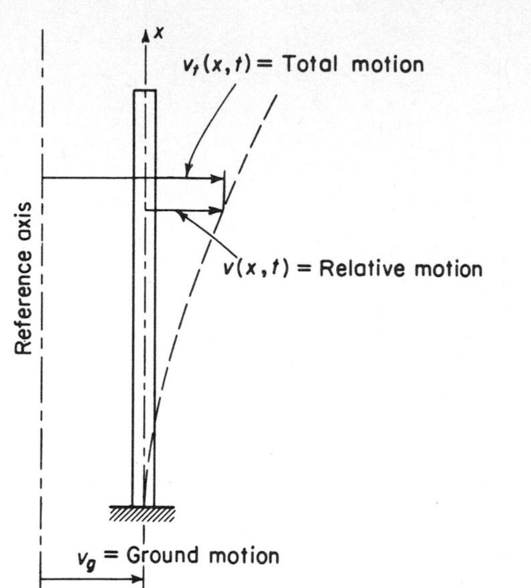

Fig. 12.8. Earthquake excitation of generalized coordinate system.

derived from the inertia force resulting from the total acceleration acting on the structure. Thus the inertia force is given by

$$f_I(x) = \mu(x)\ddot{v}_t(x) \tag{12.28}$$

But noting that the total acceleration is made up of the ground acceleration plus the relative acceleration,

$$\ddot{v}_t(x) = \ddot{v}_g + \ddot{v}(x) \tag{12.29}$$

This can be written

$$f_I(x) = \mu(x)\ddot{v}_g + \mu(x)\ddot{v}(x) \tag{12.30}$$

Then the internal virtual work associated with the inertia forces is given by

$$\begin{aligned} \delta \bar{W}_{I(\text{inertia})} &= \ddot{Y}\,\partial Y \int_0^L \mu(x)[\psi(x)]^2\,dx \\ &\quad + \ddot{v}_g\,\delta Y \int_0^L \mu(x)\psi(x)\,dx \end{aligned} \tag{12.31}$$

The second term in this expression represents the work done by the effective force; in other words, the effective generalized earthquake force may be expressed

$$P^*_{\text{eff}}(t) = -\ddot{v}_g(t) \int_0^L \mu(x)\psi(x)\,dx = -\ddot{v}_g(t)\mathcal{L} \tag{12.32}$$

where the earthquake participation factor $\mathcal{L}$ is defined as follows:

$$\mathcal{L} \equiv \int_0^L \mu(x)\psi(x)\,dx \tag{12.33}$$

Hence, the final equation of motion for a generalized coordinate system subjected to earthquake excitation may be written

$$M^*\ddot{Y} + K^*Y = P^*_{\text{eff}}(t) = -\ddot{v}_g(t)\mathcal{L} \tag{12.34}$$

in which the generalized coordinate properties are defined by Eqs. 12.33 and 12.26.

If damping were included in Eq. 12.34, the equivalent equation of motion would be

$$M^*\ddot{Y} + C^*\dot{Y} + K^*Y = P^*_{\text{eff}}(t) \tag{12.35}$$

in which C^* represents the generalized damping coefficient.

Dividing Eq. 12.35 by the generalized mass, it may be expressed

$$\ddot{Y} + 2\xi\omega\dot{Y} + \omega^2 Y = \frac{P^*_{\text{eff}}(t)}{M^*} \tag{12.36}$$

in which the symbol ξ is called the damping ratio (i.e., the ratio of the given damping coefficient to critical damping)† and is defined as follows:

$$\xi = \frac{C^*}{2M^*\omega} \tag{12.37}$$

while the symbol ω represents the circular frequency of the free vibration motion of the system and is defined as

$$\omega = \left(\frac{K^*}{M^*}\right)^{1/2} \tag{12.38}$$

12.2.1.3 Example of formulation. As an example of the formulation of the equation of motion of a system by the generalized coordinate approach, consider the cantilever column shown in Fig. 12.8. For the purpose of this discussion, it will be assumed that the flexural stiffness EI and the mass per unit length μ of the column are uniform over its full height.

The first step in the generalized coordinate formulation is the selection of an appropriate deflected shape, consistent with the prescribed boundary conditions of the system. In this case the essential geometric boundary conditions are imposed by the fixed base attachment, i.e., there must be no displacement or rotation of the elastic axis at $x = 0$. (The force conditions of zero shear and moment at the upper end are not essential and need not be satisfied by the assumed shape.) A reasonable shape which satisfies the geometric boundary conditions is given by the function

$$\psi(x) = 1 - \cos\frac{\pi x}{2L}$$

Thus, the assumed displacements may be written

$$v(x) = \left(1 - \cos\frac{\pi x}{2L}\right) Y(t)$$

where the generalized coordinate $Y(t)$ represents the displacement at the top of the column.

Using Eqs. 12.26 and 12.33, the generalized properties of this system are now found to be

$$M^* = \mu \int_0^L \left(1 - \cos\frac{\pi x}{2L}\right)^2 dx = 0.226\mu L$$

†Critical damping is defined as the least damping coefficient for which the free response of the structure is nonvibratory, i.e., for which it returns to the static position without oscillation after any excitation.

$$K^* = EI\int_0^L \left[\frac{\pi^2}{4L^2}\cos\frac{\pi x}{2L}\right] dx = \frac{\pi^4}{32}\frac{EI}{L^3} = 3.04\frac{EI}{L^3}$$

$$\mathscr{L} = \mu\int_0^L \left[1 - \cos\frac{\pi x}{2L}\right] dx = 0.364\mu L$$

Hence, the equation of motion of this undamped system subjected to a given ground acceleration $\ddot{v}_g(t)$ may be written (see Eq. 12.34)

$$(0.226\mu L)\ddot{Y} + \left(3.04\frac{EI}{L^3}\right)Y = (-0.364\mu L)\ddot{v}_g(t)$$

Alternatively, if the system is assumed to have a damping ratio of 10% critical, the equation of motion may be expressed in the form of Eq. 12.36. The circular frequency of the system is given by Eq. 12.38, as follows:

$$\omega = \left(\frac{3.04EI/L^3}{0.226\mu L}\right)^{1/2} = 3.66\sqrt{\frac{EI}{\mu L^3}}$$

Thus, the equation of motion becomes

$$\ddot{Y} + \left(0.732\sqrt{\frac{EI}{\mu L^3}}\right)\dot{Y} + \left(13.43\frac{EI}{\mu L^3}\right)Y = -1.61\ddot{v}_g(t)$$

12.2.2 Solutions for the Equations of Motion—Undamped

The preceding discussion demonstrates that the equations of motion of a single-degree-of-freedom system have the same form whether the mathematical model is based upon a lumped mass idealization or on the generalized coordinate (assumed shape) concept. It is convenient, therefore, to visualize the system as the simple frame of Fig. 12.5 in evaluating its general motion response, but it must be remembered that the analysis applies equally to any structure the deflections of which may be assumed to be represented by prescribed shape function $\psi(x)$.

The response analysis of the system will be discussed by considering first the undamped system ($C = 0$) and then indicating how damping would alter the response.

12.2.2.1 Free vibrations. The simplest form of dynamic response of any structure is the condition of free vibration that may exist in the absence of any applied excitation ($P = 0$). The equation of motion of the undamped system in this case may be written (from Eq. 12.36)

$$\ddot{v} + \omega^2 v = 0 \tag{12.39}$$

where $\omega^2 = K/M$. The solution of this differential equation is

$$v = A\sin\omega t + B\cos\omega t \tag{12.40}$$

in which the constants A and B, which define the amplitude of the free vibration response, depend upon the velocity $\dot{v}_0$ and the displacement v_0 with which the motion was initiated. Expressing the constants in terms of these

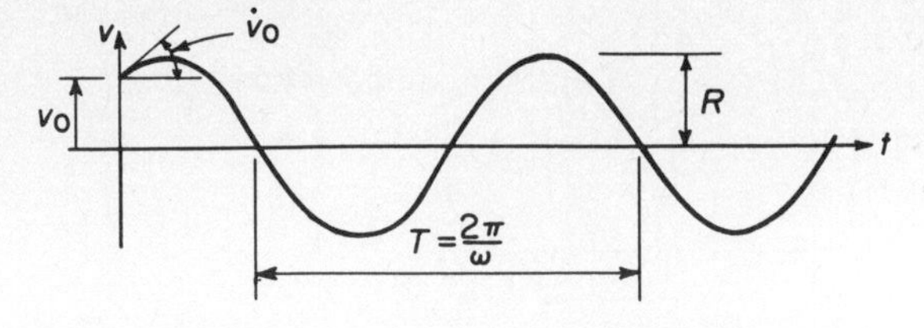

Fig. 12.9. Basic properties of free vibration motion.

initial conditions, the motion may be written

$$v = \frac{\dot{v}_0}{\omega}\sin\omega t + v_0\cos\omega t \tag{12.41}$$

This simple harmonic motion is depicted graphically in Fig. 12.9. It will be noted that the initial conditions define the displacement and slope of the response graph at time $t = 0$. Also of interest is the fact that the period T of the motion is given by

$$T = \frac{2\pi}{\omega} \tag{12.42}$$

while its amplitude R is

$$R = \sqrt{\left(\frac{\dot{v}_0}{\omega}\right)^2 + v_0^2} \tag{12.43}$$

The cyclic frequency of the motion f is the reciprocal of the period and is given by

$$f = \frac{1}{T} = \frac{\omega}{2\pi} \tag{12.44}$$

12.2.2.2 Response to impulse. An approximate expression for the response to a very short duration loading may be derived easily from the free vibration response expression. Consider a loading history $P(t)$ as shown in Fig. 12.10. If the duration of the loading t_1 is very short compared with the period of vibration of the structure T (i.e., $t_1 \ll T$) it may be assumed that

Fig. 12.10. Response to short duration impulsive load.

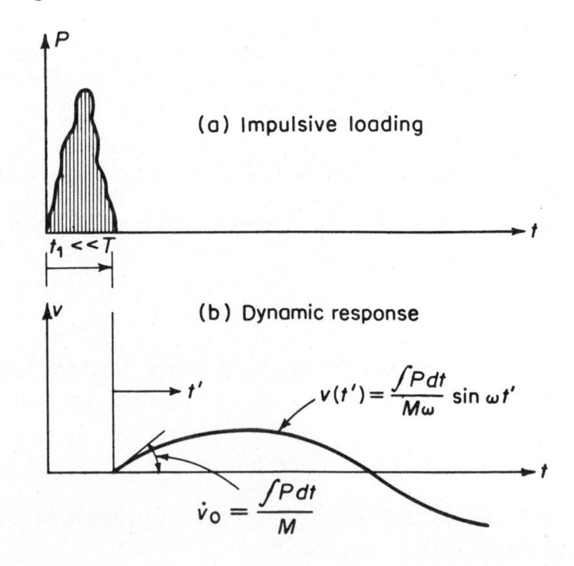

there is no significant change of displacement during the time while the load is applied, only a change of velocity $\Delta\dot{v}$. This change of velocity may be evaluated from the impulse-momentum relationship

$$M\Delta\dot{v} = \int P\,dt \tag{12.45}$$

Thus the initial conditions for the free vibration which takes place after the end of the loading are

$$v_0 \doteq 0 \qquad \dot{v}_0 \doteq \frac{\int P\,dt}{M}$$

and introducing these into Eq. 12.41 gives the response expression

$$v \doteq \frac{\int P\,dt}{M\omega} \sin \omega t \tag{12.46}$$

This response is depicted graphically in Fig. 12.10.

12.2.2.3 Example of impulse response analysis. As an illustration of the use of Eq. 12.46, consider the system shown in Fig. 12.11 and its prescribed impulsive loading. In this system the period of vibration is found to be

$$T = 2\sqrt{\frac{M}{K}} = 2\pi\sqrt{\frac{2000}{51.1(386)}} = 2 \text{ sec}$$

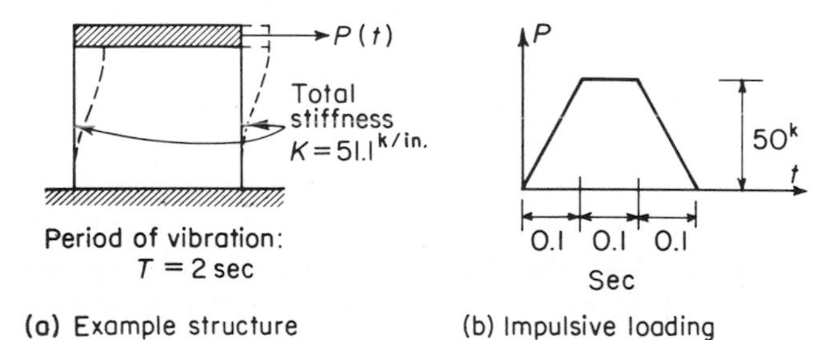

Fig. 12.11. Example of impulsive load response analysis.

where the acceleration of gravity is taken to be 386 in./sec². Thus the duration of the impulse t is quite short compared with the period. The magnitude of the impulse in this case is found to be

$$\int_0^{t_1} P\,dt = 10^{k\text{-sec}}$$

and, therefore, by Eq. 12.46 the response is given by

$$v = \frac{(10)(386)(2)}{2000(2\pi)} \sin \omega t = 0.614 \text{ in.} \sin \omega t$$

From this result it is seen that the maximum displacement is 0.614 in. and the maximum base shear force Q in the structure is given by

$$Q = kv_{\max} = 51.1(0.614) = 31.4^k$$

It is important to note that approximately this same response would be developed by any short duration impulse of $10^{k\text{-sec}}$ amplitude, regardless of the form of the loading history.

12.2.2.4 Arbitrary loading (Duhamel integral). The foregoing analysis of the response to a short duration impulse provides a convenient means for deriving an expression for the response to an arbitary loading history. Assume the given loading to be divided into a succession of very short duration impulses, as shown in Fig. 12.12, and consider one of these impulses, which ends at time τ after the beginning of the dynamic loading and is of duration $d\tau$, as shown. The magnitude of this differential impulse is $P(\tau)\,d\tau$, and the differential response which it produces may be expressed (see Eq. 12.46) as follows:

$$dv = \frac{P(\tau)}{M\omega} \sin \omega t'\,d\tau \tag{12.47}$$

where t' is the time after the end of the impulse. This may be expressed more conveniently in terms of the time after initiation of the loading, as follows:

$$t' = t - \tau \tag{12.48}$$

which leads to the following form of Eq. 12.47:

$$dv = \frac{P(\tau)}{M\omega} \sin \omega(t - \tau)\,d\tau \tag{12.49}$$

Equation 12.49 represents the portion of the dynamic response resulting from the differential impulse applied at time τ. The total response produced by the complete loading history may be obtained by superposing the effects of all the differential impulses making up the loading. Thus, by integration of Eq. 12.49, the displace-

Fig. 12.12. Response to differential impulse.

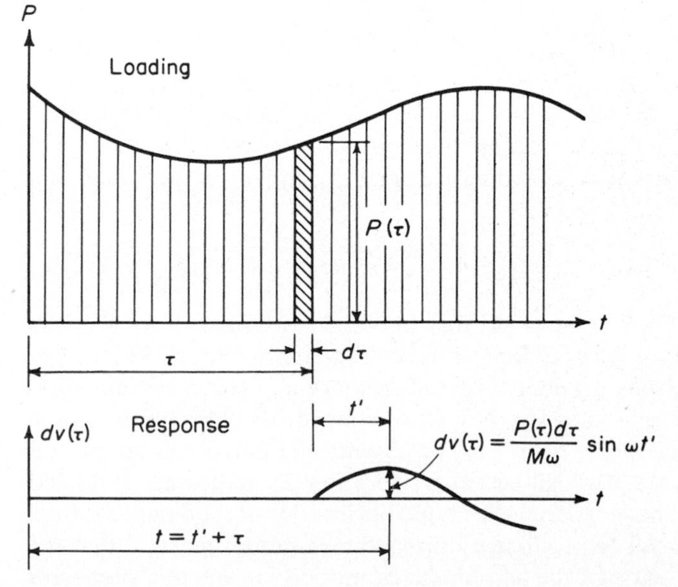

ment at time t is

$$v(t) = \int_0^t \frac{P(\tau)}{M\omega} \omega(t - \tau)\, d\tau \tag{12.50}$$

Equation 12.50, which is known as the Duhamel integral, may be used to evaluate the response of any elastic system to any given loading history; because its derivation is based on the principle of superposition it applies only to linear structures. It should be noted that it is an exact expression; the approximations that were introduced in the derivation of the impulse response expression (Eq. 12.46) became exact as the duration of the impulsive load approached zero.

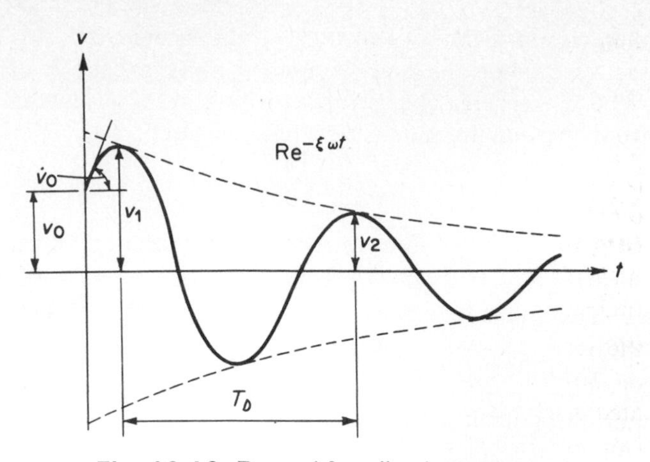

Fig. 12.13. Damped free vibration response.

12.2.3 Solution of the Equations of Motion—Damped

The undamped systems considered in the foregoing discussion are of little practical significance in that they include no energy loss mechanism. Actual structures continually lose energy during motion, and this energy loss (or damping) has an important effect in reducing the response to earthquake excitation. Damping effects may easily be incorporated into the dynamic response equation by consideration of the same three situations discussed above for the undamped system.

12.2.3.1 Free vibrations. From Eq. 12.36 it is evident that the equation of motion of a damped system in free vibration may be written

$$\ddot{v} + 2\xi\omega\dot{v} + \omega^2 v = 0 \tag{12.51}$$

The solution of this equation (for reasonable amounts of damping) may be written

$$v = e^{-\xi\omega t}[A \sin \omega_D t + B \cos \omega_D t] \tag{12.52}$$

in which

$$\omega_D = \omega\sqrt{1 - \xi^2} \tag{12.53}$$

is called the damped circular frequency. The constants A and B depend upon the initial conditions with which the system was set into motion, as with the undamped case. Expressing the constants in terms of the initial velocity and displacement, $\dot{v}_0$ and v_0 respectively, Eq. 12.52 becomes

$$v = e^{-\xi\omega t}\left(\frac{\dot{v}_0 + \xi\omega v_0}{\omega_D} \sin \omega_D t + v_0 \cos \omega_D t\right) \tag{12.54}$$

It may be noted that Eq. 12.54 is similar in form to the undamped free vibration equation (Eq. 12.40) except that the simple harmonic motion is subject to an exponential decay. A typical damped free vibration response is depicted in Fig. 12.13, with the initial conditions and the damped period of vibration T_D indicated. It should be noted that the period of vibration of the damped system does not change during the response and also that the ratio of the amplitudes of motion for any two successive cycles is constant. The natural logarithm of this ratio is called the logarithmic decrement δ and is related to the damping ratio ξ as follows:

$$\delta = \log_e \frac{v_1}{v_2} = \log_e \frac{v_n}{v_{n+1}} = 2\pi\frac{\omega}{\omega_D}\xi \tag{12.55}$$

Thus, a convenient means of evaluating the damping in any single-degree system is to set it into free vibration, measure the amplitude of motion in two successive cycles, and solve from Eq. 12.55 as follows:

$$\xi = \frac{\delta}{2\pi}\frac{\omega}{\omega_D} \doteq \frac{\delta}{2\pi} \tag{12.56}$$

In most typical structures the damping is small enough that the damped frequency is essentially the same as the undamped frequency and the approximation indicated by the right-hand term of Eq. 12.56 is satisfactory.

12.2.3.2 Response to impulse. An approximate expression for the response of a damped system to a short duration impulse may be derived exactly as for the undamped system. The initial conditions generated by the impulsive load may be assumed to be the same as before, i.e.,

$$v_0 \doteq 0 \qquad \dot{v}_0 \doteq \frac{\int P\, dt}{M}$$

Thus, the free vibration response to these initial conditions is found from Eq. 12.54 to be

$$v = \frac{\int P\, dt}{M\omega_D} e^{-\xi\omega t} \sin \omega_D t \tag{12.57}$$

12.2.3.3 Duhamel integral. The response of a damped system to arbitrary loading now may be written directly from Eq. 12.57, considering the complete loading history to be represented as a succession of differential impulses, as was done for the undamped system. By direct analogy with the derivation of Eq. 12.50 it may be seen that the Duhamel integral for a damped system is as follows:

$$v(t) = \int_0^t \frac{P(\tau)}{M\omega_D} e^{-\xi\omega(t-\tau)} \sin \omega_D(t - \tau)\, d\tau \quad (12.58)$$

12.2.4 Numerical Evaluation of the Duhamel Integral

The Duhamel integral (Eq. 12.50 or 12.58) expresses the response of a single-degree-of-freedom structure to any arbitrary loading history. The integral can be evaluated analytically only for dynamic loadings $P(\tau)$, which can be expressed mathematically. To evaluate the response to any arbitrary loading history, such as the effective loading resulting from earthquake excitation, it is necessary to apply a numerical integration process. Many numerical integration procedures are available for this purpose; however, the following method has been found to provide a simple, convenient analysis technique. Essentially the same procedure may be applied for either damped or undamped systems, only a slight modification being required to account for the damping.

12.2.4.1 Undamped. Using the following trigonometric identity,

$$\sin(\omega t - \omega\tau) = \sin \omega t \cos \omega\tau - \cos \omega t \sin \omega\tau$$

Eq. 12.50 can be rewritten in the following form:

$$v(t) = \frac{1}{M\omega}\left[\sin \omega t \int_0^t P(\tau) \cos \omega\tau\, d\tau - \cos \omega t \int_0^t P(\tau) \sin \omega\tau\, d\tau\right] \quad (12.59)$$

in which it will be noted that the trigonometric functions of t have been removed from the integral expressions. Denoting the two integrals respectively by $\bar{A}(t)$ and $\bar{B}(t)$, Eq. 12.59 may now be written

$$v(t) = \frac{1}{M\omega}[\bar{A}(t) \sin \omega t - \bar{B}(t) \cos \omega t] \quad (12.60)$$

The analysis of the dynamic response is thus reduced to the evaluation of these two integrals, and any convenient numerical process may be used for this purpose. In general it is required to compute the response at a succession of times during the course of the loading, and in this case it is easiest to proceed incrementally from one time to the next. If the loading history is divided into equal time increments $\Delta\tau$, the integrals for time t may be evaluated from the results at the previous time $t - \Delta\tau$ by simple summation, as follows:

$$\begin{aligned}\bar{A}(t) &= \bar{A}(t - \Delta\tau) + \Delta\tau P(t - \Delta\tau) \cos (t - \Delta\tau)\\ \bar{B}(t) &= \bar{B}(t - \Delta\tau) + \Delta\tau P(t - \Delta\tau) \sin (t - \Delta\tau)\end{aligned} \quad (12.61)$$

The response at time t is then given by substituting Eq. 12.61 into Eq. 12.60.

12.2.4.2 Damped. The corresponding analysis for a damped single-degree-degree-of-freedom system is given by

$$v(t) = \frac{1}{Mw_D}[\bar{A}(t) \sin \omega_D t - \bar{B}(t) \cos \omega_D t] \quad (12.62)$$

in which

$$\begin{aligned}\bar{A}(t) &= \bar{A}(t - \Delta\tau)e^{-\xi\omega\Delta\tau} + \Delta\tau P(t - \Delta\tau) \cos \omega_D(t - \Delta\tau)\\ \bar{B}(t) &= \bar{B}(t - \Delta\tau)e^{-\xi\omega\Delta\tau} + \Delta\tau P(t - \Delta\tau) \sin \omega_D(t - \Delta\tau)\end{aligned} \quad (12.63)$$

The only difference in this case (except for the usually negligible difference between the damped and undamped frequencies) is the exponential decay terms in Eq. 12.63, which cause a continual loss of energy and reduction of the response.

The summation procedure indicated in Eqs. 12.61 and 12.63 is the simplest numerical procedure possible. Greater computational efficiency can be obtained by using more refined procedures, such as the trapezoidal rule or Simpson's rule, and such refinements are generally recommended. Only slight modifications are required in the equations to represent the refined procedures, but they are omitted here for reasons of brevity.

12.2.5 Earthquake Response

The earthquake response of a general single-degree-of-freedom system may be obtained by application of the Duhamel integral; it is necessary only to recognize that the effective earthquake loading is given by the product of the mass and the ground acceleration—Eq. 12.11 (or by Eq. 12.32 using the generalized coordinate approach). Introducing Eq. 12.11 into Eq. 12.58, and neglecting the slight difference between the damped and undamped frequencies, the Duhamel integral expression for earthquake response of a damped structure may be written

$$v(t) = \frac{1}{\omega}\int_0^t \ddot{v}_g(\tau)e^{-\xi\omega(t-\tau)} \sin \omega(t - \tau)\, d\tau \quad (12.64)$$

Denoting the integral in Eq. 12.64 by the symbol $V(t)$, i.e.,

$$V(t) \equiv \int_0^t \ddot{v}_g(\tau)e^{-\xi\omega(t-\tau)} \sin \omega(t - \tau)\, d\tau \quad (12.65)$$

the earthquake response of a lumped mass system becomes

$$v(t) = \frac{1}{\omega} V(t) \quad (12.66a)$$

The corresponding expression for a generalized coordinate system is

$$v(x, t) = \psi(x) Y(t) = \psi(x)\frac{\mathscr{L}}{M^*\omega} V(t) \quad (12.66b)$$

in which $\mathscr{L}$ and M^* are defined by Eqs. 12.33 and 12.26a, respectively.

Equations 12.66a and b express the deflection

response to any given ground motion input for which the earthquake response function $V(t)$ has been evaluated. It also is of interest to determine the forces developed in the structure during the earthquake. In general these may be found most reliably in terms of the effective inertia forces, which depend upon the effective acceleration $\ddot{v}_e(t)$. The effective acceleration is defined as the product of the frequency squared and the displacement:

$$\ddot{v}_e(t) \equiv \omega^2 v(t) \tag{12.67}$$

The effective earthquake force is then given as the product of the mass and the effective acceleration:

$$Q(t) = M\ddot{v}_e(t) = M\omega^2 v(t)$$

Introducing this into Eq. 12.66a yields

$$Q(t) = M\omega V(t) \tag{12.68a}$$

Following the corresponding procedure for a generalized coordinate system, for which the effective acceleration is given by $\ddot{Y}_e(t) = \omega^2 Y(t)$, the effective inertia force per unit of length along the structure is given by

$$q(x, t) = \mu(x)\psi(x)\ddot{Y}_e(t) = \mu(x)\psi(x)\frac{\mathscr{L}}{M^*}\omega V(t) \tag{12.68b}$$

The effective earthquake force defined by Eq. 12.68b may be used exactly as a static applied load to calculate any desired internal force component in the structure, such as bending moment or shear force. Only the effective force is to be considered in such stress analysis operations—the influence of any externally applied dynamic loading which might be acting at this time is included in the effective force. Of particular interest in earthquake response analysis is the shear force developed at the base of the structure. This base shear force Q, which is representative of the total earthquake force acting in the structure, is given by the integral of all effective forces acting over the height of the structure:

$$Q(t) = \int_0^L q(x, t)\, dx = \frac{\mathscr{L}^2}{M^*}\omega V(t) \tag{12.68c}$$

12.2.6 Earthquake Response Spectra

The response at any time t of any single-degree system to earthquake excitation is defined completely by Eqs. 12.66 and 12.68; however, to obtain the entire history of forces and displacements during the earthquake obviously is a tedious computational problem. In most practical problems it is sufficient to determine only the maximum response quantities. Examination of Eqs. 12.66 and 12.68 reveals that the maximum force and displacement responses both will be given by introducing the maximum value of the response function $V(t)$ into the equations. This maximum value of the function is called the *spectral velocity* (or more accurately the spectral *pseudo*-velocity because it is not exactly equal to the maximum velocity in a damped system); it is denoted by the symbol S_v. Thus, spectral velocity,

$$S_v \equiv V_{max} \equiv \left[\int_0^t \ddot{v}_g(\tau)e^{-\xi\omega(t-\tau)} \sin \omega(t-\tau)\, d\tau\right]_{max} \tag{12.69}$$

It will be noted by examination of Eqs. 12.66a and b that the maximum displacement depends on the spectral velocity divided by the circular frequency. This ratio is called the *spectral displacement* and is denoted by S_d. Thus, spectral displacement,

$$S_d \equiv \frac{S_v}{\omega} \tag{12.70}$$

Similarly, the maximum forces are seen from Eqs. 12.68 to depend on the mass and the product of the circular frequency and the spectral velocity. This quantity is called the *spectral acceleration* (or more properly the spectral pseudo-acceleration because it is not exactly the peak acceleration value in general); it is denoted by S_a. Thus, spectral acceleration,

$$S_a \equiv \omega S_v \tag{12.71}$$

By the introduction of Eq. 12.70 into Eqs. 12.66a and b it is evident that the maximum earthquake displacement response of a structure may be written

$$V_{max} = S_d \tag{12.72a}$$

or

$$v(x)_{max} = \psi(x)Y_{max} = \frac{\mathscr{L}}{M^*}\psi(x)S_d \tag{12.72b}$$

Similarly introducing Eq. 12.71 into Eqs. 12.68a–c leads to the maximum effective earthquake forces:

$$Q_{max} = MS_a \tag{12.73a}$$

or

$$q_{max}(x) = \mu(x)\psi(x)\frac{\mathscr{L}}{M^*}S_a \tag{12.73b}$$

The corresponding maximum base shear force may be seen from Eq. 12.68c to be given by

$$Q_{max} = \frac{\mathscr{L}^2}{M^*}S_a \tag{12.73c}$$

From Eq. 12.69 it is apparent that the spectral velocity depends on three factors: (1) the characteristics of the ground motion $\ddot{v}_g(\tau)$, (2) the damping ratio of the structure ξ, and (3) the circular frequency of the structure ω. Thus, for any given earthquake input, and for any specified structure damping ratio, it is possible to determine the spectral velocity as a function of structure frequency or period $T = 2\pi/\omega$. The result of such an analysis is presented graphically in Fig. 12.14. Each of the curves on this graph was derived by evaluating the spectral velocity resulting from the earthquake acceleration history recorded at El Centro in May 1940, considering the indicated damping ratio for a succession of different periods of vibration in the range of interest. The analysis for each combination of damping and period gave a

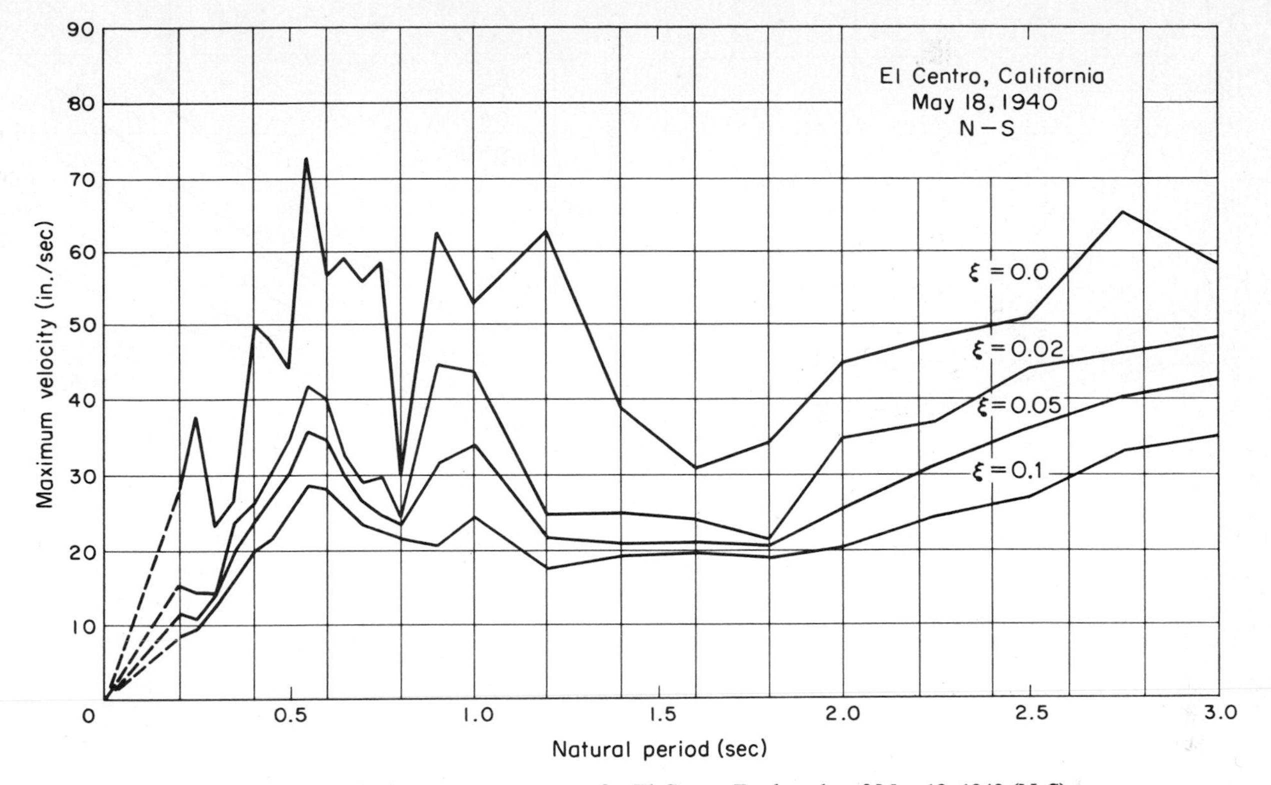

Fig. 12.14. Velocity response spectrum for El Centro Earthquake of May 18, 1940 (N–S).

single point on one curve. The complete graphs were obtained by connecting the sequence of computed points appropriately by straight lines. The graph presented in Fig. 12.14 is called the velocity response spectrum of the 1940 El Centro Earthquake motion.

The sharp peaks and valleys in the response spectrum curves shown in Fig. 12.14 are due to local resonances in the ground motion record. Such irregularities are not of fundamental significance and may be smoothed out by averaging the response spectra of a number of different earthquake records, after normalizing them to a standard intensity level. An average velocity spectrum for ground motions adjusted to the 1940 El Centro Earthquake intensity is shown in Fig. 12.15. Multiplying these average velocity response spectra by the circular frequency yields the corresponding set of acceleration response spectra, as shown in Fig. 12.16, while dividing by the circular frequency leads to the displacement response spectra of Fig. 12.17. The set of three graphs shown in Figs. 12.15 to 12.17 is a complete representation of the structural response resulting from this average earthquake motion. Because of the simple relationships between the three response quantities represented in these graphs, however, it is possible to present them all in a single graph plotted with log scales on each axis, as shown in Fig. 12.18. Each line on this graph represents velocity, acceleration, and displacement response; it is necessary only to read the values of interest from the appropriate

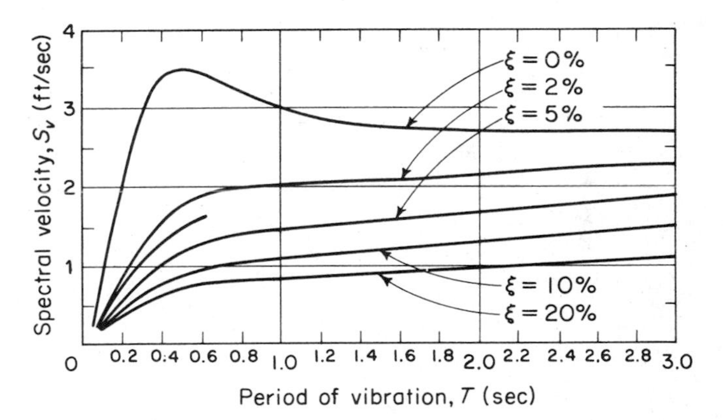

Fig. 12.15. Average velocity response spectrums, 1940 El Centro intensity. (From *U.S. Atomic Energy Commission Report TID-7024*, August 1963.)

scales, the acceleration and displacement being read from the diagonal lines. (It should be noted that the response spectra of Fig. 12.18 are plotted for an earthquake of lower intensity than the 1940 El Centro Earthquake.)

12.2.7 Examples of Earthquake Response Analysis

Earthquake response spectra such as those shown in Figs. 12.14 through 12.18 provide a direct measure of the

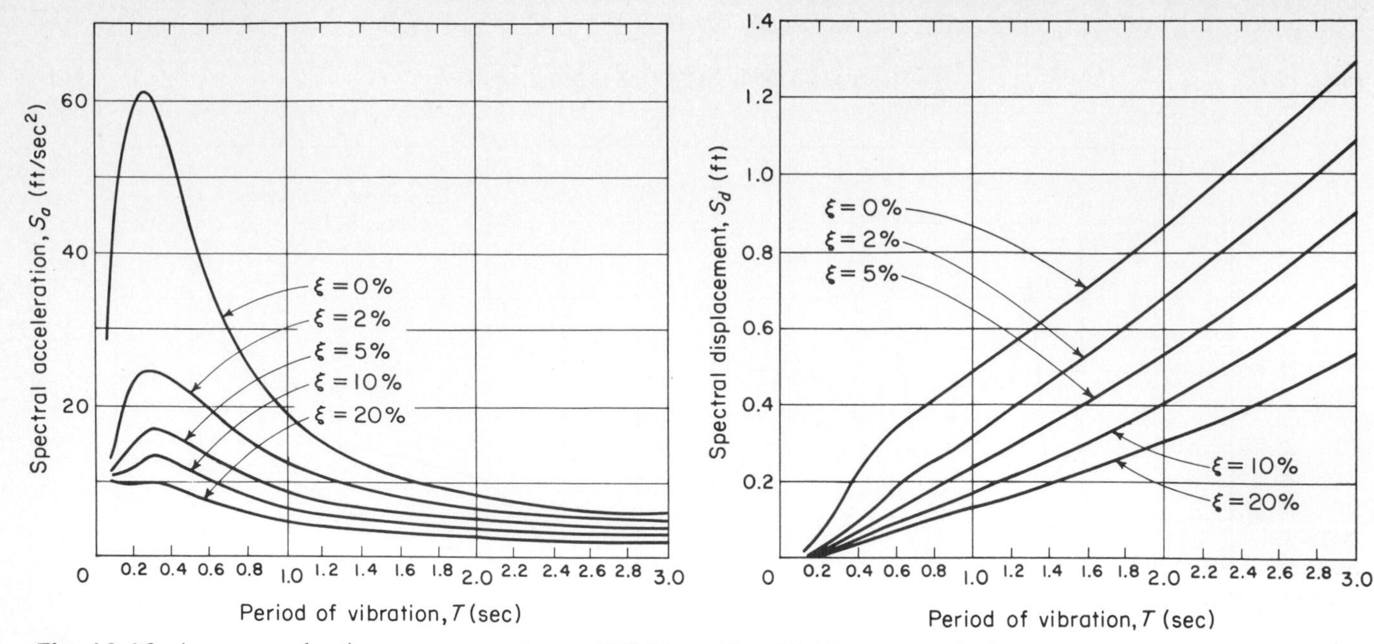

Fig. 12.16. Average acceleration response spectrums, 1940 El Centro intensity. (From *U.S. Atomic Energy Commission Report TID-7024*, August 1963.)

Fig. 12.17. Average displacement response spectrum, 1940 El Centro intensity. (From *U.S. Atomic Energy Commission Report TID-7024*, August 1963.)

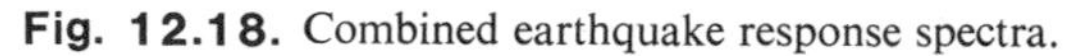

Fig. 12.18. Combined earthquake response spectra.

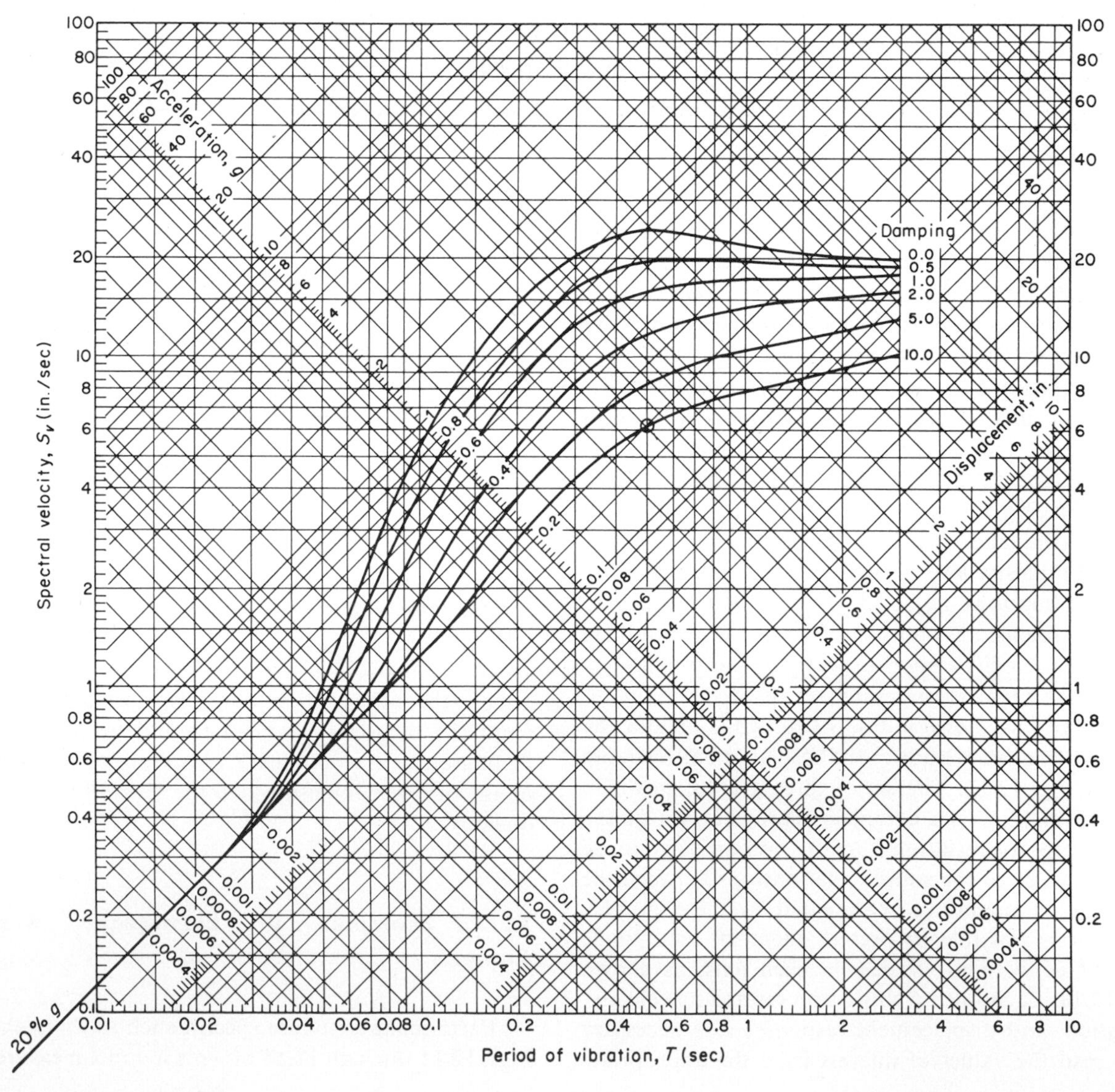

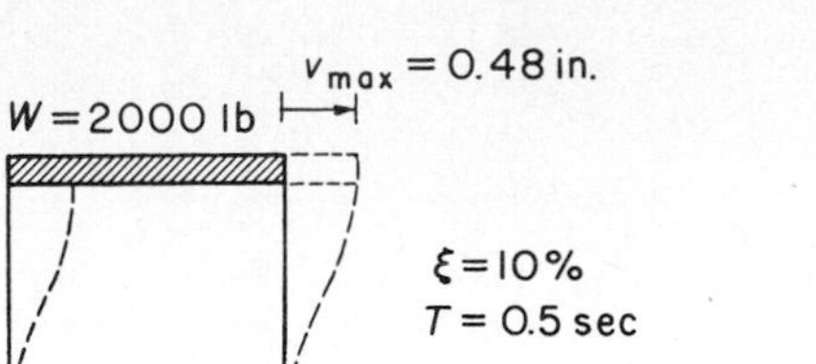

Fig. 12.19. Example, single-story frame.

response of any single-degree-of-freedom system to the earthquake motion they represent. To demonstrate how such response spectra are used in practice, three example analyses will be made using the spectral curves of Fig. 12.18. Each example structure will be assumed to have the same period of vibration, $T = 0.5$ sec, and the same damping ratio, $\xi = 10\%$. For these values, Fig. 12.18 gives the following spectral values: $S_d = 0.48$ in., $S_v = 6.0$ in./sec, $S_a = 76.0$ in./sec² ($\doteq 20\%$g).

12.2.7.1 Lumped mass system. The maximum response of the single-story frame shown in Fig. 12.19 is given directly by Eqs. 12.72a and 12.73a, with the following results:

$$v_{max} = S_d = 0.48 \text{ in.}$$

$$Q_{max} = MS_a = \frac{S_a}{g}W$$

$$= \frac{76.0}{386}(2000) = 394 \text{ lb}$$

12.2.7.2 Generalized coordinate system. The cantilever column of Fig. 12.20 is the same structure shown in Fig. 12.8, in which the deflected shape was assumed to be a cosine curve. Its generalized mass and earthquake participation factor were evaluated previously. The maximum displacement response (Fig. 12.20a) is given in this case by Eq. 12.72b as follows:

$$v_{max}(x) = \frac{\mathscr{L}}{M^*}S_d\psi(x) = \frac{0.364}{0.226}(0.48)\psi(x)$$

$$= 0.77 \text{ in. } \psi(x)$$

Thus the tip displacement is 0.77 in., and the displacements below vary as in a cosine curve. The maximum effective forces acting on the column are given by Eq. 12.73b,

$$q_{max}(x) = \frac{\mathscr{L}}{M^*}S_a\mu(x)\psi(x) = \frac{0.364}{0.226}\frac{76.0}{386}w\psi(x)$$

$$= 31.7\% \, w\psi(x)$$

where w is the weight per unit length of the column.

It is of interest to note in this case that the maximum base shear force is given by Eq. 12.73c,

$$Q_{max} = \frac{\mathscr{L}^2}{M^*}S_a = \frac{(0.364)^2}{0.226}\frac{76.0}{386}W$$

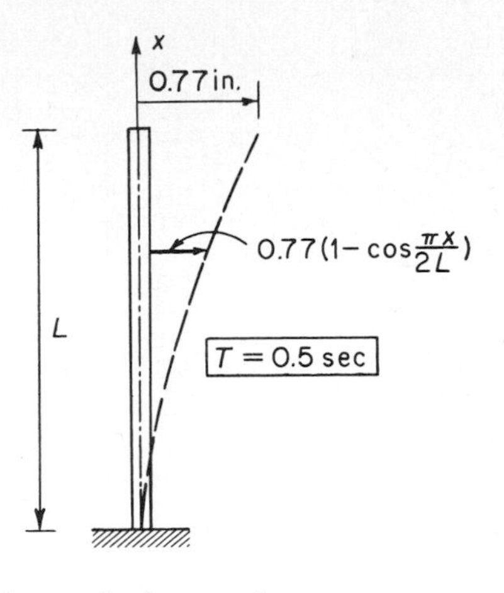

(a) Maximum displacement response

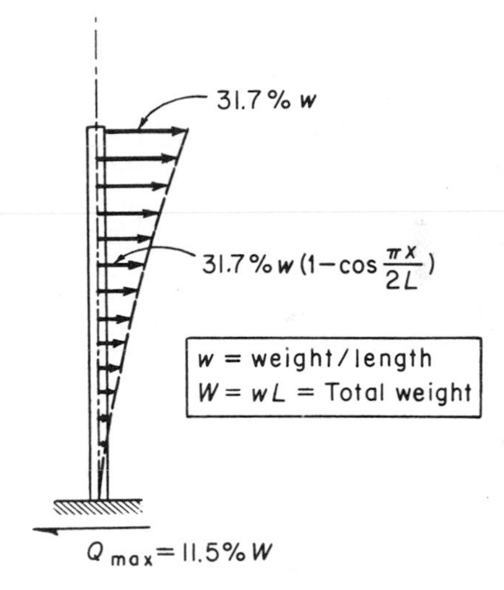

(b) Maximum earthquake forces

Fig. 12.20. Example, cantilever column.

$$= 11.5\% \, W$$

where W is the total weight of the column.

The effective inertia force can be found alternatively by distributing the base shear force over the height of the structure:

$$q_{max}(x) = Q_{max}\frac{\psi(x)\mu(x)}{\mathscr{L}} = 11.5\%\frac{w}{0.364}\psi(x)$$

$$= 31.7\% \, w\psi(x)$$

12.2.7.3 Multistory building. A typical multistory building is shown in Fig. 12.21a. As is customary in the dynamic analysis of buildings, the entire mass has been assumed to be concentrated in the floor slabs. In a complete analysis of this structure, the lateral motion of each floor slab would be an independent degree of freedom. Thus, the five-story building would have five degrees of freedom.

An approximate single-degree-of-freedom analysis, however, can be made by the generalized coordinate

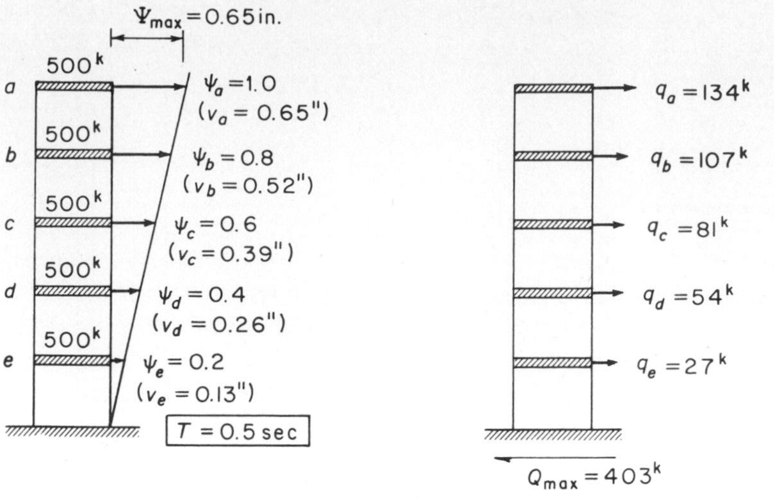

Fig. 12.21. Example, single-degree analysis of multistory building.

approach, i.e., by assuming that the lateral displacements are of a specified form. A reasonable assumption for a typical building frame is that the displacements increase linearly with height, i.e., $\psi(x) = x/L$. For this lumped mass system, the generalized mass and earthquake participation factor are given by summations that are equivalent to the integrals of Eqs. 12.26a and 12.33:

$$M^* = \sum_i M_i\phi_i^2 = \frac{1100^k}{g}$$

$$\mathscr{L} = \sum_i M_i\phi_i = \frac{1500^k}{g}$$

Thus the maximum earthquake deflection is given by Eq. 12.72b:

$$v_{i\max} = \frac{\mathscr{L}}{M^*}S_d\psi_i = \frac{1500}{1100}(0.48)\psi_i = 0.65 \text{ in. } \psi_i$$

The maximum base shear force is given by Eq. 12.73c:

$$Q_{\max} = \frac{\mathscr{L}^2}{M^*}S_a = \frac{(1500)^2}{1100}\frac{76.0}{386} = 403^k$$

Then the forces at the various story levels may be obtained by distributing the base shear force:

$$q_i = Q_{\max}\frac{M_i\psi_i}{\mathscr{L}} = \frac{403}{1500}(500)\psi_i = 134.2\psi_i$$

It is of interest to note that the base shear force computed here is 16% of the total weight of the building. The Uniform Building Code seismic force provision would give the same triangular variation of the lateral forces over the height of the structure, but for Zone 3 the base shear coefficient S would be only

$$S = \frac{0.05}{\sqrt[3]{T}} = 6.3\%$$

Thus it is evident that the earthquake intensity for the average earthquake response spectra of Fig. 12.18 is significantly greater than the maximum Code design requirement.

12.3 MULTIDEGREE-OF-FREEDOM SYSTEMS

12.3.1 Need for Multidegree Analysis

The methods described in the preceding section may be used to evaluate the response of a single-degree-of-freedom structure to any specified earthquake motions. The results may be obtained in the form of a complete time history of the stresses or displacements, by means of numerical evaluation of the Duhamel integral, or the maximum stress or displacement may be obtained directly from the response spectra of the given earthquake input. However, the results of the analysis can be representative of the actual earthquake behavior of the structure only if its motions can be defined reliably by a single displacement coordinate.

This will be the case if the mass is essentially concentrated at a single point which is constrained to move in one direction or if the arrangement of the structure is such as to permit only a single mode of displacement. In general, however, the mass of the system will be distributed throughout the structure and will be capable of displacing in many independent patterns. The dynamic response of such systems generally can be expressed realistically only by means of a number of independent displacement coordinates, and the validity of a single-degree-of-freedom approximation can be demonstrated only by making a comparative study of a multidegree system.

As was mentioned in the introduction, the equations of motion of practical structures may be discretized either by a lumped mass idealization or by assuming the displacements to be expressed as the sum of a set of prescribed displacement shapes. Only the lumped mass approach will be described here for reasons of simplicity and brevity; however, the extension of the generalized coordinate approach to the multidegree-of-freedom case is straightforward and may be advantageous in certain special cases.

12.3.2 Equations of Motion

The three-story building shown in Fig. 12.22 will be used as an example of the multidegree-of-freedom analysis procedure. As is customary in the dynamic analysis of typical buildings, the mass of the structure is assumed to be concentrated at the floor levels and to be subject to lateral displacements only. Thus the dynamic behavior of this structure is completely defined by the three story

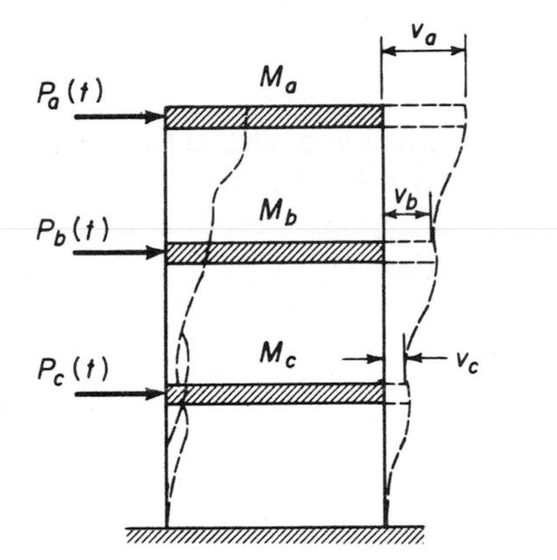

Fig. 12.22. Example, multidegree-of-freedom system.

displacements v_a, v_b, and v_c. The equation of motion of any story, then, is the equation of dynamic equilibrium of all of the forces acting on the story mass, including in general the inertia, damping, and elastic forces that result from the motion, and the externally applied force. The equations of equilibrium for the three stories might be written as follows (using symbols similar to the single-degree-of-freedom expressions):

$$\begin{aligned} F_{I_a} + F_{D_a} + F_{S_a} &= P_a(t) \\ F_{I_b} + F_{D_b} + F_{S_b} &= P_b(t) \\ F_{I_c} + F_{D_c} + F_{S_c} &= P_c(t) \end{aligned} \qquad (12.74)$$

For this lumped mass system, the inertia forces are given simply as the product of the story mass and the story acceleration:

$$\begin{aligned} F_{I_a} &= M_a \ddot{v}_a \\ F_{I_b} &= M_b \ddot{v}_b \\ F_{I_c} &= M_c \ddot{v}_c \end{aligned} \qquad (12.75)$$

which may be represented in matrix form as

$$\begin{Bmatrix} F_{I_a} \\ F_{I_b} \\ F_{I_c} \end{Bmatrix} = \begin{bmatrix} M_a & 0 & 0 \\ 0 & M_b & 0 \\ 0 & 0 & M_c \end{bmatrix} \begin{Bmatrix} \ddot{v}_a \\ \ddot{v}_b \\ \ddot{v}_c \end{Bmatrix} \qquad (12.76)$$

In a symbolic form that is applicable to systems having any number of degrees of freedom, Eq. 12.76 may be written

$$\mathbf{F}_I = \mathbf{M}\ddot{\mathbf{v}} \qquad (12.77)$$

in which $\mathbf{F}_I$ is the inertia force vector, $\ddot{\mathbf{v}}$ is the acceleration vector, and $\mathbf{M}$ is the mass matrix.

It is important to note in Eq. 12.76 that the mass matrix for a lumped mass system is of a diagonal form, i.e., the inertia force corresponding to any degree of freedom depends only on the acceleration in that degree of freedom. In general, the mass matrix for a multidegree-of-freedom system based on assumed generalized shape coordinates will not be diagonal. Therefore, it will introduce coupling between the displacement coordinates that will complicate the analysis process. The lack of mass coupling in the lumped mass system is a major advantage and is one of the principal reasons for adopting this discretization technique in practical analyses.

The elastic forces in Eq. 12.74 depend on the displacements of the system and may be expressed conveniently by means of stiffness influence coefficients:

$$\begin{aligned} F_{S_a} &= k_{aa}v_a + k_{ab}v_b + k_{ac}v_c \\ F_{S_b} &= k_{ba}v_a + k_{bb}v_b + k_{bc}v_c \\ F_{S_c} &= k_{ca}v_a + k_{cb}v_b + k_{cc}v_c \end{aligned} \qquad (12.78)$$

in which the general stiffness influence coefficient k_{ij} may be defined as the force corresponding to displacement coordinate i resulting from a unit displacement of coordinate j. In matrix form Eq. 12.78 may be expressed

$$\begin{Bmatrix} F_{S_a} \\ F_{S_b} \\ F_{S_c} \end{Bmatrix} = \begin{bmatrix} k_{aa} & k_{ab} & k_{ac} \\ k_{bb} & k_{bc} & k_{bc} \\ k_{ca} & k_{cb} & k_{cc} \end{bmatrix} \begin{Bmatrix} v_a \\ v_b \\ v_c \end{Bmatrix} \qquad (12.79)$$

or symbolically it becomes

$$\mathbf{F}_S = \mathbf{K}\mathbf{v} \qquad (12.80)$$

where $\mathbf{F}_S$ is the elastic force vector, $\mathbf{v}$ is the displacement vector, and $\mathbf{K}$ is the stiffness matrix of the structure. It has been indicated in Eq. 12.79 that the stiffness matrix $\mathbf{K}$ is not diagonal (generally): The off-diagonal terms show that elastic forces for a given coordinate depend on displacements of the other coordinates; thus there is stiffness coupling of the coordinates.

The calculation of the stiffness coefficients for any given structure is a standard problem of static structural analysis. It may be carried out by traditional procedures or by matrix methods. Stiffness matrices for large, complex systems may be obtained conveniently by digital computer analysis. The formulation of matrix structural analysis procedures is too broad a subject to be considered here, however. For the purpose of this discussion it will be assumed that the structure stiffness matrix $\mathbf{K}$ is available.

In principle, the damping forces in Eq. 12.74 could be expressed as the product of a set of damping influence coefficients multiplied by the velocities of the displacement degrees of freedom. Thus, by analogy with the expression for the elastic forces in Eq. 12.80, the damping forces could be written

$$\mathbf{F}_D = \mathbf{C}\dot{\mathbf{v}} \tag{12.81}$$

in which $\mathbf{F}_D$ is the damping force vector, $\dot{\mathbf{v}}$ is the velocity vector, and $\mathbf{C}$ is the damping matrix. In general, however, it is not practicable to evaluate the damping coefficients in the matrix $\mathbf{C}$, and damping usually is expressed in terms of damping ratios. Nevertheless, it is convenient to utilize Eq. 12.81 in the formulation of the equations of motion of a multidegree system.

Noting now that Eq. 12.74 may be expressed symbolically as

$$\mathbf{F}_I + \mathbf{F}_D + \mathbf{F}_S = \mathbf{P}(t) \tag{12.82}$$

in which $\mathbf{P}(t)$ represents the applied load vector, and substituting from Eqs. 12.77, 12.80, and 12.81, the equations of dynamic equilibrium may be written

$$\mathbf{M}\ddot{\mathbf{v}} + \mathbf{C}\dot{\mathbf{v}} + \mathbf{K}\mathbf{v} = \mathbf{P}(t) \tag{12.83}$$

Equation 12.83 represents the equations of motion of an arbitrary structural system having any number of degrees of freedom. The similarity of this matrix equation to the corresponding single-degree-of-freedom equation (Eq. 12.6) is noteworthy.

12.3.3 Vibration Mode Shapes and Frequencies

In the preceding discussion of single-degree-of-freedom systems, it is apparent that the dynamic response of the structure is dependent upon two basic factors: (1) its period of vibration T or frequency ω and (2) its assumed displacement shape $\psi(x)$ (except where the mass is assumed to be concentrated at a single point). The same factors control the response of a multiple-degree-of-freedom system. Thus, the first step in the analysis of any multidegree system must be the evaluation of its free vibration frequencies and mode shapes.

The free vibration behavior of a structure is expressed by the equations of motion adapted to the special condition of no damping ($\mathbf{C} = \mathbf{0}$) and with no applied loading ($\mathbf{P} = \mathbf{0}$); thus Eq. 12.83 becomes in this case

$$\mathbf{M}\ddot{\mathbf{v}} + \mathbf{K}\mathbf{v} = \mathbf{0} \tag{12.84}$$

However, it is known that the motions of the system in free vibration are simple harmonic. Thus, the displacement vector may be written

$$\mathbf{v} = \hat{\mathbf{v}} \sin wt \tag{12.85}$$

from which the accelerations are

$$\ddot{\mathbf{v}} = -\omega^2 \hat{\mathbf{v}} \sin wt \tag{12.86}$$

where $\hat{\mathbf{v}}$ represents the amplitude of the vibratory motion and ω is the circular frequency. Introducing Eqs. 12.85 and 12.86 into Eq. 12.84 and canceling the time variation term leads to

$$-\omega^2 \mathbf{M}\hat{\mathbf{v}} + \mathbf{K}\hat{\mathbf{v}} = \mathbf{0}$$

or

$$\mathbf{K}\hat{\mathbf{v}} = \omega^2 \mathbf{M}\hat{\mathbf{v}} \tag{12.87}$$

Equation 12.87 is a form of eigenvalue equation. Its solution may be carried out by a variety of procedures that are beyond the scope of the present discussion. Standard digital computer programs are available that will provide automatic solutions of very large eigenvalue equation systems. Thus, the solution of Eq. 12.87 need not present a major obstacle in the dynamic analysis of a multidegree system.

The complete solution of the eigenvalue equation for a system having N degrees of freedom provides a vibration frequency ω_n (or period $T_n = 2\pi/\omega_n$) and a vibration shape $\mathbf{\Phi}_n$ for each of its N modes of vibration. The mode shape vector $\mathbf{\Phi}_n$ represents the *relative* amplitudes of motion for each of the displacement components in vibration mode n. The absolute amplitude of motion is arbitrary. The vibration characteristics of an example three-story building are indicated in Fig. 12.23. This three-degree-of-freedom structure has three modes of vibration. This same information may be presented in matrix form as follows:

$$\mathbf{\Phi} = [\mathbf{\Phi}_1 \mathbf{\Phi}_2 \mathbf{\Phi}_3] = \begin{bmatrix} 1.00 & 1.00 & 1.00 \\ 0.64 & -0.60 & -2.57 \\ 0.30 & -0.67 & 2.47 \end{bmatrix}$$

$$\boldsymbol{\omega} = \langle \omega_1 \ \omega_2 \ \omega_3 \rangle = \langle 14.5 \ 31.2 \ 46.1 \rangle$$

12.3.4 Modal Equations of Motion

The vibration mode shapes of any multidegree-of-freedom system have two orthogonality properties that make possible an important simplification in the general equations of motion. The first of these properties may be expressed for a lumped mass system as follows:

$$\sum_{i=1}^{N} M_i \Phi_{in} \Phi_{im} = 0 \qquad (m \neq n) \tag{12.88a}$$

or in the more general matrix form:

$$\mathbf{\Phi}_n^T \mathbf{M} \mathbf{\Phi}_m = 0 \qquad (m \neq n) \tag{12.88b}$$

Fig. 12.23. Vibration mode shapes and frequencies of example building.

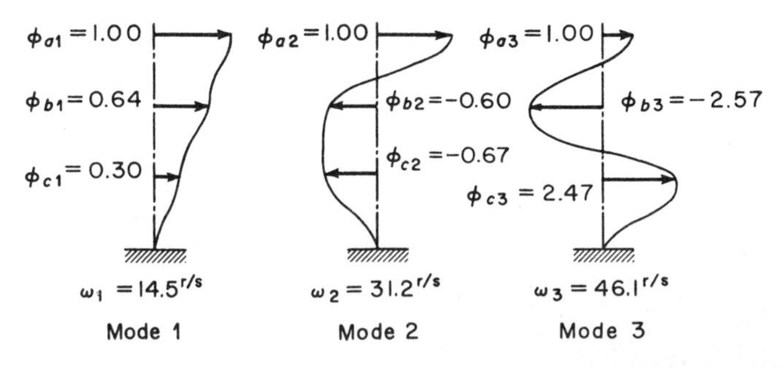

The second orthogonality property involves the corresponding relationship with respect to the stiffness matrix:

$$\mathbf{\Phi}_n^T\mathbf{K}\mathbf{\Phi}_m = 0 \qquad (m \neq n) \tag{12.89}$$

Because there are N independent vibration mode shapes for an N-degree-of-freedom system, any arbitrary displaced shape of the structure may be expressed in terms of the amplitudes of these shapes, treating them as generalized displacement coordinates. Thus, in general, any displacement v_i may be given as the sum of the contributions resulting from each mode:

$$v_i = \sum_{n=1}^{N} \Phi_{in} Y_n \tag{12.90}$$

where Y_n is the amplitude of the nth mode. In matrix form, the complete displacement vector may be expressed similarly as follows:

$$\mathbf{v} = \sum_{n=1}^{N} \mathbf{\Phi}_n Y_n = \mathbf{\Phi}\mathbf{Y} \tag{12.91}$$

in which $\mathbf{Y}$ is the generalized coordinate vector representing the vibration mode amplitudes (sometimes called the *normal* coordinates of the system).

The equations of motion of the multidegree system may be expressed in terms of the normal coordinates by differentiating Eq. 12.91 appropriately and substituting into Eq. 12.83 as follows:

$$\mathbf{M}\mathbf{\Phi}\ddot{\mathbf{Y}} + \mathbf{C}\mathbf{\Phi}\dot{\mathbf{Y}} + \mathbf{K}\mathbf{\Phi}\mathbf{Y} = \mathbf{P}(t) \tag{12.92}$$

Now this set of N simultaneous differential equations can be reduced to a single equation by multiplying by the transpose of any mode shape vector $\mathbf{\Phi}_n$. Thus,

$$\mathbf{\Phi}_n^T\mathbf{M}\mathbf{\Phi}\ddot{\mathbf{Y}} + \mathbf{\Phi}_n^T\mathbf{C}\mathbf{\Phi}\dot{\mathbf{Y}} + \mathbf{\Phi}_n^T\mathbf{K}\mathbf{\Phi}\mathbf{Y} = \mathbf{\Phi}_n^T\mathbf{P}(t)$$

which reduces to

$$\mathbf{\Phi}_n^T\mathbf{M}\mathbf{\Phi}_n\ddot{Y}_n + \mathbf{\Phi}_n^T\mathbf{C}\mathbf{\Phi}_n\dot{Y}_n + \mathbf{\Phi}_n^T\mathbf{K}\mathbf{\Phi}_n Y_n = \mathbf{\Phi}_n^T\mathbf{P}(t) \tag{12.93}$$

by virtue of the orthogonality conditions of Eqs. 12.88 and 12.89, if it is assumed that the damping matrix is such that the same type of orthogonality characteristic applies to it as well:

$$\mathbf{\Phi}_n^T\mathbf{C}\mathbf{\Phi}_m = 0 \qquad (m \neq n) \tag{12.94}$$

The normal coordinate equation of motion may be written more conveniently by introducing the following symbols for the generalized coordinate properties of each mode n:

$$\begin{aligned}
\text{Generalized mass:}\quad & M_n^* = \mathbf{\Phi}_n^T\mathbf{M}\mathbf{\Phi}_n \\
\text{Generalized damping:}\quad & C_n^* = \mathbf{\Phi}_n^T\mathbf{C}\mathbf{\Phi}_n \\
\text{Generalized stiffness:}\quad & K_n^* = \mathbf{\Phi}_n^T\mathbf{K}\mathbf{\Phi}_n \\
\text{Generalized loading:}\quad & P_n^*(t) = \mathbf{\Phi}_n^T\mathbf{P}(t)
\end{aligned} \tag{12.95}$$

Thus, Eq. 12.93 becomes:

$$M_n^*\ddot{Y}_n + C_n^*\dot{Y}_n + K_n^*Y_n = P_n^*(t) \tag{12.96}$$

A further simplification may be made by taking advantage of the fact that the generalized damping and stiffness are related to the generalized mass as follows:

$$C_n^* = 2\xi_n\omega_n M_n^* \tag{12.97a}$$

$$K_n^* = \omega_n^2 M_n^* \tag{12.97b}$$

Making use of Eqs. 12.97a and b, Eq. 12.96 may be written

$$\ddot{Y}_n + 2\xi_n\omega_n\dot{Y}_n + \omega_n^2 Y_n = \frac{P_n^*(t)}{M_n^*} \tag{12.98}$$

Equations 12.96 and 12.98 show that the equation of motion of any mode n of the multidegree-of-freedom system is exactly equivalent to the equation for a single-degree-of-freedom system. Thus, the normal coordinates (mode shapes) of a multidegree-of-freedom structure reduce its equations of motion to a set of independent equations, one for each mode of vibration. It also is of interest to note that the expressions for the generalized properties of any mode (Eqs. 12.95) are equivalent to the expressions previously defined for a single-degree-of-freedom system (Eqs. 12.26).

12.3.5 Earthquake Response Analysis

The dynamic analysis of a multidegree system by the mode superposition method is reduced to the numerical solution of Eq. 12.98 for each mode to obtain its individual contribution to the response. The total response is then obtained by superposing the modal effects, as indicated by Eq. 12.91. In the case of excitation resulting from an earthquake ground motion, the effective loads acting at the various floor levels of the structure are equal to the product of the lumped mass at the floor level and the ground acceleration, i.e., at any floor i the effective load may be written

$$P_{i_{\text{eff}}}(t) = M_i\ddot{v}_g(t) \tag{12.99}$$

Thus, the complete effective applied load vector is given by the product of the mass matrix and the ground acceleration $\ddot{v}_g(t)$, which may be expressed as follows:

$$\mathbf{P}_{\text{eff}}(t) = \mathbf{M}\hat{\mathbf{I}}\ddot{v}_g(t) \tag{12.100}$$

where $\hat{\mathbf{I}}$ represents a unit vector of dimension N. Then substituting Eq. 12.100 into Eq. 12.95 the generalized effective earthquake loading for mode n is given by

$$P_{n_{\text{eff}}}^*(t) = \mathbf{\Phi}_n^T\mathbf{M}\hat{\mathbf{I}}\ddot{v}_g(t) \equiv \mathscr{L}_n\ddot{v}_g(t) \tag{12.101}$$

in which $\mathscr{L}_n$ represents the earthquake participation factor for mode n and is defined as

$$\mathscr{L}_n = \mathbf{\Phi}_n^T\mathbf{M}\hat{\mathbf{I}} \tag{12.102}$$

Introducing Eq. 12.101 into Eq. 12.98, the earthquake equation of motion for mode n of a multidegree system becomes

$$\ddot{Y}_n + 2\xi_n\omega_n\dot{Y}_n + \omega_n^2 Y_n = \frac{\mathscr{L}_n}{M_n^*}\ddot{v}_g(t) \tag{12.103}$$

The response of the nth mode at any time t may be obtained by numerical evaluation of the Duhamel integral expression for the given earthquake motion:

$$Y_n(t) = \frac{\mathscr{L}_n}{M_n^*} \frac{1}{\omega_n} \int_0^t \ddot{v}_g(\tau) e^{-\xi_n \omega_n (t-\tau)} \sin \omega_n (t - \tau)\, d\tau$$

or, using the symbol $V_n(t)$ to represent the value of the integral at time t,

$$Y_n(t) = \frac{\mathscr{L}_n}{M_n^*} \frac{V_n(t)}{\omega_n} \qquad (12.104)$$

The complete displacement of the structure at time t is then obtained by superposing the contribution of all modes evaluated at this time, by Eq. 12.91:

$$\mathbf{v}(t) = \sum_{n=1}^{N} \mathbf{\Phi}_n Y_n(t) = \mathbf{\Phi} \mathbf{Y}(t) \qquad (12.105)$$

At this point it should be noted that an important advantage of the mode superposition procedure is that an approximate solution may be obtained by including only part of the modal contributions in the superposition. In general, the lower modes make the principal contributions to the response, and good approximations can frequently be obtained by considering only the first few modes in the analysis. On this basis, the best single-degree-of-freedom approximation generally is obtained by considering only the first mode contribution.

The earthquake forces developed in the structure may be evaluated most conveniently from its effective accelerations, which are given for each generalized coordinate as the product of its frequency squared and the displacement amplitude.

$$\ddot{Y}_{n_{\text{eff}}}(t) = \omega_n^2 Y_n(t) = \frac{\mathscr{L}_n}{M_n^*} \omega_n V_n(t) \qquad (12.106)$$

The distribution of accelerations through the structure then is of the same form as the modal displacements, i.e.,

$$\ddot{\mathbf{v}}_{n_{\text{eff}}}(t) = \mathbf{\Phi}_n \ddot{Y}_{n_{\text{eff}}}(t)$$

and the distribution of effective earthquake forces is given by the products of the local masses and local accelerations:

$$\mathbf{q}_n(t) = \mathbf{M} \ddot{\mathbf{v}}_n(t) = \mathbf{M} \mathbf{\Phi}_n \omega_n^2 Y_n(t) \qquad (12.107)$$

or, superposing the modal contributions,

$$\mathbf{q}(t) = \mathbf{M} \mathbf{\Phi} \boldsymbol{\omega}^2 \mathbf{Y}(t) \qquad (12.108)$$

Equation 12.108 represents the complete force response of any multidegree structure to a given earthquake motion. Any other desired force quantity, such as the base shear, the overturning moment, or any local stress value, may be obtained from these loads by normal static structural analysis procedures. For example, the base shear force Q_n is given by the sum of the effective earthquake forces for mode n over the height of the structure:

$$Q_n = \sum_{i=1}^{N} q_{in}(t) = \mathbf{\hat{I}}^T \mathbf{q}_n(t) = \mathbf{\hat{I}}^T \mathbf{M} \mathbf{\Phi}_n \omega_n^2 Y_n(t)$$

or, using Eqs. 12.102 and 12.104,

$$Q_n(t) = \left(\frac{\mathscr{L}_n^2}{M_n^*}\right) \omega_n V_n(t) \qquad (12.109)$$

Now, in the earthquake analysis of buildings it is convenient to designate the first term of this expression, which is a physical property of the structure depending on the mode shape and mass distribution, as its effective mass for mode n. This quantity multiplied by the acceleration of gravity g is the effective weight for mode n; it will be denoted by the symbol W_n, defined as follows:

$$W_n = \frac{\mathscr{L}_n^2}{M_n^*} g \qquad (12.110)$$

Thus, using Eqs. 12.109 and 12.110, the base shear for mode n may be written

$$Q_n(t) = \frac{W_n}{g} \omega_n V_n(t) \qquad (12.111)$$

The effective weight W_n represents the portion of the total weight of the structure that is effective in developing base shear in the nth mode. The sum of the effective weights for all modes is equal to the total weight of the structure.

If the base shear is computed by Eq. 12.111, the distribution of the modal forces through the building may be determined conveniently as follows:

$$\mathbf{q}_n(t) = \mathbf{M} \mathbf{\Phi}_n \frac{Q_n(t)}{\mathscr{L}_n} \qquad (12.112)$$

which is easily seen to be equivalent to Eq. 12.107.

12.3.6 Response Spectrum Analysis

The entire displacement and force response history of any multidegree structure is completely defined by Eqs. 12.105 and 12.108 after the modal response amplitudes have been determined as indicated by Eq. 12.104. The expressions for the response of any mode n in these equations are entirely equivalent to the expressions presented previously for the generalized coordinate analysis of a single-degree system. Thus, it is evident that the maximum response of any mode can be obtained from the earthquake response spectra by following the same procedures used for the single-degree structures.

On this basis, introducing S_{v_n}, the spectral velocity for mode n, into Eq. 12.104 leads to an expression for the maximum response of mode n:

$$Y_{n_{\max}} = \frac{\mathscr{L}_n}{M_n^*} \frac{S_{v_n}}{\omega_n} = \frac{\mathscr{L}_n}{M_n^*} S_{d_n} \qquad (12.113)$$

Then the distribution of maximum displacements in this mode is given by

$$\mathbf{v}_{n_{\max}} = \mathbf{\Phi}_n Y_{n_{\max}} = \mathbf{\Phi}_n \frac{\mathscr{L}_n}{M_n^*} S_{d_n} \qquad (12.114)$$

Similarly, the distribution of maximum effective earthquake forces in this mode (from Eq. 12.107) becomes

$$\mathbf{q}_{n_{\max}} = \mathbf{M\Phi}_n \omega_n^2 Y_{n_{\max}} = \mathbf{M\Phi}_n \frac{\mathscr{L}_n}{M_n^*} S_{a_n} \qquad (12.115)$$

or, from Eq. 12.111, the maximum base shear in mode n is

$$Q_{n_{\max}} = W_n \frac{S_{a_n}}{g} \qquad (12.116)$$

It must be recognized, however, that the maximum *total* response cannot be obtained merely by introducing the appropriate spectral response quantities into Eq. 12.105 or 12.108. These equations provide valid results only if the various modal contributions are evaluated concurrently; the modal response maxima do not occur simultaneously, in general, and thus they cannot be superposed directly to obtain the total maximum.

Consider, e.g., the base shear forces developed in the building of Fig. 12.22 when it is subjected to a specified earthquake ground motion. In each mode, the base shear force history is given by Eq. 12.111; plots of these modal responses are shown in Fig. 12.24. The total base shear force at any time may be obtained by adding the modal contributions at that time; thus, the total response history is given by the sum of the modal responses as shown in the figure.

The individual modal maximums which may be obtained from the earthquake acceleration response spectrum as indicated by Eq. 12.116 are also identified in the figure; it is quite evident from this sketch that the total maximum base shear is not given by the sum of the individual modal maximums. An approximation to the total maximum base shear, based on probability considerations, may be obtained by the so-called *root-mean-square* procedure, as follows:

$$Q_{\max} \doteq (Q_{1_{\max}}^2 + Q_{2_{\max}}^2 + Q_{3_{\max}}^2)^{1/2} \qquad (12.117)$$

The maximum value of any other response quantity, e.g., the maximum displacement of the top of the building, $v_{a_{\max}}$, may be obtained from a similar superposition of the modal maximum values:

$$v_{a_{\max}} \doteq (v_{a_{1_{\max}}}^2 + v_{a_{2_{\max}}}^2 + v_{a_{3_{\max}}}^2)^{1/2} \qquad (12.118)$$

By this procedure it is possible to get a good approximation of the earthquake response of any multidegree structure by working directly with the earthquake response spectra, without the necessity of carrying out the complete response history analysis.

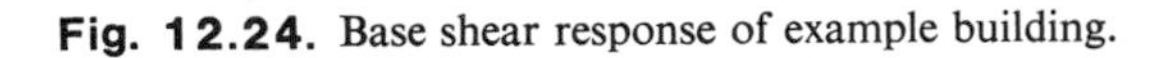

Fig. 12.24. Base shear response of example building.

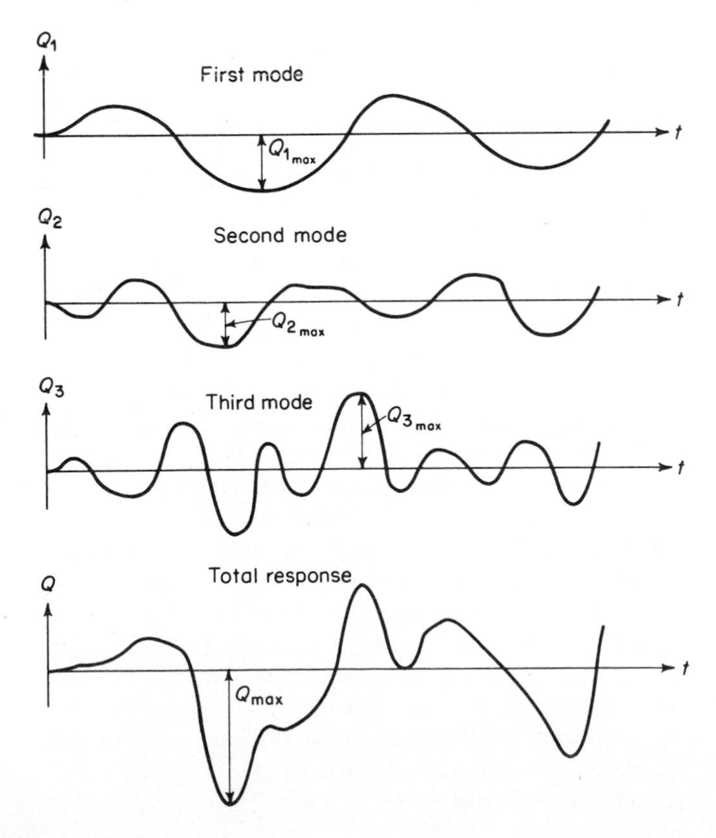

12.3.7 Approximate Vibration Analysis

The mode superposition analysis of a multidegree system may be carried out using as many modes of vibration as are desired in the analysis. If the complete eigenvalue problem (Eq. 12.87) has been solved, the entire set of mode shapes and frequencies will be available for this purpose. However, the complete solution of the eigenvalue equation for a system having many degrees of freedom is a large computational problem, and for preliminary investigations an approximate solution for the first mode shape and frequency may be adequate.

The most convenient procedure for approximating the fundamental mode frequency of any general structural system is Rayleigh's method. The basis of this method is that the maximum kinetic energy of a freely vibrating system is equal to the maximum potential energy of strain. The procedure involves first assuming the shape of the vibration mode and then equating the kinetic and potential energy developed by the system moving with that shape.

If the assumed vibrating shape of a multidegree system is represented by the dimensionless shape vector $\boldsymbol{\psi}$, the displacement vector may be expressed as

$$\mathbf{v}(t) = \boldsymbol{\psi} Y(t)$$

where $Y(t)$ represents the amplitude of the vibration. In free vibration, the maximum kinetic energy of the structure then is given by

$$\text{KE}_{\max} = \tfrac{1}{2}\omega^2 Y^2 \boldsymbol{\psi}^T \mathbf{M} \boldsymbol{\psi} \qquad (12.119)$$

and the maximum potential energy is given by

$$\text{PE}_{\max} = \tfrac{1}{2} Y^2 \boldsymbol{\psi}^T \mathbf{K} \boldsymbol{\psi} \qquad (12.120)$$

Thus, equating these two energy terms leads to the

following expression for the frequency of vibration:

$$\omega^2 \doteq \frac{\boldsymbol{\psi}^T \mathbf{K} \boldsymbol{\psi}}{\boldsymbol{\psi}^T \mathbf{M} \boldsymbol{\psi}} \qquad (12.121)$$

in which the numerator and denominator may be seen from Eq. 12.95 to be respectively the generalized stiffness and generalized mass associated with the assumed shape $\boldsymbol{\psi}$. If $\boldsymbol{\psi}$ were the exact shape of the vibration mode, Eq. 12.121 would give the exact frequency of the mode. Of greater practical importance, however, is the fact that the equation provides a good approximation of the frequency for any reasonable approximation of the mode shape.

A general method for improving an approximation of the first mode shape and frequency is to calculate the deflected shape resulting from the inertia forces associated with the original assumption. If the originally assumed displacement vector is designated $\mathbf{v}^{(0)}$, the inertia force vector resulting from a harmonic motion of that shape at the frequency ω is given by

$$\mathbf{F}_I^{(0)} = \omega^2 \mathbf{M} \mathbf{v}^{(0)} \qquad (12.122)$$

The deflections resulting from these inertia forces, which will be denoted by $\mathbf{v}^{(1)}$, are given by the force vector premultiplied by the inverse of the stiffness matrix:

$$\mathbf{v}^{(1)} = \mathbf{K}^{-1} \mathbf{F}_I^{(0)} = \omega^2 \mathbf{K}^{-1} \mathbf{M} \mathbf{v}^{(0)} \qquad (12.123)$$

(where $\mathbf{K}^{-1}$ is called the flexibility matrix of the structure).

Letting

$$\bar{\mathbf{v}}^{(1)} = \frac{1}{\omega^2} \mathbf{v}^{(1)} = \mathbf{K}^{-1} \mathbf{M} \mathbf{v}^{(0)} \qquad (12.124)$$

Eq. 12.123 may be written

$$\mathbf{v}^{(1)} = \omega^2 \bar{\mathbf{v}}^{(1)} \qquad (12.125)$$

Now an improved approximation of the potential energy of deformation may be obtained in terms of the work done by the forces $\mathbf{F}_I^{(0)}$ in moving through their displacements $\mathbf{v}^{(1)}$, i.e.,

$$\mathrm{PE} = \tfrac{1}{2} \mathbf{v}^{(1)T} \mathbf{F}_I^{(0)} = \tfrac{1}{2} \omega^4 \bar{\mathbf{v}}^{(1)T} \mathbf{M} \mathbf{v}^{(0)} \qquad (12.126)$$

The correspondingly improved expression for kinetic energy(equivalent to Eq. 12.119) is

$$\mathrm{KE} = \tfrac{1}{2} \omega^2 \mathbf{v}^{(1)T} \mathbf{M} \mathbf{v}^{(1)} = \tfrac{1}{2} \omega^6 \bar{\mathbf{v}}^{(1)T} \mathbf{M} \bar{\mathbf{v}}^{(1)} \qquad (12.127)$$

and equating these energy expressions, the frequency of vibration is found to be

$$\omega^2 \doteq \frac{\bar{\mathbf{v}}^{(1)T} \mathbf{M} \bar{\mathbf{v}}^{(0)}}{\bar{\mathbf{v}}^{(1)} \mathbf{M} \bar{\mathbf{v}}^{(1)}} \qquad (12.128)$$

Equation 12.128 provides an improved approximation of the first mode frequency; moreover, the displacements $\bar{\mathbf{v}}^{(1)}$ are a closer approximation to the first mode shape than was the original assumption $\mathbf{v}^{(0)}$. Furthermore, this same improvement process can be repeated as many times as desired to obtain still better results. Thus, by analogy with Eq. 12.124, the second improvement in the deflected shape would be given by

$$\bar{\mathbf{v}}^{(2)} = \mathbf{K}^{-1} \mathbf{M} \mathbf{v}^{(1)} \qquad (12.129)$$

while the correspondingly improved frequency expression would be

$$\omega^2 \doteq \frac{\bar{\mathbf{v}}^{(2)T} \mathbf{M} \mathbf{v}^{(1)}}{\bar{\mathbf{v}}^{(2)T} \mathbf{M} \bar{\mathbf{v}}^{(2)}} \qquad (12.130)$$

This process of successive improvement of mode shape and frequency is basically equivalent to the well-known Stodola method.

12.4 NONLINEAR EARTHQUAKE RESPONSE

12.4.1 Need for Nonlinear Analysis

The mode superposition method described above is a very effective procedure for the analysis of the response of any linearly elastic structure to any prescribed dynamic excitation. One of its most important advantages is that the same general technique may be used to obtain any desired degree of accuracy in the analysis. A simple approximate solution is provided by considering only the fundamental mode of vibration; by adding more modes the response is defined with increasing precision. Only as many modes need be considered as are required to obtain the desired accuracy—if the next higher mode makes no significant contribution to the response it may be assumed that sufficient modes have been included.

Because this method is based on the principle of superposition, however, it is evident that it may not be used in the analysis of a nonlinear structure, i.e., it is not applicable to any structure which is stressed beyond the elastic limit. On the other hand, a major earthquake can be expected to cause significant overstress in any standard multistory building which is designed by normal code procedures. For example, it was noted in the example analysis of the five-story building that the base shear force resulting from the earthquake response spectrum of Fig. 12.18 was equal to 16% of the building's weight, while the base shear coefficient given by the Uniform Building Code for this building was only 6.3%. Moreover, it was stated that the average response spectra of Fig. 12.18 represented an earthquake of lower intensity than the 1940 El Centro Earthquake. Thus, it is evident that if the building had been designed for the Code lateral force requirement, it would have been significantly overstressed by the 1940 El Centro Earthquake.

A more complete comparison of the effects of Code specified loadings with the elastic dynamic response to an actual earthquake motion is presented in Figs. 12.25–12.27. The essential dimensions and properties of a "standard" 20-story building are shown in Fig. 12.25. This frame was designed in reinforced concrete by normal procedures for the seismic loads prescribed by the Uniform Building Code; the absolute value of the member stiffnesses was adjusted to give a fundamental period of vibration of 2.2 sec.

3 @ 20 ft
Story weight = 176 k
198
198
198
226
226
226
234
234
234
284
284
284
288
288
288
314
314
314
320
19 @ 12 ft = 228 ft
15 ft

RELATIVE STIFFNESS OF COLUMNS AND GIRDERS

Ratio $(EI):(EI)_0$		
Columns		Girders
exterior	interior	
1.0	2.0	4.0
1.5	3.0	6.0
3.0	6.0	8.0
4.5	9.0	
6.0	12.0	
10.0	20.0	10.0
12.0	24.0	

Fig. 12.25. Dimensions and properties of standard building.

Lateral displacements and member forces developed in this frame by static application of the Code loads were evaluated on a digital computer using a standard building frame analysis program. In addition, the dynamic response of this building to the ground accelerations recorded at the 1940 El Centro Earthquake, shown in Fig. 12.26, was computed by the mode superposition method, using a numerical evaluation of the Duhamel integral for each mode. The complete response history of the building was evaluated; however, only the response *envelope* (i.e., the maximum value achieved by each response quantity at any time during the earthquake) will be considered here. The envelopes of lateral displacement as well as of girder and column moments are plotted in Fig. 12.27; also shown in this figure are the corresponding quantities determined from the Code seismic load analysis. Comparison of these curves shows that the dynamic response to the El Centro Earthquake is about four times greater than the Code response.

This comparison clearly shows that a building designed for Code seismic forces must be expected to suffer overstress during a major earthquake. The factor of safety provided by the actual strength of the structural materials (as compared with the design stresses) and by the participation of nonstructural components such as exterior cladding and interior partitions will not be sufficient to accommodate the observed ratio of four between elastic response and Code. Thus, it is evident that the true dynamic behavior of the structure can be determined only if allowance is made for nonlinearity (yield-

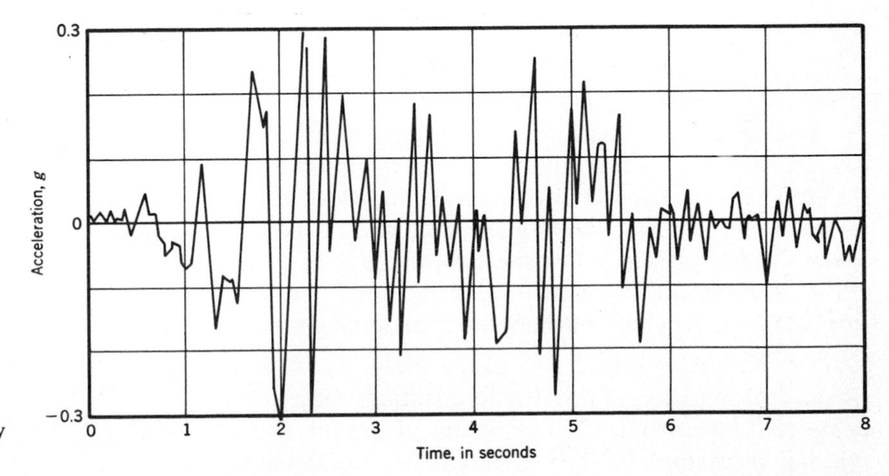

Fig. 12.26. El Centro Earthquake of May 18, 1940, N–S component.

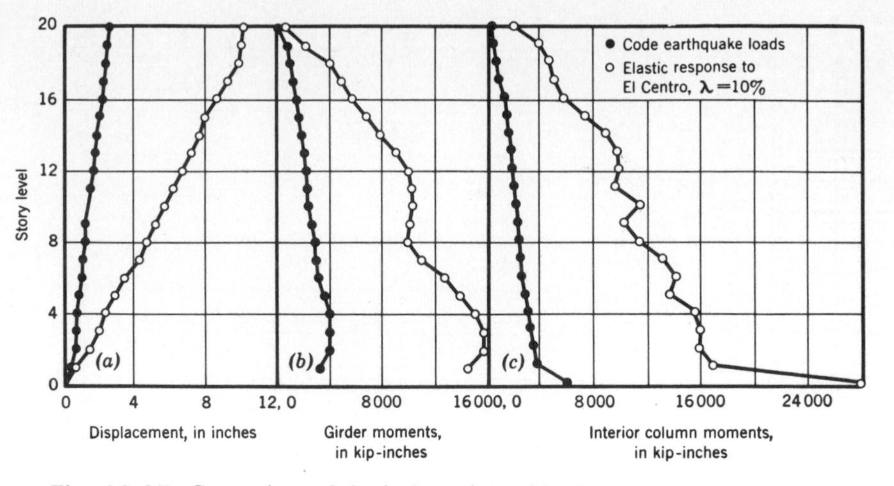

Fig. 12.27. Comparison of elastic dynamic earthquake response with code effects.

ing) in the analysis procedure. Moreover, the absolute necessity of avoiding brittle failure mechanisms (i.e., of providing a ductile structural assemblage) in the design of earthquake resistant structures is apparent.

12.4.2 Step-by-Step Analysis Procedure

The most convenient technique for evaluating the dynamic response of a nonlinear structure is by a step-by-step integration procedure. The modal decoupling, which makes possible the mode superposition analysis of linear structures, does not exist in a nonlinear system. Thus, the equations of motion must be integrated in their original form. To carry out the step-by-step analysis, the response history is divided into very short time increments. During each increment, the structure is assumed to be linearly elastic; however, between increments the properties are modified in accordance with the current condition of deformation. Thus, the nonlinear response is obtained as a sequence of linear responses of successively differing systems.

The analysis procedure involves the repeated application of the following steps for each time interval:

1. The stiffness of the structure for the time interval is evaluated, based on the state of displacement existing at the beginning of the interval.
2. Changes in displacements are computed, assuming the accelerations to vary linearly during the interval.
3. These incremental displacements are added to the displacement state at the beginning of the interval to obtain the displacements at the end of the interval.
4. Stresses are computed from the total displacements, taking account of the nonlinear material property.

The most complicated part of the analysis procedure is the determination of the structure stiffness for a given state of displacement, taking account of the prescribed yielding mechanism for the material. In general, this must be done separately for each column and girder of the structure, and then the total structural stiffness may be found by standard matrix analysis procedures. The resulting stiffness matrix expresses the static force–deflection relationship for the time increment initiated at time t as follows:

$$\mathbf{K}(t)\Delta\mathbf{v}(t) = \Delta\mathbf{P}(t) \tag{12.131}$$

The corresponding incremental equation of motion, from Eq. 12.83, may be written

$$\mathbf{M}\Delta\ddot{\mathbf{v}} + \mathbf{C}\Delta\dot{\mathbf{v}} + \mathbf{K}(t)\Delta\mathbf{v} = \Delta\mathbf{P}(t) \tag{12.132}$$

where each term represents a change of force occurring during the increment.

If it is assumed that the acceleration varies linearly during the time increment, Eq. 12.132 may be solved for the change in the displacement vector as follows: First, the change in the acceleration vector is expressed

$$\Delta\ddot{\mathbf{v}} = \frac{6}{\Delta t^2}\Delta\mathbf{v} + \mathbf{A}(t) \tag{12.133}$$

in which

$$\mathbf{A}(t) = -\frac{6}{\Delta t}\dot{\mathbf{v}}(t) - 3\ddot{\mathbf{v}}(t) \tag{12.134}$$

and Δt is the length of the time increment. Equation 12.133 is valid for a linear change of acceleration; the time t in $\mathbf{A}(t)$ refers to the beginning of the time increment. Similarly, the change in velocity vector is written

$$\Delta\dot{\mathbf{v}} = \frac{3}{\Delta t}\Delta\mathbf{v} + \mathbf{B}(t) \tag{12.135}$$

where

$$\mathbf{B}(t) = -3\dot{\mathbf{v}}(t) - \frac{\Delta t}{2}\ddot{\mathbf{v}}(t) \tag{12.136}$$

Substituting Eqs. 12.133 and 12.135 into Eq. 12.132 and rearranging leads to a pseudostatic equation similar in form to Eq. 12.131,

$$\bar{\mathbf{K}}(t)\Delta\mathbf{v} = \bar{\Delta\mathbf{P}}(t) \tag{12.137}$$

in which

$$\overline{\mathbf{K}}(t) = \mathbf{K}(t) + \frac{6}{\Delta t^2}\mathbf{M} + \frac{3}{\Delta t}\mathbf{C} \tag{12.138}$$

$$\overline{\Delta \mathbf{P}}(t) = \Delta\mathbf{P}(t) - \mathbf{MA}(t) - \mathbf{CB}(t) \tag{12.139}$$

Equation 12.137 may be solved by standard static analysis procedures for the incremental displacement vector $\Delta\mathbf{v}$, and then the total displacements at the end of the time increment are given by

$$\mathbf{v}(t + \Delta t) = \mathbf{v}(t) + \Delta\mathbf{v} \tag{12.140}$$

This step-by-step procedure is perfectly general and may be applied to systems having any number of degrees of freedom. For a single-degree-of-freedom system, the calculations may be carried out by hand for a reasonable number of time intervals. For large systems, the solution of the simultaneous equations of Eq. 12.137 is a large computational task and can be done only by digital computer. Even with a powerful computer, the complete dynamic analysis can be quite time-consuming because the set of equations must be solved for each time increment, and it may be necessary to consider as many as 100 or 200 time increments of each second of computed dynamic response. In general, the time increment should be no longer than one-tenth of the period of vibration of the highest significant vibration mode.

Another factor that must be considered in the practical usage of the step-by-step procedure is the definition of the damping matrix $\mathbf{C}$ that appears in both Eqs. 12.138 and 12.139. It is not necessary to define the damping coefficients explicitly when using the mode superposition procedure. It is much more convenient in that case to define the damping ratio ξ_n for each mode, as was noted earlier. However, this approach is not applicable to a nonlinear structure that has no true vibration modes. A useful procedure for defining the damping matrix is to assume that it consists of a linear combination of the mass and stiffness matrices,

$$\mathbf{C} = \alpha\mathbf{M} + \beta\mathbf{K} \tag{12.141}$$

in which α and β are scalar multipliers. It is evident that a damping ratio of this form will satisfy the orthogonality condition of Eq. 12.94. Moreover, by applying Eqs. 12.95 it may be shown that

$$C_n^* = \alpha M_n^* + \beta K_n^* \tag{12.142}$$

from which, using Eqs. 12.97, it is found that

$$\xi_n = \frac{\alpha}{2\omega_n} + \frac{\beta\omega_n}{2} \tag{12.143}$$

Using Eq. 12.143 it is possible to select the multipliers α and β so as to provide any desired damping ratio at any two selected frequencies, and in this way a reasonable choice may be made for the damping matrix $\mathbf{C}$.

12.4.3 Results of Nonlinear Analysis—Standard Building

The nonlinear earthquake behavior of a typical multistory building will be demonstrated by analysis of the standard building of Fig. 12.25 subjected to the 1940 El Centro Earthquake. For the purpose of this analysis, it has been assumed that the girders of the frame have yield strengths that are twice the moments produced by the static code seismic forces, while the columns have a yield strength six times their static code moments.

The results of this nonlinear analysis are compared in Fig. 12.28 with the results of a fully elastic dynamic response calculation for the same loading. The damping of the structure was assumed to be 10% of critical at the first mode frequency in the nonlinear system and in all modes of the elastic structure. It is interesting to note that the maximum displacements, shown in Fig. 12.28a, are essentially the same in both analyses, i.e., the nonlinear structure deflects only slightly more than the elastic structure.

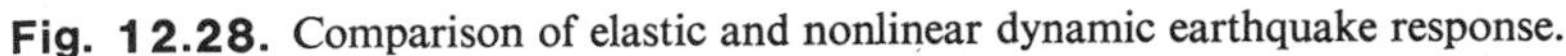

Fig. 12.28. Comparison of elastic and nonlinear dynamic earthquake response.

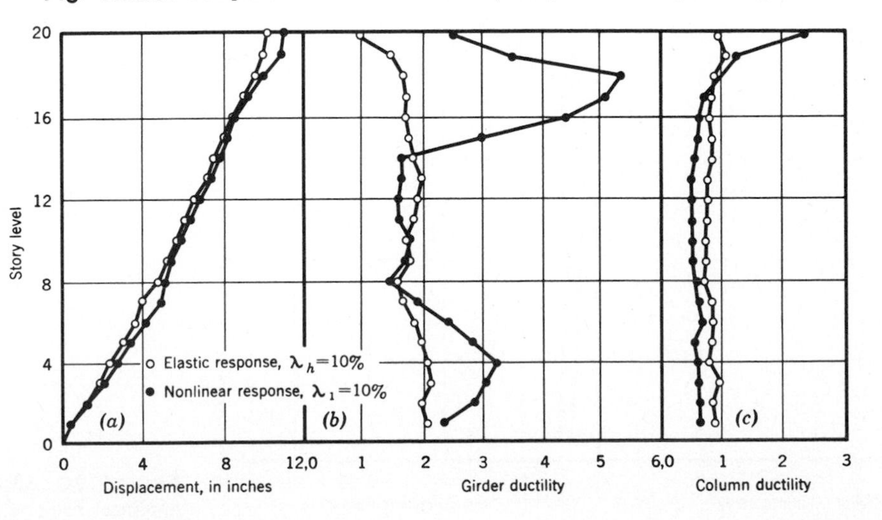

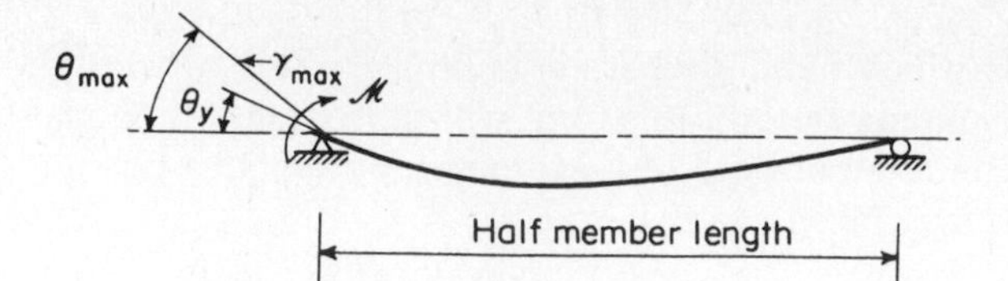

Fig. 12.29. Definition of ductility ratio.

The maximum column and girder responses, shown in Figs. 12.28b and 12.28c, are expressed in terms of the ductility ratio DR, which is defined as follows

$$\mathrm{DR} = \frac{\theta_{max}}{\theta_y} = \frac{\gamma_{max} + \theta_y}{\theta_y} \tag{12.144}$$

where θ_{max} is the maximum rotation at the end of the member and θ_y is the rotation at the initiation of yield while γ_{max} is the maximum yield angle rotation as shown in Fig. 12.29. If the member does not yield, θ_{max} is less than θ_y and the ductility ratio is less than unity. The elastic response moments in Fig. 12.28 are expressed similarly in terms of the ratio of the maximum elastic rotation to the yield rotation. The concept of a ductility factor is not actually applicable to an elastic element, of course.

Figure 12.28 shows that the girder ductility ratios in the nonlinear structure generally exceed the corresponding quantities in the elastic structure, whereas the reverse is true in the columns (except in the top stories). It is interesting to note that the girder yield requirements are greatest in the upper stories, somewhat less in the lower stories, and least in the central stories. It must be recognized that these ductility factors represent the requirements imposed on the structure by the earthquake. It is assumed that the designer has detailed each element so that it possesses the necessary ductility capacity.

12.4.4 Influence of Strength Variations

Figure 12.28 shows that in an ordinary structure designed by normal procedures the yielding is confined

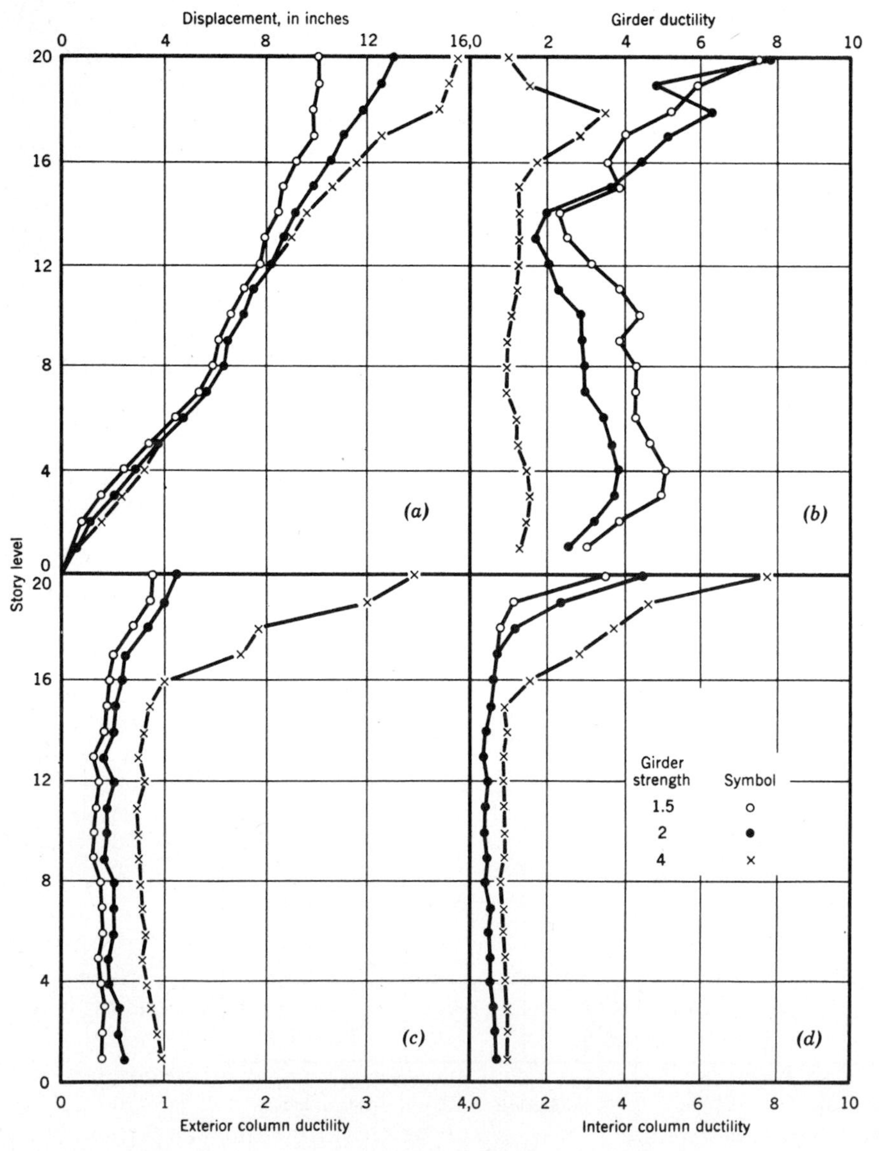

Fig. 12.30. Effect of girder strength on nonlinear response.

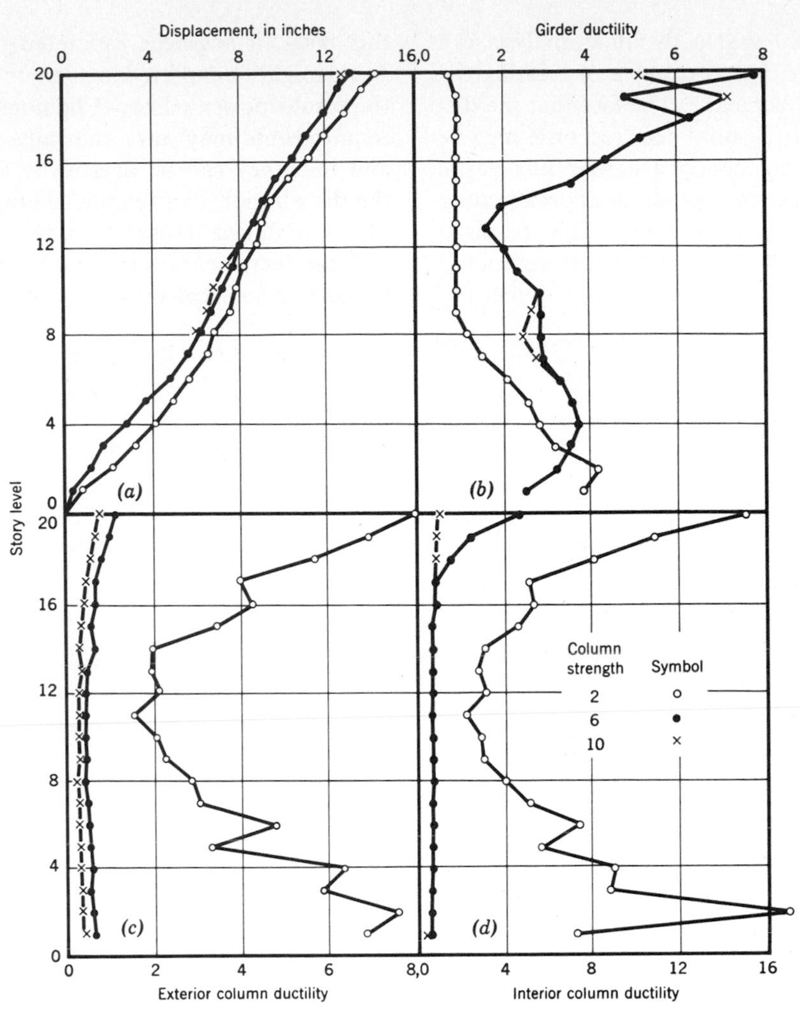

Fig. 12.31. Effect of column strength on nonlinear response.

almost entirely to the girders. The columns remain essentially elastic and actually distort less than in a completely elastic structure; thus it may be concluded that the yield energy absorbed by the girders has tended to protect the columns from overstress. By varying the strengths of the various elements, however, the yielding can be shifted from one portion of the structure to another, as is demonstrated in the following examples.

12.4.4.1 Strength of girders. To evaluate the effect of girder strength on the ductility requirements, two additional structures were analyzed and compared with the results for the standard building. The two new buildings had girder strengths of 1.5 and 4 times the design girder moments, respectively. The results of all these analyses are presented in Fig. 12.30. The maximum lateral displacements in Fig. 12.30a tend to vary inversely with the girder strengths. The reason for this somewhat surprising result becomes evident in parts b, c, and d of the figure. The building with stronger girders shows less girder yielding, which tends to force more yielding into the columns and thus to increase the lateral displacements. Reducing the girder strengths produced the opposite effect.

12.4.4.2 Strength of columns. As might be expected, the balance between column and girder yielding also can be altered by changing the column strengths. Two additional buildings were analyzed, having yield moments 2 and 10 times the design moments, respectively, with results compared to the standard building in Fig. 12.31. These figures show that increasing the column strength had little effect, which is not surprising inasmuch as the standard building showed very little column yielding. However, the reduction of column strength caused a significant reduction in girder yield and a corresponding increase in column yield. It is important to note in this case that column ductility requirements as high as 16 were computed in certain locations, which would impose a very serious design requirement on these members.

On the basis of these results, and from similar analyses of a variety of other structures, it may be concluded that ductility is an essential characteristic of earthquake resistant construction and that a dynamic analysis can give realistic results only if it is capable of treating the true nonlinear structural behavior. The fact that the displacements developed during nonlinear response may be quite similar to the displacements of a similar fully elastic structure has led to a simple procedure for approximating the earthquake ductility requirement: The required ductility ratio is assumed to be equal to the computed fully elastic force divided by the elastic limit force. On this basis, it has been estimated that ductility ratios of 4 to 6 might be expected in a major earthquake. However, the results presented above demonstrate that the ductility requirements may vary radically through the structure and that they can be drastically affected by changes in the distribution of strengths. Thus, the need for an adequate nonlinear dynamic analysis procedure is evident, and the incremental procedure described here has been found to give excellent results in earthquake analyses.

Chapter 13

Applications of Random Vibration Theory*

JOSEPH PENZIEN

Professor of Civil Engineering and Director, Earthquake Engineering Research Center, University of California, Berkeley, California

13.1 INTRODUCTION

It is the purpose of this chapter to discuss the applications of random vibration theory in earthquake engineering and to show some of the numerical results that have been obtained. This is a difficult task to perform effectively in a single chapter; especially in view of the fact that many interested readers have not had an introduction to random vibration theory previously. It is desirable, therefore, to review first the basic concepts of a random process, its characterization by probability density functions, and its effects on systems having known characteristics.

*This article was originally published in the *Bulletin of the International Institute of Seismology and Earthquake Engineering, Tokyo*, **2**: 47–69, 1965.

13.2 THE RANDOM PROCESS

A random process may be considered as an ensemble of $n(n \longrightarrow \infty)$ sample wave forms of duration s as shown in Fig. 13.1 (Crandall and Mark, 1963), where

$$^{r}x(t) \neq {}^{s}x(t) \qquad r, s = 1, 2, \ldots, n \tag{13.1}$$

By establishing probability density functions for such an ensemble, one characterizes the process so that the statistical properties of any single wave form, say $^{j}x(t)$, belonging to this ensemble will be known even though such wave form may be some future event which has not yet taken place.

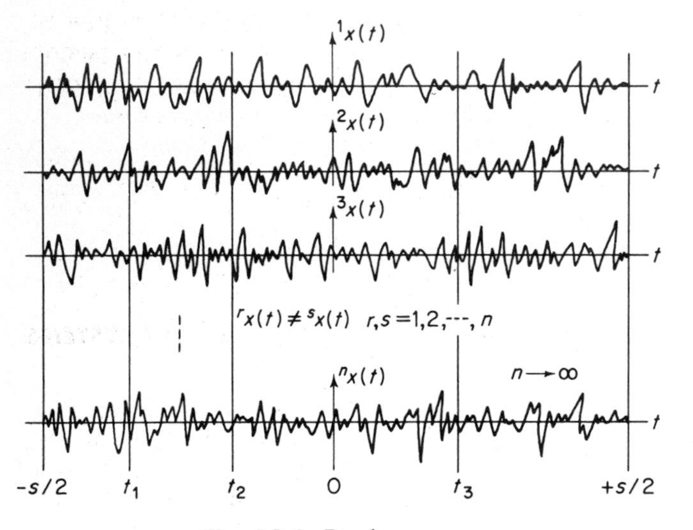

Fig. 13.1. Random process.

To completely characterize a random process, it is necessary that the probability density functions

$$p(x_1), p(x_1, x_2), p(x_1, x_2, x_3), \ldots \tag{13.2}$$

be obtained where $x_1 \equiv x(t_1)$, $x_2 \equiv x(t_2)$, etc. These probability density functions are defined such that

$$\begin{aligned} p(x_1)\, dx_1 &= Pr(x_1 < {}^{j}x_1 < x_1 + dx_1) \\ p(x_1, x_2) dx_1\, dx_2 &= Pr(x_1 < {}^{j}x_1 < x_1 + dx_1; \\ &\qquad x_2 < {}^{j}x_2 < x_2 + dx_2) \\ &\vdots \\ p(x_1, x_2, \ldots, x_n)\, dx_1\, dx_2 \ldots dx_n & \\ = Pr(x_1 < {}^{j}x_1 &< x_1 + dx_1; \\ \ldots ; x_n < {}^{j}x_n &< x_n + dx_n) \end{aligned} \tag{13.3}$$

While theoretically an infinite number of probability density functions is needed to completely define a process, it is usually sufficient in practice to use only the first two, i.e., $p(x_1)$ and $p(x_1, x_2)$.

Relying upon the Central Limit Theorem, most random processes can be considered as having Gaussian (or normal) probability distributions, i.e.,

$$p(x_1, x_2, \ldots, x_n) = \frac{1}{(2\pi)^{n/2}|S|^{1/2}} \exp\left(-\frac{1}{2}\sum_{i=1}^{n}\sum_{k=1}^{n} \frac{x_i x_k S_{ik}}{|S|}\right) \quad n = 1, 2, \ldots \tag{13.4}$$

where

$$S \equiv \begin{bmatrix} s_{11} & s_{12} & \cdots & s_{1n} \\ s_{21} & s_{22} & \cdots & s_{2n} \\ \vdots & & & \\ s_{n1} & s_{n2} & \cdots & s_{nn} \end{bmatrix} \tag{13.5}$$

$$S_{ik} \equiv \text{cofactor of } s_{ik} \tag{13.6}$$

and where s_{ik} is the ensemble average of $x_i x_k$, which can be expressed in mathematical form as

$$s_{ik} \equiv E(x_i x_k) = \frac{1}{n}\sum_{j=1}^{n} {}^{j}x_i\, {}^{j}x_k \tag{13.7}$$

The form of Eq. 13.4 assumes, provided, of course, the ensemble mean values of $x_i (i = 1, 2, \ldots)$ equal zero, that

$$E(x_i) \equiv \frac{1}{n}\sum_{j=1}^{n} {}^{j}x_i = 0 \qquad i = 1, 2, \cdots \tag{13.8}$$

One should note that all Gaussian distributions as given by Eq. 13.4 depend only upon the ensemble averages as given by Eq. 13.7. This fact is especially significant when dealing with a stationary random process since in such a case the ensemble average $E(x_i x_k)$ depends only upon the time difference $\tau \equiv (t_k - t_i)$. Thus, one can define a function of τ, namely,

$$R_x(\tau) = E(x_i x_k) \tag{13.9}$$

which is known as the autocorrelation function of the process. Having obtained this autocorrelation function, the stationary Gaussian process becomes completely defined.

Three important properties of the autocorrelation function that should be kept in mind are

$$R_x(0) = E(x^2) \equiv \sigma_x^2 \text{ (variance)} \tag{13.10}$$

$$R_x(\tau) = R_x(-\tau) \tag{13.11}$$

$$|R_x(\tau)| \leqslant R_x(0) \tag{13.12}$$

Equation 13.10 simply states that $R_x(0)$ is the mean square value of x across the ensemble; Eq. 13.11 shows the symmetry condition of the function $R_x(\tau)$ about $\tau = 0$; and Eq. 13.12 indicates that $R_x(\tau)$ can never be a diverging function but may be (and usually is) a function that decays with increasing values of τ. One should also keep in mind that the autocorrelation function as defined by Eq. 13.9 gives directly a measure of the statistical dependence of random variables $x(t + \tau)$ and $x(t)$ on each other.

Let us now consider the so-called stationary ergodic process that has the characteristic that the ensemble average of any function of x is equal to the time average (indicated by < >) of this same function as measured along any one member of the ensemble, i.e.,

$$<f(x)> = E[f(x)] \tag{13.13}$$

This being the case, a process of this type can be completely characterized by making a wave analysis of a single member of the ensemble, say member ${}^jx(t)$. This analysis is accomplished by taking the Fourier integral transform of ${}^jx(t)$ as given by the relation

$$ {}^jX(i\omega) \equiv \int_{-\infty}^{\infty} {}^jx(t)e^{-i\omega t}\,dt \tag{13.14}$$

thus separating the random wave form ${}^jx(t)$ into its harmonic frequency components. By superposition of these harmonics, the wave form ${}^jx(t)$ can be expressed by the inverse relation

$$ {}^jx(t) = \frac{1}{2\pi}\int_{-\infty}^{\infty} {}^jX(i\omega)e^{i\omega t}\,d\omega \tag{13.15}$$

Substituting Eq. 13.15 into the expression

$$\overline{x^2} \equiv <{}^jx^2> = \lim_{s\to\infty}\frac{1}{s}\int_{-s/2}^{s/2} {}^jx^2(t)\,dt \tag{13.16}$$

to obtain the mean square value of x, one obtains

$$\overline{x^2} = \int_{-\infty}^{\infty} S_x(\omega)\,d\omega \tag{13.17}$$

where

$$S_x(\omega) = \lim_{s\to\infty}\frac{|{}^jX(i\omega)|^2}{2\pi s} \tag{13.18}$$

The function $S_x(\omega)$ as defined by Eq. 13.18 is called the power spectral density function of the process. In accordance with Eq. 13.17, the area under this function represents the mean square value of x. Note that $S_x(\omega)$ is symmetric about $\omega = 0$ when ${}^jx(t)$ is a real variable.

It can be shown that the power spectral density function as defined by Eq. 13.18 and the autocorrelation function as defined by Eq. 13.9 are related to each other through the Fourier integral transforms as follows:

$$S_x(\omega) = \frac{1}{2\pi}\int_{-\infty}^{\infty} R_x(\tau)e^{-i\omega\tau}\,d\tau \tag{13.19}$$

$$R_x(\tau) = \int_{-\infty}^{\infty} S_x(\omega)e^{i\omega\tau}\,d\omega \tag{13.20}$$

Therefore, either the power spectral density function $S_x(\omega)$ or the autocorrelation function $R_x(\tau)$ can be used to completely characterize a Gaussian ergodic process.

The most common of all processes used in random theory is the so-called "white noise" process. This process is characterized by

$$S(\omega) = S_0 \quad \text{(constant)} \tag{13.21a}$$

$$R(\tau) = 2\pi S_0\,\delta(\tau) \tag{13.21b}$$

where $\delta(\tau)$ is the Dirac delta function.

13.3 EFFECT OF RANDOM PROCESS ON SYSTEMS

Consider now each wave form ${}^jx(t)$ $(j = 1, 2, \ldots, n)$ of a random process as separate inputs into a system having known transfer characteristics. Let ${}^jy(t)$ represent corresponding outputs as shown in Fig. 13.2.

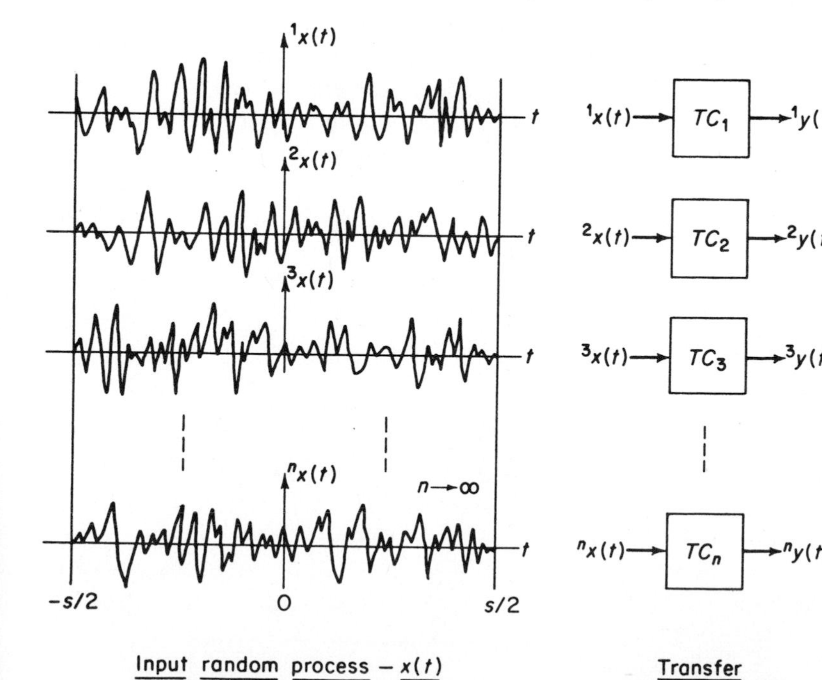

Fig. 13.2. Input-output random processes.

If the system is nonlinear it is usually necessary to first establish an input ensemble of known characteristics, to determine each member of the output ensemble by a separate deterministic analysis, and then to study the output ensemble statistically to determine its characteristics. This type of analysis represents a tremendous amount of work but can be accomplished with modern-day analog and digital computer equipment.

If the system is linear, our task is much easier as one can usually determine the output characteristic functions directly from the input characteristic functions by making use of either the unit impulse response function $h(t)$ or the complex frequency response function $H(i\omega)$. Transfer function $h(t)$ is simply the output transient response $y(t)$ as produced by a unit impulse input $x(t) = \delta(t)$, where $\delta(t)$ is the Dirac delta function while $H(i\omega)$ is the ratio of the steady state output response $y(t)$ to the harmonic input $x(t) = e^{i\omega t}$ that produces it.

In accordance with the above definition of $h(t)$ and $H(i\omega)$ and making use of the principle of superposition, it is apparent that any single deterministic output ${}^jy(t)$ is related to any arbitrary, but prescribed, input ${}^jx(t)$ by the relations

$$^jy(t) = \int_{-\infty}^{t} {}^jx(\tau)h(t-\tau)\,d\tau \qquad (13.22)$$

$$^jy(t) = \frac{1}{2\pi}\int_{-\infty}^{\infty} H(i\omega)X(i\omega)e^{i\omega t}\,d\omega \qquad (13.23)$$

where $X(i\omega)$ is given by Eq. 13.14. It can be shown that $h(t)$ and $H(i\omega)$ are Fourier transform pairs.

Now consider the entire output process ${}^jy(t)$ ($j = 1, 2, \ldots, n$). In order that one can make predictions regarding the output ${}^jy(t)$ that will result from some future input ${}^jx(t)$, one needs to establish the probability density functions as defined by the relations

$$\left.\begin{aligned} p(y_1)dy_1 &= Pr(y_1 < {}^jy_1 < y_1 + dy_1) \\ p(y_1, y_2)dy_1\,dy_2 &= Pr(y_1 < {}^jy_1 < y_1 + dy_1; \\ & \qquad y_2 < {}^jy_2 < y_2 + dy_2);\ \text{etc.} \end{aligned}\right\} \qquad (13.24a)$$

These relations correspond with those given by Eq. 13.3 for the input.

Restricting ourselves for the time being to a Gaussian ergodic input process having a zero mean value, it is correct to assume that the output process likewise will be a Gaussian ergodic process with a zero mean value. From my previous comments it follows, therefore, that all of the probability functions as given by Eq. 13.24a will be known once either the output process autocorrelation function $R_y(\tau)$ or its power spectral density function $S_y(\omega)$ has been determined.

To obtain the function $R_y(\tau)$ simply substitute Eq. 13.22 into the right-hand side of the expression

$$R_y(\tau) = E\big(y(t)y(t+\tau)\big) \qquad (13.24b)$$

which gives, upon making use of Eq. 13.9,

$$R_y(\tau) = \int_0^\infty \int_0^\infty R_x(\tau - U_2 + U_1)h(U_1)h(U_2)\,dU_1\,dU_2 \qquad (13.25)$$

where U_1 and U_2 are dummy variables for τ. $S_y(\omega)$ now can be obtained by making use of Eq. 13.19, which applies to the output process as follows:

$$S_y(\omega) = \frac{1}{2\pi}\int_{-\infty}^{\infty} R_y(\tau)e^{-i\omega\tau}\,d\tau \qquad (13.26)$$

Substituting Eq. 13.25 into Eq. 13.26 gives, after some mathematical manipulations and making use of the fact that $H(i\omega)$ is the Fourier transform of $h(t)$,

$$S_y(\omega) = |H(i\omega)|^2 S_x(\omega) \qquad (13.27)$$

Thus, it is shown that one can define the output process in this case either through the time domain, making use of Eq. 13.26, or through the frequency domain, making use of Eq. 13.27.

13.4 ELASTIC RESPONSE OF THE SINGLE-DEGREE SYSTEM TO RANDOM INPUTS

Since the deterministic dynamic response analysis of a multidegree-of-freedom linear system reduces to that of treating separately a number of single-degree systems, the engineer has been most interested in determining the dynamic response of single-degree-of-freedom systems to earthquake inputs. Therefore, let us consider now the single-degree-of-freedom elastic system shown in Fig. 13.3 when subjected to a random support acceleration $\ddot{x}_g(t)$. Usually one is interested in determining the dynamic response in terms of either the relative displacement $x(t)$ or the absolute acceleration $\ddot{x}_t(t)$. However, for low damped systems the peak relative displacement and peak absolute acceleration can be obtained approximately by simply multiplying the peak relative velocity by $1/\omega_n$ and ω_n, respectively, where ω_n is the natural undamped frequency $(k/m)^{1/2}$. Therefore, in the following discussion I shall treat $\ddot{x}_g(t)$ as the input and $\dot{x}(t)$ as the output.

It is easily shown that the transfer functions for this system are

Fig. 13.3. Single-degree-of-freedom system.

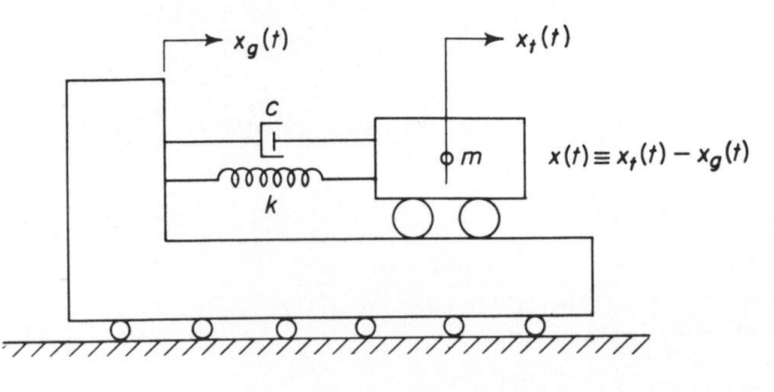

$$h_{\dot{x}}(t) = \frac{1}{p} e^{-\xi\omega_n t}(p \cos pt - \omega_n \xi \sin pt) \tag{13.28}$$

$$H_{\dot{x}}(i\omega) = \frac{i\left(\frac{\omega}{\omega_n}\right)}{\omega_n\left[\left(1 - \frac{\omega^2}{\omega_n^2}\right) + 2\xi i \frac{\omega}{\omega_n}\right]} \tag{13.29}$$

where ξ is the percent of critical damping and where p is the damped frequency which is equal to $\omega_n\sqrt{1 - \xi^2}$. The subscript $\dot{x}$ as used above simply indicates which quantity is being used as the output.

If the support acceleration $\ddot{x}_g(t)$ is considered as a Gaussian ergodic process having a known power spectral density function $S_{\ddot{x}_g}(\omega)$ or autocorrelation function $R_{\ddot{x}_g}(\tau)$, it is a simple matter to obtain $S_{\dot{x}}(\omega)$ and $R_{\dot{x}}(\tau)$ by making use of Eqs. 13.25, 13.27, 13.28, and 13.29.

Suppose, e.g., the input acceleration is represented by a "white noise" process having a constant power spectral density S_0 that corresponds with the autocorrelation function $R_{\ddot{x}_g}(\tau) = 2\pi S_0\, \delta(\tau)$. Substituting Eq. 13.29 into Eq. 13.27 for this case gives

$$S_{\dot{x}}(\omega) = \frac{\left(\frac{\omega}{\omega_n}\right)^2 S_0}{\omega_n^2\left[\left(1 - \frac{\omega^2}{\omega_n^2}\right)^2 + 4\xi^2\frac{\omega^2}{\omega_n^2}\right]} \tag{13.30}$$

while substituting Eq. 13.28 into Eq. 13.25 gives

$$R_{\dot{x}}(\tau) = \frac{\pi S_0}{2\omega_n\xi}\left(\cos p|\tau| - \frac{\xi}{\sqrt{1-\xi^2}} \sin p|\tau|\right)e^{-\omega_n\xi\tau} \tag{13.31}$$

The mean square response, i.e., $\overline{\dot{x}^2} = \sigma_{\dot{x}}^2$ (variance), is obtained most easily using the relation (see Eq. 13.10)

$$\sigma_{\dot{x}}^2 = \overline{\dot{x}^2} = R_{\dot{x}}(0) = \frac{\pi S_0}{2\omega_n\xi} \tag{13.32}$$

Following procedures similar to that above, one can easily show that

$$\sigma_x^2 = \overline{x^2} = \frac{\pi S_0}{2\xi\omega_n^3} \tag{13.33}$$

and

$$\sigma_{\ddot{x}_t^2} = \overline{\ddot{x}_t^2} = \left(\frac{1}{2\xi} + 2\xi\right)\omega_n S_0 \pi \tag{13.34}$$

The reason, of course, for interest in the mean square value of response is that it is the only quantity needed to define the probability density function $p(x)$ as given by Eq. 13.4 for $n = 1$, i.e.,

$$p[x(t)] = \frac{1}{\sqrt{2\pi}\,\sigma_x} \exp\left(-\frac{x^2}{2\sigma_x^2}\right) \tag{13.35}$$

Since the area under the constant power spectral density function for "white noise" is infinite such a process has an infinite mean square value that is impossible to experience in practice. With this observation one may at first question the extensive use of "white noise" processes in engineering studies. However, for those systems where the transfer function $|H(i\omega)|^2$ is sharply peaked at some critical value of ω, say ω_c (such as is the case at $\omega = \omega_n$ for the single-degree system treated above), and decreases rapidly with increasing frequencies beyond that critical point, one can often consider a "white noise" input with little loss of accuracy in determining the characteristics of the output processes. In such cases the response is produced primarily by the frequency components of the input that are near the critical frequency of the system. Therefore, if $S_x(\omega)$ is a reasonably slowly varying function of ω in the vicinity of the critical frequency, one can assume a "white noise" input with little loss of accuracy in the response predictions provided the constant input power spectral density $S_0 = S_x(\omega_c)$. In such cases, the output power spectral density function for relative velocity of the single-degree system treated above becomes

$$S_{\dot{x}}(\omega) = \frac{\left(\frac{\omega}{\omega_n}\right)^2 S_{\ddot{x}_g}(\omega_n)}{\omega_n^2\left[\left(1 - \frac{\omega^2}{\omega_n^2}\right)^2 + 4\xi^2 \frac{\omega^2}{\omega_n^2}\right]} \tag{13.36}$$

All of the relations derived above for stationary processes are accurate only as the duration s (see Fig. 13.3) approaches infinity. Since the engineer continually must work with processes of finite duration, one must be careful to verify the validity of making the stationary assumption in such cases. Usually the autocorrelation function of the output response is used for this purpose. Suppose, e.g., the input process $\ddot{x}_g(t)$ of the single-degree system treated above is stationary but has a finite duration s equal to 25 sec. Further assume one is interested in a system having a natural period T_n equal to 1 sec ($\omega_n = 2\pi$) and having damping equal to 5% of critical ($\xi = 0.05$). Noting that the autocorrelation function for relative velocity as given by Eq. 13.31 decays with increasing values of τ as given by the factor $e^{-\omega_n\xi\tau}$, the quantity $e^{-\omega_n\xi s}$ gives a measure of the statistical dependence of the value of $\dot{x}(t + s)$ on the value of $x(t)$ in a truly stationary process.

Evaluating $e^{-\omega_n\xi s}$ for $\omega_n = 2\pi$, $\xi = 0.05$, and $s = 25$ gives the numerical value 0.00038, thus showing very little correlation. One would conclude, therefore, that the assumption of a stationary process could be made in this case with little loss of accuracy except in the very extreme ends (or tails) of the probability distribution functions.

Since in earthquake engineering one is more interested in the probability distribution of *peak* response than in the probability distribution of response at any time t, let us consider further the character of the response $x(t)$ for a low damped single-degree system. As pointed out previously the response of such a system is due primarily to the frequencies contained in the input that are near the natural frequency of the system. As a result of this characteristic, any output function ${}^j\dot{x}(t)$ will have the appearance shown in Fig. 13.4, namely, a function

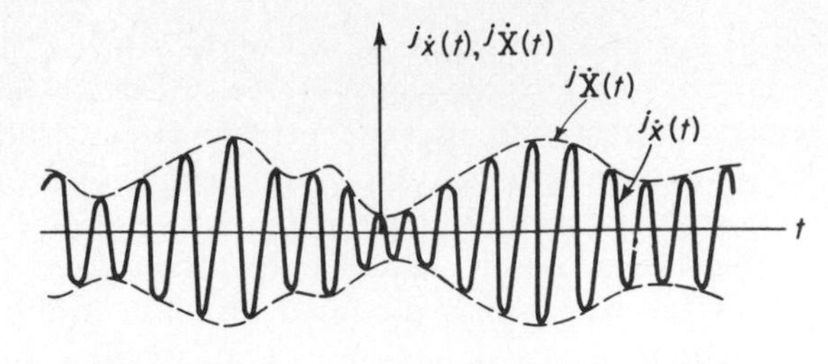

Fig. 13.4. Typical response of low damped system.

that locally appears to be nearly sinusoidal at a frequency near the natural frequency of the system and function whose amplitudes show a random "beat" characteristic. It can be shown that the probability density function $p[\dot{X}(t)]$ for the positive peak amplitude values of such systems approaches the Rayleigh distribution as given by Eq. 13.37 (see Fig. 13.5) when the percent of critical damping approaches zero

$$p[\dot{X}(t)] = \frac{\dot{X}}{\sigma_{\dot{x}}^2}\exp\left(-\frac{\dot{X}^2}{2\sigma_{\dot{x}}^2}\right) \tag{13.37}$$

For finite damping in the range common to structural systems, the Rayleigh distribution is still quite accurate as long as the process is stationary. One should, of course, realize that as damping approaches zero all mean square values of response approach infinity under the stationary assumption. Also one should note that the autocorrelation decay function $e^{-\omega_n\xi\tau}$ does *not* decay for zero damping. Therefore, the minimum duration s for which the stationary assumption can be made in practice increases very rapidly as $\xi \longrightarrow 0$.

The Rayleigh distribution, as given by Eq. 13.37 and which is plotted in Fig. 13.5, indicates that under the stationary assumption infinite response is possible even though the probability of its occurrence approaches zero. In practice, of course, processes can never be completely stationary ($s \longrightarrow \infty$); therefore, the "tail ends" of the Gaussian and Rayleigh probability distributions

Fig. 13.5. Rayleigh probability distribution of envelope $\dot{X}(t)$.

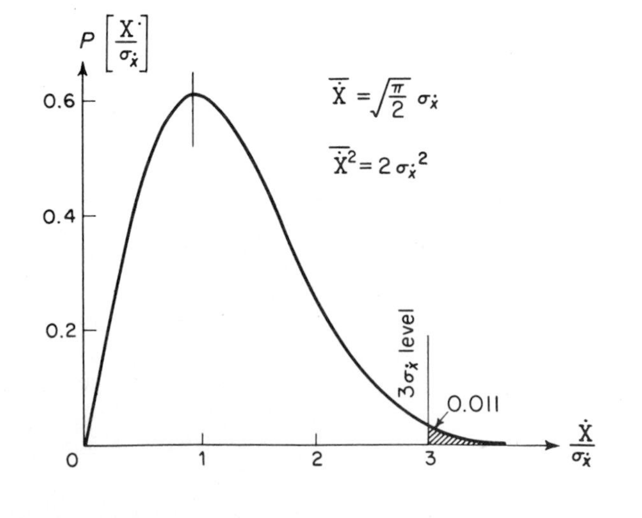

are usually considerably in error since finite maximum response can be expected. It is common practice in many fields of engineering to assume this maximum response at the 3 σ level.

Caughey and Stumpf (1961) have treated the above single-degree system when subjected to a stationary random input of *finite* duration s and having an arbitrary (but slowly varying in the vicinity of ω_n) power spectral density function. These investigators show the mean square relative displacement response $\sigma_x^2(t)$ is given by the relation

$$\sigma_x^2(t) \cong \frac{\pi S_{\ddot{x}_g}(\omega_n)}{2\xi\omega_n^3}\left\{1 - \frac{e^{-2\omega_n\xi t}}{p^2} \cdot \left[p^2 + \frac{(2\omega_n\xi)^2}{2}\sin^2 pt + \omega_n p\xi \sin 2pt\right]\right\} \tag{13.38}$$

where time t is here measured from the start of the input. For low damping the oscillatory terms of Eq. 13.38 are negligible, thus permitting the mean square response to be evaluated by the expression

$$\sigma_x^2(t) \cong \frac{\pi S_{\ddot{x}_g}(\omega_n)}{2\xi\omega_n^3}[1 - e^{-2\omega_n\xi t}] \tag{13.39}$$

Equation 13.39 differs from Eq. 13.33 only by the presence of the square bracket term that shows the rate at which the stationary value of σ_x^2 is reached. Applying L'Hospital's rule to Eq. 13.39, the limiting value of $\sigma_x^2(t)$ as $\xi \longrightarrow 0$ is found to be

$$\sigma_x^2(t) \cong \frac{\pi t S_{\ddot{x}_g}(\omega_n)}{\omega_n^2} \qquad \xi = 0 \tag{13.40}$$

Mean square relative velocities for the above case are obtained approximately by multiplying Eqs. 13.38–13.40 by ω_n^2.

13.5 APPLICATIONS IN EARTHQUAKE ENGINEERING

One of the most significant contributions to the field of earthquake engineering was the introduction of the idea of the earthquake response spectrum, which was first introduced by M. A. Biot, E. C. Robinson, G. Housner, and others. These spectra simply give the *maximum* response reached by the simple one-degree-of-freedom system, as shown in Fig. 13.3, when subjected to support accelerations $\ddot{x}_g(t)$ corresponding to the measured ground accelerations of past strong-motion earthquakes. Many such response spectra can be found in a report by Alford, Housner, and Martel (1951).

The relative velocity response spectrum as given by the approximate relation

$$V(\xi, T_n) = \left|\int_0^t \ddot{x}_g(\tau)e^{-2\pi\xi(t-\tau)/T_n}\sin\frac{2\pi}{T_n}(t-\tau)\,d\tau\right|_{\max} \tag{13.41}$$

is most commonly used since for low damped systems

the relative displacement response spectrum and approximate absolute acceleration spectrum can be obtained by simply multiplying the right-hand side of Eq. 13.41 by $(T_n/2\pi)$ and $(2\pi/T_n)$, respectively. T_n is the undamped period $2\pi/\omega_n$.

Using the eight components of the four strongest ground motions recorded to date, Housner normalized each accelerogram to a common intensity level, and by averaging the velocity spectra resulting therefrom he obtained what is now commonly known as Housner's standard velocity spectra (see Fig. 13.6; Housner, 1959; Housner, 1960). Even though the number of sample functions (eight) may be small in this case, one can still think of Housner's standard velocity spectra as giving the mean value of maximum (or peak) response due to earthquakes of a given intensity level $I(\xi)$. Housner has defined this intensity level as

$$I(\xi) = \int_{0.1}^{2.5} V(\xi, T_n)\, dT_n \tag{13.42}$$

which is generally considered a good measure of the expected damage to structures having fundamental periods in the range $0.1 < T_n < 2.5$.

In the above-mentioned studies dealing with response spectra, deterministic analyses were made using prescribed (measured) ground accelerations. Let us now consider nondeterministic analyses using ground motion inputs that are described in a statistical sense only.

It has been indicated by numerous investigators that ground acceleration as caused by earthquakes may be considered as being composed of a series of velocity impulses of random amplitude v_i, time spacing t_i, and wave shape (Housner, 1947; 1955; Goodman and Rosenblueth, 1955; Thompson, 1959; Rosenblueth, 1956; Rosenblueth and Bustamante, 1962). If the ensemble average $E[v_i^2]/E[t_i]$ of such wave forms remains constant $(2k)$ with time, a "white noise process" will result as $t_i \longrightarrow 0$. This suggests that, if the average velocity pulse spacing in actual earthquake accelerograms is considerably smaller than the natural period of any particular structure being considered, a "white noise" ground acceleration process may be assumed. It can be shown that the constant power spectral density S_0 for this process is related to the above ensemble average as follows:

$$S_0 = \frac{1}{\pi}\frac{E(v_i^2)}{E(t_i)} \tag{13.43}$$

Using an analog computer, Bycroft (1960) studied the possibility of using a "white noise" process to represent earthquake ground motion of a given intensity level. In these studies, Bycroft noted the maximum velocity response values resulting from 20 separate bursts of "white noise" input of 25 sec duration each. It was necessary, of course, in these studies, to limit the input band width having constant power spectral density to the range of 0–35 cps. To compare his average maximum response values with Housner's velocity spectra, Bycroft normalized his results to that power spectral density S_0 that would give full agreement with Housner's results for $T_n = 3$ sec and $\xi = 0.20$. This normalization criterion resulted in a value of S_0 equal to 0.75 ft²/sec³ [0.375, if both positive and negative frequencies are used in the function $S(\omega)$]. A further normalization of these same results so that they may be compared with Housner's *standard* velocity spectra requires that $S_0 = 0.0063$ ft²-rad/sec³ (when both + and − frequencies are considered). As shown in Fig. 13.6a, Bycroft's results would seem to indicate that "white noise" is a reasonable simulation of earthquake excitation.

Now let us attempt to predict maximum response directly by analytical methods when assuming the ground motion as a "white noise" stationary process of infinite duration $(s \longrightarrow \infty)$. Certainly, when observing any single earthquake accelerogram, one immediately gets the impression that the stationary assumption is invalid. It may be suggested, however, that since for typical values of damping ξ and earthquake duration s the statistical dependence of the value of $x(t+s)$ on the value of $x(t)$ is very low, as previously shown, one can justify placing many accelerograms of the same intensity end to end, thus establishing a truly stationary process. One should be cautious at this point in such reasoning, however, since any statistical dependence of the above two variables even though small can and does have large effects on the extreme tail ends of the probability density functions. Also, if an input process is established in this manner, one needs to consider the validity of basing the power spectral function of this stationary process on analyses of individual accelerograms.

Let us now proceed on the basis of a stationary "white noise" input of intensity $S_0 = 0.0063$ ft²-rad/sec³, which is the normalized intensity used in comparing Bycroft's results with Housner's standard velocity spectra. If one considers using the $3\,\sigma_{\dot{x}}$ level as given by Eq. 13.32 and restated again by Eq. 13.44 for design purposes, as is common practice in other fields of engineering, one would like to know how such results compare with Housner's standard velocity spectra:

$$3\sigma_{\dot{x}} = \left(\frac{9S_0T_n}{4\xi}\right)^{1/2} \tag{13.44}$$

Equation 13.44 is plotted using dashed lines in Fig. 13.6b where easy comparisons can be made with Housner's standard velocity spectra. You will note reasonable agreement in the short period range for damping even as low as 2% of critical. In the high period range reasonable agreement is seen for 10% damping; however, considerable differences are found to exist as damping decreases. Zero damping, of course, gives infinite response by this approach.

If one attempts to account for finite duration effects

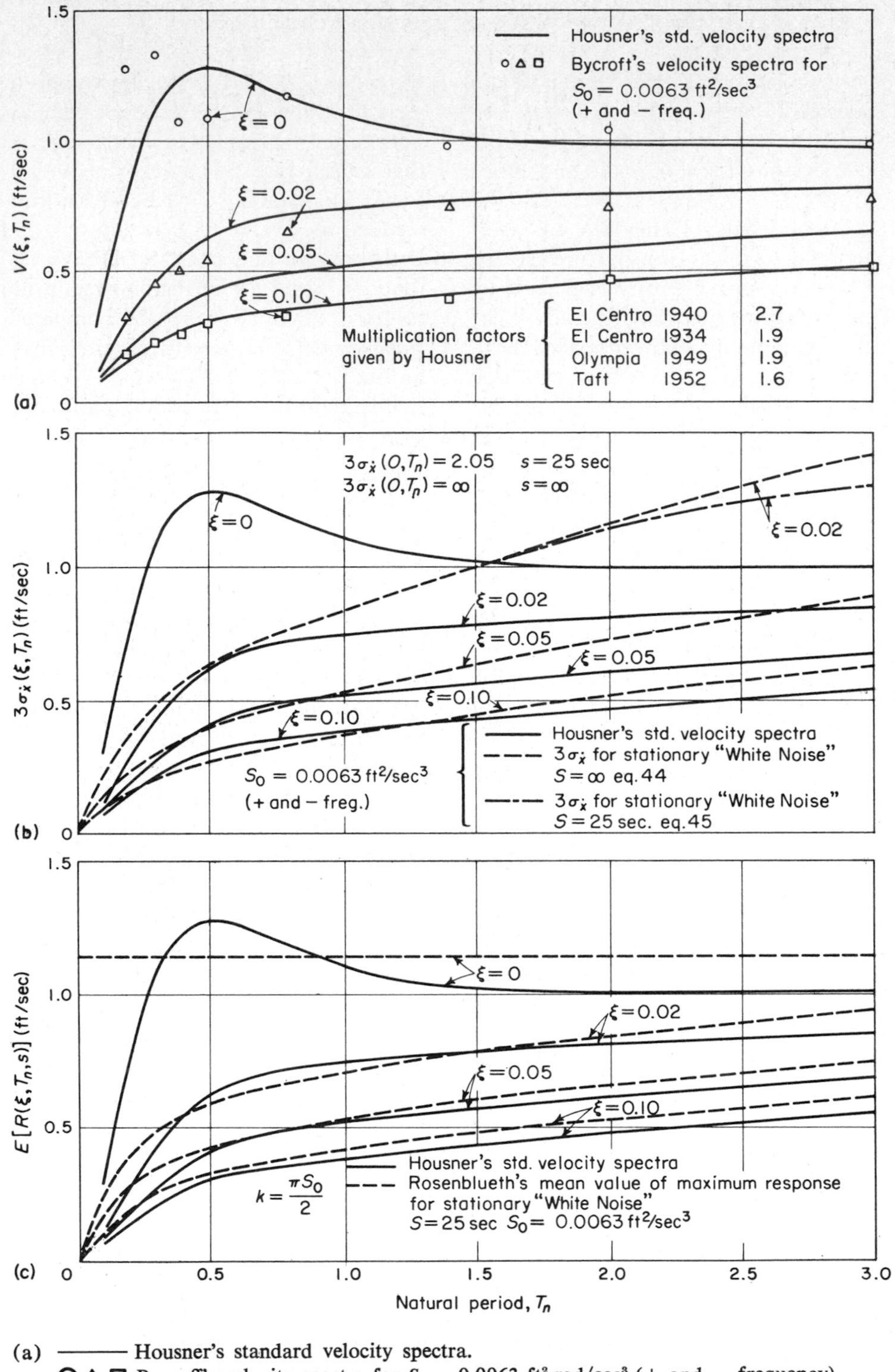

(a) ——— Housner's standard velocity spectra.
○ △ □ Bycroff's velocity spectra for $S_0 = 0.0063$ ft²-rad/sec³ (+ and − frequency).

Multiplication factors given by Housner { El Centro 1940 2.7; El Centro 1934 1.9; Olympia 1949 1.9; Taff 1952 1.6 }

(b) $3\sigma_{\dot{x}}(0, T_n) = 2.05$; $s = 25$ sec; $3\sigma_{\dot{x}}(0, T_n) = \infty$; $s = \infty$. ——— Housner's standard velocity spectra.

$S_0 = 0.0063$ ft²-rad/sec³ (+ and − frequency) { - - - - $3\sigma_{\dot{x}}$ for stationary "white noise," $s = \infty$ Eq. 13.44; — · — $3\sigma_{\dot{x}}$ for stationary "white noise," $s = 25$ sec Eq. 13.45 }

(c) ——— Housner's standard velocity spectra.
- - - - Rosenblueth's mean value of maximum response for stationary "white noise," $s = 25$ sec; $S_0 = 0.0063$ ft²-rad/sec³.

Fig. 13.6. Earthquake response spectra.

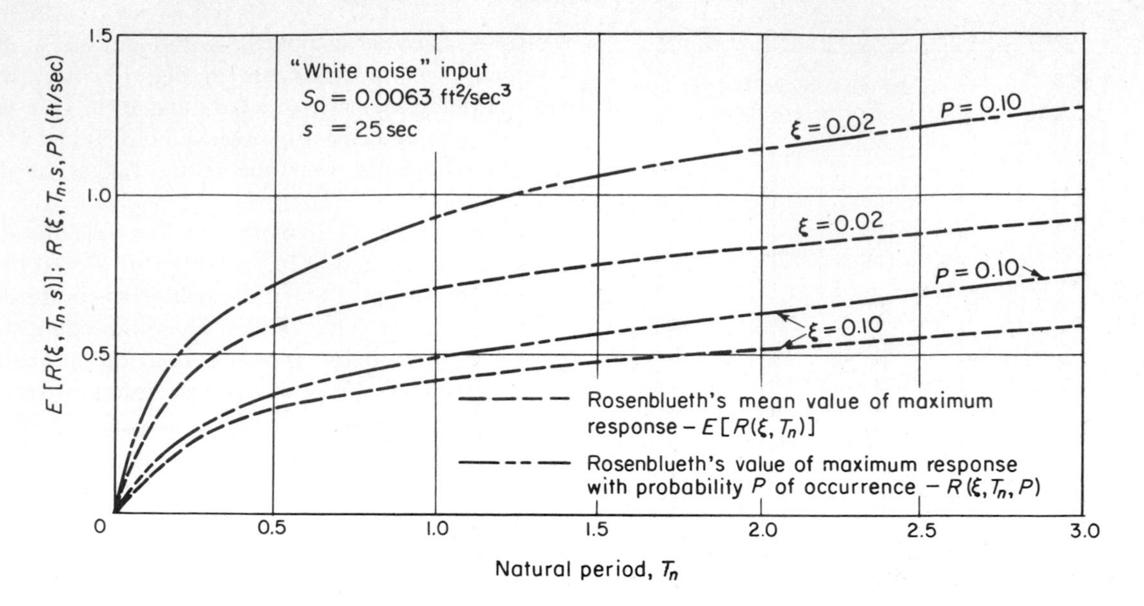

"White noise" input: $S_0 = 0.0063$ ft²/sec³; $s = 25$ sec.
Rosenblueth's mean value of maximum response—$E[R(\xi, T_n)]$.
Rosenblueth's value of maximum response with probability P of occurrence—$R(\xi, T_n, P)$.

Fig. 13.7. Earthquake response spectra.

by using the Caughey–Stumpf equation in a similar manner, i.e., using Eq. 13.39 written in the form of Eq. 13.45

$$3\sigma_{\dot{x}} = \left(\frac{9S_0T_n}{4\xi}\right)^{1/2}(1 - e^{-4\pi\xi s/T_n})^{1/2} \qquad (13.45)$$

one obtains the results as shown by the dash-dot-dash line in Fig. 13.6b for $s = 25$ sec. One will note some difference between Eqs. 13.44 and 13.45 at the long period end for the case of $\xi = 0.02$. However, these same two equations can be assumed to coincide for all practical purposes over the entire period range ($0 < T_n < 3$) for $\xi = 0.05$ and 0.10. In the case of $\xi = 0$, Eq. 13.44 yields an infinite response while Eq. 13.45 yields $3\,\sigma_{\dot{x}} = 2.05$ ft²/sec³ (see Eq. 13.40). Thus, it appears that the differences noted between the $3\,\sigma_{\dot{x}}$ level and Housner's standard spectra do not result nearly as much from the *apparent* effects of making the stationary assumption as they do from the fact that the $3\,\sigma_{\dot{x}}$ level is too high when considering cases of low damping and long periods. In other words, the assumed Rayleigh distribution of peak values becomes considerably in error at the "tail end" for such cases. Actually, the finite duration effects are involved in this distortion of the Rayleigh distribution.

A more accurate method of predicting the maximum or peak response when concerned with the longer periods and lower damping ratios and when assuming a "white noise" input process of finite duration s is the method presented by Rosenblueth and Bustamante (1962; also, Rosenblueth, 1964). These investigators, using $r(t)$ defined by the relation

$$r(t) \equiv e^{-\xi\omega_n t}\{[px(t)]^2 + [\dot{x}(t) + \xi\omega_n x(t)]^2\}^{1/2} \qquad (13.46)$$

as a measure of response and using two-dimensional random walk concepts and diffusion analogies, derived probability density functions for *peak* response that permit the numerical evaluations of the functions

$$E[R(\xi, T_n, s)] \quad \text{and} \quad R(\xi, T_n, s, P) \qquad (13.47)$$

where $E[R(\xi, T_n, s)]$ is defined as the ensemble average of peak response $|r(t)|_{\max} \equiv R(\xi, T_n, s)$, while $R(\xi, T_n, s, P)$ is defined as that value of peak response associated with probability P of being exceeded. Note that for very low damped systems one can assume

$$|\dot{x}(t)|_{\max} \cong |r(t)|_{\max} \qquad (13.48)$$

Rosenblueth and Bustamante used their theory in establishing the graphical data shown in Figs. 13.8 and 13.9.

Using the relation as given by Rosenblueth and Bustamante for ensemble average of peak response in the case of no damping ($\xi = 0$), i.e., using

$$E[R(0, T_n, s)] = 2.348\,(ks)^{1/2} = 2.348\left(\frac{\pi S_0 s}{2}\right)^{1/2} \qquad (13.49)$$

along with the results of Figs. 13.8 and 13.9, one can obtain the numerical values of $E[R(\xi, T_n, s)]$ and $R(\xi, T_n, s, P)$ by the equations

$$E[R(\xi, T_n, s)] = 2.348\,\beta_e\left(\frac{\pi S_0 s}{2}\right)^{1/2} \qquad (13.50)$$

and

$$R(\xi, T_n, s, P) = \beta\,R(0, T_n, s, P) \qquad (13.51)$$

Equation 13.50 is plotted in Fig. 13.6 where the

results can be compared with Housner's standard velocity spectra and again in Fig. 13.7 for two cases of damping, namely, $\xi = 0.02$ and 0.10. One will note quite good agreement between Eq. 13.50 and Housner's standard spectra over the entire ranges of damping and period.

Equation 13.51 is plotted in Fig. 13.7 for $P = 0.10$ and $\xi = 0.02$ and 0.10. A value of $P = 0.10$ simply means that 1 out of every 10 earthquakes in the ensemble will produce a value of $R(\xi, T_n, s)$ exceeding the value shown. It is quite apparent that the probability of exceeding a certain peak response decreases very rapidly with peak response. Further studies of such plots using additional values of P can provide one with a fairly good picture of the probability aspects of peak response

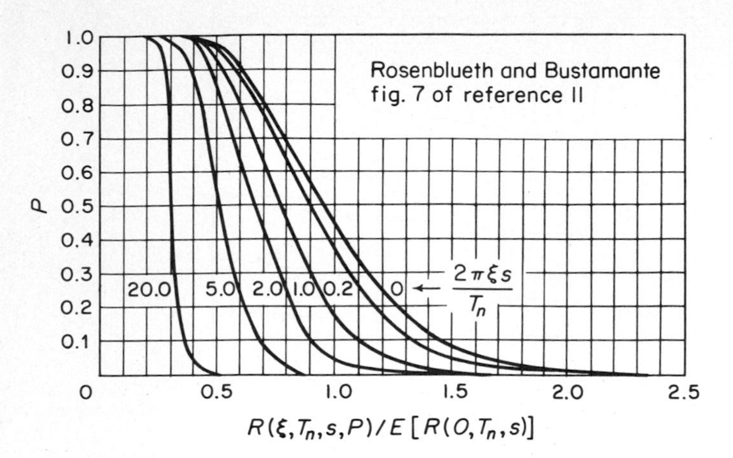

Fig. 13.8. Probability distribution of peak response.

Fig. 13.9. Rosenblueth and Bustamante (1962) β factors.

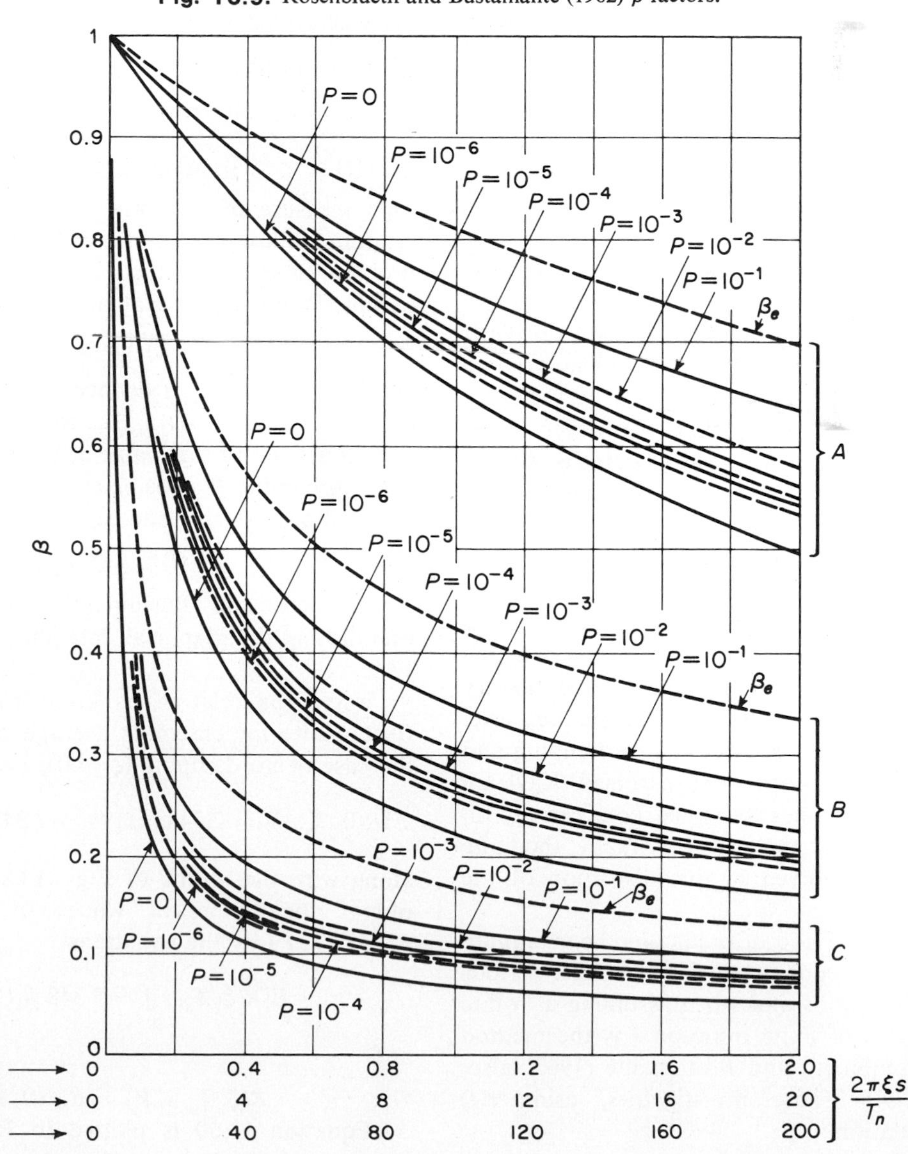

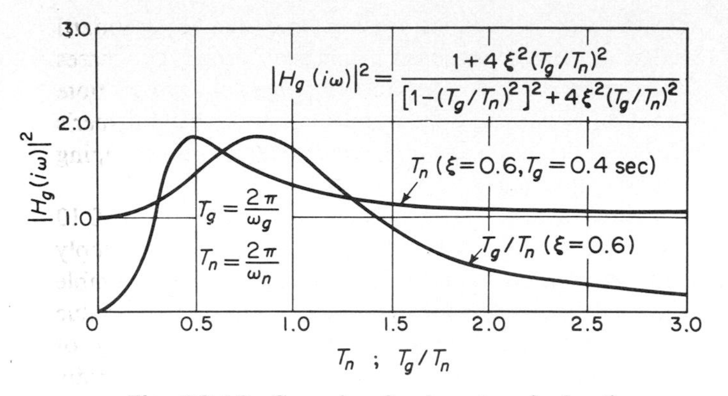

Fig. 13.10. Ground surface layer transfer function.

under the assumption of a "white noise" input of finite duration.

The above discussion on the applications of random vibration theory in earthquake engineering assumes ground motion can be represented by a stationary "white noise" process. Finite duration effects have been considered in this case.

Let us now turn our attention to the question regarding the validity of using a flat or constant power spectral density function. Since we are interested in the response of low damped systems, we need not be concerned about the intensity of the power spectral density function for frequencies considerably above the frequency range of our structures being considered; however, for frequencies within the frequency range of these structures the variation of intensity with frequency is very important.

Ground measurements show that the assumption of a flat spectrum is considerably better at bedrock level than it is at the surface of the weaker, less stiff overburden layers. Analyses of many accelerograms show that the power spectral density functions obtained therefrom, on the basis of an integration over the complete duration of each individual accelerogram separately, all have similar characteristics, namely, (1) they are all highly oscillatory in a random fashion, (2) smooth curves drawn through these oscillatory functions peak at frequencies ω_g (or periods T_g) usually in the range $0.2 < T_g < 1.0$, and (3) these smooth functions decay rapidly with increasing frequencies. Kanai and Tajimi have suggested that such smooth curves are approximately given by the function (Kanai, 1957; Tajimi, 1960; Tajimi, 1964)

$$S_{\ddot{x}_0}(\xi_g, \omega_g, \omega) = \frac{1 + 4\xi_g^2(\omega/\omega_g)^2}{[1-(\omega/\omega_g)^2]^2 + 4\xi_g^2(\omega/\omega_g)^2} S_0 \quad (13.52)$$

where ξ_g and ω_g may be thought of as ground damping factor and predominant frequency, respectively, and where S_0 is a constant power spectral density function (see Fig. 13.10). It has been suggested by H. Tajimi (1964) that for firm ground conditions ω_g be taken as $5\pi(T_n = 0.4 \text{ sec})$ and S be taken as 0.6. He cautions, however, that these values should be adjusted to fit local ground conditions. It is apparent that only after carefully studying the power spectral density functions of many strong-motion accelerograms can these factors be established for the various ground conditions commonly encountered in engineering practice. One should note that the entire factor on the right-hand side of Eq. 13.52 that is multiplied by S_0 to obtain $S_{\ddot{x}_g}$ is actually the transfer function $|H_{\ddot{x}_t}(i\omega)|^2$ for a single-degree-of-freedom system having a natural frequency equal to ω_g and a damping ratio equal to ξ_g and when support acceleration $\ddot{x}_g(t)$ is the input and absolute acceleration $\ddot{x}_t(t)$ is the output.

Power spectral density functions for N–S and E–W components of the 1940 El Centro ground motion as recorded are shown in Figs. 13.11a and b. These functions, which were published by Ravara (1965), are

Fig. 13.11. Power spectral density functions.

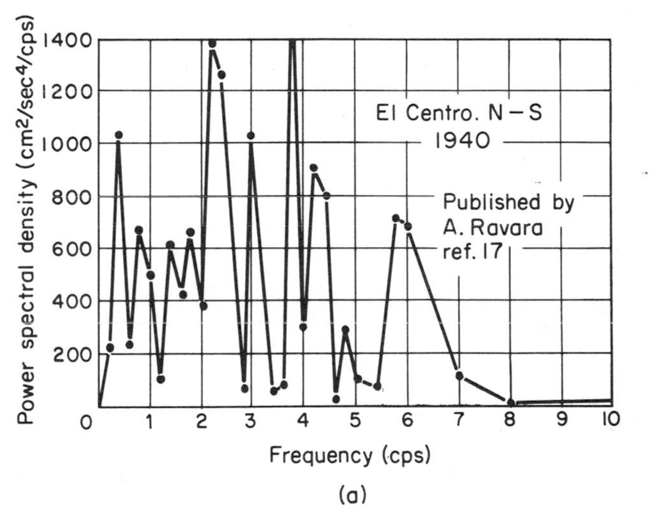

(a)

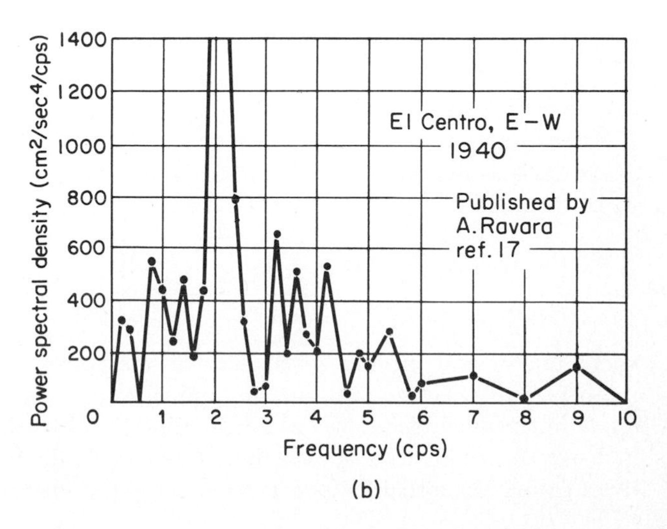

(b)

shown here as examples of how typical power spectral density functions appear.

Once a smooth power spectral density function $S_{\ddot{x}_0}(\xi_g, \omega_g, \omega)$ has been established for a given site, whether it be of the form of Eq. 13.52 or otherwise, one can approximate the random response of a single-degree-of-freedom system to such a stationary input by assuming "white noise" as the input as previously discussed, but now the constant power spectral density S_0 must be considered as a function of ω_n, i.e., one should use a constant $S_{\ddot{x}_0}(\xi_g, \omega_g, \omega_n)$. The justification of making this simplification lies in the fact that response is due primarily to the frequencies in the input that are near the natural frequency ω_n of the system.

One should always keep in mind that the smooth function $S_{\ddot{x}}(\xi_g, \omega_g, \omega_n)$ is essentially a *mean* spectral density function. While one may be able to establish the probability density function for maximum response using this mean spectral density function as the specified input, one cannot at the present time predict the probability density function for maximum response due to actual future earthquakes. In other words, to establish such a probability density function it is not sufficient to use just a *mean* power spectral density function in establishing it but, in addition, one must consider the joint probability density function

$$p[S_{\ddot{x}}(\xi_g, \omega_g, \omega_n), \omega_n] \tag{13.53}$$

which, of course, does not exist at the present time due to lack of strong ground motion statistical information.

All applications of random theory as previously discussed herein have made use of a stationary process that was somewhat justified on the basis of the fact that the quantity $e^{-\omega_n \xi t}$, which appears in the response autocorrelation function, decays rapidly with t for typical values of ξ and ω_n. However, since earthquakes are not stationary in the strict sense, the response probability density functions obtained on the basis of the stationary assumption are admittedly in error to a certain degree. The error increases as one moves into the "tail" regions of these functions. Since maximum or peak response is of greatest interest to the engineer, accuracy even in these "tail" regions is important.

There have been some attempts to study earthquake response using nonstationary input processes. Bogdanoff, Goldberg, and Bernard (1961) assumed an input process of the form

$$\left.\begin{aligned} {}^j\ddot{x}_g(t) &= \sum_1^n t a_j e^{-\alpha_j t} \cos(\omega_j t + \phi_j) \qquad & t > 0 \\ {}^j\ddot{x}_g(t) &= 0 & t \leqslant 0 \end{aligned}\right\} \tag{13.54}$$

where the a_j and ω_j are given sets of real positive numbers with $\omega_1 < \omega_2 < \cdots < \omega_n$ and $\phi_1, \phi_2, \ldots, \phi_n$ are n independent random variables uniformly distributed over the interval 0–2π. Using such an input these investigators generated output processes for the linear single-degree-of-freedom system and made statistical studies of them to establish probability density functions for response. Due to the lack of strong ground motion data it is, of course, difficult at this time to establish the validity of using the above-described nonstationary input process (Eqs. 13.54).

Before ending this general discussion a few words should be said about applying random theory to nonlinear systems. It is usually necessary to first establish an input process of known characteristics, to determine each member of the output process by a separate deterministic analysis, and then to study the output process statistically to determine the necessary probability density functions. Of course, all of the problems previously discussed regarding defining a proper ground-acceleration process still remain to be solved in this case. Jennings (1963, 1965) has conducted extensive studies on simple yielding structures using a stationary, Gaussian input process having a power spectral density function as determined from the average undamped velocity spectrum of linear systems. Goldberg, Bogdanoff, and Sharpe (1964) also have studied some simple nonlinear systems but using a nonstationary input process as defined by

$$\left.\begin{aligned} {}^j\ddot{x}_g(t) &= \sum_1^n t a_j e^{-\alpha_j t} \cos(\Omega_j t + \phi_j) \qquad & t > 0 \\ {}^j\ddot{x}_g(t) &= 0 & t < 0 \end{aligned}\right\} \tag{13.55}$$

where $\Omega_1, \Omega_2, \ldots, \Omega_n$ are considered as independent random variables which are independent of random variables $\phi_1, \phi_2, \ldots, \phi_n$. Both these sets of random variables are considered to have uniform probability distributions. Bycroft (1960) has also studied nonlinear systems.

REFERENCES

Alford, J. L., G. W. Housner, and R. R. Martel (1951). *Spectrum Analyses of Strong-Motion Earthquakes*, Pasadena: California Institute of Technology, Earthquake Research Laboratory.

Bogdanoff, J. L., J. E. Goldberg, and M. C. Bernard (April 1961). "Response of a Simple Structure to a Random Earthquake-Type Disturbance," *Bull. Seism. Soc. Am.*, **51** (2), 293–310.

Bycroft, G. N. (April 1960). "White Noise Representation of Earthquakes," *Proc. Am. Soc. Civil Engr.* **86** (EM2), 1–16.

Caughey, T. K., and H. J. Stumpf (December 1961). "Transient Response of a Dynamic System Under Random Excitation," *J. Appl. Mech.* **28** (4), 563–566.

Crandall, S. H., and W. D. Mark (1963). *Random Vibration in Mechanical Systems*, New York: Academic Press.

Goldberg, J. E., J. L. Bogdanoff, and D. R. Sharpe (February

1964). "The Response of Simple Nonlinear Systems to a Random Disturbance of the Earthquake Type," *Bull. Seism. Soc. Am.*, **54** (1), 263–276.

Goodman, L. E., E. Rosenblueth, and N. M. Newmark (1955). "Aseismic Design of Firmly Founded Elastic Structures," *Trans. Am. Soc. Civil Engr.*, **120**, 728–802.

Housner, G. W. (January 1947). "Characteristics of Strong-Motion Earthquakes," *Bull. Seism. Soc. Am.*, **37** (1), 9–31.

Housner, G. W. (July 1955). "Properties of Strong Ground Motion Earthquakes," *Bull. Seism. Soc. Am.*, **45** (3), 197–218.

Housner, G. W. (October 1959). "Behavior of Structures During Earthquakes," *Proc. Am. Soc. Civil Engr.*, **85** (EM4), 109–129.

Housner, G. W. (1960). *Design of Nuclear Power Reactor Against Earthquakes, Proceedings of the Second World Conference on Earthquake Engineering*, Vol. I, Tokyo.

Jennings, P. C. (July 1963). *Response of Simple Yielding Structures to Earthquake Excitation*, Pasadena: California Institute of Technology, Earthquake Engineering Research Laboratory.

Jennings, P. C. (1965). *Response of Yielding Structures to Statistically Generated Ground Motion, Proceedings of the Third World Conference on Earthquake Engineering*, Auckland and Wellington, New Zealand.

Kanai, K. (June 1957). "Semi-empirical Formula for the Seismic Characteristics of the Ground," *Bull. Earth. Res. Inst., Univ. Tokyo*, **35**, 309–325.

Ravara, A. (1965). *Spectral Analysis of Seismic Actions, Proceedings of the Third World Conference on Earthquake Engineering*, Auckland and Wellington, New Zealand.

Rosenblueth, E. (June 1956). *Some Applications of Probability Theory in Aseismic Design, World Conference on Earthquake Engineering*, Berkeley, California.

Rosenblueth, E., and J. I. Bustamante (June 1962). "Distribution of Structural Response to Earthquakes," *Proc. Am. Soc. Civil Engr.* **88** (EM3), 75–106.

Rosenblueth, E. (October 1964). "Probabilistic Design to Resist Earthquakes," *Proc. Am. Soc. Civil Engr.* **90** (EM5), 189–219.

Tajimi, H. (July 1960). *A Statistical Method of Determining the Maximum Response of a Building Structure During an Earthquake, Proceedings of the Second World Conference on Earthquake Engineering*, Vol. II, Tokyo and Kyoto, Japan.

Tajimi, H. (1964). *Introduction to Structural Dynamics*, Tokyo, Japan: Corona-sha.

Thomson, W. T. (January 1959). "Spectral Aspect of Earthquakes," *Bull. Seism. Soc. Am.*, **49**, 91–98.

Chapter 14

Soil–Pile Foundation Interaction

JOSEPH PENZIEN

Professor of Civil Engineering and
Director, Earthquake Engineering Research Center,
University of California, Berkeley, California

14.1 INTRODUCTION

The improvement of modern traffic systems requires in certain localities that major bridges be constructed across coastal estuaries which, when located in active seismic regions, present an extremely difficult design problem due to the poor foundation conditions usually present. Under such conditions, the bridge piers are normally supported on piles driven through a deep layer of relatively weak material to a penetration in the underlying firm material deemed sufficient to carry the superimposed loads by friction or by point bearing. Since structural types such as buildings also may be supported in this manner, the seismic response investigation reported in this chapter can be considered a case study of the seismic effects on pile-supported structures.

The California State Division of Highways is presently (1968) designing a bridge approximately 3000 ft in length to be erected across Elkhorn Slough, which is located near Monterey, California. At this bridge site a surface layer of saturated soft to firm gray silty clay is present that varies in depth from approximately 50 ft at the proposed bridge abutment locations to approximately 120 ft midway between these locations. A very shallow layer of water covers the clay layer, and a well-compacted sand lies below the clay layer. In order to aid the State Division of Highways in carrying out a rational design of a bridge-pile structural system for this site, the investigation reported herein was undertaken to determine the interaction effects between the bridge structure and its supporting piles and between the supporting piles and the clay medium when subjected to a prescribed strong-motion earthquake excitation.

14.2 METHODS OF ANALYSIS

14.2.1 General

The purpose of this section is to present a general method for investigating the seismic effects on bridges that are supported on long piles extending through deep, sensitive clays. The analysis is separated into two parts; (1) the determination of the dynamic response of the clay medium alone when excited through its lower boundary by a prescribed horizontal seismic motion and (2) the determination of the interaction of the entire structural system, including the piles, with the moving clay medium. Application of the method to a specific bridge structure now being designed by the California State Division of Highways is presented in considerable detail.

14.2.2 Dynamic Response of Clay Medium

The first part of the general analysis is the determination of the dynamic response of the clay medium alone without any bridge structure being present. Since the deformations produced in this medium by a horizontal excitation are essentially pure shear, the real system will be represented by the idealization shown in Fig. 14.1. It is necessary in this case to use a discrete parameter system due to the nonlinearities and hysteresis effects involved in the shear stress–strain relations for the clay medium. Thus, the basic idealized system consists of a series of discrete masses connected by nonlinear linkages that resist relative lateral shearing deformation.

Since the clay medium is assumed to be of infinite extent in the horizontal plane and of constant depth, the discrete properties of the mathematical model may be based on a column of clay having a unit cross-

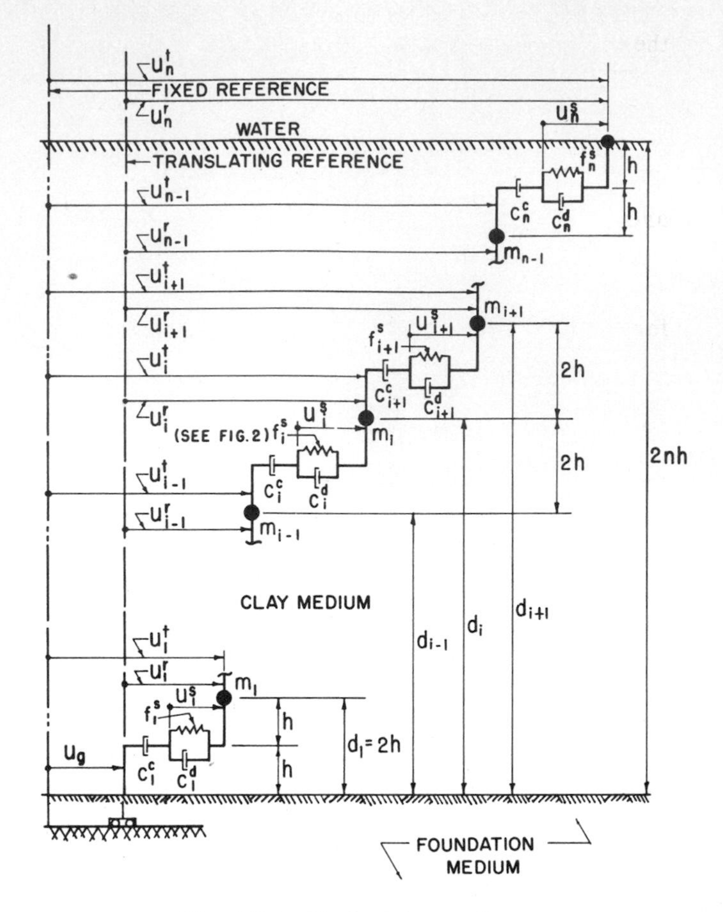

Fig. 14.1. Idealized clay medium.

sectional area and a height equal to the depth of the clay layer. Using this column as the physical system, its mass will be lumped at discrete points uniformly spaced, as shown in Fig. 14.1. Thus,

$$m_1 = m_2 = \cdots = m_{n-1} = \frac{2\gamma h}{g} \tag{14.1a}$$

$$m_n = \frac{\gamma h}{g} \tag{14.1b}$$

in which γ is the unit weight of clay medium, g denotes the acceleration of gravity, and $2h$ is the vertical spacing of the discrete masses. The number of discrete masses (n) is selected of sufficient magnitude so that the response of the idealized model to any prescribed strong-motion earthquake input will adequately represent the response of the continuous system.

Each of the linkages connecting adjacent pairs of masses consists of a bilinear hysteresis-type spring having the idealized force-displacement characteristics shown in Fig. 14.2 and a nonlinear dashpot placed in parallel that in turn is placed in series with a second nonlinear dashpot. These three linkage components represent the elasto-plastic, damping, and creep properties, respectively, of the clay medium. Based on the discrete parameter model (Fig. 14.1), the nonlinear coupled differential equations of motion that define the dynamic response of the clay medium are the following:

$$m_i\ddot{u}_i^r + c_i^d\dot{u}_i^s - c_{i+1}^d\dot{u}_{i+1}^s + f_i^s h_i\left(\frac{t}{T}\right) - f_{i+1}^s h_{i+1}\left(\frac{t}{T}\right) = -m_i\ddot{u}_g \quad (14.2)$$

and

$$c_i^c(\dot{u}_i^r - \dot{u}_{i-1}^r - \dot{u}_i^s) - c_i^d\dot{u}_i^s - f_i^s h_i\left(\frac{t}{T}\right) = 0 \quad (14.3)$$

for $i = 1, 2 \ldots, n$, in which

$$c_i^c \equiv [c_i^{co} + c_i^{c'}|\dot{u}_i^r - \dot{u}_{i-1}^r - \dot{u}_i^s| + c_i^{c''}(\dot{u}_i^r - \dot{u}_{i-1}^r - \dot{u}_i^s)^2]f_i\left(\frac{t}{T}\right) \quad (14.4)$$

and

$$c_i^d \equiv [c_i^{do} + c_i^{d'}|\dot{u}_i^s|]g_i\left(\frac{t}{T}\right) \quad (14.5)$$

Note that the time dependent coefficients c_i^c and c_i^d ($i = 1, 2, \ldots, n$) as defined by Eqs. 14.4 and 14.5 and the spring force terms given in Eqs. 14.2 and 14.3 include the dimensionless time functions $f_i(t/T)$, $g_i(t/T)$, and $h_i(t/T)$, respectively. These functions will be defined so that they equal unity at time zero and decay with time thereafter in such a way that changes in clay characteristics due to remolding can be included in the dynamic analysis. These changes in characteristics are undoubtedly dependent on the entire past history of response. However, the appropriate history dependency is extremely difficult to establish. The dimensionless time functions as well as the constants appearing in Eqs. 14.4 and 14.5 can be determined only by extensive laboratory tests performed on clay samples taken from the bridge site. A discussion of these specific tests and their interpretation toward establishing the preceding characteristics of the clay medium are presented in a subsequent section of this chapter.

In the proposed analysis, the clay medium is excited through its lower boundary by a prescribed horizontal acceleration ($\ddot{u}_g$) of the foundation medium, which is assumed to be time dependent only, i.e., space variations in the horizontal direction of motion are ignored. This excitation is represented by the term ($-m_i\ddot{u}_g$) on the right side of Eq. 14.2. Ground accelerations as recorded by the Coast and Geodetic Survey, U.S. Department of Commerce (USCGS) for several past strong-motion earthquakes are used as the prescribed inputs. Solutions of the coupled differential Eqs. 14.2 and 14.3 are obtained assuming zero initial conditions and using known step-by-step methods of analysis.

Fig. 14.2. Force-displacement characteristics of springs.

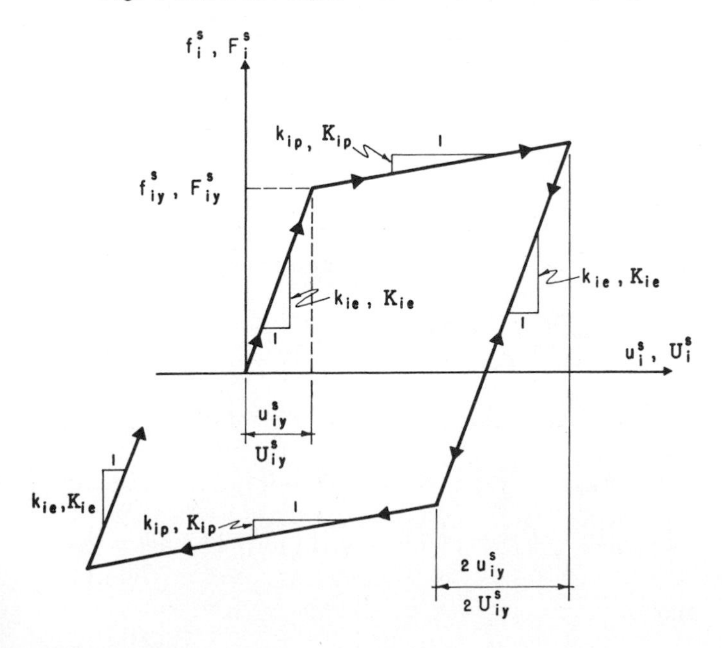

14.2.3 Dynamic Response of the Bridge Structural System

The specific bridge structural system used in this investigation is shown schematically in Fig. 14.3. This system includes the bridge itself, the supporting piers, and the piles. The dynamic response of this entire bridge structural system in the direction of the longitudinal axis of the bridge, which results from its interaction with the clay and foundation media, is determined using the idealized discrete parameter system shown in Fig. 14.4. This idealized system characterizes a typical longitudinal section of the bridge structure that is located between two transverse expansion joints.

Since there are two distinct types of pile groupings in the bridge structural system being considered, namely, a single row grouping and a multiple row grouping, it is necessary that the lumped mass M^d representing the rigid deck girder mass be connected to the foundation medium by two separate but parallel discrete parameter systems having somewhat different properties. The discrete parameters that characterize this idealized structural system are the following:

1. A single mass M^d representing the mass of the bridge deck and girders.
2. Masses m_i^b ($i = 1, 2, \ldots, m - 1$) representing the mass of the piers.
3. Stiffness influence coefficients k_{ij}^b, $k_{i\theta}^b$ ($i, j = 1, 2, \ldots, m$) representing the elastic force characteristics of the piers.
4. Stiffness influence coefficients $k_{\theta j}^b$, $k_{\theta\theta}^b$ ($j = 1, 2, \ldots, m$) representing the elastic moments that are applied to the pile caps by the piers.
5. Linear viscous dashpots having constant coefficients C_i^b ($i = 1, 2, \ldots, m$) that represent damping present in the piers and bridge deck.
6. Mass M_n representing the mass of a pile cap that has rotary moment of inertia equal to I_n.

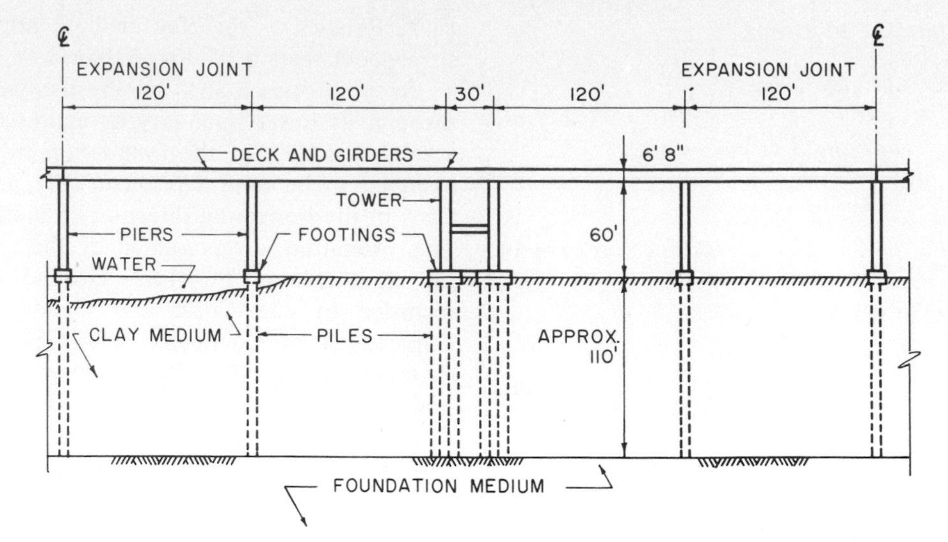

Fig. 14.3. Bridge structural system.

7. A linear rotational spring (spring constant K_θ) representing the elastic moment resistance to a rotation of the pile cap at the tower as produced by the resulting axial forces developed in the pile grouping.

8. Linear viscous dashpot having constant coefficient C_θ that represents damping due to rotational motion of the pile cap.

9. Stiffness influence coefficients k^a_{ij}, $k^a_{i\theta}$ ($i = 1, 2, \ldots, n$; $j = 1, 2, \ldots, n-1$) representing the elastic force characteristics of the piles.

10. Stiffness influence coefficients $k^a_{\theta j}$, $k^a_{\theta\theta}$ ($j = 1, 2, \ldots, n-1$) representing the elastic moments that are applied to the pile caps by the piles.

11. Bilinear hysteresis-type springs representing the interaction forces F^s_i (excluding damping) between the clay medium and piles that have characteristics similar to those represented in Fig. 14.2.

12. Masses M_i ($i = 1, 2, \ldots, n-1$) representing the mass of the piles that include an effective mass of the clay medium (M^e_i: $i = 1, 2, \ldots, n-1$) resulting from the relative motion caused by interaction.

13. Nonlinear time dependent viscous dashpots having the variable coefficients C^c_i ($i = 1, 2, \ldots, n$) that represent the creep properties of the clay medium resulting from its interaction with the piles.

14. Nonlinear time dependent viscous dashpots having the variable coefficient C^d_i ($i = 1, 2, \ldots, n$) that represent the damping characteristics of the clay medium resulting from the elastic deformations produced by interaction with piles.

15. Axial forces S_i ($i = 1, 2, \ldots, m$) representing the vertical imposed loads carried by the piers, and axial forces T_i ($i = 1, 2, \ldots, n$) representing the vertical imposed loads carried by the piles.

Referring to Fig. 14.4, the idealized model will be excited simultaneously at its base by the prescribed horizontal acceleration $\ddot{u}_g$ of the foundation medium and at each level i ($i = 1, 2, \ldots, n$) by a known support acceleration time history u^t_i that represents the acceleration time history of the clay medium at level i assuming no interaction with the bridge structure (see Fig. 14.1). The coupled nonlinear differential equations of motion that govern the response of this idealized bridge structural system are the following:

$$[M_i]\{\ddot{U}^r_i\} + [C^d_i]\{\dot{U}^s_i\} + [F^s_i] + [k^a_{ij}]\left\{U^r_i - \frac{d_i}{d_n}U^r_n\right\} + \left(\theta - \frac{U^r_n}{d_n}\right)\{k^a_{i\theta}\} + [T_{ij}]\}U^r_i\} = -\ddot{u}_g[M_i]\{I\} + [M^e_i]\{\ddot{u}^t_i\} \tag{14.6}$$

$$[m^b_i]\{\ddot{U}^b_i + \ddot{U}^r_n\} + [C^d_i]\{\dot{U}^b_i\} + [k^b_{ij}]\{U^b_i\} + \theta\{k^b_{i\theta}\} + [S_{ij}]\{U^b_i\} = -\ddot{u}_g[m^b_i]\{I\} \tag{14.7}$$

$$M_n\ddot{U}^r_n + C^d_n\dot{U}^s_n + F^s_n + < k^a_{nj} >\left\{U^r_i - \frac{d_i}{d_n}U^r_n\right\} + k^a_{n\theta}\left(\theta - \frac{U^r_n}{d_n}\right) + < k^b_{0j} >\{U^b_i\} + \frac{T_n}{d_n - d_{n-1}}U^r_{n-1} - \left(\frac{T_n}{d_n - d_{n-1}} + \frac{S_i}{e_i}\right)U^r_n + \frac{S_1}{e_1}(U^r_n + U^b_i) = -\ddot{u}_g M_n + M^e_n\ddot{u}^t_n \tag{14.8}$$

$$M^d(\ddot{\bar{U}}^b_m + \ddot{\bar{U}}^r_n) + \bar{C}^d_m\dot{\bar{U}}^b_m + < \bar{k}^b_{mj} >\{\bar{U}^b_i\} + < \bar{\bar{k}}^b_{mj} >\{\bar{\bar{U}}^b_i\} + \bar{\theta}\bar{k}^b_{m\theta} + \bar{\bar{\theta}}\bar{\bar{k}}^b_{m\theta} + \frac{\bar{S}_m(\bar{U}^b_{m-1} - \bar{U}^b_m)}{\bar{e}_m - \bar{e}_{m-1}} + \frac{\bar{\bar{S}}_m(\bar{\bar{U}}^b_{m-1} - \bar{\bar{U}}^b_m)}{\bar{\bar{e}}_m - \bar{\bar{e}}_{m-1}} = -\ddot{u}_g M_d \tag{14.9}$$

$$I_n\ddot{\theta} + < k^b_{\theta j} >\{U^b_i\} + k^b_{\theta\theta}\theta + < k^a_{\theta j} >\left\{U^r_i - \frac{d_i}{d_n}U^r_n\right\} + k^a_{\theta\theta}\left(\theta - \frac{U^r_n}{d_n}\right) + K_\theta\theta + C_\theta\dot{\theta} = 0 \tag{14.10}$$

$$[C^c_i]\{\dot{U}^r_i - \dot{U}^s_i\} - [C^d_i]\{\dot{U}^s_i\} - [F^s_i] = [C^c_i]\{\dot{u}^t_i - \dot{u}_g\} \tag{14.11}$$

and

$$\bar{U}^r_n + \bar{U}^b_m = \bar{\bar{U}}^r_n + \bar{\bar{U}}^b_m \tag{14.12}$$

Matrix Eq. 14.6 represents the horizontal force equilibrium equations of motion for masses $M_1, M_2, \ldots, M_{n-1}$ of pile group 1, thus giving $(n - 1)$ individual equations. Similarly, Eq. 14.7 represents the horizontal force equilibrium equations of motion for masses $m_1^b, m_2^b, \ldots, m_{m-1}^b$, therefore giving an additional $(m - 1)$ individual equations. The horizontal force equilibrium equation of motion for masses M_n is given by Eq. 14.8. Similar equations to those given by Eqs. 14.6, 14.7, and 14.8 must be written for corresponding masses in pile group 2. In addition, the equation of motion for mass M^d must be written as given by Eq. 14.9. Note that in this equation (also in Eq. 14.12) a bar is placed above those quantities representing pile group 1 and a double bar is placed above those quantities representing pile group 2.

The horizontal forces included in the above equations are inertia, damping, interaction, transverse shear, net horizontal components of axial loads, and effective seismic loads. Equation 14.10 is the moment equilibrium equation of motion for mass M_n. In addition to these equilibrium equations of motion, it is necessary that force balance be maintained throughout each individual linkage system representing the pile–clay interaction characteristics. The required equation that satisfies this condition for pile group 1 is matrix Eq. 14.11. A similar equation, of course, must be written for pile group 2. The final equation which must be written to satisfy the condition of continuity of horizontal displacement at mass M^d for pile groups 1 and 2 is given by Eq. 14.12.

The preceding differential equations of motion contain numerous time dependent coefficients that permit nonlinearities to be present in the creep and damping characteristics of the clay medium and also permit changes in these characteristics due to remolding. These variable coefficients are defined in a similar manner to those given by Eqs. 14.4 and 14.5 and may be represented as follows:

$$C_i^c = [C_i^{co} + C_i^{c'}|\dot{U}_i^r - \dot{U}_i^s - \dot{u}_i^r| + C_i^{c''}(\dot{U}_i^r - \dot{U}_i^s - \dot{u}_i^r)^2]F_i\left(\frac{t}{T}\right) \quad (14.13)$$

and

$$C_i^d = [C_i^{do} + C_i^{d'}|\dot{U}_i^s|]G_i\left(\frac{t}{T}\right) \quad (14.14)$$

The dimensionless time functions as well as the constants appearing in Eqs. 14.13 and 14.14 will be determined by considering the results of laboratory tests performed on clay samples taken from the bridge site. Many of these functions and constants are of identical form and magnitude, respectively, in the final analysis of the general problem. All of the previously coupled differential equations of motion representing the bridge structural system are solved numerically, assuming zero initial conditions.

14.2.4 Evaluation of Discrete Parameters of Bridge Structural System

14.2.4.1 Mass M^d. This quantity represents the total mass of the barrier railings, deck slab, girders, and diaphragms between the transverse expansion joints

Fig. 14.4. Idealized structural system.

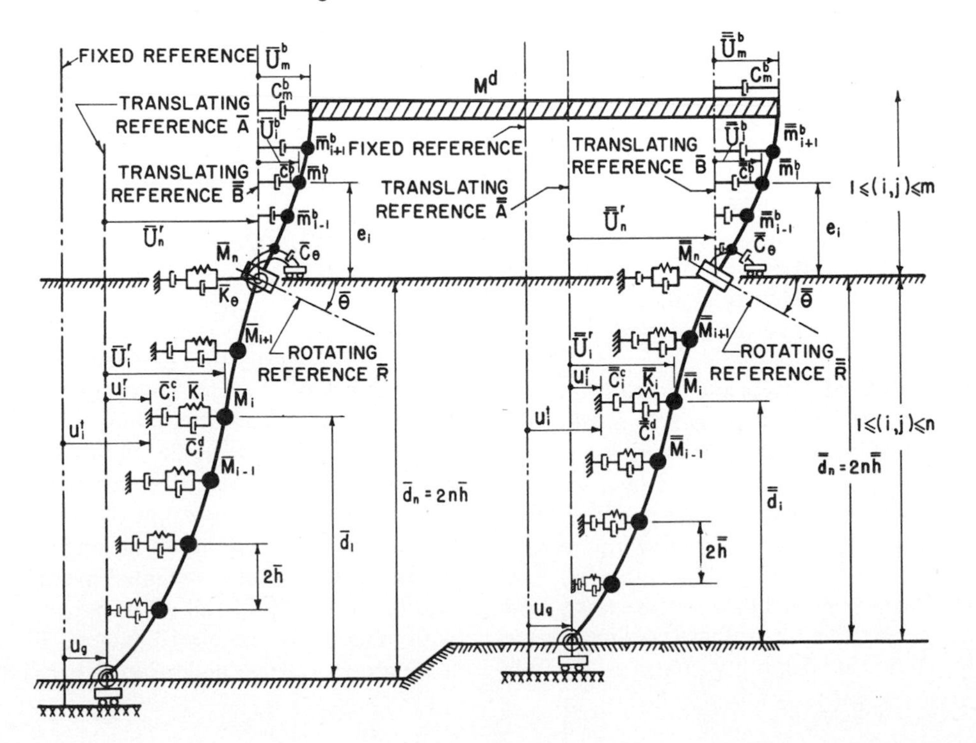

(510 ft), plus the mass of the piers over the height $(e_m - e_{m-1})/2$.

14.2.4.2 Masses $\bar{m}_i^b$ and $\bar{\bar{m}}_i^b$. These masses are defined as follows:

$$\bar{m}_i^b = 2\frac{W^p}{g}\left(\frac{\bar{e}_{i+1} - \bar{e}_{i-1}}{2}\right) \tag{14.15a}$$

$$\bar{\bar{m}}_i^b = 3\frac{W^p}{g}\left(\frac{\bar{\bar{e}}_{i+1} - \bar{\bar{e}}_{i-1}}{2}\right) \tag{14.15b}$$

in which W^p denotes the weight of single pier per foot of height.

14.2.4.3 Stiffness influence coefficients $\bar{k}_{ij}^b$ and $\bar{\bar{k}}_{ij}^b$ $(i, j = 1, 2, \ldots, m, \theta)$. These coefficients are obtained by standard methods of structural analysis, i.e., by inverting a flexibility coefficient matrix representing the pier elastic properties. In the evaluation of the flexibility coefficient matrix, for the case of longitudinal motion, the top of the pier is allowed to translate but not rotate, whereas the bottom is assumed to be pinned.

14.2.4.4 Damping coefficients $\bar{C}_i^b$ and $\bar{\bar{C}}_i^b$. These coefficients define viscous damping forces for the bridge superstructure that are assumed proportional to the velocity of their respective masses relative to the base of the pier. All damping coefficients are based on a percentage of critical damping of the system.

14.2.4.5 Masses $\bar{M}_n$ and $\bar{\bar{M}}_n$. These masses represent the total mass of the pile caps for pile groups 1 and 2, respectively, plus a contribution from the piers to a height of $e_1/2$ and a contribution from the piles to a depth of $(d_n - d_{n-1})/2$.

14.2.4.6 Rotary mass moments of inertia $\bar{I}_n$ and $\bar{\bar{I}}_n$. These terms represent the combined mass rotary moments of inertia of the pile caps for groups 1 and 2, respectively.

14.2.4.7 Spring constant $\bar{K}_\theta$. Because of the multiple row grouping of the piles at the tower piers there will be an elastic resistance to rotation of the pile cap due to the axial forces developed in the piles as a result of this rotation. This resistance to rotation is expressed by the rotational spring constant $\bar{K}_\theta$.

14.2.4.8 Damping coefficients $\bar{C}_\theta$ and $\bar{\bar{C}}_\theta$. The rotational damping of the pile caps is due to the dissipation of energy resulting from the axial loads that are induced in the piles and piers as caused by a rotation of the pile cap, plus a contribution to the frictional resistance of the clay medium to this same rotation.

14.2.4.9 Stiffness influence coefficients $\bar{k}_{ij}^a$ and $\bar{\bar{k}}_{ij}^a$ $(i, j = 1, 2, \ldots, n, \theta)$. These stiffness coefficients are obtained by inverting the flexibility matrices for pile groups 1 and 2, respectively. In determining the flexibility coefficients, pinned end conditions are specified for the piles.

14.2.4.10 Bilinear hysteresis spring forces $\bar{F}_i^s$ and $\bar{\bar{F}}_i^s$. These forces are assumed to be related to their respective displacements $\bar{U}_i^s$ and $\bar{\bar{U}}_i^s$, as shown in Fig. 14.2.

The fundamental expression used in the valuation of the elastic spring constants $\bar{K}_{ie}$ and $\bar{\bar{K}}_{ie}$ is the following Mindlin equation, which gives the x component of displacement as produced by a single concentrated force P located at any arbitrary point (0, 0, c) within an isotropic half-space and acting in the x direction:

$$u_x(x, y, z) = \frac{P(0, 0, c)}{16\pi(1 - \nu)G}\left\{\frac{3 - 4\nu}{R_1} + \frac{1}{R_2} + \frac{2cz}{R_2^3} + \frac{4(1 - \nu)(1 - 2\nu)}{R_2 + z + c} + x^2\left[\frac{1}{R_1^3} + \frac{3 - 4\nu}{R_2^3} - \frac{6cz}{R_2^5} - \frac{4(1 - \nu)(1 - 2\nu)}{R_2(R_2 + c + z)^2}\right]\right\} \tag{14.16}$$

in which ν equals Poisson's ratio, c equals the z distance of the load below the surface xy boundary plane, and

$$R_1^2 \equiv x^2 + y^2 + (z - c)^2 \qquad R_2^2 \equiv x^2 + y^2 + (z + c)^2$$

It is assumed in the dynamic analysis that no transfer of water takes place in the saturated clay medium during the earthquake. Therefore, it is reasonable to use a Poisson's ratio ν of $\frac{1}{2}$. Thus $G = E/3$, which reduces Eq. 14.16 to

$$u_x(x, y, z) = \frac{3P(0, 0, c)}{8\pi E}\left[\frac{1}{R_1} + \frac{1}{R_2} + \frac{2cz}{R_2^3} + x^2\left(\frac{1}{R_1^3} + \frac{1}{R_2^3} - \frac{6cz}{R_2^5}\right)\right] \tag{14.17}$$

or, in cylindrical coordinates,

$$u_x(r, \theta, z) = \frac{3P(0, 0, c)}{8\pi E}\left\{\frac{1}{[r^2 + (z - c)^2]^{1/2}} + \frac{1}{[r^2 + (z + c)^2]^{1/2}} + \frac{2cz}{[r^2 + (z + c)^2]^{3/2}} + r^2 \cos^2 \theta\left[\frac{1}{[r^2 + (z - c)^2]^{3/2}} + \frac{1}{[r^2 + (z + c)^2]^{3/2}} - \frac{6cz}{[r^2 + (z + c)^2]^{5/2}}\right]\right\} \tag{14.18}$$

Due to a singularity that exists at the point of application of the load (0, 0, c), deflections must be considered at finite distances from this point. Since the displacements of interest are those at the boundaries of the piles, i.e., $r = B$, this singularity does not cause difficulty. From Eq. 14.18 it can be seen that the deflection at a constant radius r is a function of the angle θ. Therefore, it is necessary to obtain a weighted average deflection in the x direction at this radius using an averaging process with respect to the coordinate y. Thus, for radius B, a weighted average deflection is defined as

$$u_x(B, z) = \int_0^B u_x(B, \theta, z)dy = \int_0^{\pi/2} u_x(B, \theta, z) \cos \theta d\theta \tag{14.19}$$

Substituting Eq. 14.18 into Eq. 14.19 gives

$$u_x(B, z) = \frac{3P(0, 0, c)}{8\pi E}\left\{\frac{1}{R_1} + \frac{1}{R_2} + \frac{2cz}{R_2^3} + \frac{2}{3}B^2\left[\frac{1}{R_1^3} + \frac{1}{R_2^3} - \frac{6cz}{R_2^5}\right]\right\} \tag{14.20}$$

in which

$$R_1 \equiv [B^2 + (z - c)^2]^{1/2} \qquad R_2 \equiv [B^2 + (z + c)^2]^{1/2}$$

When subjected to a displacement in the elastic range, a pile encounters a continuously distributed loading over its length as caused by the resistance of the clay medium. This loading can be approximated by assuming that the intensity of loading is uniformly distributed along the length of the pile within each height interval $2h$ but that the intensity varies from one interval to the next. The general expression for the weighted average deflection at radius B as caused by a uniformly distributed loading acting over the height of one interval is obtained by substituting the intensity of loading $p(0, 0, \bar{c} \pm h)$ between points $(\bar{c} - h)$ and $(\bar{c} + h)$ for the concentrated load $P(0, 0, c)$ in Eq. 14.20 and integrating with respect to c over this interval. Thus, there is obtained

$$\begin{aligned} u_x(B, z)_{\text{av}} = \frac{3p(0, 0, \bar{c} \pm h)}{8\pi E}\Bigg\{&\sinh^{-1}\frac{\bar{c} + h - z}{B} - \sinh^{-1}\frac{\bar{c} - h - z}{B} + \sinh^{-1}\frac{\bar{c} + h + z}{B} - \sinh^{-1}\frac{\bar{c} - h + z}{B} \\ &+ \frac{2}{3B^2}\left[\frac{B^2(\bar{c} + h) - 2B^2z + (\bar{c} + h)z^2 + z^3}{[B^2 + (\bar{c} + h + z)^2]^{1/2}} - \frac{B^2(\bar{c} - h) - 2B^2z + (\bar{c} - h)z^2 + z^3}{[B^2 + (\bar{c} - h + z)^2]^{1/2}}\right] \\ &- \frac{2}{3}\left[\frac{z - (\bar{c} + h)}{[B^2 + (\bar{c} + h - z)^2]^{1/2}} - \frac{z - (\bar{c} - h)}{[B^2 + (\bar{c} - h - z)^2]^{1/2}}\right] \\ &+ \frac{4}{3}\left[\frac{B^2z + (\bar{c} + h)z^2 + z^3}{[B^2 + (\bar{c} + h + z)^2]^{3/2}} - \frac{B^2z + (\bar{c} - h)z^2 + z^3}{[B^2 + (\bar{c} - h + z)^2]^{3/2}}\right]\Bigg\} \end{aligned} \tag{14.21}$$

in which $\bar{c}$ is the vertical distance from the boundary surface to the midheight of the interval over which the load acts.

The Mindlin theory permits one to characterize completely an elastic half-space. However, to use this theory in the determination of the discrete elastic spring constants as previously defined for the idealized bridge structural system, it is necessary that numerous simplifying assumptions be made so that the problem at hand is amenable to solution. To recognize the appropriate

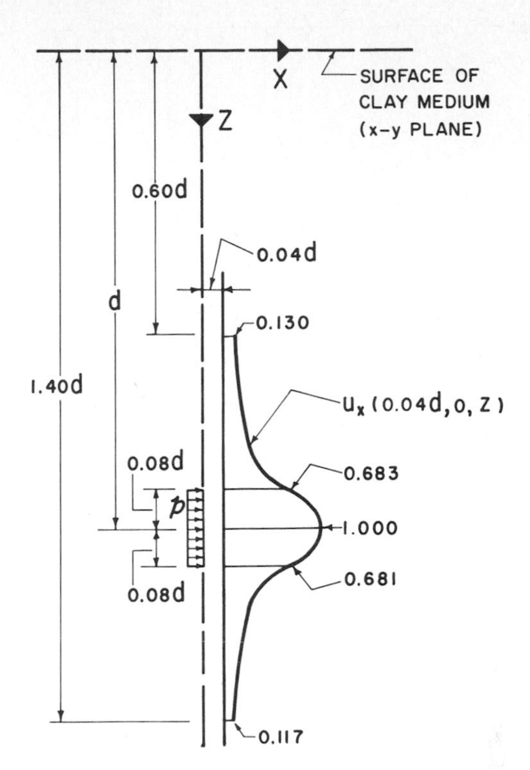

Fig. 14.5. Variation of horizontal displacement along a vertical axis for uniformly distributed load of intensity p.

simplifying assumptions that can be made without altering appreciably the basic characteristics of the structural system, one must understand the basic characteristics of the previously mentioned theory of the half-space. To demonstrate one very important characteristic, Eq. 14.21 has been plotted in dimensionless form in Figs. 14.5 and 14.6 for $h/\bar{c}$ ratios of 0.08 and 1.00, respectively. Figure 14.5 demonstrates the fact that the displacement $u_x(B, z)_{\text{av}}$ decays very rapidly with distance from the loaded region in the vertical direction, i.e., along the z axis. Similar rapid decay rates in the displacement field are also present along the x and y axes. This same rapid decay phenomenon also may be seen in Fig. 14.6 at the base of the loaded region. It is important to note, for subsequent considerations, that displacements $u_x(B, z)_{\text{av}}$ produced by the loaded conditions of Fig. 14.6 are reasonably constant over most of the loaded region except for a very narrow region near its base, i.e., at $z = 2c$, and for a very narrow region near the surface, i.e., at $z = 0$.

From the preceding analysis of the plots shown in Figs. 14.5 and 14.6, one may conclude that (1) the displacement of a point along the axis of loading within an elastic half-space is produced primarily by loading that is present in the immediate vicinity of the point being considered and (2) the displacement of this point is not influenced greatly by its vertical position in the half-space, i.e., its z coordinate, unless it is located very near the surface. Due to these characteristics of the

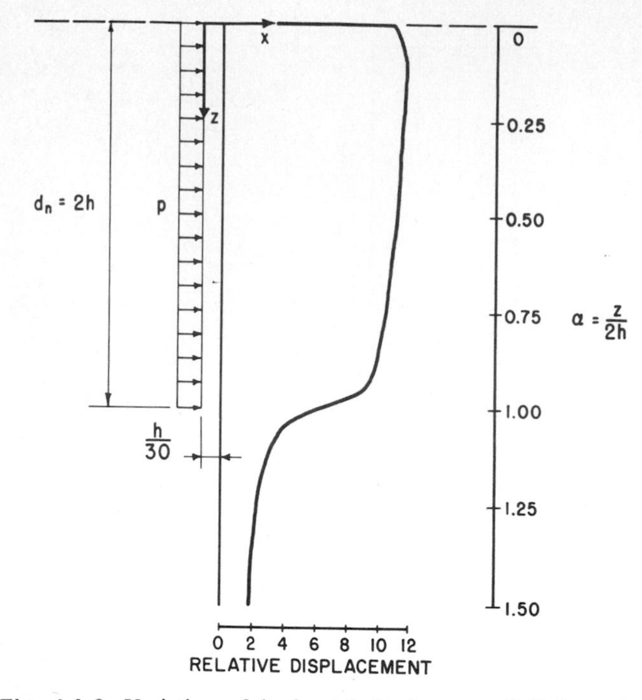

Fig. 14.6. Variation of horizontal displacement field for uniformly distributed load for the case of $\bar{c} = h$.

half-space, one can approximate the pile–clay medium interaction effects by a method similar to the classical "beam on elastic foundation" theory. This theory assumes that the reaction force per unit length along a foundation at a given location can be represented by the expression ku in which u is the deflection of the foundation and k is a constant usually called the modulus of the foundation. This constant, therefore, is the reaction per unit length when the deflection is unity. The simple assumption that the continuous reaction intensity at a point along the foundation is proportional to the deflection at that point only is commonly known as the Winkler assumption and has been proven to be very satisfactory in practical cases. Therefore, it is considered appropriate for the pile–clay medium interaction problem being considered in this investigation.

Consider now the determination of a Winkler continuous spring constant $k'(z)$ that the clay medium provides for a single pile of length d_n at depth z. This spring constant is easily obtained by applying the uniform line loading of intensity p as shown in Fig. 14.6 to the half-space and determining the resulting deflections $u_x(B, z)_{av}$ as shown in this same figure. By definition of the spring constant, one may state that

$$k'(z) \equiv \frac{p}{u_x(B, z)_{av}} \tag{14.22}$$

in which $u_x(B, z)_{av}$ is given by Eq. 14.21. Substituting Eq. 14.21 into Eq. 14.22 gives

$$k'(z) = \frac{8\pi E(z)}{3}\left\{\text{from Eq. 14.21 with } \bar{c} = h = \frac{d_n}{2}\right\}^{-1} \tag{14.23}$$

Note that Eq. 14.21 was derived for an elastic half-space in which the modulus E does not vary with location. Due to the rapid decay phenomenon previously discussed, this expression can be applied, however, with little loss of accuracy to a clay medium where the modulus E varies slowly with depth provided the constant E is replaced by $E(z)$. Thus, Eq. 14.23 reflects this substitution.

Consider now the determination of an appropriate Winkler continuous spring constant, say $k''(z)$, for a pile group of length d_n consisting of Q individual piles. For typical spacings of piles within a group some interaction effects between individual piles are present and should be accounted for in determining $k''(z)$. Due to these interaction effects between piles within a group, one can state the following inequality, i.e.,

$$k''(z) \leq Qk'(z) \tag{14.24}$$

For the case where pile spacings are very large, no appreciable interaction effects would be present, and as a result the pile group spring constant $k''(z)$ would simply be equal to $Qk'(z)$.

The loading that will now be considered as applied to the clay medium for the determination of $k''(z)$ is similar to that shown in Fig. 14.6 except that this same line loading of intensity power depth d_n will be applied to the clay medium along each of the intended axes of piles within the group. It is necessary here to determine the horizontal x component of displacement of the clay medium along each axis and average these displacements for any level z.

The entire displacement field (x component) for a single line loading of the type defined previously is obtained by substituting $p\,dc$ for $P(0, 0, c)$ in Eq. 14.17 and integrating with respect to c over the interval $0 < c < d_n$. This results in the expression

$$\begin{aligned} u_x(x, y, z) = \frac{3p}{8\pi E(z)}\Bigg\{&\sinh^{-1}\frac{d_n + z}{(x^2 + y^2)^{1/2}} \\ &+ \sinh^{-1}\frac{d_n - z}{(x^2 + y^2)^{1/2}} \\ &+ \frac{1}{x^2 + y^2}\Bigg[-\frac{2x^2zd_n(d_n + z)}{[x^2 + y^2 + (d_n + z)^2]^{3/2}} \\ &+ \frac{4x^2z^2(d_n + z)}{(x^2 + y^2)[x^2 + y^2 + (d_n + z)^2]^{1/2}} \\ &+ \frac{x^2(d_n + z) - 2z[y^2 + z(d_n + z)]}{[x^2 + y^2 + (d_n + z)^2]^{1/2}} \\ &+ \frac{x^2(d_n - z)}{[x^2 + y^2 + (d_n - z)^2]^{1/2}} \\ &+ \frac{2z[(x^2 + y^2)y^2 - (x^2 - y^2)z^2]}{(x^2 + y^2)(x^2 + y^2 + z^2)^{1/2}}\Bigg]\Bigg\} \end{aligned} \tag{14.25}$$

in which the x and y coordinates have their origin at the point of application of the load.

It should be noted that Eq. 14.21 represents a weighted average displacement of points on a circle of radius B that lies on a horizontal plane at level z and whose center is at the location of the horizontal load, whereas Eq. 14.25 represents the displacement at an arbitrary point (x, y, z) in which $(x^2 + y^2)^{1/2} > B$. In using the latter mentioned expression, it will be assumed that the deflection of point (x, y, z) that is some distance removed from the points of application of the loading adequately represents the weighted average deflection of points on a circle of radius B whose center is located at this same point.

Using Eqs. 14.21 and 14.25, one can obtain the horizontal x component of displacement of any vertical axis (intended pile axis) within the clay medium at level z due to a single-line static loading of intensity p applied to the clay medium over the interval $0 < z < d_n$. By moving this same line loading to each of the Q intended pile axes within the group and using Eqs. 14.21 and 14.25 in each case, one obtains $Q \times Q$ displacement functions, each of which may be denoted as $u_{rs}(z)$, where $r, s = 1, 2, \ldots, Q$. Thus, an individual displacement function $u_{rs}(z)$ is defined as the horizontal displacement of the clay medium along axis r due to the line loading that is placed along axis s. Referring to Eqs. 14.21 and 14.25, these displacement functions may be expressed in equation form as follows: For the case $r \neq s$

$$\begin{aligned} u_{rs}(z) = \frac{3p}{8\pi E(z)}\Bigg\{&\sinh^{-1}\frac{d_n + z}{(R_{rs})^{1/2}} + \sinh^{-1}\frac{d_n - z}{(R_{rs})^{1/2}} \\ &+ \frac{1}{R_{rs}}\Bigg[-\frac{2x_{rs}^2 z d_n(d_n + z)}{[R_{rs} + (d_n + z)^2]^{3/2}} \\ &+ \frac{4x_{rs}^2 z^2(d_n + z)}{R_{rs}[R_{rs} + (d_n + z)^2]^{1/2}} \\ &+ \frac{x_{rs}^2(d_n + z) - 2z[y_{rs}^2 + z(d_n + z)]}{[R_{rs} + (d_n + z)^2]^{1/2}} \\ &+ \frac{x_{rs}^2(d_n - z)}{[R_{rs} + (d_n - z)^2]^{1/2}} \\ &+ \frac{2z[R_{rs}y_{rs}^2 - (x_{rs}^2 - y_{rs}^2)z^2]}{R_{rs}[R_{rs} + z^2]^{1/2}}\Bigg]\Bigg\} \end{aligned} \tag{14.26}$$

in which $R_{rs} \equiv x_{rs}^2 + y_{rs}^2 \geq B^2$ and x_{rs} and y_{rs} are the x and y coordinates, respectively, of axis r when the origin of these coordinates coincides with the s axis.

$$r = 1, 2, \ldots, Q; \qquad s = 1, 2, \ldots, Q \qquad r \neq s$$

and Q is the number of intended piles in the group under consideration. For the case $r = s$,

$$\begin{aligned} u_{ss}(z) = \frac{3p}{8\pi E(z)}\Bigg\{&\sinh^{-1}\frac{d_n + z}{B} + \sinh^{-1}\frac{d_n - z}{B} \\ &+ \frac{4}{3}\frac{B^2 z + z^2(d_n + z)}{[B^2 + (d_n + z)^2]^{3/2}} \\ &+ \frac{2}{3B^2}\Bigg[\frac{B^2(d_n - 2z) + z^2(d_n + z)}{[B^2 + (d_n + z)^2]^{1/2}} \\ &+ \frac{B^2(d_n - z)}{[B^2 + (d_n - z)^2]^{1/2}} - \frac{z^3 - B^2 z}{[B^2 + z^2]^{1/2}}\Bigg]\Bigg\} \end{aligned} \tag{14.27}$$

It is now necessary to obtain the average displacement of all Q axes within the pile group, say $u(z)_{av}$, which by definition may be expressed as

$$u(z)_{av} = \frac{1}{Q}\sum_{r=1,2,\ldots}^{Q}\ \sum_{s=1,2,\ldots}^{Q} u_{rs}(z) \tag{14.28}$$

Using Eqs. 14.26 and 14.27, Eq. 14.28 may be written as

$$\begin{aligned} u(z)_{av} = \frac{3p}{8\pi E(z)Q}\Bigg[&\sum_{r=1,2,\ldots}^{Q}\ \sum_{s=1,2,\ldots}^{Q}(1 - \delta_{rs}) \\ &\{\text{from Eq. 14.26}\} + \sum_{s=1,2,\ldots}^{Q}\{\text{from Eq. 14.27}\}\Bigg] \end{aligned} \tag{14.29}$$

where the Kronecker delta term δ_{rs} is defined as

$$\delta_{rs} \equiv \begin{cases} 0 & r \neq s \\ 1 & r = s \end{cases}$$

The Winker spring constant $k''(z)$ for the pile group is now defined as

$$k''(z) = \frac{Qp}{u(z)_{av}} \tag{14.30}$$

Substituting Eq. 14.29 into Eq. 14.30 gives

$$k''(z) = \frac{8\pi Q^2 E(z)}{3}(\text{from Eq. 14.29})^{-1} \tag{14.31}$$

The bridge structural system between expansion joints as shown in Fig. 14.3 contains two multiple row pile groups directly under the tower and three single row groups. Using Eq. 14.31, the spring constant $k''(z)$ for one multiple row group and one single row group may be obtained. Denoting these functions as $k_1''(z)$ and $k_2''(z)$, respectively, the entire bridge structural system under consideration will experience under longitudinal motion a lateral foundation modulus of $2k_1''(z)$ for the combined multiple row pile groupings and $3k_2''(z)$ for the combined single row pile groupings.

Considering that the dynamic analysis is based on a discrete parameter system as shown in Fig. 14.4, the continuous spring systems just determined must be lumped into concentrated springs. Thus, lumping the continuous springs over each interval of height $2h$ gives the following elastic spring constants K_{ie} for the concentrated spring at level i that characterize the system shown in Fig. 14.4:

$$\bar{K}_{ie} = 4hk_1''(z_i) \tag{14.32a}$$

$$\bar{\bar{K}}_{ie} = 6hk_2''(z_1)\text{: } i = 1, 2, \ldots, n - 1 \tag{14.32b}$$

and

$$\bar{K}_{ne} = 2hk_1''(z_n) \tag{14.33a}$$

$$\bar{\bar{K}}_{ne} = 3hk_2''(z_n) \tag{14.33b}$$

The yield values of the interaction spring forces $\bar{F}_{iy}^s$ and $\bar{\bar{F}}_{iy}^s$ as represented by F_{iy}^s in Fig. 14.2 are determined by multiplying the ultimate lateral bearing pressure q_f that the clay medium is capable of exerting on a pile by their respective total projected pile areas, i.e.,

$$\bar{F}_{iy}^s = 4q_f Bh\bar{Q} \tag{14.34a}$$

$$\bar{\bar{F}}_{iy}^s = 4q_f Bh\bar{\bar{Q}} \tag{14.34b}$$

in which $\bar{Q}$ and $\bar{\bar{Q}}$ represent the total number of piles represented by pile groups 1 and 2, respectively. Previous investigations indicate that the ultimate lateral bearing pressure q_f is approximately equal to $7.5c_s$ in which c_s is the soil cohesive strength or ultimate shear strength as measured by a standard triaxial test. Since the above interaction yield force values are based on ultimate shear strength of the soil, the slopes $\bar{K}_{ip}$ and $\bar{\bar{K}}_{ip}$ of the force-displacement relations, as represented by K_{ip} in Fig. 14.2, are considered to be essentially zero. However, small but finite values are actually assigned to these quantities in the numerical analysis for stability reasons.

14.2.4.11 Masses $\bar{M}_i$ and $\bar{\bar{M}}_i$. The idealized discrete masses $\bar{M}_i$ and $\bar{\bar{M}}_i$, as represented by M_i in Fig. 14.4, are obtained by summing the total mass of the piles (or pile groups) being represented in interval i and adding an effective mass of the clay medium resulting from the relative motion caused by interaction. Thus, it may be stated that

$$M_i = M_i^p + M_i^e \qquad i = 1, 2, \ldots, n-1 \quad (14.35)$$

where M_i^p represents the combined pile masses while M_i^e represents contributions from the clay medium.

Since the effective mass M_i^e is entirely the result of interaction effects, it is convenient to separate this motion (Fig. 14.7) into two distinct types as shown in Figs. 14.8 and 14.9: one type where the pile group is forced to displace with the moving clay medium without interaction and a second type where the pile group is forced to interact with an undisturbed clay medium.

In Figs. 14.7–14.10, three reference lines (OA'', OA', and OA) are used to locate the structural system relative to the foundation medium. Line OA'' represents points along a vertical axis within the clay medium in its completely undisturbed state and also represents the axis of the pile group in its initial position. Line OA' represents the displaced position of the line OA'' due to the

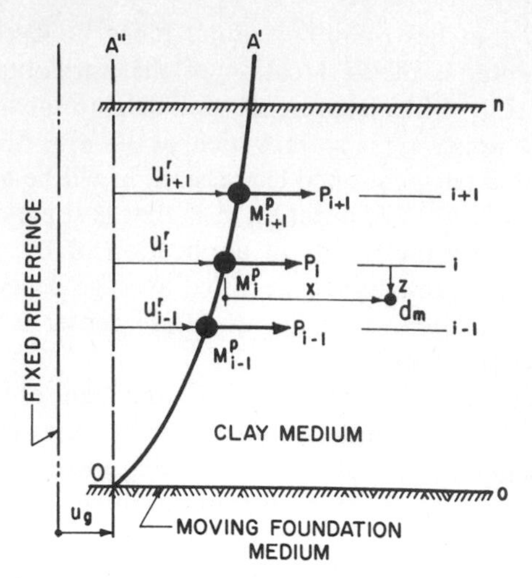

Fig. 14.8. Forced motion of structural system with no interaction.

dynamic response of the clay medium with no interaction being present while line OA is the displaced position of OA'' with interaction present.

It is quite apparent that there exists a set of dynamic loads which, if applied to the entire pile–pier–bridge structural system, would produce motion resulting in no interaction with the clay medium. This type of motion is represented in Fig. 14.8, where only the applied loads P_i on the pile masses are shown. It is easily established that these loads are represented by the expression

$$P_i = M_i^p(\ddot{u}_g + \ddot{u}_i^r) + < k_{ij}^a >\left\{u_i^r - \frac{d_i}{d_n}u_n^r\right\} + \left(\theta' - \frac{u_n^r}{d_n}\right)k_{i\theta}^a + < T_{ij} >\{u_i^r\} \quad (14.36)$$

Since no interaction is present in this case, its contribution to M_i^e is zero.

To determine the true motion of the real structural system shown in Fig. 14.7, one must combine the motion

Fig. 14.7. True motion of structural system.

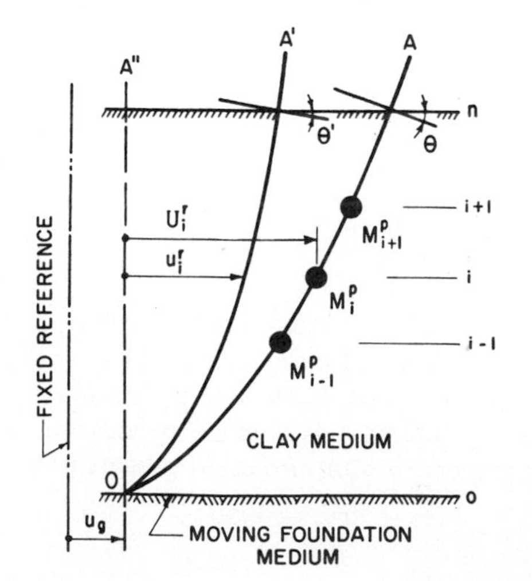

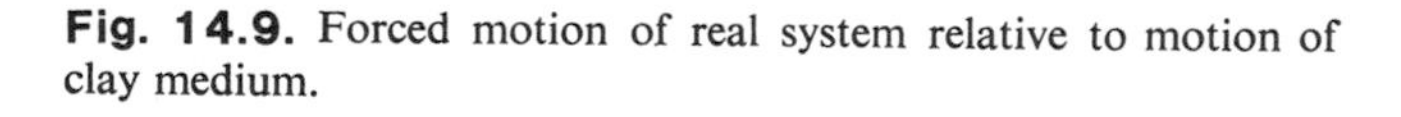
Fig. 14.9. Forced motion of real system relative to motion of clay medium.

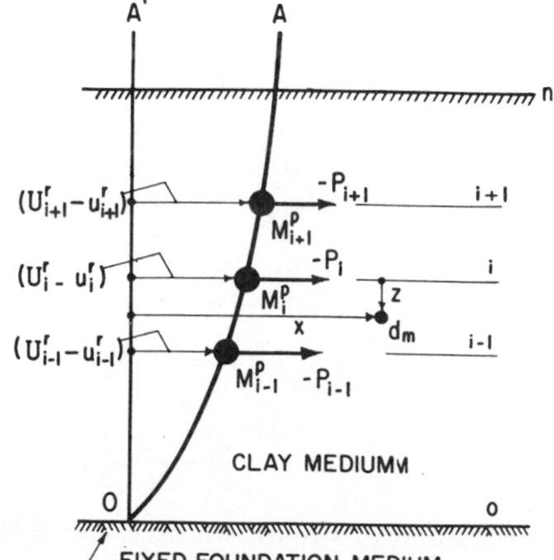

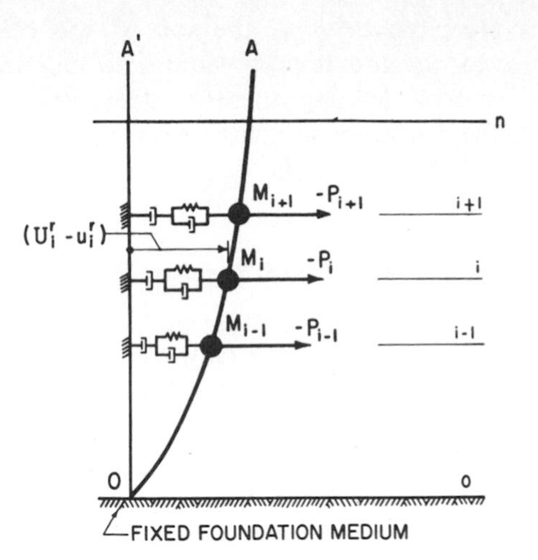

Fig. 14.10. Motion of idealized system relative to motion of clay medium.

of Fig. 14.8 with that of Fig. 14.9. This latter case properly accounts for all interaction effects, thus making it possible to determine M_i^e.

The motion of the real system in Fig. 14.9 must be exactly the same as the idealized system shown in Fig. 14.10. To be equivalent it is a necessary condition that the total kinetic energy in any interval i be the same in each case. This condition results in the equation

$$M_i^e \cong \int_{\bar{z}_i-h}^{\bar{z}_i+h} \int_{-\infty}^{\infty} \int_{-\infty}^{8} (\psi_u^2 + \psi_v^2 + \psi_w^2)\rho(x, y, z)\, dx\, dy\, dz \tag{14.37}$$

in which $\psi_u = \psi_u(x, y, z)$, $\psi_v = \psi_v(x, y, z)$ and $\psi_w(x, y, z)$ represent the displacement fields within the clay medium in the x, y, and z directions, respectively, when subjected to the interaction forces of all piles within an intended pile group but normalized so that the relative displacement $(U_i^r - u_i^r)$ equals unity.

14.2.4.12 Creep coefficients C_i^c. The coefficients that represent creep effects in the clay medium due to its interaction with a pile group is determined from an approximate strain rate field for each layer i as produced by the distributed interaction forces within that group over layer i. Since creep will take place primarily near the piles where the stresses are high, the effects on creep in layer i due to those interaction forces in neighboring layers will be neglected. The strain rate associated with any given stress level will be determined from triaxial tests.

For purposes of determining the creep coefficient in layer i, the elastic stress field due to F_i^s will be assumed. A study of this stress field along the x axis shows that the major principal axis is oriented reasonably close to the x axis; thus, the stress quantity $(\sigma_x - \sigma_y)$ may be compared directly with the triaxial deviator stress to obtain an approximate measure of the creep rate present at any given point. After observing the nature of the overall stress field in the vicinity of a pile group, it appears that the $(\sigma_x - \sigma_y)$ stress distribution along a single axis can be used to adequately predict the creep displacement rate of a pile group due to a given interaction loading condition. This particular x axis is oriented in the direction of motion in a horizontal plane that is located at midheight of the particular interval i being investigated. The origin of this x-y coordinate system is coincident with the centroidal axis of the pile group. The resulting $(\sigma_x - \sigma_y)$ stress distribution along this x axis is antisymmetric with respect to the origin. Therefore, one need consider only the distribution of stress along the positive side of this axis. From triaxial test data a relationship between strain rate and deviator stress can be obtained. Having this relationship and the $(\sigma_x - \sigma_y)$ stress distribution along the x axis for level i, a curve showing strain rate along this axis can be determined. The area under this curve represents the velocity of the pile group at level i due to creep effects. Thus,

$$C_i^c \equiv \frac{F_i^s}{\dot{U}_i^r - \dot{U}_i^s} = \frac{F_i^s}{\int_B^{\infty} \dot{\epsilon}_x(x; F_i^s)dx} \tag{14.38}$$

14.2.4.13 Damping coefficients C_i^d. It is suggested that the coefficients that represent damping effects be evaluated by equating the rate of energy dissipation resulting from elastic deformations (creep effects excluded) of the real system in their respective i intervals to the rate of energy dissipation in the damping dashpots at corresponding locations in the idealized model.

A measure of the energy dissipation per unit volume for the clay medium can be determined from free vibration tests performed on clay samples. By definition the rate of energy dissipation per unit volume of a cylindrical test specimen is

$$E_{Di} = c_i^* \frac{\dot{u}_1^2}{AL} = c_i^* \frac{(\epsilon_i L)^2}{AL} \tag{14.39}$$

in which A is the cross-sectional area and L is the height of the test specimen. The total rate of energy dissipation in the clay medium within interval i is found by integrating this expression over the volume of the interval under consideration.

The total rate of energy dissipation in the damping dashpot at level i in the idealized model can be expressed in terms of the velocity of the generalized coordinate $\dot{U}_i^s$ as follows:

$$(E_{Di})_{\text{total}} = C_i^d(\dot{U}_i^s)^2 \tag{14.40}$$

For the evaluation of the damping coefficients, the time dependent stress field is assumed to be of the form

$$\sigma_x(x, y, z, t)_i = U_i^s(t)\kappa_i(x, y, z) \tag{14.41}$$

in which $\kappa_i(x, y, z)$ is a stress influence function and is the generalized expression for the stress field that corresponds to a unit displacement at the intended location of the pile group. Taking the time derivative of

Eq. 14.41 and dividing by the modulus of elasticity for that level yields an expression for strain rate that can be substituted into Eq. 14.39. Equating the total rates of energy dissipation as given by Eqs. 14.39 and 14.40, an expression for the creep dashpot coefficient at level i is obtained:

$$C_i^d = \frac{c_i^* L}{A(E_i)^2} \int_{v_i} [\kappa_i(x, y, z)]^2 dv \qquad (14.42)$$

14.2.4.14 Axial forces $\bar{S}_i$, $\bar{\bar{S}}_i$, $\bar{T}_i$, and $\bar{\bar{T}}_i$. These forces represent the portions of the total dead wieght of the bridge structural system that are carried by their respective members at level i. The horizontal components t_i and t_{i+1} of the axial forces T_i and T_{i+1}, respectively, are approximated by the relations

$$\begin{aligned} t_i &= \frac{T_i}{L_i}u_{i-1} - \frac{T_i}{L_i}u_i \\ t_{i+1} &= \frac{T_{i+1}}{L_{i+1}}u_i - \frac{T_{i+1}}{L_{i+1}}u_{i+1} \end{aligned} \qquad (14.43)$$

Thus, the resulting horizontal component of force at mass m_i is the difference Δt_i of the preceding components, i.e.,

$$\Delta t_i = \frac{T_i}{L_i}u_{i-1} + \left(\frac{T_i}{L_i} + \frac{T_{i+1}}{L_{i+1}}\right)u_i - \frac{T_{i+1}}{L_{i+1}}u_{i+1} \qquad (14.44)$$

Written in matrix form for $i = 1, 2, \ldots, n - 1$, Eq. 14.44 may be designated simply as

$$\{\Delta t_i\} = [T_{ij}]\{u_i\} \qquad (14.45)$$

14.3 DETERMINATION OF SOIL PROPERTIES

14.3.1 General

For purposes of analysis in this investigation it was assumed that the shear stress deformation characteristics of the clay could be represented by the model shown in Fig. 14.11. The elements in this model are (1) a non-linear elastic spring with hysteresis characteristics to represent the immediate deformation characteristics of the soil structure under cyclic loading, (2) a viscous dashpot in parallel with the spring to represent internal damping within the soil, and (3) a viscous element in series with the spring–dashpot combination to represent the creep behavior of the soil. It was necessary, therefore, to develop test procedures to determine the soil parameters defining these characteristics and to establish the variation in these parameters throughout the depth of the *in situ* clay layer.

Fig. 14.11. Proposed model representing stress–deformation characteristics of clay under short-term cyclic loading.

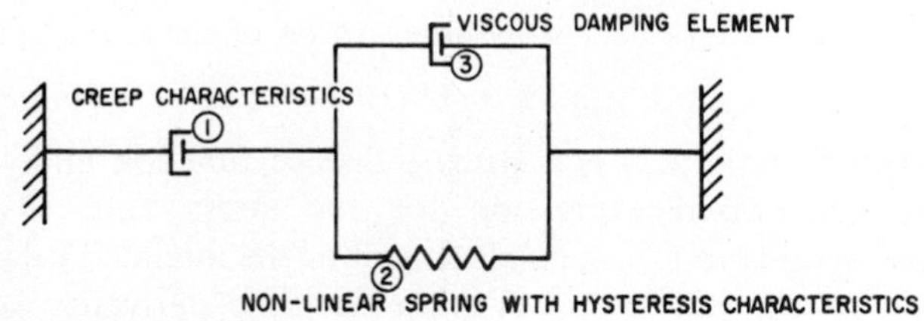

14.3.2 Boring and Field Testing Program

The characteristics of the clay were investigated by means of three borings designated PS-1, PS-2, and PS-3 made along the axis of the proposed bridge location as shown in Fig. 14.12. The logs of these borings are shown in the soil profile along the bridge site, Fig. 14.13.

Undisturbed samples were recovered from the borings at the elevations shown in Fig. 14.14 for use in the laboratory testing program. In boring PS-1, these were obtained in 18 in. lengths by means of 6-in. diameter Shelby tubes. In borings PS-2 and PS-3 the undisturbed samples were taken at more frequent intervals in 30 in. lengths using 2.8-in. diameter Shelby tubes.

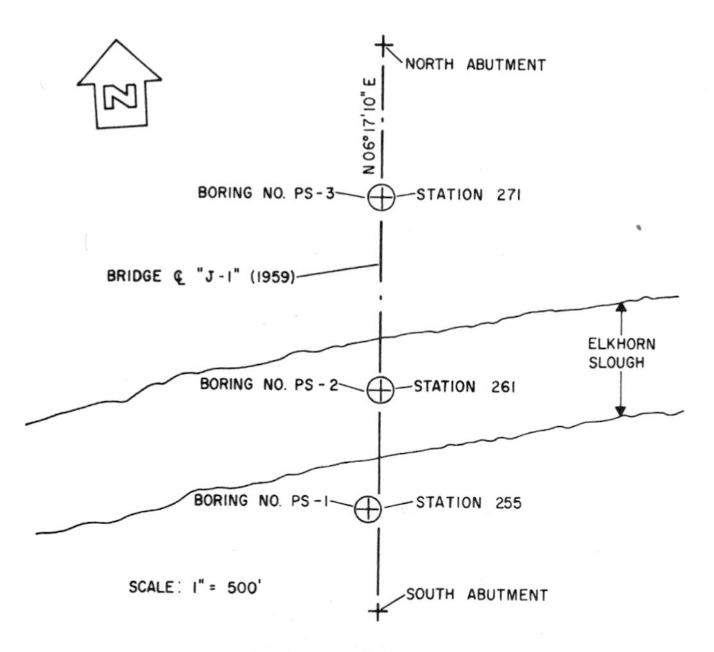

Fig. 14.12. Plan of borings.

In all cases the tubes were forced into the soil by hand or by hydraulic jack except for the bottom 15 ft of boring PS-2 where the tubes were driven by hammer blows. The ends of the tubes were sealed with wax as soon as practicable after recovery to prevent moisture loss from the soil prior to testing.

Adjacent to each boring made to obtain undisturbed samples, a supplementary boring was made, within about 5 ft, in which field vane shear tests were made at 5 ft intervals. The results of these tests are presented in Fig. 14.14.

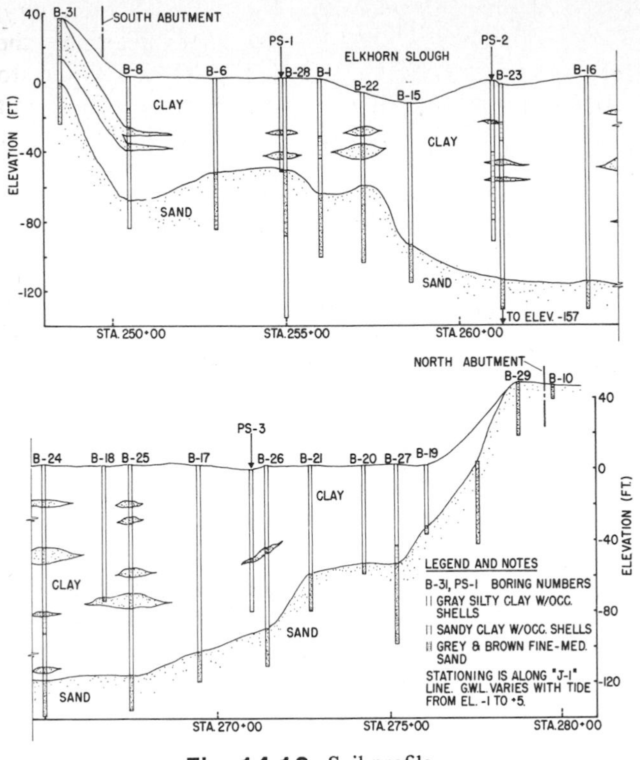

Fig. 14.13. Soil profile.

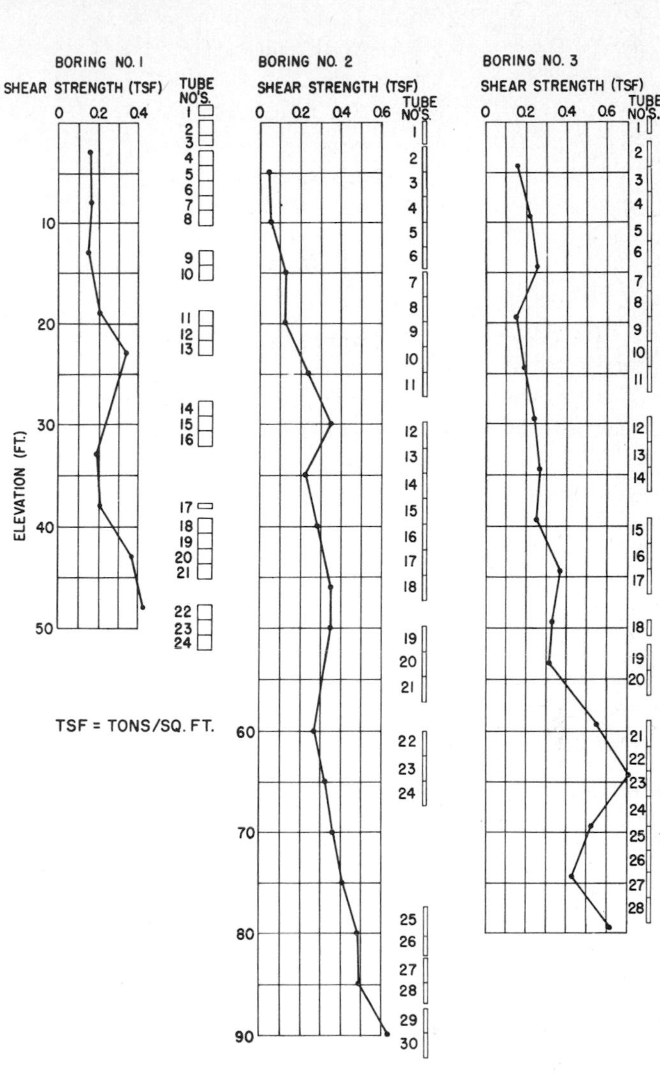

Fig. 14.14. Vane shear strengths and sample elevations.

Fig. 14.15. Summary of results of the vane shear tests.

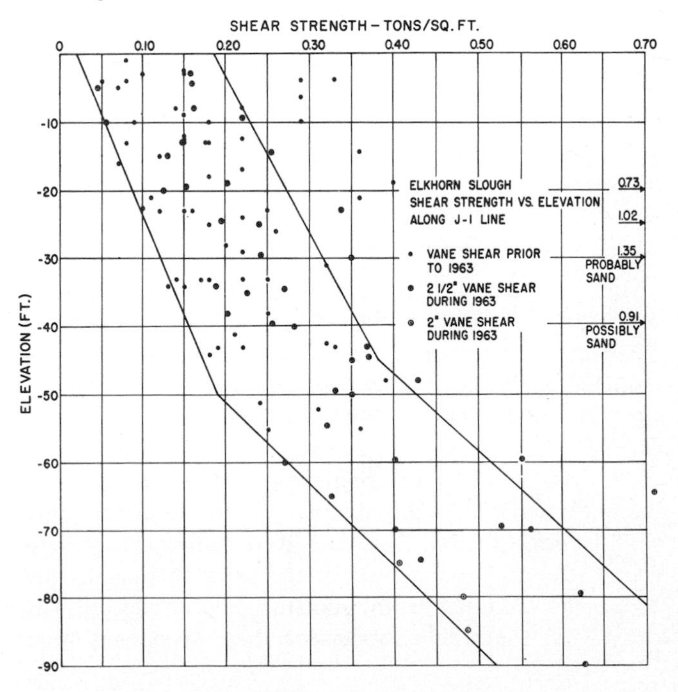

14.3.3 Soil Conditions

The clay underlying the proposed site for the Elkhorn Slough bridge is a soft-to-firm gray, silty clay containing many seams and thin layers of silt and silty sand up to 5 ft in thickness. The thickness of the clay deposit varies from about 50 ft at the bridge abutments to about 120 ft in the center portion of the bridge span.

Atterberg limits of typical samples of the clay are liquid limit = 76 and plastic limit = 22, and the water content varies from an average of about 80% at elevation −30 to an average of about 50% at elevation −75.

Corresponding to the decrease in water content with depth, the soil strength shows a general increase with depth as illustrated by the vane shear strength data in Fig. 14.14. A summary of all vane test data obtained at the bridge site, both in the borings made for the present investigation and in previous investigations, is presented in Fig. 14.15. Although there is considerable scatter in the strength values at any elevation, the average shear strength increases from about 0.1 tons/ft^2 at elevation 0 to about 0.6 tons/ft^2 at elevation −90. The envelope enclosing most of the strength data is shown in Fig. 14.15.

14.3.4 Laboratory Strength Tests

In conjunction with the laboratory test program to determine the soil deformation characteristics, a number

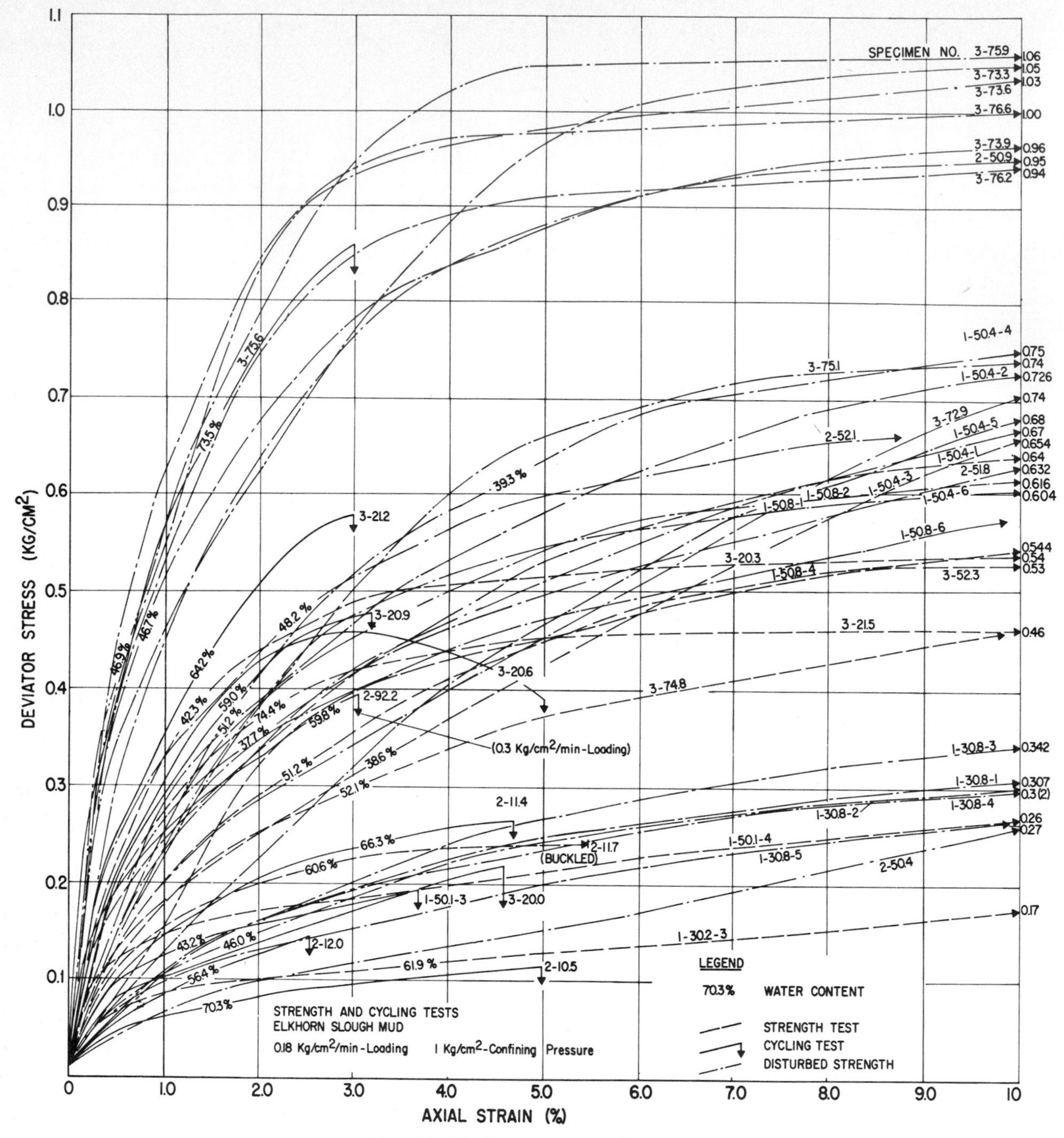

Fig. 14.16. Static stress–strain curves.

of unconsolidated, undrained, triaxial compression tests were conducted using a confining pressure of $10\,kg/cm^2$. The tests were performed on 1.4-in. diameter specimens trimmed from the undisturbed samples, and the stress vs strain relationships obtained in these tests are shown in Fig. 14.16.

A number of other specimens used for determining damping and creep characteristics were deformed only slightly during these tests and were subsequently subjected to compression tests in the same manner as the specimens trimmed from undisturbed samples. It was considered that the strengths of these specimens were only slightly affected by the prior testing procedures and provided a reasonable indication of the strengths of the undisturbed samples in the laboratory. Thus the stress vs strain relationships for these specimens are also shown in Fig. 14.16.

The shear strengths for all these samples, taken in each case as one-half the compression strength, are plotted vs sample depth in Fig. 14.17. As in the case of the vane test data, the laboratory strength tests show considerable variations in strength for samples at any one elevation but a general increase in strength with depth.

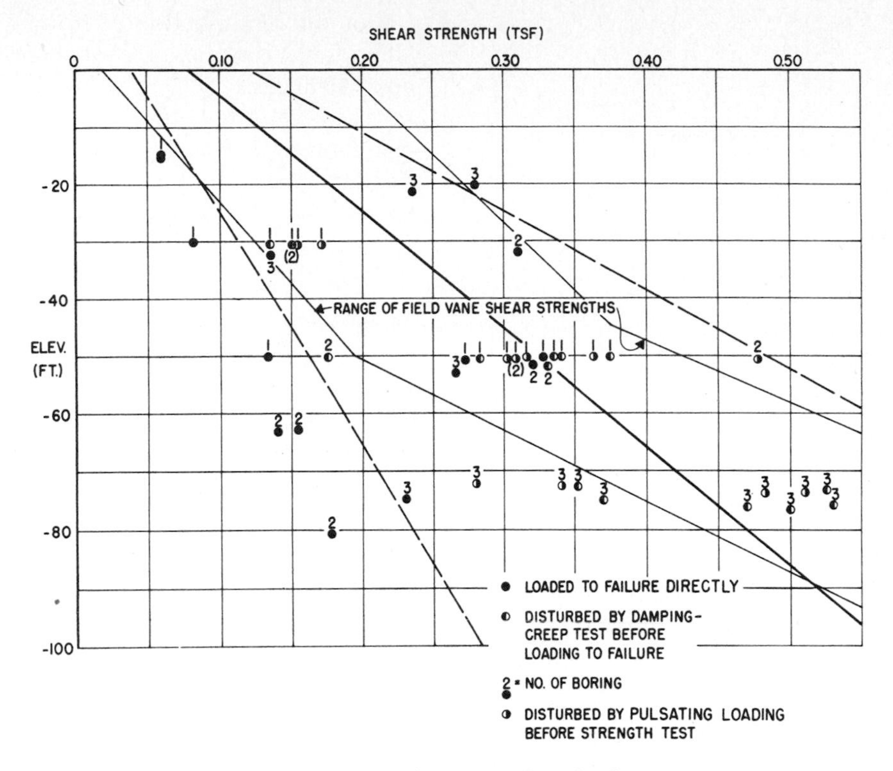

Fig. 14.17. Shear strength vs depth.

The envelope for the field vane shear strength values is also plotted in Fig. 14.17.

Based on the results of the laboratory and field tests it would appear that the shear strength vs depth relationship for the clay deposit can be reasonably represented as a uniform increase from an average value of 0.08 ton/ft² at elevation 0 to 0.52 ton/ft² at elevation −90, indicated by the solid line in Fig. 14.17, together with a possible variation of ±50% from these values as indicated by the dashed lines. These relationships will be used later as a basis for determining the variation of deformation parameters throughout the depth of the deposit.

14.3.5 Determination of Shear Properties from Compression Tests

The analysis of the response of a clay layer overlying a rigid stratum or the interaction between the clay layer and piles driven into it requires a knowledge of the shear stress vs strain characteristics of the clay. In laboratory test programs it is more convenient to determine the stress vs strain relationships for soils in axial compression. However, for saturated clays subjected to short-term stress applications, deformations occur at constant volume and thus Poisson's ratio will be equal to 0.5. Thus, it is possible to determine numerical relationships between principal stresses and strains in compression and shear with sufficient accuracy by means of the expressions

$$\tau = \frac{\sigma_1 - \sigma_3}{2} = \frac{\sigma_d}{2} \tag{14.46}$$

and

$$\gamma_s = (1 + \nu)\epsilon_1 \tag{14.47}$$

where τ = principal shear stress; σ_d = principal compression stress difference, $\sigma_1 - \sigma_3$; σ_1 = major principal stress in compression test; σ_3 = minor principal stress in compression test; γ_s = principal shear strain; ϵ_1 = major principal compression strain; and ν = Poisson's ratio. Substituting $\nu = 0.5$ in the above expressions gives

$$\tau = 0.5\sigma_d \tag{14.48}$$

$$\gamma_s = 1.5\epsilon_1 \tag{14.49}$$

Thus the shear stress vs strain behavior for the clay can readily be determined from the compression test data.

14.3.6 Dynamic Tests

The creep, damping, and dynamic elasticity characteristics of soil specimens were determined by dynamic tests, using the apparatus illustrated in Fig. 14.18. A specimen of clay, 1.4 in. in diameter and about 3.5 in. high was fitted with a lucite cap and base and mounted on a rigid slab. A rigid mass M was then lowered gently onto the cap of the specimen until 10% of its weight

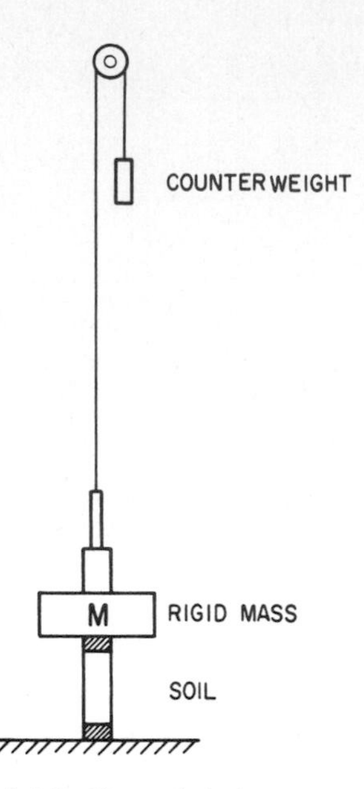

Fig. 14.18. Dynamic test arrangement.

was supported by the specimen and the remaining 90% was supported by the pulley and counterweight shown in the figure. At a given instant the wire attaching the rigid mass to the counterweight was severed, releasing the mass onto the cap of the specimen. The accelerations of the cap of the sample due to this sudden load application were recorded by an accelerometer mounted on the cap and deformations, by a linear variable differential transformer attached to the cap. Variation of acceleration and deformation with time were recorded by a Honeywell Visicorder capable of recording at frequencies up to 1000 cps.

Fig. 14.19. Idealizations of dynamic test results.

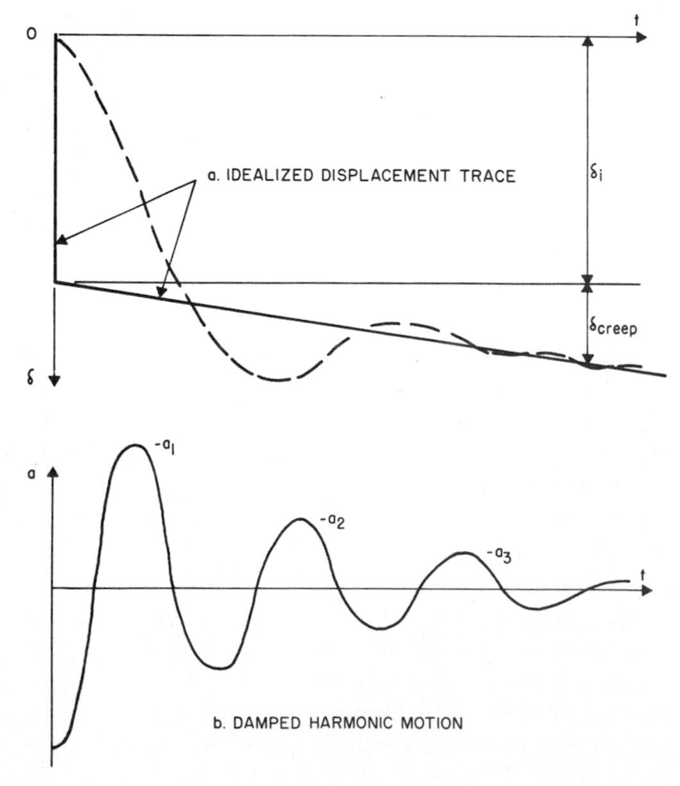

The acceleration–time curve has the general form of the damped harmonic motion shown in Fig. 14.19. For this type of motion the ratio of actual damping to its critical value λ is determined by the relationship

$$\lambda = \frac{1}{2\pi}\log\frac{a_1}{a_2} = \frac{1}{2\pi}\log\frac{a_2}{a_3} = \cdots = \frac{1}{2\pi}\log\frac{a_n}{a_{n+1}} \quad (14.50)$$

where a_1, a_2, a_3, etc. are the amplitudes of successive peaks of the response curve. Thus, an average value of λ could be readily determined from the acceleration record for each test. However, this procedure is valid only for reversible elastic systems, and values of λ can be determined only for samples subjected to low stress levels where this type of behavior is evidenced.

The deformation vs time relationship provides a convenient means for determining the creep and dynamic elastic characteristics of the samples. By extrapolating the displacement curve to zero time an instantaneous displacement δ_1 for the specimen may be determined, and subsequent deformations can be attributed to creep of the sample (see Fig. 14.19). Thus the progressive increase in creep movement over any desired length of time can be readily determined. The stress causing this creep is equal to the counterweight divided by the average area of the sample since creep due to the seating load already has occurred.

For the present program it was found that the rate of creep did not vary significantly over the first 10 sec, and an average rate was reasonably representative of the sample behavior over this period of time. Thus, from any one test an average rate of creep and the stress inducing it could be determined. By repeating the test on similar specimens but using masses of different magnitudes a relationship between average creep rate and applied stress can be determined as shown in Fig. 14.20 for 10 psi ultimate strengths.

Finally, the dynamic stress–strain characteristics of the soil can be obtained in two ways as follows:

1. In each test the average stress on the specimen and the instantaneous elastic deformation δ are determined. By conducting a series of tests on identical specimens but using rigid masses of different magnitudes, a series of corresponding pairs of values of stress and elastic deformation can be determined and plotted to establish a dynamic elastic stress–strain relationship, as shown, e.g., in Fig. 14.21.

2. Due to the fact mass M is large compared with the mass of the soil sample, this system will act essentially as a single-degree-of-freedom system, i.e., no elastic waves will propagate up and down the sample during the dynamic test. Under this condition the strains will be nearly uniform throughout the specimen, thus giving displacements that vary essentially linearly from zero at

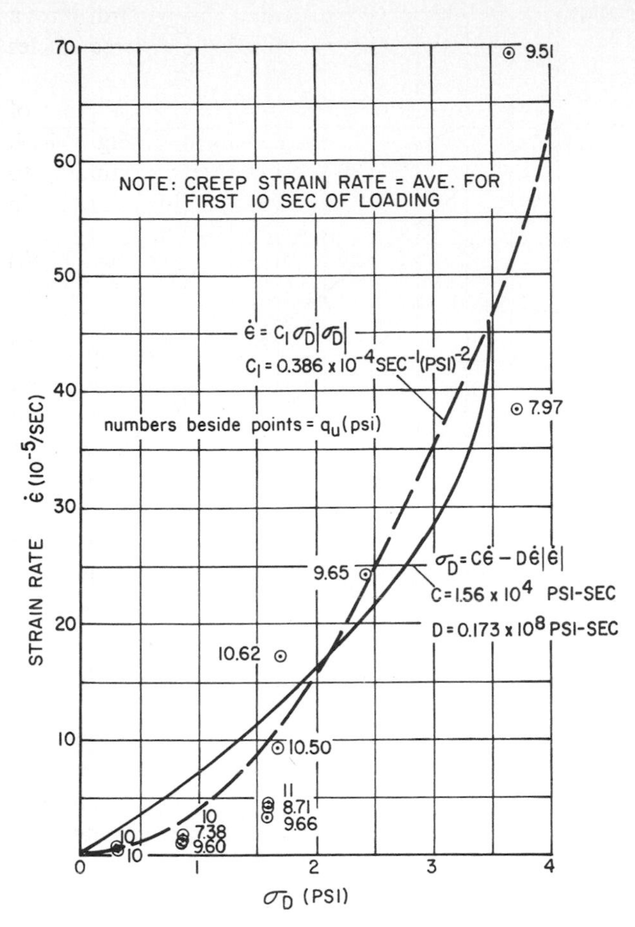

Fig. 14.20. Creep strain rate vs stress; samples of 10 psi strength.

the sample base to a maximum value at the sample top. The generalized single-degree-of-freedom system therefore consists of a mass equal to the rigid mass plus a contribution from the soil sample equal to one-third the sample mass. The spring constant of this generalized system, of course, is the spring constant of the soil sample itself. Since the accelerations of the mass at the top of the specimen are known, the resultant forces acting on the spring can be computed (as mass time acceleration), and hence the force developed at the top of the specimen at any time can be determined. The corresponding average strains of the sample at the same times can be determined from the observed record of sample displacement. Thus, by plotting stress vs strain at corresponding times, dynamic stress–strain curves under these conditions are obtained that are similar to the average dynamic curve shown in Fig. 14.21.

14.3.7 Damping Characteristics

Following the procedure previously described, the damping ratio λ developed in a number of dynamic tests on samples of different strengths was determined. The magnitude of the rigid mass used in these tests was restricted to a value such that the maximum applied stress would not exceed 15% of the sample strength in order that response would be controlled by the initial portion of the stress–strain relationship where reversible elastic behavior is evidenced.

Values of λ for the various specimens ranged from 0.08 to 0.12 and appeared to be relatively independent of strength or elevation of the sample. Since the analyses are not sensitive to minor variations in damping coefficients, the average value of λ equal to 0.094 would seem to be approximate for the entire deposit of clay at the Elkhorn Slough bridge site.

14.3.8 Creep Characteristics

Tests were conducted to determine the relationships between average rate of creep over a period of 10 sec following stress application and applied stress for three series of samples having compressive strengths of the order of 4 psi, 10 psi, and 15 psi. The results of these tests for the 10 psi material are presented in Fig. 14.20.

It was found that average creep rate varied considerably with the magnitude of the applied stress and with sample strength. However, for samples of any given strength, the relationship between average creep rate and applied stress could be represented with reasonable accuracy by a parabolic relationship.

In representing the effects of creep in the analysis of clay–pile interaction, it is convenient to utilize creep strain rate as the dependent variable and applied stress as the independent variable. Thus, a parabola of the form

$$\dot{\epsilon} = C_1\sigma_0|\sigma_0| \tag{14.51}$$

Fig. 14.21. Stress–strain relations; samples of 10 psi strength.

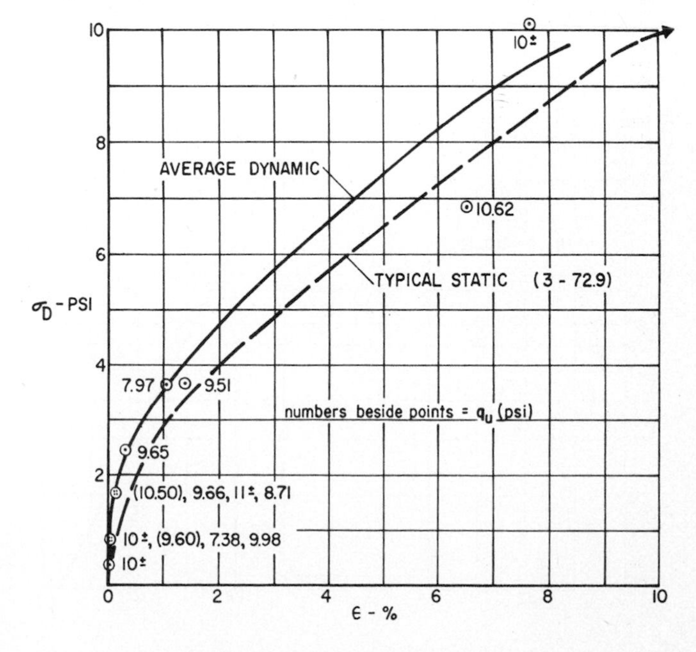

as shown in Fig. 14.20 was used to provide a satisfactory means of representing the soil behavior. Values of C_1 that best represent the test data can be readily determined for samples of different strengths and plotted as shown in Fig. 14.22. Actually, it is necessary to know the variation in the creep coefficient C_1 with depth in the soil mass. Since the average variation in strength with depth has already been established (Fig. 14.17), the relationship between the coefficient C_1 and elevation can be readily deduced as shown in Fig. 14.23. The range of values shown in this figure reflects the range in strength values of $\pm 50\%$ at any elevation in the soil mass.

In studying the response of the clay layer alone it is convenient to include creep effects by considering stress as the dependent variable and creep strain rate as the independent variable. For this purpose the above equation might be expressed as

$$\sigma_0 = \left(\frac{|\dot{\epsilon}|}{C_1}\right)^{1/2} \text{ sign } \dot{\epsilon} \tag{14.52}$$

The evaluation of σ_0 using this type of equation on a computer involves considerably more time than that

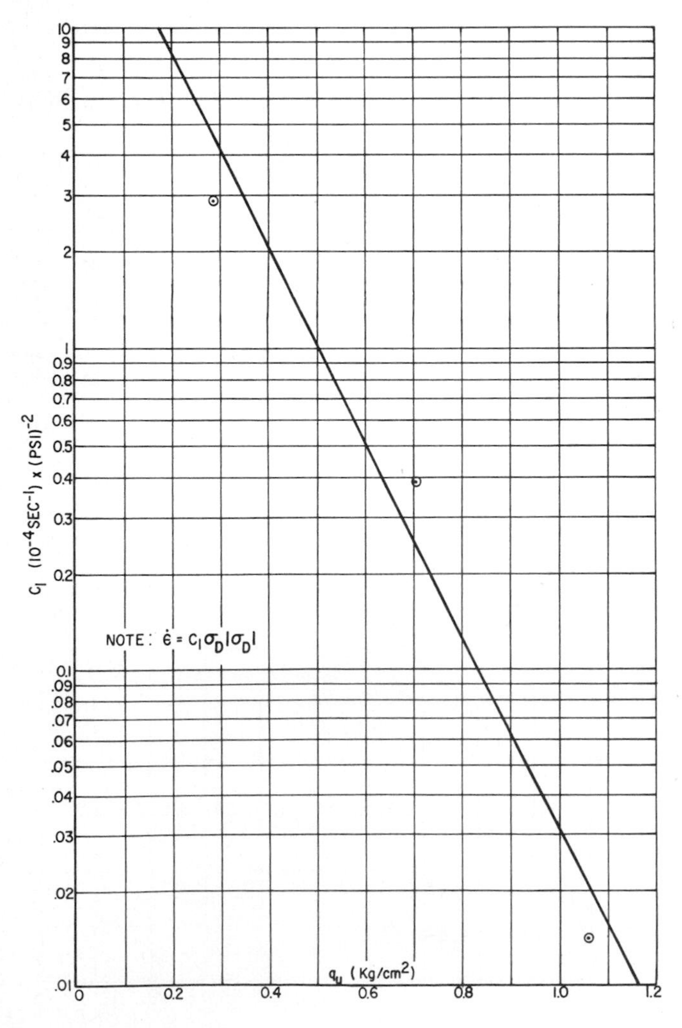

Fig. 14.22. Creep coefficient C_1 vs strength.

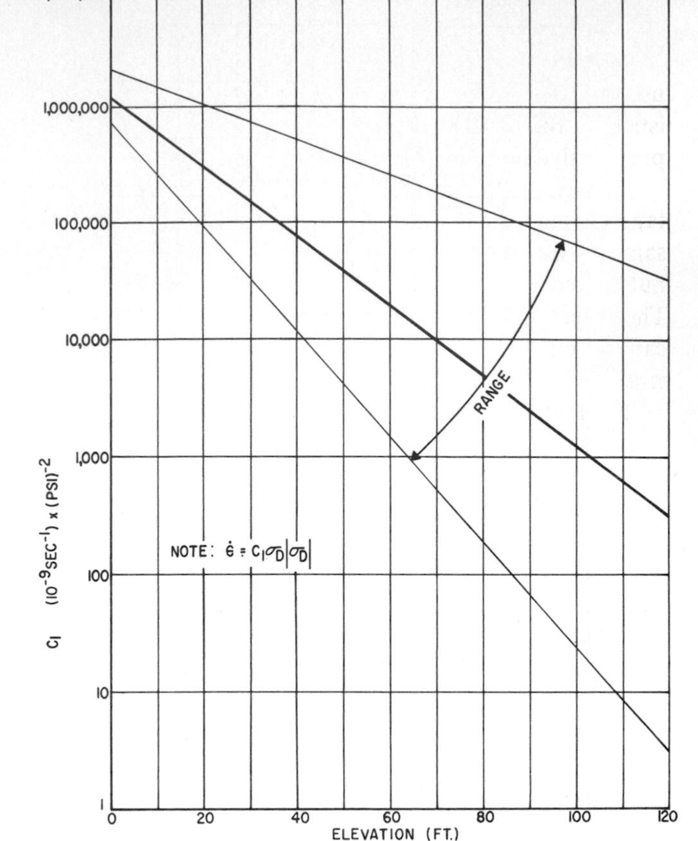

Fig. 14.23. Creep coefficient C_1 vs elevation.

required by using an equivalent equation of the form

$$\sigma_0 = C\dot{\epsilon} - D\dot{\epsilon}|\dot{\epsilon}| \tag{14.53}$$

which also can be used to represent the test data with an acceptable degree of accuracy. Thus, in order to simplify the computer operations, test data also have been represented by equations of the latter form, and appropriate values of the constants C and D, for soil samples of different strengths, have been determined. Using the relationships between strength and elevation shown in Fig. 14.17, the corresponding variation of the coefficients C and D with elevation can be easily obtained.

14.3.9 Variation in Creep and Damping Characteristics During Cyclic Loading

In order to investigate the possible variation in creep and damping characteristics of the clay during an earthquake, values of damping ratio λ and creep coefficient C were determined for specimens that previously had been subjected to a number of stress cycles causing axial strains as large as 10%. It was found that the resulting changes in λ and C were insufficient to warrant the inclusion of this effect in the analysis.

14.3.10 Dynamic Stress–Strain Characteristics

Finally, the data from the impact–creep tests were used to determine the dynamic stress–strain characteristics of the samples by each of the methods described previously.

Figure 14.21 shows the relationships between instantaneous elastic deformation and average stress for eight samples having compressive strengths of about 10 psi but tested using rigid masses of different magnitudes. The average relationship drawn through these points can be considered to characterize the dynamic stress–strain relationship for samples of this strength.

Similar data for tests on samples having compressive strengths of 4 and 15 psi were obtained.

Also plotted with the dynamic stress–strain relationship in Fig. 14.21 is the stress vs strain relationships for samples of the same strength tested under static loading conditions using a rate of stress application of about 0.2 kg/cm²/min. It may be seen that in each case the dynamic stress–strain relationship is slightly steeper than corresponding static relationship, reflecting the additional strains due to creep that can occur during the slower rate of loading.

In the analysis of the response of the soil deposit to earthquake-induced motions is was found desirable to characterize the stress–deformation relationship of the soil at low strains. This was done by defining a secant modulus E_o as the ratio of axial stress to axial strain at a strain level of 0.35%. Within this range in the dynamic loading tests the soil exhibits little or no hysteresis effects and may be considered to be reversibly elastic. The secant modulus determined in this way from the results of a dynamic loading test is designated E_{od}.

A similar secant modulus E_{os} at an axial strain of 0.35% also can be determined from the stress–strain relationship obtained under static or slow loading conditions.

From a comparison of dynamic and static stress vs strain relationships such as those shown in Fig. 14.21 it was found that the dynamic modulus E_{od} was greater than the corresponding static modulus E_{os} by amounts varying from 24 to 74%, with an average value of about 50%. This result was subsequently used to convert static load test data to corresponding dynamic test values.

14.3.11 Stress–Strain Relationships During Cyclic Loading

Examination of the dynamic elastic stress–strain relationships for samples revealed that it would be necessary to incorporate a nonlinear spring with hysteresis characteristics in the model of soil behavior. To determine quantitative values for the properties of such a spring required the determination of the inelastic behavior of the soil during cyclic loading in which the directions of shear stresses and strains were repeatedly reversed. Because of difficulties in making such determinations under dynamic loading conditions and in view of the fact that dynamic and static stress–strain relationships did not appear to be too different, it was decided to conduct the cyclic loading tests under static loading conditions, which permitted accurate determination of applied stresses and corresponding strains, and subsequently to modify these results appropriately to determine probable dynamic behavior.

Accordingly, a number of special unconsolidated, undrained triaxial compression tests were conducted on specimens, 1.4 in. in diameter and 3.5 in. high, trimmed from the undisturbed samples. In these tests the piston of the triaxial compression cell was attached to the specimen cap so that the axial stress on the specimen could be increased or decreased, as desired, throughout the test. The rate of stress application was about 0.2 kg/cm²/min.

In a typical test, the axial stress was increased progressively until the axial strain was about 3%, at which stage the axial stress was restored to its initial value. By pulling on the piston the axial stress was then gradually decreased until the axial extension was about 3% and the stress again was restored to its original hydrostatic condition. It should be noted that because of the initial ambient pressure of 1.0 kg/cm² acting on the specimen, the axial stress remained compressive throughout the entire loading cycle. Usually this procedure was repeated until about 10 cycles of loading had been applied.

A typical relationship between the change in axial stress and the corresponding axial strains obtained in such a test is shown in Fig. 14.24. In general the stress–strain relationship for the second cycle of loading differed considerably from that during the initial loading, but subsequent changes in further cycles were relatively small. Furthermore, it appeared that when variations in cycles were introduced—such as limiting one or more cycles to axial compression only or to reduced amounts of axial extension—the general form of the stress–strain relationship was maintained.

For purposes of analysis it is necessary to idealize the relationships shown in Fig. 14.24. When strains are small, even during cyclic loading, the stress–deformation relationships will be controlled by the initial part of the first loading cycle and can be represented with a satisfactory degree of accuracy by a straight line. For axial strains up to about 0.6% this relationship can be approximated by the secant modulus determined at an axial strain of 0.35%. This modulus, for tests conducted under static loading conditions, has been designated by E_{os}.

When high strains are induced during the stress cycles, the modulus E_{os} will apply for only the first part

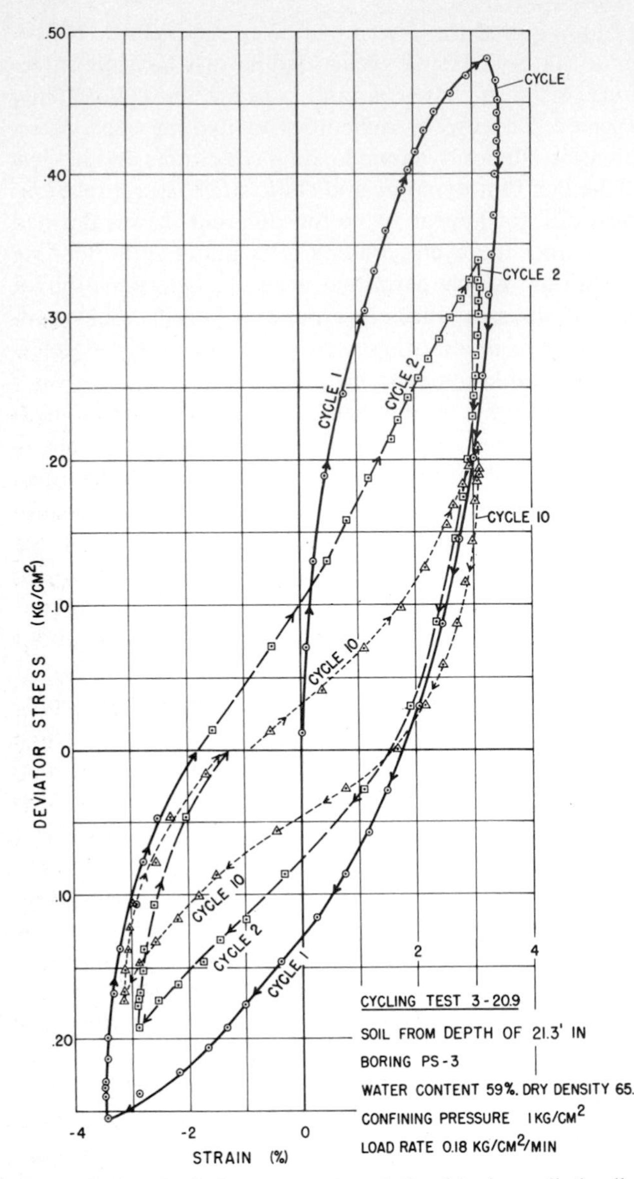

Fig. 14.24. Typical stress–strain relationship in cyclic loading test.

of the first cycle and the behavior of the soil is better characterized by the average of reasonably consistent stress–strain relationships developed during cycles 2–10, as shown in Fig. 14.24. This behavior may be idealized using the standard bilinear hysteresis loops. In this representation, the first cycle is ignored and the best parallelogram fitting the stress–strain relationships for subsequent cycles and conforming to the following mathematically consistent rules is established. Behavior during initial loading is assumed to be controlled by the modulus E_{1s}, until a limiting strain ϵ_{ys} is reached, at which stage further deformations are determined by the modulus E_{2s}. When the loading is reversed, behavior is again determined by the modulus E_{1s} until a strain change equal to $2\epsilon_{ys}$ has occurred, at which stage further deformations are determined by E_{2s}. On reloading, the modulus E_{1s} is again operative until a strain change of $2\epsilon_{ys}$ has occurred and the modulus E_{2s} begins to control the behavior. Similar behavior is developed during subsequent loading cycles.

By this means, the cyclic stress–strain relationships of the clay under static loading conditions are represented by the moduli E_{os}, E_{1s}, E_{2s}, and the yield strain ϵ_{ys}. Following the procedure described above, values for these parameters have been determined for a number of samples, of different strengths, subjected to cyclic loading tests. Although the samples used for these tests were not loaded to failure, their probable strength values could be estimated with a high degree of accuracy from the knowledge of their stress–strain behavior during initial loading, and the general form of such relationships could be obtained in the large number of strength tests previously conducted.

Thus, the values of E_{os}, E_{1s}, E_{2s}, and ϵ_{ys} can be obtained as a function of the soil strength. These moduli have been found to vary considerably with sample strength, but in each case the data can be represented by a linear relationship. However, the magnitude of ϵ_{ys} is essentially constant, regardless of the strength of the specimens, with an average value of 0.6%.

Using the average relationship between soil strength and elevation established previously (Fig. 14.17), the above moduli can be readily converted to show the variation of moduli with elevation in the clay layer.

Finally, it is necessary to determine the parameters determining the elastic behavior of the soil under dynamic cyclic loading conditions. A comparison of the dynamic and static stress–strain relationships during the first loading cycle indicates considerable similarity in form with the exception that the dynamic relationship is much steeper at low strains. (It was previously shown that the dynamic modulus at low strains is about 50% higher than the static modulus at the same strain.) Thus, it appears reasonable to approximate the dynamic deformation behavior of the soil by parameters E_{od}, E_{1d}, E_{2d}, and ϵ_{yd} corresponding to those utilized for describing the cyclic loading behavior under static stress conditions (E_{os}, E_{1s}, E_{2s}, and ϵ_{ys}), with the following relationships:

$$E_{od} = 1.5E_{os} \qquad E_{1d} = 1.5E_{1s}$$

$$E_{2d} = E_{2s} \qquad \epsilon_{yd} = \epsilon_{ys}$$

Based on these relationships, the variations in parameters describing the dynamic elastic behavior of the soil at different elevations were obtained.

14.4 COMMENTS ON PROPOSED METHOD OF ANALYSIS

It is apparent that it is impossible to obtain a completely rigorous solution to the general problem under consideration. However, it is believed that an approximate solution sufficiently accurate for design purposes can be obtained by the methods set forth in the preceding section of this chapter.

To justify the methods of analysis that have been used herein, several of the major assumptions that have been made should be recalled. First, it has been assumed that the elastic stress and displacement fields within the clay medium can be adequately defined by a static theory, i.e., the Mindlin theory. This assumption can be justified when the characteristic wave length in the clay medium, which in this case is the shear wave length (λ_s), is long compared with the horizontal distance (D_h) across the zone of major influence resulting from interaction. This shear wave length is calculated using the relation

$$\lambda_s = \frac{1}{f}\sqrt{\frac{G}{\rho}} \tag{14.54}$$

in which G is the shear modulus of clay medium, ρ denotes the mass density of clay medium, and f refers to the predominant frequency of the forcing function. It is believed that this condition is reasonably well satisfied for the general problem being considered and, therefore, the use of a static theory in defining the stress and displacement fields is considered satisfactory. It also should be noted that satisfying this condition implies that the inertia forces of masses $\bar{M}_i$ and $\bar{\bar{M}}_i$ are small compared with the interaction spring forces and the elastic shear forces in the piles. Dynamic response calculations show this latter statement to be true. Thus, the dynamic response calculations are quite insensitive to reasonably large changes in the values assigned to the effective clay masses $\bar{M}_i^e$ and $\bar{\bar{M}}_i^e$.

The second major assumption regarding the pile–clay medium interaction was that which uncoupled the interaction springs as used in the idealized system (Winkler assumption). It was only after considerable effort had been spent on a coupled system that the decision was made to simplify the problem by using uncoupled springs. The coupled system led to an extremely complicated method of analysis that required a great deal more effort in carrying out the solution and did not seem to be justified.

Due to the transient nature of earthquake ground motion and its relatively short duration, creep effects are small. Therefore, the proposed approximate method of determining creep coefficients is considered satisfactory.

The effects that all of the various simplifying assumptions have on the accuracy of the general solution could be examined at great length. However, in justifying the use of these assumptions it must be remembered that all soil properties of the clay medium have very wide variations, i.e., of the order of 100 or 200%. Therefore, inaccuracies introduced in the analysis by the various assumptions made are considered acceptable.

14.5 BRIDGE STRUCTURAL SYSTEMS INVESTIGATED

The objective of the general investigation reported herein has been to develop methods for predicting seismic effects on bridges that are supported on piles extending through deep, sensitive clays and to apply these methods to the bridge structure presently being designed by the California State Division of Highways for erection across Elkhorn Slough. This bridge structure is shown schematically in Fig. 14.3 and is also shown in considerably more detail in Fig. 14.25. It should be realized that such design details are preliminary and may not necessarily correspond with the final design.

While the above-described structure served as the basic structure being analyzed in the general investigation, a single variation in this structure was assumed for additional analysis. This variation consisted of changing the outside diameter of all piles from 54 to 36 in. and the inside diameter from 44 to 26 in. However, the number and arrangement of piles and the superstructure design were assumed to remain unchanged.

For analysis purposes the mass of the bridge superstructure was lumped at the bridge deck level, three intermediate levels on the piers, and at the pile cap level. This lumping of mass corresponds to $m = 4$ in Fig. 14.4. Damping in the superstructure was assumed to be 5% of critical ($\xi = 0.05$) in its fundamental mode for all analyses performed.

The depth of the clay layer varies considerably along the longitudinal axis of the proposed bridge site and has a maximum depth of approximately 120 ft. Therefore, it seemed most appropriate for analysis purposes to assume the clay medium as an infinite layer horizontally having a depth of 120 ft. Two different sets of properties were used for this medium in analyzing each of the above bridge structural types. These two sets of properties, which will be referred to subsequently as Clay System Numbers 1 and 2, were established on the basis of tests performed on San Francisco Bay "mud" and Elkhorn Slough clay, respectively.

The number of samples tested in establishing the properties of Clay System Number 1, i.e., San Francisco Bay mud, was very small as these tests were performed early in the program to aid in developing experimental techniques before Elkhorn Slough samples were available. A large number of samples, however, was later tested in establishing the properties of Clay System Number 2, i.e., Elkhorn Slough clay. Since Clay System Number 2 represents the actual bridge site conditions,

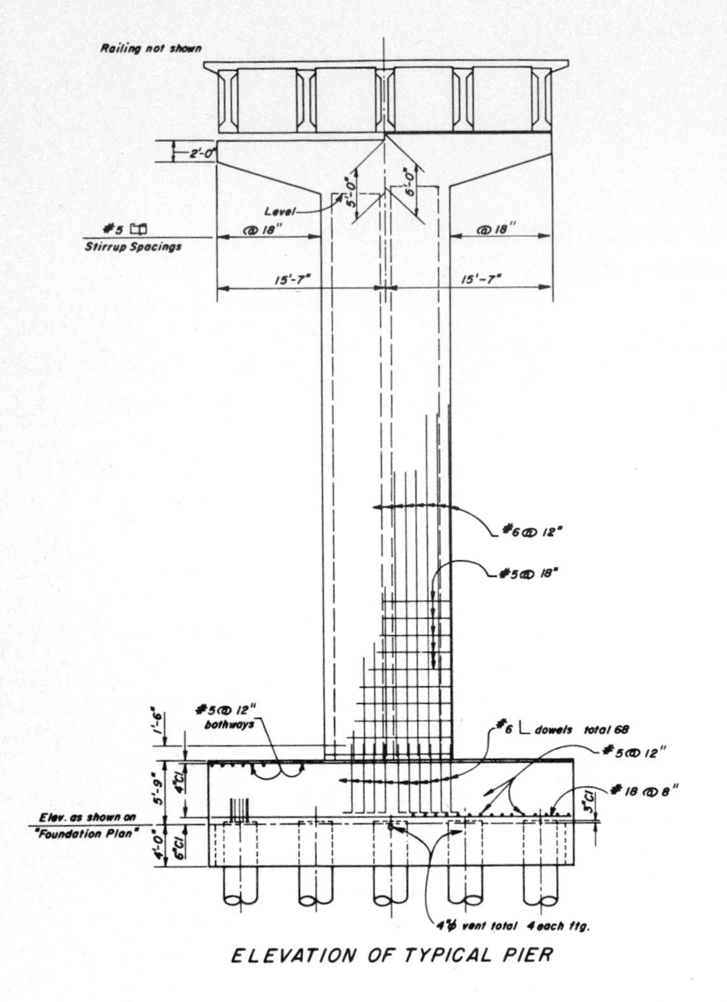

ELEVATION OF TYPICAL PIER

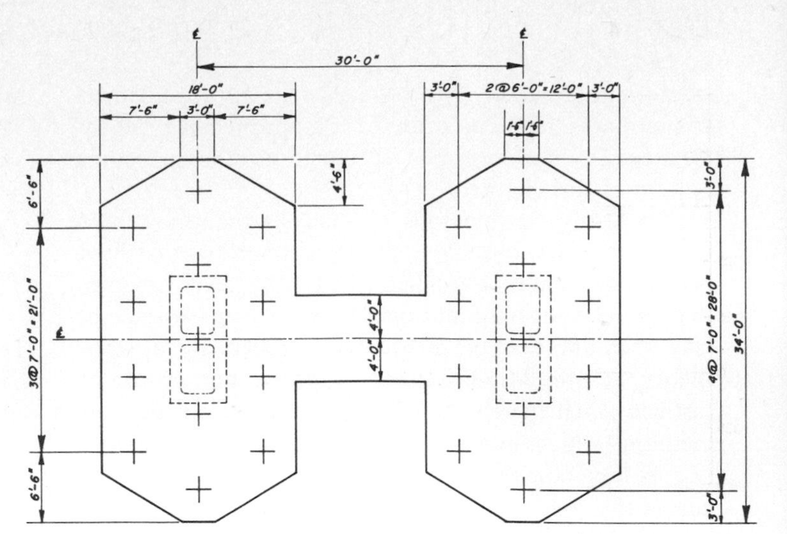

PLAN OF PILE CAP FOR THE TWO MULTIPLE ROW GROUPINGS

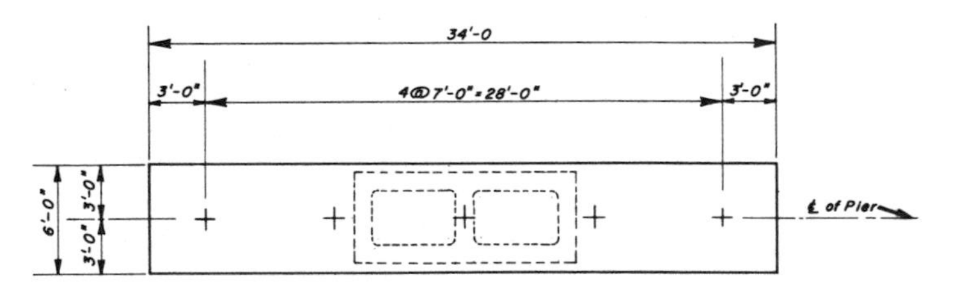

PLAN OF PILE CAP FOR SINGLE ROW GROUPING

Fig. 14.25. Details of bridge structure.

the dynamic response studies based on this clay medium are of principal importance. However, it is considered worthwhile to present the results based on Clay System Number 1 to illustrate some differences in response that can be expected with certain specific changes in clay medium properties.

The properties of Clay System Number 2 were presented in some detail in the preceding section of this report; therefore, no further discussion of these properties is necessary. However, it is necessary at this point in our discussion to point out the basic differences between the properties of Clay System Numbers 1 and 2. The important differences as they affect the dynamic response of the bridge structural systems relate to the stress–strain relations. The response studies using Clay System Number 1 were based on only the two moduli E_{1d} and E_{2d}, which varied with depth z as given by the relations

$$E_{1d}(z) = 142 + 8.9z$$
$$E_{2d}(z) = 60.9 - 0.45z \qquad (14.55)$$

where z is depth in feet and the moduli are given in pounds per square inch. Comparing these moduli with those given for Clay System Number 2, relatively small differences were noted for E_{1d}, while significant differences were noted for E_{2d}. However, the greatest and most significant difference in the stress–strain properties of these two clay systems is their difference in yield strain ϵ_{yd}. This normal yield strain equals 0.003 for Clay System Number 1 and 0.006 for Clay System Number 2, which corresponds to shear strains γ_{yd} of 0.0045 and 0.009, respectively. Since both clay systems have essentially the same moduli E_{1d} with depth, it is quite apparent that the above yield strains represent a much weaker material for Clay System Number 1 as compared with Clay System Number 2. Other properties such as creep and damping are similar.

For analysis purposes, the mass of the clay medium was lumped at 18 equally spaced levels, i.e., $n = 18$, as shown in Fig. 14.4.

14.6 DISCUSSION OF RESULTS

14.6.1 Total Acceleration of Clay Medium

The recorded acceleration of the N–S component of the 1940 El Centro Earthquake was used as the prescribed horizontal acceleration $\ddot{u}_g(t)$ at the base of the

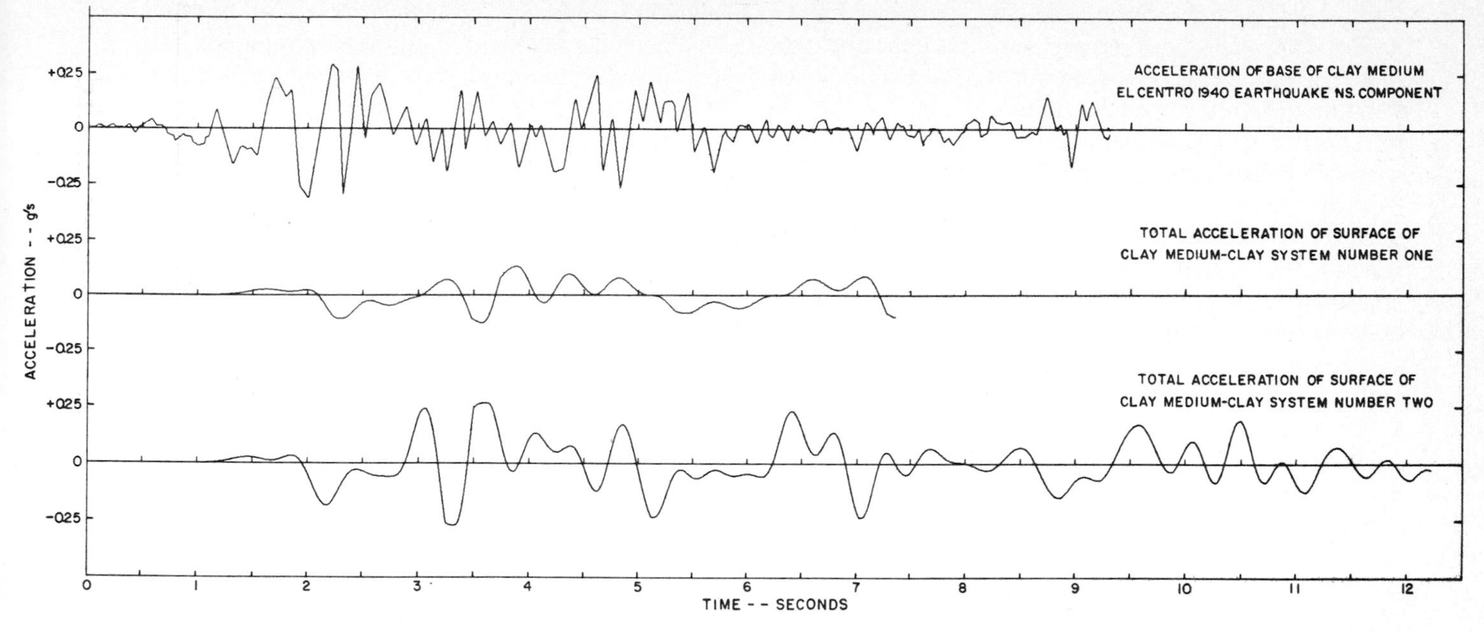

Fig. 14.26. Acceleration vs time curves at base and surface of clay mediums.

clay medium. Approximately 9 sec of this acceleration–time function is shown in Fig. 14.26, and its standard absolute acceleration response spectrum as defined by the relation

$$AA(\omega) = \left|\sqrt{1-\xi^2}\,\omega \int_0^t \ddot{u}_g(t)e^{-\xi\omega(t-\tau)} \cdot\left[\left(1-\frac{\xi^2}{1-\xi^2}\right)\sin\omega\sqrt{1-\xi^2}(t-\tau) + \frac{2\xi}{\sqrt{1-\xi^2}}\cos\omega\sqrt{1-\xi^2}(t-\tau)\right]d\tau\right|_{\max} \tag{14.56}$$

is shown in Fig. 14.27. The absolute acceleration given by Eq. 14.56 is simply the absolute value of the maximum total acceleration that the mass of a single-degree-of-freedom system will experience when excited through its support by the prescribed ground acceleration $\ddot{u}_g(t)$. The spectral values obtained in accordance with this definition are plotted in Fig. 14.27 vs the undamped natural frequency ω for five different values of the damping ratio ξ, i.e., $\xi = 0$, 0.02, 0.05, 0.10, and 0.20. Note that the absolute acceleration response spectral curves approach an asymptotic value, with increasing frequency ω, which corresponds with the peak acceleration in the ground motion. In the case of the N–S component of the El Centro Earthquake, this peak acceleration is approximately 0.33 g.

Also shown in Fig. 14.26 are the time histories of total horizontal acceleration produced at the surface of Clay System Numbers 1 and 2 when excited at their base by the N–S component of the El Centro Earthquake. Absolute acceleration response spectra for these acceleration functions are shown in Fig. 14.27.

The above clay surface acceleration time histories have been calculated considering the response of the clay medium alone, i.e., the bridge structural system is assumed not to be present.

Comparing the three acceleration–time functions and

Fig. 14.27. Absolute acceleration response spectra vs undamped natural frequency.

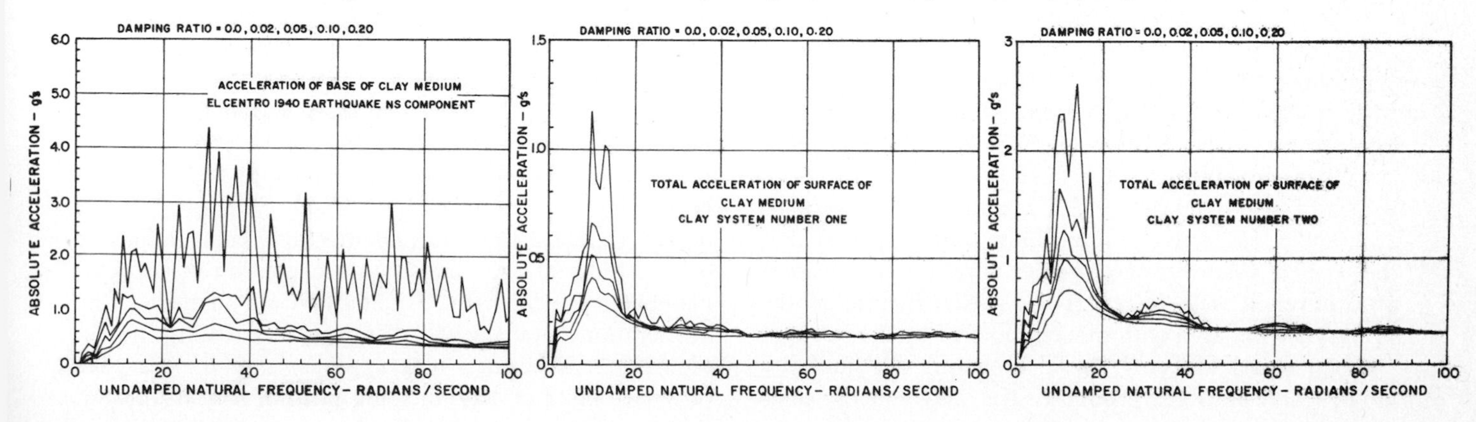

their absolute acceleration response spectra as shown in Figs. 14.26, and 14.27, respectively, it is quite apparent that the surface acceleration functions contain much lower frequency components than are contained in the base acceleration function. This fact shows the effectiveness with which a soft clay medium filters out the higher frequency components. The predominant circular frequencies contained in the surface motion of both clay systems are, as shown by their response spectra, in the approximate range of 8–15 rad/sec. Since the stiffnesses of both clay systems are approximately the same, one could expect this peaking of response spectra in approximately the same frequency range.

Using the response spectra of Fig. 14.27 to indicate magnitude of seismic forces that would be developed in linear elastic structures if subjected to base accelerations corresponding to the three acceleration functions given in Fig. 14.26, it is apparent that these forces would be comparable for damped structures when considering the effects of the El Centro motion vs the effects of the surface motion of Clay System Number 2. (Reasonable damping factors for reinforced concrete bridge structures are in the approximate range of 0.05–0.15). However, the response spectra for Clay System Number 1 would indicate considerably lower seismic forces than those indicated by the response spectra for either Clay System Number 2 or the N–S component of El Centro. This large reduction for Clay System Number 1 is undoubtedly due to the relatively large amount of inelastic action which takes place during the response of this weaker system. Very little inelastic action will take place, however, during the response of Clay System Number 2 due to its higher yield strength.

The maximum peak acceleration reached during the period of the earthquake input is noted to be approximately 0.30 g for the surface motion of Clay System Number 2 and 0.13 g for the surface motion of Clay System Number 1.

14.6.2 Total Longitudinal Acceleration of Bridge Deck

The total horizontal acceleration time history for the bridge deck is shown in Fig. 14.28 for the three different clay–pile systems analyzed: (1) Clay System Number 1, 54 in. OD piles; (2) Clay System Number 2, 54 in. OD piles; and (3) Clay System Number 2, 36 in. OD piles.

These acceleration time histories show maximum peak values of approximately 0.6, 1.2, and 1.2 g for systems 1, 2, and 3, respectively. Again, it is quite apparent that the weaker Clay System Number 1 produces considerably lower seismic forces in the superstructure of the bridge as compared with those produced by the stronger Clay System Number 2. Note, however, the similarity of the accelerations produced by Clay–Pile Systems Numbers 2 and 3. This similarity shows the relatively small influence that pile stiffness has on the response of the superstructure. As will be shown subsequently, relatively small clay–pile interaction displacements result during the earthquake. Therefore, the piles are forced to move generally with the moving clay medium. In other words the displacements of the piles are controlled to a much larger degree than the clay–pile interaction forces.

It is somewhat difficult, by observing the acceleration time histories of Fig. 14.28, to isolate the predominant frequencies contained therein. However, it is estimated that a Fourier analysis of the two similar wave forms representing Clay System Number 2 would show predominant frequencies in a rather narrow band near a frequency of 2 cps. This observation undoubtedly reflects a response of the superstructure primarily in its fundamental mode of vibration, which has been calculated to be 1.7 cps on the assumption of completely fixed piers at the location of the pile caps. Since the piles cannot provide full fixity as assumed, the correct fundamental frequency of the superstructure should be somewhat greater than 1.7. Using a frequency of say 1.8 cps ($\omega = 11.3$ rad/sec) and a damping factor ξ of 0.05, the absolute acceleration spectral value given in Fig. 14.27 for Clay System Number 2 is approximately 1.25 g. This acceleration level is in very close agreement with the peak acceleration of 1.2 g observed for the bridge deck. Realizing that the above prediction of a peak acceleration of 1.25 g using the response spectrum of Clay System Number 2 neglects the effects of interaction between piles and clay medium while the observed peak of 1.2 g includes these interection effects, one again concludes that such interaction effects have a relatively small effect on the forces developed in the superstructure. Therefore, if one is interested only in seismic effects on the superstructure, the surface motion of the clay medium as determined with no bridge structure present could be used as a direct input into the base of the bridge piers. This analysis, of course, would not provide an indication of the forces and deformations produced in the piles.

It is of interest to compare the above peak acceleration (1.2 g) of the bridge deck when Clay System Number 2 is a part of the overall system with the peak acceleration that would be produced if the El Centro ground motion was the prescribed motion at the base of the piers. This latter peak acceleration can be obtained from the acceleration spectrum shown in Fig. 14.27 for the N–S component of El Centro. Using a damping factor of 0.05 and the estimated fundamental circular frequency of 11.3 rad/sec, one obtains an acceleration of approximately 0.9 g, which is considerably lower than the 1.2 g acceleration obtained with the presence of Clay System Number 2. Therefore, it is apparent that the presence of a deep layer of clay, in some cases, if it has sufficient

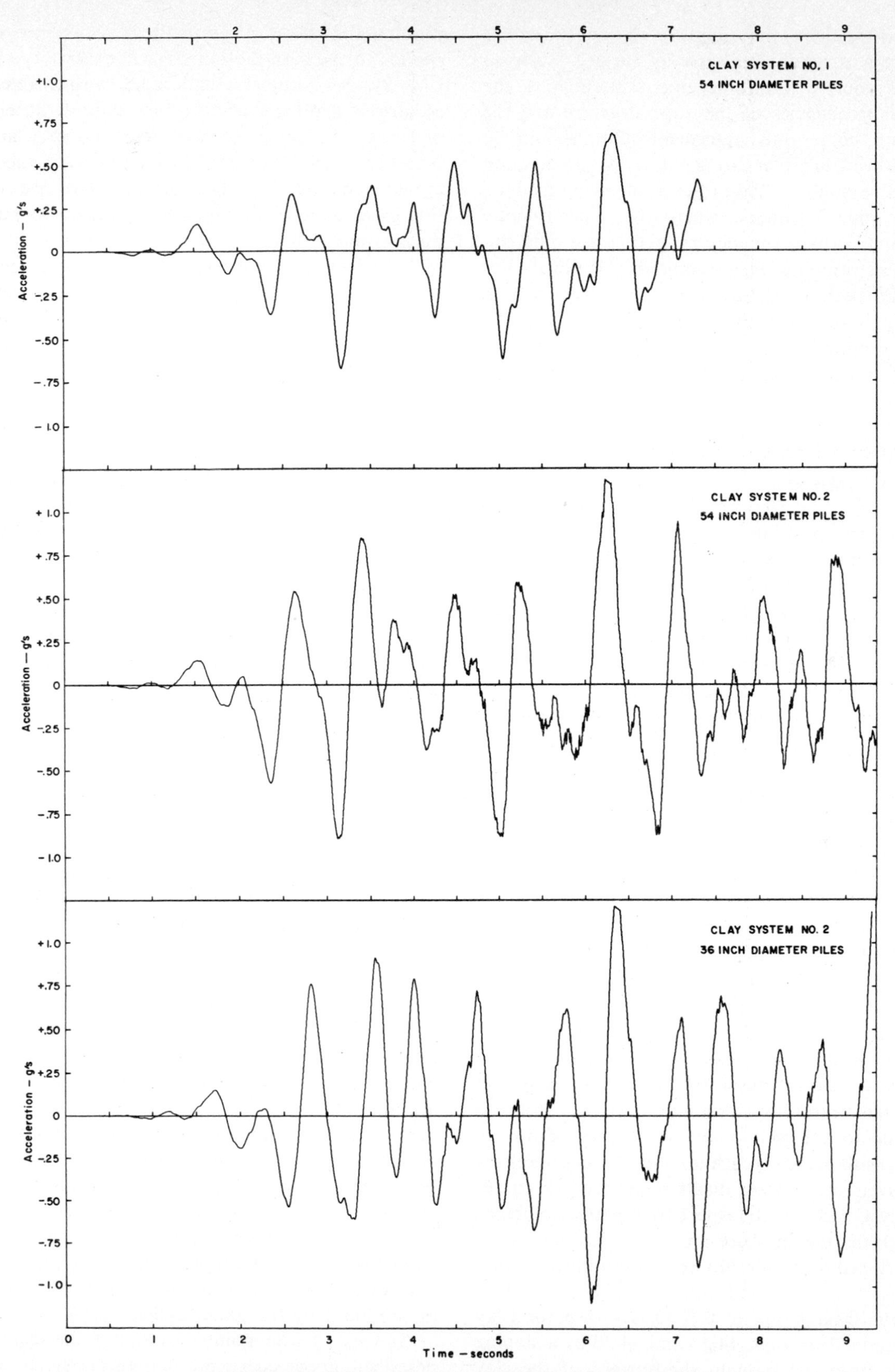

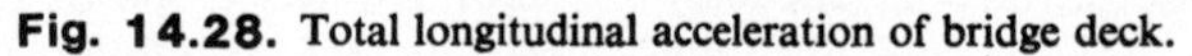
Fig. 14.28. Total longitudinal acceleration of bridge deck.

strength and elasticity, may increase the seismic forces produced in a structure built on its surface. Such an increase, of course, would become noticeable as the fundamental frequencies of the superstructure and the clay medium come into agreement. One should be careful, however, in generalizing the above observation of increased response in the presence of a clay medium because in other instances, where the superstructure frequency and perhaps the clay properties are different, a decrease in response easily could be obtained. For example, the peak acceleration of the bridge deck is considerably lower than the peak acceleration of the El Centro ground motion when the clay medium is represented by Clay System Number 1.

14.6.3 Deflected Shapes of Clay and Interaction Systems

The deflected positions of the clay medium as a function of depth with no bridge structure present and the deflected positions of the piles as a function of depth with the entire bridge structure present have been determined at times $T = 0, 0.5, 1.0, \ldots, 8.0$ sec. Due to lack of space, only example results as shown in Figs. 14.29 and 14.30 will be presented herein. All curves representing clay medium displacements are identified by the letter C, and all curves representing pile displacements are identified by the letter P. The vertical line in each case is the fixed reference line from which both types of displacements are measured. This reference has been established along the axis of the pile group at time $T = 0$. Therefore, both curves labeled C and P coincide with this reference line at time $T = 0$.

Only the first 8 sec of response are reported herein as this initial time period contains the critical response. No additional information of value would be presented if the response at later times were included.

The displacement u_g corresponding with the El Centro acceleration function as prescribed in the analysis is shown in each case in Figs. 14.29 and 14.30. It should be noted that these displacement values are not too accurate due to the well-known difficulties of double integrating, with accuracy, a highly oscillatory acceleration. However, the relative displacements of all curves identified by C and P with respect to a vertical reference axis through the moving base are quite accurate, as the above-mentioned difficulties do not arise in their evaluation.

The yield shear strain of 0.0090 is shown for Clay System Number 2 on Figs. 14.29 and 14.30 by a sloping dashed line from the base to the surface of the clay medium at $T = 2.0$ sec. By comparing the absolute value of slopes along all curves labeled C with corresponding slopes of the dashed lines representing yield strain, one can easily see the amounts and locations of yielding that takes place in the clay medium as it responds to the base motion. Such comparisons show that Clay System Number 1 undergoes considerable yielding at various depths, thus absorbing considerable amounts of energy. However, such comparison for Clay System Number 2 shows (Figs. 14.29 and 14.30) rather small amounts of yielding. The yielding that does occur in this case takes place in more restricted locations and during much shorter periods of time than in the case of Clay System Number 1. Thus, Clay System Number 2 responds with much more of the character of an elastic system than does Clay System Number 1. This basic difference is the reason for the differences in accelerations produced by these two media, as previously noted in Fig. 14.26.

Examining now the interaction displacements, i.e, the horizontal displacements between curves C and P, one generally observes these displacements to be small in comparison with the total displacements as measured from a vertical reference through the moving base. However, these interaction displacements are not insignificant as they have an appreciable effect on the curvatures produced in the piles. The pile yield radius of curvature is indicated on each of Figs. 14.29 and 14.30. This radius of curvature is based on the classical linear flexure theory and has been calculated on the basis of a maximum flexure strain of 1.00×10^{-3} in./in. This strain corresponds to a concrete modulus of elasticity of 4.5×10^6 psi and a maximum concrete flexure stress of 4.5×10^3 psi. One will note that in general the responses in Figs. 14.29 and 14.30 indicate maximum pile curvatures that are of the same order of magnitude as their yield curvatures. Comparing the actual curvatures produced in the 36 in. OD piles with those produced in the 54 in. OD piles, one finds that they are considerably larger. However, the ratios of the actual maximum curvatures for these two sizes are about the same as the ratio of their yield curvatures. Therefore, it is difficult to say whether one pile size is more critical than the other. If one carried this reasoning to smaller and smaller pile diameters, of course, one would reach a point where the curvatures in the piles coincide with the curvatures represented by the clay displacement curves C. In such a limiting case the curvatures would be controlled entirely by the response of the clay medium and therefore any further reduction in pile diameter would indeed represent a reduction in the maximum flexural stress produced. Note that the flexural stresses as referred to here do not reflect the direct stress as produced by the axial loads.

It needs to be pointed out that the displacements described above were plotted automatically and that because the computer calculated these displacements only at the 18 discrete levels, the plotter cannot accurately show these displacements at levels between these discrete locations. Since the plotter operates on a linear

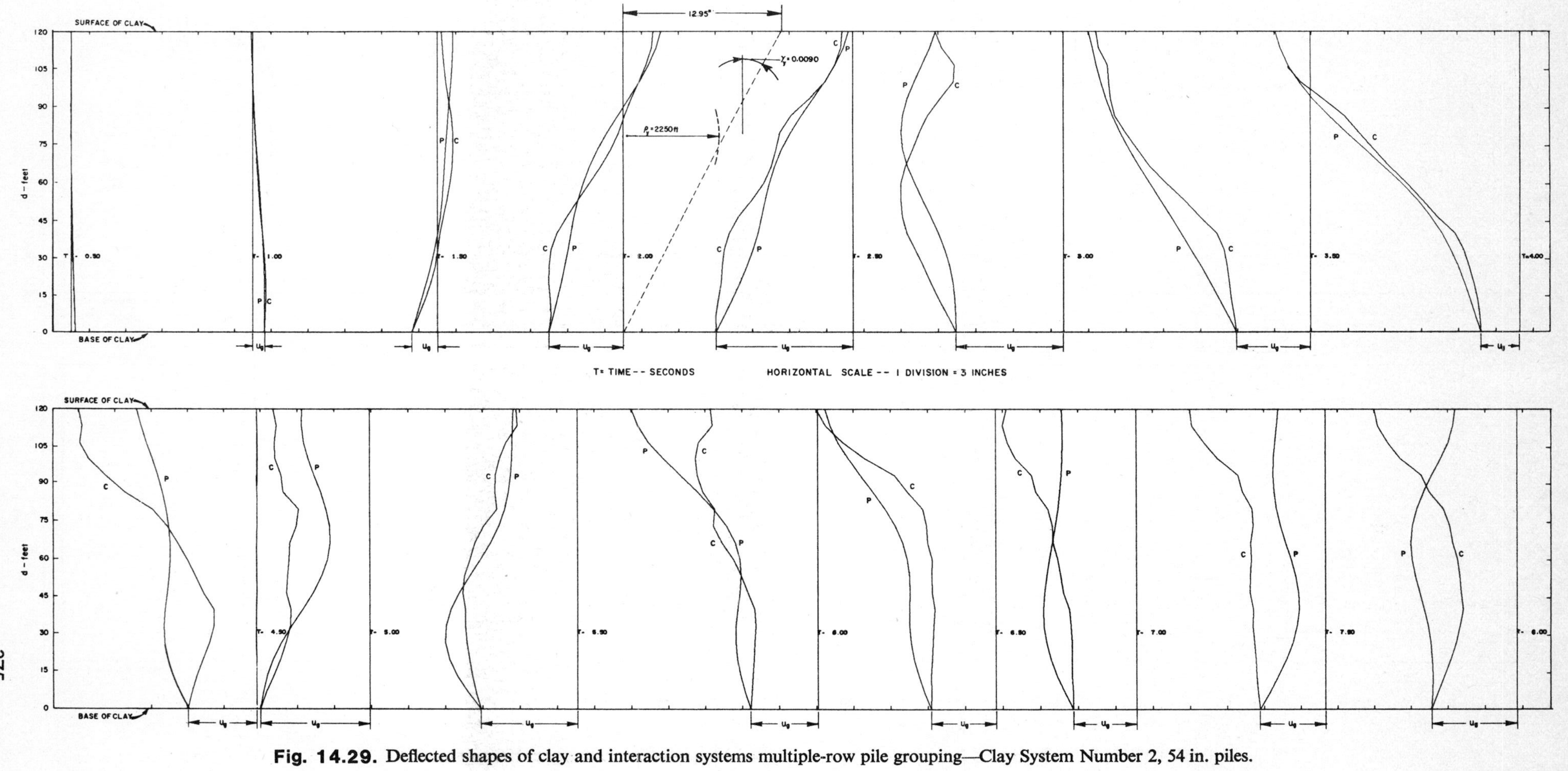

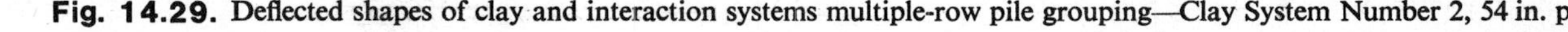
Fig. 14.29. Deflected shapes of clay and interaction systems multiple-row pile grouping—Clay System Number 2, 54 in. piles.

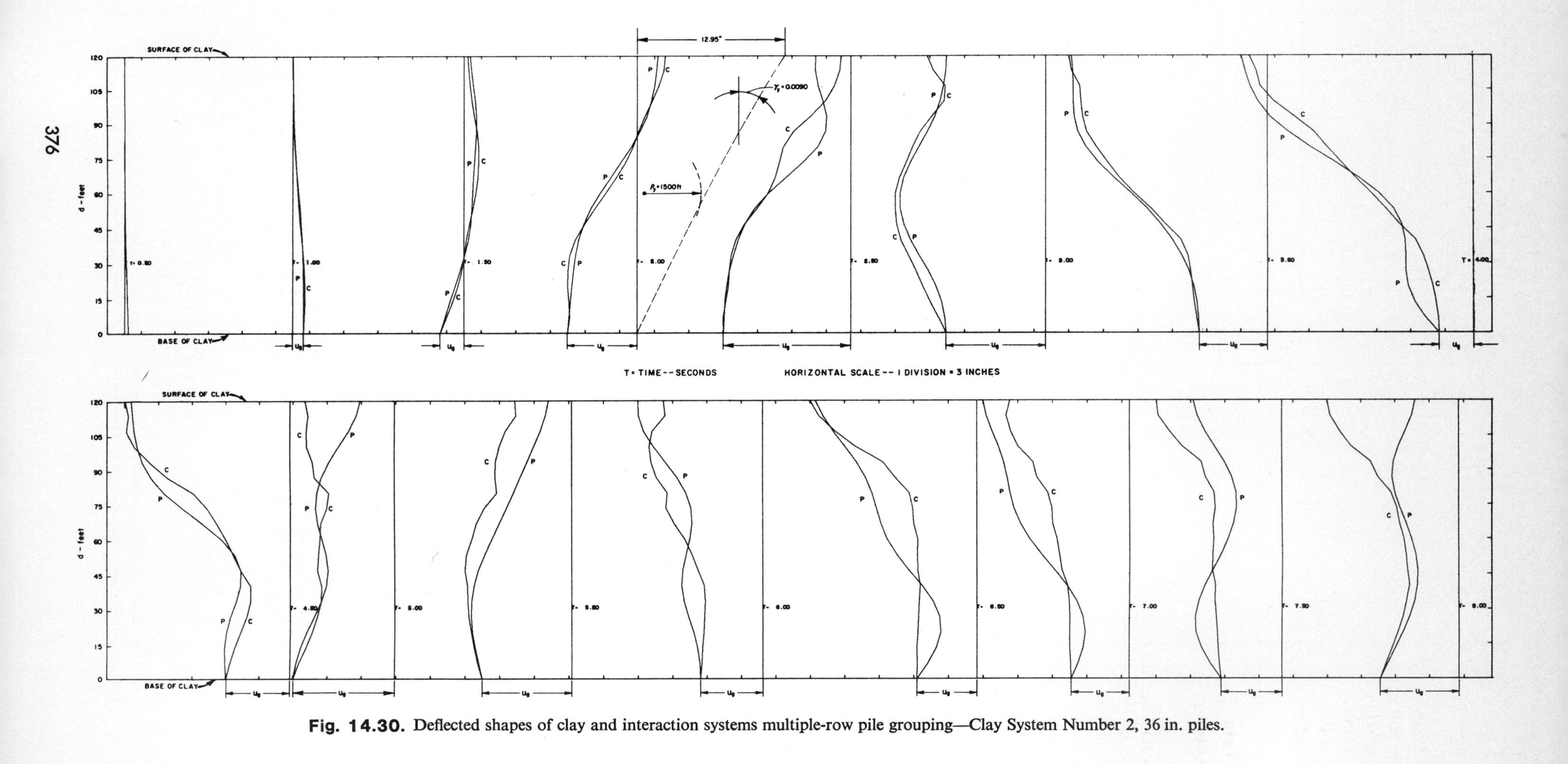

Fig. 14.30. Deflected shapes of clay and interaction systems multiple-row pile grouping—Clay System Number 2, 36 in. piles.

variation basis between points, some of the largest curvatures as shown by the above-described plots are too large. When interpreting these data, one should visualize smooth curves drawn through the 19 data points ($n = 0, 1, 2, \ldots, 18$) rather than the straight line segments between points as produced by the plotter.

14.6.4 Total Force Distribution Along Pile Groupings

The total force intensity per unit of vertical dimension for the entire pile grouping is shown as a function of vertical location in the clay medium in Figs. 14.31 and 14.32 for the 54 and 36 in. diameter, multiple row pile grouping, respectively. These force intensity plots need little explanation as one readily sees that they vary through the depth of the clay medium as one would expect from an examination of the corresponding interaction displacements shown in Figs. 14.29 and 14.30. However, one very significant observation should be made. Note that the maximum force intensities are very much below the so-called "yield intensity" given on each figure. This yield intensity is that force intensity at which the pile grouping would start to cut its way through the clay medium. Since the force intensities could have reached their largest maximum values at times intermediate between those shown in Figs. 14.31 and 14.32, the computer program was written so that the maximum positive and negative values as selected from those calculated at 0.0025-sec intervals could be obtained. These maximum force intensities along with the times at which they occurred are shown in Fig. 14.33. Again, it is quite apparent that the maximum interaction force intensities developed were far below the yield intensities.

One would conclude from the above observation that standard size piles will never cut their way through a deforming clay medium. Remember that the above force intensities are sufficient to produce curvatures of the order of magnitude of the yield curvatures. As far as the author is aware, this conclusion is not in conflict with field observations of embankment failures where piles have been in place.

14.6.5 Lower Boundary Condition of Piles

A pinned boundary condition has been assumed for all piles at the base of the clay medium. This boundary condition is in reality incorrect as such piles must be driven sufficiently into the base medium to develop the necessary point-bearing capacity. If the base medium below the clay layer has much higher stiffness and strength characteristics than the clay medium and if the piles are driven an appreciable distance into this medium, a rather bad situation is created at the interface of these two media. In other words, the large discontinuity in soil shear strains at this point would likely cause curvatures in the piles that far exceed their yield values. Thus, in effect plastic hinges would be developed in the piles at this location. The designer should recognize this situation and design the piles in this region so that vertical load-carrying capacities are not lost when such yielding occurs.

If the penetration distances of the piles into the base medium are sufficiently small, such pile failures at this location would not occur since local soil failures would occur instead.

It should be recognized that assuming pinned lower boundary conditions in the general analysis does not lead to any general error in response of the entire system. Such errors involved are only in the immediate vicinity of the base of the piles.

14.6.6 Stability of Piles

The idealized structural model shown in Fig. 14.4 represents the interaction effects between piles and clay medium by linkages, each of which consists of a damping dashpot in parallel with a nonlinear interaction spring that in turn are placed in series with a creep dashpot. All creep dashpot coefficients calculated by the methods previously presented were sufficiently high for the clay systems studied so that the creep displacements developed during the short period of the prescribed earthquake were small compared with the total interaction displacements $U_i^r - u_i^r$. In fact, they have been found to be sufficiently small during this short transient period so that they could have been neglected without any great loss in accuracy of the general solution.

Under static loads of long duration the above-mentioned interaction linkages, of course, would give no lateral support to the piles because of the presence of the creep dashpots. Therefore, while the idealized model used is a satisfactory one in calculating the transient response during an earthquake, it could not be used in the static case.

The clay medium does possess a certain permanent elasticity even though it may be small. This permanent elasticity could be represented in the model by an elastic spring placed in parallel with the creep dashpot. This spring, of course, would be much less stiff than the spring placed in parallel with the damping dashpot. Therefore, in view of the large creep dashpot coefficients this additional spring would transfer very little of the total interaction load during the transient period of an earthquake but would transfer all of the interaction load under static conditions.

Considering now the lateral stability of a single pile

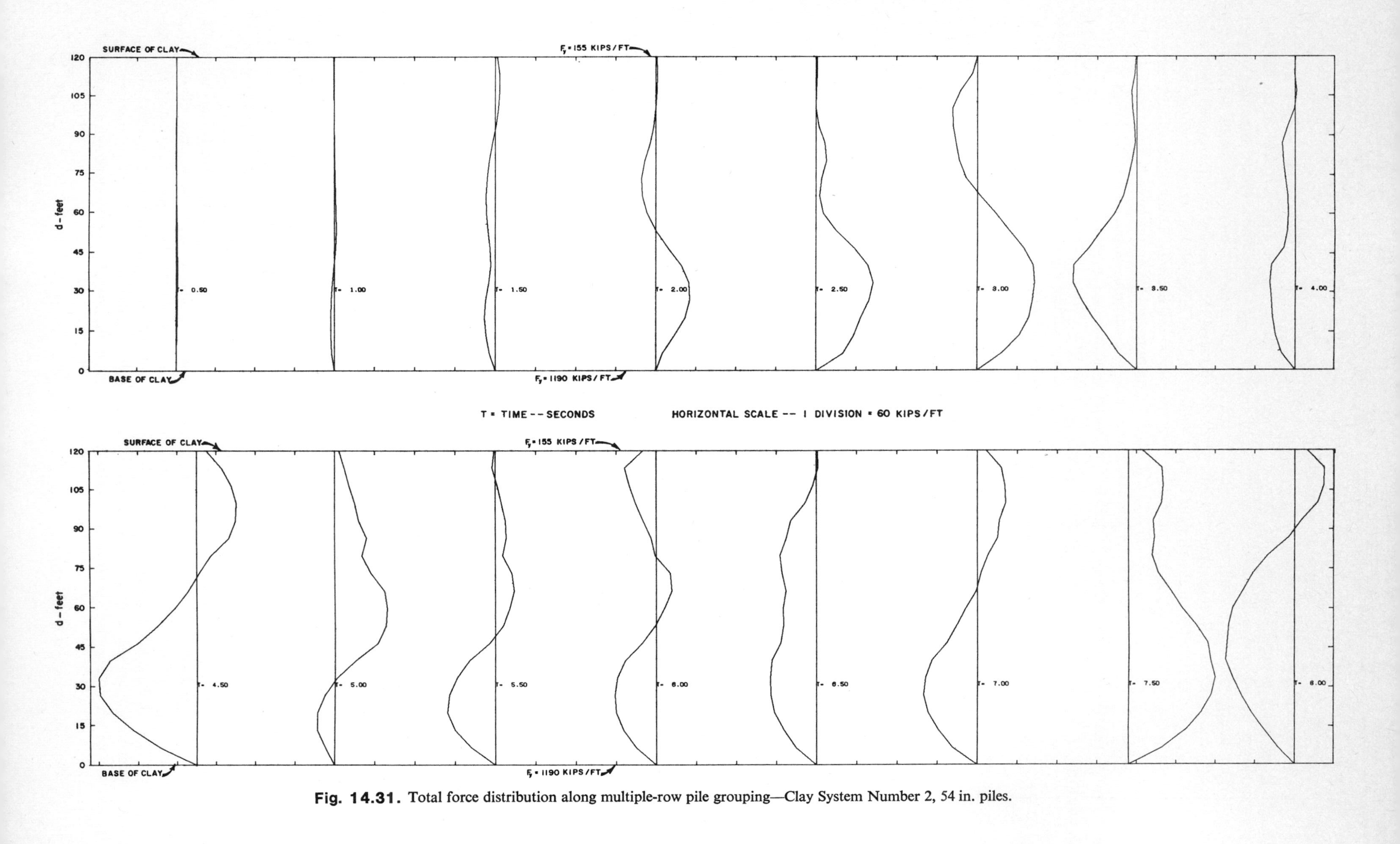

Fig. 14.31. Total force distribution along multiple-row pile grouping—Clay System Number 2, 54 in. piles.

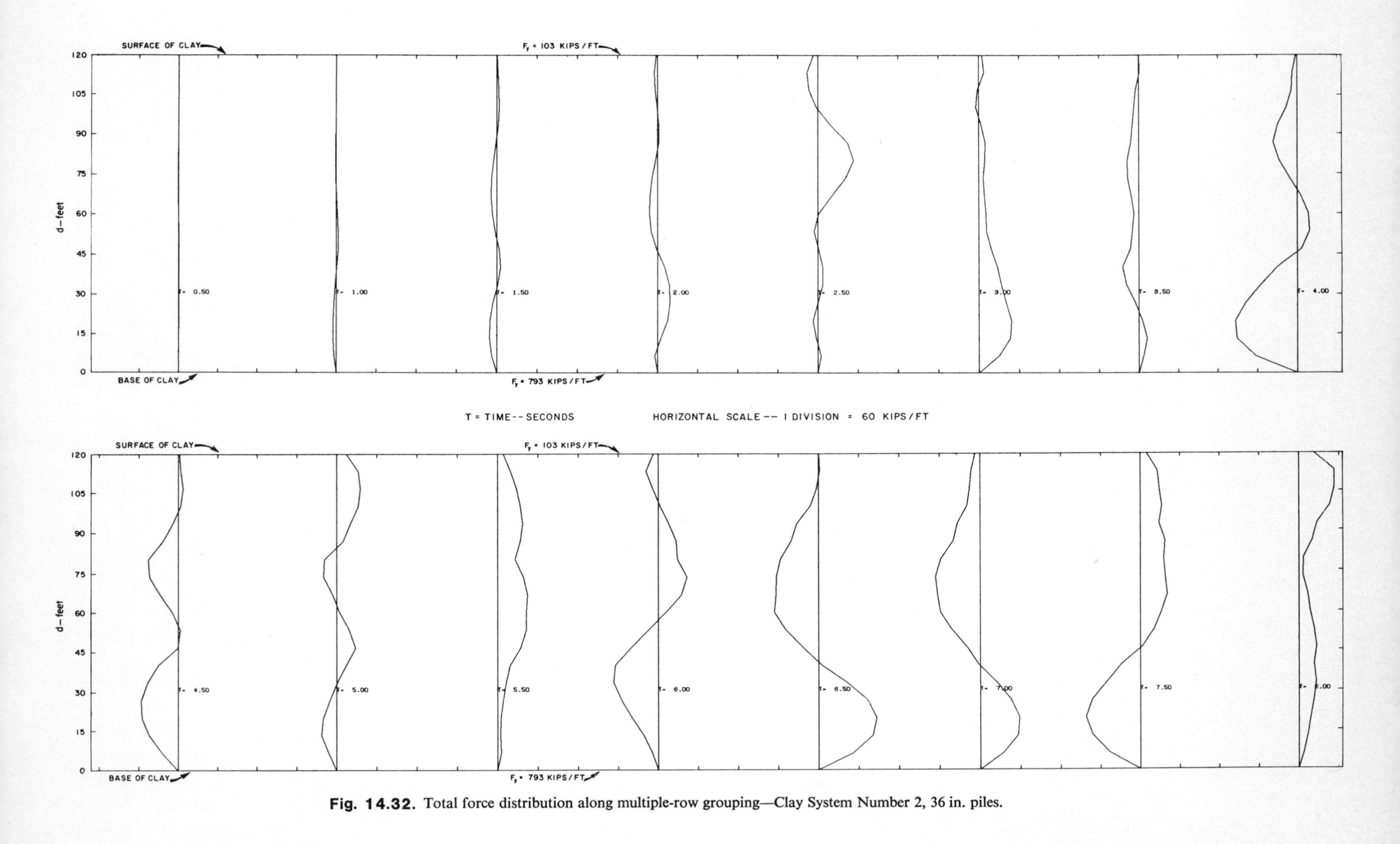

Fig. 14.32. Total force distribution along multiple-row grouping—Clay System Number 2, 36 in. piles.

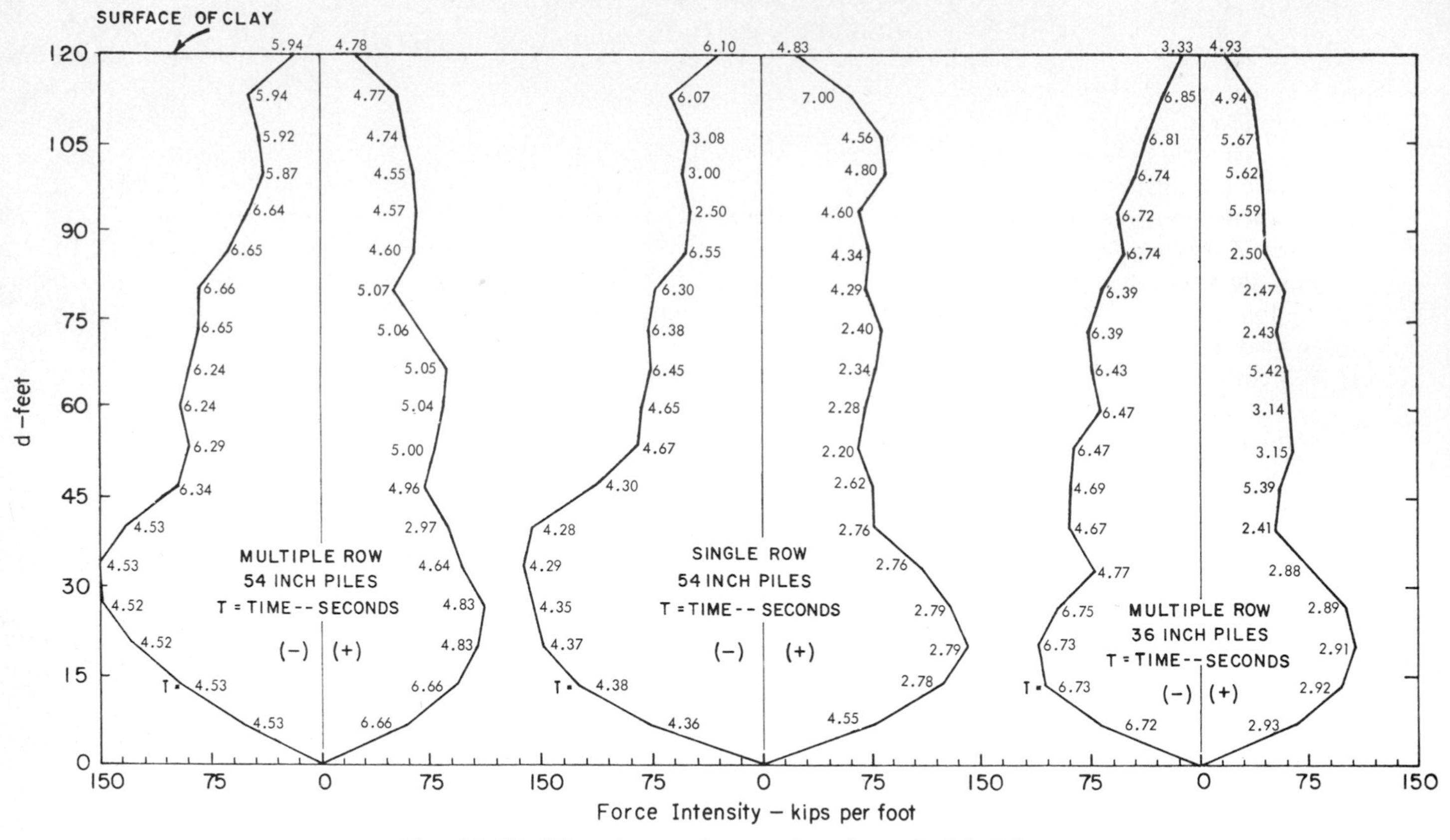

Fig. 14.33. Dimensions and properties of standard building.

under static conditions, one needs to know the elastic spring constant k, i.e., the interaction force per unit of length per unit of lateral displacement, that the clay medium can provide permanently in giving lateral support to the pile. No attempt has been made to establish this spring constant in the general investigation reported herein. The experienced soil engineer, however, can give the designer some basis for establishing this constant.

Once the above spring constant has been established, the flexural stiffness EI of the pile can be selected to be of sufficient magnitude that the vertical loads can be carried without lateral stability being a problem. The designer will need to recall in this case the theory of buckling of a uniform beam on an elastic foundation. If such a beam is of infinite length, has a flexural stiffness EI, and is supported by a Winkler-type elastic foundation having a uniform spring constant k, it can be shown that the critical axial load P is given by the expression

$$P_{cr} = 2\sqrt{kEI} \tag{14.57}$$

and that the beam buckles as a sine wave having full wave lengths λ as given by the relation

$$\lambda = 2\pi\left(\frac{EI}{k}\right)^{1/4} \tag{14.58}$$

This theory shows that even a very flexible foundation, relatively speaking, is very effective in shortening the wave length λ and thus increasing the critical load.

14.7 CONCLUSIONS AND RECOMMENDATIONS

The basic theory presented in this chapter provides a rational approach for investigating seismic effects on bridges that are supported on long piles extending through deep, sensitive clays. Solutions based on this theory can be obtained by digital computer with sufficient accuracy so that the dynamic behavior of the entire bridge structural system, including the piles, is adequately defined for design purposes.

Before applying the above general theory to a specific structural system, it is necessary that all required clay medium properties be established. Based on the results of the soil testing program reported herein, it appears that the basic idealized model selected for this material can provide an adequate representation of its stress–deformation characteristics during the period of an earthquake. The testing procedures utilized in the investigation appear to provide an adequate means for determining these clay medium properties.

The application of the above methods in determining the dynamic response of the proposed Elkhorn Slough bridge structural system has produced results on which the following specific conclusions and recommendations are based:

1. A deep clay layer can be expected to greatly filter the higher frequency components of a typical earthquake

acceleration input at its base before such accelerations reach the surface. However, the lower frequency components that are near the fundamental shear mode frequency of the clay layer are likely to be amplified if the clay system has sufficient strength. In such cases structures built on the surface and having fundamental frequencies that match or nearly match the fundamental frequency of the clay layer will experience greater peak response than if excited directly by the earthquake acceleration. The proposed Elkhorn Slough bridge shows a somewhat greater peak response in this respect when the N–S component of El Centro Earthquake acceleration is the prescribed input, i.e., the peak bridge deck acceleration is approximately 1.2 g when considering the entire clay–pile–bridge superstructure system and is approximately 0.9 g when considering only the bridge superstructure system.

2. The bridge superstructure, including attachments to piles, should be designed with full recognition of the importance of providing ductility so that large amounts of energy can be absorbed during the period of a very strong earthquake.

3. The deformations that could be expected in the clay medium at the Elkhorn Slough site, if subjected at its base to an earthquake similar to that recorded at El Centro, would produce curvatures in the piles of the same order of magnitude as their yield curvatures. Such piles therefore should be designed so that they can withstand a considerable amount of inelastic deformation without losing their vertical load-carrying capacity.

4. It is quite apparent that standard size piles will never "cut" their way through a moving clay medium of the Elkhorn Slough type. Rather, such piles will be forced to deform essentially with the clay medium and will be given only relatively small relief by the interaction displacements. This type of behavior means that considerably more control is placed on pile curvatures than on pile moments. Therefore, standard or possibly somewhat smaller than standard diameter piles would have an advantage over the larger diameter piles as far as flexural stresses are concerned. Of course, one must use a larger number of smaller size piles than larger size piles because of their lower vertical load-carrying capacities.

5. If the piles are driven to a considerable depth in the highly compacted sand layer just below the clay medium, very large curvatures should be expected to develop in the piles at the interface of these two layers during a strong earthquake. In such a case the piles should be designed with the necessary ductility in this region so that their vertical load-carrying capacities are maintained.

6. Further investigation is recommended to establish the existing "permanent elastic" moduli for the Elkhorn Slough clay medium that can be used to study the lateral stability of the piles under static conditions. Lateral stability of the piles is, of course, not a problem during the short period of transient excitation produced by an earthquake.

7. Since the phase relations of the dynamic response of the bridge deck will differ from one section to the next, adequate separation should be provided in the expansion joints so that one section of bridge deck will not "pound" against the adjacent sections during the period of a strong earthquake.

It should be fully recognized that the specific analytical results obtained and conclusions drawn therefrom in this study apply only to the specific structural systems analyzed when subjected to an excitation corresponding with the N–S component of the 1940 El Centro Earthquake. While these results and conclusions are considered extremely helpful when designing similar structures for future earthquakes, one always must recognize the many parameters involved that could differ appreciably, thus producing significant changes in the dynamic response characteristics of such systems.

ACKNOWLEDGMENTS

The material presented in this chapter has been taken from a report submitted to the California State Division of Highways entitled *Seismic Effects on Structures Supported on Piles Extending Through Deep Sensitive Clays* by R. A. Parmelee, J. Penzien, C. F. Scheffey, H. B. Seed, and G. R. Thiers, SESM 64-2, Institute of Engineering Research, University of California, Berkeley, August 1964. The author therefore wishes to acknowledge and express his sincere thanks to the above individuals who as a team conducted the general investigation reported herein.

The author also wishes to express his appreciation to Mr. A. Shah, Mr. M. Venkatesan, and Dr. I. King, who wrote the digital computer programs; to Mr. K. L. Lee, who participated in the soil testing program; to Dr. V. Jenschke, who developed the response spectra program; and to Dr. A. Chopra and Mr. G. Wang who participated in the generation of basic response data and in the preparation of material for the above-mentioned report.

Special thanks are also expressed to Mr. John Kozak, formerly of the California State Division of Highways, for his advice and encouragement on all phases of the program.

Finally, the author expresses his appreciation to the California State Division of Highways and the U.S. Bureau of Public Roads, whose financial support made this investigation possible.

Chapter 15

Earth Slope Stability During Earthquakes*

H. BOLTON SEED

Professor of Civil Engineering
University of California
Berkeley, California

15.1 INTRODUCTION

Evaluation of the degree of stability or potential instability of earth slopes is a major problem in seismically active regions of the world. However, it is only in recent years that soil engineers in the United States have directed any significant attention to this problem.

Possibly one reason for this previous lack of attention was some doubt in the profession as to whether a problem really existed. Few cases of major slides developing during earthquakes have found their way into soil mechanics literature in the English language, and while geological reports and foreign publications contain many reports of such incidents, they have usually occurred far away from the United States so that there was some tendency to consider them peculiar to other countries and more a matter of interest than concern. However, catastrophic

*This chapter appeared originally as "Slope Stability During Earthquakes" in the *Journal of the Soil Mechanics and Foundations Division, ASCE,* (SM4), 299–323, July 1967.

Fig. 15.1. Failure of Sheffield Dam in Santa Barbara Earthquake of 1925.

slope failures during recent earthquakes, particularly those in Alaska in 1964, have done much to dispel any doubts that existed on this score. Those engineers who witnessed the complete devastation in the slide area at Turnagain Heights, Anchorage (Seed and Wilson, 1967), the complete disappearance of dock facilities and the boat harbor at Seward (Shannon, 1966), or the many other slides that occurred during the Alaskan Earthquake, never again will fail to consider the awesome possibility of earthquake-induced landslides. Hopefully, enough of this experience will be recounted in various reports to impress the importance of such considerations on other soil engineers and geologists.

A reappraisal of earlier failures also serves as a sobering reminder of landslide possibilities. Figure 15.1 shows a view of the Sheffield Dam in Santa Barbara, California, that failed during an earthquake in 1926; Fig. 15.2 shows failures along the banks of the Solfatara Canal during the El Centro, California, Earthquake of 1940; and Fig. 15.3 shows a view of a small part of a slide (Duke and Leeds, 1963) involving some 30 million yd^3 of soil which moved vertically about 60 ft and laterally about 1000 ft during the Chilean Earthquake of 1960.

Fig. 15.2. Disruption of banks of Solfatara Canal in El Centro Earthquake of 1940.

These illustrations leave little doubt that the problem exists. In fact, various conditions leading to slope instability during earthquakes might be listed as follows (Seed, 1966):

1. Flow slides caused by liquefaction of cohesionless soils, e.g., Seward, 1964; Valdez, 1964; Kenai Lake, 1964; Kansu Province, 1920; and San Francisco, 1957.
2. Slides caused by liquefaction of thin seams or lenses of sand, e.g., Fourth Avenue Slide, Anchorage, 1964; L-Street Slide, Anchorage, 1964; and Government Hill Slide, Anchorage, 1964.
3. Slides in clay deposits facilitated by liquefaction of sand lenses, e.g., Turnagain Heights, Anchorage, 1964, and Rinihue, Chile, 1960.

Fig. 15.3. Slide near Lake Rinihue in Chilean Earthquake of 1960.

4. Slumping of fills on good foundation materials, e.g., Hebgen Dam, 1957; San Francisco, 1906; Chile, 1960; and Kern County, California, 1952.
5. Collapse and cracking of fills on poor soil foundations, e.g., Napier, 1931; Southern California, 1940; Chile, 1960; Portage, Alaska, 1964; and Niigata, Japan, 1964.

It is the responsibility of the soil engineer to recognize and guard against conditions that may lead to any of these eventualities. Recognition of the problems is an important first step in developing satisfactory solutions for their avoidance. However, this step must be supplemented by other bases, involving qualitative and quantitative assessments of the conditions involved, in appraising the degree of safety of an existing slope or a proposed design. Let us now look at some of the tools that have been used for this purpose.

15.2 PAST PRACTICE IN THE EVALUATION OF SLOPE STABILITY DURING EARTHQUAKES

One of the primary tools used to minimize the danger of slope failures during earthquakes has been the exercise of judgment, based on past experience, to recognize and avoid potentially unstable conditions. It has long been recognized that loose, saturated sands are particularly vulnerable to liquefaction during earthquakes, leading to flow slides or unstable foundation conditions for overlying sloping deposits. The potential dangers associated with the effect of earthquakes on highly sensitive clays also have been recognized; Miller and Dobrovolny (1959), e.g., anticipated the possibility of landslides in the Anchorage area long before they were induced by a major earthquake. Careful judgment, utilizing geological and soil engineering information, undoubtedly will continue to be a major tool for assessing slope stability during earthquakes for many years to come.

At the same time, guidance in the exercise of judgment can be obtained by appropriate analyses of embankment stability. Past practice and most current practice in the analysis of embankment stability against earthquake forces involve the computation of the minimum factor of safety against sliding when a static, horizontal force of some magnitude is included in the analysis. The analysis is treated as a static problem, and the horizontal force is expressed as the product of a seismic coefficient k and the weight W of the potential sliding mass. If the factor of safety approaches unity, the section is generally considered unsafe, although there is no generally recognized limit for the minimum acceptable factor of safety. In effect, the dynamic effects are replaced by a static force, and the approach therefore might be termed a a pseudostatic method of analysis.

15.3 SELECTION OF SEISMIC COEFFICIENT IN PSEUDOSTATIC ANALYSIS

One of the major problems facing the engineer using this type of approach is that of selecting the value of the seismic coefficient to be used for design purposes. Methods used include (1) the adoption of empirical rules, (2) the assumption of rigid body response, and (3) the use of viscoelastic response analyses. A detailed account of the different meanings associated with the term "seismic coefficient" and methods of selecting values for analysis purposes has been presented by Seed and Martin (1966). The following summary is excerpted from that paper.

15.3.1 Use of Empirical Values

Most engineers in the United States who adopt a pseudostatic method of seismic stability analysis adopt some empirical value for the design seismic coefficient; typically this lies in the range of 0.05–0.15. While there is good reason to adopt some value of this type as a means of differentiating between the seismicity and foundation conditions at different sites or for studying the advantages of different sections, there appears to be no published justification for using values in the range of 0.05–0.15 as a basis for selecting or approving final design slopes. It appears that continued use of these empirical values has given them some semblance of an authoritative design criterion—yet no one seems to know why the values should have been selected in the first place.

Because no reasonable basis for adopting seismic coefficients of this order of magnitude (0.05–0.15) for the design of earth slopes has so far been advanced, the validity of their use cannot be evaluated. Few major dams have been subjected to major earthquakes, so there is little experience to serve as a guide. Since present practice has not really been tested by a major earthquake, there is no way of knowing whether embankments analyzed on this basis will be found to be adequate or inadequate. It is entirely possible that empirical values on the order of 0.1 or 0.15 may lead to safe design conditions in many cases, but until some means of judging their validity is developed, their use must be considered of questionable value in the design procedure.

It is interesting to note that, whereas the design seismic coefficient is typically on the order of 0.1 in the United States, somewhat higher values are used in Japan (Japan Society of Civil Engineers, 1960). The design seismic coefficients given in the design criteria for earth dams established in 1957 by the Japanese National Committee on Large Dams, and commonly followed in Japan, range from 0.12 to 0.25 depending on the location of the dam, the type of foundation, and the possible downstream effects of damage caused by an earthquake.

15.3.2 Rigid Body Response Analyses

If an embankment is assumed to behave as a rigid body, the accelerations will be uniform throughout the section and equal at all times to the ground accelerations. Thus it is sometimes argued that the design seismic coefficient should be equal to the maximum ground acceleration. The main limitations of this approach are:

1. Although low, stiff embankments or those in narrow canyons may respond essentially as rigid structures, there is considerable evidence from field tests in which actual dams have been subjected to forced vibrations by means of large shaking machines that all earth dams do not behave as rigid bodies but respond in different ways to any given series of imposed motions.

2. The maximum acceleration will be developed in an embankment for only a short period of time, so that the deformation resulting from it may be small. Although it will be supplemented by deformations produced by other accelerations and inertia forces developed during the earthquake, there is no reason to believe that their combined effects will be equivalent to those produced by

applying an inertia force, corresponding to the maximum acceleration in the embankment, as a static force, i.e., as if it were acting for an unlimited period of time.

In view of these limitations, there is nothing to guide the design engineer on the appropriateness of using a seismic coefficient corresponding to the maximum ground acceleration in a pseudostatic analysis of seismic stability or on how this value might be appropriately modified for different structures and ground motions.

15.3.3 Elastic Response Analyses

The deficiencies in the use of empirical rules or the assumption of rigid body response have led a number of investigators to propose the use of elastic response solutions for the determination of design seismic coefficients. In effect, the embankment is considered to consist of a series of infinitely thin horizontal slices, the slices being connected by linearly elastic shear springs and viscous damping devices, and the response at different levels resulting from a uniformly distributed base motion is determined.

The first analysis using this type of approach was made by Mononobe, Takata, and Matumura (1936) for the case of an infinitely long embankment with a symmetrical and homogeneous cross section resting on a rigid foundation. Similar analyses since have been developed for other conditions. Hatanaka (1955) extended the analysis to include the case of a triangular elastic wedge in a rectangular canyon and considered the variation of horizontal response over both the length and height of the dam. He showed that when the length of the dam is about four times the height, the influence of end restraint has negligible effect on the natural frequencies of vibration and the magnitude of response in the central region, and thus the use of an analysis based on the assumption of infinite length can be considered sufficiently accurate for all practical purposes.

Subsequently, Ambraseys (1960) developed solutions for the case of a truncated wedge of uniform modulus and for a symmetrical wedge of uniform modulus resting on an elastic layer of infinite thickness. More recently, Rashid (1961) analyzed the response of a symmetric wedge in which the shear modulus increased as the cube root of the depth, a variation more appropriate for dams composed of cohesionless materials.

Despite the availability of these solutions, the only specific suggestions for their use in design have been made by Ambraseys (1960b), who proposed that the lateral forces acting on a dam during an earthquake be expressed by a static force with magnitude determined by a seismic coefficient evaluated in one of the following ways: (1) The seismic coefficient at any depth should be taken as the square root of the sum of the squares of the seismic coefficients for peak response in the first four modes, i.e.,

$$\bar{k}(y) = \left[\sum_{n=1}^{n=4} \{k_n(y)\}^2\right]^{1/2} \tag{15.1}$$

or (2) the seismic coefficient at any depth should be taken as the maximum value at that depth for any one of the modal distributions, i.e.,

$$k(y) = [k_n(y)]_{\max} \tag{15.2}$$

Both these expressions represent an attempt to estimate the maximum seismic coefficient distribution acting at any instant during the earthquake. Similar recommendations were subsequently advocated by Krishna (1962). Application of these approaches leads to a distribution of seismic coefficient varying from a maximum value at the crest of the embankment to zero at the base.

Used in this manner the dynamic response analyses are utilized only for the purpose of determining the maximum inertia forces likely to be developed in an embankment at any instant during an earthquake, and these values are incorporated in a pseudostatic analysis of stability. This approach suffers from the following limitations:

1. The selection of the maximum seismic coefficient at any depth for any mode leads to a conservative estimate of the inertia forces developed.

2. The use of the computed seismic coefficients in a static stability analysis leads to the conclusion that if the factor of safety becomes equal to unity the embankment will fail. Actually the computed inertia forces are dynamic in nature and the factor of safety may be reduced below unity for a short period or a series of short periods without significantly large deformations developing.

3. Any response analysis that considers the dam as a series of slices suffers from the following limitations: (a) The analysis is based on the assumption that the response of the embankment to the ground motion is controlled only by the shearing action developed between horizontal slices—however, the work of Ishizaki and Hatakeyama (1962) and Clough and Chopra (1966) indicates that both horizontal and vertical compressive and tensile deformations within an earth dam, resulting from horizontal ground motions, can contribute significantly to the overall dynamic stress pattern and (b) the analysis is developed only for horizontal ground shaking and does not include any consideration of the effects of ground motions in a vertical direction.

4. The approach assumes elastic deformation in the soil mass with energy being dissipated in terms of viscous damping. However, soil deformations under higher stresses are inelastic and some energy is dissipated by hysteretic damping. Thus, considerable care is required in selecting equivalent moduli and viscous damping factors for use in the analysis in order to obtain meaningful estimates of embankment response.

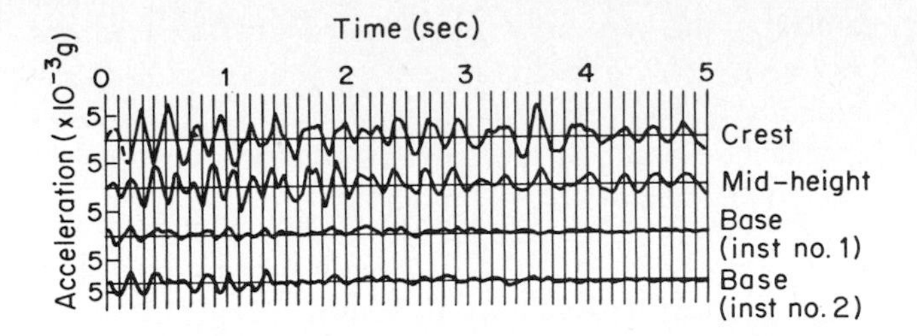

Fig. 15.4. Observed response of Sannokai Dam. (From Okamoto, Hakuno, Kato, and Kawakami, 1965.)

Despite these limitations there is a substantial body of evidence to show that embankments do respond to earthquake ground motions as deformable bodies.

In recent years several dams, about 100 ft high, have undergone field tests in which they have been subjected to forced vibrations using large shaking machines (Keightley, 1963, 1964; Martin and Seed, 1966). The response of the dams clearly has shown that peak accelerations are developed at certain characteristic frequencies and, furthermore, the response is in good agreement with that predicted by viscoelastic response analyses. This type of analysis also indicates that the accelerations induced in an embankment by an earthquake increase with height above the base. This behavior is in good accord with that observed by seismographs installed at the base and crest of the Cachuma Dam in California during a small earthquake in 1957 (Ambraseys, 1960a) and in measurements of the response of the Sannokai Dam in Japan to several small earthquakes (Okamoto, Hakuno, Kato, and Kawakami, 1965). The accelerations recorded at the base, midheight, and crest of this dam during an earthquake are shown in Fig. 15.4. The form of these results is in excellent agreement with the computed distribution of acceleration determined by a viscoelastic response analysis of a 100-ft high embankment shown in Fig. 15.5.

Fig. 15.5. Computed response of 100-ft high dam to El Centro Earthquake.

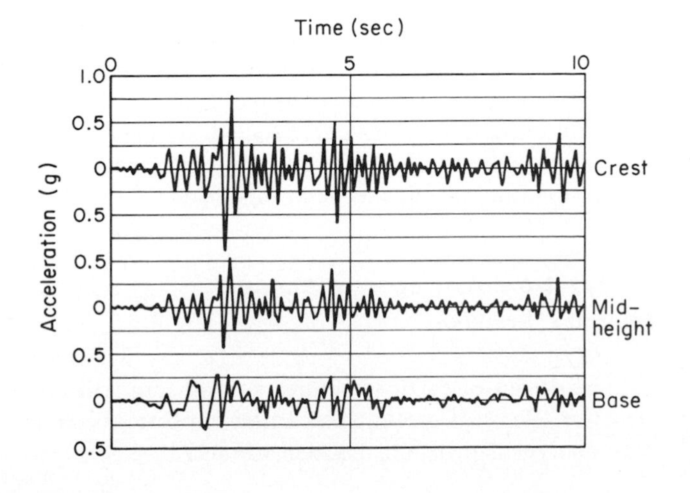

Thus, it would appear that viscoelastic response analyses can provide a reasonable approach for assessing the dynamic forces induced in an embankment by an earthquake, but the merits of the approach are not used to full advantage by incorporating the maximum dynamic forces, as static forces, in pseudostatic methods of analysis.

15.3.4 Comparison of Methods for Determining Seismic Coefficients

In view of the different approaches currently used for determining seismic coefficients for use in pseudostatic analyses for design purposes, it is interesting to compare the values obtained by the various methods. This is

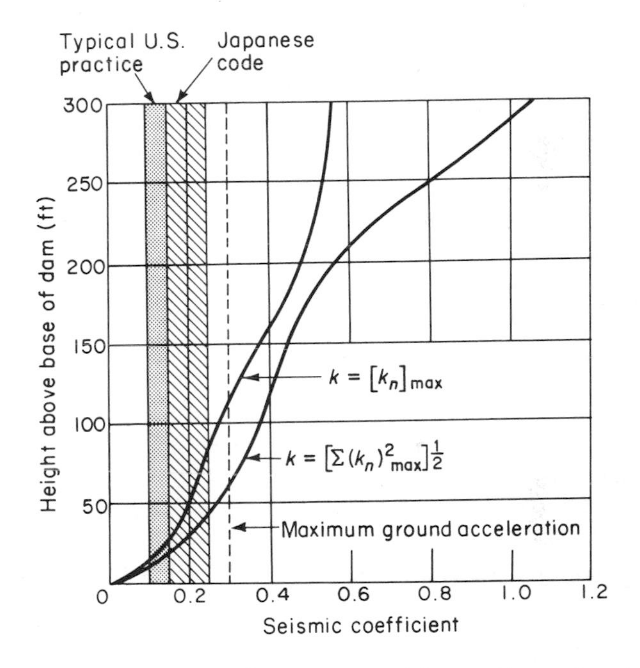

Fig. 15.6. Seismic coefficients suggested for use in pseudostatic analysis.

best illustrated by a numerical example (Seed and Martin, 1966). Suppose values are determined for a long dam, with a homogeneous section 300 ft high, constructed of compacted soil having a shear wave velocity of 1000 ft/sec (a value typical of modern cohesive alluvial fills) and subjected to the N–S component of the El Centro Earthquake of 1940, for which the maximum ground acceleration was about 0.3 g.

Values determined by the following methods are plotted in Fig 15.6: (1) an empirical design coefficient of 0.10–0.15 (typical United States practice); (2) a design coefficient in the range of 0.15–0.25 (Japanese earth dam code requirement); (3) a design coefficient equal to the maximum ground acceleration, corresponding to the assumption of rigid body response; and (4) Ambraseys' recommended values, determined by an elastic response analysis for 20% critical damping, using Eqs. 15.1 and 15.2.

It is apparent that an engineer has a wide choice in selecting a seismic coefficient for use in design. The wide range of values and opinions currently in use is a serious limitation of present practice and presumably reflects the uncertainty of engineers concerning this aspect of embankment design.

15.3.5 Limitations of Past Practice in Evaluating Slope Stability

While the use of previous experience and pseudostatic methods of analysis can provide some guidance in evaluating slope stability during earthquakes, their use is seriously limited by available knowledge. At the present time (1967) there are very few cases of slope failures where soil conditions have been sufficiently well established to provide anything other than a qualitative guide in extrapolating the experience to new sites. In fact, one of the most urgent needs in the development of improved methods of evaluating slope stability is the provision of quantitative information on soil properties at sites where major failures have occurred to provide a basis for evaluating the usefulness of proposed methods of approach.

At the same time, pseudostatic methods of analysis incorporating seismic coefficients on the order to 0.1 do not serve to explain the few cases of failure for which detailed studies have been made. Three major slope failures in Anchorage occurred by sliding near the surface of a layer of soft, sensitive clay; a pseudostatic analysis indicates that failure would develop at the base of the layer. This type of analysis would not explain the fact that failure did not develop at all until about 2 min after the start of the ground motions. Finally, the method would not account for the failure of the Sheffield Dam in Santa Barbara during an earthquake in 1926. Because pseudostatic approaches fail to provide a reasonable evaluation of slope behavior in several of the few well-defined case histories available for study, their usefulness must be considered limited to assuring some degree of increased conservatism in slope selection on a purely empirical basis.

15.4 NEW DEVELOPMENTS IN ANALYSIS OF SLOPE STABILITY DURING EARTHQUAKES

During the past few years a number of factors have combined to provide an improved understanding of slope stability during earthquakes. These include the accumulation of significant field data concerning slope failures of this type, new concepts in analysis procedures, new analytical tools for assessing slope response, new methods of soil testing, and new proposals for design approaches. Some of the more significant developments are summarized below.

15.4.1 Field Studies of Unstable Slopes

During the Alaskan Earthquake of 1964 a number of slope failures occurred that since have been the subject of intensive investigations by soil engineers and geologists. As a result they provide the best-documented case histories of earthquake-induced landslides available to date.

Excellent descriptions of the soil conditions in the landslide areas and of the slide development in Anchorage have been prepared by Shannon and Wilson (1964) and Hansen (1965), and similar descriptions of the slides at Valdez and Kenai Lake have been presented by Coulter and Migliaccio (1966) and McCulloch (1966). In addition, a description by Shannon (1966) of the major slide at Seward and a detailed analysis by Seed and Wilson (1967) of the Turnagain Heights landslide in Anchorage have been presented.

As a result of these detailed studies a new appreciation has been gained of (1) the mechanics of slide movements leading to the formation of grabens; (2) the complex mechanism by which slide movements during earthquakes may retrogress large distances behind a slope; (3) the significance of liquefaction of sand layers and lenses in facilitating landslides during earthquakes; (4) the potential danger of highly sensitive clay soils in contributing to slope failures during earthquakes; and (5) the nature of submarine slides caused by liquefaction of cohesionless materials.

Similar studies of a major slide that occurred near Lake Rinihue in Chile (Davis and Karzulovic, 1961) during an earthquake in 1960 and the soil liquefaction at Niigata, Japan (Seed and Idriss, 1967), during an earthquake in 1964 also have provided important information on field behavior.

Such studies provide not only valuable experience records and a means for establishing behavior patterns but also a means for evaluating the merits of new concepts and approaches. In this way detailed studies of slide movements in earthquakes can make a major contribution to the fund of knowledge in this field.

15.4.2 New Concepts in Slope Stability Analyses

The design of earth structures to withstand the destructive effects of earthquakes safely constitutes a complex analytical problem. Sudden ground displacements during earthquakes induce large inertia forces in embankments. As a result any one slope of an embankment will be subjected to inertia forces that alternate in direction many times during an earthquake, and it is necessary to determine the effects of these pulsating stresses, superimposed on the initial dead load stresses, on the embankment configuration.

During an earthquake the inertia forces in certain zones of an embankment may be sufficiently large to drop the factor of safety below unity a number of times, but only for brief periods of time. During such periods permanent displacements will occur, but the movement will be arrested when the magnitude of the acceleration decreases or is reversed. The overall effect of a series of large but brief inertia forces may well be a cumulative displacement of a section of the embankment, but once the ground motions generating the inertia forces have ceased, no further deformation will occur unless there has been a marked loss in strength of the soil.

Thus the magnitude of the deformations that develop will depend on the time history of the inertia forces, and a logical method of design requires (1) a determination of the variation of inertia forces with time and (2) an assessment of the embankment deformations induced by these forces.

The important concept that the effects of earthquakes on embankment stability should be assessed in terms of the deformations they produce, rather than the minimum factor of safety developed, was first proposed by N. M. Newmark (1963). Methods of analysis based on this concept subsequently have been presented by Newmark (1965) and Seed (1966a). Both of these methods presume a knowledge of the time history of the inertia forces acting on an embankment during the earthquake.

Deformation approaches of this type are considerably more involved than the conventional method of stability analysis for static loading conditions or pseudostatic methods of analysis for slope stability during earthquakes, but they are clearly necessary if meaningful assessments of the effects of earthquakes on slopes are to be made.

A detailed consideration of the limitations of the pseudostatic method of analysis also has led to further suggestions for improved analytical procedures (Seed and Martin, 1966). It may be seen from the preceding summary that attempts to determine a static seismic coefficient to incorporate in a pseudostatic analysis procedure leave much to be desired. Empirical methods of selecting a coefficient have little rational basis, are based on past precedents rather than past experience and, while they may sometimes be adequate for their intended purpose, may also mislead the engineer into a false sense of security. Seismic coefficients representing the maximum inertia force to which a dam might be subjected, whether determined by rigid body or viscoelastic response analyses, can be justified only if the designer accepts the concept that any deformation, no matter how small, constitutes failure of the dam (a concept that is both unnecessary and uneconomical). Attempts to determine a static force that will be equivalent in its deforming effects to those of the earthquake ground motions are both unnecessary and unachievable, because they necessarily must be based on a previous evaluation of the dynamic response and deformation of the embankment that in itself, is the objective of the analysis.

In view of these facts, it would appear that the complexity of the analytical problem is in fact increased by attempts to represent the earthquake effects by a static seismic coefficient and that a simpler, more rational approach could be developed by determining, instead, dynamic seismic coefficients that represent the time history of inertia forces to which the embankment will be subjected by any given ground motion. In fact, evaluation of such forces is a necessary prerequisite to any rational approach to the problem of predicting embankment deformations during earthquakes.

In addition to the foregoing arguments, there is a growing body of evidence to show that soil strength mobilized during earthquakes may be quite different from that determined under static or transient loading conditions and is a function of the entire time history of the stresses developed during the earthquake. Typical examples of this evidence are:

1. The failure of saturated sands causing large slides in the recent Alaskan Earthquake (Shannon and Wilson, 1964) and foundation failures in the subsequent Niigata Earthquake (Seed and Idriss, 1967). No failures were produced in these sands by relatively high stresses applied statically by footings or by steep slopes, but complete failure by liquefaction occurred under relatively flat and gently sloping areas as a result of modest stress increases produced by the earthquakes.

2. Failure of saturated sands causing sliding in the Alaskan Earthquake occurred after 1 or 2 min of severe ground shaking. No significant deformations would have developed in many of the slide areas if the duration of the earthquake had been only 30–60 sec. Thus, the duration of pulsating stress applications or the number of pulses developed is a significant factor that must be considered, as well as the applied stresses, in determining whether failure will occur.

3. Studies of slope failures in the Alaskan Earthquake indicated that saturated sensitive clays also may lose strength under pulsating loading conditions.

4. Laboratory tests show that soil strength under cyclic loading conditions is quite different from that mobilized under static loading and is a function of the number of stress cycles as well as the stress intensity (Seed and Chan, 1966; Seed and Lee, 1966).

Since both field and laboratory evidence demonstrate the major influence of the time history of stress applications on soil behavior, it would appear that failure to consider this factor would be a serious deficiency of any design procedure.

Recognition of the importance of predicting embankment deformations rather than a transient factor of safety, of the significance of the time history of stresses rather than the maximum stress developed, and of the necessity of considering the possible reduction in strength of soils during earthquakes represent significant advances in analytical approaches aimed at the investigation of slope stability during earthquakes.

15.4.3 Improved Procedures for Evaluating Slope Response to Earthquakes

In keeping with the foregoing concepts, new procedures have been proposed for evaluating the response of embankments and slopes to earthquake ground motions. The developments have been made possible largely through the availability of computers for making the detailed computations required.

To provide information on the variation of inertia forces with time in an embankment, Seed and Martin (1966) proposed the use of the shear slice approach to determine the entire time history of accelerations and stresses developed during the period of significant ground motions. A comparison of the acceleration distribution in a dam determined in this way with that recorded on the Sannokai Dam in Japan is shown in Figs. 15.4 and 15.5. The good agreement in the form of the computed and observed values is readily apparent. On this basis, it was found possible to express the stresses developed at different sections of an embankment in terms of dynamic seismic coefficients (Fig. 15.7) and to represent the results for embankments of different heights and materials subjected to a given base motion in a simplified chart form (Table 15.1). It may be noted in Fig. 15.7 that the values of the seismic coefficients increase with increasing elevation of the potential sliding mass within the body of the embankment, and in Table 15.1 that the seismic coefficients vary with the height of the embankment and the material characteristics. They also would vary with the nature of the earthquake ground motions. This type of information provides the necessary

Fig. 15.7. Values of average seismic coefficient for 100-ft high embankment subjected to El Centro Earthquake: v_s = 1000 ft/sec; 20% critical damping.

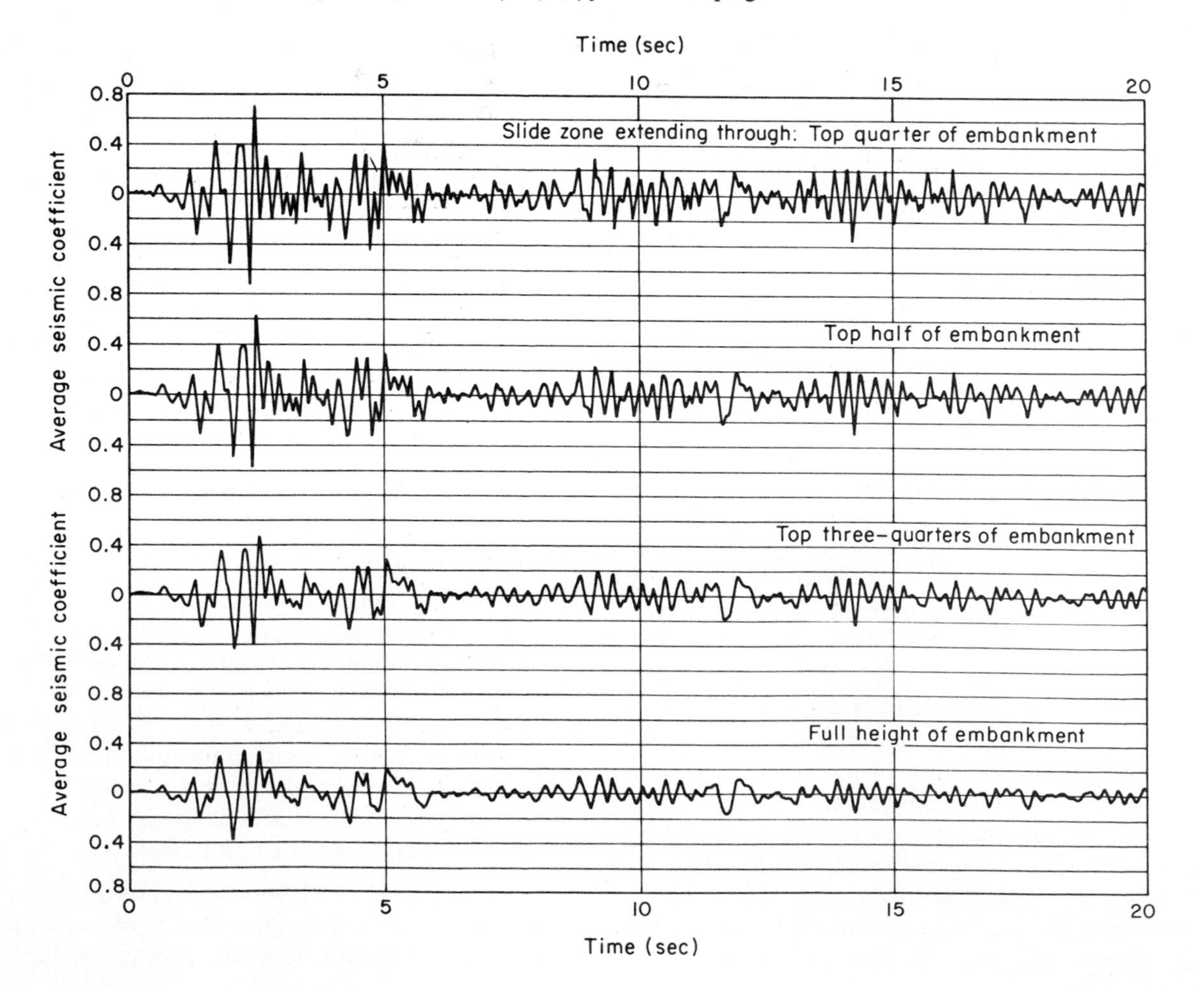

Table 15.1. RESULTS FOR EMBANKMENTS OF DIFFERENT HEIGHTS AND MATERIALS SUBJECTED TO EL CENTRO (1940) BASE MOTION

	v_s = 300 ft/sec Height of dam, ft			v_s = 1000 ft/sec Height of dam, ft		
Equivalent seismic force series	100	300	600	100	300	600
Number of significant force cycles	10	5	3	15	12	7
Predominant frequency of force cycles	1.25	0.4	0.3	3.3	1.25	0.7
Equivalent maximum seismic coefficient operative over different portions of embankment: Top quarter	0.35	0.20	0.10	0.40	0.36	0.24
Top half	0.30	0.15	0.07	0.35	0.28	0.16
Top three-quarters	0.22	0.10	0.04	0.30	0.22	0.11
Full height	0.16	0.08	0.03	0.25	0.16	0.08
Natural period of embankment, sec	0.87	2.61	5.22	0.26	0.78	1.57

basis for the analysis of deformations and for the planning of laboratory test procedures.

Recognizing the limitations of the shear slice procedure, however (only shear modes considered, no vertical ground motions introduced, etc.), efforts were made to develop improved analyses. As a result, Clough and Chopra (1966) introduced the finite element approach for evaluating the response of embankments constructed of linear viscoelastic materials. In this approach, the embankment is considered to consist of an assemblage of elements, interconnected at a finite number of nodal points as illustrated in Fig. 15.8. An internal displacement distribution, selected to satisfy certain required conditions, is assumed in each element. With the aid of this displacement distribution it is possible to determine the stiffness properties for each element. The stiffness of the complete assemblage of elements is obtained by adding the appropriate stiffness components of the individual elements surrounding each nodal point. The stiffness relationship for the complete assemblage then can be used in the derivation of the equations needed for the evaluation of the response of the system. The analysis yields the complete time history of displacements, velocities, accelerations, strains, and stresses at the nodal points in the system. In addition, the dynamic seismic coefficients denoting the net inertia force acting on any section of the embankment at different instants of time can be determined. Computer programs can be written to perform the computations and to plot the time history of different response values or the distribution of any response characteristic at any instant of time.

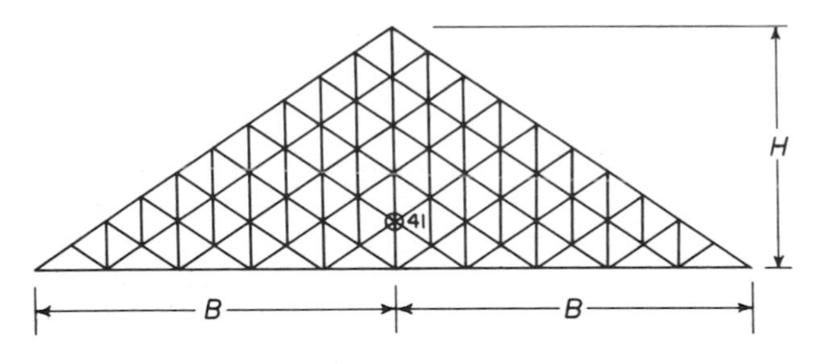

Fig. 15.8. Finite element idealization of earth embankment.

Typical results are shown in Figs. 15.9–15.13. Figure 15.9 shows the first six mode shapes and frequencies for a symmetrical embankment 300 ft high with side slopes of 1 on $1\frac{1}{2}$. It is readily apparent that although the first mode resembles a pure shear distortion of the type assumed in the shear slice analysis, vertical and rocking motions are clearly involved in the higher modes. The time history of stresses at a point on the center line of the embankment, 60 ft above the base, when the embankment is subjected to the horizontal and vertical components of ground motions recorded in the El Centro Earthquake of 1940 are shown in Fig. 15.10, and stress contours in the embankment at an instant 2.25 sec after the start of ground shaking are shown in Fig. 15.11. Similar results can be readily obtained for other points in the embankment or times during the earthquake.

Expression of results of this type in terms of seismic coefficients (Seed and Martin, 1966; Chopra and Clough, 1965) is illustrated by the data presented by Chopra and Clough in Fig. 15.12 for a 300-ft embankment with side slopes of 1 on $2\frac{1}{2}$ subjected to the same earthquake. The locations of four wedges within the embankment are shown in the upper part of Fig. 15.12, and the time histories of horizontal inertia forces acting on these wedges, expressed in terms of dynamic seismic coefficients, are shown in the lower part of the figure. The increase in values of the seismic coefficients with increasing elevation of the wedge within the embankment is again readily apparent. Also shown in Fig. 15.12 are values of the seismic coefficient computed by the shear slice method (shown by dots on the plots for wedges 2 and 3). It may be seen that in this case the values do not differ greatly from those determined by the finite element approach, indicating that the shear slice analysis often may be adequate for seismic coefficient determination.

The significance of the vertical component of the base motion (Chopra and Clough, 1965) is illustrated by the computed values of horizontal and vertical seismic coefficients shown in Fig. 15.13 for a 300-ft high embankment with side slopes of 1 on $2\frac{1}{2}$ subjected to the ground

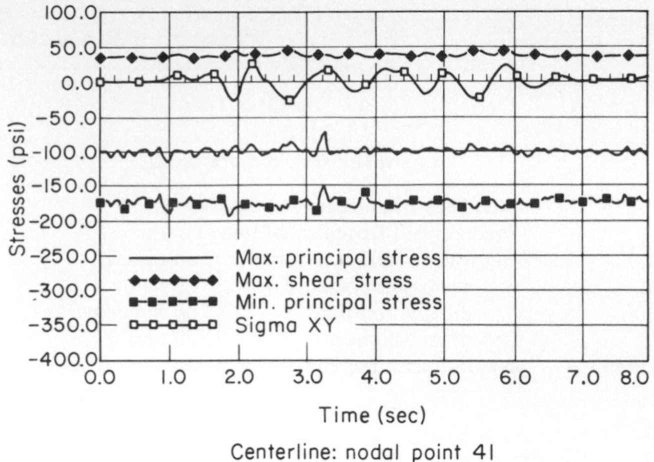

Fig. 15.10. Time history of stresses at point on center line of embankment.

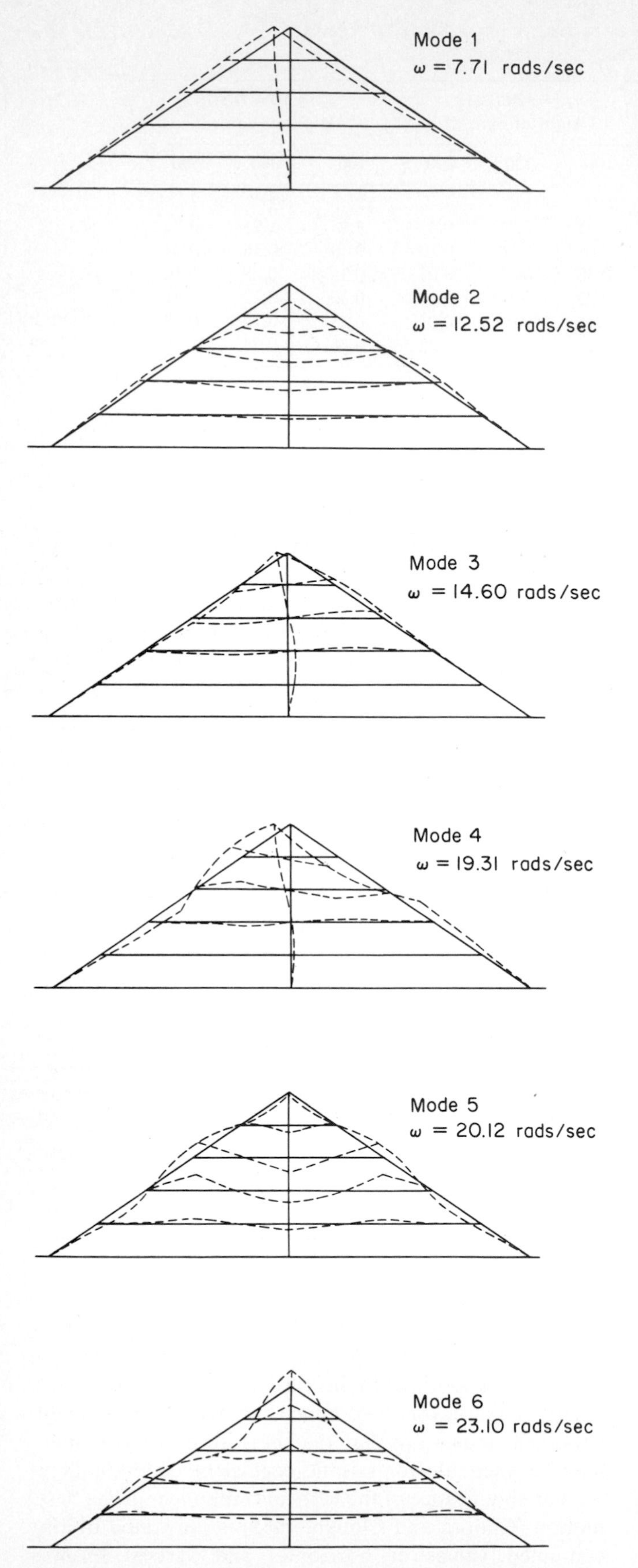

Fig. 15.9. Natural modes (1–6): shapes and circular frequencies; dam side slopes $1\frac{1}{2}$: 1. (From Clough and Chopra, 1966.)

motions recorded at Taft in the Kern County, California, Earthquake of 1952. It may be seen that the inclusion of the vertical components of the base motion has virtually no influence on the horizontal seismic coefficients but does affect considerably the vertical seismic coefficients.

It is interesting to note the much higher frequency of the inertia forces corresponding to the vertical motions. A similar form of response was observed by accelerometers located on the crest of the El Infiernillo Dam in Mexico during an earthquake in 1965 (Marsal, 1967).

Based on this approach Finn and Khanna (1966) more recently have made a preliminary evaluation of the influence of a soil foundation and sloping cores on the response of embankments. However, techniques for utilizing finer mesh representations of embankments and their foundations are required for more detailed studies of the effects of these factors.

Similar analyses have been made by Idriss and Seed (1966) to evaluate the response of a bank of soil. In this case the problem is complicated by the necessity of extending the finite element mesh a sufficient distance from the area of the slope for an accurate response picture to be obtained. It has been shown that the array of elements shown in Fig. 15.14 provides an adequate system for many purposes and that the computed stresses in a bank of stiff clay subjected to the El Centro Earthquake of 1940 are as shown in Fig. 15.15. Maximum values of the dynamic seismic coefficients for different potential sliding masses also were evaluated. As for embankments, the magnitude of the seismic coefficients expressing the horizontal inertia forces was found to decrease with increasing depth of the potential sliding mass below the crest of the slope. Details of the analysis and its applications have been presented (Idriss and Seed, 1966). A similar study of rock slopes has been described by Finn (1966).

Analyses capable of handling traveling wave base

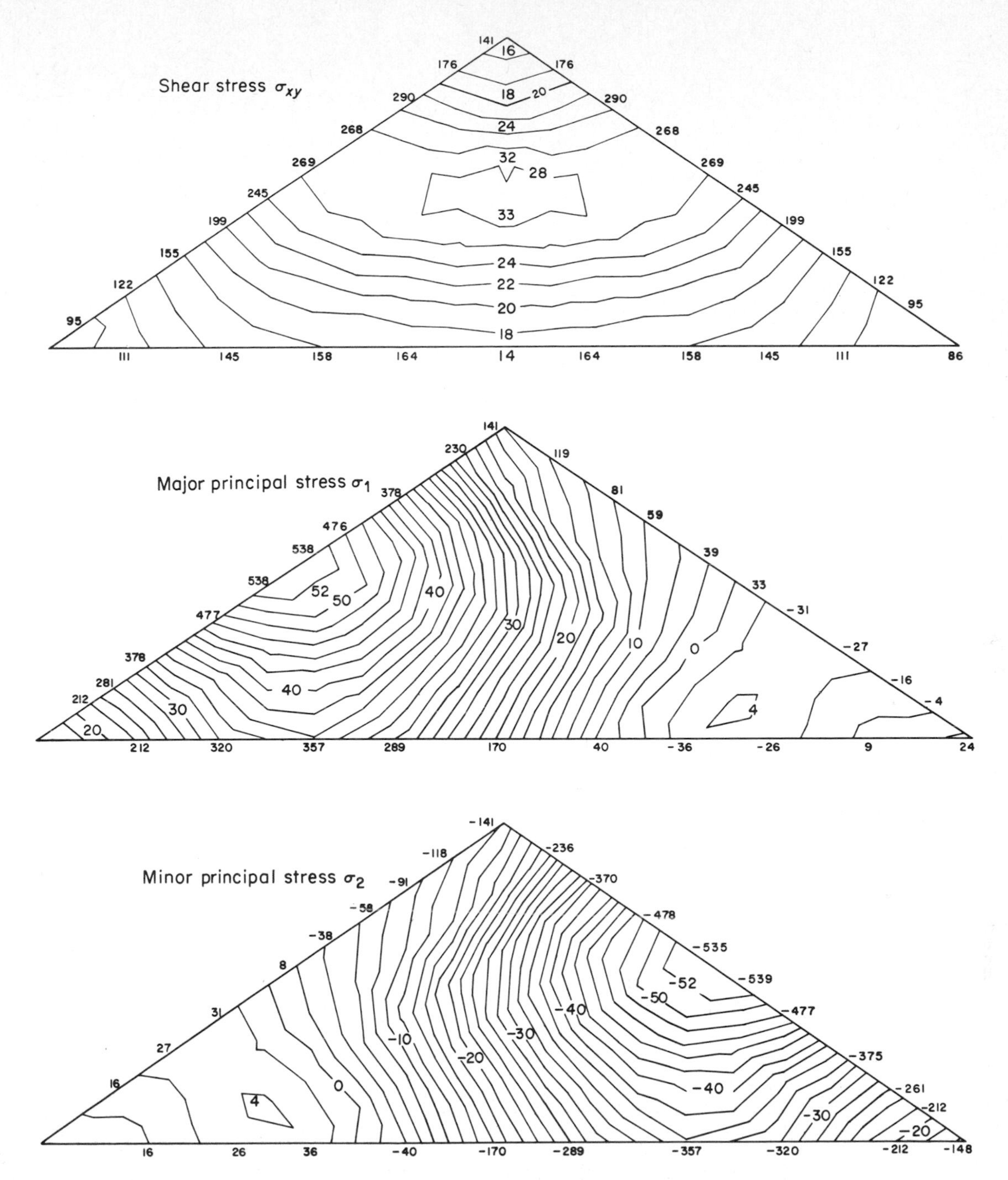

Fig. 15.11. Stress contours at $t = 2.25$ sec: dam height 300 ft; slopes $1\frac{1}{2}:1$; subjected to El Centro 1940 Earthquake (N–S component). (From Clough and Chopra, 1966.)

motions also have been developed, and efforts are now being directed to the solution of problems considering the nonlinear stress–strain characteristics of the soils comprising slopes and embankments.

15.4.4 Improvements in Soil Testing Procedures

The past several years also have seen the development of improved methods of testing soils to evaluate the material characteristics required for design and analysis purposes.

Vibratory equipment (Wilson and Dietrich, 1960; Hall and Richart, 1963; Hardin, 1965) has been developed to determine the dynamic moduli and damping characteristics of soils under small amplitude motions and cyclic loading triaxial compression (Seed and Lee, 1966; Kondner and Krizek, 1965; Parmalee et al., 1964; Martin, 1965; de Graft Johnson, 1965; Taylor and Hughes, 1965), and simple shear tests have been used to measure the stress–strain relationships of soils for a range of

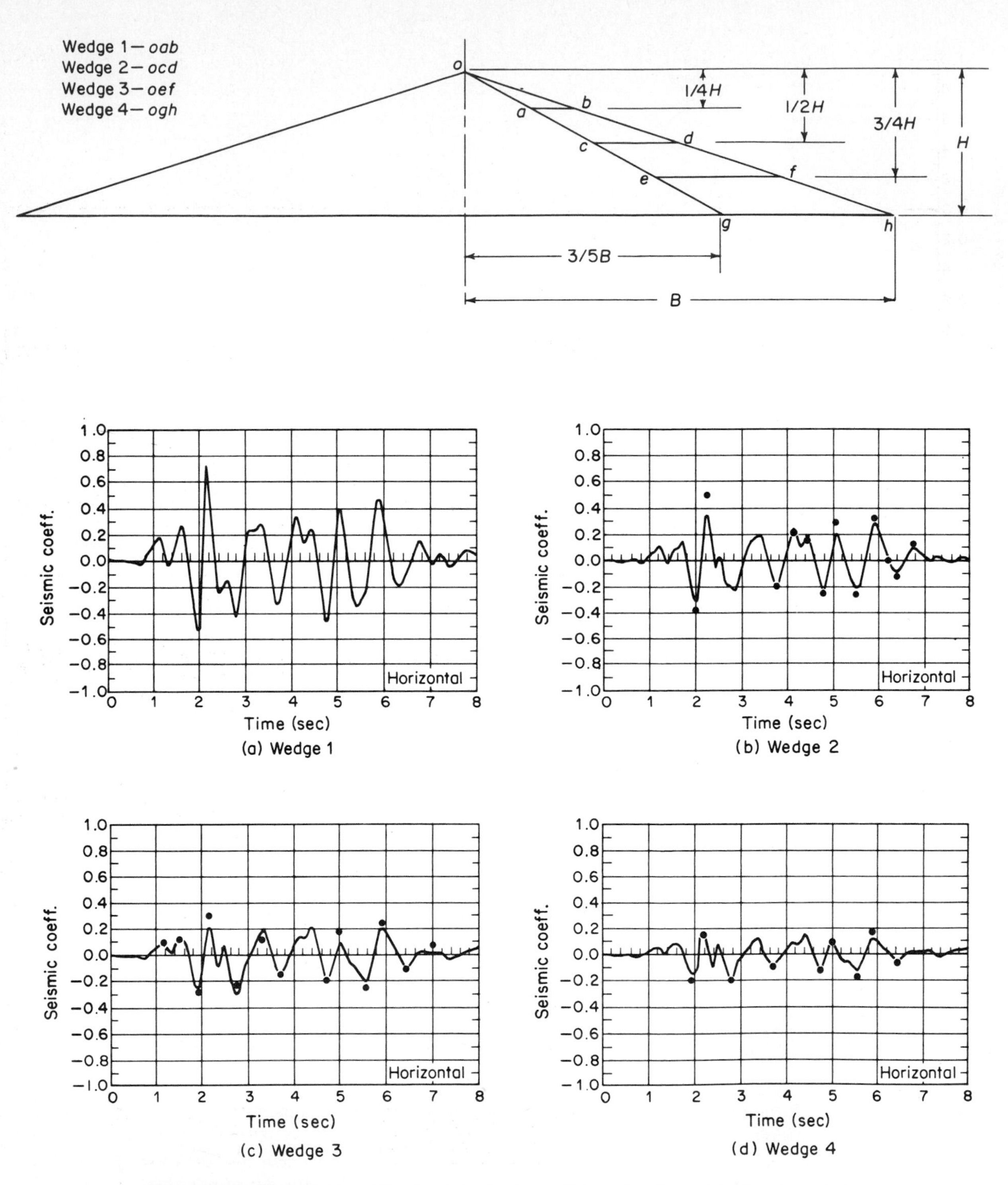

Fig. 15.12. Time history of horizontal seismic coefficients for dam with side slopes 3:1. (From Chopra and Clough, 1965.)

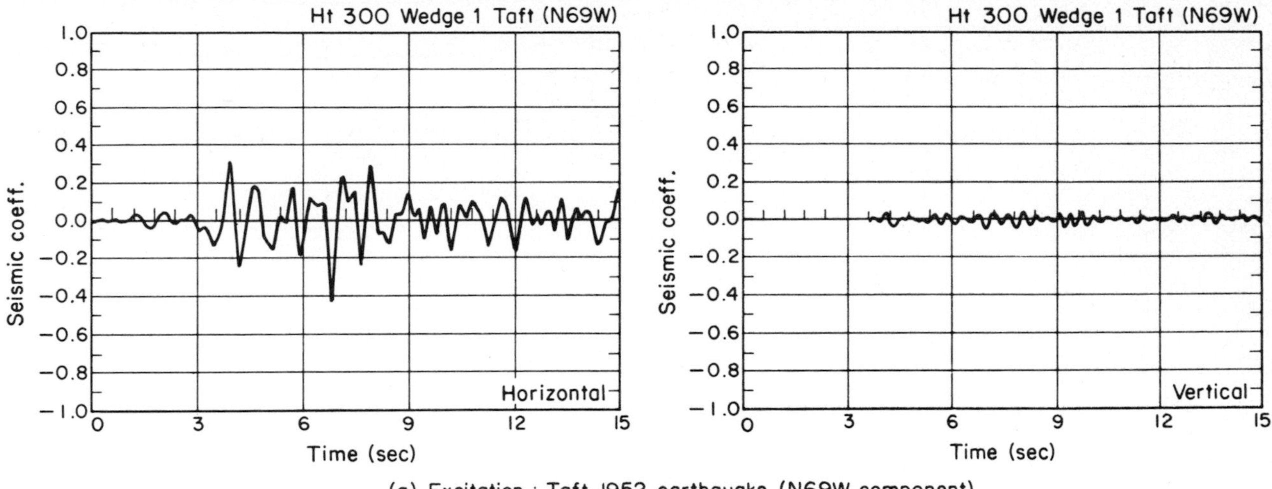

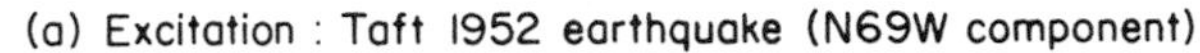
(a) Excitation : Taft 1952 earthquake (N69W component)

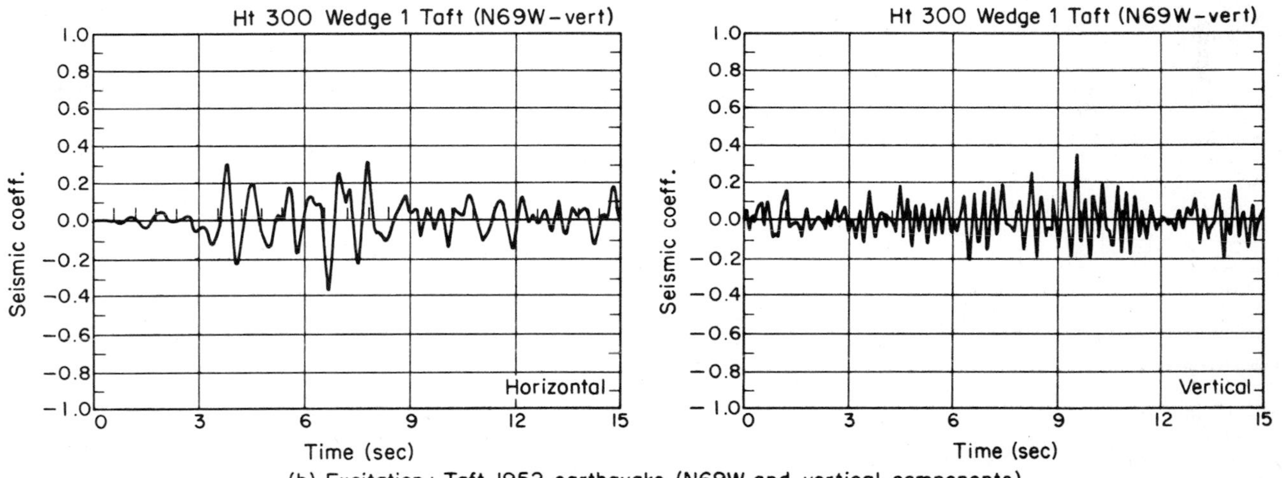

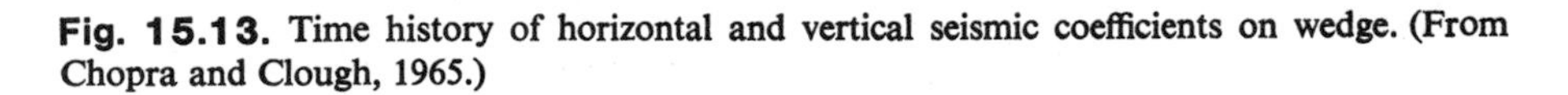
(b) Excitation : Taft 1952 earthquake (N69W and vertical components)

Fig. 15.13. Time history of horizontal and vertical seismic coefficients on wedge. (From Chopra and Clough, 1965.)

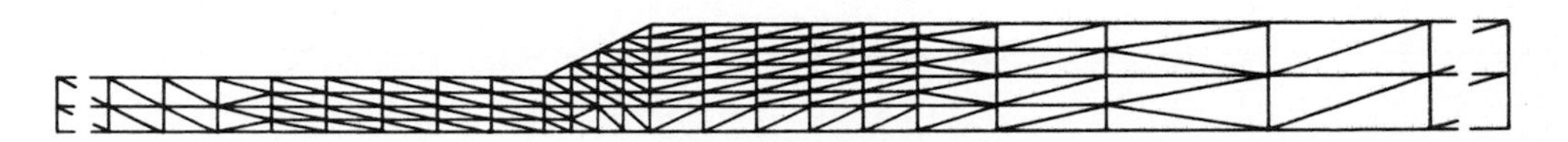

Fig. 15.14. Finite element idealization of earth bank.

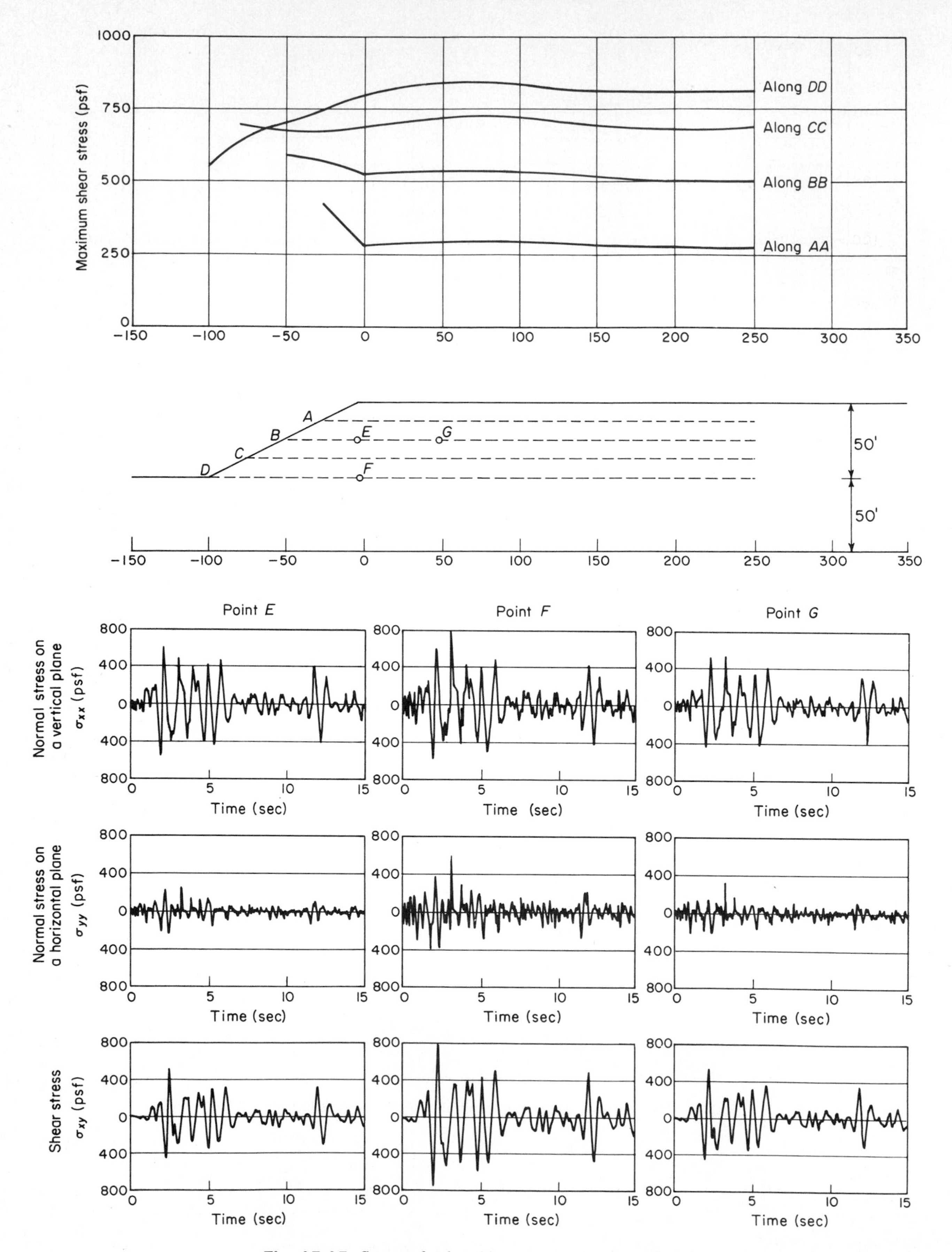

Fig. 15.15. Stresses developed in earth bank during earthquake.

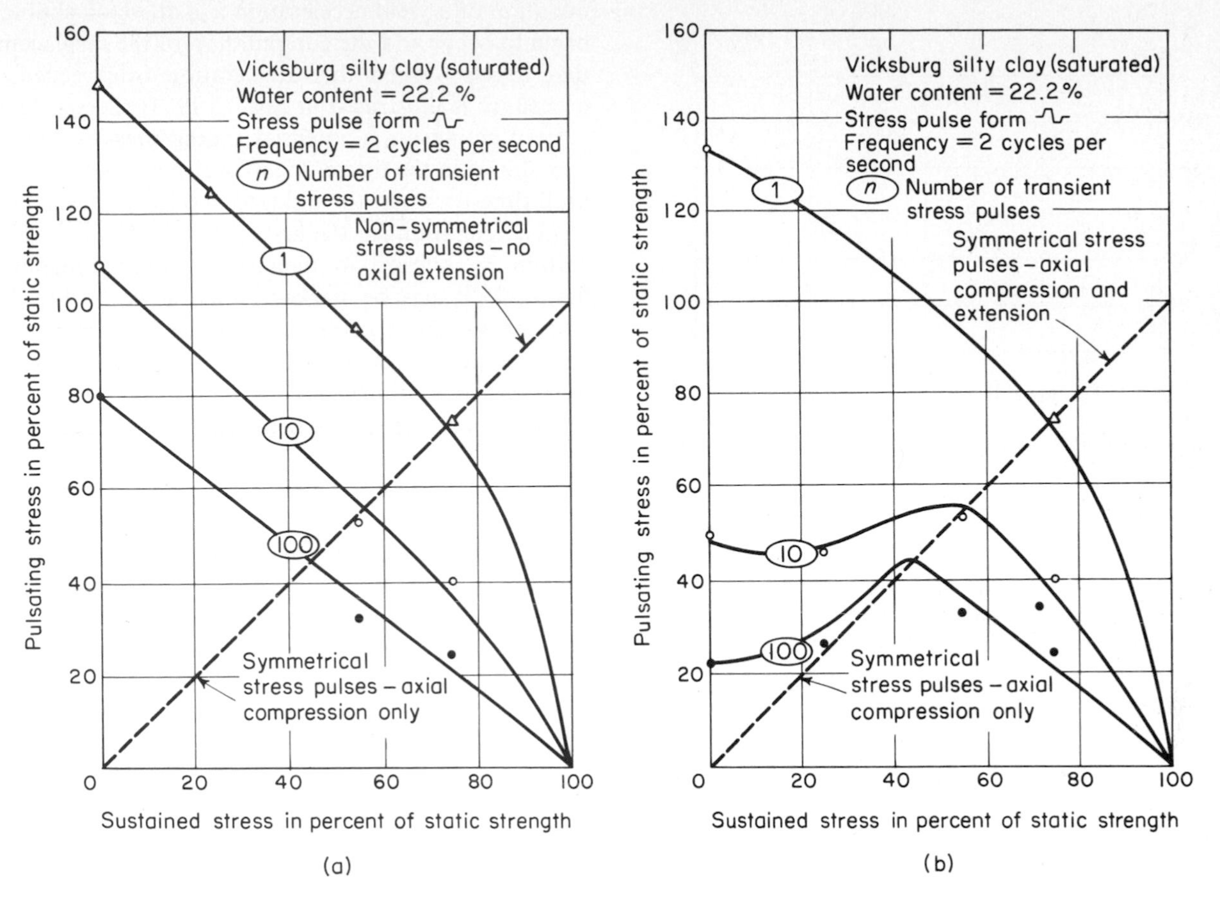

Fig. 15.16. Combinations of sustained and pulsating stresses causing failure: Vicksburg silty clay, one- and two-directional loading.

amplitudes that might be expected to develop during strong earthquakes. Using these techniques it has been possible to determine equivalent moduli and viscous damping parameters for use in the response analyses described previously. In addition, studies have been made of the strength of saturated sands at different densities and under a range of confining pressures (Seed and Lee, 1966; Lee and Seed, 1967).

For compacted clays, the combinations of sustained stress and pulsating stress that will cause failure have been shown to depend on the nature of the loading conditions (reversing or nonreversing stresses), the soil type, the frequency and duration of the pulsating stresses, and the form of the stress pulse (Seed and Chan, 1966). Typical test data are shown in Fig. 15.16. Combinations of sustained and pulsating stresses causing different strains have also been determined. For a practical range of initial stress conditions in earth embankments this relationship often can be expressed as a relationship between total stress (sustained plus pulsating) and total strain for a given number of stress cycles, as shown in Fig. 15.17. For numbers of cycles between 10 and 100 at a frequency of 2 cps it appears that such relationships for compacted and insensitive clays will not differ greatly from the stress–strain relationships for the soils determined by conventional undrained test procedures (Seed and Chan, 1966; Ellis and Hartman, 1967; see Fig 15.17). However for sensitive clays, under

Fig. 15.17. Relationships between total stress and total strain under pulsating load conditions. San Francisco Bay mud.

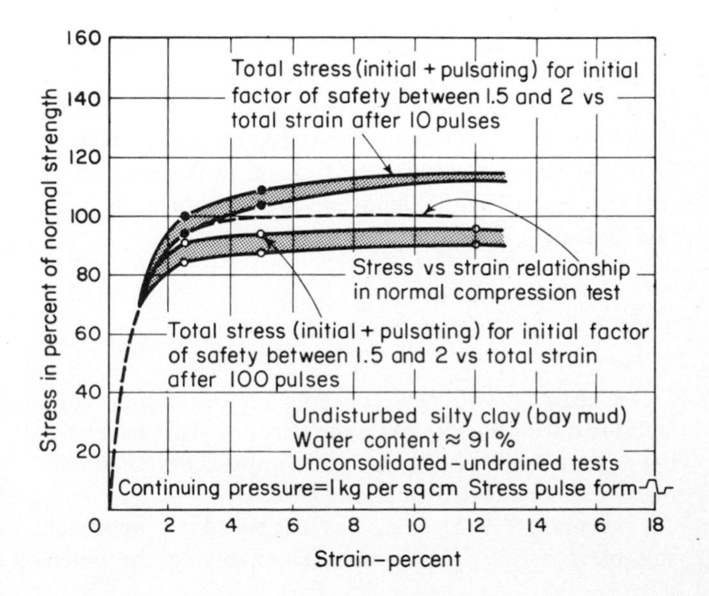

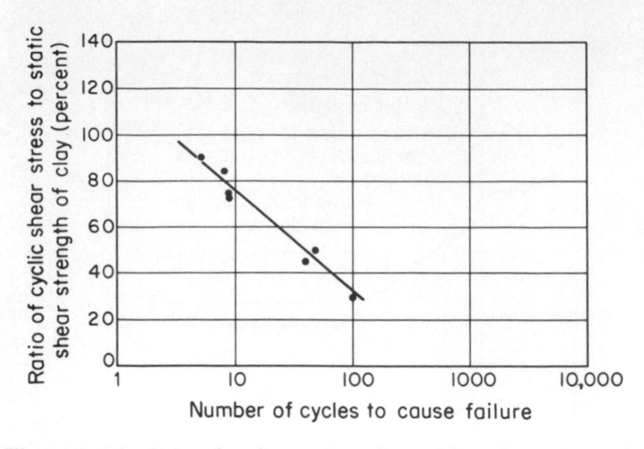

Fig. 15.18. Strength of samples of sensitive silty clay under cyclic loading conditions.

cyclic loading conditions representative of those developed under a level ground surface, failure may occur at total stress levels significantly less than the static strength of the soil (Seed and Chan, 1964; Thiers, 1965), as illustrated by the test data for Bootlegger Cove clay from Anchorage, Alaska, shown in Fig. 15.18.

Determination of the stability of embankments and earth slopes under earthquake loading requires a knowledge of the strength of samples of soil that have been consolidated under anisotropic stress conditions representing those existing before the earthquake and tested under pulsating loading conditions.

Consolidation of test specimens under anisotropic stress conditions is extremely important in dealing with saturated cohesionless materials, because the strength of these soils under pulsating loading conditions increases considerably with increasing values of the principal stress ratio during consolidation (Lee and Seed, 1966). Test data for saturated cohesive soils do not appear to be influenced to anywhere near the same extent by changes in this parameter.

Investigations of soil behavior under cyclic loading conditions can provide significant improvements in analysis of previous slope failures and evaluation of the stability of proposed slopes. The use of test data of this type in the design of canal banks has been described by Ellis and Hartman (1967), and its use in the analysis of the Turnagain Heights landslide in Anchorage, by Seed and Wilson (1967).

15.4.5 New Design Methods

In keeping with new concepts for evaluating slope stability during earthquakes, the past few years have seen the advancement of new analytical approaches for assessing embankment deformations (1965).

Newmark (1965), e.g., has suggested an approach suitable for rigid plastic materials involving the determination of a yield acceleration $k_y g$ at which sliding will begin to occur and the computation of the displacements that develop when this acceleration is exceeded. The procedure is illustrated in Fig. 15.19. If the acceleration pattern acting on a potential sliding mass is similar to that shown in the figure, then no displacement will occur until time t_1, when the induced acceleration reaches the yield acceleration for the first cycle k_{y1}. If the yield acceleration is assumed to remain constant throughout the first cycle, it may be marked off as shown on Fig. 15.19, and the variation in velocity of the sliding mass may be computed by integration over the shaded area. The velocity will continue to increase until time t_2, when the acceleration again drops below the yield value, and the velocity is finally reduced to zero at time t_3, as the direction of acceleration is reversed. The rate of displacement of the sliding mass may then be computed by integration of the velocity vs time relationship, as shown in Fig. 15.19.

This approach has been used successfully to predict the surface displacements of banks of dry, cohesionless soils subjected to known series of base motions (Goodman and Seed, 1966; Bustamente, 1965). However, for soils in which pore pressure changes develop as a result of the shear strains induced by the earthquake, determination of appropriate values of the yield acceleration becomes extremely difficult. Furthermore, for some types of soil, no well-defined yield acceleration exists and displacements take place over a wide range of accelerations.

Fig. 15.19. Integration of accelerograms to determine downslope displacements.

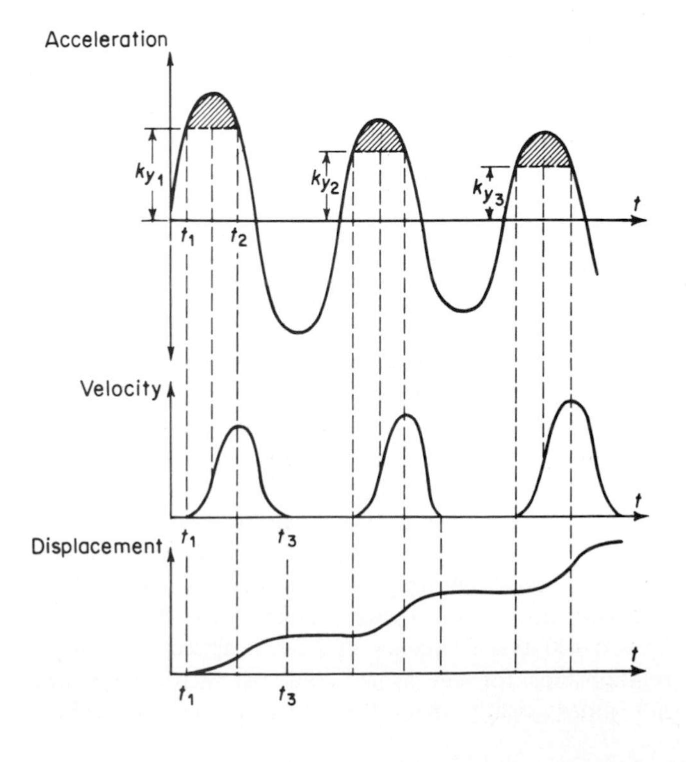

Accordingly, an alternative approach has been suggested (Seed, 1966a) based on (1) a determination of the stresses acting on soil elements within an embankment both before and during an earthquake, (2) subjecting typical soil samples in the laboratory to the same sequence of stress changes experienced by corresponding elements in the field and observing the resulting deformations, and (3) estimating the deformations of the slope from the observed deformations of the soil elements comprising it. The method thus gives consideration to the time history of forces developed in the embankments or slopes during an earthquake, the behavior of the soil under simulated earthquake loading conditions, and the desirability of evaluating embankment deformations rather than a factor of safety. It would appear that for embankments constructed of saturated soils this method provides a more tractable approach at the present time.

Special problems in connection with slope stability arise in the design of earth dams. A consideration of design philosophy in dealing with such structures has been presented by Sherard (1967).

15.5 CONCLUSIONS

The present state of the art with regard to the stability of earth slopes during earthquakes might be summarized as follows:

1. There can be no doubt that major and catastrophic slope failures have occurred during medium and large earthquakes, and the development of reliable methods for anticipating and preventing such failures is a major cause for concern to the soil engineer working in seismically active areas.

2. Qualitative assessments of slope stability during earthquakes often can be made on the basis of experience and judgment, but there is an insufficient backlog of well-defined case histories to provide a basis for quantitative evaluations.

3. Pseudostatic methods of analysis provide a means for comparing the merits of different embankment sections and for assuring increased conservatism in design sections on an empirical basis. However, there is little to guide the design engineer in selecting an appropriate value for the seismic coefficient and, furthermore, this method of analysis is inadequate to explain the mechanics of a considerable number of embankment failures. Thus, the method leaves much to be desired.

4. In recent years considerable progress has been made in the development of new concepts in earthquake-resistant design of embankments; new and greatly improved methods of analyzing the response of embankments to earthquakes; new methods of soil testing to determine those characteristics controlling soil behavior during earthquakes; new methods of design incorporating the foregoing developments; and new interest in the documentation and analysis of actual slope failures, thereby providing a basis for evaluating the merits of the new analytical, experimental, and design procedures. These developments provide a framework for evaluating previous failures and thereby offer the possibility for a more meaningful categorization of experience and an improved guide to engineering judgment in the evaluation of slope stability during earthquakes.

5. The current interest in the problem augurs well for continued research and improvements in design techniques. Much progress can be expected in the next 5 years, and it is not difficult to foresee the time when satisfactory solutions to many of our earthquake-resistant design problems involving soils will be available.

ACKNOWLEDGMENT

The author gratefully acknowledges the permission of the Water Department, City of Santa Barbara, the Bureau of Reclamation, and C. M. Duke for permission to reproduce the damage photographs in Figs. 15.1, 15.2 and 15.3, respectively.

REFERENCES

Ambraseys, N. N. (1960a). *On the Seismic Behaviour of Earth Dams, Proceedings of the Second World Conference on Earthquake Engineering*, Tokyo, Japan.

Ambraseys, N. N. (1960b). *The Seismic Stability of Earth Dams, Proceedings of the Second World Conference on Earthquake Engineering*, Tokyo, Japan.

Bustamente, J. (January 1965). *Dynamic Behaviour of Non-Cohesive Embankment Models, Proceedings of the Third World Conference on Earthquake Engineering*, New Zealand.

Chopra, A. A., and R. W. Clough (November 1965). "Earthquake Response of Homogeneous Earth Dams," *Report*, Berkeley, California: Soil Mechanics and Bituminous Materials Laboratory, University of California.

Clough, R. W., and A. K. Chopra (April 1966). "Earthquake Stress Analysis in Earth Dams," *J. Engr. Mech. Div., ASCE*, **92** (EM2), 197–211.

Coulter, H. W., and R. R. Migliaccio (1966). "Effects of the Earthquake of March 27, 1964, at Valdez, Alaska," *Geological Survey Professional Paper 542-C*, Washington, D.C.: U.S. Department of the Interior.

Davis, S., and J. K. Karzulovic (1961). "Deslizamientos en el valle del rio San Pedro Provincia de Valduia Chile," *Publication, No. 20*, Anales de la Facultad de Ciencias Fisical y Matematicos, Santiago, Chile: University of Chile, Institute of Geology.

Duke, C. M., and D. J. Leeds (February 1963). "Response of Soils, Foundations, and Earth Structures to the Chilean Earthquakes of 1960," *Bull. Seism. Soc. Am.*, **53** (2).

Ellis, W., and V. B. Hartman (July 1965). "Slope Stability During Earthquakes, San Luis Unit, California," *J. Soil Mech. Found. Div., ASCE*, **93** (SM4).

Finn, W. D. L. (May 1966). "Static and Seismic Analyses of Slopes," *Soil Mechanics Series No. 4*, University of British Columbia.

Finn, W. D. L., and J. Khanna (March 1966). "Dynamic Response of Earth Dams," *Soil Mechanics Series No. 3*, Department of Civil Engineering, University of British Columbia, Canada.

Goodman, R. E., and H. B. Seed (March 1966). "Earthquake-Induced Displacements in Sand Embankments," *J. Soil Mech. Found. Div., ASCE*, **92** (SM2).

deGraft Johnson, J. W. (1965). "Dynamic Response of Clay Under Axial Cyclic Loading," thesis presented to the University of California, at Berkeley, California, in 1965, in partial fulfillment of the requirements for the degree of Doctor of Philosophy.

Hall, J. R., and F. E. Richart, Jr. (November 1965). "Dissipation of Elastic Wave Energy in Granular Soils," *J. Soil Mech. Found. Div., ASCE*, **89** (SM6).

Hansen, W. R. (1965). "Effects of the Earthquake of March 27, 1964 at Anchorage, Alaska," *Geological Survey Professional Paper 542-A*, Washington, D.C.: U.S. Department of the Interior.

Hardin, B. O. (January 1965). "The Nature of Damping in Sands," *Jour. Soil Mech. Found. Div., ASCE*, **91** (SM1).

Hatanaka, M. (December 1955). "Fundamental Considerations on the Earthquake Resistant Properties of the Earth Dam," *Bulletin No. 11*, Disaster Prevention Research Institute, Japan: Kyoto University.

Idriss, I. M., and H. B. Seed (April 1966). "The Response of Earth Banks During Earthquakes," *Report*, Berkeley, California: Soil Mechanics and Bituminous Materials Laboratory, University of California.

Ishizaki, H., and N. Hatakeyama (February 1962). "Considerations of the Vibrational Behavior of Earth Dams," *Bulletin No. 52*, Disaster Prevention Research Institute, Japan: Kyoto University.

Japan Society of Civil Engineers (1960). "Earthquake Resistant Design for Civil Engineering Structures, Earth Structures and Foundations in Japan," *Report*.

Keightley, W. O. (July 1963). "Vibration Tests of Structures," Pasadena, California: Earthquake Engineering Research Laboratory, California Institute of Technology.

Keightley, W. O. (September 1964). "A Dynamic Investigation of Bouquet Canyon Dam," Pasadena, California: Earthquake Engineering Research Laboratory, California Institute of Technology.

Kondner, R. L., and R. Krizek (January 1965). *Dynamic Response of Cohesive Soils for Earthquake Considerations, Proceedings of the Third World Conference on Earthquake Engineering*, New Zealand.

Krishna, J. (November 1962). *Earthquake Resistant Design of Earth Dams, Proceedings of the Earthquake Symposium*, Roorkee, India: Roorkee University.

Lee, K. L., and H. B. Seed (July 1966). "Strength of Anisotropically Consolidated Samples of Saturated Sand under Cyclic Loading Conditions," *Report*, Berkeley, California: Soil Mechanics and Bituminous Materials Laboratory, University of California.

Lee, K. L., and H. B. Seed (January 1967). "Cyclic Stress Conditions Causing Liquefaction of Sands," *J. Soil Mech. Found. Div., ASCE*, **93** (SM1).

Marsal, R. J. (July 1967). "Performance of El Infiernillo Dam, 1963–66," *J. Soil Mech. Found. Div., ASCE*, **93** (SM4).

Martin, G. R. (1965). "The Response of Earth Dams to Earthquakes," thesis presented to the University of California, Berkeley, California, in 1965, in partial fulfillment of the requirements for the degree of Doctor of Philosophy.

Martin, G. R., and H. B. Seed (October 1966). "An Investigation of the Dynamic Response Characteristics of the Bon Tempe Dam, California," *Report*, Berkeley, California: Soil Mechanics and Bituminous Materials Laboratory, University of California.

McCulloch, D. S. (1966). "Slide Induced Waves, Seiching and Ground Fracturing Caused by the Earthquake of March 27, 1964 at Kenai Lake, Alaska," *Geological Survey Professional Paper 543-A*, Washington, D. C.: U.S. Department of the Interior.

Miller, R. D., and E. Dobrovolny (1959). "Surficial Geology of Anchorage and Vicinity, Alaska," *Bulletin 1093*, Washington, D. C.: U.S. Geologic Survey.

Mononobe, N., A. Takata, and M. Matumura (1936). *Seismic Stability of the Earth Dam, Proceedings of the Second Congress on Large Dams*, Washington, D. C.

Newmark, N. M. (1963). "Earthquake Effects on Dams and Embankments," presented at the October 7–11, 1963, ASCE Structural Engineering Conference, San Francisco, California.

Newmark, N. M. (June 1965). "Effects of Earthquakes on Dams and Embankments," *Geotechnique*, **XV** (2).

Okamoto, S., M. Hakuno, K. Kato, and F. Kawakami (1965). *On the Dynamical Behaviour of an Earth Dam During Earthquake," Proceedings of the Third World Conference on Earthquake Engineering*, New Zealand.

Parmalee, R. J., J. Penzien, C. F. Scheffey, H. B. Seed, and G. A. Thiers (August 1964). "Response of Structures Supported on Long Piles Through Deep Sensitive Clay," *Report No. SESM-64-2*, Berkeley, California: University of California, Structural Engineering and Structural Mechanics Laboratory.

Rashid, Y. R. (1961). "Dynamic Response of Earth Dams to Earthquakes," Graduate Student Research Report, Berkeley, California: University of California.

Seed, H. B. (January 1966a). "A Method for Earthquake-Resistant Design of Earth Dams," *J. Soil Mech. Found. Div., ASCE*, **92** (SM1), 13–41.

Seed, H. B. (September 1966b). Soil Stability Problems Caused by Earthquakes," *Report*, Berkeley, California: Soil Mechanics and Bituminous Materials Laboratory, University of California.

Seed, H. B., and C. K. Chan (1964). "Pulsating Load Tests on Samples of Clay and Silt from Anchorage, Alaska," *Report on Anchorage Area Soil Studies, Alaska, to U.S. Army Engineer District, Anchorage, Alaska*, Seattle, Washington: Shannon and Wilson.

Seed, H. B., and C. K. Chan (March 1966). "Clay Strength under Earthquake Loading Conditions," *J. Soil Mech. Found. Div., ASCE*, **92** (SM2), 53–78.

Seed, H. B., and I. M. Idriss (May 1967). "Analysis of Soil Liquefaction: Niigata Earthquake," *J. Soil Mech. Found. Div., ASCE*, **93** (SM3), 83–108.

Seed, H. B., and K. L. Lee (November 1966). "Liquefaction of Saturated Sands During Cyclic Loading," *J. Soil Mech. Found. Div., ASCE*, **92** (SM6), 105–134.

Seed, H. B., and G. R. Martin (May 1966). "The Seismic Coefficient in Earth Dam Design," *J. Soil Mech. Found. Div., ASCE*, **92** (SM3), 25–58.

Seed, H. B., and S. D. Wilson (July 1967). "The Turnagain Heights Landslide, Anchorage, Alaska," *J. Soil Mech. Found. Div., ASCE*, **93** (SM4), 325–353.

Shannon, W. L. (1966). "Slope Failures at Seward, Alaska," presented at the August 22–26, 1966, ASCE Soil Mechanics and Foundations Division Conference on Stability and Performance of Slopes and Embankments, Berkeley, Calif.

Shannon and Wilson, Inc. (1964). *Report on Anchorage Area Soil Studies, Alaska, to U.S. Army Engineer District, Anchorage, Alaska*, Seattle, Washington.

Sherard, J. L. (July 1967). "Some Considerations in Earth Dam Design," *J. Soil Mech. Found. Div., ASCE*, **93** (SM4).

Taylor, P., and J. Hughes (January 1965). *Dynamic Properties of Foundation Subsoils as Determined from Laboratory Tests, Proceedings, of the Third World Conference on Earthquake Engineering*, New Zealand.

Thiers, G. R. (1965). "The Behavior of Saturated Clay Under Seismic Loading Conditions," thesis presented to the University of California, at Berkeley, California, in 1965, in partial fulfillment of the requirements for the degree of Doctor of Philosophy.

Wilson, S. D., and R. J. Dietrich (1960). "Effect of Consolidation Pressure on Elastic and Strength Properties of Clay," *Proceedings*, June 13–17, ASCE Research Conference on Shear Strength of Cohesive Soils, at Boulder, Colorado.

Chapter 16

Current Trends in the Seismic Analysis and Design of High-Rise Structures

NATHAN M. NEWMARK

Professor of Civil Engineering
University of Illinois
Urbana, Illinois

16.1 INTRODUCTION

The design of a structure is a process of synthesis, as contrasted with the analysis for given loadings or environmental conditions. In the design of a building to resist earthquake motions, the designer works within certain constraints, such as the architectural configuration of the building, the foundation conditions, the nature and extent of the hazard should failure or collapse occur, the possibility of an earthquake, the possible intensity of earthquakes in the region, the cost or available capital for construction, and similar factors. In the light of available information, the designer chooses the materials to be used, the method of construction, and the design concepts. He may choose to use steel, aluminum, concrete, masonry, or a combination of materials. He may select a frame with rigid connections, a frame with bracing, a structure carrying lateral forces primarily by deep walls or "shear walls," or a combination of these elements.

Whatever the choice, the designer must have some basis for selection of the strength and the proportions of the building and of the various members in it. The required strength depends on factors such as the intensity of earthquake motions to be expected, the flexibility of the structure, and its ductility or reserve strength before damage occurs. Because of the interrelations among flexibility and strength of a structure and the forces generated in it by earthquake motions, the dynamic design procedure must take these various factors into account. The ideal to be achieved is one involving appropriate flexibility and energy-absorbing capacity, permitting the earthquake displacements to take place without unduly large forces being generated. To achieve this end, control of the construction procedures and appropriate inspection practices are necessary. The attainment of the ductility required to resist earthquake motions must be emphasized.

In the material presented here a general description of the response of relatively simple systems to earthquake motions is presented, first for elastic behavior and then for inelastic behavior. Generalizations are made about the relation of the response of multidegree-of-freedom systems to simple systems, both in the elastic and the inelastic range. Consideration is given to the general nature of the provisions of current building codes for earthquake-resistant design. Based on the relations between the results of theoretical analyses and current design provisions, estimates are made of the required ductility for the earthquake-resistant design of buildings. Finally, some comments are made about the design of actual buildings and their behavior in earthquakes.

16.2 RESPONSE OF SIMPLE STRUCTURES TO EARTHQUAKE MOTIONS

A series of structures of varying size and complexity is shown in Fig. 16.1, corresponding to a simple, relatively compact machine anchored to a foundation in Item 1, a simple bent or frame in Item 2, a more complex frame in Item 3, multistory buildings of 15 stories in Item 4, and of 40 stories in Item 5, an elevated water tank in Item 6, and a suspension bridge responding either laterally or vertically in Item 7. The period of vibration T or the frequency of vibration f in the fundamental mode of vibration is indicated for each of these structures.

Each of the structures shown in Fig. 16.1 could be represented by a simple oscillator consisting of a single mass supported by a spring and a dashpot, as shown in Fig. 16.2. The relation among the circular frequency of vibration ω, the natural frequency f, and the period T is given by the following equation in terms of the spring constant k and the mass m:

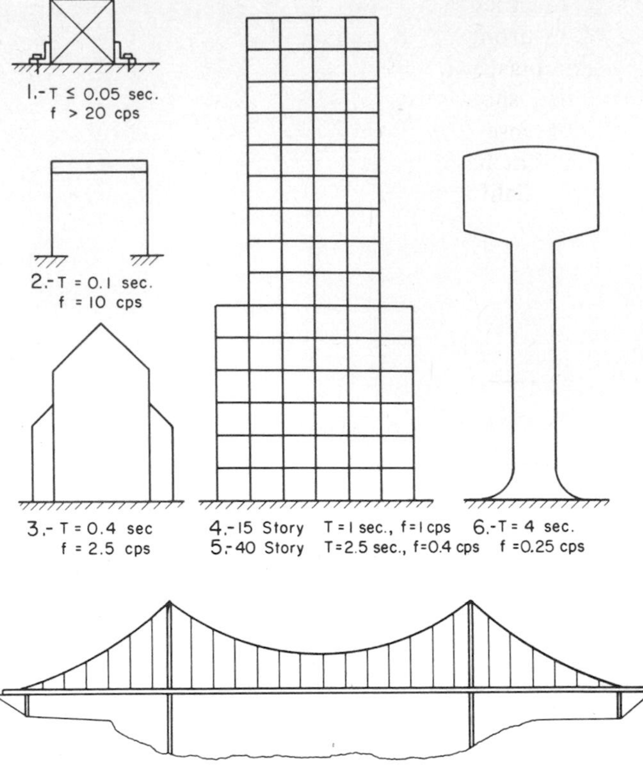

Fig. 16.1. Structures subjected to earthquake ground motions.

$$\omega^2 = \frac{k}{m} \tag{16.1}$$

$$f = \frac{1}{T} = \frac{\omega}{2\pi} = \frac{1}{2\pi}\sqrt{\frac{k}{m}} \tag{16.2}$$

The effect of the dashpot is to produce damping of free vibrations or to reduce the amplitude of forced vibrations in general. The amount of damping is most conveniently considered in terms of the proportion of critical damping β, which for most practical structures is relatively small (in the range of 0.5 to 10 or 20%) and does not appreciably affect the natural period or frequency of vibration.

The simple system in Fig. 16.2 can be used to represent the various modes of vibration of a multidegree-of-freedom system. For the time being, however, we shall consider only the fundamental mode of vibration of the multidegree-of-freedom systems in Fig. 16.1, as represented by the single-degree system in Fig. 16.2.

When the base of the system in Fig. 16.2 moves with respect to time, the mass is set into motion also and strains are induced in the spring. The motion of the base may be described by giving the displacement as a function of time or, equally as descriptive of the motion, the time history of the velocity of the base or the time history of its acceleration. Strong-motion earthquake accelerations with respect to time have been obtained for a number of earthquakes. Ground motions from other sources of disturbances, such as quarry blasting, nuclear blasting, etc. are also available and show many of the same characteristics. The most intense long duration strong-motion earthquake record that has been recorded so far is that of the El Centro, California, earthquake of May 18, 1940. The recorded accelerogram for that earthquake, in the north–south component of horizontal motion, is shown in Fig. 16.3. On the same figure are shown integration of the ground acceleration to give the variation of ground velocity with time and also the integration of velocity to give the variation of ground displacement with time. These integrations require base-line corrections of various sorts, and the magnitude of the maximum displacement may vary, depending on how the corrections are made. The maximum velocity is relatively insensitive to the corrections, however. For this earthquake, with the integrations performed as shown in Fig. 16.3, the maximum ground acceleration is 0.32 g, where g is the acceleration of gravity, the maximum ground velocity is 13.7 in/sec, and the maximum ground displacement is 8.3 in. These three maximum values are of particular interest because they help to define the response motions of the various structures considered in Fig. 16.1 most accurately, if all three maxima are taken into account.

For the ground motions in Fig. 16.3 or any other type of ground motion that might be considered, the response

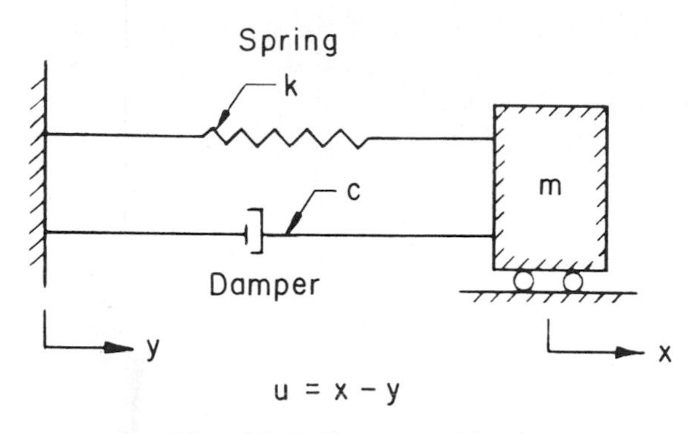

Fig. 16.2. System considered.

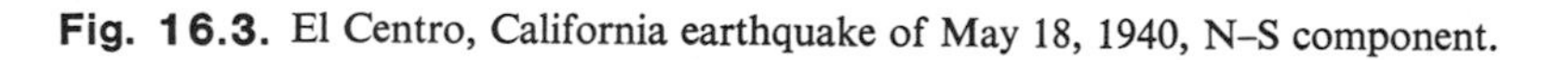

Fig. 16.3. El Centro, California earthquake of May 18, 1940, N–S component.

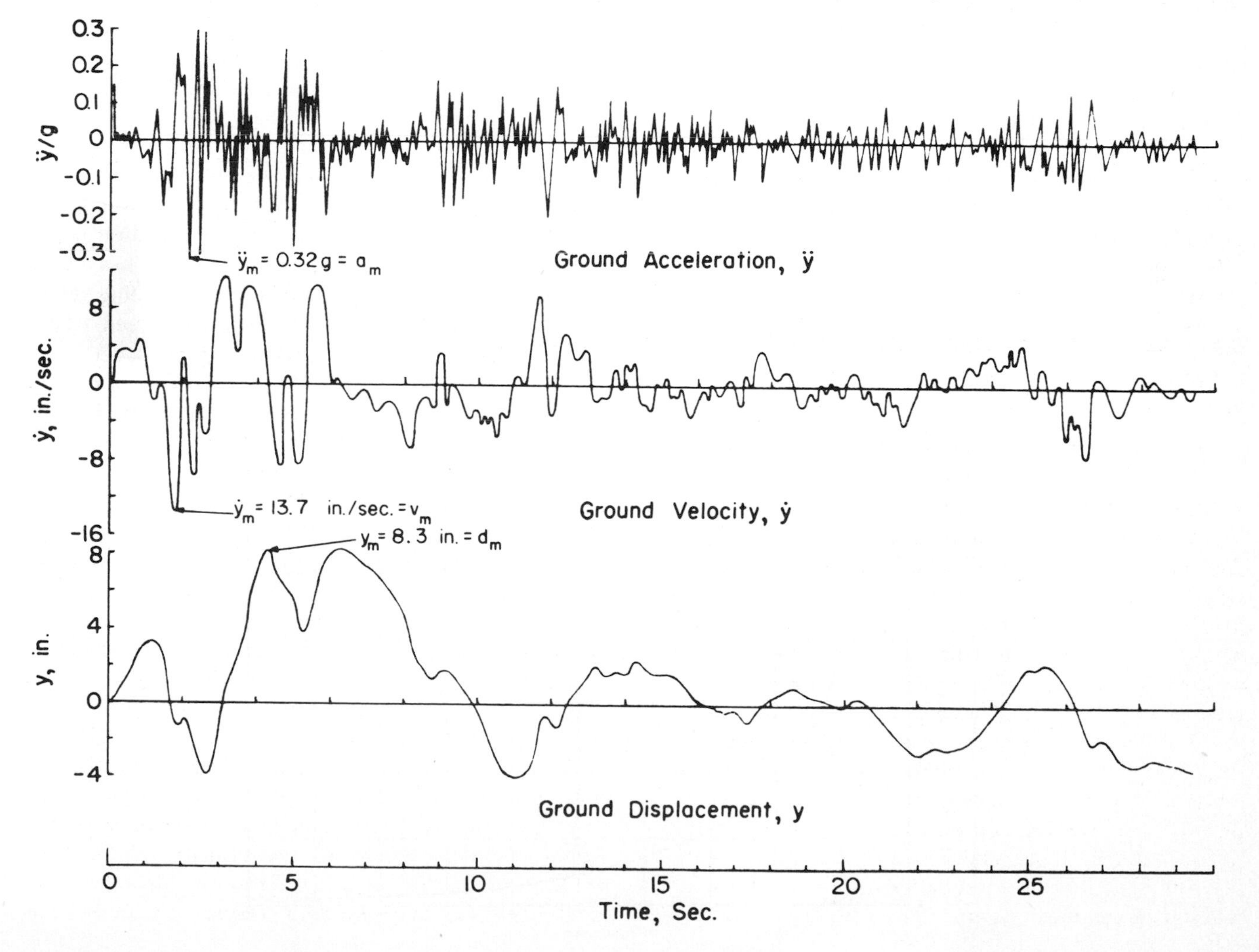

of the simple oscillator shown in Fig. 16.2 can be readily computed as a function of time. The maximum values of the response of this oscillator are of particular interest. These maximum values might be stated in terms of the maximum strain in the spring in Fig. 16.2, $u_m = D$ or, alternatively, the response can be stated as the maximum spring force or the maximum acceleration of the mass that is related to the maximum spring force directly when there is no damping or it may be stated by a quantity having the dimensions of velocity, which gives a measure of the maximum energy absorbed in the spring. This quantity, designated the pseudovelocity, is defined in such a way that the energy absorption in the spring is $\frac{1}{2}mV^2$. The relations among the maximum relative displacement of the spring D, the pseudovelocity V, and the pseudoacceleration Ag, which is a measure of the force in the spring, are as follows:

$$V = \omega D \tag{16.3}$$

$$Ag = \omega V = \omega^2 D \tag{16.4}$$

The pseudovelocity V is nearly equal to the maximum relative velocity for systems with moderate or high frequencies, but it may differ considerably from the maximum relative velocity for very low frequency systems. The pseudoacceleration A is exactly equal to the maximum acceleration for systems with no damping and is not greatly different from the maximum acceleration for systems with moderate amounts of damping, over the whole range of frequencies from very low to very high values. Typical plots of the response of the system as a function of period or frequency are called response spectra. Response spectrum plots are shown in Fig. 16.4 for acceleration and for relative displacement, for a system with a moderate amount of damping, subjected to an input similar to that shown in Fig. 16.3. This arithmetic plot of maximum response is simple and convenient to use. However, a somewhat more useful plot, which indicates at one and the same time the values for D, V, and A, is indicated in Fig. 16.5. This has the virtue that it also indicates more clearly the extreme or limiting values of the various parameters defining the response.

For Fig. 16.5, the frequency is plotted on a logarithmic scale. Since the frequency is the reciprocal of the period, the logarithmic scale for period would have exactly the same spacing of the points, or in effect the plot would be turned end for end.

The pseudovelocity is plotted on a vertical scale, also logarithmically. Then on diagonal scales, as indicated on the figure, along an axis that extends upward from right to left are plotted values of the displacement, and along an axis that extends upward from left to right is plotted the pseudoacceleration, in such a way that any one point defines for a given frequency the displacement D, the pseudovelocity V, or the pseudoacceleration Ag. Points are indicated in Fig. 16.5 for the seven structures of Fig. 16.1, plotted at the fundamental frequencies for the structures considered. One can read directly from the plot the response values. Some further interpretation is needed for the response of a multidegree-of-freedom system, however. A more detailed explanation of these points is given in Blume, Newmark, and Corning (1961).

The typical shape of the response spectrum shown in Fig. 16.5 is characteristic of the response for almost any type of input. A wide variety of motions have been considered by Newmark and Veletsos (1964) and Veletsos, Newmark, and Chelapati (1965), ranging from simple pulses of displacement, velocity, or acceleration of the ground through more complex motions such as those arising from nuclear blast detonations and for a variety of earthquakes as taken from available strong-motion

Fig. 16.4. Arithmetic plots of response.

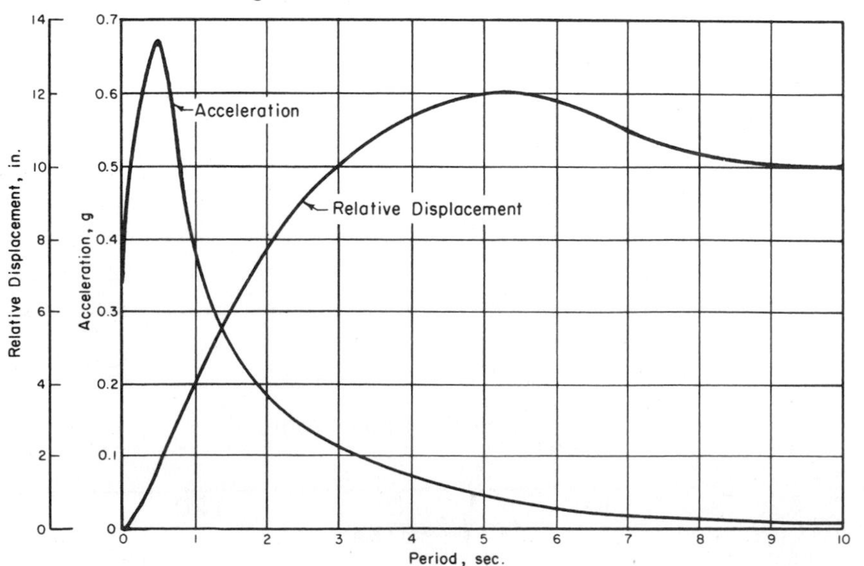

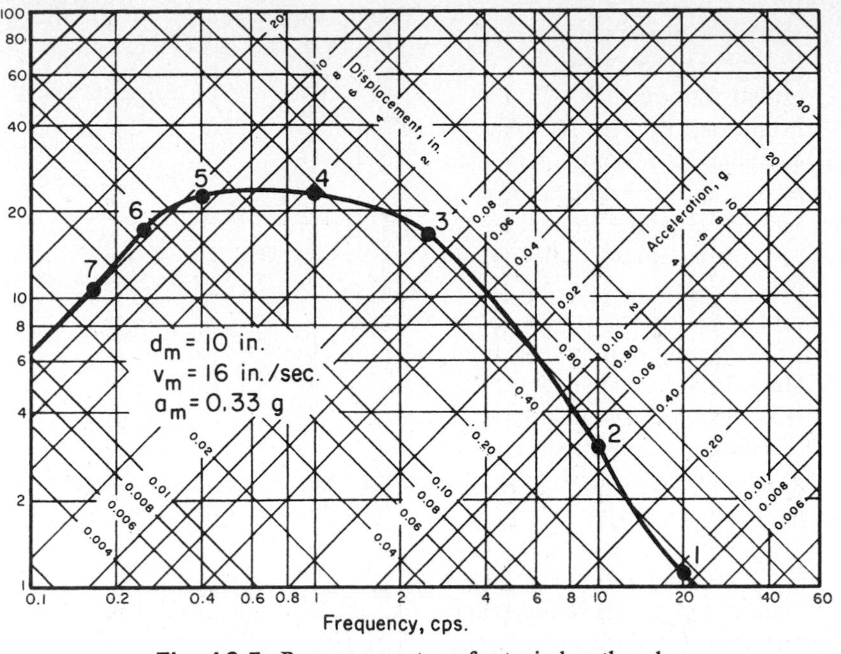

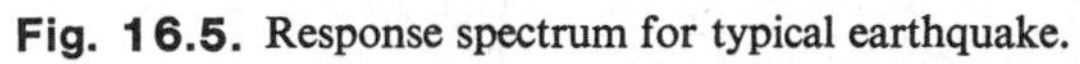

Fig. 16.5. Response spectrum for typical earthquake.

Fig. 16.6. Deformation spectra for elastic systems subjected to the El Centro earthquake.

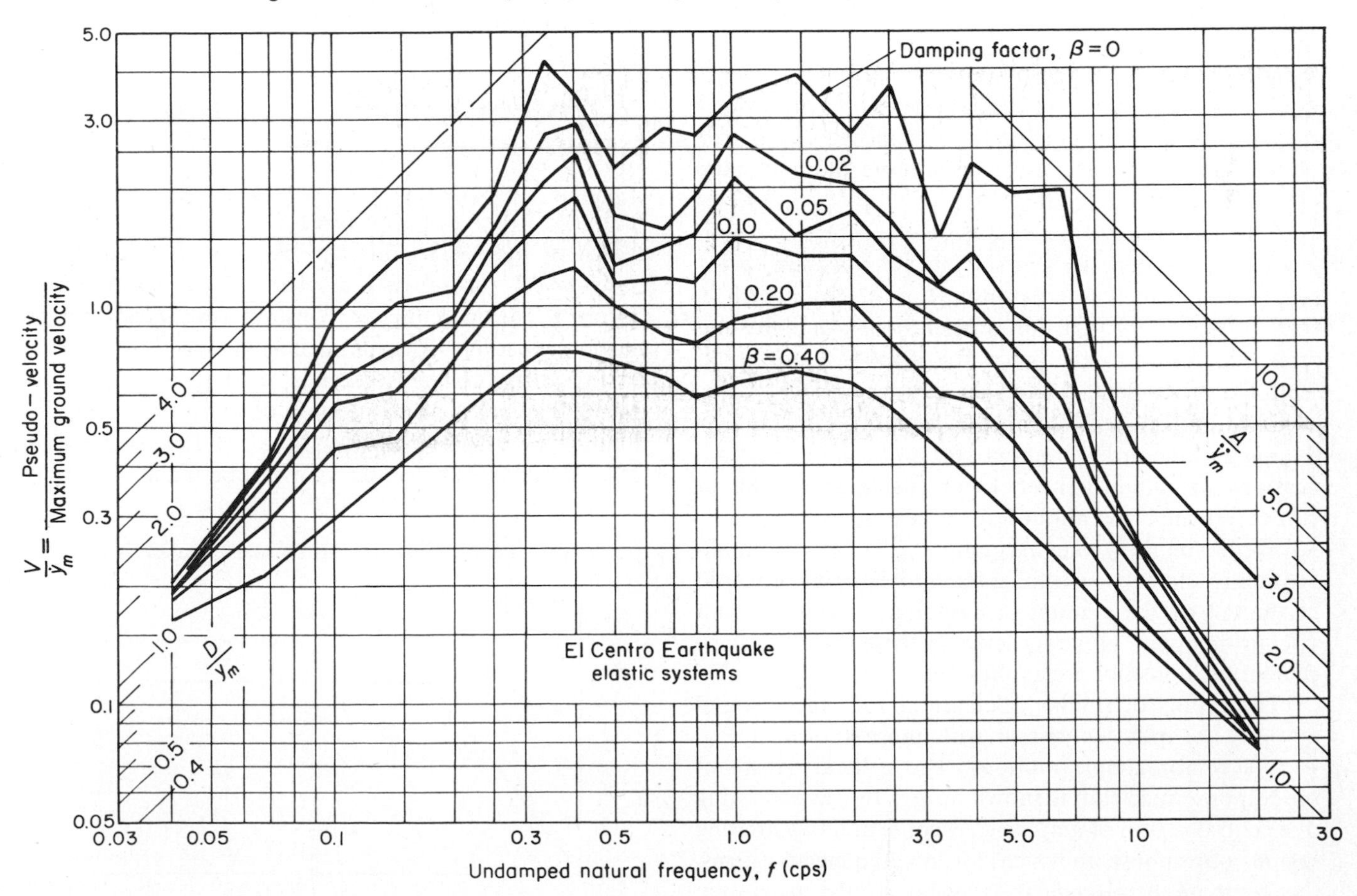

records. Typical of all of these is the response spectrum shown in Fig. 16.6 for the same El Centro earthquake, the motion records for which are given in Fig. 16.3. The response spectrum for small amounts of damping is much more jagged than indicated by Fig. 16.5, but for the higher amounts of damping the response curves are smooth. The scales are chosen in this instance to represent the amplifications of the response relative to the ground-motion values of displacement, velocity, or acceleration.

The spectrum shown in Fig. 16.6 is typical of response spectra for nearly all types of ground motion. It is noted that on the extreme left of Fig. 16.6, corresponding to very low frequency systems, the response for all degrees of damping approaches an asymptote corresponding to the value of the maximum ground displacement. A low frequency system corresponds to one having a very heavy mass and a very light spring. When the ground is moved relatively rapidly, the mass does not have time to move, and therefore the maximum strain in the spring is precisely equal to the maximum displacement of the ground.

On the other hand, for a very high frequency system, the spring is relatively stiff and the mass is very light. Therefore, when the ground is moved, the stiff spring forces the mass to move in the same way the ground moves, and the mass must therefore have the same acceleration as the ground at every instant. Hence, the force in the spring is that required to move the mass with the same acceleration as the ground, and the maximum acceleration of the mass is precisely equal to the maximum acceleration of the ground. This is shown by the fact that all of the lines on the extreme right-hand side of the figure approach as an asymptote the maximum ground acceleration line.

For intermediate frequency systems, there is an amplification of motion. In general the amplification factor for displacement is less than that for velocity, which in turn is less than that for acceleration. Amplification factors indicated by the figure are of the order of about 3.5 for displacement, 4.2 for velocity, and about 9.5 for acceleration for the undamped system in Fig. 16.6. For damping of the order of about 10% critical, these amplifications are slightly over 1 for displacement, about 1.5 for velocity, and about 2 for acceleration.

For an infinitely long harmonic oscillation, the amplification would become infinite for zero damping or would be limited by the amount of damping for systems with viscous damping. However, even in these cases the same general relationships are applicable.

The results of similar calculations for other ground motions are quite consistent with those shown in Fig. 16.6, even for simple motions. The general nature of the response spectrum is shown in Fig. 16.7 as consisting of a central region of amplified response and two limiting regions of response, in which, for low frequency systems, the response displacement is equal to the maximum

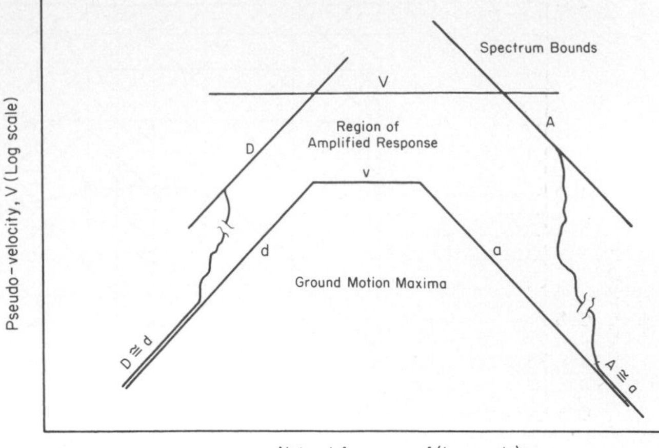

Fig. 16.7. Typical tripartite logarithmic plot of response spectrum bounds compared with maximum ground motions.

ground displacement. For high frequency systems, the response acceleration is equal to the maximum ground acceleration. For damping of the order of about 5 to 10% critical, the amplification factors for displacement, velocity, and acceleration are only slightly over 1, 1.5, and 2.0, respectively, for a wide variety of earthquake and ground shock motions. These amplification factors increase quite rapidly, however, as the damping decreases, and they decrease relatively slowly as the damping factor increases above the values of 5 to 10%.

16.3 RESPONSE SPECTRA FOR INELASTIC SYSTEMS

A typical inelastic spring force-displacement relation is shown in Fig. 16.8. This can be approximated by an elasto-plastic relation as indicated, with an elastic initial region, a plastic ceiling of constant resistance, and an elastic unloading. The unloading is considered to be elastic until yielding is reached in the opposite direction. For equal yield values in either direction, calculations of the response of the system of Fig. 16.2 for an elasto-plastic resistance function can be made. A variety

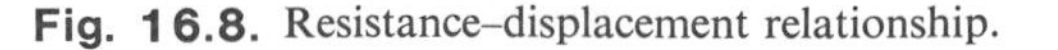

Fig. 16.8. Resistance–displacement relationship.

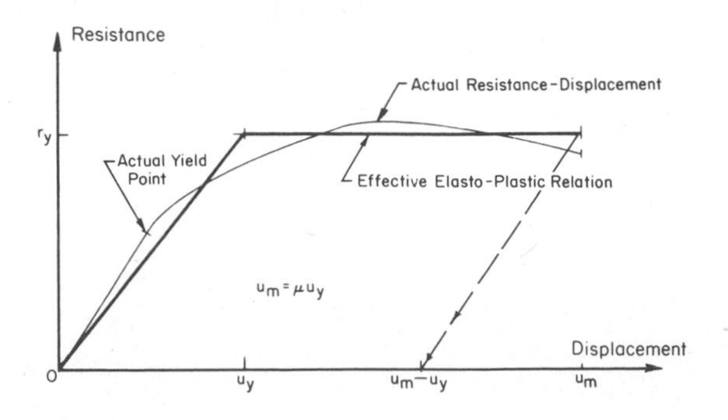

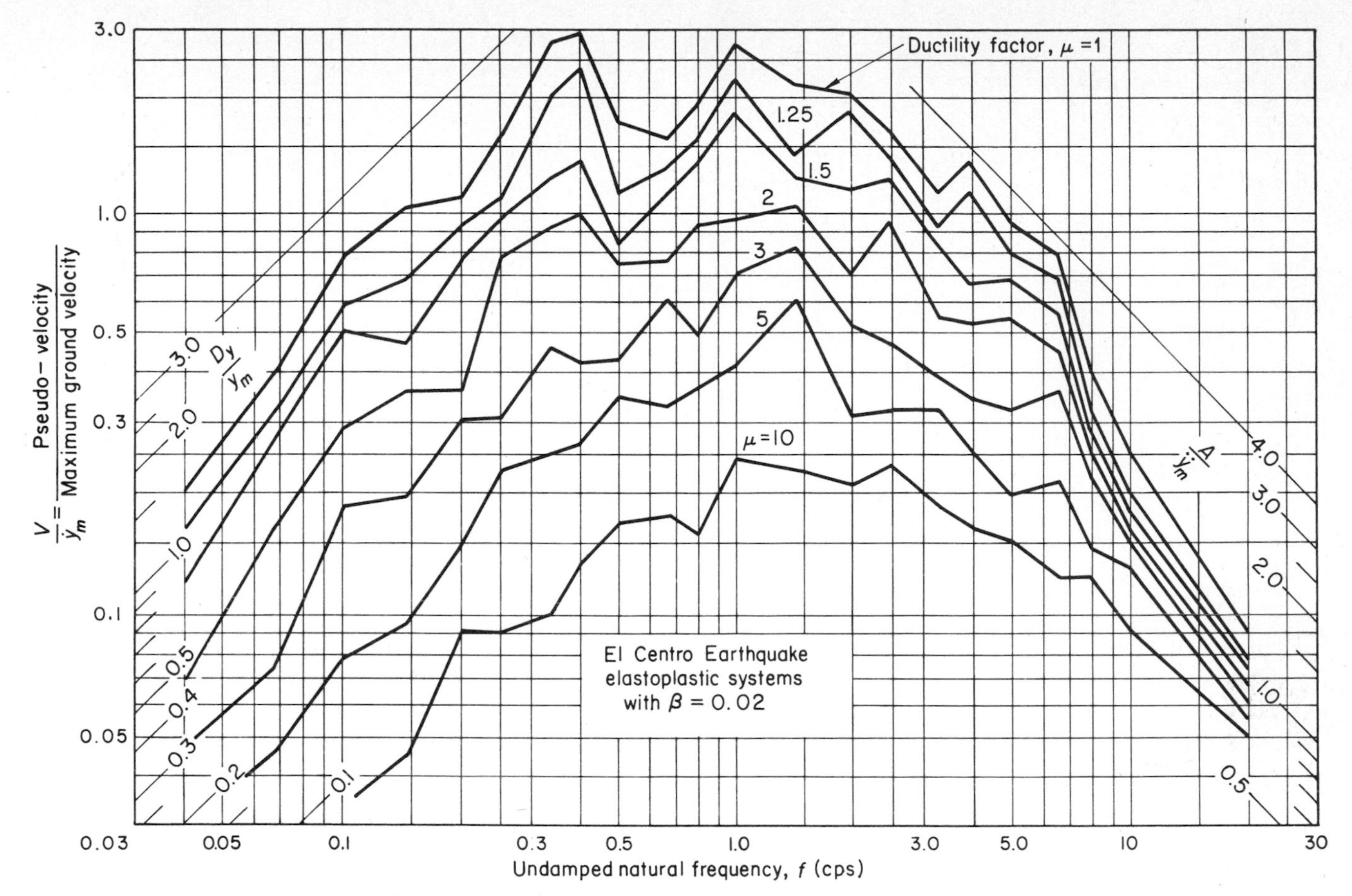

Fig. 16.9. Deformation spectra for elasto-plastic systems with 2% critical damping subjected to the El Centro earthquake.

of such calculations have been made and are reported by Newmark and Veletsos (1964) and Veletsos and Newmark (1960). It is interesting and instructive to plot the results of such calculations on a chart similar to the tripartite response spectrum charts of Figs. 16.5 and 16.6. This can be done directly for the elasto-plastic system, for constant values of ductility factor μ, which is the ratio of the maximum relative displacement of the spring to the yield point value of displacement. However, the plot can be made only in terms of the elastic component of displacement, in which case the accelerations are properly presented, or alternatively for the total displacement, in which case the accelerations are not properly presented. Since the former case is most convenient, this is what has been used as a basis for the chart shown in Fig. 16.9, which is presented also for the El Centro earthquake, for elasto-plastic systems having varied amounts of yielding, but with a damping factor of 2% of critical in the elastic range of the response. Ductility factors ranging from 1 (elastic behavior) up to 10 are shown in the figure. The total displacements can be obtained directly from the figure by multiplying the displacement components by the value of μ, the ductility factor, assigned to each curve. It is noted in Fig. 16.9 that the elastic components of displacement vary roughly inversely as the ductility factor for the left-hand side of the chart or for low or even intermediate frequency systems and that the accelerations are nearly the same for all high frequency systems. This is consistent with the observations made earlier, for low frequency systems, that the total spring displacement is equal to the maximum ground displacement, and this is true regardless of the nature of the spring or force displacement curve. For high frequency systems, the acceleration of the mass must be the same as the ground and, therefore, the acceleration must be the same as the ground acceleration regardless of the characteristics of the spring.

The results obtained in Fig. 16.9, and from similar calculations for other earthquake and ground shock motions, are approximated in Fig. 16.10. In the left-hand side of this figure it is apparent that the total displacement is the same for the elasto-plastic response as for an elastic system; and for the right-hand side, the acceleration is the same for the elasto-plastic system as for the elastic system. In the intermediate region, one can approximate the results by use of the relationship that the energy is the same for the elasto-plastic system as for an elastic

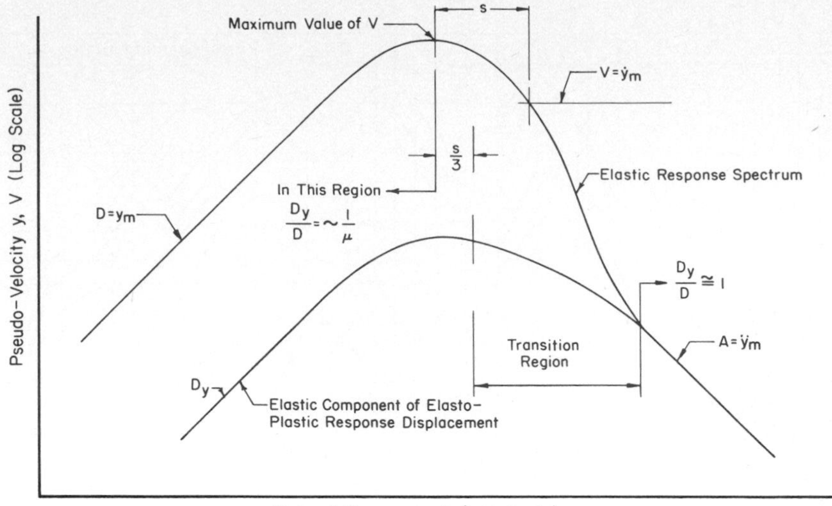

Fig. 16.10. Approximate design rule for construction of deformation spectra for elasto-plastic systems.

system having the same frequency. These observations lead to further generalizations that have been verified by additional calculations some of which are reported by Newmark and Veletsos (1964), but others are still under study. The generalizations may be stated as follows:

1. For low frequency systems, the total displacement for the inelastic system is the same as for an elastic system having the same frequency.
2. For intermediate frequency systems, the total energy absorbed by the spring is the same for the inelastic system as for an elastic system having the same frequency.
3. For high frequency systems, the force in the spring is the same for the inelastic system as for an elastic system having the same frequency.

A number of stress-strain curves are shown in Fig. 16.11, and the above-mentioned rules are indicated in that figure. In the region where displacement is preserved, the forces or accelerations vary in accordance with the ordinates to the curve of stress or force relative to strain

Fig. 16.11. Comparison of strains for equal displacement, energy, or force.

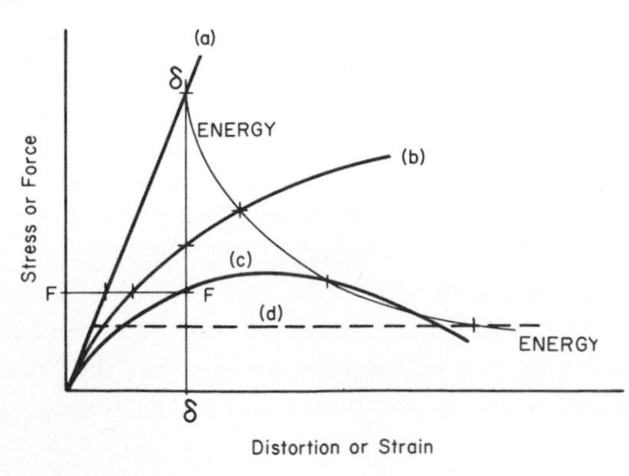

at a constant strain ordinate. In the region where energy is preserved, both the displacement and the force vary in such a way as to keep the area under the various curves the same for all of the curves shown. In the region where force is preserved, the displacement varies in accordance with the displacement abscissas for a constant value of stress.

Shown in Fig. 16.11 are lines corresponding to (a) an elastic resistance, (b) an inelastic resistance having the same initial slope, (c) an inelastic resistance showing a maximum and a decay beyond the maximum, and (d) an elasto-plastic resistance, as indicated by the dashed line.

However, Fig. 16.11 indicates that the line for a force greater than the yield point force for the elasto-plastic curve will never intersect the elasto-plastic curve and will therefore give an infinite displacement. This is not the case under actual conditions. Limits that are more realistic can be obtained if one plots the response spectrum in terms of total relative displacement for the inelastic curves, as indicated in Fig. 16.12. Here the lower set of lines corresponding to $\bar{D}$, $\bar{V}$, and $\bar{A}$ are for an elastic condition. The curves $\bar{D}_1$, $\bar{V}_1$, $\bar{A}_1$ are drawn for an inelastic force-displacement relation in which displacement is preserved for $\bar{D}_1$, energy is preserved for $\bar{V}_1$, and acceleration is preserved for $\bar{A}_1$. Similar curves are shown for $\bar{V}_2$ and $\bar{A}_2$ at somewhat higher levels of inelastic behavior. Curves for other levels also can be drawn for $\bar{V}_3$ and $\bar{A}_3$, for example.

At a frequency such as f_a, all of the displacements are bounded by point a, and the displacement limit is the same for all of the inelastic curves considered.

At a frequency such as f_b, the displacement limit corresponds to point b_0 on $\bar{V}$, and a greater displacement corresponding to point b_1 on $\bar{V}_1$. However, one cannot reach $\bar{V}_2$ without crossing the line corresponding to $\bar{D}$. Hence, the upper bound of displacement is given by $\bar{b}$.

At a frequency such as f_c, the displacement bound is given by c_0 on line $\bar{A}$, c_1 on line $\bar{A}_1$, but by c_2 on $\bar{V}_2$, which intersection is reached before the line reaches $\bar{A}_2$. Similarly, the upper bound is $\bar{c}$ on line $\bar{D}$, which is below $\bar{V}_3$.

In other words, for inelastic systems, for low frequencies, displacement is preserved. For intermediate frequencies, energy is preserved, except that the displacement cannot be greater than the *displacement bound* for the elastic response spectrum. Furthermore, for high frequency systems, force (or acceleration) is preserved, provided that the energy absorbed is not greater than the *energy bound* for the elastic spectrum and also that the displacement is not greater than the *displacement bound* for the elastic response spectrum.

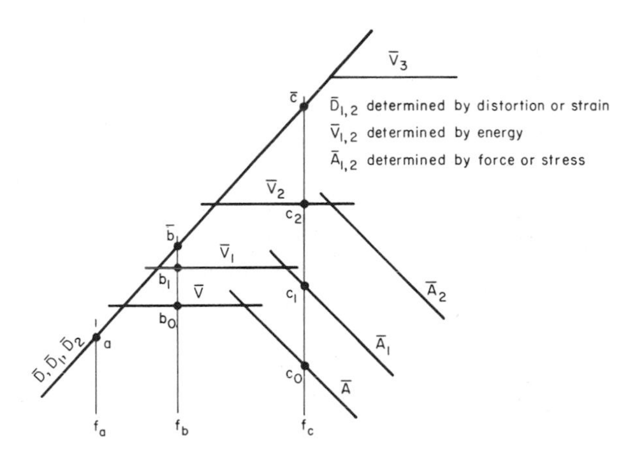

Fig. 16.12. Response spectrum displacement limits.

It should be pointed out that the relative values to be used here for the elastic response spectrum should be those corresponding to about 5 to 10% of damping; otherwise the acceleration and velocity bounds will be too high.

16.4 MULTIDEGREE-OF-FREEDOM SYSTEMS

A multidegree-of-freedom system has a number of different modes of vibration. For example, the shear beam shown in Fig. 16.13 has a fundamental mode of lateral oscillation as shown in Fig. 16.13b, a second mode as shown in Fig. 16.13c, and a third mode as shown in Fig. 16.13d. Each of these modes can be considered to vibrate independently, with participation factors as defined in the usual way, and as described in detail by Blume, Newmark, and Corning (1961). A response spectrum for a multidegree-of-freedom system can be drawn for a particular system, as a function of the fundamental frequency of the system, in much the same way as a response spectrum for a single-degree-of-freedom system is drawn. We may do this so as to define dis-

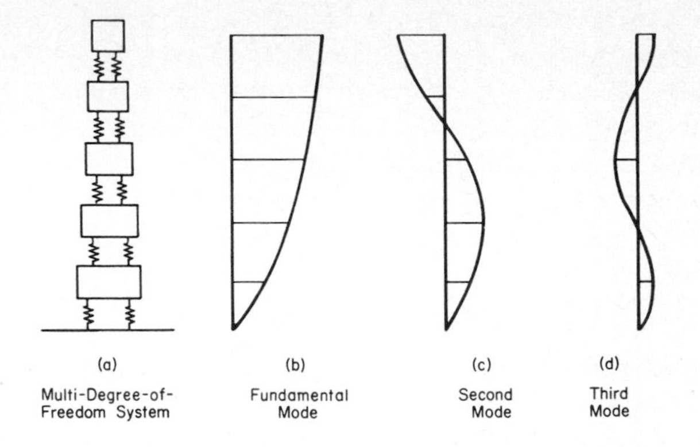

Fig. 16.13. Modes of vibration of shear beam.

placement bounds D', V', and A', for the multidegree-of-freedom system, which are drawn in such a way that they can be used instead of D, V, and A to give the response values desired for the multidegree-of-freedom system when the fundamental frequency is used to define the frequency for the response spectrum value for the multidegree-of-freedom system. These curves then involve the participation factors and modal responses for the various modes. The relationships are shown schematically in Fig. 16.14. Further exploration of these concepts is underway and is presented in somewhat more detail in Newmark *et al.* (December 1965).

For horizontal motions of the base of a structure founded on a firm foundation without rocking, the participation factors of the various modes can be selected by proper choice of the modal values, as indicated in Blume, Newmark, and Corning (1961), so as to make participation factors unity. When the modal values are so chosen, a particular response parameter at a particular point in the structure, α, has values for each of the modes designated by α_n (for the nth mode). The quantity α may be an acceleration of a particular mass, a strain at a particular point, a moment at a particular joint

Fig. 16.14. Tripartite logarithmic response spectrum plot for single- and multidegree-of-freedom systems.

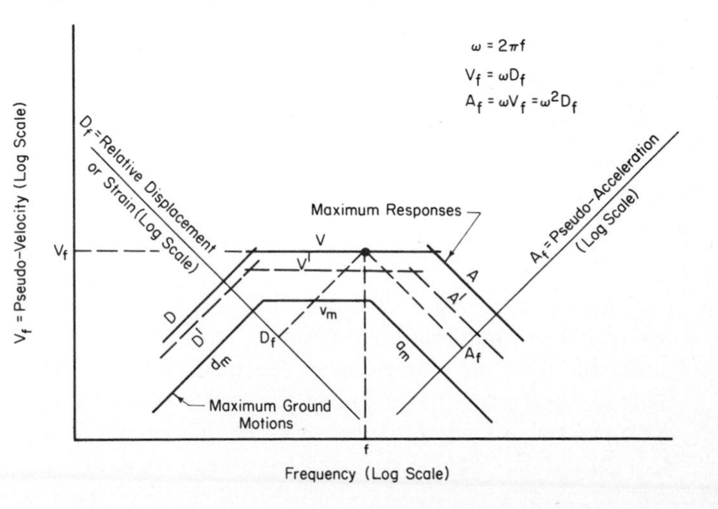

in a particular member, a shear in a particular story, displacement of a particular node or joint, etc. If for every frequency f_n of the structure there is defined by the response spectrum a relative displacement response value of D_n, then because the various modal maxima cannot occur simultaneously, an upper bound to the particular response quantity is given by the following relation:

$$\bar{\alpha} \leq \sum |\alpha_n D_n| \qquad (16.5)$$

This equation indicates that the actual response quantity $\bar{\alpha}$ is less than or equal to the sum of the absolute values of all of the modal response values, each of which is equal to the response value for that mode, α_n, multiplied by the spectrum displacement value for that mode, D_n, provided that the participation factor for the nth mode is unity.

Relations equivalent to Eq. 16.5 can be stated in terms of the other response spectrum parameters as follows:

$$\bar{\alpha} \leq \sum \left|\alpha_n \frac{V_n}{\omega_n}\right| \qquad (16.6)$$

$$\bar{\alpha} \leq \sum \left|\alpha_n \frac{A_n g}{\omega_n^2}\right| \qquad (16.7)$$

These relationships of course are applicable only to an elastic system.

The summations in Eqs. 16.5 through 16.7 give absolute upper bounds to the response quantities. It is shown in Goodman, Rosenblueth, and Newmark (1955), e.g., that the probable value of the response parameter α_p is equal approximately to the square root of the sum of the squares of the modal values, as indicated by Eq. 16.8.

$$\alpha_p \cong \sqrt{\sum (\alpha_n D_n)^2} \qquad (16.8)$$

Equations corresponding to 16.8 can be written involving V_n or $A_n g$, of course.

Along a line where D_n is constant, Eq. 16.5 assumes a simpler form. Similarly along lines where V_n or A_n are constant, Eqs. 16.6 and 16.7 assume also simpler forms. For example, if $D_n = D =$ constant, one has the result

$$\bar{\alpha} \leq D \sum |\alpha_n| \qquad (16.9)$$

If $\bar{\alpha}$ is set equal to D', then one has the result

$$\frac{D'}{D} \leq \sum |\alpha_n| \qquad (16.10)$$

Similarly one can compute V' and A' as follows for the cases where V_n or A_n is constant, respectively:

$$\frac{V'}{V} \leq \sum \left(\frac{\omega_1}{\omega_n}\right)|\alpha_n| = \sum \left(\frac{f_1}{f_n}\right)|\alpha_n| \qquad (16.11)$$

$$\frac{A'}{A} \leq \sum \left(\frac{\omega_1^2}{\omega_n^2}\right)|\alpha_n| = \sum \left(\frac{f_1}{f_n}\right)^2|\alpha_n| \qquad (16.12)$$

Equations 16.10, 16.11, and 16.12 give a procedure by which the multidegree-of-freedom spectrum can be plotted from the single-degree spectrum values. A different spectrum will be drawn for each response quantity desired of the multidegree-of-freedom system.

A number of comparisons have been made for the exact responses of multidegree-of-freedom systems for various ground motions, compared with the computations of upper bounds from equations such as 16.5, 16.6, and 16.7 or of probable values from equations such as 16.8.

Such calculations are reported by Newmark *et al.* (December 1965) and Jennings and Newmark (1960), e.g,. Jennings and Newmark (1960) have shown that for systems with a small number of degrees-of-freedom, say 4 or less, the true response for an earthquake motion is very nearly equal to but slightly less than the sum of the absolute values of the modal maxima. For a large number of degrees-of-freedom, say 12 or more, the true response is very nearly equal to the square root of the sum of the squares of the modal maxima. For systems with an intermediate number of degrees-of-freedom, the true response is generally about midway between these two values. In Figs. 16.15 and 16.16 there are shown calculations of response for two different 5-degree-of-freedom systems; in Fig. 16.15 the system corresponds to uniform masses and springs; and in Fig. 16.16 the system has varying springs but uniform masses, in both cases subjected to a particular ground shock, as recorded in event Aardvark, a nuclear detonation. The responses are given for the relative story deflections in each of the five stories of the 5-degree-of-freedom shear beam. Both the sum of the absolute values of the modal maxima and the square root of the sum of the squares are plotted in the figure, which compares the true spring distortion (computed for the actual ground motion input by a step-by-step integration procedure) with the approximate spring distortion computed from the true response spectrum. A value of $s_a = 1.00s$ would correspond to the approximate value equal to the exact value. It is noted that in every case the sum of the absolute values of the modal maxima lies above this value, as it should, and the square root of the sum of the squares value lies below, although there is no reason why in some cases some of these values should not lie above the line $s_a = 1.00s$. The general rule described previously would indicate, for a 5-degree-of-freedom system, that one should use the average of the two values so computed. It appears that this is borne out by the results plotted in Figs. 16.15 and 16.16 also. However, errors less than about 40% in general and in most cases not much greater than about 20% would be involved if one were to use either of the two approximations for this particular case.

Similar calculations for other input motions and somewhat different structures lead to the same general conclusions, which appear to be valid except where the spacing of the frequencies is such that several modes having nearly the same frequency are involved.

It appears, therefore, that we can use the results of modal calculations to infer the responses for actual earthquakes by use of the response spectrum. This may introduce errors in some cases, but the errors are relatively small. One must keep in mind the fact that the

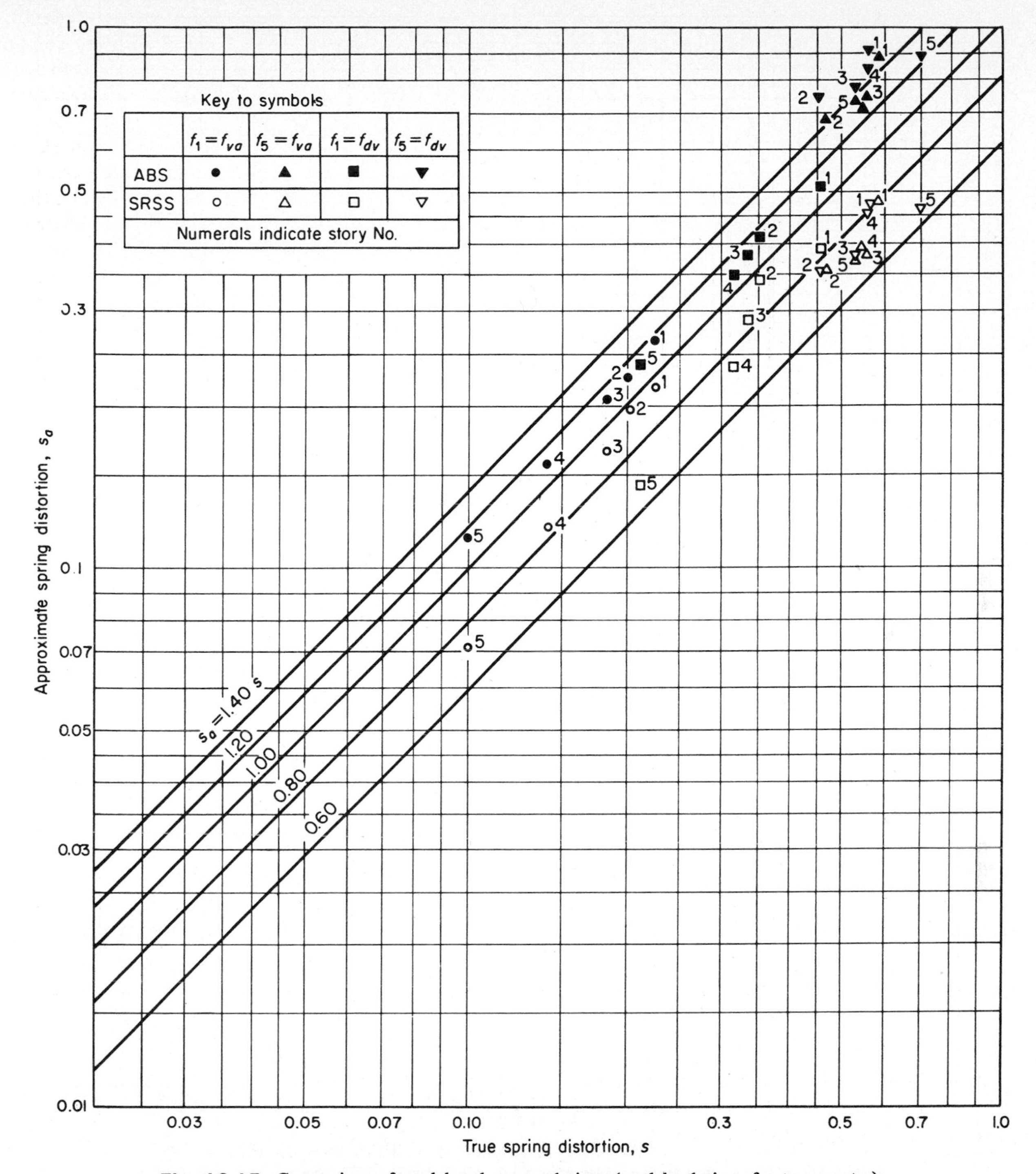

Fig. 16.15. Comparison of modal and exact solutions (modal solutions for true spectra); system 5A; ground shock from nuclear detonation, event Aardvark.

average of the responses for a large number of earthquakes would be more nearly consistent with the results arising from the use of a smooth spectrum value and would be more nearly representative of probable values than the values for a particular single earthquake.

Calculations have been made for a number of cases of inelastic multidegree-of-freedom systems to give their responses to particular earthquakes. Such calculations are reported by Newmark *et al.* (December 1965) for a 2-degree-of-freedom system and by Penzien (1960), Clough, Benuska, and Wilson (1965), and Heer (1965) for a 5-degree system. Although generalizations cannot be readily made because of the sparseness of the data, it does appear that the generalizations made herein, and outlined in Fig. 16.12, are applicable as bounds for a multidegree-of-freedom system, in general, provided that the response spectra for the multidegree-of-freedom system is determined as indicated in Fig. 16.14.

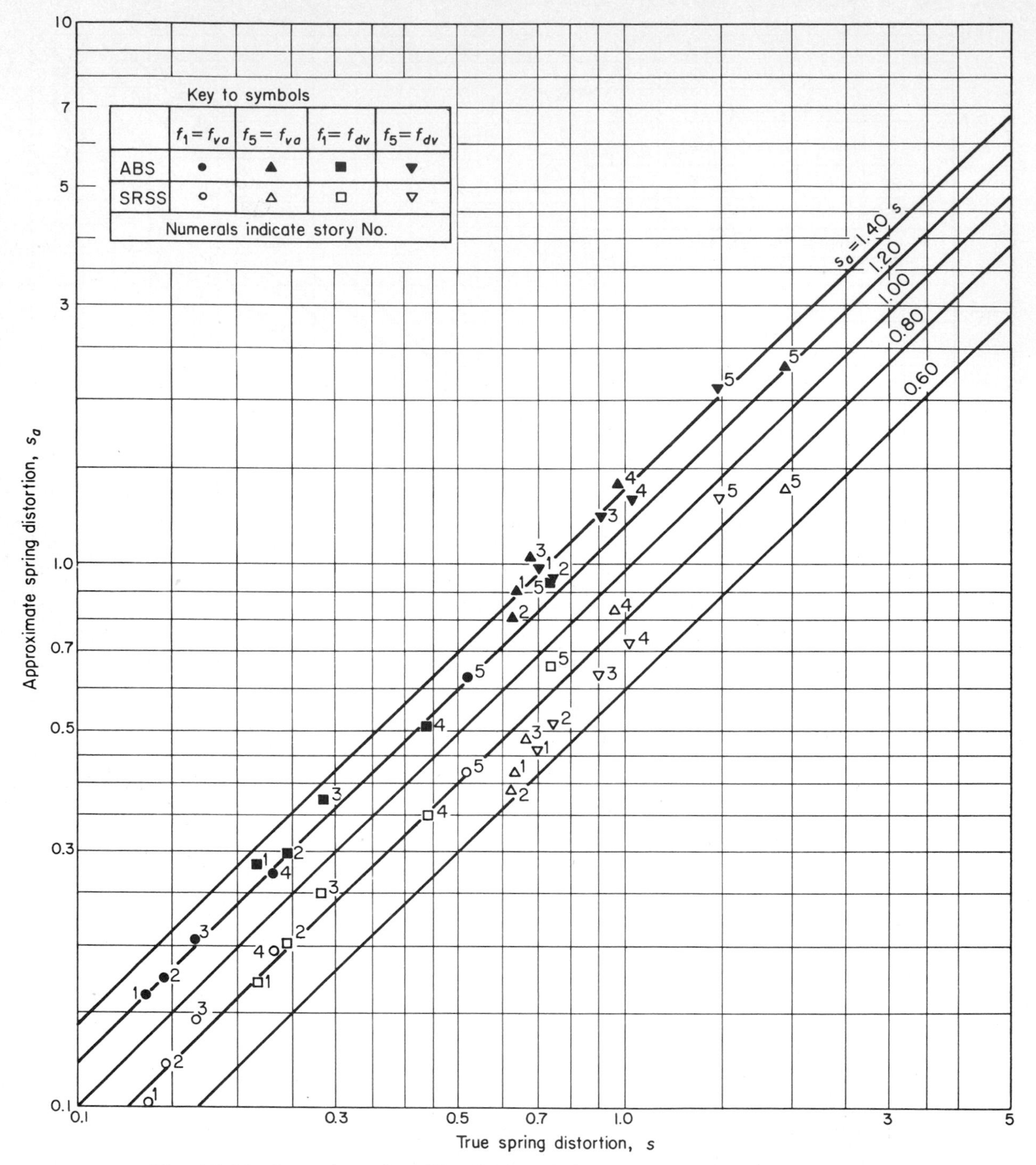

Fig. 16.16. Comparison of modal and exact solutions (modal solutions for true spectra); system 5D; ground shock from nuclear detonation, event Aardvark.

16.5 RESULTS OF ELASTIC ANALYSES FOR TALL BUILDINGS

Shears and overturning moments in a number of tall buildings have been computed and are reported in various references. A particularly interesting comparison has been shown by Bustamante (1965). In using the data from Bustamante, however, one should keep in mind the fact that the response spectrum used was one which corresponded to a maximum velocity response of about 1.2 in./sec or roughly about one-sixteenth the El Centro response spectrum. The calculations when so interpreted indicate that the Uniform Building Code values for base shear are from one-half to one-sixth those computed by the more exact analysis and that the shears at higher elevations of the building generally increase relative to those given by the Uniform Building Code. Overturning moments are also somewhat higher than those given by

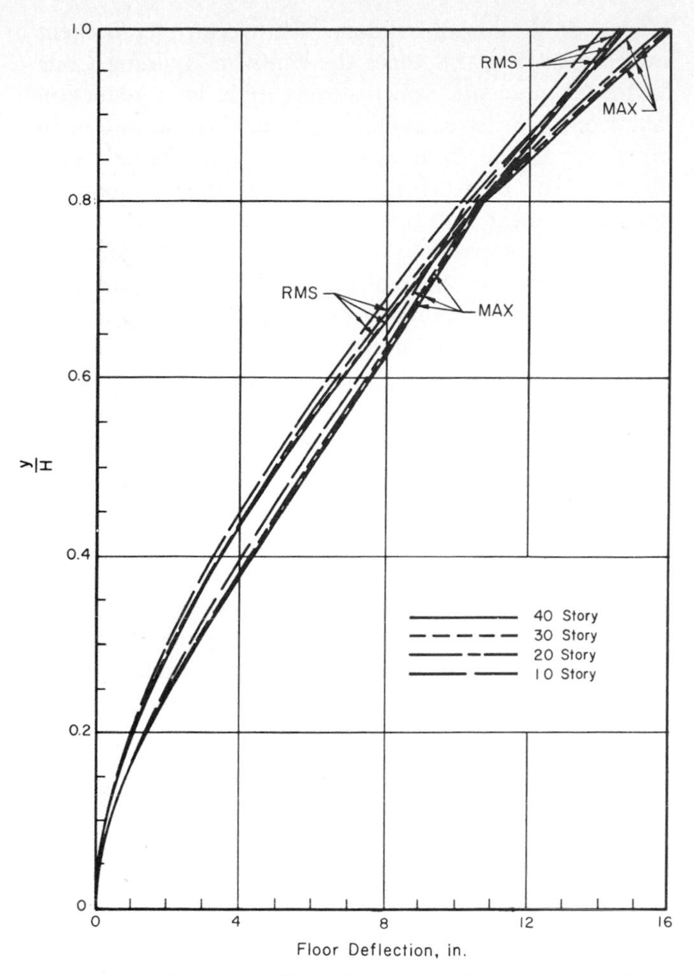

Fig. 16.17. Maximum floor deflections: 40-, 30-, 20-, and 10-story flexural buildings; $T_0 = 3$ sec., $D = 10$ in., $V = 20$ in. per sec., $A = 0.667\,g$.

the Code, but the upper stories generally have a more conservative moment relative to the base value.

A new series of calculations is reported herein. These calculations were made by S. J. Fenves, using the high-speed digital computer in the Department of Computer Science at the University of Illinois. Calculations were made for a series of buildings having 40, 30, 20, and 10 stories, with parameters chosen so that the fundamental periods of vibration for all of these were either 3 or 1 sec. The response spectrum used was one having a displacement bound of 10 in., a velocity bound of 20 in./sec, and an acceleration bound of 0.667 *g*. This is very nearly the response spectrum for the El Centro earthquake for elastic conditions. Two types of buildings were considered: a flexural building, corresponding to a shear wall structure of uniform properties over the height, and a "shear beam" building, corresponding to a frame structure. Deflections of the flexural building are shown in Fig. 16.17. Values are given for the square root of the sum of the squares, designated by RMS on the figure, and for the sum of the absolute values of the modal maxima, designated by MAX.

The maximum story shears are shown in Fig. 16.18. The Uniform Building Code values, for a coefficient $K = 1$ but multiplied by a factor of 1.5 to account for the difference between working stress and yield point, are shown on the figure for comparison. It is seen that the base shear computed for these buildings is about 3 to 3.5 times the Uniform Building Code value. Near the top of the building the values are from 5 to 6.5 times the Uniform Building Code values. These comparisons are for the RMS values. The MAX values are considerably higher but are not considered reasonable to use for the multistory building. This is an indication that for the building described and for the El Centro earthquake a ductility factor of the order of 3 is required at the base and about 5 near the top of the building, in order that the Uniform Building Code lead to a design which is adequate. Unless these ductility factors are provided for by the details of construction and inspection, an earthquake of intensity equal to the El Centro earthquake would produce serious consequences in the building considered.

Fig. 16.18. Maximum story shears: 40-, 30-, 20-, and 10-story flexural buildings; $T_0 = 3$ sec., $D = 10$ in., $V = 20$ in. per sec., $A = 0.667\,g$.

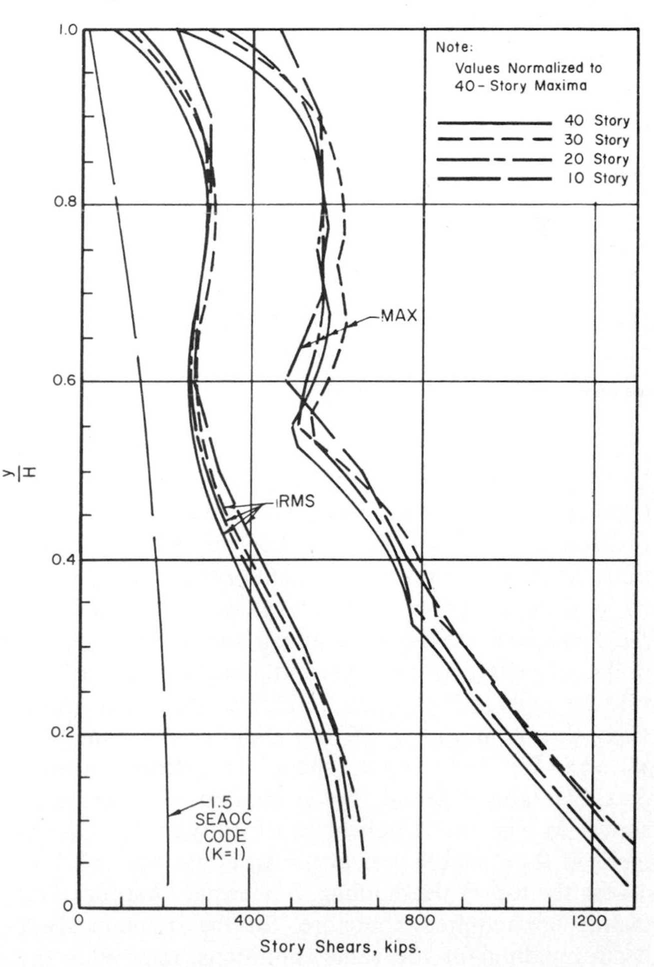

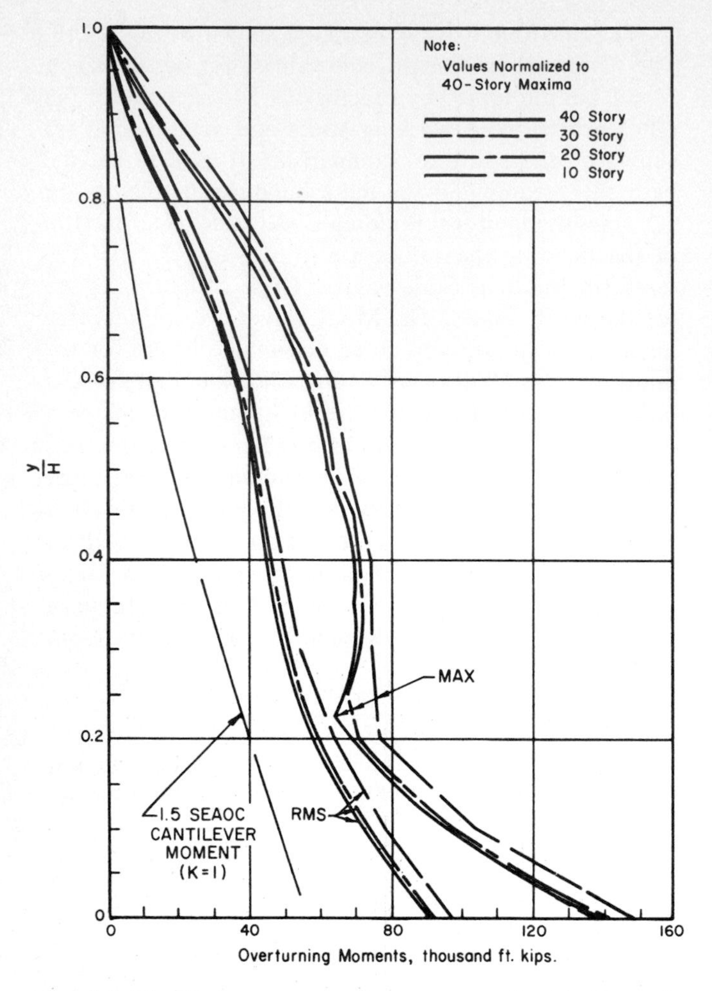

Fig. 16.19. Maximum overturning moments: 40-, 30-, 20-, and 10-story flexural buildings; $T_0 = 3$ sec, $D = 10$ in., $V = 20$ in. per sec, $A = 0.667\,g$.

A similar comparison is made for overturning moments in Fig. 16.19. Here the comparisons are made against the cantilever moments rather than the design overturning moments by the Uniform Building Code. Near the base of the building the RMS values are 1.6 to 1.8 times the Uniform Building Code values for overturning moments. These are about one-half the values, relatively, of the base shears and correspond to a reduction factor for overturning moment of the order of about 50% of the cantilever moment values corresponding to the RMS base shears. For comparison, the shear beam building for the same heights and periods leads to results that are given in Figs. 16.20, 16.21, and 16.22. The deflections shown in Fig. 16.20 are slightly less than those shown in Fig. 16.17, but the shape of the deflection curve is considerably different. The maximum story shears, as shown in Fig. 16.21, indicate a ductility factor requirement of 2 or slightly less at the base and about 2.5 to 4 near the top of the building. Somewhat lower ductility factors are required, therefore, for the frame or shear beam building for the same conditions, relative to the flexural or shear wall building. The overturning moment values are 1.7 to 1.8 times the Uniform Building Code cantilever moments, which corresponds to a reduction factor of the order of about 0.85 to 0.9 instead of 0.5. In other words, the shear beam building or frame has a much greater overturning moment, relatively, than the flexural or shear wall building.

Further comparisons are given in Figs. 16.23, 16.24, and 16.25 for the shear beam building for a period of 1.0 sec for the same number of stories. This is, of course, an unreasonable type of structure even to consider for 40 stories, but not for 10 stories. However, it is of interest to note that the ductility factors required for shear for these buildings range from about 3.7 to 4 at the base and are about the same at the top, although for the 10-story building they do go up to slightly larger values. The reduction factor for overturning moment is about 0.95 to 1.0, however, or in other words the overturning moment is nearly equal to the cantilever moment.

Based on these results and other analytical studies, it is concluded that a reasonably conservative design basis for a building involves a response spectrum approach,

Fig. 16.20. Maximum floor deflections: 40-, 30-, 20-, and 10-story shear buildings; $T_0 = 3$ sec, $D = 10$ in., $V = 20$ in. per sec, $A = 0.667\,g$.

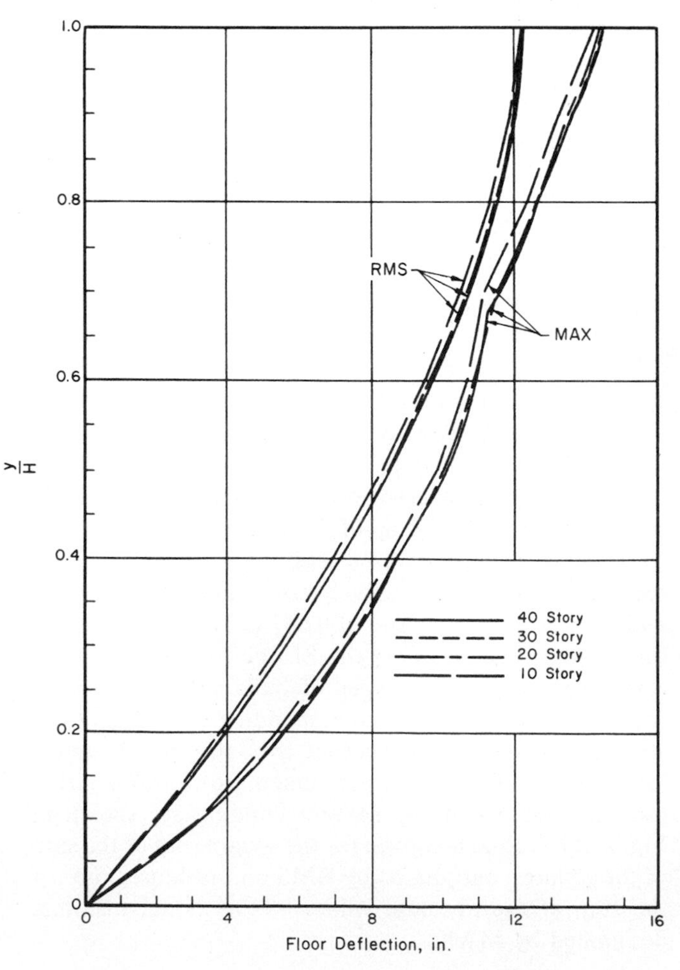

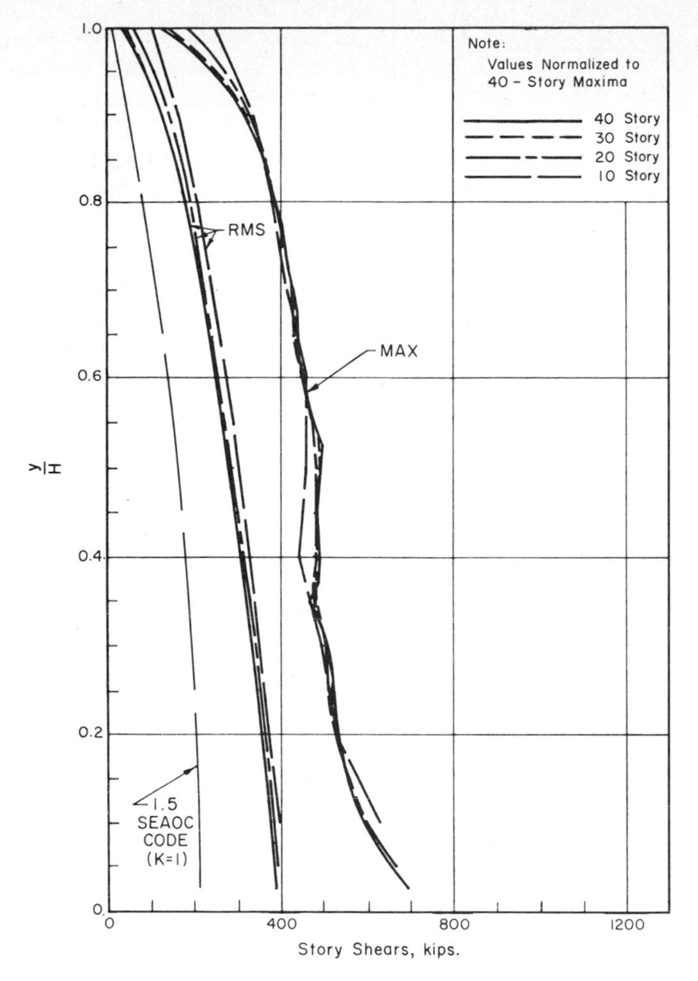

Fig. 16.21. Maximum story shears: 40-, 30-, 20-, and 10-story shear buildings; $T_0 = 3$ sec, $D = 10$ in., $V = 20$ in. per sec, $A = 0.667\,g$.

but with the use of a reduced ground motion, corresponding to a selected value of ductility factor that can be mobilized by the method of construction chosen. The method of selecting the response spectrum to use in such an analysis is indicated in Fig. 16.26. The trapezoidal set of lines designated by the legend "ground motion" corresponds to the maximum values of ground displacement, velocity, and acceleration. The elastic spectrum, designated by the symbol $\mu = 1$, for displacement and acceleration (D and A) represents slightly amplified values, corresponding to an elastic response spectrum for the ground motion considered. The curve marked D for $\mu = 5$ is the displacement spectrum for a ductility factor of 5, and the curve marked A for $\mu = 5$ is the acceleration or force spectrum for the same conditions. These are drawn so as to conserve displacement on the left-hand side, force on the right-hand side, and energy in the central part. An elastic analysis made for the reduced acceleration spectrum therefore would correspond to the ductility values derived for the conditions described.

The relations between the various bounding lines in Fig. 16.26, for an elasto-plastic resistance function, are

Table 16.1 Ratios of elasto-plastic to elastic response spectrum values in various ranges

Quantity conserved	Elasto-plastic relative to elastic response: Total displacement	Acceleration
Displacement	1	$\dfrac{1}{\mu}$
Energy or velocity	$\dfrac{\mu}{\sqrt{2\mu - 1}}$	$\dfrac{1}{\sqrt{2\mu - 1}}$
Force or acceleration	μ	1

relatively simple to compute. A summary of the values so computed is given in Table 16.1. The table lists, for each of the quantities that can be conserved (displacement, energy or velocity, and force or acceleration), the ratio between the elasto-plastic and the purely elastic response values for total displacement or for acceleration as a function of the ductility factor μ. For example, when $\mu = 5$, the tabulated values indicated are as follows: Along a constant displacement line, the displacement is

Fig. 16.22. Maximum overturning moments: 40-, 30-, 20-, and 10-story shear buildings; $T_0 = 3$ sec, $D = 10$ in., $V = 20$ in. per sec, $A = 0.667\,g$.

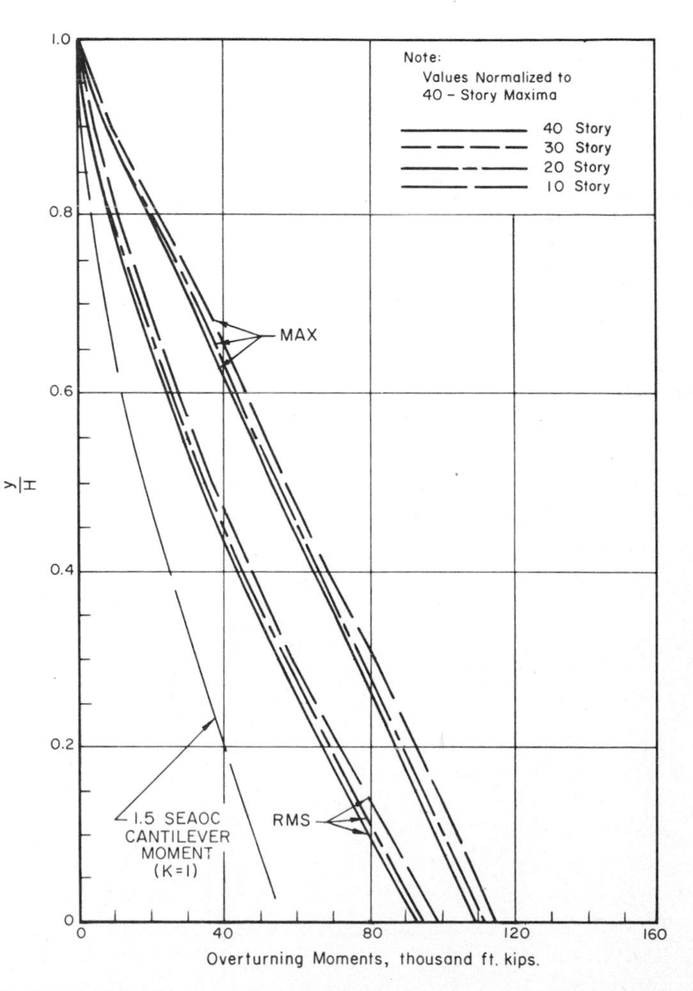

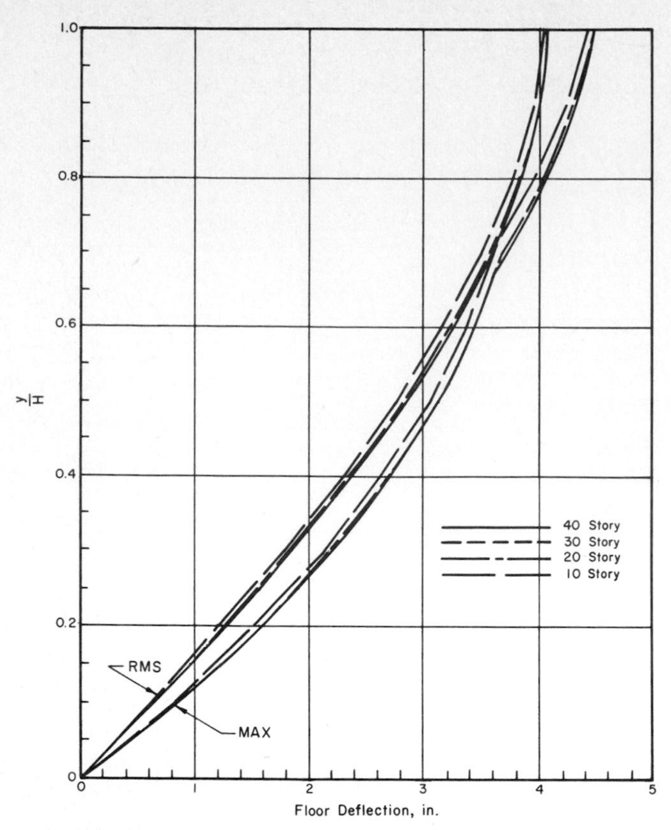

Fig. 16.23. Maximum floor deflections: 40-, 30-, 20-, and 10-story shear buildings; $T_0 = 1$ sec, $D = 10$ in., $V = 20$ in. per sec, $A = 0.667\,g$.

Fig. 16.24. Maximum story shears: 40-, 30-, 20-, and 10-story shear buildings; $T_0 = 1$ sec, $D = 10$ in., $V = 20$ in. per sec, $A = 0.667\,g$.

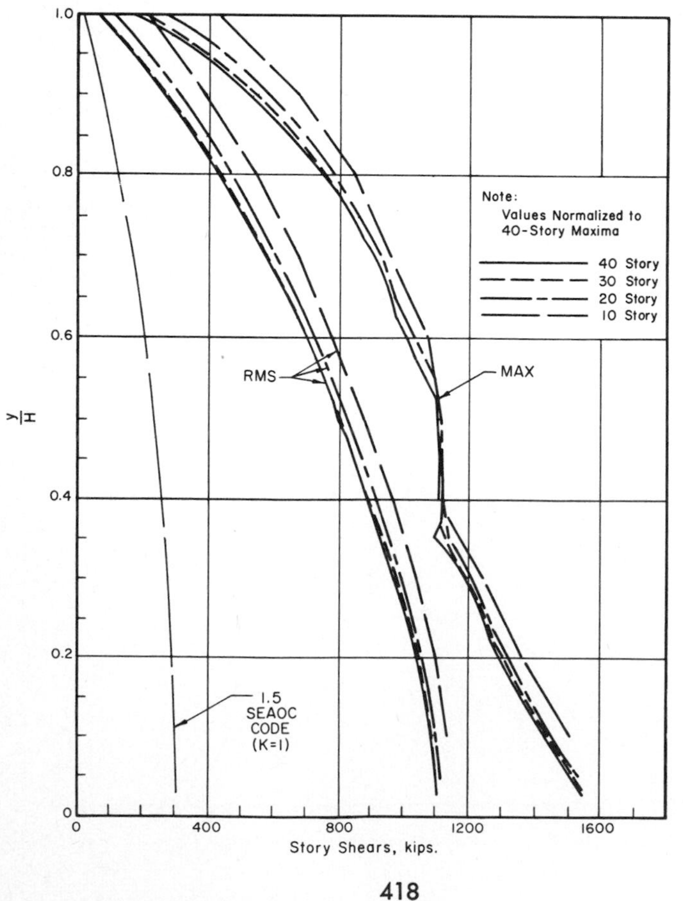

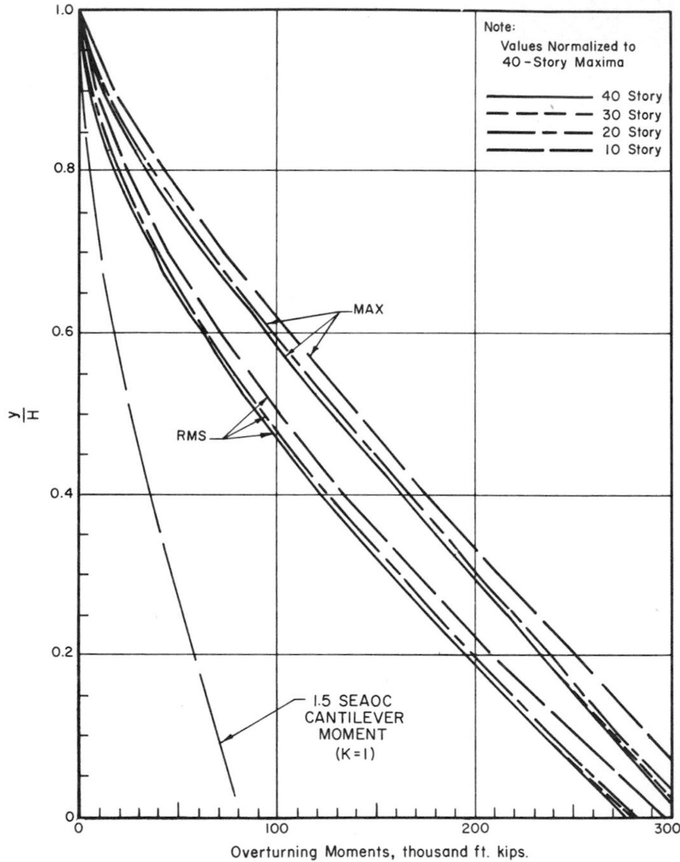

Fig. 16.25. Maximum overturning moments: 40-, 30-, 20-, and 10-story shear buildings; $T_0 = 1$ sec, $D = 10$ in., $V = 20$ in. per sec, $A = 0.667\,g$.

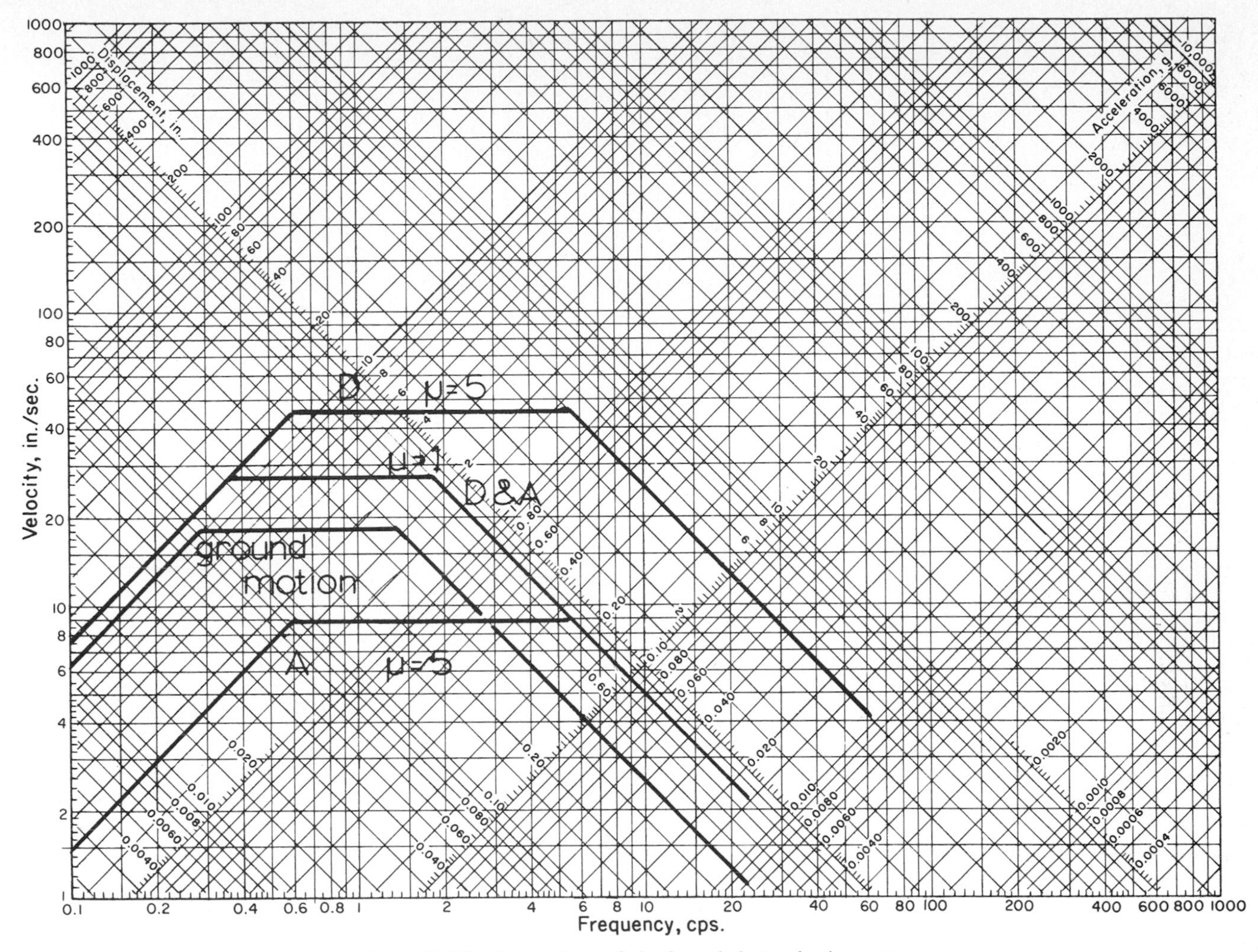

Fig. 16.26. Comparison of elastic and elasto-plastic spectra.

the same and the acceleration is one-fifth as much for the elasto-plastic spectrum as for the elastic spectrum. Along a constant velocity line, the displacement is five-thirds as great and the acceleration one-third as great for the elasto-plastic spectrum compared with the elastic spectrum. Finally, along a line of constant acceleration, the displacement is five times as great and the acceleration value is the same as the value for elastic response.

16.6 DESIGN OF COMPOSITE 41-STORY BUILDING

An example of the use of these concepts in the design of an actual building is illustrated next. The particular building chosen was designed originally as a shear wall structure. However, in reviewing the design, I felt that it was desirable to add more flexibility in the lower stories and recommended that these be made of steel frame construction so as to permit greater flexibility and energy-absorbing capacity there. The lower six stories therefore were made of steel-frame construction and the upper part was to be designed to be either a concrete shear wall or a braced steel frame. The deflection for the composite building, for a Zone 2 earthquake-response spectrum, is shown in Fig. 16.27. This is on the upper bound of Zone 2 and is about three-quarters of the El Centro earthquake response spectrum. The building has a period of about 2.3 sec.

The maximum story shears are shown in Fig. 16.28. For the response spectrum selected, the ductility factor required is about 2 or slightly less. Indicative of this is the comparison shown in Fig. 16.29, where the spectrum is one which corresponds to a ductility factor of about 2 relative to that used in Fig. 16.28. Here the design values are slightly less than those corresponding to the Uniform Building Code, when the latter is increased by the ratio of yield point to working stress values. It is concluded that this building will be adequate for a ductility factor of 2 for the exposure considered, namely, about three-quarters of El Centro, and will have adequate resistance

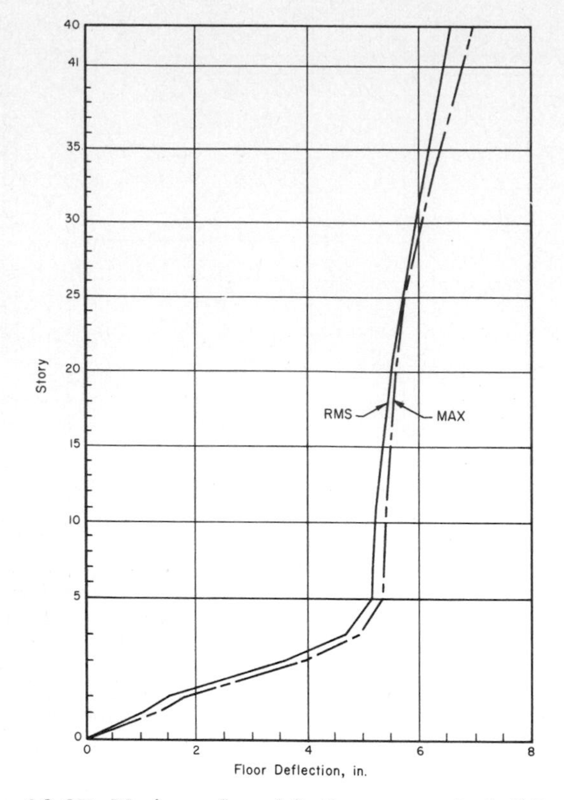

Fig. 16.27. Maximum floor deflections: composite building; $T_0 = 2.324$ sec, $D = 6$ in., $V = 15$ in. per sec, $A = 0.5\,g$.

Fig. 16.28. Maximum story shears: composite building; $T_0 = 2.324$ sec, $D = 6$ in., $V = 15$ in. per sec, $A = 0.5\,g$.

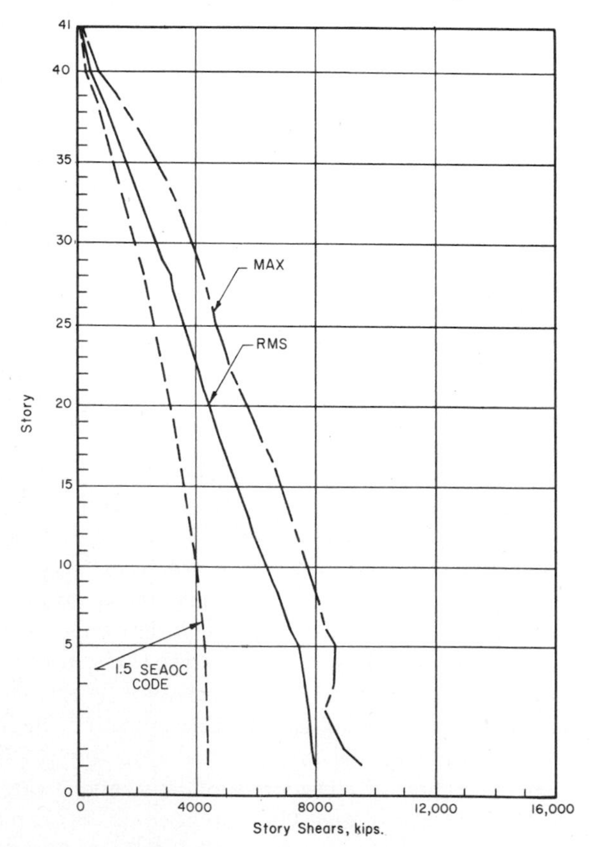

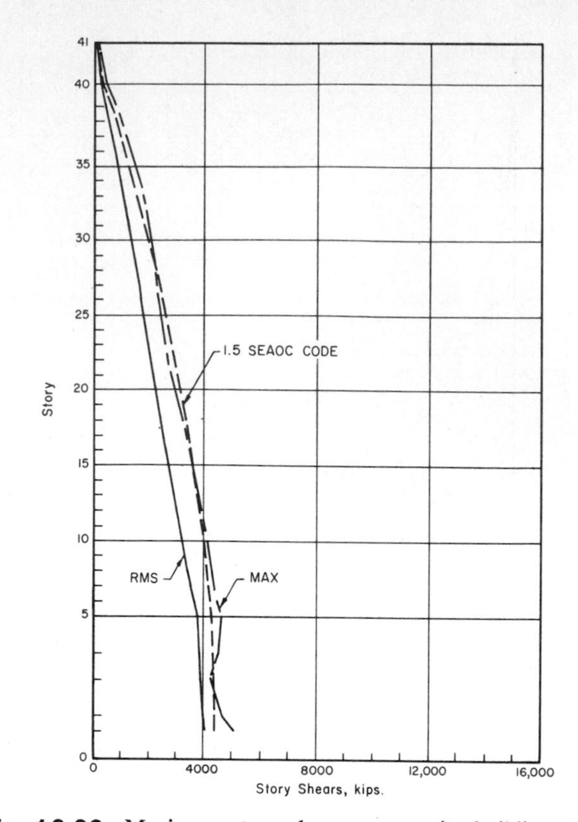

Fig. 16.29. Maximum story shears: composite building; $T_0 = 2.324$ sec, $D = 3$ in., $V = 7.5$ in. per sec, $A = 0.375\,g$.

Fig. 16.30. Maximum overturning moments: composite building; $T_0 = 2.324$ sec, $D = 6$ in., $V = 15$ in. per sec, $A = 0.5\,g$.

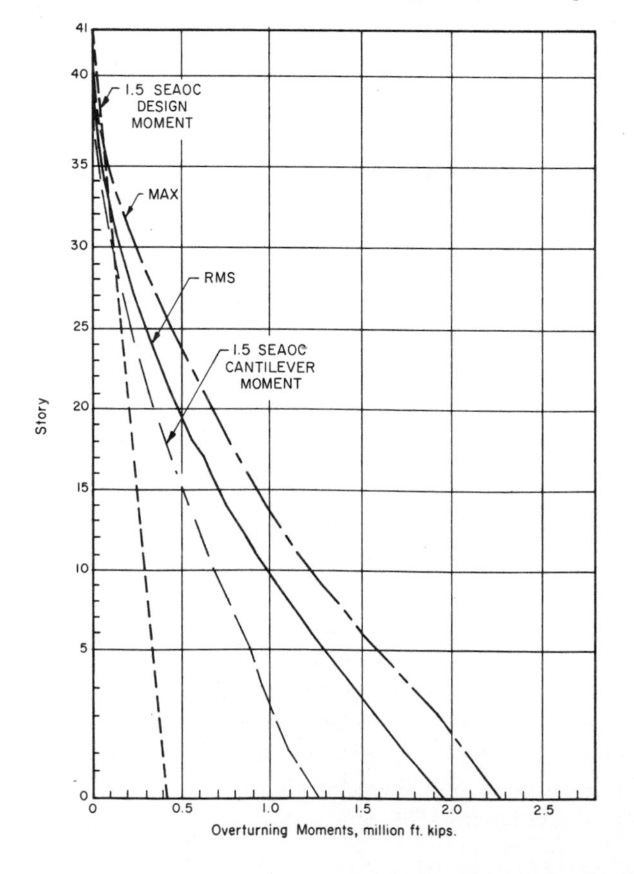

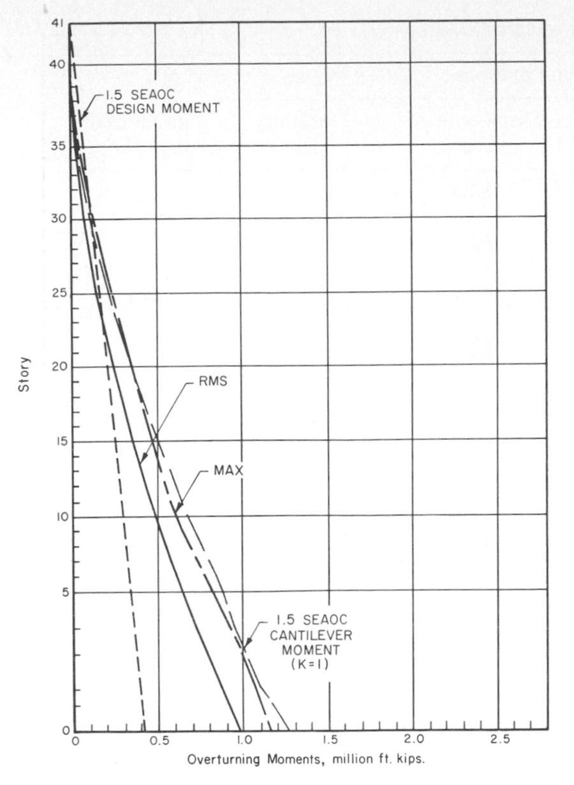

Fig. 16.31. Maximum overturning moments: composite building; $T_0 = 2.324$ sec, $D = 3$ in., $V = 7.5$ in. per sec, $A = 0.375\,g$.

to cope with a stronger earthquake without danger of collapse. The purpose of selecting a relatively conservative ductility factor of 2 in this case was to avoid expensive repairs for the expected earthquake hazard.

The overturning moments for the same building are shown in Fig. 16.30. The overturning moment is considerably higher than that given by the Uniform Building Code, and the design was made so as to be consistent with the computed values rather than with the Code values. The overturning moment corresponding to the reduced spectrum is shown in Fig. 16.31. It appears that the reduction in cantilever moment corresponding to the Uniform Building Code value for this building is almost twice as great as from the analysis.

16.7 SPECIAL CONSIDERATIONS

In regions where unusual types of ground motions can be expected because of oscillations of the soil over deeply buried rock, modifications to the response spectrum must be considered. This is particularly essential in places like Mexico City, where amplification of ground motions in the range of periods from about 2 to 2.5 sec occurs because of the natural frequency of the bowl of soft soil on which most of Mexico City is founded. An example of a building designed for the special conditions in Mexico City has been given by Zeevaert and Newmark (1956). This building, the Latino Americana Tower, was designed for a base shear of the order of 500 metric tons, corresponding to an earthquake of Modified Mercalli Intensity VIII but taking into account the amplification of motions corresponding to the natural period of vibration of the soil on which the building rests. During the construction the building was modified to provide for a heavy television tower at the top. Shortly after the construction was completed, a major earthquake occurred, corresponding to the intensity for which the building was designed. Recording instruments had been installed in the building, on which records were obtained during the earthquake. The values recorded were almost precisely those that had been considered in the design as being consistent with the probable values corresponding to Eq. 16.8 herein.

In regions where substantial vertical earthquake accelerations occur, or in some cases for unusual types of construction that may have a different resistance to motion in one direction than in another, a peculiar phenomenon similar to "pumping" may be encountered. This is best illustrated by an analysis described by Newmark (1965). Consider, for example, a mass sliding under a constant force, having a frictional resistance against the sliding surface. The friction coefficient may be characterized by the coefficient N. If one considers the effect of the gravity action on the sliding mass, the force required to produce downhill sliding is less than that required to produce uphill sliding. Two extreme cases of rigid-plastic resistance, corresponding to the condition shown in Fig. 16.32, are considered in the calculations described in the following. These involve a symmetrical resistance corresponding to a mass sliding horizontally, with the same resistance in either direction, and an unsymmetrical resistance, corresponding to a frictional resistance against sliding in one direction and an infinite resistance against sliding in the other direction.

Calculations were made for these two conditions for the four earthquakes described in Table 16.2. For these four earthquakes the values of acceleration and time scale were modified so as to give a maximum acceleration for all of the earthquakes of 0.5 g and a maximum ground

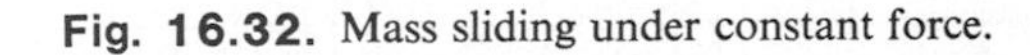
Fig. 16.32. Mass sliding under constant force.

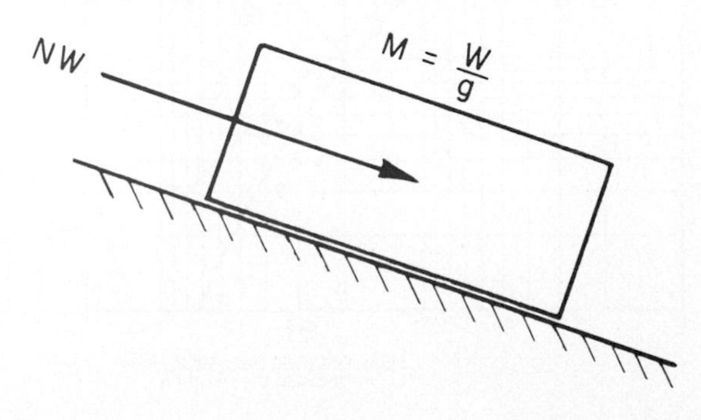

Table 16.2. EARTHQUAKES CONSIDERED IN ANALYSIS

Earthquake	Maximum ground motions				
	Acceleration, g	Velocity, in./sec	Displacement, in.	Duration, sec	Normalized* displacement, in.
Ferndale 12/21/54, N45E	0.205	10.5	8.26	20	27.7
Eureka 12/21/54, S11W	0.178	12.5	10.0	26	51.2
Olympia 4/13/49, S40W	0.210	8.28	9.29	26	20.5
El Centro 5/18/40, N–S	0.32	13.7	8.28	30	25.5

*Normalized to give acceleration = 0.50 g and velocity = 30 in./sec.

velocity of 30 in./sec. The maximum displacements for the four earthquakes range from 20.5 to 51.2 in. The purpose of normalizing the earthquakes was to obtain a more consistent set of data for strong earthquake conditions.

The case of symmetrical resistances is shown in Fig. 16.33, where the maximum displacement of the sliding mass relative to its support is plotted against the ratio of the resistance coefficient to the maximum acceleration of 0.5 g. The envelope of the plotted points corresponds to a condition in which the maximum energy of the mass, corresponding to the quantity $\frac{1}{2}mV^2$, is absorbed by the resistance multiplied by the distance over which sliding occurs. A slight correction to this energy, to account for the fact that sliding does not occur unless the acceleration exceeds the resistance, is indicated by the lower curve. In no case does the maximum displacement exceed the maximum ground displacement for the condition of symmetrical resistance.

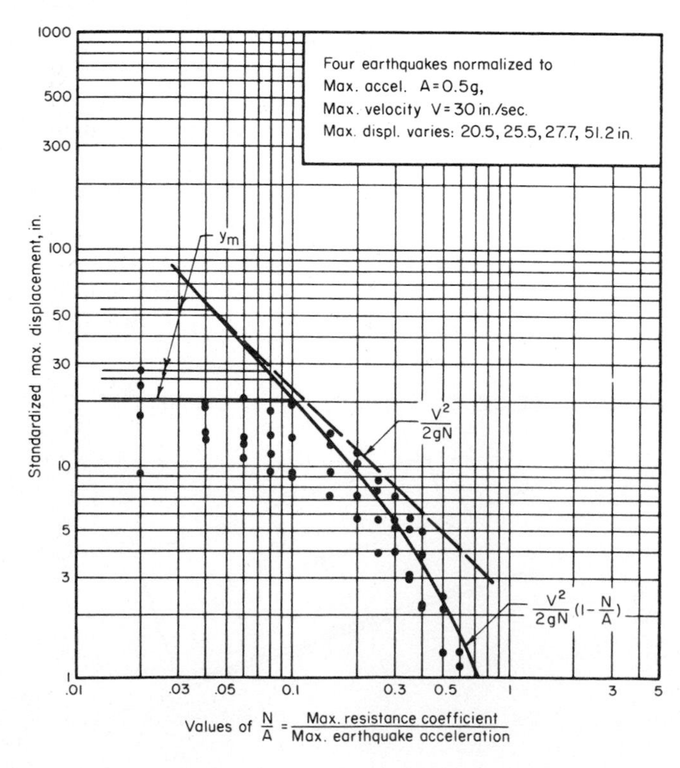

Fig. 16.33. Standardized displacement for normalized earthquakes (symmetrical resistance).

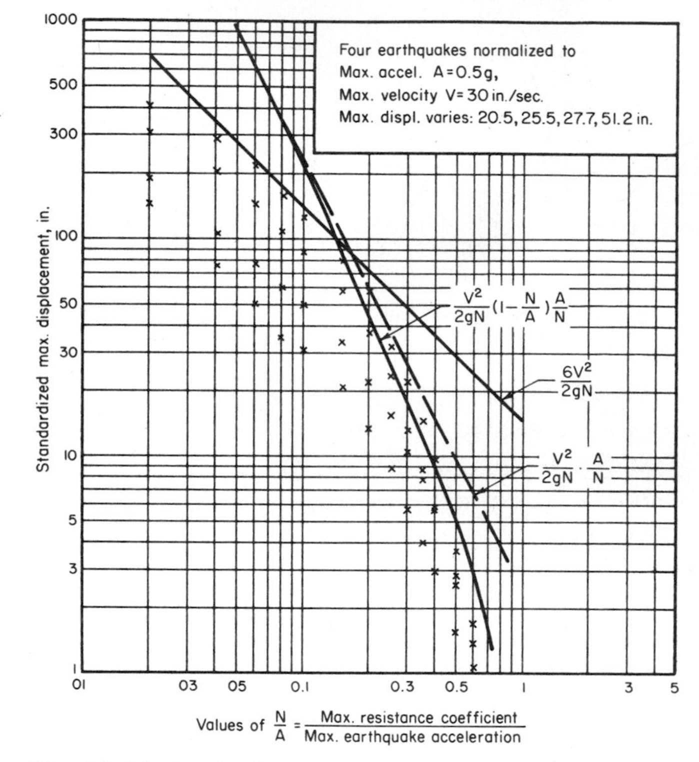

Fig. 16.34. Standardized displacement for normalized earthquakes (unsymmetrical resistance).

Figure 16.34 summarizes the displacement for unsymmetrical resistance. Here the displacements are almost 6 times as great as those in Fig. 16.33 for the maximum condition. In other words, the earthquake considered corresponds to a condition of something like 6 pulses for unsymmetrical resistance contrasted with only 1 effective pulse for symmetrical resistance. The factor 6 is no doubt dependent upon the duration of the earthquake motions. It is probably proportional to the square

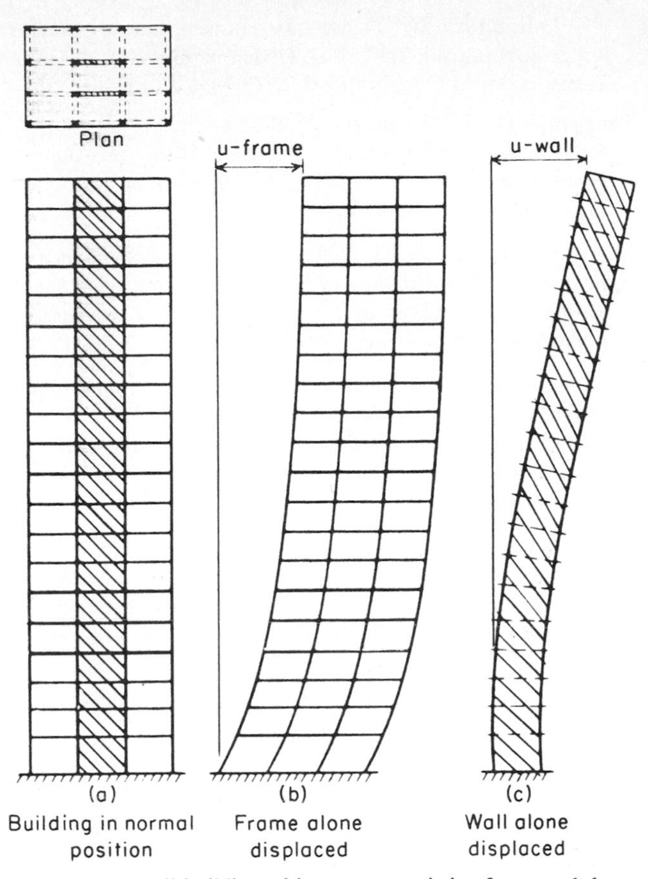

Fig. 16.35. Tall building with moment-resisting frame and shear walls in center interior bay.

root of the total duration. This factor is consistent with a duration of about 30 to 40 secs, corresponding to the duration of the earthquakes for which the calculations were made.

Similar calculations were made with elasto-plastic resistances having different yield points in the two directions. As soon as the difference in yield point was more than just nominal in value, the "pumping" appeared to be almost as great as for the rigid plastic resistance reported in Figs. 16.33 and 16.34. Hence, it is concluded that substantial increases in displacement can occur under conditions where yielding occurs for vertical motions or where the conditions of yielding or of coupling with adjacent structures introduce a substantially greater resistance to deflection in one direction than in another. Care must be taken to avoid such conditions or to provide adequately for them.

In buildings which have a combination of a frame and a shear wall to resist earthquakes, consideration must be given to the interaction of the different types of construction and their different ductilities and flexibilities. The difference in pattern of displacement of a frame and a shear wall is shown in Fig. 16.35. When a building contains these two elements, the partition of shear between them must be such as to produce equal deflections of the two elements. Because of the difference in the shape of curves, it appears that the shear wall will take more than the total shear near the base but will be restrained relatively in the opposite direction by the frame in the upper part of the structure. Provision for the inter-reaction between the two elements must be made if the structure is to behave properly. Moreover, consideration of the change in configuration and energy-absorbing capacity, if the shear wall fails during the course of the deflection of the structure, must be taken into account in assessing the overall behavior of the composite structure.

16.8 CONCLUDING REMARKS

For earthquakes of the intensities experienced in regions subjected to strong earthquakes, energy-absorbing capacity and ductility are essential to permit deformations to occur beyond the range of linear behavior or the range of ordinary working stresses. It is possible to design modern tall buildings to resist earthquakes with an adequate margin of safety. The margin that can be achieved is a function of the price one is willing to pay as a sort of insurance against normally expected earthquake intensities and the degree of damage one is willing to permit in an extraordinarily severe earthquake.

The general philosophy is proposed that a building should suffer little, if any, damage in the intensity of earthquake that might normally be expected several times during its life in order to avoid expensive repairs. However, the building should have an adequate reserve capacity against collapse should an extreme earthquake occur at any time.

REFERENCES

Blume, J. A., N. M. Newmark, and L. H. Corning (1961). *Design of Multi-Story Reinforced Concrete Buildings for Earthquake Motions*, Chicago, Illinois: Portland Cement Association.

Bustamente, J. I. (1965). *Seismic Shears and Overturning Moments in Buildings*, *Proceedings of the Third World Conference on Earthquake Engineering*, New Zealand.

Clough, R. W., K. L. Benuska, and E. L. Wilson (1965). *Inelastic Earthquake Response of Tall Buildings*, *Proceedings of the Third World Conference on Earthquake Engineering*, New Zealand.

Goodman, L. E., E. Rosenblueth, and N. M. Newmark (1955). "Aseismic Design of Firmly Founded Elastic Structures," *Trans. ASCE*, **120**, 782–802.

Heer, J., *Response of Inelastic Systems to Ground Shock* (1965). Ph.D. Dissertation, University of Illinois.

Jennings, R. L., and N. M. Newmark (1960). *Elastic Response of Multi-Story Shear-Beam-Type Structures Subjected to Strong Ground Motion, Proceedings of the Second World Conference on Earthquake Engineering*, Vol. II, Tokyo.

Newmark, N. M. (1965). "Effects of Earthquakes on Dams and Embankments," Fifth Rankine Lecture, *Geotechnique*, Institution of Civil Engineers, London.

Newmark, N. M., and A. S. Veletsos (June 1964). *Design Procedures for Shock Isolation Systems of Underground Protective Structures, Vol. III, Response Spectra of Single-Degree-of-Freedom Elastic and Inelastic Systems*, Report for Air Force Weapons Laboratory by Newmark, Hansen, and Associates under subcontract to MRD Division, General American Transportation Corporation, RTD TDR 63-3096.

Newmark, N. M., W. H. Walker, A. S. Veletsos, and R. J. Mosborg (December 1965). *Design Procedures for Shock Isolation Systems of Underground Protective Structures, Vol. IV, Response Spectra of Two-Degree-of-Freedom Elastic and Inelastic Systems and Vol. V, Response Spectra of Multi-Degree-of-Freedom Systems*, Report for Air Force Weapons Laboratory by Newmark, Hansen, and Associates under subcontract to MRD Division, General American Transportation Corporation, RTD TDR 63-3096.

Penzien, J. (1960). *Elasto-Plastic Response of Idealized Multi-Story Structures Subjected to a Strong Motion Earthquake, Proceedings of the Second World Conference on Earthquake Engineering*, Vol. II, Tokyo.

Veletsos, A. S., and N. M. Newmark (1960). *Effect of Inelastic Behavior on the Response of Simple Systems to Earthquake Motions, Proceedings of the Second World Conference on Earthquake Engineering*, Vol. II, Tokyo.

Veletsos, A. S., N. M. Newmark, and C. V. Chelapati (1965). *Deformation Spectra for Elastic and Elasto-Plastic Systems Subjected to Ground Shock and Earthquake Motions, Proceedings of the Third World Conference on Earthquake Engineering*, New Zealand.

Zeevaert L., and N. M. Newmark (1956). *Aseismic Design of Latino Americana Tower in Mexico City, Proceedings of World Conference on Earthquake Engineering*, Berkeley, California.

Chapter 17

Design of Earthquake-Resistant Structures—Steel Frame Structures

HENRY J. DEGENKOLB

Structural Engineer and President
H. J. Degenkolb & Associates
Consulting Engineers
San Francisco, California

17.1 INTRODUCTION

Current building code forces specified for providing earthquake resistance in building structures are based principally on the observation of the performance of tall buildings that have been subjected to major earthquakes. This elementary approach is still necessary at the present time because the theoretical design factors obtained from the presently available ground motion measurements are much too high for the usual elastic methods of analysis. While great progress has been made on analyses based on elasto-plastic performance, these studies have not progressed to the point where they can be effectively used in the design office for the average building. It is also true, unfortunately, that there are absolutely no ground records available at the locations of major damage or maximum motion for the truly major earthquakes such as the 1906 San Francisco Earthquake. Thus it is necessary to study the buildings that have performed well in the past in

Table 17.1. Buildings in San Francisco with a complete steel frame supporting all wall and floor loads*

Building	Stories	Approx. area (sq ft)	Columns	Floors
Claus Spreckels	19	5,600	Z	Concrete
New Chronicle	16	5,600	Pl & A	Tile
Merchants Exchange	14	25,000	Pl & A	Concrete
St. Francis Hotel	14	22,000	Z	Concrete
Flood Building	12	40,000	Z	Tile
Hamilton Hotel	12	3,100	Channel	Concrete
Mutual Savings	12	5,000	Z	Concrete
Alexander Hotel	11	4,700	Channel	Concrete
Kohl	11	9,000	Channel	Concrete
Shreve	11	9,000	Z	Concrete

*Eleven others listed from 6 to 10 stories. "Damage to steel frames almost negligible." Cracked partitions, walls, tile, stonework. Shifting stone. Based on data from Galloway *et al.* (1907).

order to determine why they were satisfactory and from these studies to determine the requirements for future designs.

The primary experience with the performance of tall structures in major earthquakes has been in the San Francisco Earthquake of 1906. This earthquake has been estimated to be of magnitude $8\frac{1}{4}$ on the Richter Scale of magnitude. Downtown San Francisco is about 9 mi from the San Andreas fault where the rupture occurred. Tables 17.1 and 17.2 are taken from the ASCE Committee Report on the effects of that earthquake on buildings (Galloway, Couchot, Snyder, Derleth, and Wing, 1907). According to this committee report, the buildings in Table 17.1, those with complete steel frames, performed well. Note that this group of buildings includes one 19-story, one 16-story, and several 14-story buildings as well as lower structures.

The buildings in Table 17.2, which had masonry bearing walls with interior steel frames were reported by the committee to have suffered more damage by earthquake and fire than those in Table 17.1.

Table 17.2. Buildings in San Francisco with steel frames supporting floors but brick and stone bearing walls supporting themselves and some supporting outer bays of floors*

Building	Stories	Approx. area (sq ft)	Columns	Floors
Crocker	11	10,000	Phoenix	Tile
Mills	11	21,000	Z	Tile
Mutual Life	9	5,000	Z	Tile
Old Chronicle	9	—	Cast iron	Tile
Union Trust	9	8,400	Channel	Tile

*Seventeen others listed from 2 to 7 stories. Based on data from Galloway *et al.* (1907).

It is interesting to compare the magnitude of well-known earthquakes and the distance to high-rise construction:

San Francisco 1906	$8\frac{1}{4}$ magnitude	Downtown 9 mi from fault
Mexico City, July 1957	7.5 magnitude	Epicenter 170 mi from Mexico City, 60 mi from Acapulco
Alaska 1964	8.4 magnitude	Anchorage 80 mi from fault
El Centro 1940	7.1 magnitude	No tall buildings near

From this comparison, it can be seen that the 1906 San Francisco Earthquake has provided the most severe test of high-rise construction to date.

In Anchorage, Alaska, there were three 14-story buildings, all seriously damaged in the 1964 earthquake, and one 14-story building in Whittier, also damaged (Wood, 1967).

Because of the essentially satisfactory performance of the tall structures in San Franscisco in a nearby major earthquake, it becomes important to study the characteristics of those buildings to determine and evaluate building code requirements.

All of the taller buildings in downtown San Francisco in 1906 used structural steel frames; most of them used semirigid connections. It also is interesting to note that the least damaged of the 14-story buildings in Anchorage was a steel-framed structure with concrete shear walls.

In studying the performance of these steel structures, three basic structural elements are involved: (1) beam-column connections, (2) bending of structural shapes, and (3) column compression and bending. Other elements contributing to the structural performance were the floor slabs and the masonry walls. The floor slabs, acting as diaphragms to distribute the loads, were not damaged by the earthquake although they suffered materially in the fire. The masonry walls undoubtedly furnished a large portion of the support in some cases, but at critical locations, such as at the first floor of most buildings, they were omitted so that all or most of the structural resistance had to be taken by the steel frame.

17.2 BEAM-COLUMN CONNECTIONS

The first of the three major structural elements to be considered are the beam-column connections since these were the weakest elements of these older structures. The connections used were generally of the type that now would be classified as Series A web connections with top and bottom clip angles, generally $\frac{3}{8}$in. thick. Because of the past use and excellent performance of these connections, it is somewhat interesting and informative to see how they have performed under test. These con-

Table 17.3. MOMENT TESTS ON WEB SHEAR CONNECTIONS

		Number of rivets*				
Test	Member, in. I	Number of vertical rows	Web R	Rivets in outstanding legs	Ultimate moment, in. kips	Maximum recorded rotation
1	6	1	2	2	23.5	0.030
2	8	2	4	4	135	0.024
3	8	2	4	8	159.5	0.030
4	12	3	3	6	403	0.026
5	12	3	6	12	529	0.022
6	18	5	5	10	1225	0.010
7	18	5	10	20	1300	0.007

*All rivets 7/8 in.; all angles 3/8 in. There was no failure and no definite yield point. Angles deformed badly without visible distress in rivets or beams. There were several repetitions of load. $\theta = 0.030$ corresponds to 12 ft 6 in. high story drift of 4 1/2 in. Based on data from Rathbun (1936).

nections have great ductility but very little rigidity, and they are usually assumed to act as hinged connections. At large deformations they can have a surprising amount of moment strength. During earthquake loading conditions, it is obvious that these connections must be loaded in the plastic range. Unfortunately, most of the available tests give very little data in this region since they were performed to produce data on rigidity so that allowances for joint slip could be made in the various elastic analyses based on slope deflection. Occasionally they were carried to some load approaching maximum, but little data is given on deformations, since at that time the research workers were concerned with verifying an adequate factor of safety based on loads.

A series of connection tests was performed and described by J. C. Rathbun (1936) in the transactions of ASCE. Table 17.3 gives the results for web connections without clip angles while Table 17.4 gives the test results for clip-angle moment connections. Unfortunately, since the tests were run for studying the elastic properties of the joints, the deformation readings were not taken at high strains. From the rotation data given and the ultimate resisting moment, it can be assumed that very large rotations were achieved before the ultimate load, and this assumption is substantiated by the pictures included in the report. In order to indicate scale, a rotation of 0.007 would correspond to a story drift of 1 in. in a 12 ft 6 in. high story—the drift caused by rotations of the joint itself, not including the bending deformations in columns or girders. The ability of this type of connection to hold together and deform without collapse may be

Table 17.4. MOMENT TESTS ON TOP AND BOTTOM CLIP ANGLES*

Test	Tension rivets	Ultimate moment, in. kips	Maximum recorded θ at moment
8	4 (2 rows)	580	0.006 at 268
9	6 (2 rows)	665	0.010 at 430
10	8 (2 rows)	1055	0.014 at 560
11	4 (2 rows)	Not given	0.007 at 560
12	8 (2 rows)	Above 1026 did not fail	0.010 at 750

*All tests on 12 in. I beam, 7/8 in. rivets, and 3/8 in. angles. All tests had four rivets from angle to each flange. From pictures, large deformations were observed but not given. Based on data from Rathbun (1936).

Fig. 17.1. Bolted moment-resisting steel joint.

largely responsible for the reputation of steel frames to perform well in earthquakes. If this connection can be assumed to be "plastic" under vertical load and then to perform "elastically" under lateral loads, this type of connection using thicker angles usually can resist a lateral load of about 1% g in moderate height buildings of about 10 stories.

As the requirements of the building codes regarding lateral forces became more severe and as buildings in earthquake regions became taller, it was necessary to provide stronger moment connections than could be designed with top and bottom clip angles. The most commonly used connection uses the split tees top and bottom of the beam and has been used in most recent buildings designed for major earthquake forces except where all welded construction has been used. Figure 17.1 shows a representative connection of this type for a major steel-framed building in the San Francisco Bay area. This type of connection has great strength and very great ductility when properly designed. Table 17.5 shows some tension tests on split tees using rivets as the tension connection between column and tee. These tests were reported in Toriggino and Cope's (1933) discussion of Berg's (1933) paper in the transactions of ASCE.

It is interesting to note, that in these 14 tests, ranging up to a flange stress ultimate tension of 294 kips requiring 4–1$\frac{1}{4}$ in. tension rivets, the separation of the "tee" from the column goes up to over 1$\frac{1}{4}$ in. with an average deformation of 0.80 in. In relating this data to ductility, drift, or total building deformation, this average "tee" deformation of 0.80 in. corresponds to a story drift of 5 in. using 24 in. deep girders in 12 ft 6 in. high stories.

Other tests have been made on connection deformation of split tees, principally in the elastic range. C. R. Young and K. B. Jackson (1934) conducted a series of tests to compare welded and riveted rigidity. The welded types of connections tested in this report are not used for large lateral forces. Figure 17.2 shows the typical tee connection tested, using an 18WF47 girder connected to a 12WF110 column. The data provided by the test report does not give the joint deformations at the ultimate loads, but from the lower portion of the curves shown in Fig. 17.3 and the ultimate loads, it is obvious that the connections had very great ductility. The tests also are interesting since they were among the first to recognize the importance of reversed loadings. The tests included five reversed loads while the sixth loading was carried to ultimate failure.

In Rathbun's series of tests, split tee connections were included, again to determine the elastic properties of connections, so deformation data is lacking at higher loads. The available information is presented in Table 17.6. Again, by comparing the ultimate moments with the deformations at the largest moments for which the

Fig. 17.2. Connection test set-up. (Taken directly from Young and Jackson, 1934.)

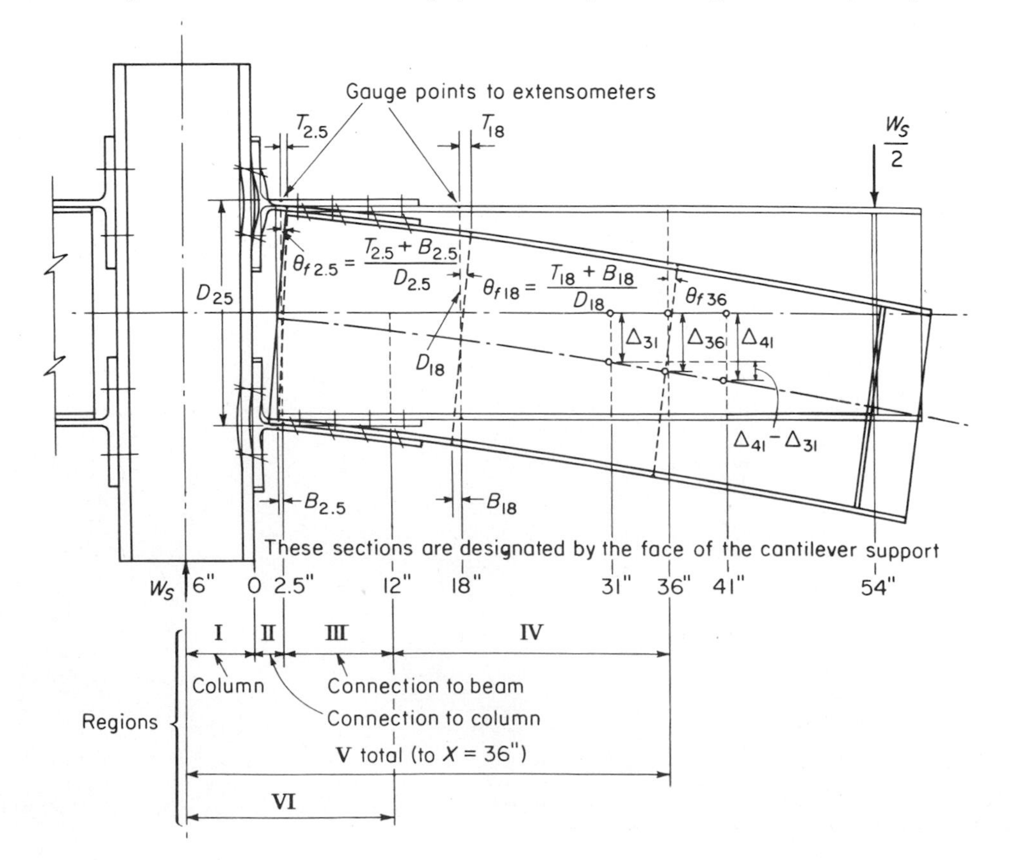

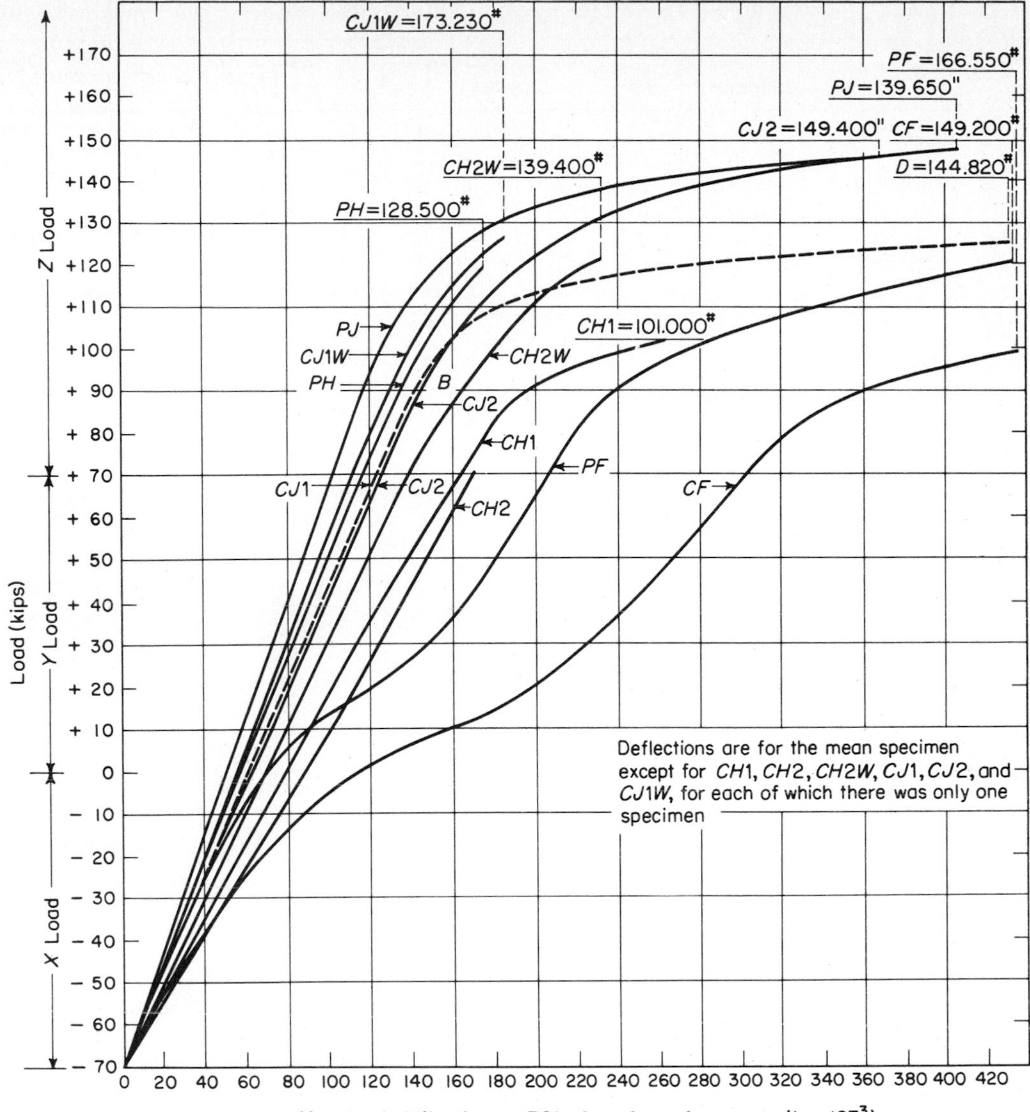

Fig. 17.3. Deflection curves and ultimate loads. (Taken directly from Young and Jackson, 1934.)

Table 17.5. Tests for Mills Tower, San Francisco, 1930*

Test	"Tee"	Tension rivets, in.	Ultimate load, kips	Deformation at ultimate, in.	Failure
1	24I100	4 $\frac{3}{4}$	123	0.46	Rivet necking
2	24I100	4 $\frac{3}{4}$	120	0.48	Rivet necking
3	24I100	4 $\frac{7}{8}$	176	0.60	Rivet necking
4	24I100	4 $\frac{7}{8}$	181.5	0.60	Rivet necking
5	24I100	4 1	206.5	0.82	Rivet necking
6	24I100	4 1	212.0	0.65	Rivet necking
7	36B260	4 $1\frac{1}{8}$	209.0	1.46	Rivet necking
8	36B260	4 $1\frac{1}{8}$	205.0	0.80	Head snapped off
9	36B260	4 $1\frac{1}{4}$	253.4	0.90	Head snapped off
10	36B260	4 $1\frac{1}{4}$	251.0	0.95	Head snapped off
11	36B300	4 $1\frac{1}{8}$	205.4	0.74	Head sheared
12	36B300	4 $1\frac{1}{8}$	237.0	1.05	Rivet necking
13	36B300	4 $1\frac{1}{4}$	293.6	0.94	Head snapped off
14	36B300	4 $1\frac{1}{4}$	281.6	0.60	Head snapped off

Deformation tabulated

*Average deformation 0.80 in. Deformation is separation of flange from column at C.L. of web. 0.80 in. corresponds to 12 ft 6 in. high story drift of 5 in. using 24 in. deep girders. Deformations at ultimate *many* times (10 to 20) that at yield. Based on data from Berg (1933).

Table 17.6. Moment connection test, split tee*

Test	Beam	Tee	Shear rivets Number	Shear rivets Size	Tension rivets Number	Tension rivets Size	Ultimate M (in. kips)	Failure	Deformations§ θ at M (in. kips)
13	12-in. I	15I99	6	$\frac{7}{8}''$	4	$\frac{7}{8}''$	1845	Tension R	0.007 at 1350
14	12-in. I	15I99	6	$\frac{7}{8}''$	8	$\frac{7}{8}''$	1640	‡	0.0045 at 1600
15	16-in. I	24I105.9	8	1″	8	1″	5700	Shear R	0.007 at 3400
16	22-in. I	30I240	10	1″	8	1″	8500	‡	0.007 at 4600
17	22-in. I	30I240	16	1″	16	1″	8530	‡	0.005 at 8000
18	16-in. I	24I105.9†	8	1″	8	1″	‡	Column deformed	0.007 at 3400

*From Rathbun (1936).
†Fastened to flange of 14 WF 167.
‡Test failed to reach ultimate.
§For comparison, $\theta = 0.007$ corresponds to 12 ft 6 in. high story drift of 1.05 in.

deformations are presented, it can be seen that the connections must have been very ductile.

One of the problems in designing split "tee" connections concerns the "prying" action under the tension bolt as discussed in several of the references. In designing the structure for the International Building in San Francisco, the edge margins of the tension bolts were reduced to the minimum possible, eliminating the "prying" action and causing the flanges of the tee to bend as simple cantilevers. Twelve samples of the tees were tested (Cooper and Errera, 1960) in six back-to-back tests using enough packing between tees to simulate the thickness of the columns. This was done in order to provide the proper relationship between grip of bolts and the bending properties of the tee flange. The results of these tests are given in Table 17.7. It is interesting to note that the minimum elongation was 0.98 in. on Specimen T83 where a tension bolt failed. This minimum elongation corresponds to a story drift of about 5 in. for a 30 in. deep girder in 12 ft 6 in. story height, again neglecting column and girder bending. The tests indicate that the minimum ratio or ultimate deformation to yield deformation was greater than 20.

Figure 17.4 shows the typical failure of the test specimens while Fig. 17.5 shows the use of these welded tees in a typical connection in the International Building. Figure 17.6 shows the stress–strain curves of the various tests in this series.

Fig. 17.4. Failure of test specimen. (From Cooper and Errera, 1960.)

Table 17.7. Tests for the International Building, 1960*

Test	Stem "t," in.	Flange plate	Tension bolts Number	Tension bolts Size	Design	Loads in kips Yield	Loads in kips Ultimate	Ultimate deformation, in.
T28	$\frac{3}{8}$	$14 \times 1\frac{1}{2}$	4	$\frac{7}{8}''$	79	124	248.5	1.30
T18	$\frac{1}{2}$	$14 \times 1\frac{5}{8}$	4	1″	103	158.2	327	1.54
T77	$\frac{5}{8}$	$16 \times 1\frac{1}{2}$	8	$\frac{7}{8}''$	158	270	522	1.46
T83	$\frac{7}{8}$	$16 \times 1\frac{5}{8}$	8	1″	206	359.5	706	0.98
T44	1	16×2	8	$1\frac{1}{8}''$	260	444	825	1.84
T129	$1\frac{1}{4}$	$16 \times 2\frac{1}{4}$	8	$1\frac{1}{4}''$	330	561	1017	1.61

*Tension tests on welded, stress-relieved "tees" with extra high strength, chrome molybdenum steel tension bolts. All failures except T83 were through net section of Stem Pls. T83 failed through fracture of bolts. Data from Cooper and Errera (1960).

Fig. 17.5. Bolted and welded connection, International Building.

More recent tests by Douty and McGuire (1965) have been reported studying the performance of high-strength bolted moment connections in the plastic range. These tests separated the action of the stem of the tee "splicing" the beam from the flange action of the tee involving "prying" action on the bolts in tension. Figure 17.7 shows the typical stress–strain relationship for the "stem splice" action. In this test unspliced 10WF21 and 16WF36 were compared to similar beams spliced with various connections. The maximum developed moments in the spliced beams were slightly greater than the theoretical plastic moment, which in turn was based on the gross section of the beams. The gross sections were about 23% greater than the actual net section. In these tests of the plain unspliced beams local buckling started at about 1.13 times the yield moment and at about six times the yield deflection. In the spliced beams, the splice plate reinforced the flange of the beam against buckling so that the ultimate deflection reached seven to eight times the yield deflection. In the tests where beams were connected to 14WF150 column flanges, the ultimate moment was 1.3 to 1.55 times the plastic moment, but the maximum deformations are not reported.

Also included in the Douty and McGuire paper are data on tests where connection plates are welded to ends of beams with the plates bolted to the column as shown in Fig. 17.8. This type of connection has not been used extensively in high-rise construction but has been used to a certain extent in lower structures. If properly designed, ratios of ultimate deformation to elastic deformation of seven or more may be achieved.

There have been relatively few tests on welded con-

Fig. 17.6. Test curves from International Building tests at Lehigh. (Taken directly from Cooper and Errera, 1960.)

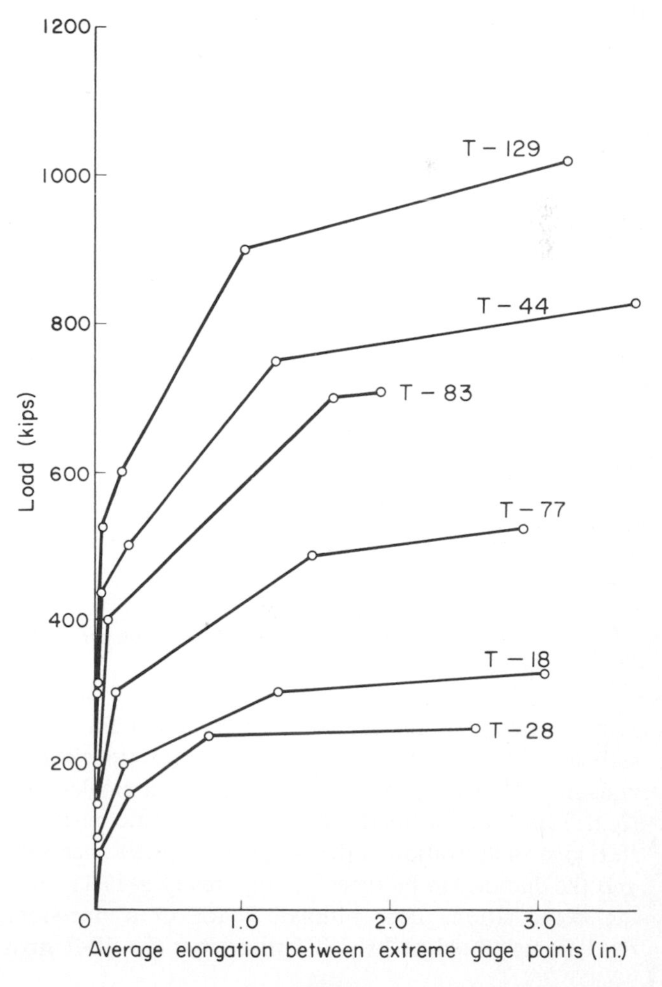

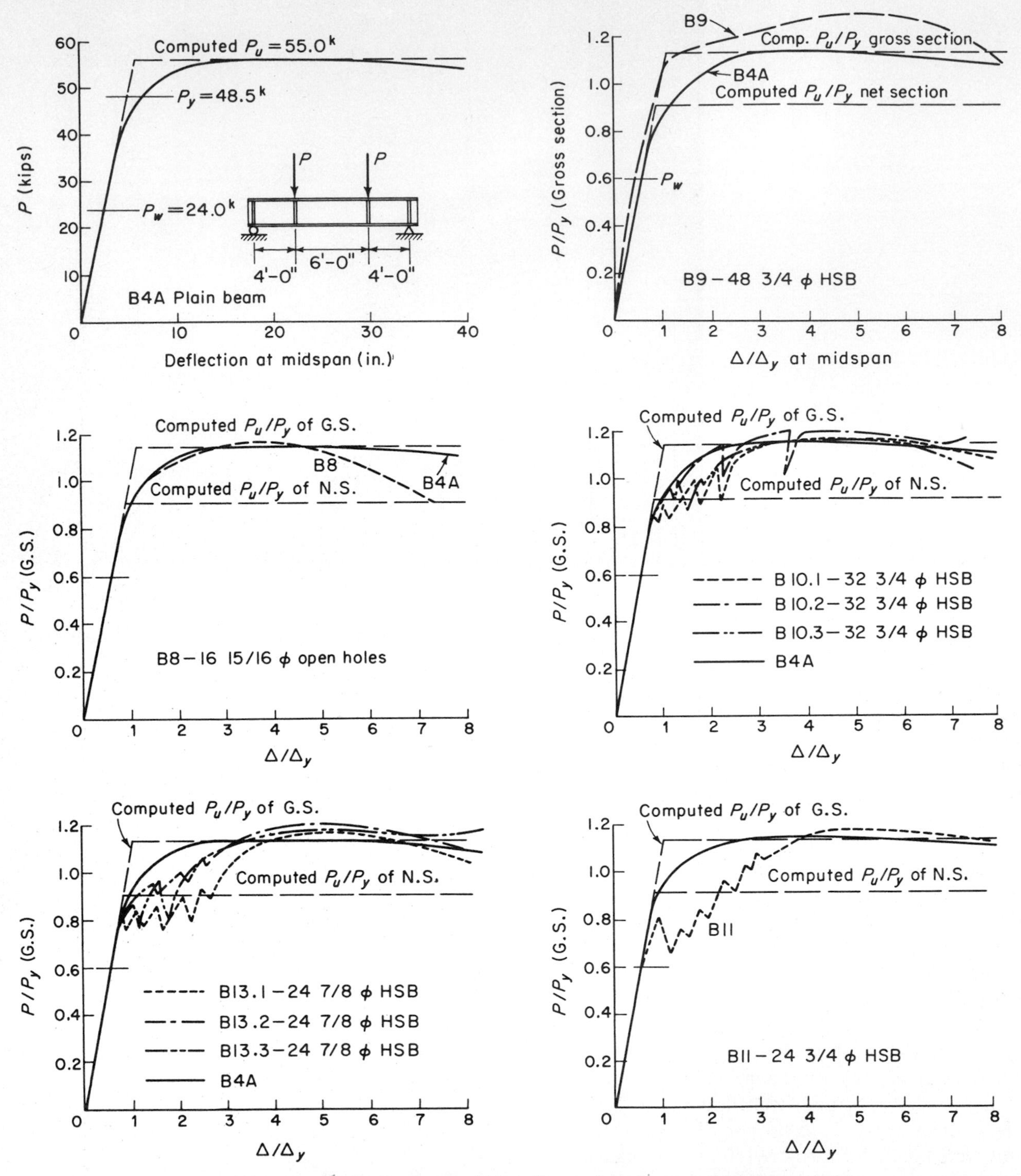

Fig. 17.7. Load deflection curves. (Taken directly from Douty and McGuire, 1965.)

nections designed for high-rise construction for which sufficient data are available to evaluate their postelastic behavior. However, the tests performed on single-story rigid frames to formulate the rules for plastic design in steel give an indication of the design procedures necessary and the ductility to be expected. Figures 17.9–17.11 show the deformations near ultimate loads of a one-story rigid frame tested at Lehigh University. The 10-ft high column is still carrying maximum load although it had deflected 7 in. from its vertical position.

There are several items to consider in the design of steel connections, especially with regard to the internal actions of the joint. Consider the moment resisting joint shown in the upper left portion of Fig. 17.12. Tees are provided to develop the full moment capacity of the beam. The web of the tee must have enough net

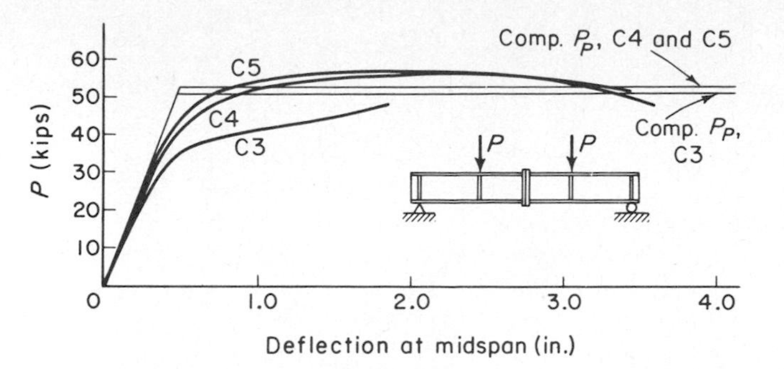

Fig. 17.8. Load deflection curves. (Taken directly from Douty and McGuire, 1965.)

Fig. 17.9. Vertical loading on steel frames. (Taken directly from Beedle, 1954.)

Fig. 17.10. Vertical and lateral loading on steel frames. (Taken directly from Beedle, 1954.)

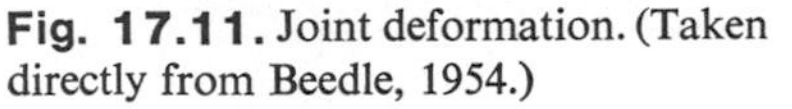
Fig. 17.11. Joint deformation. (Taken directly from Beedle, 1954.)

section and a sufficient number of bolts acting in shear to develop the chord stress T. This stress T is then transferred through the flange of the tee in bending and then through tension bolts to the column. In considering the flange of the tee and the tension bolts, there are two alternate conditions of design as indicated in portions A of Fig. 17.12. The more usual condition, shown on the left, has generally been assumed in the past since tees traditionally have been made by cutting portions of rolled sections. With the sections available for large moments, the thickness of the tee flange is limited, and consequently a point of inflection (P.I.) has to be assumed between the tension bolts and the web of the tee. In order to assume the points of inflection, prying forces C must also be assumed, so that the tension force in the tension bolts is larger than the flange stress T by the amount of prying force C. In actual connections, this force C may range from 25 to 50% of T. An alternate

Fig. 17.12. Stresses on steel connections.

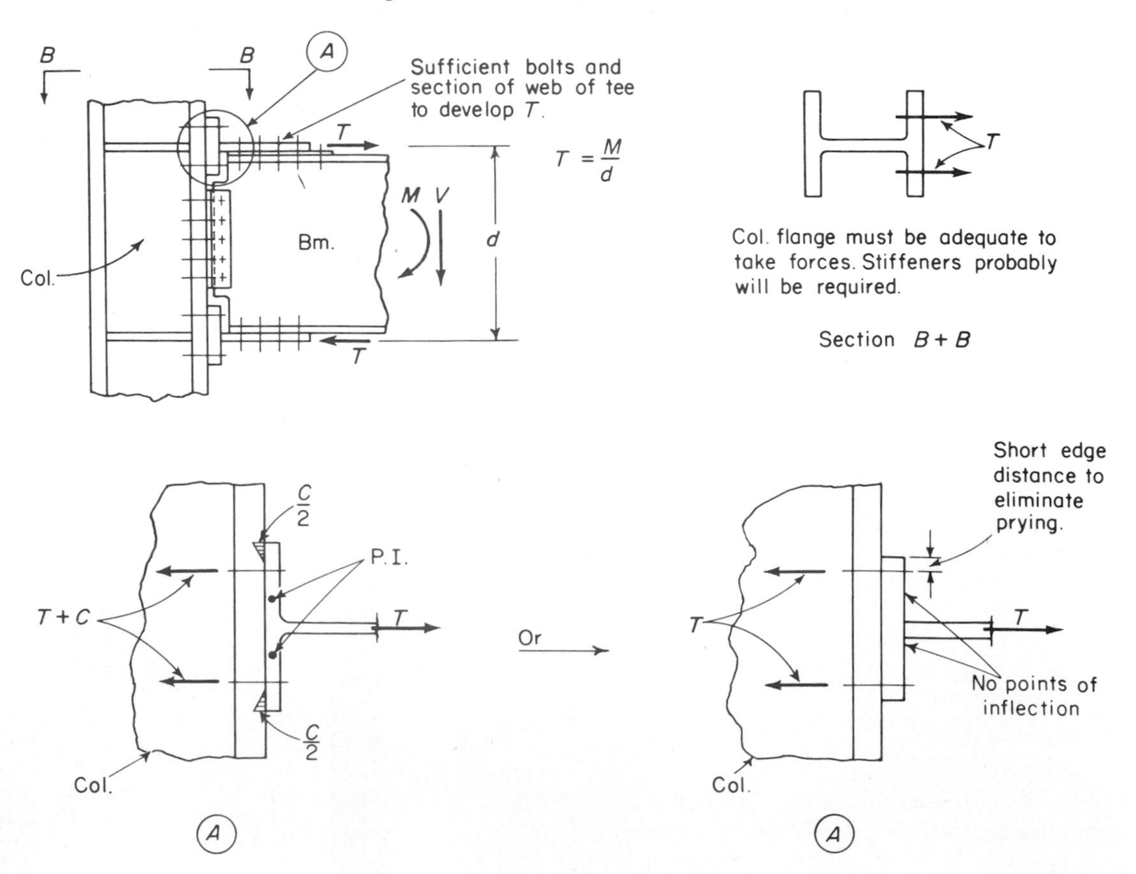

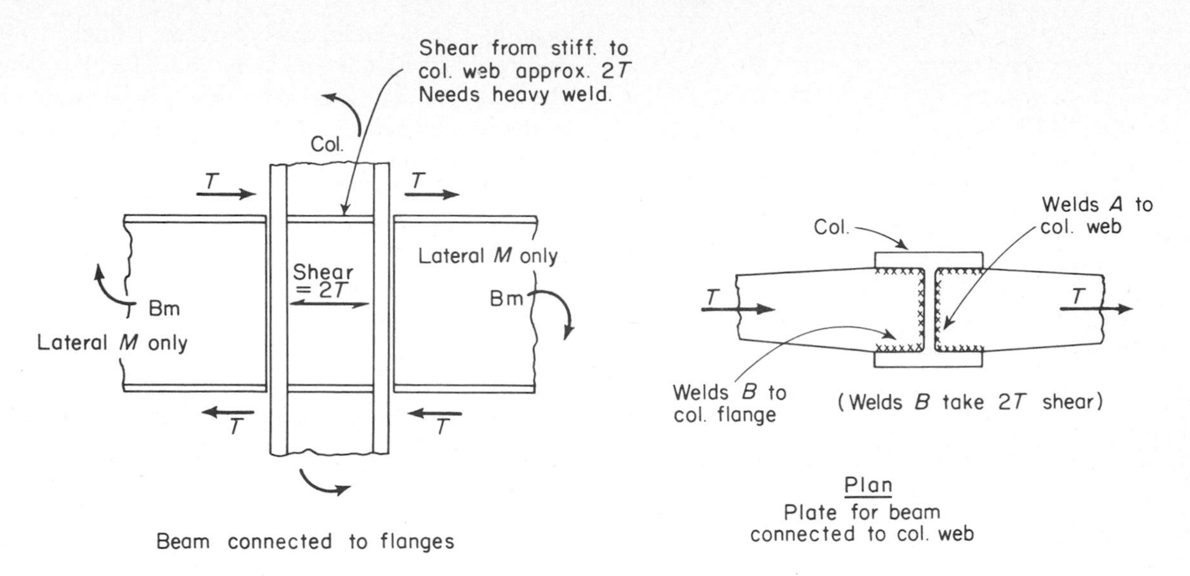

Fig. 17.13. Stresses on steel connections.

condition of design may be as shown in the right-hand detail *A* of Fig. 17.12. If the edge distance of the tension bolts is kept short, no prying forces can be developed, and hence there can be no point of inflection in the flange of the tee. The flange of the tee, then, must be designed as a simple cantilever from the web, requiring more thickness of flange than the first alternate but permitting smaller tension bolts. For high-rise buildings in areas subjected to major earthquakes, this flange thickness may be greater than available in rolled sections so that welded built-up tees may be required.

If a view is taken of the column in plan as shown in Section *B*, it can be seen that there may be additional prying forces as the column flange tends to bend. In the past these forces often have been overlooked. However, with the large forces associated with the earthquake stresses in high-rise structures and the general necessity to provide moment connections in the opposite direction (beams connected to web of column), stiffeners usually will be required to reduce column flange bending to acceptable levels. Therefore, prying forces in this plane usually do not control. It might be mentioned at this time that although all welded steel construction may eliminate the need for this type of column-beam connection in many cases, it is still a commonly used connection, especially where coverplated columns must be used because of heavy loads.

In Fig. 17.13 two loading conditions are shown that inexperienced designers frequently overlook. At the left the shear condition in the column web between the beam flanges is shown. Regardless of whether this is a bolted or welded joint, the stiffener plates have tension at one side and compression on the other (considering only the lateral loading portions of the beam moments) so that the shear on the weld to the column web is measured by twice the beam flange stress. Similarly, the shear force in the column web is approximately twice the beam flange stress.

When the column is bending in the weak direction as shown in plan, only the welds to the flanges of the column are fully effective and again the total welding to the flanges must take twice the beam flange stress.

17.3 BENDING OF BEAMS AND GIRDERS

The second factor that influences the overall performance of a steel-framed structure is the bending capacity of the beams and girders. In steel, failure is usually by instability and in order to perform satisfactorily in a major earthquake, the moment-rotation curve must develop a long plateau. Ultimate failure usually is not caused by lateral buckling but is triggered by local buckling. A typical relationship between moment and rotation is shown in Fig. 17.14. In order to have an adequately ductile situation, the proportions of the beam and its physical surroundings must be such that a plateau as shown by the solid line is developed rather than a falling off of the load as indicated by the dashed line. It must be remembered in the following discussion that the tests and references used here are those available from the many tests performed to develop safe design criteria for plastic design with A7 steel and were not performed to investigate earthquake design criteria, so some interpretation is necessary and some points of difference in action must be pointed out.

The first considerations in evaluating bending ductility are the proportions of the basic cross section of the beam. Because local buckling usually triggers the ultimate failure load, the proportions of the section as shown in Fig. 17.15 must be such that local buckling will occur only at large strains. Fortunately, in

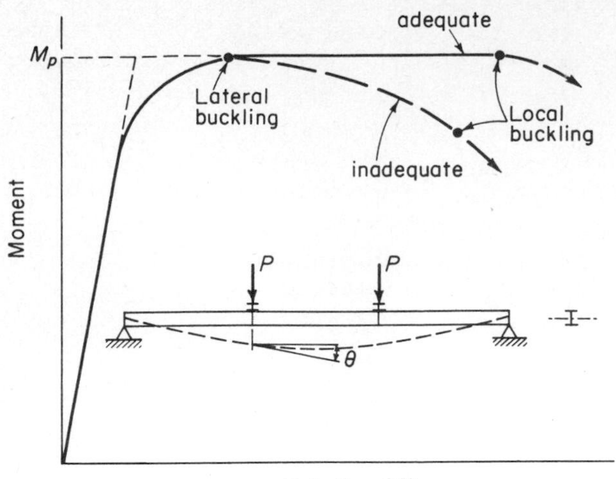

Fig. 17.14. Beam M–θ relationships. (Taken directly from Lee, Ferrara, and Galambos, 1963.)

developing plastic design in steel, much research (Welding Research Council—ASCE Joint Committee, 1961) has been devoted to this phase of the problem, and it is found that most rolled shapes are satisfactory. In order to develop adequate ductility b/t must be less than 17 and d/w must be less than 70 for A7 steel. Lower figures probably would be required for high strength steel, and research is now taking place to determine the critical ratios.

Lateral buckling for the beam as a whole under earthquake conditions depends on various factors such as the (1) shape of the moment curve, (2) end restraint of the flanges, (3) type of connection detail, (4) spacing of lateral bracing, (5) stiffness of lateral bracing, and (6) attachment of intermediate floor beams. Not all of these have been tested but comparisons can be made with tests that have been performed on similar factors.

In the plastic design of steel frames, the spacing of the lateral bracing beams forms one of the prime parameters in evaluating the buckling tendencies of members in bending. Tests were performed at Lehigh University and reported by Lee and Galambos (1962) to determine the basic requirements. When the critical section is subjected to a uniform moment and with the critical central section restrained only by the side spans, it was found that ratios of critical length L to radius of gyration about the weak axis of r_y of 45 or less gave ϕ/ϕ_y of 10 or greater. It will be noted from Fig. 17.16 that the tension flange is not restrained and the compression flange is restrained only by a knife edge and the rigidity of the adjacent span.

Now, in earthquake loadings it will be seen that the moment curve is not uniform, as conducted in the tests, but varies on a triangular basis from maximum at the column to minimum near midspan. It seems reasonable to expect that there is less of a tendency to buckle with a decreasing moment than with a uniformly maximum moment. This is confirmed in a report by Lee, Ferrara, and Galambos (1963) where test P4 with a $60r_y$ side span and $40r_y$ center span had a ϕ/ϕ_y ratio of 12.1 as compared to test P3 with $40r_y$ on all spans which had a ϕ/ϕ_y ratio of 13.3.

The amount of restraint also would affect the tendency to buckle. In the tests the restraint for the critical center

Fig. 17.15. D/t and b/t relationship.

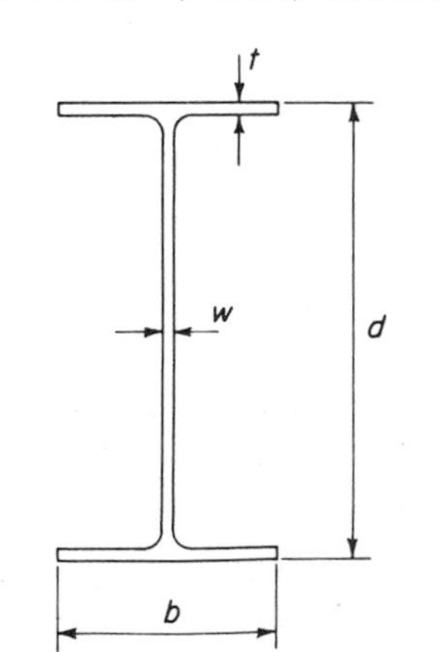

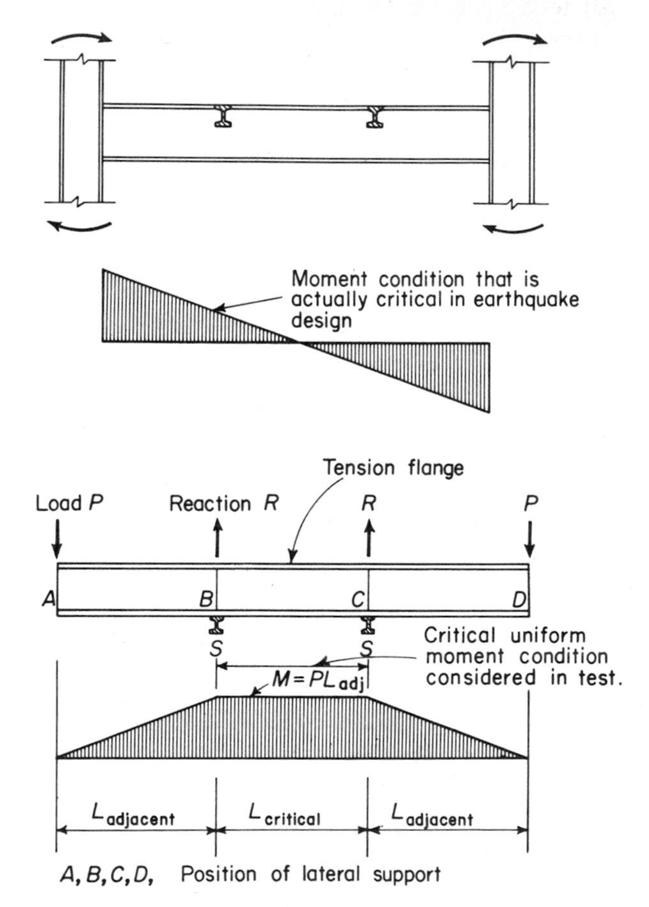

Fig. 17.16. Moment and restraint relationship.

section is provided only by the lateral stiffness of the adjacent portion. In the earthquake loading, the restraint at the point of maximum load—at the column connection—is provided by the rigidity of the column in torsion plus the other beams framing into the column. The type of connection detail at the column also would materially affect the tendency to buckle. Where moment plates are used to connect the beam flange to the column, the plates add additional stiffness laterally and also reinforce the flanges against local buckling, again indicating a greater provision in normal buildings against local buckling than is reflected in the tests for plastic design.

In the last reference (Lee, Ferrara, and Galambos, 1963) tests were run to determine various methods of attachment of the lateral beams to the main beam, as shown in Fig. 17.17. It was found that the tendency to buckle depended only slightly on the stiffness of the lateral beams. All tests were run at a ratio of $L/r_y = 40$, and all connections were made to the compression flange only. It was found that it was most important to hold the beam in position but that it was not necessary to try to prevent rotation. Stiffeners running half the depth of the beam as compared to full height stiffeners did not affect the results. These tests also indicated the following:

1. Lateral buckling started at about three times yield rotation—it varied in the individual tests from 1.5 to 5.3—but the section did not unload when buckling occurred.

2. Local buckling generally started from about 8 to 12 times yield rotation with an average of 11.5. Each test exhibited at least 10 times yield rotation before serious unloading occurred. A typical curve is shown in Fig. 17.18.

3. The onset of local buckling varied throughout the tests, but it generally occurred when the compression flange strain was from 12 to 23 times the elastic limit strain.

In all of the above conditions it must be considered that in actual high-rise buildings under earthquake loadings the top flange is usually continuously supported by the floor construction using either concrete slabs or metal deck. The intermediate floor beams are fastened to the main girder by Series A or equivalent connections, which extend down to the region of the bottom flange. The tests were run on the more severe condition of the tension flange completely unbraced and the compression flange restrained only by a knife edge or by the purlin connections shown in Fig. 17.17. Considering all conditions it would be most reasonable to expect that there would be less tendency to buckle in the actual building than in the tests so that expected yield ratios of at least 10 would be a conservative estimate of performance.

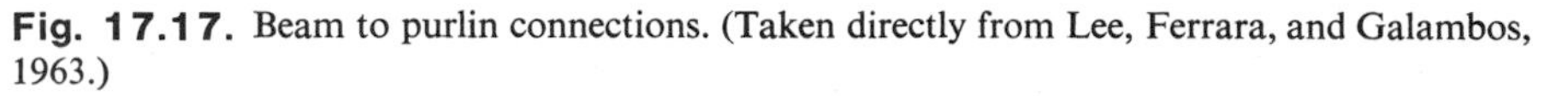

Fig. 17.17. Beam to purlin connections. (Taken directly from Lee, Ferrara, and Galambos, 1963.)

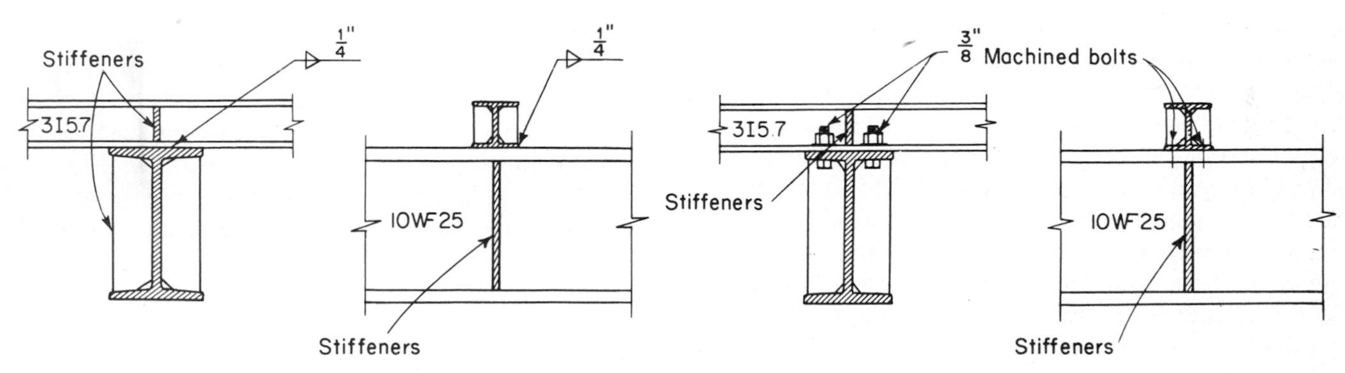

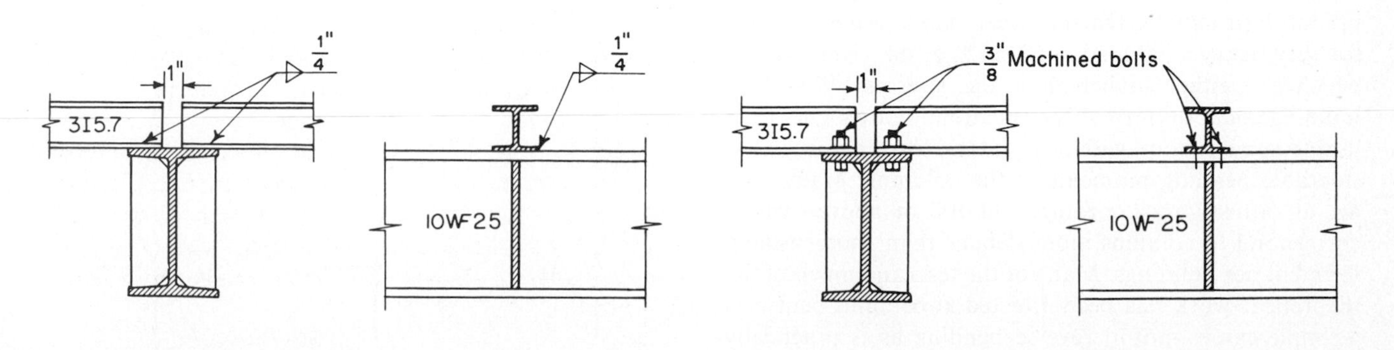

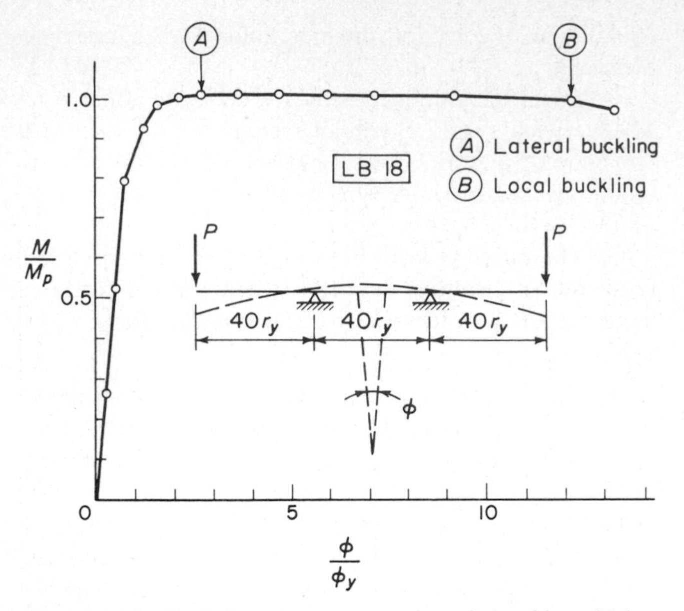

Fig. 17.18. Typical moment–curvature relationships. (Taken directly from Lee, Ferrara, and Galambos, 1963.)

17.4 COLUMNS

Columns are the last of the three basic units of construction in steel frame structures that must be considered. Over the years much work has been done on column research, but much of this has been done with axial loadings with little or no eccentricity to determine basic column formulas. It is only recently that extensive work has been performed on beam columns where large bending moments are applied at the same time as the axial loads. Much of the early work was done for aircraft design. That pertaining to major structural work has been done for the purpose of developing criteria for plastic design in steel but does not extend into the ranges of load combinations customary in providing for earthquake resistance in major steel construction. In high-rise buildings the general range of usable column loadings is in a fairly narrow range, and many of the tests are too far out of this range to provide much usable information. For example, many tests are at light axial loadings at 10–12% of the yield stress (3000–4000 lb/in.^{2}), which is much less than the customary usable loadings found in practical structures. On the other hand, some tests are for very heavy axial loads—50–65% of the yield stress—which are either higher than the stresses allowed in building codes or normal specifications or are higher than usable in a structure with large lateral loads causing considerable bending moments in the columns. Many tests are at rather long L/r ratios—90–100 or above—which correspond to columns more slender than those usually found in tier buildings. Many of the tests and much of the theoretical work has been directed at columns bent into a simple curve—not in reverse bending as is practically universal under lateral loadings. Finally, most tests were stopped when the columns buckled under the mistaken assumption of many engineers that buckling is synonymous with failure. Actually, most usable steel columns have considerable strength in the postbuckling range, sustaining loads greater than the usual design loads at very large deformations. In most structural design, strength in this range is assumed to be meaningless so it rarely has been investigated. In earthquake-resistant design this postbuckling strength is a major factor in the actual strength of the structure and so must be considered in establishing design loads, allowable stresses, and general design criteria.

Most of the usable work in this field has been performed at Lehigh University, and most of the data presented here will summarize the work of Galambos and Lay (1962). The tests that have been performed to date on columns combining axial load with bending moments consider bending only in one plane and that along the strong axis of the column. No tests to date consider bending on the weak axis.

In considering the various parameters presently known to affect the toughness or ductility or post-elastic behavior of steel column sections, we can itemize the following:

1. There are the usual properties of the cross section, i.e., area, moment of inertia about each axis, section modulus, and the derived property radius of gyration. One parameter, similar to that of beams, is the ratio of maximum applied moment M to the yield moment M_y of the section in pure bending without axial load.
2. There is the length of the column which, together with the radius of gyration, determines the L/r ratio.
3. The ratio of applied axial load P to the yield point axial load on a short column P_y must be related to the applied moment M on the column considered.
4. The sign and proportion of the bending moment applied at each end of the column must be considered as illustrated in Fig. 17.19. A moment applied to only one end of a hinged column is defined as $\beta = 0$. Opposite

Fig. 17.19. Definition of β.

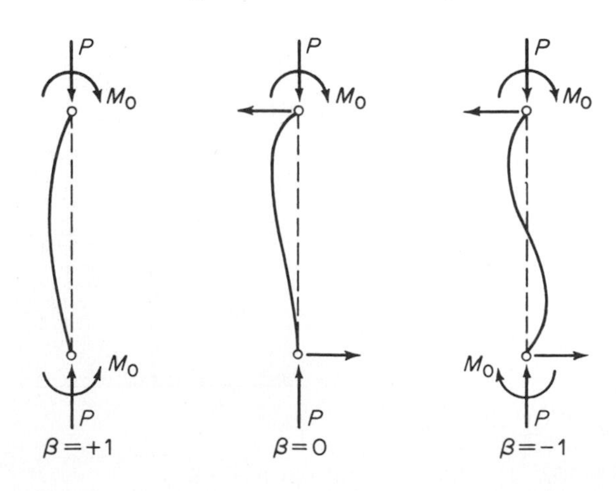

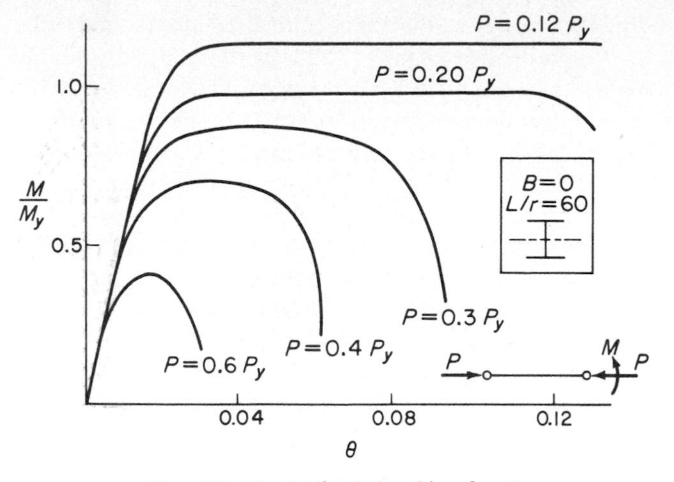

Fig. 17.20. M–θ relationship; $\beta = 0$.

equal moments at each end of a column—moments resulting in a single curvature without a point of contraflexure—are defined as having $\beta = +1$. Equal moments having the same direction of rotation and causing an S-shaped curve with a point of contraflexure in the center are defined as $\beta = -1$.

5. The relationship of D/w and b/t as defined for beams determines the possibility of the occurrence of local buckling. In the lower portions of tall tier buildings, it is improbable that local buckling will be of major importance.

6. Finally, the torsional properties of the section in relation to other parameters will determine the possibility of torsional failure. This has been only partially investigated, and design methods so far proposed are not suitable for use in a design office. Again, it is improbable that the sections ordinarily used in major tier buildings will be influenced to a major extent by torsional flexibility.

Because of the many parameters, the influence of each will be discussed separately.

Figure 17.20 shows the effect of the ratio of load P to axial yield load P_y to the moment-rotation capacity for the single unique case where $\beta = 0$ (one end moment only), $L/r = 60$, and with torsional effects and local buckling effects neglected. Note that lightly loaded columns have long plateaus, indicating high ductility. Heavy loads have short plateaus, indicating quick unloading and low ductility. In ordinary high-rise construction subjected to earthquake the zone of greatest interest is in the range of $P/P_y = 0.20$ to $P/P_y = 0.35$. In the ultimate range, columns with small P/P_y ratios and $\beta = 0$ or $\beta = -1$ are most affected by local buckling and torsional effects.

Figure 17.21 shows the effect of the end restraint. The condition where $\beta = +1$ (single curvature) gives the shortest plateau, but fortunately in earthquake-resistant design this is not the usual condition. The condition when $\beta = 0$ (moment on one end only) gives a moderately long plateau. The conditions for $\beta = +1$ to $\beta = 0$ have been the most researched conditions since they are of major importance in the plastic design of steel frames. Earthquake loadings usually produce reverse curvature conditions that provide the longest plateau and consequently the highest rotation capacity. Because research on earthquake effects has lagged behind that of plastic design, little laboratory data is available on this loading condition. As given in Figs. 17.20 and 17.21, the rotation plotted is directly in radians and is not relative to the yield rotation.

Figure 17.22 defines the rotation constant R_c as used here and in the primary reference (Galambos and Lay, 1962). Note that the commonly used ductility factor μ would be one larger than R_c (ductility factor $= R_c + 1$). With the definition presented here it also can be seen that the capacity of the column at the end point ($0.95\ M$) is still much larger than the design capacity of the column, and a portion of the column capacity therefore has been neglected.

The effect of the end moment ratio β can be seen from the tabulation of test data in Table 17.8. It can be seen that there is little data available. However, what little data is available indicates larger R_c values for negative β

Fig. 17.21. M–θ relationship; various β.

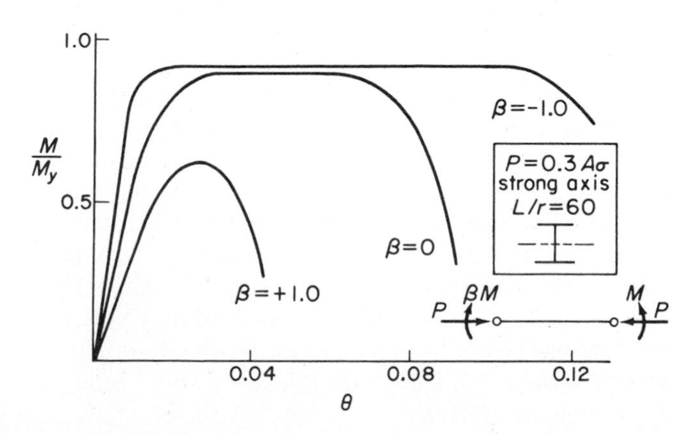

Table 17.8. BEAM-COLUMN TESTS

Test	Member	l/r	P/p_y	β	R_c
T26	4WF13	84	0.12	+1.0	0.90
T23	4WF13	83	0.11	0	1.85
T24	4WF13	83	0.12	−0.52	(5.6)*
T29	4WF13	84	0.13	−1.0	(1.95)*
T20	4WF13	56	0.12	+1.0	1.15
T17	4WF13	56	0.12	−0.5	1.70
T12	8WF31	55	0.12	+1.0	1.2
T13	8WF31	55	0.12	0	(3.0)*
T4	8WF31	55	0.12	−0.5	(1.7)*

*Tests not carried to ultimate. Information concerning effect of β from information given in Lee, Ferrara, and Galambos (1963).

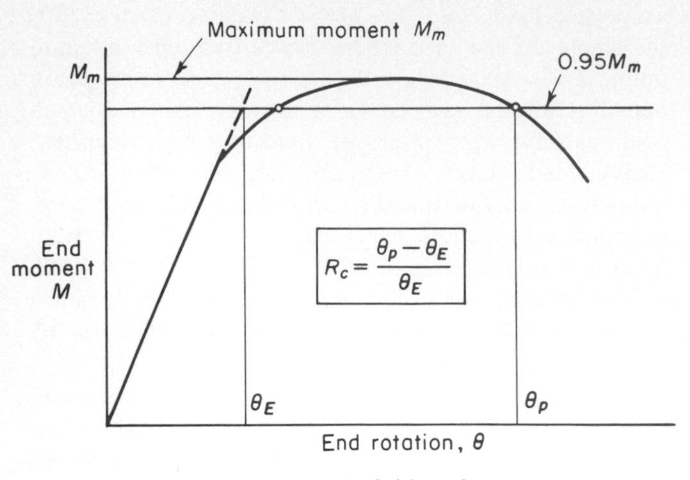

Fig. 17.22. Definition of R_c.

conditions, even though most tests were not carried to ultimate values.

Local buckling and torsion effects reduce the strength of the column as indicated in Fig. 17.23. In sections where local buckling occurs, the shape and value of the moment-rotation curve remain the same as in compact sections up to the point where the flange buckles locally and as shown in Figs. 17.21 and 17.22. When local buckling is a factor, the length of the plateau may be shortened when it is quite long as in the upper curve, but a shorter plateau may be unaffected as indicated in the lower curve. In other words, local buckling acts as a cut-off point for the moment- rotation curve, shortening the plateau in some cases. Limitations due to torsion are similar but are much more complicated.

The previous figures have shown the actual moment-rotation capacity in terms of absolute rotation θ, in radians. With the addition of additional parameters, it becomes more convenient to observe the various effects on rotation capacity R_c, thereby combining the critical information from an entire curve to one point. In doing so, however, it must be borne in mind that the value of R_c is as defined in Fig. 17.22 only. Point θ_E is probably higher than the yield rotation since it is located at 95% of the maximum moment. Point θ_P is smaller than the usable capacity of the column, as stated earlier, since it also is based on the 95% M value. However, at present, this is the best information available.

Fig. 17.23. Effect of local buckling. (Taken directly from Galambos and Lay, 1962.)

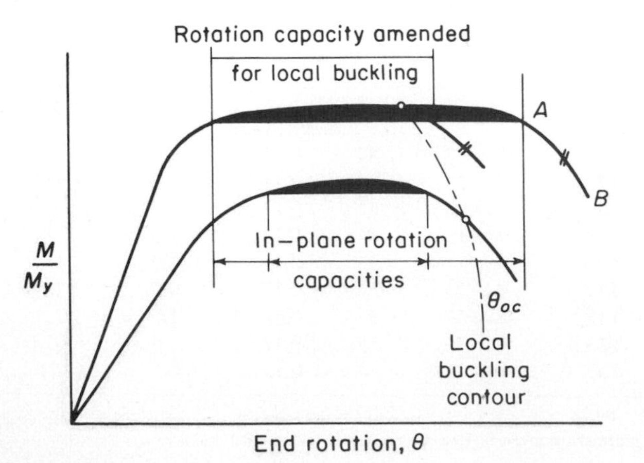

Figure 17.24 shows the relationship of load to rotation capacity for the case of $\beta = 0$. To the lower right of the dotted line labeled "local buckling zone" is the area where the column must be investigated for local buckling effects. Those columns with compact sections probably would follow the curves shown. Those sections with wide thin flanges and webs subject to local buckling would show a drop-off from the curves shown. The portion at the upper left of the dotted line labeled "Lateral Torsional Buckling Zone" is the zone where torsional effects may become important. Note that this is generally

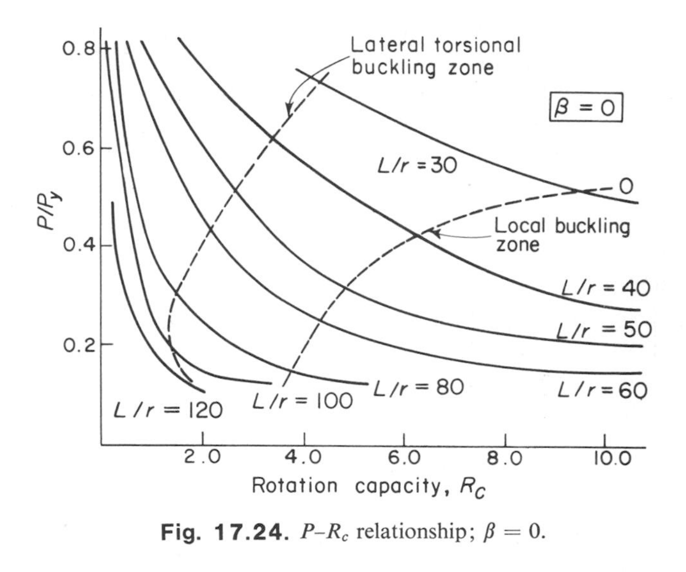

Fig. 17.24. P–R_c relationship; $\beta = 0$.

the zone where the P/p_y are greater than found in high-rise buildings and where the L/r ratios of the columns are usually higher than found in practical tier buildings.

Figure 17.25 is the similar plot for the single curvature condition where $\beta = +1.0$. The moment rotation capacities are substantially less than those found in Fig. 17.24, where $\beta = 0$.

The corresponding data for $\beta = -1.0$ is not available. However, when considering the great increase in rotation capacities for $\beta = 0$ (Fig. 17.24) over that found for $\beta = +1.0$ (Fig. 17.25) it would be expected that a substantial increase in rotation capacity might occur in going from $\beta = 0$ to $\beta = -1.0$, especially when considering the information in Fig. 17.21 and Table 17.8. Even with $\beta = +1.0$, and the usual case in a practical building where $P/p_y = 0.30$ and $L/r = 40$, R_c would be about 2.0 ($\mu = 3.0$) while with $\beta = 0$, R_c would be in the 8–10 range or greater.

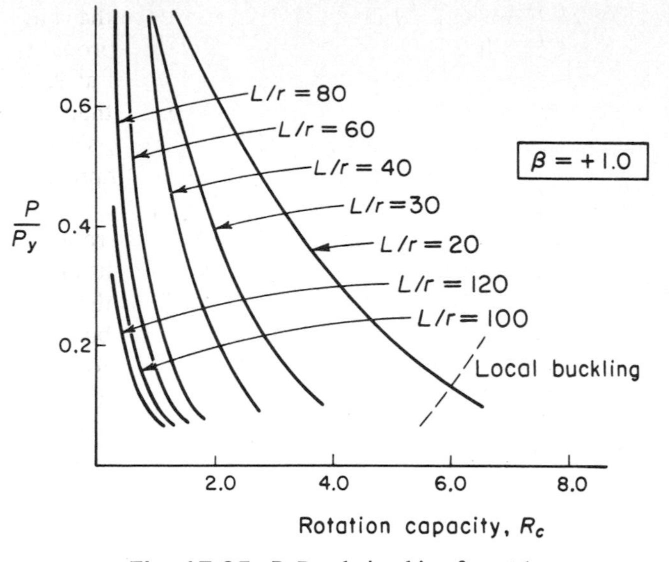

Fig. 17.25. P–R_c relationship; $\beta = +1$.

The effect of torsion is very complicated and is not in a form that is practical for use in a design office. Theoretical studies have been made by Galambos and Lay (1962) on the performance, including torsion and local buckling, on an 8WF31. These are shown in Fig. 17.26 for the condition $\beta = 0$ and in Fig. 17.27 for the condition $\beta = +1$. Again, no data is available on the condition where $\beta = -1.0$.

The 8WF31 section is not compact, has a high b/t value, and is torsionally quite flexible, so the curves shown for the 8WF31 are probably quite conservative as compared to most columns. Even so, Fig. 17.26 indicates that in the range of interest for tier buildings subject to earthquake loadings, R_c factors of 3–5 may be expected even for the case where $\beta = 0$. Conversely, of course, in any design where high L/r's, high P/p_y ratios, or single curvature bending ($\beta > 0$) is encountered, the designer should proceed with caution and should consider that less ductility in the columns must be expected in that portion of the structure.

Fig. 17.26. P–R_c relationship; $\beta = 0$ with effect of local buckling.

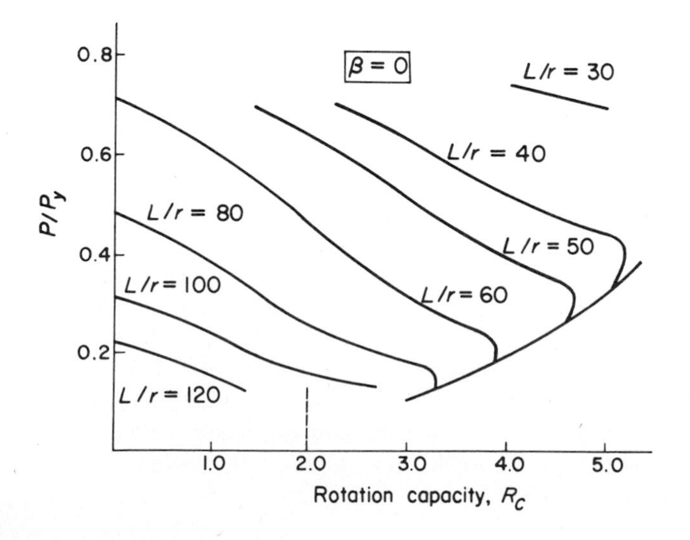

It also is true for most high-rise structures that in the lower two-thirds or so of the building columns are much stronger and stiffer than the beams and girders, so hinges will probably not occur in the columns and the required ductility is less than might be anticipated. Most recent studies in dynamic loadings indicate that hinges probably will occur in the beams and girders rather than in the columns, especially in the lower portions of the structure. However, any changes in building codes or design specifications, increasing the permitted loads on columns as compared to beams, could upset these findings and columns would have to be investigated more thoroughly.

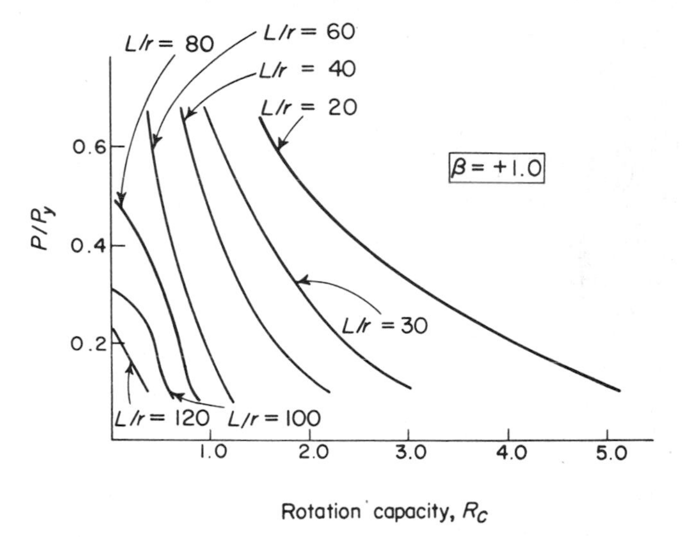

Fig. 17.27. P–R_c relationship; $\beta = 1$ with effect of local buckling.

17.5 REPETITIVE LOADINGS

One of the more important aspects of earthquake-resistant construction has only recently been generally recognized. In the vast majority of older testing a single load application was provided. In a very few cases, several cycles of loading were used, but the number of cycles was always minimal. The 1964 Alaskan Earthquake emphasized the importance of the length of time of shaking on the performance of the structures. The major damage to at least several structures in Anchorage occurred after several minutes of motion. In recognition of this, it becomes important to know the performance of the various structural elements under several or many cycles of loading. There has been no agreement among engineers as to the number of cycles required. If the major motion in Anchorage lasted from 3 to 7 min

Fig. 17.28. Photo from Galambos and Lay, 1962.

(according to various observers) and if the predominant period was 1 sec, then from 150 to 400 cycles of load could be deduced, although certainly not all or probably not even the majority of these cyclic loadings would be at maximum strain. Even a smaller earthquake of, say, 30 sec duration on ground having a dominant period of $\frac{1}{4}$ sec would require about 100 cycles of satisfactory performance—but again these probably would not be all at maximum strain.

The recent work in this field has been performed at the University of California under the guidance of Egor Popov (Bertero and Popov, 1965). The pioneering work was a test on 4 × 4 × 13 lb beams under heavy alternating loads producing strains from 1 to $2\frac{1}{2}$%, as shown in Table 17.9. The b/t value of the flanges was 10.5. Tests were run from full positive strain to full negative strain, causing local buckling of the flanges and finally fracture. Figure 17.28 shows one of the beams with typical flange buckling and the start of the fracture.

Table 17.9. NUMBER OF CYCLES TO OBTAIN LOCAL BUCKLING OR FRACTURE AS A FUNCTION OF THE CONTROLLING CYCLIC STRAIN*

	Number of cycles to cause	
Strain, %	Local buckling	Fracture
1	70	650
1.12	20	240
1.3	12	125
2	3	50
2.5	1	16

*Average yield stress of steel 41,000 psi. Data from Bertero and Popov (1965).

The average yield stress of the steel was 41,000 psi. It will be noted, then, that yield-point strain is 0.14% and that a 1% strain indicates that the beam has been strained to seven times the yield-point strain. Later work, now under way at the University of California, is on details and connections of various types including welded and bolted connections and various proportions of beams and columns.

17.6 ANALYSIS

The analysis of the high-rise steel frame has been thoroughly covered in the standard texts. There are several factors, however, that should receive emphasis. Most engineers seem to feel that the so-called exact methods of analysis, sych as slope deflection, moment distribution, and many other variations of these and other methods, are more accurate than the approximate methods. For low or even medium height buildings, this may be true. However, the basic assumptions for all of these methods neglect member shortening or lengthening. If a tall structure is analyzed by a so-called exact method, it will be found that the outside columns will shorten or lengthen a considerable amount as compared to adjacent columns, thereby invalidating the entire analysis as shown in Fig. 17.29. The approximate cantilever method that accounts for this relative column length change is much more accurate than the so-called exact methods. The use

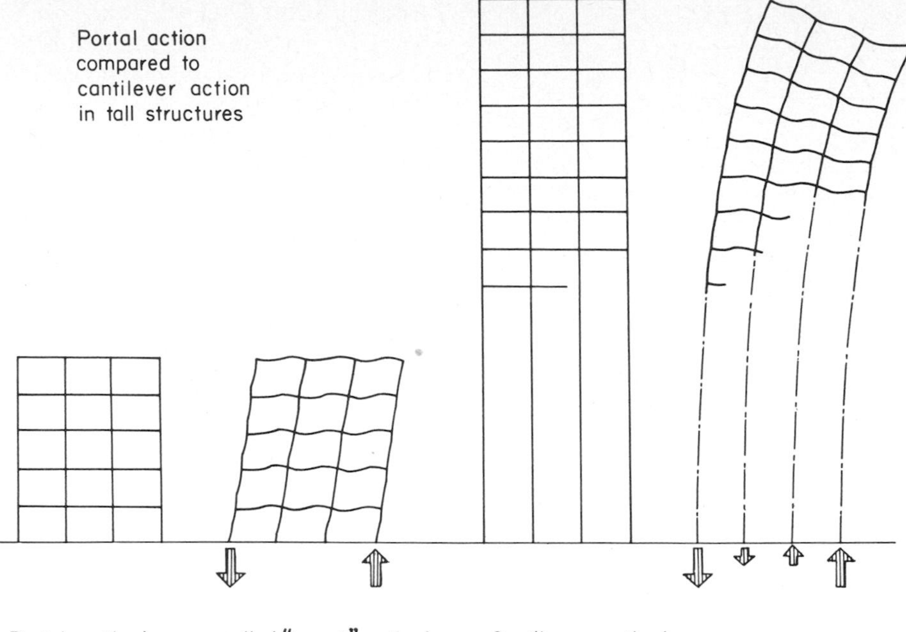

Fig. 17.29. Approximate vs "exact" analysis.

of computer programs that include axial deformations in the frame analysis eliminate this problem.

A second factor in analysis concerns the relative rigidities of dissimilar elements in high-rise construction. The standard texts and all basic approaches relate relative rigidities to the distribution of known forces between the various resisting elements. As given in most building codes, the assumed forces are generally related to the fundamental period of the structure. As shown in Fig. 17.30, we can consider the case of two resisting elements in a tall building—one a slender shear wall as shown on the left and the other a structural frame as shown on the right. For purposes of discussion we will not consider

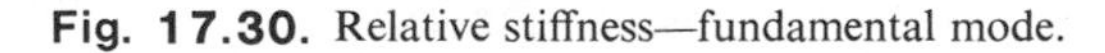

Fig. 17.30. Relative stiffness—fundamental mode.

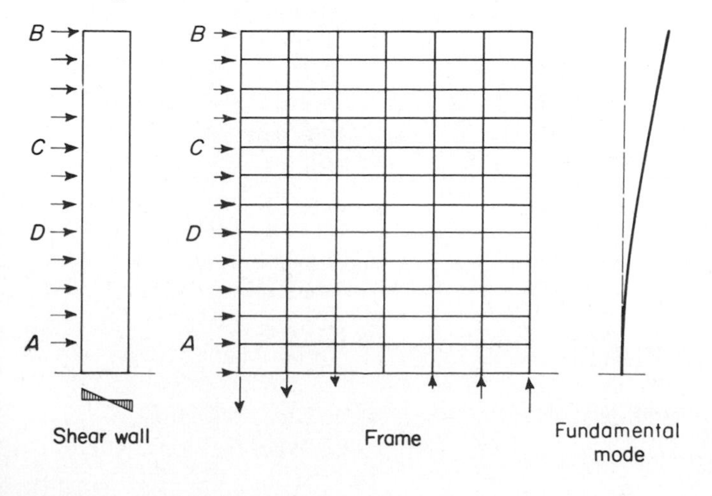

shear wall–frame interaction that has become recognized recently in the technical journals.

For the fundamental mode of vibration, and for the proportions of frame members usually found in practical structures, a load at a lower floor such as at A would go mostly into the shear wall. At this location the shear wall is much stiffer than the frame since shear effects are of the greatest importance. When considering foundation yielding, shear, column, and girder deformations in both bending and axial loads, it will be found that for loads at B, the frame is much stiffer than the slender shear wall. The deflections of the shear wall and frame for all loads—those at A, B, C and D—relate to the reactions at the ground and the rigidities, of course, are inversely proportional to the deflections.

From any given dimensions and the physical constants of the structural materials and the foundations, a fairly accurate calculation can be made for the relative rigidity of the shear wall as compared to frame. It will vary from floor to floor, and the loads at various floors will interact in a highly redundant and complex manner, but with modern computers this is a minor detail—very definite calculations can be made with considerable assurance as to their reliability. However, it was assumed that the structure will vibrate in the fundamental mode. When we examine the higher modes of vibration, we find that the relative rigidities completely change, as shown in Fig. 17.31. Loads C and D now do not span to the ground as they formerly did but tend to partially cancel each other, and the shear wall stiffness greatly increases as compared to the frame. This means that the shear wall is taking a

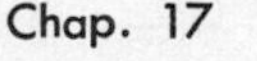

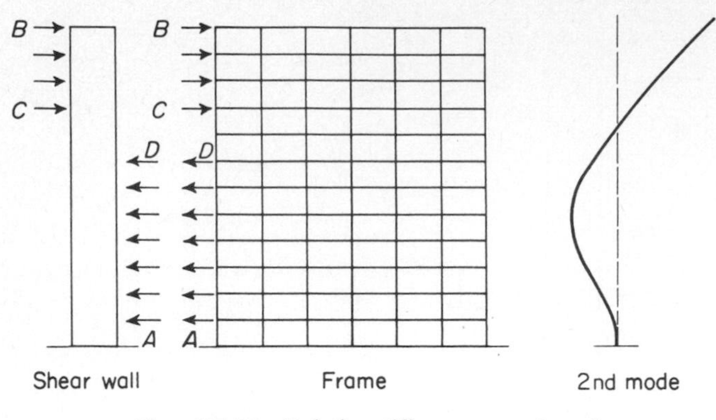

Fig. 17.31. Relative stiffness—second mode.

much larger proportion of the total load than was assumed for the calculations based on the fundamental load. It is not necessary to review the higher modes of vibration, because they would only further illustrate the same effect.

The importance of this observation concerns the fact that even if the *total* lateral load on the structure is known, the *distribution* of that load to the varying resisting elements is unknown unless they are very similar to each other. The structural engineer's final design of member sizes and strengths depends on the loads and the *distribution* of loads to the various elements.

There is no known way in which the effects of the varying rigidities can be incorporated in static design calculations unless the proportion of load that can be assigned to each of the modes of vibration is known. The only mathematical solution is to perform a dynamic analysis. For a valid dynamic analysis, a known ground motion is necessary. The type of ground motion will greatly influence the relative rigidity of the various resisting elements. If the response spectrum shows that long-period motions predominate, relative rigidities based on the fundamental mode may be appropriate, but if the ground motion is rich in short period vibrations, the basic relative rigidities may vary greatly from that assumed. Here the *lack* of ground motion records again is critical. There are no available ground motion records available in the large cities where high-rise construction is prevalent. Ground motion records for sites subjected to short-period vibrations are completely unacceptable for sites where long-period vibrations may predominate even for calculating the relative rigidities of various resisting elements.

A third major factor in analysis relates to the interrelation of shear walls and frames. Much of the dynamic research that has been attempted and published considers the shear wall and the frame as two separate resisting elements. Some of the studies have progressed to the stage where the two elements are tied together at each floor so that deflections at each floor must be equal. However, even this is rarely, if ever, the case in an actual structure. The shear wall is interconnected with the frame and changes the relative rigidities and the actions of the various individual components to a major—and generally critical—degree. Figure 17.32 briefly illustrates this

Fig. 17.32. Shear wall and frame acting together.

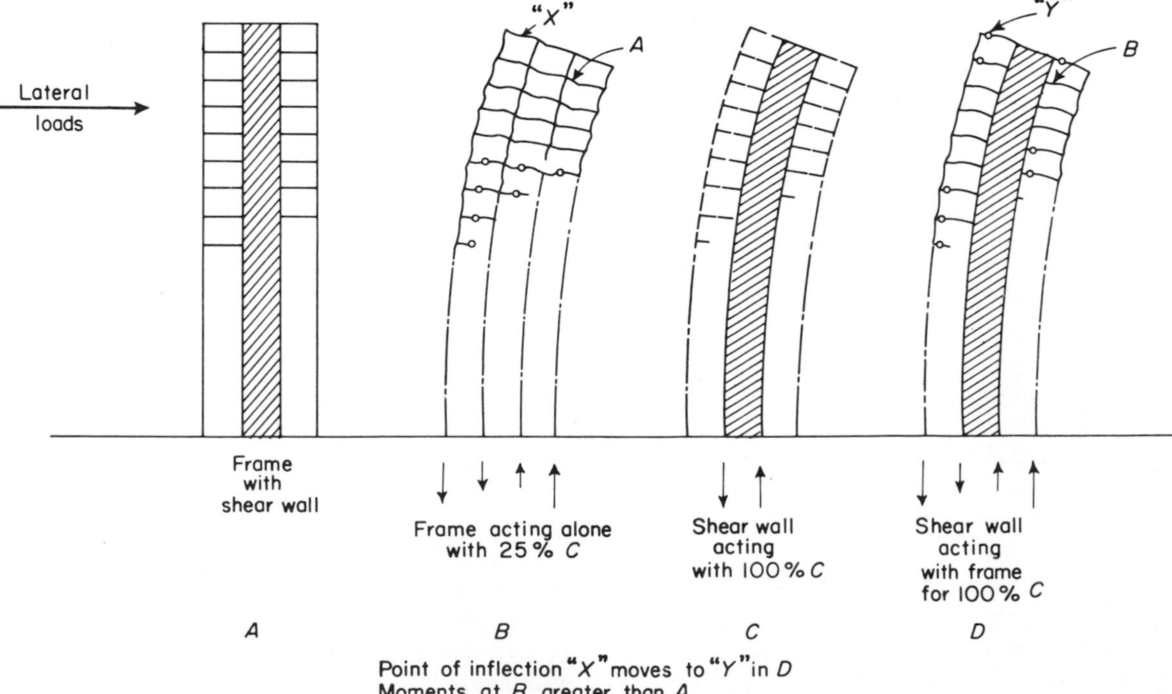

effect. If a structure, as in (a), consists of a shear wall in a structural frame, the design usually considers the frame to act alone with 25% of the code required force, as in (b).

Simultaneously, the shear wall, usually assumed to act alone, as in (c), is required to take 100% of the code required force without regard to its effect on the adjacent frame member. The condition shown in (d) is closer to the true action. The total lateral force is applied to the *system*. Because of the large rigidity of the wall, the points of inflection of the girders shift from the approximate midpoint of the girders (x) toward the outside columns (y) and the girder end moment adjacent to the shear wall becomes much larger. A large proportion of the overturning stresses are shifted from the shear wall itself to the outside columns. It will be found that the condition at (d)—shear wall and frame acting together—will produce the critical stresses in major portions of the frame.

17.7 DIAPHRAGMS

A major element that has escaped critical observation in modern steel-framed high-rise construction is the floor system acting as a diaphragm. Most research and analysis directed to metal deck diaphragms relates to low one-story buildings, developing allowable shear values and exterior chord requirements. The more critical situations in high-rise construction relate to openings, cores, and location of resisting elements.

To demonstrate certain conditions, consider the connections between the floor diaphragm and shear walls of a hypothetical structure. Figure 17.33 shows a typical steel-framed office floor 120 feet square, with all lateral forces resisted by the 20- by 24-ft elevator core. The height of the building is immaterial in this illustration, and for simplicity the torsional stresses as specified by code are neglected. The dead load on the floor is 1500 kips, and the reaction at the core is 10%g as required by code. Note that this is not related to the base shear. Beams are spaced 8 ft o.c. with a metal deck and $2\frac{1}{2}$-in. thick concrete floor fill over. The steel framing around the core is not recommended, as will become clear later.

Consider first the lateral loads in the north–south direction, with forces on the structure going toward the south. The shear stresses in the portion to the east (right) of Section 1-1 are 48 ft/120 ft × 150 kips = 60 kips for a shear loading of 500 lb/linear ft and a unit shear in the concrete of 16.7 psi. Considering the tension across. Section 2-2, there is a force of 50 ft/120 ft × 150 kips = 62 kips. Since the total resistance of this force is at points D and B, there is a tension of 31 kips at B. Correspondingly, there is a compression of 31 kips at points A and C. With steel framing as shown, there are steel beams at B and D to take this tension. If moment-resisting connections are used from beams to columns,

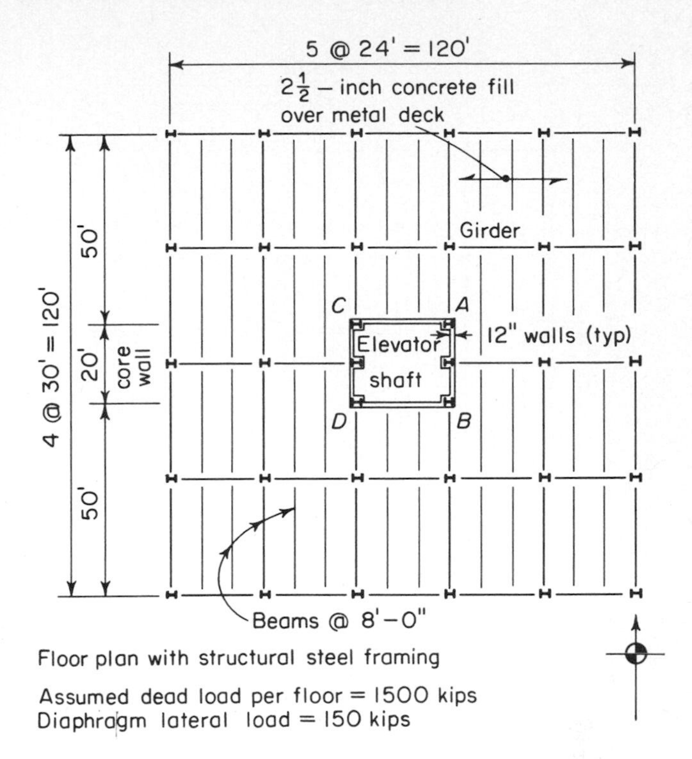

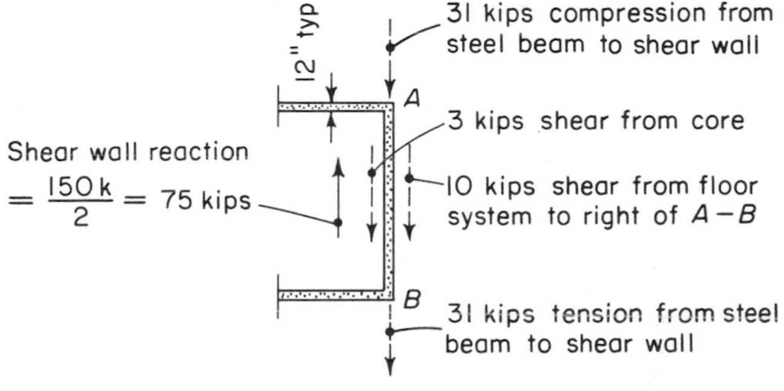

Fig. 17.33. Plan of steel-frame building—north/south lateral.

this tension is a small increase in design, but if shear connections only are used, this 31-kip tension may require some special connection at these corners. In order to handle the shears from the slab to wall between A and B, shear key or other devices should be used. Incidentally, the 50-ft length of beam connecting to point B must be connected to the concrete floor fill for a total shear of 31 kips, and this detail is frequently overlooked when these shear transfers are high.

Figure 17.34 shows the same floor, but it now is loaded with the lateral force going in the westerly (left) direction. The tension on Section 3-3 is 48 ft/120 ft × 150 kips = 60 kips, 30 kips of which are to connect in tension at the shear walls at C and D, but there is no steel member at that point to take the required tension. Along Section 4-4, the shear is 50 ft/120 ft × 150 kips = 62.5 kips, with shear of 520 lb/linear ft or 17 psi shear in the concrete. Again it is noted that no chord has been

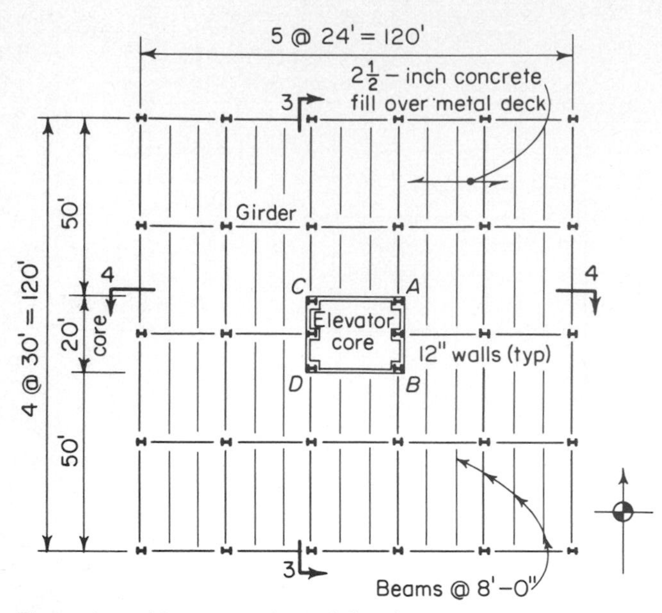

Floor plan with structural steel framing

Assumed dead load per floor = 1500 kips
diaphragm lateral load = 1500 x 10% = 150 kips.

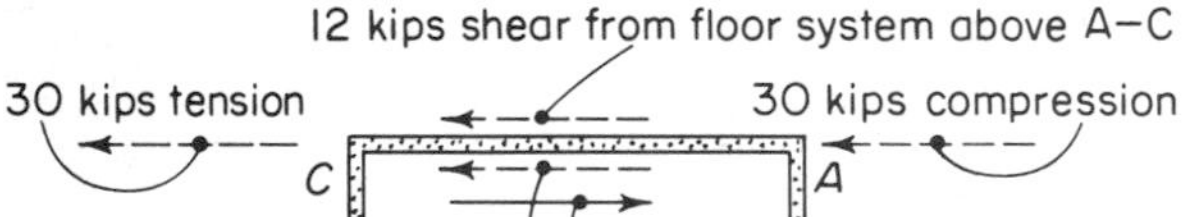

Shear wall reaction = 75 kips
3 kips shear from core

Stresses delivered to shear wall *AC*

Fig. 17.34. Plan of steel frame building—east/west lateral. (Used by Henry J. Degenkolb in previous work; see Wood, 1967.)

provided at *C* or *A* to transfer the 30-kip loads from the diaphragm to the shear walls. Figure 17.35 shows two ways of providing these required chords. On the upper left is the condition where steel framing is used, which would be preferred for this style of construction. A structural steel tie is provided for the design load of 30-kip tension. It must be anchored to the column and girder on the far side with the design forces to transfer adequately the load into the concrete. It must extend far enough to the left to transfer the shears through the diaphragm to such a point that the adjacent parallel girders can take the load. One advantage of this type of framing is the ability of the steel column to take the overturning chord stresses for the cantilever shear wall.

If a steel column is not provided in the core corners, as shown at point *A* to the right in Fig. 17.35, the required diaphragm tie can be provided with reinforcing steel in the slab. In this event the thickness of the slab must be checked to ensure adequate bar coverage. Again, the bars must extend far enough to transfer the tensions to adjacent girders or to typical slab reinforcing. While this detail may seem to be simpler and cheaper than the one shown for point *C*, the vertical chord steel for the shear walls conflicts with the carrying girder and the two Number 7 bars and often makes this detail impractical.

The illustrations used here are for a typical case where the core is near the center of the building. Where the core is at one end of the building as shown in the lower part of Fig. 17.35, the chord stresses at *X* and *Y* become much greater and of primary importance in the transfer of stresses to the shear walls.

Note that in earthquake-resistant design, as in all structural design, a complete, continuous stress path must be provided from the origin of the load to its final point of resistance, recognizing the relative rigidities of the various elements.

17.8 SUMMARY

In order to arrive at some basis for the design of our tall steel structures to resist earthquakes, the engineering profession is forced to adopt one of two approaches or some combination of them. One approach is to take an assumed ground motion, run a dynamic analysis on the structure, use engineering knowledge to assign the loads to their proper resisting elements, and then proportion the members and their connections to resist these loads. This can be done on either an elastic basis or on an elasto-plastic basis. This approach has certain difficulties. First, we do not have a ground motion record of a strong or great earthquake on the type of foundation material

Fig. 17.35. Steel detail. (Used by Henry J. Degenkolb in previous work; see Wood, 1967.)

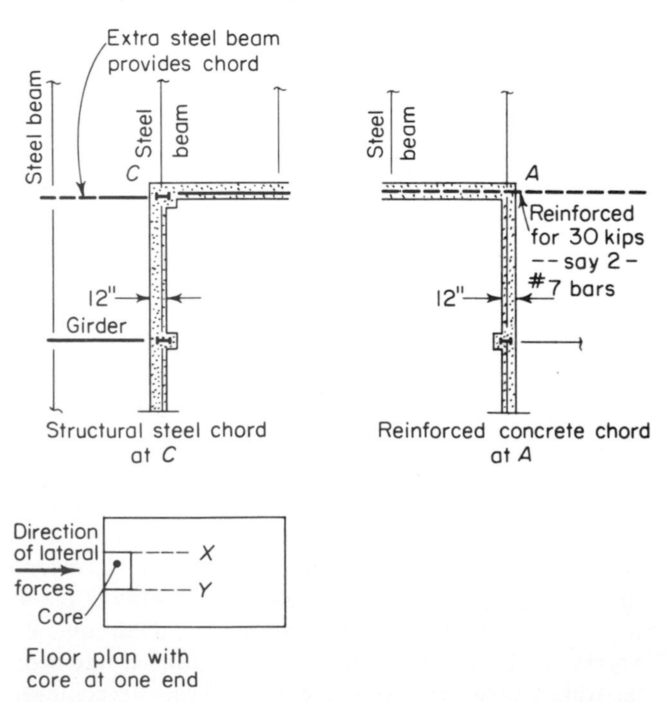

on which our big buildings are located. Second, while much research has been performed on the response and various parameters that govern response, all of it has been done on nice, regular, simplified elements. There has been little or no work done on models, either physical or mathematical, that resemble an actual building including its torsional motions, its variation in stiffness or loads throughout its vertical height, its discontinuous resisting elements, and many other actual practical factors. Third, there has been no work on the relative rigidities of differing elements under various dynamic loadings. Finally, there are great gaps in our knowledge relating to the physical performance of the structural elements under earthquake loadings. Such work that has been done at present on regular idealized models of buildings indicates that even with the moderate El Centro Earthquake (1940) motion, and under our present code prescribed forces, the full ductility ratio capacity of our structural materials is required—or may even be exceeded.

The other approach on which to base our designs is on the performance of our structures in the past, and this has been the basis for our code forces because major buildings in San Francisco up to 19 stories in height did survive the 1906 earthquake successfully. These tall structures that performed well had steel frames with very ductile connections and possibly members. It is important, therefore, to review their actual deformation capacities and to try to relate them to future designs. With the deformation capacities found from tests and the usual proportions of members to be found in high-rise buildings designed to resist major earthquake forces and with the types of connections commonly used in those buildings, properly designed to resist code-required forces, it is probable that ductility ratios of at least 8–10 or more on the frame as a whole will result.

However, the type of buildings we are building now do not even remotely resemble the structures on which our codes are based. In place of the integral masonry walls of the past that cracked and absorbed energy but still provided stability, we now have clean, bare frames with loose, nonstructural curtain walls, partitions that are deliberately kept free of the structure to prevent damage, lightweight metal decks, and a completely different type of structure. It is probably completely erroneous to extrapolate the successful performance of these old structures into a basis of design of our present, completely different structures—at least as the only approach.

It is necessary, therefore, that the design of connections, chords in diaphragms, shear transfers, and all other details and the proportioning of the various members be properly performed to assure a structure of sufficient ductility. It is of equal importance that this careful design and detailing be carried through into the actual field construction. This can be assured only through careful, thorough, knowledgeable field inspection, utilizing all the modern techniques of inspection now available.

REFERENCES

Beedle, L. S. (1954). "Recent Tests of Rigid Frames," talk given to Structural Engineers Association of California (unpublished).

Berg, U. T. (1933). "Wind Bracing Connection Efficiency," *Trans. ASCE*, **98**, 709–770 (see discussions by Gould, Cope, Young, Huntington).

Bertero, V. V., and E. P. Popov (February 1965). "Effect of Large Alternating Strains on Steel Beams," *Proc. ASCE*, **91**, 1–12.

Cooper, P. B., and S. J. Errera (August 1960). *Static Tension Tests of Structural Tee Joints*, Lehigh University Institute of Research, Fritz Engineering Laboratory Report No. 200.60.345.

Douty, R. T., and W. McGuire (April 1965). "High Strength Bolted Moment Connections," *Proc. ASCE*, **91**, 101–128.

Galambos, T. V., and M. G. Lay (May 1962). *End-Moment End-Rotation Characteristics for Beam-Columns*, Lehigh University, Institute of Research, Fritz Engineering Laboratory Report No. 205A.35.

Galloway, J. D., M.C. Couchot, C.H. Snyder, C. Derleth, Jr., and C. B. Wing (1907). "The Effects of the San Francisco Earthquake of April 18, 1906–Report of Committee on Fire and Earthquake Damage to Buildings," Appendix B, *Trans. ASCE*, **59**, 223–244.

Lee, G. C., A. T. Ferrara, and T. V. Galambos (March 1963). *Experiments on Braced Wide-Flange Beams*, Lehigh University, Institute of Research, Fritz Engineering Laboratory Report No. 205 H.6.

Lee, G. C., and T. V. Galambos (February 1962). "Post Buckling Strength of Wide Flange Beams," *Proc. ASCE*, **88**, 59–78.

Rathbun J. C. (1936). "Elastic Properties of Riveted Connections," *Trans. ASCE*, **101**, 524–596.

Toriggino, A., and E. L. Cope (1933). "Discussion of 'Wind Bracing Connection Efficiency,'" *Trans. ASCE*, **98**, 756–761.

Welding Research Council–ASCE Joint Committee, (1961). "Commentary on Plastic Design in Steel," Chap. 6, *ASCE Manual No. 41*.

Wood, F. J. (ed.) (1967). *The Prince William Sound, Alaska, Earthquake of 1964 and Aftershocks*, Vol. II, Washington, D.C.: U.S. Department of Commerce, Coast and Geodetic Survey.

Young, C. R., and K. B. Jackson (1934). "The Relative Rigidity of Welded and Riveted Connections," *Can. J. Res.*, **11** (1), 62–100(July); **11** (2), 101–134 (August).

Chapter 18

Design of Earthquake-Resistant Poured-in-Place Concrete Structures

JOHN A. BLUME

President, John A. Blume & Associates, Engineers
San Francisco, California

18.1 INTRODUCTION

Many aspects of the design of poured-in-place concrete structures to resist earthquakes apply to structures of any material or combination of materials or to various methods of construction. This chapter discusses certain of these aspects that may not be covered elsewhere in this book, and it emphasizes others that apply directly to reinforced concrete buildings and other structures constructed by placing a concrete mix in forms representing the final configuration and location of the overall structure. These forms contain nonstressed reinforcing bars. Precast, prestress, and poststress construction are covered in the following chapter.

Great strides have been made in recent decades in the fields of earthquake engineering and structural dynamics. These advances have been made possible by increased general interest in, and support of, intensive studies of earthquake problems. This interest in turn has resulted from continuing earthquakes and resulting disasters in many parts of the world coupled with a growing fund of knowledge and data, significant new concepts, and improved equipment with which to work such as the high-speed digital computer. Some of the concepts and techniques are so new as not to be generally understood or to be generally applied in design. Some recently have been included in building code provisions for earthquake resistance. Although there is much more to be learned in the future, if what is known now were to be generally applied in the design and construction of poured-in-place concrete structures, the risk associated with the future earthquake shaking of such structures would be reduced to a very low level.

18.2 BASIC CONCEPTS

There are certain basic concepts that must be recognized in order, first, to understand the nature of the problem and, second, to properly cope with the problem in design. These have to do with the earthquake demands, the work capacity of the structure, and the fact that poured-in-place reinforced concrete may be designed to be ductile and have great work capacity prior to failure.

18.2.1 Earthquake Demands

The demands of the earthquake on the structure are uncertain as to time, frequency of occurrence, and intensity. Assuming for the moment that lateral forces represent the earthquake loading, it is not the case that the design forces prescribed by seismic codes are the maximum forces to which a structure in an active seismic area may be subjected. The actual forces—so long as the structure remains elastic—may be as much as several times the code specified forces during severe earthquake demand (Blume, 1960a,b). Whether a particular structure would ever be subjected to such great forces depends not only on its structural and dynamic characteristics but upon compound probabilities involving the earthquake location relative to the structure; the earthquake energy; the depth of the earthquake focus; the media of the source, the propagation path, and of the point of reception of the earthquake waves; the duration of shaking; the frequency composition of the disturbing waves; and other factors. No other phase of structural loading includes such a wide variation of possible demands. Although equivalent forces long have been used for floor live loading, for wind, for train or truck loading on bridges, etc., these demands do not exceed the accepted equivalent design forces by any such amount as is possible in the earthquake problem.

This leads to philosophical questions and judgments such as the following: What should the design basis be? What is the probability of what loading? What probability should be the basis for design? One simple approach is to consider the average frequency of occurrence for a specific area based upon the data available. At best this rate of return method is a poor vehicle for sparse real data. Assume that a certain type and intensity of earthquake demand has occurred an average of once in 100 years, based upon a short 200-year history. This provides no real answer as to when the next demand will be—it could occur tomorrow, or several hundred years from now. There are, of course, more sophisticated approaches to the problem of risk (beyond the scope of this chapter), but they all suffer from lack of adequate data. It may be convenient to take refuge in a small probability—of, say, that associated with 2 or 3 standard deviations—as being of no concern or as an "accepted risk." It is not logical, however, to consider a once-in-50-year possibility as being of no concern during the design of a 40-year life structure—the "once" may occur in the first, or in any, year of the structure's life.

The most sensible approach is to design on some reasonable basis, to recognize the uncertain nature of the demands, and to provide all the reserve capacity that can be mobilized in a severe future emergency but at little or no extra cost in initial construction cost. Of course, unusual risks should have additional design considerations and refinements.

18.2.2 Capacity of the Structure

If the demand should exceed the capacity, failure results. Although the demand—with few exceptions—is a random variable beyond the control of the designer, the mean capacity is to a large extent in his hands. A basic concept, not yet generally understood, is that energy absorption or work capacity is a more meaningful measure of earthquake capacity of most structures than strength alone (Blume, 1960a; Housner, 1956). It is to be noted, however, that in the elastic or linear range of structural response, energy capacity is proportional to the applied force. The problem is that the code lateral forces may be exceeded during actual earthquakes and the unit deformations associated with the yield point —or its equivalent—may be greatly exceeded. The performance of the structure in such cases depends upon its properties beyond the initial yield point. If there is no ductility and if there is no alternate stress path, the structure collapses under continued earthquake demands of similar or greater amount. On the other hand, ductility and reserve inelastic energy capacity and/or alternate and stronger stress paths may be available and may save the structure. Although seismic codes specify lateral forces rather than work capacity, some attempts have been made to allow for the value of ductility (Structural Engineers Association of California, 1966; International Conference of Building Officials, 1967) and to

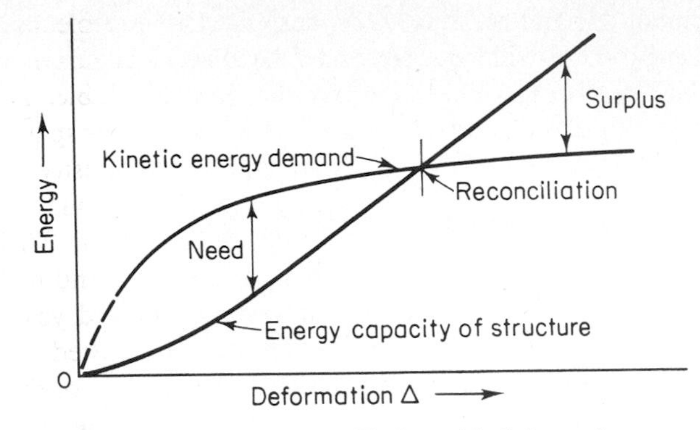

Fig. 18.1. Energy reconciliation with deformation.

include the parameter of deformation as well as strength. More extensive procedures are available to more rationally provide for energy capacity (Blume, 1960b; Blume, Newmark, and Corning, 1961).

It can be demonstrated that earthquake response depends upon the natural periods and damping of a structure as well as the frequency content and time history of the earthquake motion. In the short period portion of a spectral response diagram, acceleration (or force based upon Newton's second law) is the most meaningful criterion. In the long period portion, displacement is most meaningful. In the remaining portion—and generally the most significant band for most natural building periods—velocity is the most significant index. Of course, acceleration or displacement can be converted to velocity at any frequency with the generally justified assumption of harmonic response motion. Since kinetic energy (demand) is related to the square of velocity, it follows that work capacity under deformation on the basis of the conservation of energy is a most meaningful parameter for buildings. Figure 18.1 indicates the relationship of energy demand and work capacity under increasing distortion.

18.2.3 Ductile vs Ordinary Concrete

The most important point in this chapter is that poured-in-place reinforced concrete structures may be essentially brittle or they may have great inelastic work capacity, depending upon the amount and the disposition of the reinforcing steel relative to the concrete sections. New design procedures are required to ensure ductility and energy absorption, the importance of which can be inferred from the uncertainty of the earthquake demand previously outlined. Because of the significance of ductility and some confusion as to terms and procedures, the familiar electrical designations DC and AC are adopted here, with appropriate meanings as follows: DC refers to "ductile concrete," which is so designed as to ensure that in flexural members shear failure and compression failure in the concrete cannot occur prior to stretching of the tensile bars, and in compression members shear failure cannot occur and any concrete that fails in compression will be confined. AC refers to "any (other) concrete" that may or may not be subject to failure as described for DC, depending upon several conditions not checked or controlled in the design process.

The ultimate strength method of design, ACI 318 Part IV-B (American Concrete Institute, 1965), is not to be confused with design for DC concrete. Structures designed by either the working stress method, ACI 318 Part IV-A, or by the ultimate strength method may be DC or AC, depending upon conditions not controlled unless the initial design (by either method) is subsequently checked and refined as indicated by DC procedures.

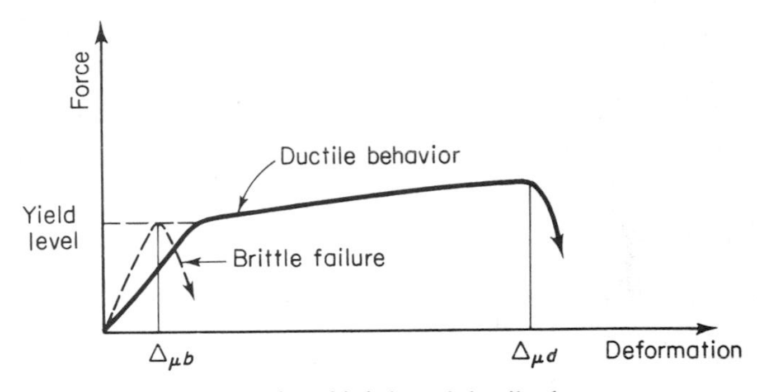

Fig. 18.2. Capacity of brittle and ductile elements.

The capacity of a structure—based upon the logical criterion of ability to absorb earthquake energy rather than on strength alone—is vastly increased with DC as compared to brittle material. Figure 18.2 indicates the relative work capacities of two hypothetical elements or structures, each with the same yield point strength. The area under the curves to Δ_{ub} and to Δ_{ud} respectively represents work capacity, which is obviously much greater—along with the probability of survival in a strong earthquake—for the ductile element.

18.3 THE EARTHQUAKE PERFORMANCE OF POURED-IN-PLACE REINFORCED CONCRETE STRUCTURES

In considering the earthquake performance of any structure, it is essential to evaluate the particular nature of the earthquake exposure and of the characteristics of the exposed structures in the light of current knowledge. Broad generalizations can be, and have been, very misleading. There are countless possible earthquake disturbances, and yet relatively very few severe shocks have occurred in densely populated areas. However, within the last few decades there have been great losses in many countries, including Chile, Colombia, Ecuador, Greece, Iran, Italy, Japan, Mexico, Morocco, New Zealand,

Peru, Russia, Turkey, Venezuela, Yugoslavia, and the United States.

There has been damage to buildings of all materials, and some buildings of all consistent materials have remained undamaged. Detailed analyses in the light of current knowledge provide answers to the apparent anomalies, however, and add to the fund of knowledge. Insofar as poured-in-place concrete structures are concerned, a great deal of inadequate design and construction has been exposed. In most cases, where damage occurred, there was no specific earthquake-resistant design and, in some cases, the aseismic provisions were inadequate and/or were inadequately executed. Considering the design and construction in the light of today's knowledge, the performance was predictable. Unfortunately, no buildings of modern DC design have been exposed—in fact, very few have yet been designed—and so that record does not yet exist. There is every reason to believe, based upon damaged buildings that did not conform to both good seismic design practice and to DC requirements, that (1) both are needed and (2) if both are provided along with good construction the results would be quite satisfactory. It is essential that this new concept—DC—be added to the designer's functions and responsibilities. Modern seismic codes provide for this (Structural Engineers Association of California, 1966; International Conference of Building Officials, 1967; Blume, Newmark, and Corning, 1961. Note: Subsequently these will be referred to as SEAOC, UBC, and BNC, respectively).

18.4 FORCE-DEFORMATION CHARACTERISTICS

A scale plot of force vs translation or of applied moment vs rotation reveals a tremendous amount of information about the earthquake resistance of a member, assembly, story, or whole structure. Not only is the strength shown, but also the deformation and the relationship of strength and deformation at various levels of loading. The slopes of the plot and the changes in the slopes are most meaningful. For example, a negative slope beyond which there is no positive slope or zero slope indicates failure under continued loading. The area under the curve (the integral) indicates the sum of stored elastic energy in the linear range and the capacity to do work in the inelastic range. Figure 18.3 indicates various force–deformation characteristics under slow, or "static," loading. These depend upon materials, connection details, and combinations of materials and elements.

An important consideration is that there are various types of loading, failure, and unit stress and that the shape of the force–deformation curve varies not only

Fig. 18.3. Common types of inelasticity: (a) nonlinear, softening; (b) elasto-plastic; (c) bilinear, softening; (d) "plateau" resistance.

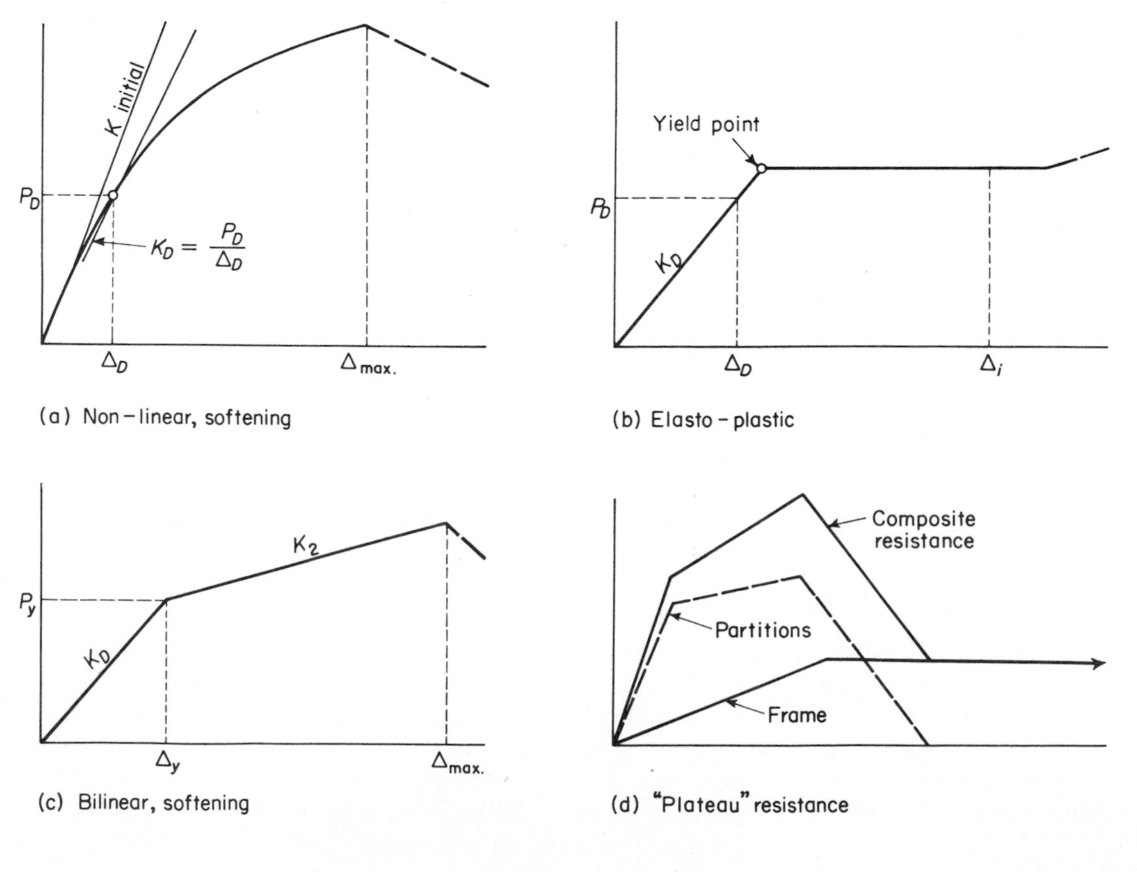

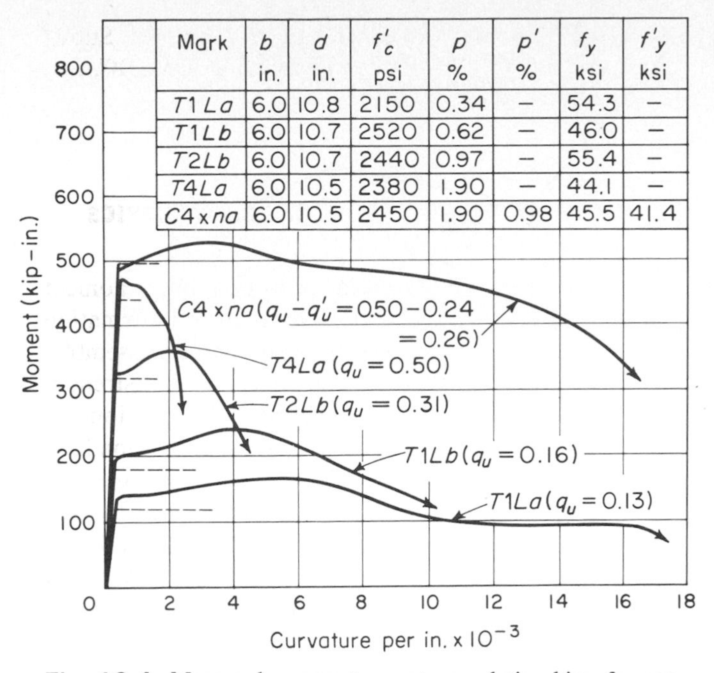

Mark	b in.	d in.	f'_c psi	p %	p' %	f_y ksi	f'_y ksi
T1La	6.0	10.8	2150	0.34	—	54.3	—
T1Lb	6.0	10.7	2520	0.62	—	46.0	—
T2Lb	6.0	10.7	2440	0.97	—	55.4	—
T4La	6.0	10.5	2380	1.90	—	44.1	—
C4xna	6.0	10.5	2450	1.90	0.98	45.5	41.4

Fig. 18.4. Measured moment–curvature relationships for sections subjected to bending only. (Gaston, Siess, and Newmark, 1952; Blume, Newmark, and Corning, 1961.)

with materials but with the type of stress and failure. A slender mild steel bar has a different response to pure tension than it does to compression in which long column action, or buckling, would be the governing condition. A concrete beam failing in shear has a much different response than one failing by stretching of the tensile reinforcing bars. An entire story of a building with alternate stress paths would have different characteristics than a single member or a single joint of a member in that story.

Under certain stress conditions, force–deformation characteristics are different under rapid or dynamic loading than under slow or static loading. Mild steel in tension is generally stronger under rapid than under slow loading. However, it is often difficult and costly to obtain dynamic data. Moreover, the differences generally are not great. For these reasons plus the fact that the error is on the side of safety, static test data may be used to assess characteristics under earthquake response. Another important point is that the number of load cycles induced by earthquakes is so small compared with the number of cycles necessary to cause fatigue failure that fatigue failure need not be considered in the design of an earthquake-resistant building (BNC, 1961). Of course, this assumes that adequate repairs are made between earthquakes.

Figure 18.4 indicates the moment–curvature characteristics of reinforced concrete test beams with various degrees of ductility determined by the amount of the reinforcing steel (BNC, 1961; Gaston, Siess, and Newmark, 1952). Only the very ductile characteristics are permitted under ductile concrete (DC) design procedures. The reasons for this are obvious from the poor ductility and low work capacity of the nonductile specimens T4La and T2Lb in Figure 18.4 These specimens are overreinforced and thus the concrete fails in compression. The double reinforcing of specimen C4xna greatly improves the characteristics.

18.4.1 Ductility

It is convenient to define ultimate ductility μ_u as the ratio of the maximum deformation prior to failure to the deformation at initial yield. Although this term is generally used in connection with elasto-plastic behavior, it can be applied to any inelastic behavior pattern if failure deformation is defined as that deformation beyond which the slope of the force-deformation plot remains negative. Figure 18.5 indicates the terms and conditions for several typical cases. Ductility μ may be considered as a demand ductility or the ratio of the required deformation under any demand i to that at the yield point.
Thus

$$\mu_u = \frac{\Delta_u}{\Delta_y} \tag{18.1}$$

$$\mu_i = \frac{\Delta_i}{\Delta_y} \tag{18.2}$$

where Δ_y = the yield point deformation; Δ_u = the deformation beyond which point the slope of the force–deformation curve is always negative; Δ_i = the deformation under condition or trial i; and μ_i, μ_u = ductility factors as defined above.

It is important to note that the maximum deformation Δ_u may be established by direct stress, by secondary stress, by buckling, by local failure, or by whatever condition or combination of conditions—for the particular element or structure under consideration—leads to the minimum value of Δ_u. Where there is indeterminacy or there are multiple stress paths in a structure or element such as a building story, there may be local failure at one or several points, but the overall force–deformation characteristic may not be at the initial yield. Of course, the element or overall Δ_u is established at whatever point an overall failure mechanism first develops as defined above.

Ductility also may be measured in rotational units, such as the ratio of ultimate member curvature to curvature at yield.

18.4.2 Energy Absorption Capacity

Ductility alone is not adequate for energy reconciliation. There could be a large value of μ associated with a small force, and the area under the curve would be inadequate. At the other extreme, a very great force

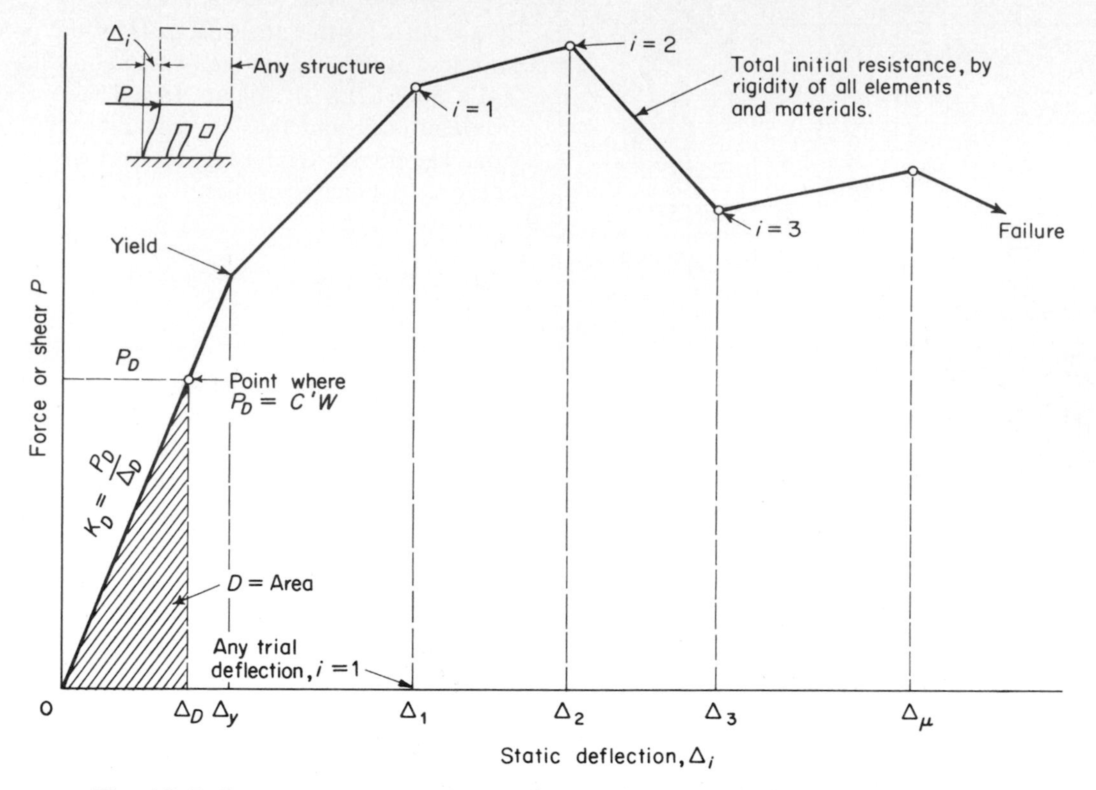

Fig. 18.5. Reserve-energy geometry, general case. (Blume, Newmark, and Corning, 1961.)

could develop sufficient energy in the elastic range or at a value of μ between 0 and 1. In most buildings, however, μ_u should greatly exceed 1 under normal code design conditions in order that the structure be able to mobilize adequate work capacity in the event of a severe earthquake. The reserve energy technique (Blume, 1960b; BNC, 1961), discussed subsequently, provides for both energy and force development under seismic code conditions and normal design terminology.

Unfortunately, a great many laboratory tests of materials, members, and of assemblies of members were not extended into the inelastic range of deformation. Thus there is a wealth of data on ultimate loads and yield point unit stresses but a relatively small amount on μ_u and inelastic work capacity. It is only in recent years that the inelastic range has been of interest to engineers as a means of reconciling earthquake performance with structural characteristics (Blume, 1960a; 1960b; Housner, 1956; BNC, 1961). A compilation of references on inelastic test data on various materials has been developed (Blume, 1962). An intensive study of the inelastic behavior of reinforced concrete members and of means of providing ductility and energy-absorbing capacity was sponsored by the Portland Cement Association. This work culminated in a book (BNC, 1961) that first provided data for DC design; it will be referred to from time to time in this chapter. Considerable testing has been done in the inelastic behavior of reinforced concrete members, joints, and assemblies. It has been clearly shown that reinforced concrete can be designed to have ductile characteristics, and the 1967 Uniform Building Code (International Conference of Building Officials, 1967) requires ductility for tall, and certain other, buildings. Some of the test data will be provided herein.

18.4.3 Inelastic Behavior of Reinforced Concrete

Figure 18.6 is a force–deformation diagram for a test beam in the PCA series conducted at the University of Illinois (BNC, 1961). This specimen had a μ_u value of about 16. The μ values for the loads numbered 14, 58, and 90 are 1, 2.5, and 8 respectively. Failure beyond point 153 occurred as the compression reinforcement buckled. Figure 18.7 is for a similar test beam but with equal positive and negative reinforcement that was subjected to load reversals as indicated by the arrows.

In a more recent test series conducted by the Portland Cement Association to specifications outlined by the seismic committee of the Structural Engineers Association of California, the ductility of various types of joints was demonstrated. Figure 18.8 indicates a specimen and the loading and reaction points, and Figure 18.9 gives the force–deformation diagram for test specimen VA under cycled loading (Bennett *et al.*, 1965). The ductility

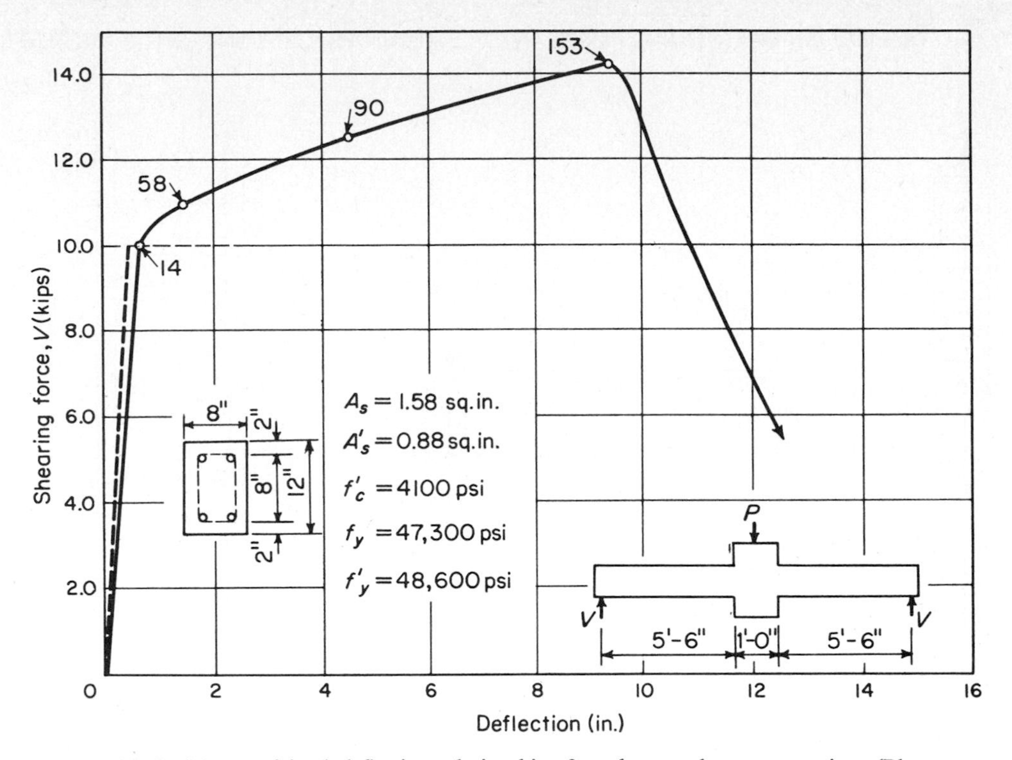

Fig. 18.6. Measured load–deflection relationships for a beam-column connection. (Blume, Newmark, and Corning, 1961.)

Fig. 18.7. Measured load–deflection relationships for a beam-column connection subjected to reversals of load. (Blume, Newmark, and Corning, 1961.)

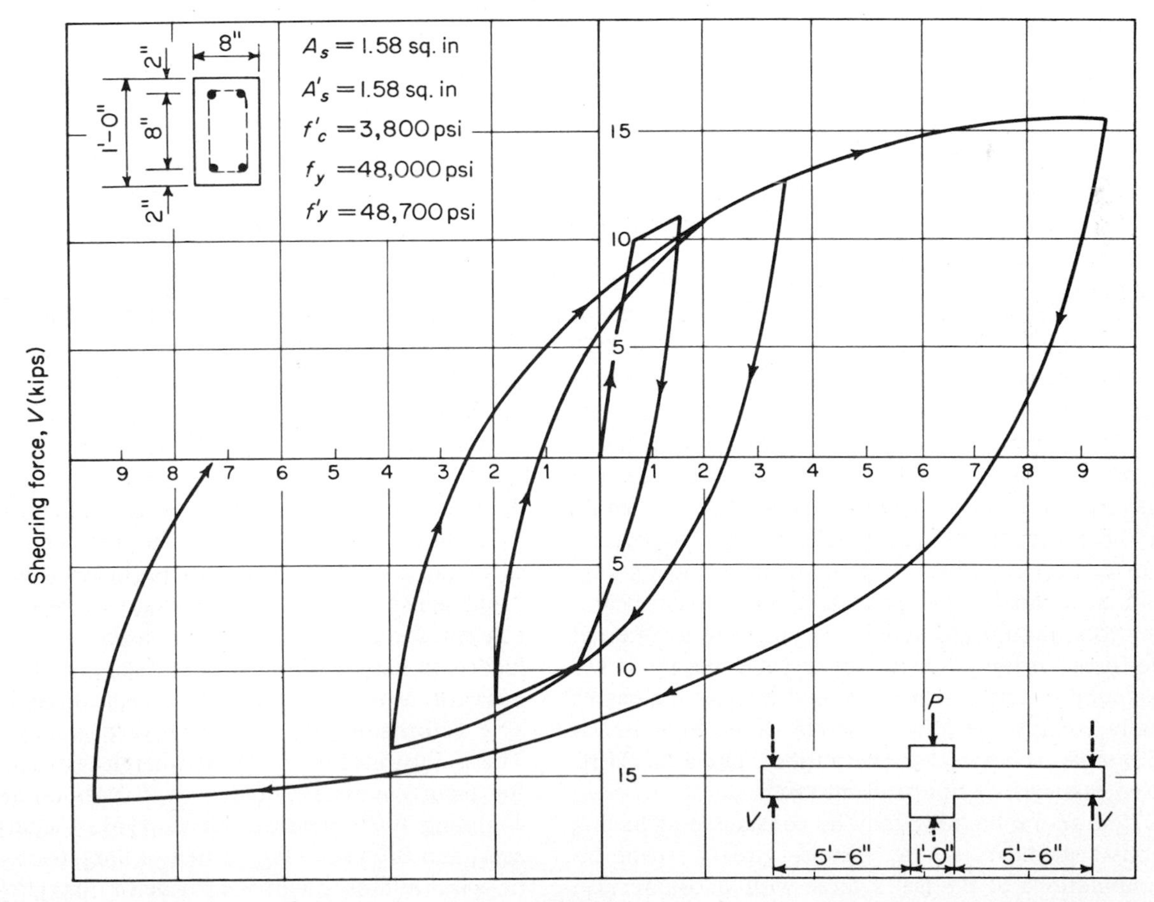

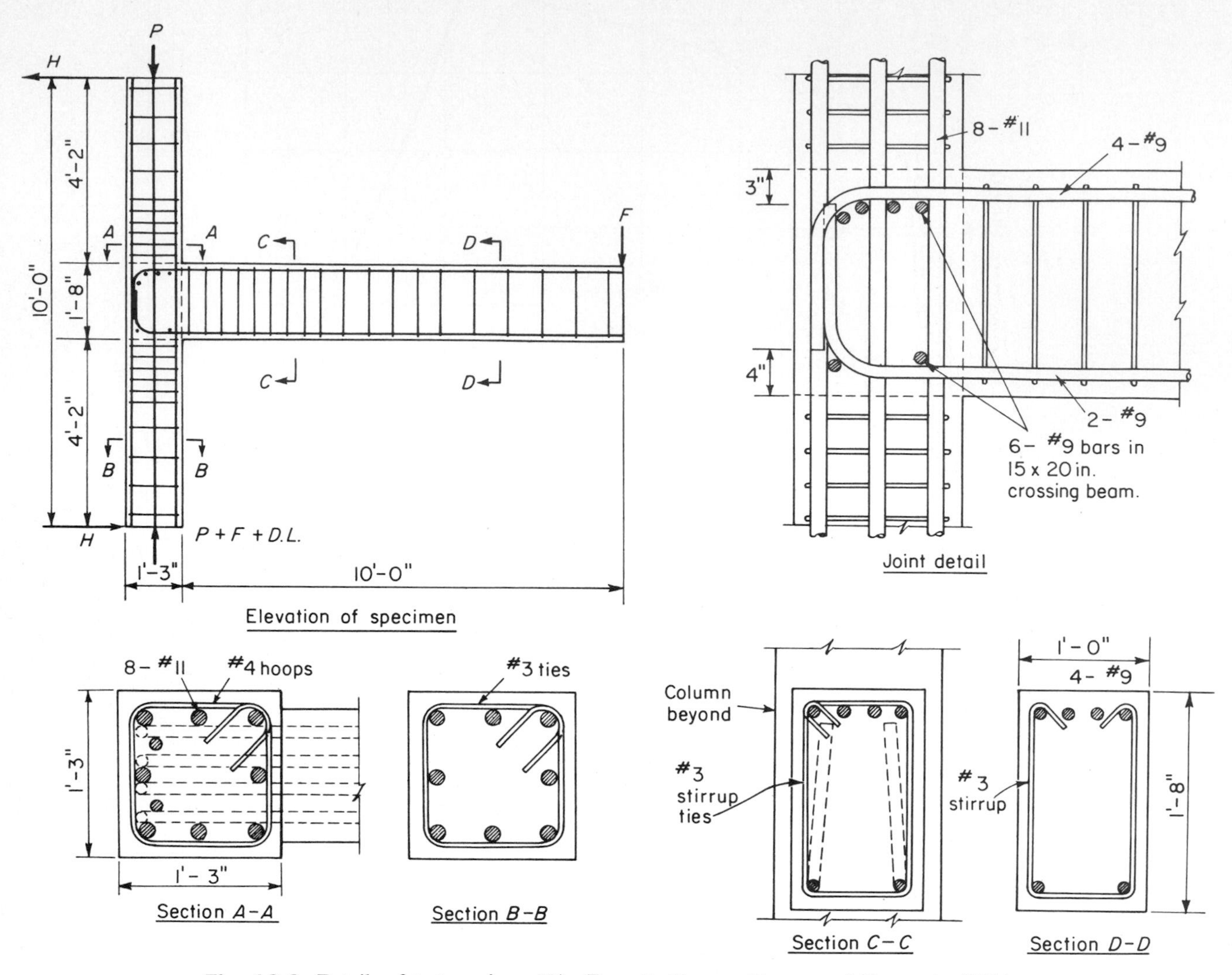

Fig. 18.8. Details of test specimen VA. (Bennett, Hanson, Parme, and Sbarounis, 1965.)

is good in spite of several specimen details that are not in accord with the best recommended practice for DC. Other specimens without the joint stub (representing a typical spandrel beam normal to an exterior span test beam) had somewhat less ductility. The tests, intended to be severe and to reveal any weaknesses in laterally unrestrained exterior joints of a concrete frame, indicated again that ductility and energy-absorbing capacity can be provided even at unfavorable joints and under severe column axial loading. A free-standing exterior beam-column joint requires some joint hoops to resist internal tensile forces in the joint. Although such free joints are rarely found in actual structures, and intersecting beams generally confine the joint concrete, some joint hoops are desirable as an added precaution. The 1967 UBC requirements specify hoops in all joints.

A rigid frame building may be considered as having horizontal members (slabs, beams, girders, spandrels or combinations of the latter three with slabs), vertical members such as columns and walls, and joints. The horizontal members are subjected to moment and shear and a minor amount of axial force and their loading is basically flexural in nature. They can be designed so as to have great ductility and energy absorption capacity.

Columns are basically subjected to compressive axial forces, although considerable moment and shear also may be developed especially under earthquake motion. In addition to interaction of moment and axial load, as in normal concrete design problems, the matter of "confinement" must be included in consideration of column force–deformation characteristics. If material loaded in axial compression is restrained from bulging outward, or is confined, the load capacity and the allowable deformation before failure are greatly increased. This is illustrated in Fig. 18.10, wherein test data are shown for plain concrete cylinders with various amounts of confining fluid pressure (BNC, 1961; Richart, Brandtzaeg, and Brown, 1928). In lieu of fluid, transverse hoop or spiral reinforcement is provided in design practice to confine the core concrete. The "shell" concrete, that

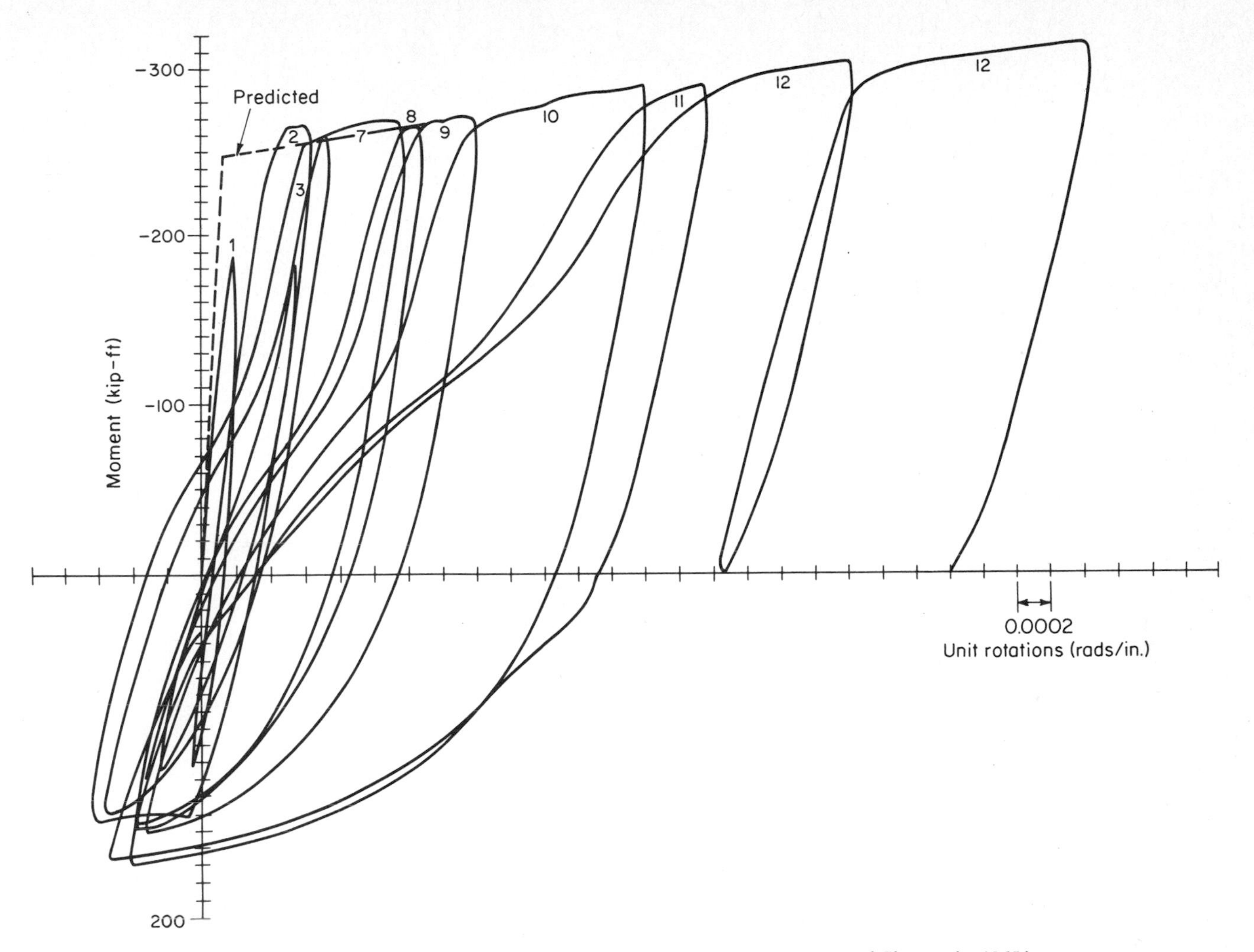

Fig. 18.9. Test results for specimen VA. (Bennett, Hanson, Parme, and Sbarounis, 1965.)

Fig. 18.10. Stress–strain curves from compression tests of confined concrete cylinders. (Blume, Newmark, and Corning, 1961.)

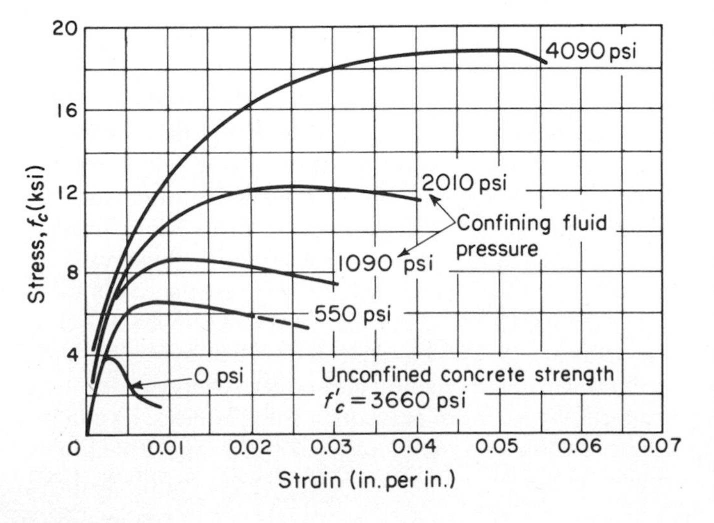

outside the transverse reinforcement, is of course of no value under severe loading conditions. Thus for any given column section, the ductility available under axial force and moment will depend upon the amount of axial force and of moment (interaction phenomena) and also upon the capacity of the transverse reinforcement to confine the core concrete. A confined concrete column section under axial load and moment has much greater ductility at a given axial load below the balance point than does an unconfined section.

Figure 18.11 shows theoretical interaction curves for a specific concrete section, and Fig. 18.12 indicates the ductility ratio for curvature for the same section in the confined and in the unconfined state and under various column load ratios (BNC, 1961). In general, columns have less ductility than horizontal members or they are relatively more costly if designed to have equal ductility. For this and for other reasons it is often desirable to proportion a structure so that "hinges" will develop in the horizontal rather than in the vertical members under

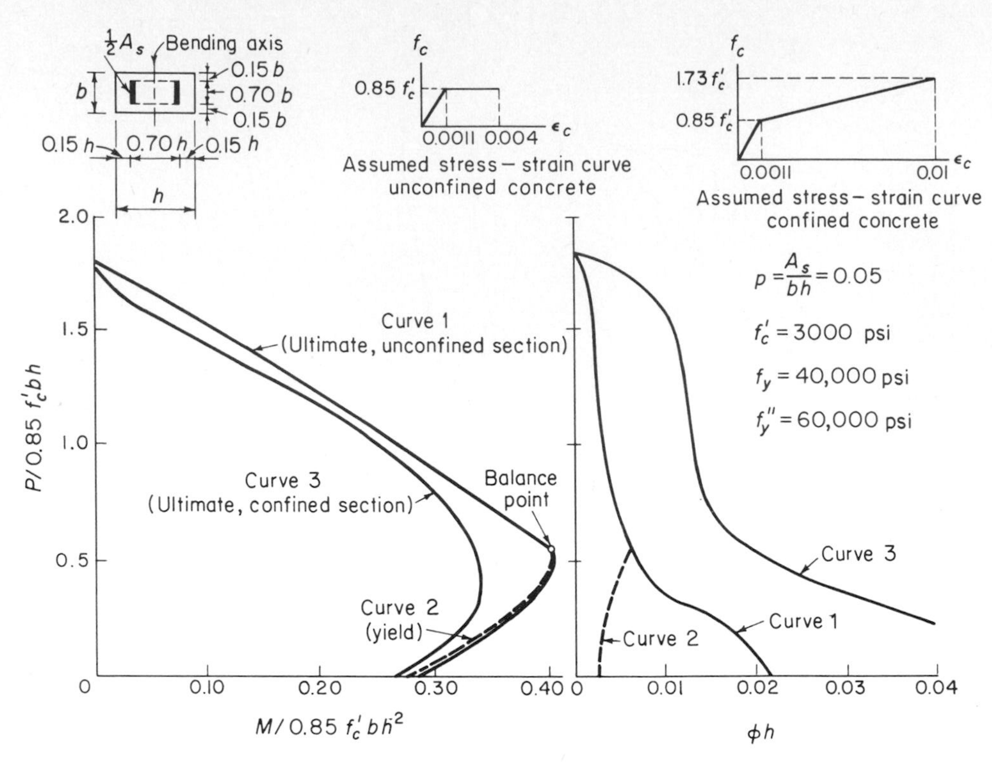

Fig. 18.11. Interaction curves for a rectangular section having a confined concrete core. (Blume, Newmark, and Corning, 1961.)

severe lateral loading. This is not difficult to do under normal building spans, and it is required for ductile concrete in the 1967 Uniform Building Code.

Beams and girders are made ductile by so reinforcing them that concrete shear failure and compression failure in flexure are impossible. The desired mode of failure is stretching of the tensile reinforcement, so the members are purposely underreinforced and some compression steel is usually provided. Figure 18.13 shows rotational ductility for two values of f_y and for d'/d ratios of 0.1. The ratio of compression to tensile steel is the parameter. These ductility determinations are explained in detail in BNC (1961). As an example, if q_u were 0.13 and f_y were 40,000 psi with $d'/d = 0.1$, the ratio ϕ_u/ϕ_y would be 16 with no compression steel. However, if q_u were 0.24, p'/p would have to be 0.5 to provide the same ductility.

The joints are subject to moment, shear, and axial forces, and sometimes torsion as well, delivered by the columns and the intersecting horizontal members. Although the internal stresses are complex, they can be resolved by rational methods if required. Empirical data from test programs indicate that intersecting horizontal members and continuous bars strengthen the joints adequately under most conditions so that hinge development occurs outside the joint. Exterior joints require transverse ties to resist internal diagonal tension, and it is good practice to provide some ties in interior joints as well. The code requirements for ductility conservatively specify more joint hoops than would be essential in most cases (SEAOC, 1966; International Conference of Building Officials, 1967). Thus reinforced concrete may be made to have adequate ductility and reserve energy absorption capacity by means of preventing shear failures, by confining any compression failures, and by forcing local failures under severe emergency loading to occur mainly by local

Fig. 18.12. Variation of ϕ_u/ϕ_y for tied columns of unconfined and confined concrete in respect to axial load. (Blume, Newmark, and Corning, 1961.)

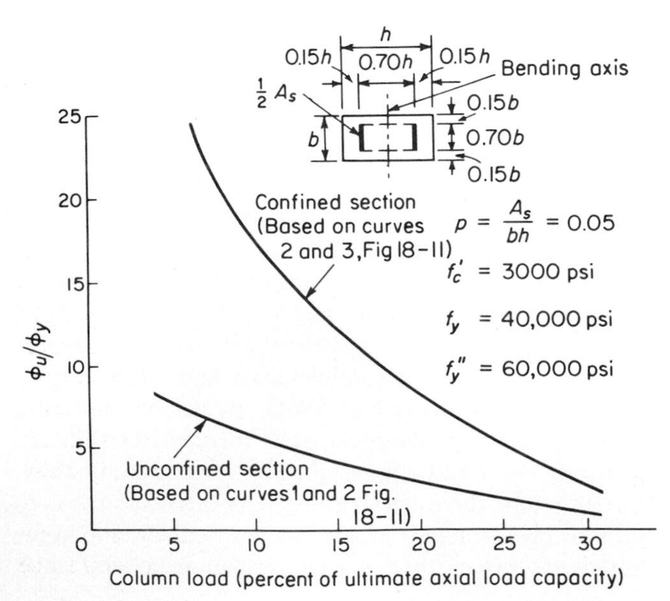

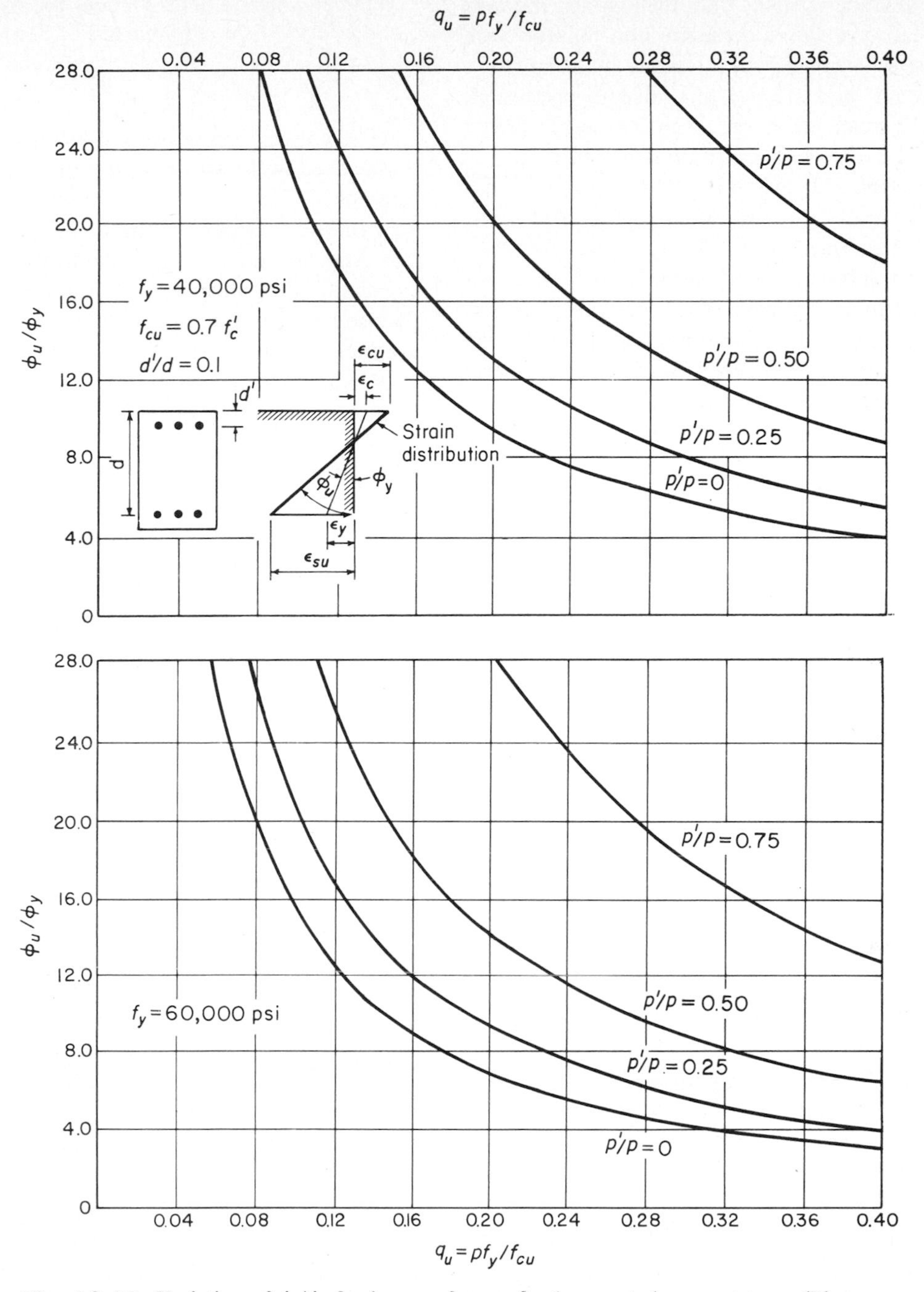

Fig. 18.13. Variation of ϕ_u/ϕ_y for beams of unconfined concrete in respect to q_u. (Blume, Newmark, and Corning, 1961.)

stretching of tensile steel from moment in horizontal members. The means of doing this will be considered in the following sections.

18.5 DESIGN REQUIREMENTS AND OPERATIONS

Historically, specific design for earthquake resistance is recent—within the last few decades—and the design requirements and operations cover a wide spectrum. A great many existing buildings, of all materials, have had no earthquake-resistant design per se. Some of these structures, however, may possess considerable seismic resistance because of their geometry, because of designed wind resistance, or because of other fortunate but happenstance conditions. Other buildings may be very poor risks indeed. Hopefully, these latter structures will not be exposed to severe earthquake demands in the remaining years of their useful life.

Buildings designed under earthquake codes of one type or another and with various amounts of lateral force are generally much better risks than their nonearthquake predecessors. However, even these are not immune from risk and damage because (1) the seismic building codes basically were not, and are not, intended to prevent all damage and (2) much more has been learned in recent years and decades about the earthquake problem and of how to deal with it. For example DC only recently has been introduced and brought into code requirements. Many pre-DC buildings are of the type that may not require DC characteristics and some others may accidently, and fortunately, possess the desirable properties. Another reason for possible damage is that code provisions have not always been completely or properly applied, in both design and construction, as is necessary. In the future, however, none of these matters should be left to chance.

The requirements for ductile concrete (DC) were first proposed by Blume, Newmark, and Corning 1961. In 1966 the Seismology Committee of the Structural Engineers Association of California (which developed the seismic provisions of the 1961 and 1964 editions of the Uniform Building Code, 1960) issued a revised document (1966) that includes two new proposed code sections: 2630, "Ductile Moment-Resisting Space Frames," and 2631, "Concrete Shear Walls and Braced Frames." Section 2630 with some minor exceptions and additions has the same requirements for DC as Chapter 6 of BNC (1961). Together with these new sections, the SEAOC 1966 revision of "Recommended Lateral Force Requirements" includes a revised section, 2313(j), as follows:

> (j) Structural Systems. 1. Buildings more than one hundred and sixty feet (160′) in height shall have ductile moment-resisting space frames capable of resisting not less than 25% of the required seismic force for the structure as a whole. All buildings designed with a horizontal force factor (K) of 0.67 or 0.80 shall have ductile moment-resisting space frames.
>
> Moment-resisting space frames and ductile moment-resisting space frames may be enclosed by or adjoined by more rigid elements which would tend to prevent the space frame from resisting lateral forces where it can be shown that the action or failure of the more rigid elements will not impair the vertical and lateral load resisting ability of the space frame.
>
> 2. Construction. The necessary ductility for a ductile moment-resisting space frame shall be provided by a frame of structural steel conforming to ASTM A-7, A-36 or A-441 with moment-resisting connections, or by a reinforced concrete frame complying with Section 2630 of this Code.
>
> Shear walls in buildings where $K = 0.80$ shall be composed of axially loaded bracing members of ASTM A-7, A-36 or A-441 structural steel; or reinforced concrete bracing members or walls conforming with the requirements of Section 2631 of this Code. Reinforced concrete shear walls and reinforced concrete braced frames for all buildings shall conform to the requirements of Section 2631 of this Code.

The entire 1966 SEAOC Sections 2630 and 2631 are reproduced in the appendix to this chapter for convenient reference. These SEAOC recommended provisions are now in effect as part of the building code in the City of Los Angeles and also in the County of Los Angeles. The International Conference of Building Officials accepted the provisions at their October 1966 convention, and these are included in the 1967 edition of the Uniform Building Code. Reference should be made to applicable codes for the precise wording. These long-needed provisions for ductile concrete are to prevent brittle earthquake failures of the buildings in which the requirements are met. It would be well indeed if at some future time all poured-in-place concrete buildings met such requirements unless there is positive assurance that the yield point in critical members would never be exceeded by earthquake demands. In this connection it must be noted that the relatively short history of earthquake motion and damage cannot be expected to represent extreme conditions that may have some probability of future occurrence in various areas.

18.5.1 Diaphragm Action

One of the initial basic steps is to determine or to specify the effectiveness of the floor and roof systems as diaphragms. The term, which is somewhat of a misnomer because stiffness normal to the plane is of less importance than that in the plane of the floor or roof, applies to the in-plane plate action of the horizontal system and its ability to deliver forces to the vertical frame elements and walls. If the horizontal plate should be infinitely rigid, forces would be delivered to the vertical elements in proportion to their relative rigidities. This is a common assumption, especially with concrete floor systems that are quite rigid. At the other extreme, the horizontal plate system may be so flexible due to combined flexural, shear, and possibly local torsional deformations under lateral forces as to have no significant distribution capabilities. There are, of course, various degrees of effectiveness between these limits. In any case the diaphragm condition must be established and consistently maintained in the lateral force calculations (U.S. Departments of the Army, Navy, and Air Force, 1966). Concrete floor or roof systems that are cast in place should normally be treated as rigid diaphragms.

18.5.2 Relative Rigidity

Another very important concept in earthquake-resistant design is that of relative rigidity. It is beyond the scope of this chapter to cover this subject in detail, and it is not necessary because of available reference material (BNC, 1961; U.S. Departments of the Army, Navy, and Air Force, 1966; Portland Cement Association, 1955; Plummer and Blume, 1953). If elements of a structure are interconnected by a rigid element, the system is in parallel and moves as a unit if the force is applied to the connecting element. Since all the interconnected elements move the same amount, their forces in the elastic range must be proportional to their relative rigidities in the entire system. Stiffness is the reciprocal of flexibility.

It is most significant that stiffness and strength are different considerations. An element may be very stiff but have less strength than a more flexible element. Obviously, the stiffer element would attract more force and fail first.

It is essential in most cases to consider shear as well as flexural participation. Another important point is the effective resistance against rotation of the top and bottom of a vertical element. The common assumption of the "fixed-but-guided" condition for a vertical element is often not justified by the rigidity and/or the strength of the horizontal members.

Connected elements are in series when they carry the same load but have different deflections. In this case the flexibility of the system is the sum of the flexibilities of the individual elements and the combined stiffness is the reciprocal of the sum of the flexibilities. On the other hand, a system in parallel has the same deformation for all elements and the forces carried by each vary. The total stiffness of a symmetrical group in parallel is the sum of the stiffnesses of the individual elements in the group.

Should an element yield, it will carry for all additional deformations only its yield level force (plus any additional stiffness contribution, often assumed to be zero as in an elastoplastic system) and the remaining elements will carry the remainder of the force or shear in accordance with their relative stiffnesses. If a brittle element fails completely, the entire burden is then transferred to any remaining elements in the system.

If shear walls are combined with moment-resisting frames, code requirements may require special analysis. In any event, the designer must carefully consider geometric compatibility of the frame and the walls under lateral forces to avoid unwanted failures. In tall buildings, the shear wall may actually be so slender that flexure is the dominant freedom in the "shear" wall. In the lower stories the wall will tend to be more rigid than the frame, and in the upper stories the opposite situation may exist. In such cases a rigorous computer program with proper geometric compatibility may be the best design aid.

Although relative rigidity is basic to all structural analysis, in the earthquake design problem it assumes unusual significance because of the need to consider walls, wall elements, piers, spandrels, and core units (about vertical openings in a building), as well as ordinary frame members.

There are, of course, many other problems that are generally identified in earthquake code provisions or in textbooks. These would include torsion, overturning moment, individual elements, and appendages. One item not specified generally is the method of frame analysis for the lateral forces. Often a "proper" or an "accepted" analysis is required. In this connection it is to be noted that many modern buildings with long spans and low story heights have such flexible girders (or column-connected beams) compared to the columns that approximate methods of frame analysis may not be applicable, especially in the lower and in the upper stories (BNC, 1961). Instead of points of inflection occurring at mid-height or some other arbitrary height in the first story, e.g., it may not occur anywhere in the story. These conditions should be reviewed and corrections applied as necessary or a rigorous analysis should be employed (BNC, 1961; Portland Cement Association, 1964, 1965; Blume, 1967a).

18.5.3 Design Steps

There are several operations or steps in current design practice. In earthquake-resistant design there are additional steps, some required under code specifications and some beyond normal code requirements. The latter may not be followed for typical structures and normal risks but they should be, as additional precautions, for special or unusual structures or for unusual risks of earthquake exposure or of structure occupancy or function. For example, a nuclear power plant normally would have special dynamic analyses for earthquake motion in addition to meeting all applicable seismic code requirements.

It is very desirable that the engineer responsible for the earthquake resistance of a building collaborate with others in the design operation early in the planning and preliminary design stage. He may, with his experience and understanding of the seismic design problem, offer valuable contributions affecting the basic concept of the structure. He may be able to increase the seismic resistance or to reduce its cost or both. For example, he may suggest structural symmetry of columns and walls that will eliminate torsion (except, of course, the "accidental" torsion required in design by current codes) and torsional coupling. He may also provide trial values of story heights and column sizes that will be most effective in not only earthquake matters but—even more important to many—in compatibility of building functions and reduced costs of design and construction.

It is usually essential that the design proceed on a step-by-step basis and that a preliminary seismic analysis be conducted before the story heights and important member sizes are finally established. In addition, it is desirable to determine the probable drift or lateral motion under wind forces as well as earthquake design forces. Although flexibility (with long periods of vibration) results in minimum seismic forces under code design requirements, it may lead to excessive lateral deformation and human perception under frequent wind storms. However, this condition is not typical for concrete structures of ordinary proportions.

One of the first steps is to select the optimum concrete strength, or strengths, as well as to determine the concrete unit weight and the grades of reinforcing steel to be used. The codes must be considered in this step as well as good practice and sound judgment. Then the floor-framing system is planned to be consistent with the column layout, architectural requirements, and costs. Provision for lateral force resistance must be made even at this stage to avoid unnecessary repeated trial designs. Trial member sizes are determined, usually by approximate methods.

It must be determined whether the design process is to be conducted under working stress design, ACI 318, IV-A or the ultimate strength method, ACI 318, IV-B (American Concrete Institute, 1965). The new seismic codes give some preference to the latter—with, of course, changed U factors from ACI 318—although working stress design is not prohibited. (See SEAOC Section 2630 in the appendix to this chapter.) Some engineers believe that the IV-B procedures lead to slightly undersized columns for seismic exposures and suggest a U factor of 2.10 for such members.

Another early decision involves the type of framing and the corresponding K-factor under the seismic code. Will there be a complete moment-resisting frame, a partial moment-resisting frame with walls, a box system, or some other system? The answer depends upon the height, code requirements, building layout and function, judgment, and other factors. If the building is under the 1967 UBC, or equivalent requirements, and there is a space frame to resist all the required lateral force, then K is 0.67, and there must be ductile concrete (DC). The dual bracing system, which provides $K = 0.80$ with the partial moment-resisting frame, has three possible systems of design as shown in the code, but again DC shall be provided and the shear walls shall comply with the special shear wall provisions. The above is true regardless of the building height if $K = 0.67$ or 0.80. However, if the building is more than 160 ft in height the above two types of frame are the only ones permitted. If the building is 160 ft or less in height, K may be 0.67, 0.80, 1.00, or 1.33. The latter is for the box system.

The ductile and shear wall provisions are new and will be treated further in the next section. The other code requirements, with few exceptions, have been in effect for several years and have been discussed in some detail (BNC, 1961; SEAOC, 1960; U.S. Departments of the Army, Navy, and Air Force, 1966).

18.5.4 Special Risk Design

Special risk structures such as nuclear plants or tall, narrow buildings may be further analyzed after initial design by code specifications and their seismic resistance improved by various methods not required by building codes (Blume, 1965). Approximate dynamic analysis for particular earthquake exposure may be conducted with the use of spectral response diagrams (BNC, 1961; Hudson, 1956). In such cases individual modes are considered with their participation factors for whatever condition is under consideration and then the modal responses are combined by root-mean-square or other approximations (BNC, 1961; Clough, 1962). The sum of the absolute modal responses is generally considered as somewhat conservative while it is believed that under some conditions the root-mean-square approximation may be nonconservative.

A more complex but also more accurate method for a given earthquake condition is to subject the entire dynamic system with all of its natural modes to the entire time history of a real earthquake or of a specified artificial earthquake. The procedure requires computer aid (Clough, Wilson, and King, 1963).

The problem with either the spectral or the time-history method is that the computed elastic response provides accelerations usually several times the normal code seismic coefficients. Figure 18.14 indicates some spectral response accelerations for particular damping values. The values are typically greatly in excess of the building code design coefficients. This anomaly can be explained in many cases by several factors, including the difference between allowable design unit stresses and yield unit stresses, by noncalculated elements such as partitions or filler walls, possibly by the design providing some excess material, and above all by the energy absorption capacity in the inelastic range if the structure is ductile. Any part of the anomaly still not explained must rest upon the chance that the earthquake may not occur in the expected intensity (risk) and upon damage and risk of collapse should the intensity be experienced. These matters have been discussed at length by Blume (1960a; 1960b) and BNC (1961).

Approximate analysis procedures are available to consider the amount of ductility or the reserve energy capacity required for specific situations. A common assumption, and one that may be somewhat conservative for ductile reinforced concrete that has bilinear softening characteristics, is elastoplastic behavior beyond the yield point. Under reversals and cycling, however, some deterioration and loss of initial stiffness take place (BNC,

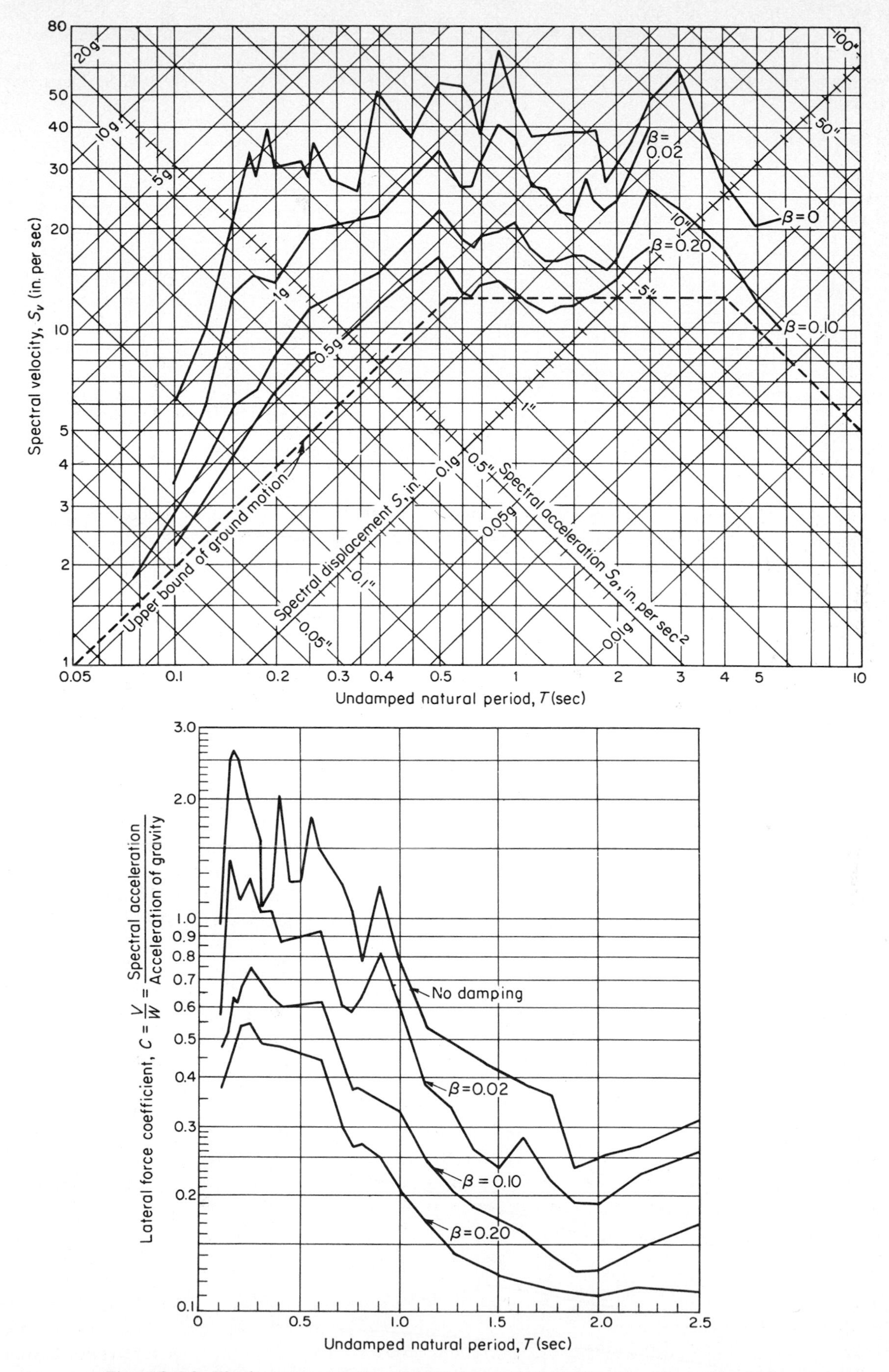

Fig. 18.14. Elastic response spectra, 1940 El Centro, N–S, Earthquake. (Blume, Newmark, and Corning, 1961.)

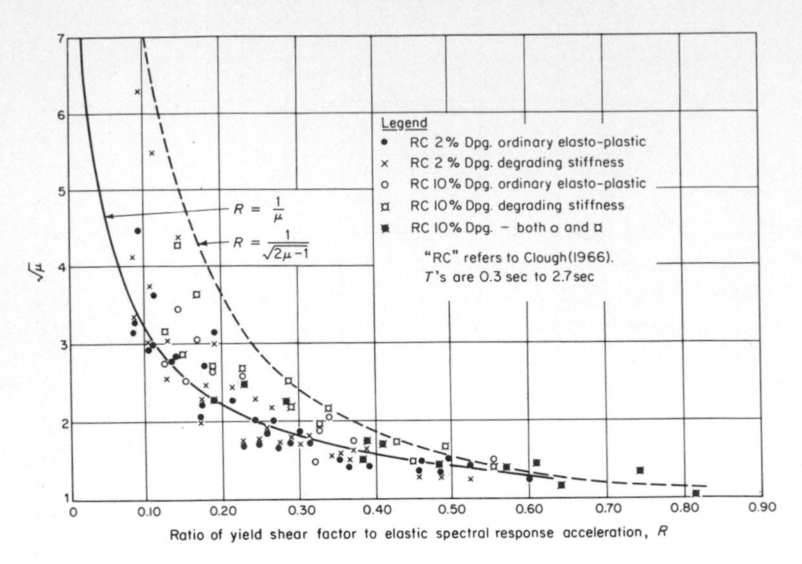

Fig. 18.15. Ductility vs. strength to demand ratio, El Centro, 1940, N–S. (Blume, 1967c.)

1961; Clough, 1966), and this roughly tends to cancel the conservative aspect of the elastoplastic assumption. The required reserve energy reduction factor R, for most cases of single mass elastoplastic systems, falls between β/μ and $\beta/\sqrt{2\mu - 1}$ (BNC, 1961). In view of deterioration or degrading and other factors, β/μ may be nonconservative and $\beta/\sqrt{2\mu - 1}$ is a realistic upper bound, as shown in Fig. 18.15 (Clough, 1966; Blume, 1967c).

Thus

$$R\alpha = ZKC \tag{18.3}$$

$$R = \frac{\beta}{\sqrt{2\mu - 1}} \tag{18.4}$$

and

$$\mu_u = \frac{\beta^2\alpha^2}{2(ZKC)^2} + \frac{1}{2} \tag{18.5}$$

where R = reserve energy reduction coefficient; Z = seismic zone factor; K = seismic code factor for type of building; C = design coefficient for base shear V; μ_u = ductility factor as in Eq. 18.1; α = elastic spectral response acceleration, generally for 5% damping for concrete buildings; and β = ratio of design unit stress to yield unit stress. A single-mass system is used here.

Assuming as an example that $Z = K = 1$, $\beta = 0.6$, and $C = 0.10$, the following values of μ_u are obtained for the α values shown:

α:	0.2	0.4	0.6	0.8	1.0	1.2 g units
μ_u:	1.2	3.4	7.0	12.0	18.5	26.4

All of these μ_u values are well into the ductile range except 1.2, which is close to the yield value of 1.0 and might be developed by a nonductile or brittle material. It is apparent that for the probable range of spectral acceleration α, in severe earthquake, considerable ductility is required under code design shears. An energy safety factor would

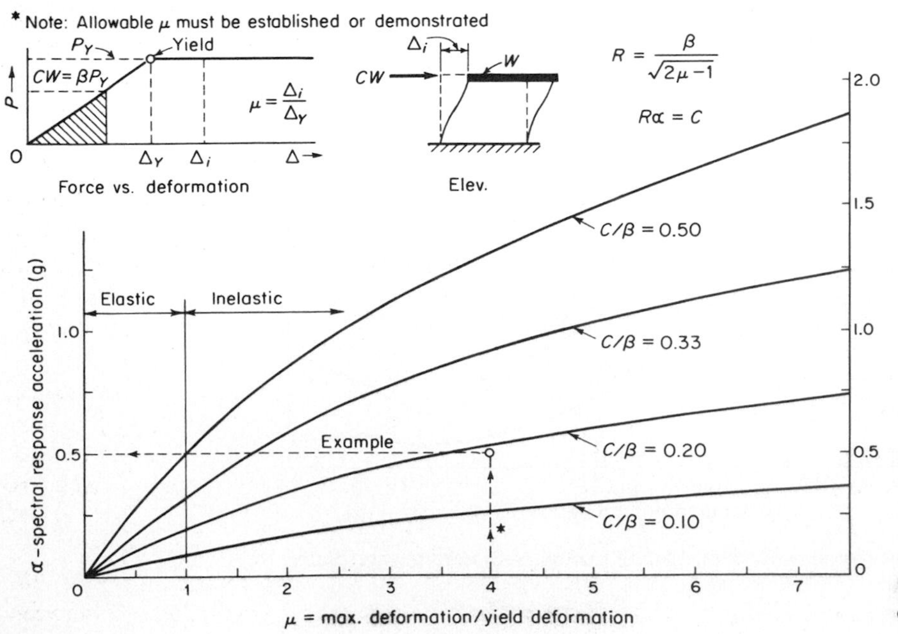

Fig. 18.16. Reserve energy data for elastoplastic system. (Blume, 1967c.)

require an increase in the above μ_u values that merely provide energy reconciliation.

Figure 18.16 (Blume, 1967c) indicates the relationships of μ and α with the convenient parameter C/β. In the example shown a μ value of 4 develops energy capacity for $\alpha = 0.5g$ if C/β is 0.19. In code terms, C may be ZKC.

The above elastoplastic assumption is actually a special case of the Reserve Energy Technique (Blume, 1960b; BNC, 1961), which is general and provides for any type of inelastic system. It enables one to estimate deformation and damage for any spectral exposure. Application to multistory buildings involves assumptions as to the effective mobilization of energy capacity from story to story and, of course, is only an approximation until that complex subject is better understood. A square root assumption has been employed as a judgment factor at this stage of knowledge, i.e., where there are 16 stories above the level under consideration, the effective energy mobilization is considered to be one-fourth of the summation of the story energy capacities.

It is apparent that most buildings require ductility and energy-absorption capacity as well as reasonable strength—or else great elastic strength—to survive major earthquake emergencies. The next section provides data on the code requirements for ductile reinforced concrete.

18.6 DUCTILE CONCRETE

Existing publications on the design of ductile concrete are BNC (1961) and Sections 2630 and 2631 of SEAOC (1966), included in the appendix to this chapter. The former is in narrative form with theory, figures, tables, test data, and examples, and the latter is in code terminology. There are a few minor variations or additions in the code document as compared to the former reference. The SEAOC reference should be employed in case of conflict since it is the pattern for official code requirements such as the 1967 Uniform Building Code, which also should be consulted.

DC—essential for good performance of concrete frame structures in severe earthquakes—entails additional design steps and new responsibilities for design engineers. It may be thought of as a new dimension—that of deformation or ductility and work capacity—added to the traditional design dimension of strength. As for all new and different procedures, it will no doubt seem somewhat awkward at first. However, experience and design aids soon should make DC design a part of routine design operations, although the design effort will be increased over that for conventional methods. At first some may prefer to design according to traditional strength requirements and then to provide for DC as a check, or final, procedure. It is suggested, however, that concrete dimensions be established for basic DC requirements prior to detailed reinforcing selection in order to minimize the total work effort and further that with experience DC design be attempted directly.

In general, concrete sections may tend to be slightly larger than under previous design methods, especially columns, and main reinforcing bar areas may be somewhat less. Web reinforcement and column transverse reinforcement tend to be greater in total volume especially near, and in, the joints. As experienced concrete designers know, there need be no extra cost, and perhaps a saving as well as a better structure, with slightly greater concrete sections and the planned re-use of forms. In other words, small concrete girder, beam, and/or column sections may lead to excessive (and costly) steel percentages and congestion, with great placing difficulty for the bars and for the concrete. The results can be poor as well as costly. All too often the relatively minor savings in net floor space, or in building usable volume, obtained by using minimum column sections with maximum percentages of longitudinal reinforcing steel, are of academic rather than of real value to the owner. Economic studies that include the cost of concrete in place, of formwork and shoring with allowance for re-uses, and of reinforcing bars of various types in place indicate that the minimum concrete section often is not the least costly member for a given function. Moreover, in view of DC requirements, the minimum concrete section generally has the least value in ductility and ultimate shear capacity and therefore is the least desirable for earthquake resistance. Structural design is an art, especially for seismic forces—it will continue to be so under DC requirements.

18.6.1 Horizontal Members

The 1966 SEAOC Section 2630(d) specifies that the reinforcement ratio p for negative moment in a horizontal member at the face of columns shall not exceed 0.025 or $0.46 f'_c p'/f_y p$, whichever value is least. BNC (1961) has the same 0.025 maximum value, but its second requirement is specified as

$$q_u - q'_u \leqq 0.25 \tag{18.6}$$

where $q_u = pf_y/f_{cu}$; $q'_u = p'f'_y/f_{cu}$; $p = A_s/bd$, of normal design significance; $p' = A'_s/bd$, of normal design significance; f_y = yield stress of the tension reinforcement; f'_y = yield stress of the compression reinforcement; and $f_{cu} = 0.7f'_c$ for $f'_c \leqq 5000$ psi, and $f_{cu} = 1500$ psi $+ 0.4f'_c$ for $f'_c > 5000$ psi.

The 1966 SEAOC code also specifies that f_y for flexural members shall not exceed 40,000 psi and that the positive moment capacity at the face of columns shall not be less than 50% of the negative moment capacity provided. Assuming that $p'/p = 0.5$ and $f_y = f'_y = 40{,}000$ psi, values of p are obtained as shown in Table 18.1.

It can be seen in Table 18.1 that for the given conditions and limitations there is very little difference between

Table 18.1. Values of p, q_u, and ϕ_u/ϕ_y for $f_y = f'_y = 40{,}000$ psi

f'_c, psi	SEAOC $p_{max} = \dfrac{0.46 f'_c p'}{f_y p}$	SEAOC p_{max} value	BNC p_{max} based upon $q_u - q'_u = 0.25$	BNC p_{max} value	q_u SEAOC	q_u BNC	ϕ_u/ϕ_y† SEAOC	ϕ_u/ϕ_y† BNC
			$p'/p = 0.50$					
3000	0.0173	0.0173	0.0262*	0.025	0.33	0.47	11	8
4000	0.0230	0.0230	0.0350*	0.025	0.33	0.36	11	10
5000	0.0288*	0.025	0.0437*	0.025	0.29	0.29	13	13
			$p'/p = 0.75$					
3000	0.0260*	0.025	0.0525*	0.025	0.47	0.47	17	17
4000	0.0345*	0.025	0.0700*	0.025	0.36	0.36	20	20
5000	0.0432*	0.025	0.0874*	0.025	0.29	0.29	26	26

*Exceeds the 0.025 limitation.
†Data obtained from Blume, Newmark, and Corning (1961).

Anchorage length beyond column face

$$= \frac{A_s f_y}{1.5 u_\mu \Sigma_0};$$

or 24" minimum.

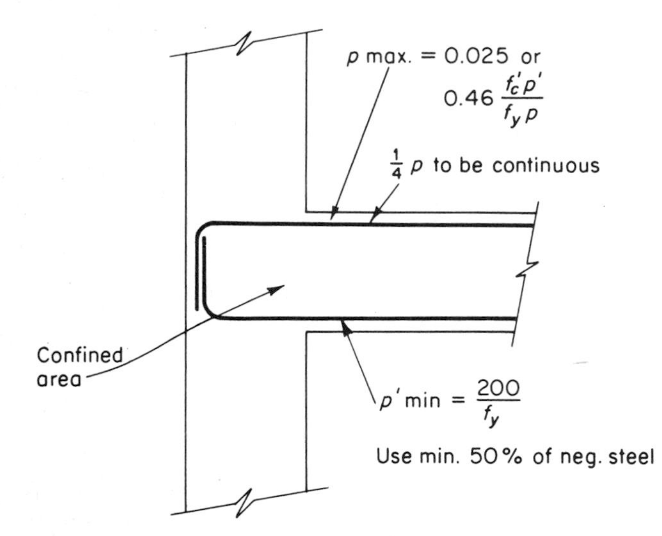

Use min. of 2 bars top and bottom

No splices within one column width or within 2 x beam depth from face of column.

No splices within tension area or reversing stress area unless concrete is confined.

Fig. 18.17. Beam longitudinal reinforcement. (From Structural Engineers Association of California, 1966 and Blume, 1967b.)

SEAOC (1966) and BNC (1961). For $p'/p = 0.50$ and $f'_c = 3000$ psi, the resulting rotational ductilities are 11 and 8 respectively. They are 11 and 10 for $f'_c = 4000$ psi, and the two values are equal for each of the other conditions tabulated. These rotational ductilities were obtained from Fig. 18.13. The 0.025 maximum value of p governs in all but two of the cases shown.

The effectiveness of bottom bars at the column is noteworthy. As p'/p increases, the ductility increases out of proportion. However, the minimum ductility of 11 should be more than adequate in most cases. It is to be noted though that p' may have to be more than the minimum value in many cases because of net moment requirements under earthquake reversals.

Figure 18.17 indicates SEAOC requirements for longitudinal beam or girder reinforcement. Figure 18.18

Fig. 18.18. Beam shear design. (From Structural Engineers Association of California, 1966 and Blume, 1967b.)

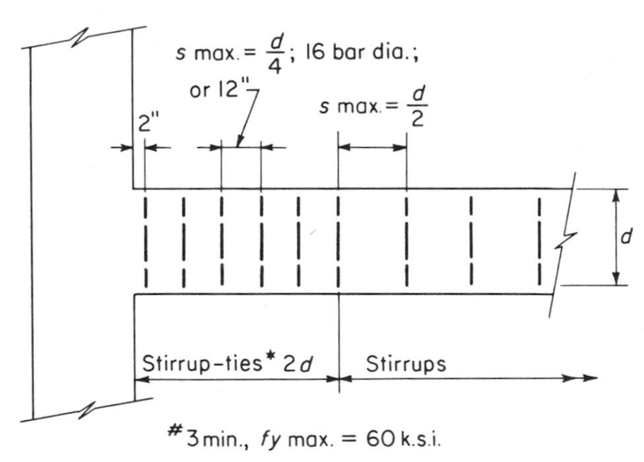

* Required each end; where M_u developed; where required compression steel occurs.

1.4(D+LL)

M_u^A L M_u^B

$$V_u = \frac{M_u^A + M_u^B}{L} + 1.4\, V_{D+L}$$

$v_c = 0.3\phi F_{sp}\sqrt{f'_c}$; $F_{sp} = 6.67$ for normal aggregates.

v_u max. $= 10\phi\sqrt{f'_c}$; $v_u = \dfrac{V_u}{bd}$

$V_u - V_c = V'_u = \dfrac{A_v \phi f_y d}{s}$; $V_c = v_c bd$

illustrates the requirements for beam shear design and transverse reinforcement. Figure 18.19 shows the types and functions of transverse reinforcement under ductile concrete requirements (Blume, 1967b). The confinement function, which is associated with ductile concrete, can be satisfied only with hoops or spirals. The latter are the same, however, as normally used in spiral columns. As shown, a hoop or spiral also can qualify to perform the other function: to act as a tie to prevent buckling of longitudinal bars. It is important to differentiate these various types of reinforcement and their functions in ductile concrete design.

In addition to reinforcement requirements, there are limitations on concrete dimensions under the SEAOC requirements that affect both horizontal members and columns. Some of these may prove in time to be conservative and unduly restrictive in design. Figure 18.20 illustrates dimensional limitations that must be considered in all phases of design under SEAOC (1966). These are especially significant in the early planning and preliminary stages.

The design of beams and girders under ultimate strength procedures may be greatly simplified by various techniques. An approximate method suggested by Zweig (1965) is particularly useful in selecting bar sizes. Figure

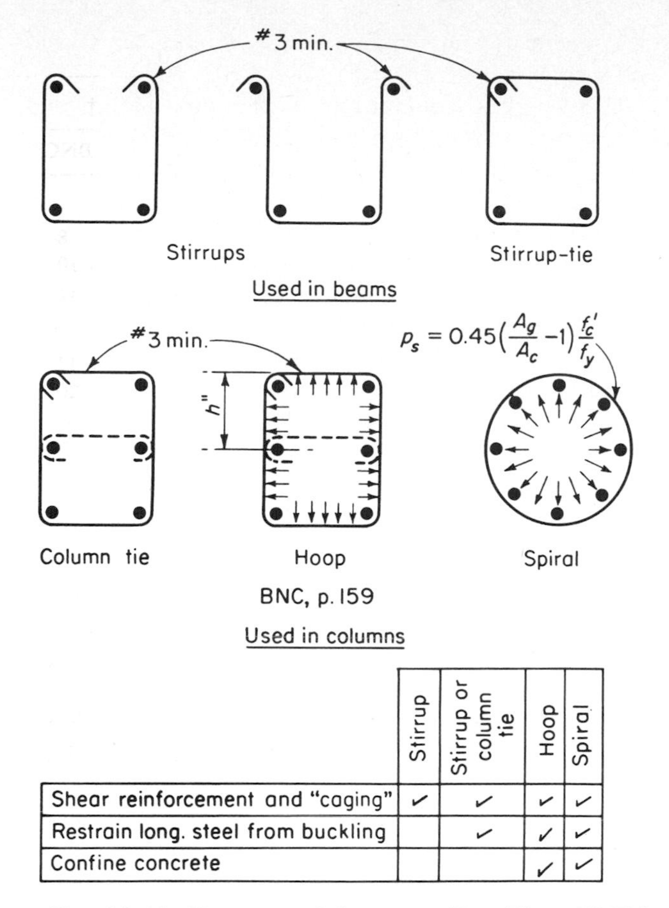

	Stirrup	Stirrup or column tie	Hoop	Spiral
Shear reinforcement and "caging"	✓	✓	✓	✓
Restrain long. steel from buckling		✓	✓	✓
Confine concrete			✓	✓

Fig. 18.19. Transverse reinforcement. (From Blume, 1967b.)

Fig. 18.20. Frame dimensional limitations. (From Structural Engineers Association of California, 1966 and Blume, 1967b.)

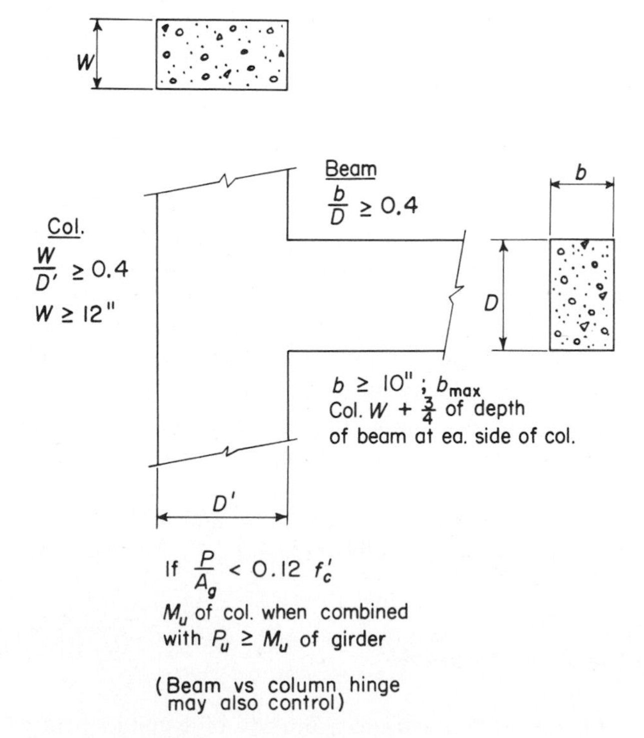

Fig. 18.21. USD for bending. (From Blume, 1967b and Driskell, 1967.)

Reference: A. Zweig, A.C.I. Jour. Feb. 1965

1. $\bar{A}_s = \frac{M_u}{ad}$ (Not $M_u' = \frac{M_u}{\phi}$)

2. $\bar{p} = \frac{\bar{A}_s}{bd}$ — M_u must be in ft-kips, d and b in inches

3. $A_s = \bar{A}_s\,(1 + \bar{p}c)$

$a = 0.075\, f_y$; f_y in ksi units

Values of c: f_c', psi	f_y in ksi: 40	50	60
3000	7.86	9.83	11.80
4000	5.90	7.38	8.85
5000	4.72	5.90	7.08

Example: $M_u = 523^{ft-k}$, $d = 31.3"$, $b = 24"$

$f_c' = 5000$ psi, $f_y = 40$ ksi,

$a = 3.00$, $c = 4.72$

$$\bar{A}_s = \frac{M_u}{ad} = \frac{523}{3 \times 31.3} = 5.57^{\square"}$$

$$\bar{p} = \frac{\bar{A}_s}{bd} = \frac{5.57}{24 \times 31.3} = 0.00741$$

$$A_s = \bar{A}_s\,(1 + \bar{p}c) = 5.57[1 + (0.00741)(4.72)] = 5.76^{\square"}$$

Use 4 - #11; $A_s = 6.24^{\square"}$; $M_u \approx 566^{ft-k}$

Check by ACI stress block gives $M_u = 560^{ft-k}$

18.21 provides the equations, constants for various f'_c and f_y values, and an example of the procedure (Blume, 1967b; Driskell, 1967).

18.6.2 Columns

Columns must be given several considerations under DC design. Many requirements are new, e.g., the column bars must be spliced in the midhalf of the column. The minimum percentage of steel is 1% and the maximum 6%. The maximum yield value of the bars is 60,000 psi. Bars may be ASTM A-15, A-408, or A-432. The most practicable way to design the columns is with the aid of interaction diagrams such as in ACI SP-7 (Everard and Cohen, 1964).

Figure 18.22 illustrates the requirements for column spirals or hoops above and below the joint and a minimum requirement through exterior joints and, as modified, other joints. These portions plus the joint per se constitute the confined length of the column. Generally—but this should be verified—the confinement steel is more than adequate for column shear requirements that also must be satisfied.

Fig. 18.22. Column steel above and below joint (spiral or hoops). (From Structural Engineers Association of California, 1966 and Blume, 1967b.)

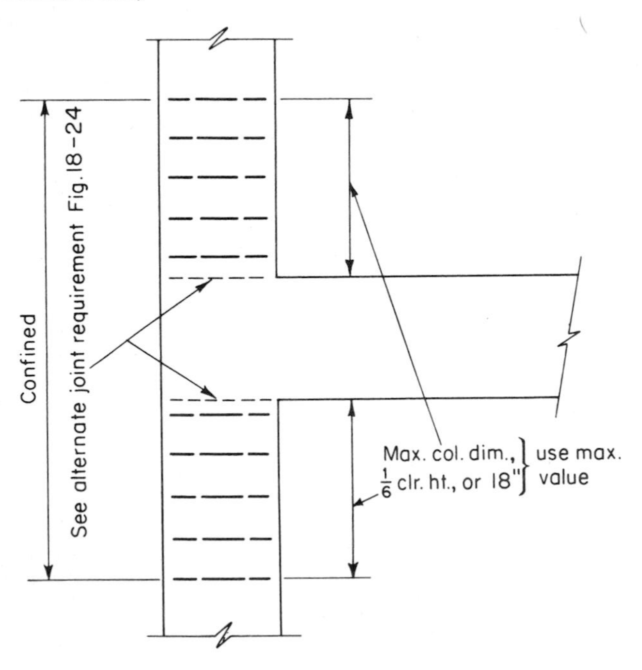

For spirals:

$$p_s = 0.45\left(\frac{A_g}{A_c} - 1\right)\frac{f'_c}{f_y}$$

But not less than required for A_v per Fig. 18 – 23

For hoops:

$2 \times p_s$; max. spacing = 4" o.c.

Check span of hoops, h''

The length of column outside the joint and beyond the confined portion also must meet the shear requirements. These, of course, are consistent with the ductile concrete objectives—to prevent shear failure in preference to tensile yielding of reinforcing steel. Figure 18.23 indicates the column shear design procedure. Column ties, as per Figure 18.19, may be used only in the length between the limits of the confined concrete.

The joint, of course, is part of the horizontal member as well as the column. It not only must be confined in whole or in part according to SEAOC '66, but in exterior columns especially the joint must be checked for local shear. Figure 18.24 illustrates an exterior joint and the forces that cause a concentration of shear in the joint. Whichever governs, confinement or this local joint shear, controls the joint transverse bar design.

Interior joints are subject to an amount of transverse reinforcement that depends upon the relative dimensions of the intersecting horizontal members and the column. If there are members framing into four sides of the column, as is the typical case, the depth of joint is defined by the depth of the shallowest member (SEAOC, 1966). The transverse steel requirement is established by the relative width of the beams and their associated column faces. If each beam at the joint is at least three-fourths as wide as the associated column face, and if the extension of column corners beyond the beam sides do not exceed 4 in., the amount of transverse reinforcing in the joint may be one-half of that required by confinement (Fig. 18.22) or joint shear (Fig. 18.24). Otherwise, the full amount of transverse steel is required in the joint (SEAOC, 1966). Obviously, the designer must carefully select his member sizes to avoid this penalty requirement.

The importance of joint integrity during severe earthquake response is basic. Should a joint fail, the capacity of all members framing into the joint is reduced or eliminated. On the other hand, a hinge in a horizontal member, or even in a column, is discrete. Beyond this point, a joint failure would tend to be in shear or diagonal tension and thus to be of a nonductile nature. However, in consideration of the performance of test specimen VA (Fig. 18.9), which had no transverse joint reinforcement and only three intersecting beams, two of which were stubs, it would seem that the code joint requirements are quite conservative as well as restrictive in geometric proportions. Time and additional test programs may lead to code changes.

18.6.3 An Example

Figure 18.25 illustrates a portion of a building frame designed to the ductile concrete requirements of the codes (SEAOC, 1966; International Conference of Building

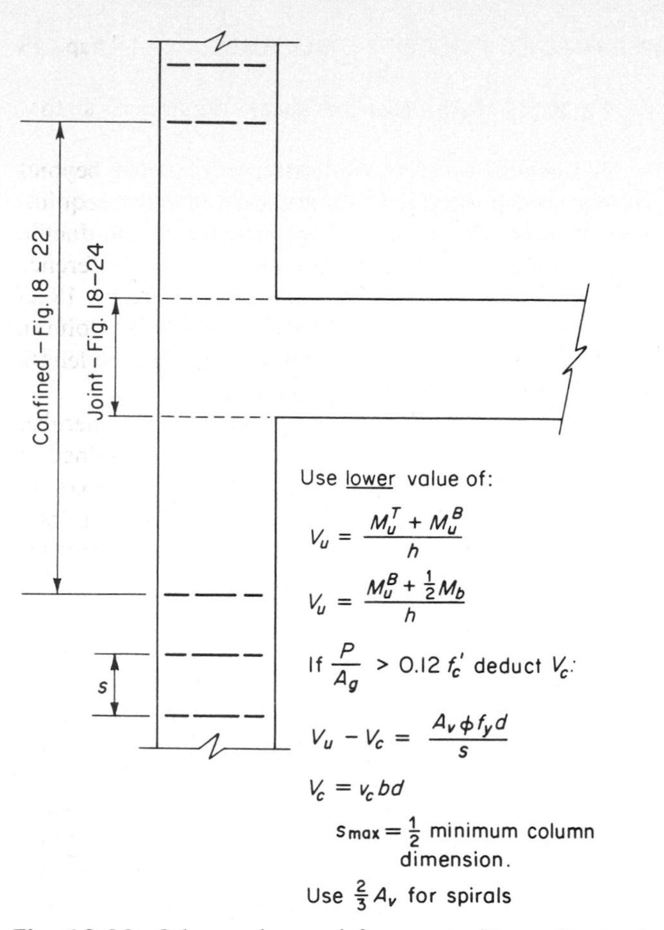

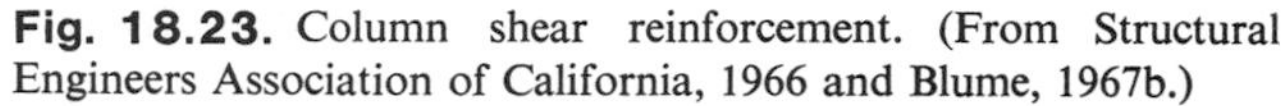

Fig. 18.23. Column shear reinforcement. (From Structural Engineers Association of California, 1966 and Blume, 1967b.)

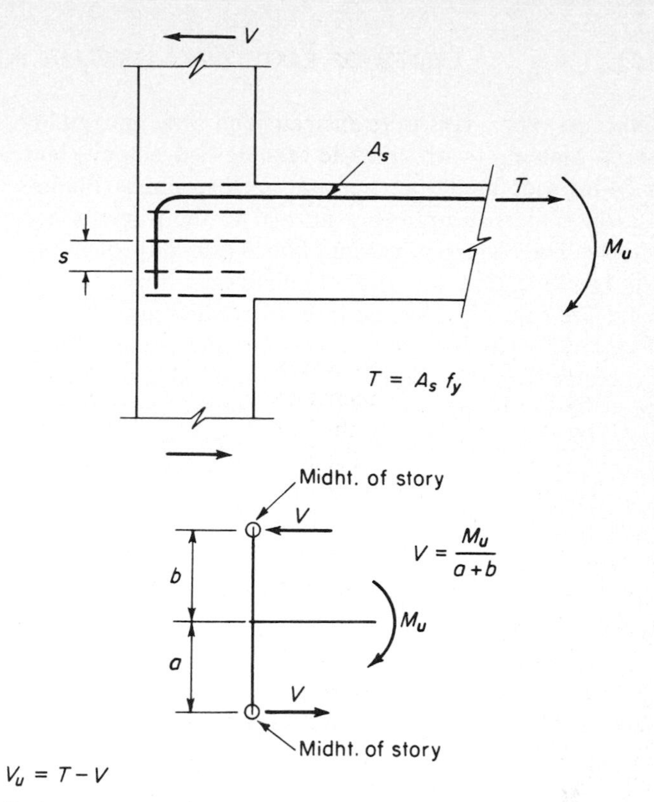

Fig. 18.24. Shear check within exterior joint. (From Blume, 1967b and Driskell, 1967.)

Fig. 18.25. Portion of a ductile frame. (Blume, 1967b.)

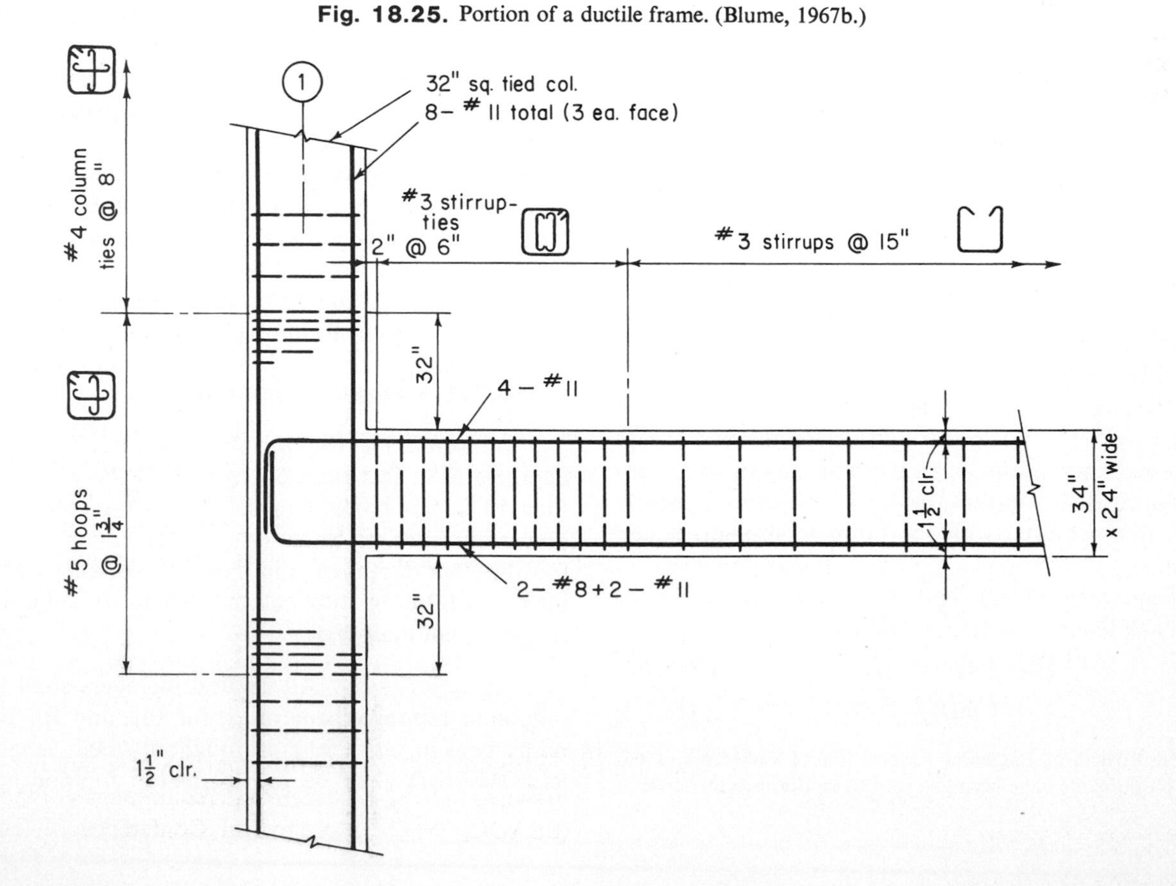

Officials, 1967). This is an exterior joint, low in a multistory building, with a 32-in. square-tied column and a 24-in. wide by 34-in. deep intersecting girder (Blume, 1967b). The spandrel beam normal to the frame is not shown. The Number 5 column hoops extend through the joint and for 32 in. into the column in each story (the column size controls over the three alternate requirements, Fig. 18.22). The bottom beam reinforcing at the column is greater than the minimum amount of 50% of the top (Fig. 18.17) because of the net moment requirements from dead load and earthquake moment reversals. The beam span in this case is 20 ft center to center of columns and the story height is 12 ft above and 15 ft below the floor level.

Concrete strength f'_c is 5000 psi, normal density aggregates are used, column bars and ties are ASTM A-432 ($f_y = 60{,}000$ psi), and all other bars are ASTM A-15 ($f_y = 40{,}000$ psi). The basic column requirements were $P_u = 2780^K$ and $M_u = 690'^K$. However, these were subsequently increased under ACI 318, Section 9-4, to 2880^K and $715'^K$ respectively. In addition, under the code the column must be able to force the hinge into the beam.

Ductile concrete is a new vehicle for aseismic design that is expected to have a considerable impact upon the construction industry and to lead to vastly increased earthquake resistance. Designers of buildings and other structures intended for seismic areas—active or potentially active—are faced with the responsibility of utilizing the DC concept in their design operations.

18.7 APPENDIX*

18.7.1 Ductile Moment-Resisting Spaces Frames (Sec. 2630, SEAOC, 1966)

18.7.1.1 General

1. Design and construction of cast-in-place, monolithic reinforced concrete framing members and their connections in ductile moment resisting space frames shall conform to the requirements of A.C.I. Building Code, A.C.I. 318, and all the requirements of this section.

2. All lateral load-resisting frame members shall be designed by the ultimate strength design method except that the working stress design method may be used provided that it is shown that the factor of safety is equivalent to that achieved with the ultimate strength design method.

3. Equations (15-2) and (15-3), A.C.I. 318, for earthquake loading shall be modified to:

$$U = 1.40(D + L + E) \tag{18.7.1}$$

$$U = 0.90D + 1.25E \tag{18.7.2}$$

* From Structural Engineers Association of California, 1966; paragraph numbers have been changed to conform to this book.

18.7.1.2 Definitions

1. Confined concrete. Concrete which is confined by closely spaced special transverse reinforcements which are provided to restrain the concrete in directions perpendicular to the applied stresses.

2. Special transverse reinforcement. Spirals, stirrup ties, or hoops provided to restrain the concrete to make it qualify as confined concrete.

3. Stirrup ties or hoops. Continuous reinforcing steel of not less than a No. 3 bar bent to form a closed hoop which encloses the longitudinal reinforcing and the ends of which have a standard 135 degree bend with a 10 bar diameter extension.

18.7.1.3 Physical requirements for concrete and reinforcing steel

1. Concrete. The minimum specified 28-day strength of the concrete, f'_c, shall be 3000 lb/in.^{2}.

2. Reinforcement. All longitudinal reinforcing steel shall be new billet-steel bars (ASTM A-15, A-408, A-432). For flexural members reinforcing steel complying with ASTM standards A-15 or A-408 shall be either structural grade or intermediate grade, and the specified yield strength, f_y, shall not exceed 40,000 lb/in.^{2}. A-408 reinforcing bars when bent more than 10 degrees shall meet the 90 degree bend test requirement for A-15 reinforcing steel except that the pin about which the specimen is bent shall have a diameter of eight times the bar diameter. For columns, the specified yield strength of the vertical reinforcing steel, f_y, shall not exceed 60,000 lb/in.^{2}. Grades of steel other than those specified in the design shall not be used.

Where reinforcing steel is to be welded, a chemical analysis of the steel shall be provided. The welding procedure shall be as set forth in the American Welding Society's publication, AWS D12.1 "Recommended Practice for Welding Reinforcing Steel, Metal Inserts and Connections in Reinforced Concrete Construction."

18.7.1.4 Flexural members

1. General. Flexural members shall not have a width-depth ratio of less than 0.4, nor shall the width be less than 10 in. nor more than the supporting column width plus a distance on each side of the column of three-fourths the depth of the flexural member. Flexural members framing into columns shall be subject to a rational joint analysis.

2. Reinforcement. All flexural members shall have a minimum reinforcement ratio, for top and for bottom reinforcement, of $200/f_y$ throughout their length. At least two bars shall be provided both top and bottom.

The reinforcement ratio, p, shall not exceed 0.025 nor 0.46 $f'_c p'/f_y p$, whichever is least, for negative moment at the face of columns, and the positive moment capacity at such locations shall be not less than 50% of the negative moment capacity provided. A minimum of one-fourth of the larger amount of the negative reinforcement required at either end shall continue throughout the length of the beam.

3. *Splices.* Tensile steel shall not be spliced by lapping in a region of tension or reversing stress unless the region is confined by stirrup ties. Splices shall not be located within the column or within a distance of twice the member depth from the face of the column. At least two stirrup ties shall be provided at all splices.

4. *Anchorage.* Flexural members framing into only one side of a column, in any vertical plane, shall have top and bottom reinforcement extending to the far face of a confined concrete region [Section 18.7.1.5(4)], terminating in a standard 90 degree hook. Length of required anchorage shall be computed beginning at the near face of the column. Length of anchorage in confined regions shall be determined by:

$$L = \frac{A_s f_y}{1.5 u_u \sum o} \qquad (18.7.3)$$

including hook and vertical extension, but not less than 24 in. Main flexural reinforcement shall be capable of being anchored, without horizontal offsets, within the confined column core when framing into only one side of a column.

5. *Web reinforcement.* Vertical web reinforcement of not less than No. 3 bars shall be provided in accordance with the requirements of Chapter 17, A.C.I. 318, except that:

(a) Maximum

$$V_u \geq \frac{M_u^A + M_u^B}{L} + 1.4 V_{D+L} \qquad (18.7.4)$$

where M_u^A and M_u^B are ultimate moment capacities of opposite sense at each end of the member and V_{D+L} is the simple span shear.

(b) Stirrups shall be spaced at no more than $d/2$ throughout the length of the member.

(c) Stirrup ties, at a maximum spacing of not over $d/4$, 16 bar diameters or 12 in., whichever is least, shall be provided in the following locations:

1. At each end of all flexural members. The first stirrup tie shall be located not more than 2 in. from the face of the column and the last, a distance of at least twice the member depth from the face of the column.
2. Wherever ultimate moment capacities may be developed in the flexural members under inelastic lateral displacement of the frame.
3. Wherever required compression reinforcement occurs in the flexural members.

18.7.1.5 Columns subject to direct stress and bending

1. *Dimensional limitations.* The ratio of minimum to maximum column thickness shall not be less than 0.4 nor shall any dimension be less than 12 in.

2. *Vertical reinforcement.* The reinforcement ratio, p, in tied columns shall not be less than 0.01 nor greater than 0.06.

3. *Splices.* Lap splices shall be made within the center half of column height, and the splice length shall not be less than 30 bar diameters or 16 in. Continuity may also be effected by welding or by approved mechanical devices provided not more than alternate bars are welded or mechanically spliced at any level and the vertical distance between these welds or splices of adjacent bars is not less than 24 in.

4. *Special transverse reinforcement*

(a) Special transverse reinforcement shall be provided through the joint and in that portion of a column over a length equal to the maximum column dimension, or one sixth the clear height of the column, but not less than 18 in. from either face of the joint.

EXCEPTION: Special transverse reinforcement of one-half the amount otherwise required by Subsections 4 and 5 shall be required within the joint, determined by the depth of the shallowest framing member, where such members frame into all four sides of a column and whose width is at least three-fourths the column width. When a corner, unconfined by flexural members of a tied column, exceeds 4 in., the full special transverse reinforcement shall be provided through the joint.

(b) Where spiral reinforcement is used to satisfy the requirement for special transverse reinforcement, it shall not be less than that required in Section 913 of A.C.I. 318.

(c) The volume of transverse reinforcement provided by hoops shall not be less than two times that required for spiral reinforcement. Supplementary cross ties may be included as part of the volume of transverse reinforcement up to 25% of the total volume. These supplementary ties must have the standard hook and must engage the exterior hoop and vertical bar. For hoop type reinforcement, the maximum unsupported length h'', shall not exceed:

$$h'' = \frac{2A''_{sh} f''_{yh}}{p'' a f''_y} \qquad (18.7.5)$$

where a is the center-to-center spacing of hoops; p'' is as required for spiral reinforcement, A.C.I. 318, but not less than 0.008 for cold-drawn wire, 0.010 for hard-grade bars, or 0.012 for intermediate grade bars; A''_{sh} is the area of cross section of transverse hoop; f''_y is the useful limit stress of transverse spiral reinforcement, to be taken as the yield strength for intermediate and hard-grade steel and as the stress corresponding to strain of 0.005 for cold-drawn wire of high strength steel with an indefinite yield strength; and f''_{yh} is the useful limit stress of hoop reinforcement defined in the same way as f''_y.

Where the length h'' is less than the length of a side of a rectangular or square column, a sufficient number of overlapping hoops may be provided in order to avoid exceeding the limiting value of h'' as given above. The center-to-center spacing of spiral or hoops shall not exceed 4 in. The minimum size of reinforcing steel for hoops shall be No. 3 bars. Additional ties on column bars, in addition to the perimeter hoops, may be needed to meet shear requirements.

5. *Beam-column joint analysis.* The transverse reinforcement through the joint shall be proportioned according to the requirements of Subsection 4. The transverse reinforcement thus selected shall be checked according to the provisions set forth in Subsection 6, with the exception that the V_u acting on the joint shall be equal to the maximum shears in the joint computed by a rational analysis taking into account the column shear and the concentrated shears developed from the forces in the beam reinforcement at a stress assumed at f_y.

EXCEPTION: The provisions of this subsection shall be modified in accordance with the provisions of Subsection 4 for those cases where confinement of the joint is effected by beams framing into all four sides.

6. *Column shear.* The transverse reinforcement in columns subjected to bending and axial compression shall satisfy the following requirement:

$$A_v f_y \frac{d}{s} = V_u - V_c \qquad (18.7.6)$$

where V_u is the maximum ultimate shear on the column due to earthquake, computed as:

$$\frac{M_u^B + \frac{1}{2}M_b}{h} \qquad (18.7.7)$$

but not more than

$$\frac{M_u^T + M_u^B}{h} \qquad (18.7.8)$$

where M_u^T and M_u^B are the ultimate moment capacities of the column at the top and bottom, respectively, under design earthquake axial load at the locations indicated; h is the clear height of the column; and M_b is the maximum sum of the moment capacities of the beams framing into the top connection; for instance, the sum of negative moment capacity of one beam and positive moment capacity of the other at the faces of the column. The factor $\frac{1}{2}$ shall be omitted if only one column frames into the top connection. $V_c = v_c bd$, where v_c shall be in accordance with Section 1701 of A.C.I. 318, except that v_c shall be considered zero when $P/A_g < 0.12f'_c$.

s = spacing, $\leq \frac{1}{2}$ minimum column dimension;
d = effective depth of section;
A_v = total cross-sectional area of special transverse reinforcement in tension within a distance s, except that two-thirds of such area shall be used in the case of circular spirals.

7. *Design limitations.* At any beam-column connection where $P/A_g \geq 0.12f'_c$ the total ultimate moment capacity of the column, at the design earthquake axial load, shall be greater than the total ultimate moment capacity of the beams, along their principal planes at that joint.

EXCEPTION: Where certain beam-column connections at any level do not comply with the above limitations, the remaining columns and connected flexural members shall comply and further shall be capable of resisting the entire shear at that level accounting for the altered relative rigidities and torsion resulting from the omission of elastic action of the nonconforming beam-column connections.

Where $P/A_g < 0.12f'_c$, the column shall further conform to the requirements for flexural members.

8. *Effective column length.* All columns shall have their effective length for design determined in accordance with A.C.I. 318, Section 915(d), with design determined in accordance with Section 916.

18.7.1.6 Inspection. For buildings designed under this section, a specially qualified inspector under the supervision of the registered professional engineer responsible for the structural design shall provide continuous inspection of the placement of the reinforcement and concrete, and shall submit a certificate indicating compliance with the plans and specifications.

18.7.2 Concrete Shear Walls and Braced Frames (Sec. 2631, SEAOC, 1966)

18.7.2.1 General

1. Design and construction of earthquake reinforced concrete shear walls and reinforced concrete braced frames subjected primarily to axial stresses for all buildings shall conform to the requirements of the A.C.I. Building Code, A.C.I. 318, and all the requirements of this section.

2. Shear walls and braced frames shall be designed by the ultimate strength design method except that the

working stress design method may be used provided that the factor of safety in shear and diagonal tension is equivalent to that achieved with the ultimate strength design method.

3. Equations (15-2) and (15-3), A.C.I. 318, for earthquake loading shall be modified to:

$$U = 1.4(D + L + E) \quad (18.7.9)$$

$$U = 0.9D + 1.25E \quad (18.7.10)$$

provided further that twice the U value set forth above shall be used in calculating shear and diagonal tension in buildings without a 100% moment resisting space frame.

18.7.2.2 Braced frames. Reinforced concrete members of braced frames subjected primarily to axial stresses in buildings with a ductile moment-resisting space frame shall have special transverse reinforcing as set forth in Section 18.7.1.5(4) throughout the full length of the member. Tension members shall additionally meet the requirement for compression members.

18.7.2.3 Shear and diagonal tension-ultimate strength design

1. The nominal ultimate shear stress in a concrete wall shall be computed by

$$v_u = \frac{V_u}{bd} \quad (18.7.11)$$

The shear stress v_u shall not exceed

$$v_u = \left(0.8 + 4.6\frac{H}{D}\right)\phi\sqrt{f'_c} \quad (18.7.12)$$

where H is the total height to which the shear wall extends in the structure, and D is the width of the wall in the direction of the shear force.

The value for v_u shall not exceed $10\phi\sqrt{f'_c}$ for H/D ratios greater than 2 and $5.4\phi\sqrt{f'_c}$ for H/D ratios less than one.

2. The total shear carried by a reinforced concrete wall shall be determined in accordance with the following:

$$V_u = v_c bd + V'_u \quad (18.7.13)$$

3. The shear stress carried by the concrete shall not exceed

$$v_c = \left(3.7 - \frac{H}{D}\right)2\phi\sqrt{f'_c} \quad (18.7.14)$$

The maximum value for v_c shall not exceed $5.4\phi\sqrt{f'_c}$ for H/D ratios less than one and $2\phi\sqrt{f'_c}$ for H/D ratios greater than 2.7.

When the structural lightweight concrete is used, the limiting value of v_c as set forth above shall be multiplied by the factor of $0.15 f_{sp}$.

The area of shear reinforcement required to resist the portion of the shear V'_u shall be computed by

$$A_v = \frac{V'_u s}{\phi f_y d(H/D - 1)} \quad (18.7.15)$$

but in no case will the reinforcement be less than required in Section 2202 A.C.I. 318 or by

$$A_v = \frac{V'_u s}{\phi f_y d} \quad (18.7.16)$$

18.7.2.4 Vertical boundary members for shear walls

1. Special vertical boundary elements shall be provided at the edges of concrete shear walls in buildings whose lateral force resisting system is as described for $K = 0.80$ in Table 23C. These elements shall be composed of concrete encased structural steel elements of ASTM A-7, A-36, or A-441 or shall be concrete reinforced as required for columns in Section 18.7.1.5 with special transverse reinforcement as described in Section 18.7.1.-5(4) for the full length of the element.

2. The boundary vertical elements and such other similar vertical elements as may be required shall be designed to carry all the vertical stresses resulting from the wall loads in addition to tributary dead and live loads and from the horizontal forces as prescribed in Section 2313. Horizontal reinforcing in the walls shall be fully anchored to the vertical elements.

3. Similar confinement of horizontal and vertical boundaries at wall openings shall also be provided unless it can be demonstrated that the unit compressive stresses at the opening have a load factor two times that indicated in Section 18.7.2.1(1).

REFERENCES

American Concrete Institute (1965). *ACI Standard Building Code Requirements for Reinforced Concrete* (ACI 318–63), Detroit, Michigan.

Bennett, W. B., J. M. Hanson, A. L. Parme, and J. A. Sbarounis (September 1965). *Laboratory Investigation of Reinforced Concrete Beam-Column Connections Under Lateral Loads*, Chicago: Portland Cement Association.

Blume, J. A. (1960a). "Structural Dynamics in Earthquake Resistant Design," *Trans. ASCE*, **125**, 1088–1139.

Blume, J. A. (July 1960b). *A Reserve Energy Technique for the Design and Rating of Structures in the Inelastic Range, Proceedings of the Second World Conference on Earthquake Engineering*, Tokyo, Japan.

Blume, J. A. (1962). *The Earthquake Resistance of California School Buildings—Additional Analyses and Design Implications*, Division of Architecture, Sacramento: California State Printing Division, Documents Section.

Blume, J. A. (1965). *Earthquake Ground Motion and Engi-*

neering Procedures for Important Installations Near Active Faults, Proceedings of the Third World Conference on Earthquake Engineering, New Zealand.

Blume, J. A. (1967a). "A Structural–Dynamic Analysis of an Earthquake Damaged Fourteen-Story Building," *The Prince William Sound, Alaska, Earthquake of 1964 and Aftershocks*, Vol. II, Part A, U.S. Department of Commerce, Environmental Science Services Administration, Publication 10-3.

Blume, J. A. (May 1967b). *Design of a High-Rise Building Under the New Code Requirements for Ductile Reinforced Concrete*, Session No. 8, American Society of Civil Engineers Seminar on Buildings in Earthquake and Wind, Seattle, Washington.

Blume, J. A. (May 1967c). *Earthquake Demands Related to Code Lateral Force Coefficients and Ductility*, Session No. 4, American Society of Civil Engineers Seminar on Buildings in Earthquake and Wind, Seattle, Washington.

Blume, J. A., N. M. Newmark, and L. H. Corning (1961). *Design of Multistory Reinforced Concrete Buildings for Earthquake Motions*, Chicago, Illinois: Portland Cement Association.

Clough, R. W. (July 1962). "Earthquake Analysis by Response Spectrum Superposition," *Bull. Seism. Soc. Am.*, **52** (3).

Clough R. W., E. L. Wilson, and I. P. King (August 1963). "Large Capacity Multistory Frame Analysis Programs," *J. Am. Soc. Civil Eng.*, ST 4.

Clough, R. W. (October 1966). "Effect of Stiffness Degradation on Earthquake Ductility Requirements," *Report No. 66-16*, Structural Engineering Laboratory, Berkeley: University of California.

Driskell, J. J. (1967). *Design of a 30-Story Reinforced Concrete 100% Moment-Resisting Frame to Meet Ductility Requirements*, American Concrete Institute Seminar, Los Angeles.

Everard, N. J., and E. Cohen (1964). *Ultimate Strength Design of Reinforced Concrete Columns*, Interim Report of ACI Committee 340, ACI Publication SP-7, Detroit, Michigan: American Concrete Institute.

Gaston, J. R., C. P. Siess, and N. M. Newmark (December 1952). *An Investigation of the Load-Deformation Characteristics of Reinforced Concrete Beams Up to the Point of Failure*, Civil Engineering Studies, Structural Research Series No. 40, Urbana: University of Illinois.

Housner, G. W. (1956). *Limit Design of Structures to Resist Earthquakes, Proceedings of the First World Conference on Earthquake Engineering.*

Hudson, D. E. (1956). *Response Spectrum Techniques in Engineering Seismology, Proceedings of the First World Conference on Earthquake Engineering.*

International Conference of Building Officials (1967). *Uniform Building Code.*

Plummer, H. C., and J. A. Blume (1953). *Reinforced Brick Masonry and Lateral-Force Design*, Washington, D. C.: Structural Clay Products Institute.

Portland Cement Association (1955). *Analysis of Small Reinforced Concrete Buildings for Earthquake Forces*, 4th ed.

Portland Cement Association (1964). "Design Constants for Rectangular Long Columns," *Advanced Engineering Bulletin No. 12*, Chicago.

Portland Cement Association (1965). "Capacity of Restrained Eccentrically Loaded Long Columns," *Advanced Engineering Bulletin No. 10*, Chicago.

Richart, F. E., A. Brandtzaeg, and R. L. Brown (November 1928). "A Study of the Failure of Concrete Under Combined Compressive Stresses," *Bulletin No. 185*, Urbana: University of Illinois Engineering Experiment Station.

Structural Engineers Association of California, Seismology Committee (1960). *Recommended Lateral Force Requirements and Commentary.*

Structural Engineers Association of California, Seismology Committee (1966). *Recommended Lateral Force Requirements.*

U.S. Departments of the Army, Navy, and Air Force (March 1966). *Seismic Design for Buildings*, TM 5-809-10; NAVDOCKS P-355; AFM 88-3.

Zweig, A. (February 1965). "Ultimate Strength Design for Bending by Iteration," *J. Am. Concrete Inst.*, **62**.

Chapter 19

Prestressed and Precast Concrete Structures

T. Y. LIN

Professor of Civil Engineering
University of California
Berkeley, California

19.1 INTRODUCTION

Basic principles underlying the design of prestressed and/or precast concrete structures are the same as those for conventional in-place concrete structures. For prestressed structures, the new factors are the use of high-tensile steel and high-strength concrete and the stressing of one against the other, all of which introduce magnitudes, details, and behavior different from conventional concrete structures. For precast structures, the methods of joinery and their reserve strength and ductility present special problems not always encountered in poured-in-place concrete. Only the different and new features are emphasized in this chapter. Readers are advised to refer to other sections of this book for principles and methods common to earthquake design of structures in general and concrete structures in particular.

Prestressed concrete differs from conventionally reinforced concrete in the introduction of internal stresses by tensioning the steel and holding it against the con-

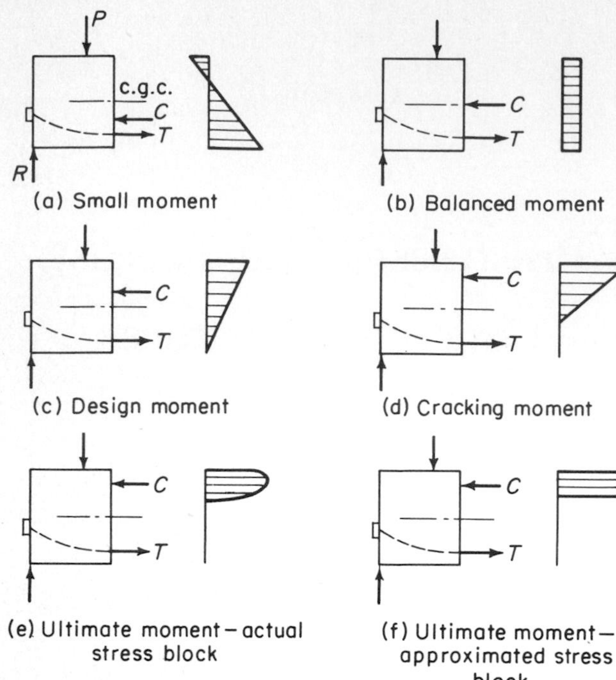

Fig. 19.1. Stress blocks on beam sections corresponding to different magnitudes of external moment.

crete. When there is no external moment acting on a statically determinate member, the center of tension T in the steel coincides with the center of compression C in the concrete. As external moment is applied, the center of compression C moves away from the center of tension T, thus creating an internal resisting moment. Different locations of C yield different stress distributions (Fig. 19.1) ranging from the elastic to the ultimate. Readers are referred to a treatise (Lin, 1963) on prestressed concrete for details and fundamentals. Only portions related to earthquake design are discussed in this chapter.

19.2 MATERIAL PROPERTIES

Three types of steel most frequently used for prestressing are wires, strands, and bars. Typical stress–strain curves for these are shown in Fig. 19.2. It is noted that the elastic strain in each case is only a very small portion of the total strain, indicating a ductile material good for earthquake resistance. While the excess ductility is not as high as structural grade steel, it is seldom that prestressed steel can be stressed to rupture unless the structure is already seriously deformed to the range of total collapse.

Fig. 19.2. Typical stress–strain curve for prestressing steels.

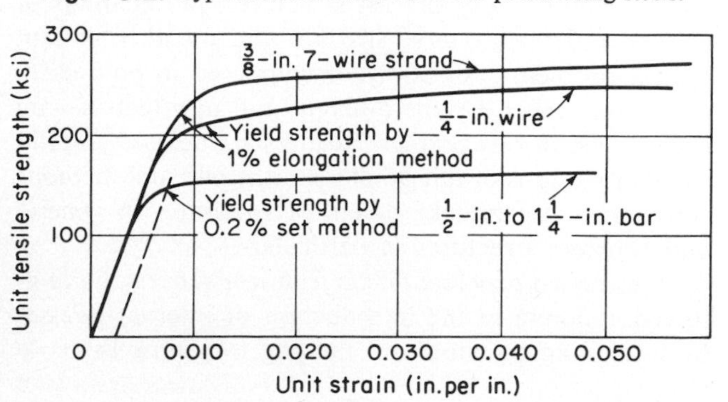

High-strength concrete has typical stress–strain curves, as shown in Fig. 19.3. Its ductility decreases somewhat as the strength goes up. However, the excess energy absorption capacity in the plastic range is still many times the elastic energy under the normal working stress.

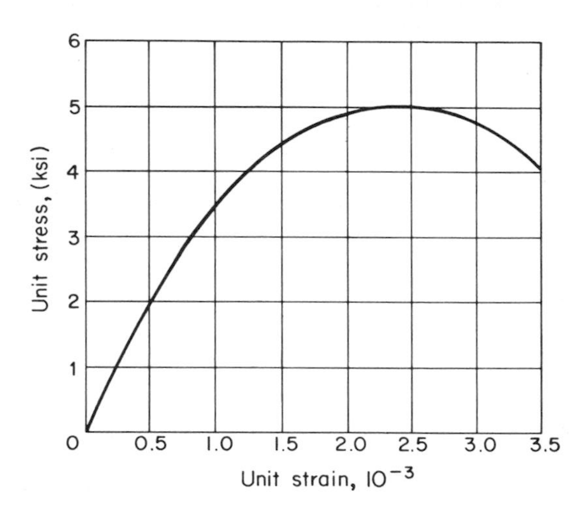

Fig. 19.3. Typical stress–strain curve for 5000-psi concrete.

Prestressed steel and concrete work together through bond and anchorages. In the case of bonded construction, a broken bond or a ruptured tendon is localized and seldom transmitted to the end anchorages. For unbonded construction, integral action between steel and concrete is assured only by the end anchorages. Fortunately, stress variation in an unbonded tendon is of a smaller range, and there is little chance of any breakage in the tendon proper. Some typical end anchorages are shown in Fig. 19.4.

19.3 ELASTIC BEHAVIOR

Because earthquakes act on buildings with varying and alternating directions, a member that is eccentrically prestressed may seem to be incapable of resisting moment applied in an unfavorable direction. For example, the simple beam shown in Fig. 19.5 prestressed with a parabolic cgs (centroid of steel) would seem to have little resistance for upward loadings produced by earthquakes. In actuality, however, this is not true, because under the action of dead load a prestressed member

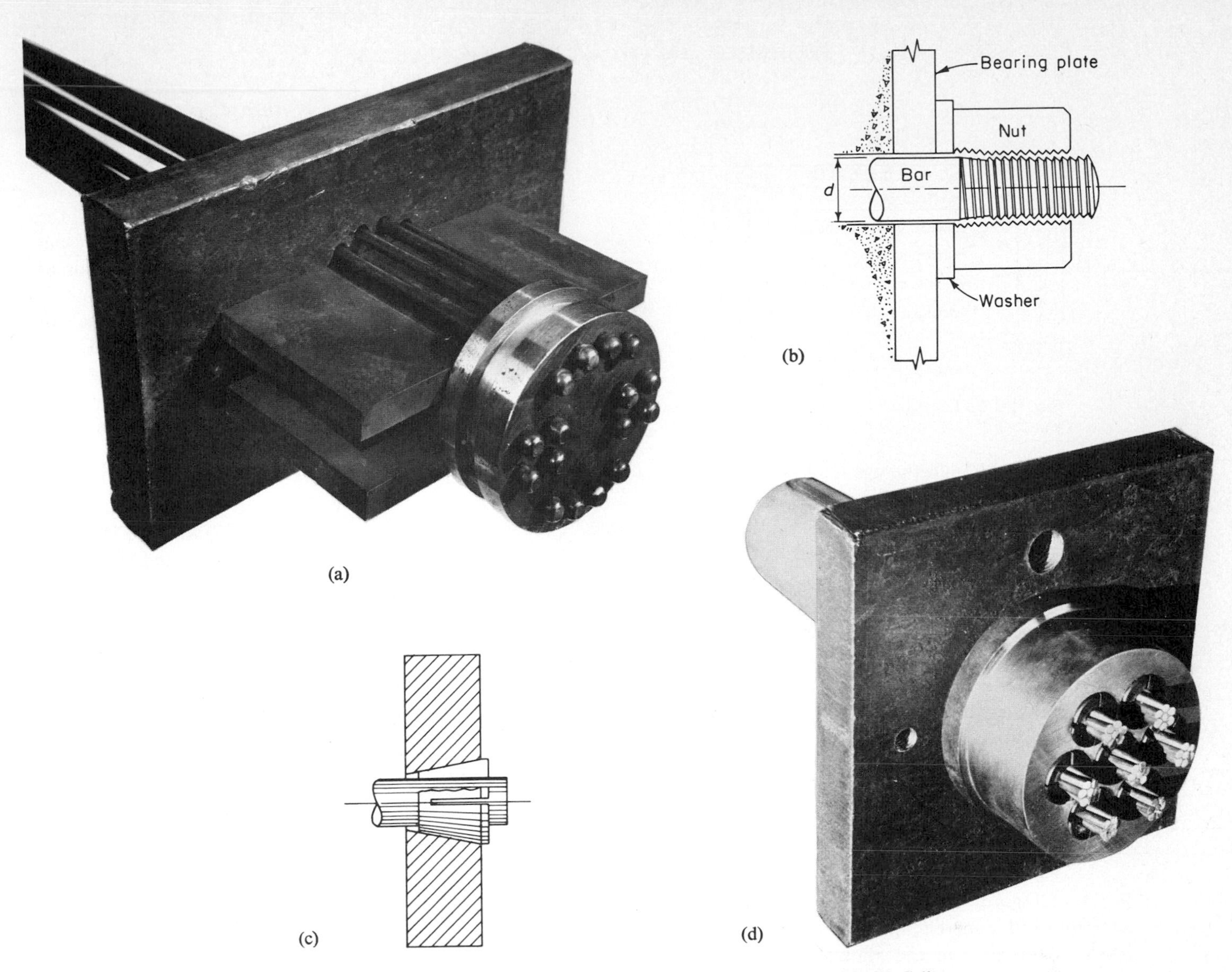

Fig. 19.4. (a) End anchorage, Prescon system; (b) end anchorage for the Lee-McCall or Stressteel system; (c) wedge for anchoring rods; (d) anchorage for seven strands.

would usually possess a line of pressure (the *C*-line) close to the cgc (centroid of concrete section). In fact, in an ideal case, such as shown in Fig. 19.5, the upward component of the tendons balances the gravity action of the dead load, and the *C*-line will pass exactly through the cgc. Then any additional load would simply move the *C*-line away from the cgc, and the beam is able to resist a sizable loading whether upward or downward.

Sometimes a prestressed member is subjected to the moment reversal of earthquake loadings with its *C*-line already quite far away from the cgc; in this case, the resisting capacity is relatively low and additional non-prestressed steel may be required. This is similar to a reinforced concrete member under flexure when reinforcement may be required on both sides in order to provide sufficient resistance to reversal under earthquake actions.

Fig. 19.5. A *C*-line location in a simple beam.

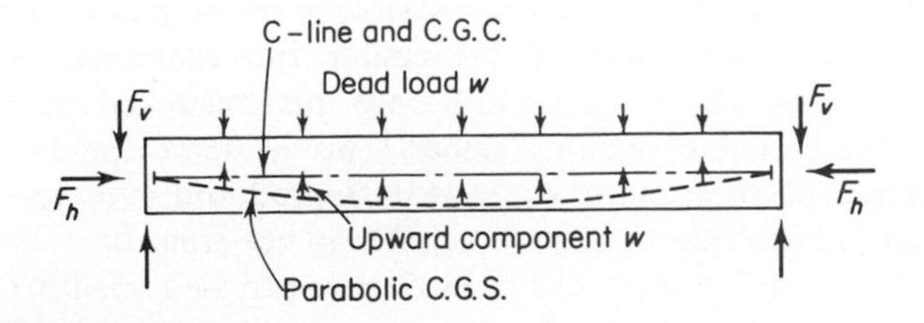

The elastic behavior of a prestressed rigid frame is shown in Fig. 19.6. An ideal design for earthquake resistance is again to locate the *C*-line through the cgc before the application of earthquake forces. For example, the cgs in the beam should be located to supply an upward component to balance the dead load so that we will have a *C*-line in the beam coinciding with the cgc. If the horizontal component of the prestress F_1 should act with an eccentricity e_1 at an end of the beam, that moment F_1e_1 should be counterbalanced with another moment supplied by the tendons in the columns, such that $F_2e_2 = F_1e_1$. If this is done the columns will be under concentric prestress with the *C*-lines coinciding

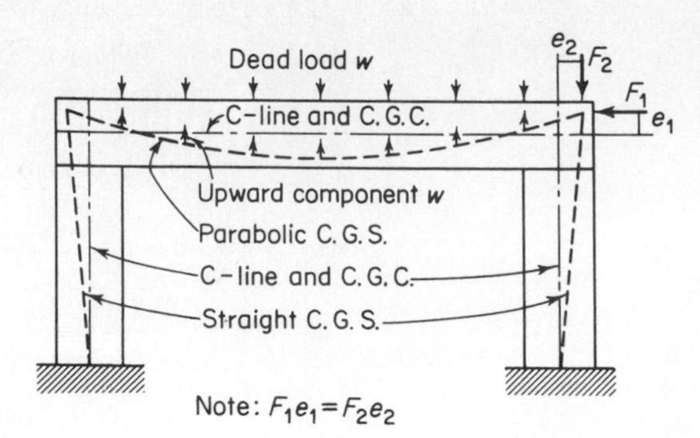

Fig. 19.6. Location of a *C*-line in a rigid frame.

with the cgc although the cgs is physically located away from the cgc.

In an actual building it is not always possible to achieve the ideal location indicated in Fig. 19.6. But if the *C*-line is not too far from the cgc, the elastic capacity of the members to resist moment reversal can be sufficient to meet code requirements and the action of moderately heavy earthquakes. It is emphasized that the resistance of the prestressed members to moment reversal is not indicated by the physical location of the cgs but by the location of the *C*-line before the application of earthquake forces.

19.4 ALLOWABLE STRESSES AND LOAD FACTORS

For buildings of conventional materials, such as steel or reinforced concrete, most building codes permit a one-third increase in the allowable stresses when considering earthquake loadings. This increase is justified because of the infrequent occurrence and the short duration of the design earthquakes. It is also convenient because this procedure avoids the necessity of reproportioning all members for earthquake effects. While the above reasoning is equally applicable to prestressed concrete, the value of one-third increase cannot be directly applied.

Consider the case of allowable tension for extreme fiber stresses in a prestressed concrete member. A one-third increase for zero tension would still culminate in zero tension; a one-third increase for an allowable tension of 400 psi would add to the moment capacity by only approximately 5%. Consider the case of allowable tension in prestressed steel, which is usually set at 0.60 f'_s; a one-third increase would raise this to 0.80 f'_s, which is clearly much too high. Hence, a straight one-third increase for allowable stresses cannot be specified when considering earthquake effects in prestressed concrete.

In order to preserve the original intent of the one-third increase in allowable stresses without encountering the technical difficulties mentioned above, it is suggested that the loads, shears, and moments be modified by a reduction factor of $\frac{3}{4}$ when considering earthquake effects while the allowable stresses remain unchanged. This method, sometimes used for earthquake design of conventional structures, is indeed the sensible solution for prestressed concrete. This method simply permits the overloading of all parts by one-third of the normal loadings when considering earthquake effects. It assumes that all members would be able to carry a one-third increase in load for a short duration without signs of distress; this is a valid assumption for prestressed concrete, as in the case of other conventional materials when designed according to the usual standards.

When ultimate strength method is used for seismic design of prestressed concrete, most building codes specify the same load factors as for conventional materials. The 1967 Uniform Building Code (International Conference of Building Officials, 1967) and the Building Code Requirements for Reinforced Concrete (ACI, 1963) both call for a load factor of

$$U = 1.25(D + L + W)$$

in which W = earthquake effects.

This, when compared to the requirement of $U = 1.5D + 1.8L$ indicates, for loadings including earthquake load, an increase of about one-third in allowable stresses or a reduction factor of about 3/4 as suggested above. Where gravity load effects counteract earthquake effects, only 0.9 of the dead load is considered and the earthquake load is increased by the factor of 1.1. Thus,

$$U = 0.9D + 1.1W$$

19.5 ENERGY ABSORPTION CAPACITY AND MOMENT-CURVATURE RELATIONSHIPS

While the static earthquake load specified by building codes together with the allowable stresses or load factors considered above will usually yield satisfactory designs, the energy absorption capacity of the resisting members becomes an important criterion when considering unusually heavy earthquakes. This is especially significant for tall buildings, where the code forces are relatively low and the effects of higher modes of vibration become more pronounced.

The ductility of prestressed concrete is dependent on (1) the tensile strength and elongation of the prestressed steel and (2) the compressive strength and shortening of concrete. Before the cracking of concrete, the entire concrete section absorbs the energy with the strain energy of steel sharing only a small part of the work. After cracking, however, the steel participates fully in the energy absorption, while only the uncracked part of the concrete remains active. This postcracking behavior of prestressed concrete is indeed quite similar and comparable to that of reinforced concrete.

The energy absorption capacity of prestressed steel is typically described in Fig. 19.7. During the process of prestressing a sizable amount of energy is stored in the steel, although part of that energy is lost as a result of the loss of prestress. Under the application of external loading and previous to cracking, the change in steel strain is quite small, and the energy is essentially absorbed by the concrete. After cracking, however, the reserve energy capacity of steel is exceedingly high. It is noted that the steel seldom can be stretched to its ultimate strain, which has a 4% specified minimum but generally extends to more than 6% for the ASTM A416 seven-wire strands now prevalent in the United States.

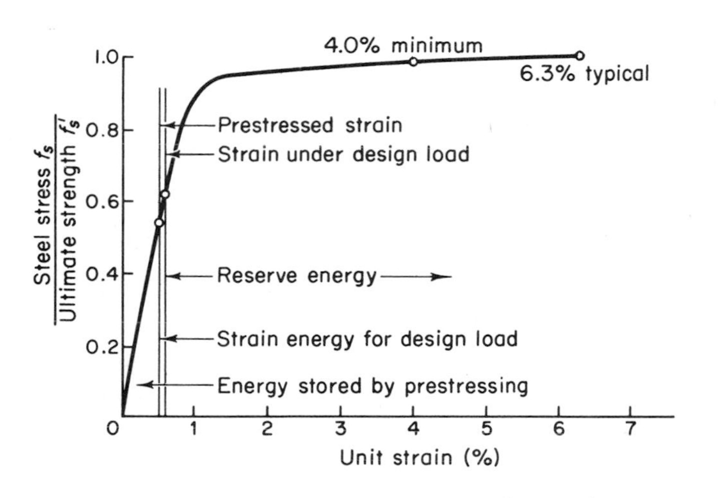

Fig. 19.7. Stress-strain diagram for 7-wire strand.

Hence, there is no question of the sufficiency of the prestressed steel to absorb an unusual earthquake, e.g., 4–5 times heavier than the design earthquake. Although the ultimate strain of 4–6% is much lower than that for reinforcing bars, or for structural steel, the ductility is already more than enough to meet the requirements and can seldom if ever be fully utilized in any case.

To illustrate the energy absorption capacity of prestressed concrete beams, two typical examples are shown in Fig. 19.8—one for a post-tensioned beam B with a tendon bonded to the concrete and two nonprestressed steel bars in addition and another for an unbonded beam U with two greased tendons (Caulfield and Patton, 1963). Both beams were loaded at the third points, and the middle third part was measured for moment–curvature relationships. Because moment M multiplied by curvature ϕ is a measure of the energy stored in the beam, the area under the curve before cracking is a measure of the elastic energy while the remaining area indicates the energy absorption capacity in the plastic range. It can be shown that for beam U, the ratio of the plastic energy absorption capacity to the elastic capacity is approximately 22, while that for beam B is about 70. In both cases, there is plenty of reserve capacity. The bonded beam B, with lower percentage of steel and some nonprestressed bars, exhibited especially high ductility.

While curves in Fig. 19.8 are typical of under-reinforced slender beams failing in flexure, beams with short shear spans and thin webs may fail in shear, especially if no web reinforcement is provided. The two lower

Fig. 19.8. Moment-curvature relationship for prestressed beams.

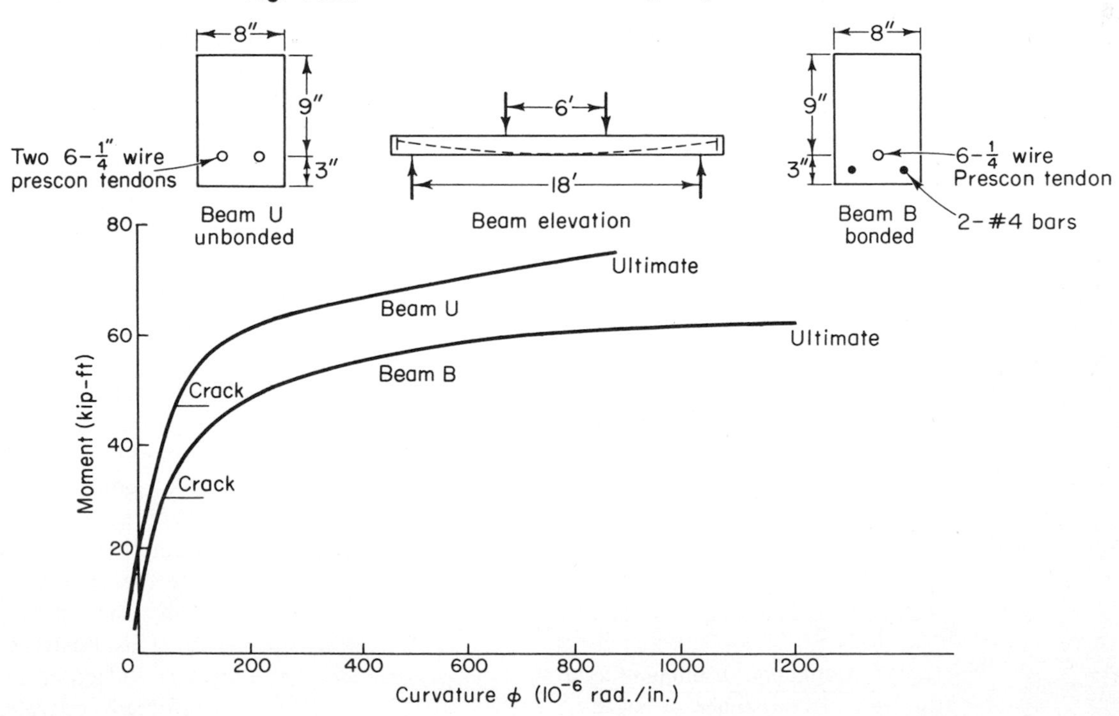

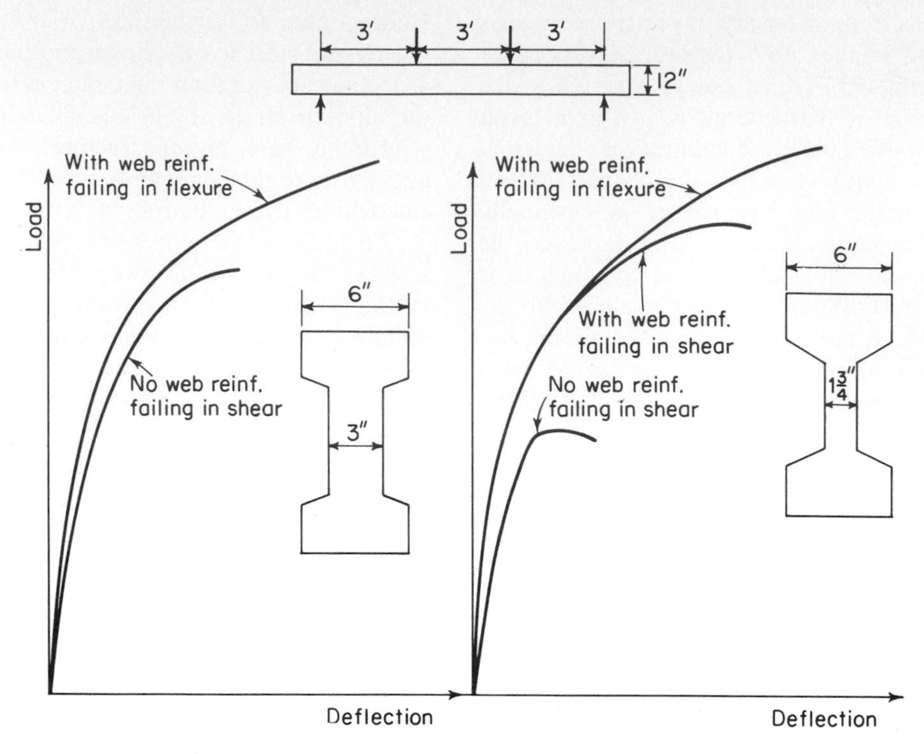

Fig. 19.9. Ductility of beams with and without web reinforcement.

curves in Fig. 19.9 indicate the lack of ductility of such beams and clearly bring out the necessity for web reinforcement in these cases (MacGregor, Sozen, and Siess, 1960). The two top curves, on the other hand, confirm the ductility of beams with proper web reinforcement. Generally speaking, the amount of web reinforcement called for by the 1963 ACI Code will be sufficient to assure a reasonable amount of ductility for the purpose of earthquake resistance because it ensures flexural failure to occur before shear failure.

Figure 19.10 gives the results of some tests for rectangular beams without web reinforcement subjected to high shear in combination with moment (Walther and Warner, 1958). The low shear span to depth ratio of 2 was the main reason for the low ductility. It will be noticed that the ductility, measured by the ratio of

Fig. 19.10. Ductility of beams under combined bending and shear.

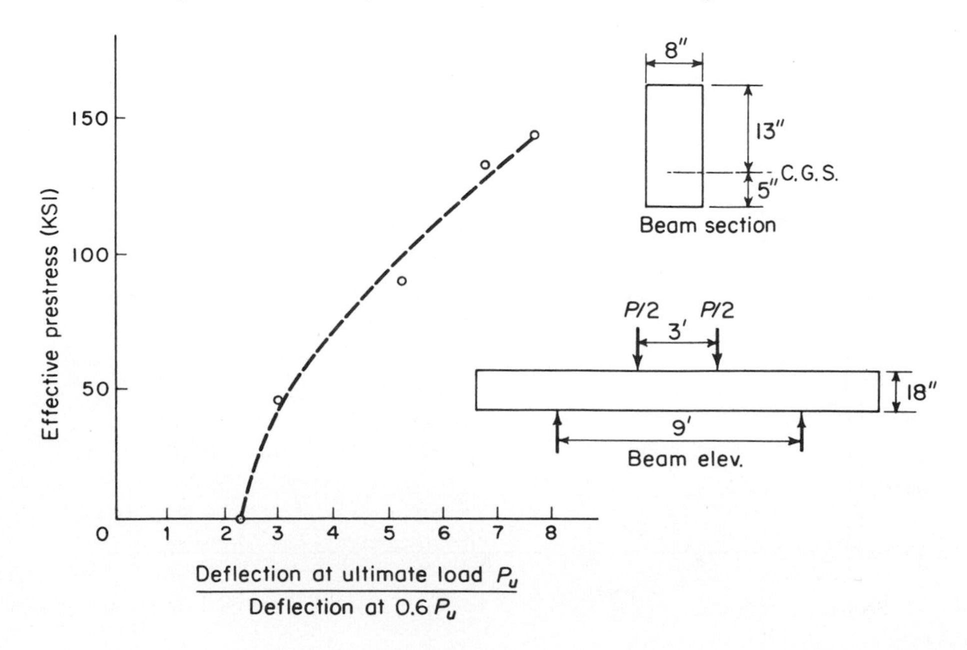

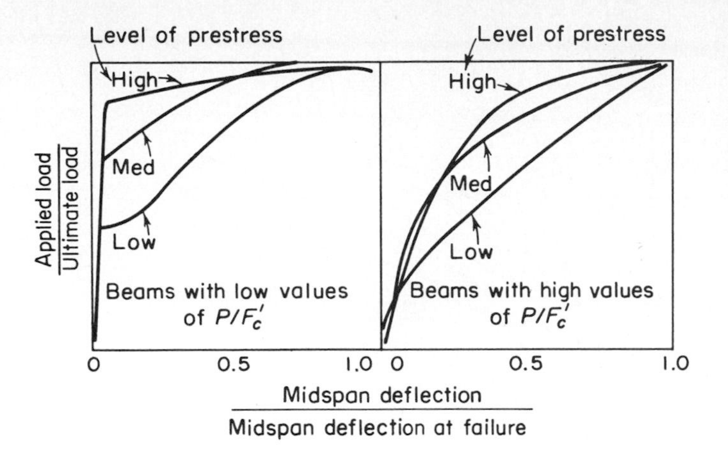

Fig. 19.11. Load-deflection curves for beams failing in flexure.

the deflection at ultimate load P_u to that at 0.6 P_u ranged from 2 to 8, depending on the extent of the effective prestress in the steel, indicating that higher prestress results in higher ductility. This is also confirmed by tests at the University of Illinois (Fig. 19.11), which indicated higher resilience for higher levels of prestress (Warwaruk, Sozen, and Siess, 1962).

Three rigid frames of prestressed concrete tested at the University of California (Fig. 19.12) exhibited load deflection characteristics as in Fig. 19.13. (Tests were conducted by Jakub Mames, Silesian Technical University, Gliwice, Poland.) Two of these frames had prestressed columns and one had reinforced columns without prestressing. The postcracking ductility of these frames indicated their capacity for energy absorption.

The energy absorption capacity of prestressed beams subjected to external moment producing compression on the precompressed side is not too well known. However, from available data it can be said that the use of nonprestressed reinforcement on the opposite side will greatly increase the strength and the ductility of the beam (Scordelis, Lin, and May, 1957).

The energy absorption potential of prestressed concrete (as with other materials) when subjected to alternating loads is not fully described by its ductility potential, i.e., the ratio of total deformation to yield deformation. The entire load-deformation history must be taken into account. Unloading must be considered as well as loading.

Fig. 19.12. Dimensions of rigid frame No. 1 for Fig. 19.9.

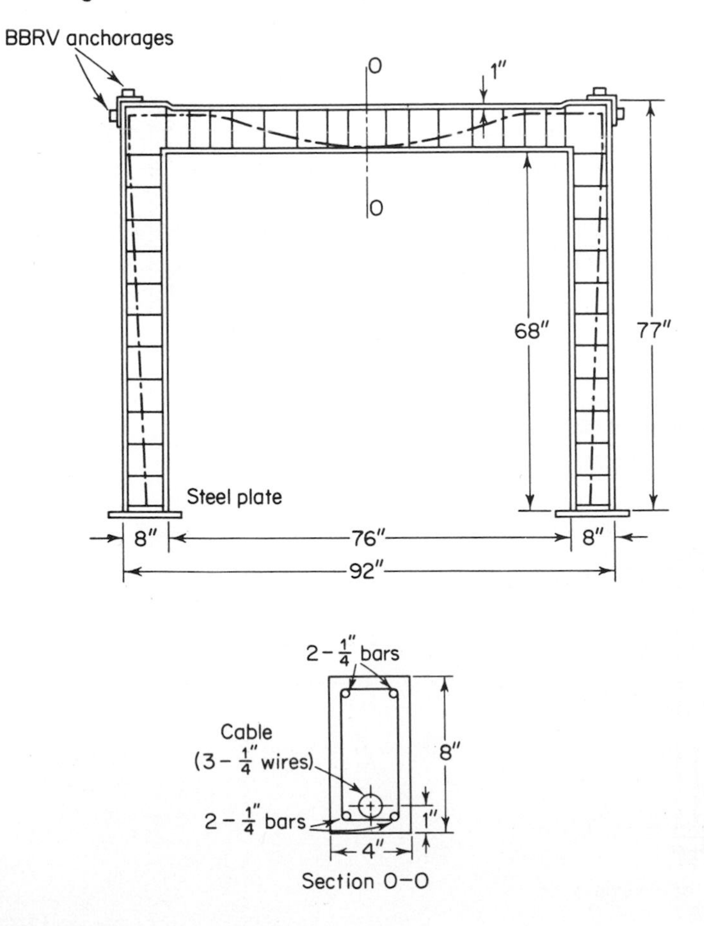

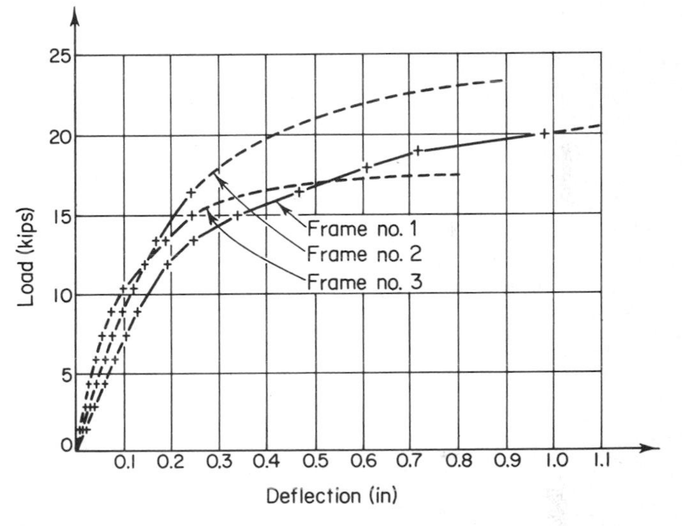

Fig. 19.13. Mid-span deflection of prestressed concrete rigid frames.

The Prestressed Concrete Engineering Association of Japan published the following double-tee test (Fig. 19.14) in their tentative draft on "General Principles of Earthquake Resistant Design of Prestressed Concrete Structures." This test points out that prestressed concrete may have less than a full ideal hysteresis property and, thus, absorb less energy than assumed in the typical calculation of elasto-plastic earthquake response. It is noted that for loads bringing the steel just past yielding good elastic recovery is obtained, while for extreme deformations recovery characteristics are similar to reinforced concrete beams.

Kiyoshi Nakano of the Building Research Institute, Japan, has reported on three prestressed concrete portal

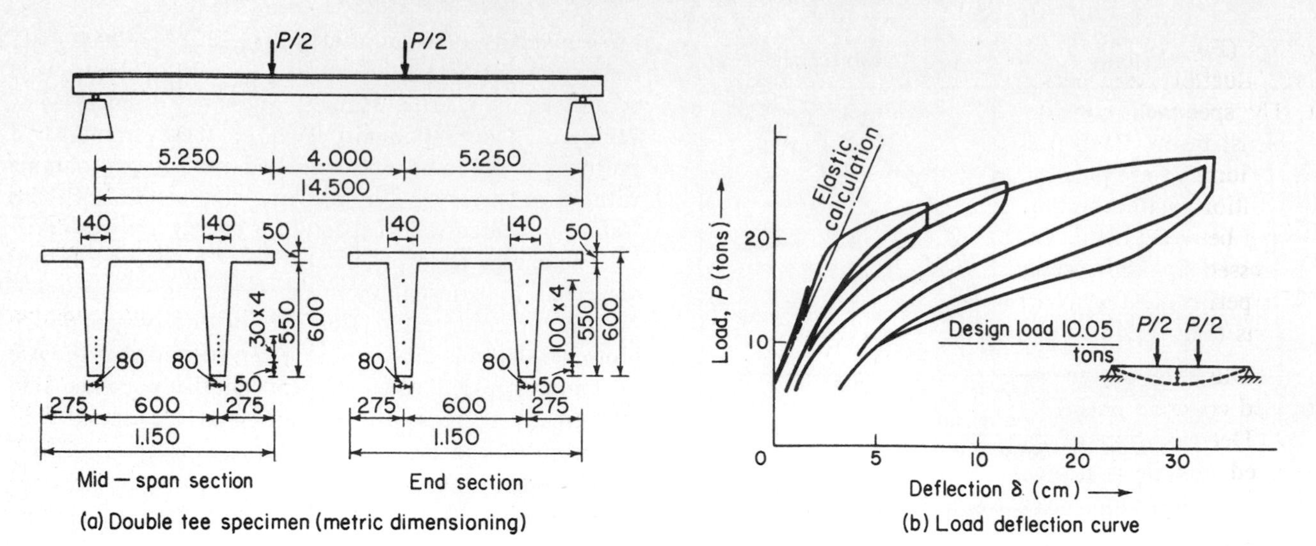

(a) Double tee specimen (metric dimensioning)

(b) Load deflection curve

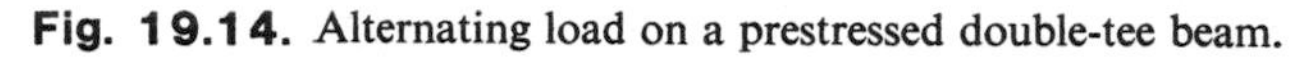
Fig. 19.14. Alternating load on a prestressed double-tee beam.

Fig. 19.15. Precast prestressed portal frames.

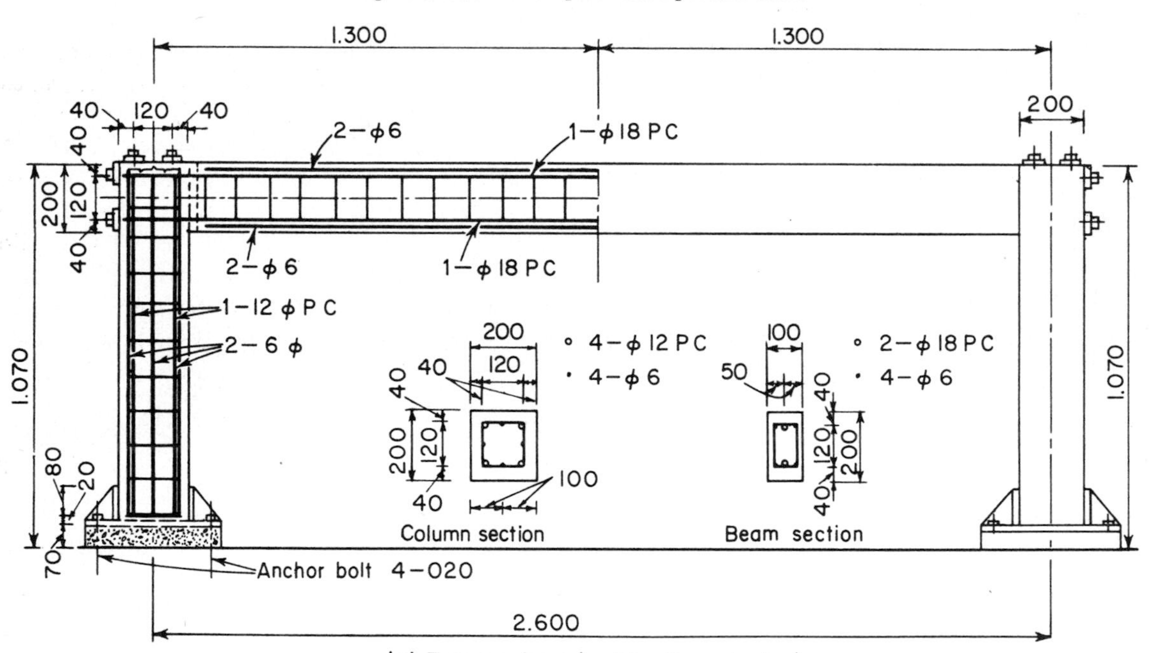

(a) Test specimen (metric dimensioning)

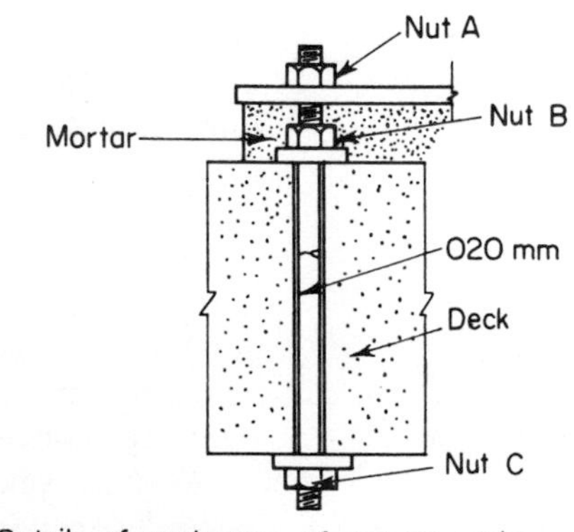

(b) Details of anchorage of column ends

frames (Fig. 19.15) in which resistance against lateral force, ductility, and period of vibration was examined.

The specimens consisted of two precast columns and a precast beam. Each frame was assembled as follows: The columns were prestressed, and the beam was placed in position. After casting and hardening of the joint mortar between the beam and column, the beam was prestressed by high-strength steel rods. As a result of the experiment, Dr. Nakano drew the following conclusions (Fig. 19.16):

1. Ultimate design method is applicable for prestressed concrete portal frame.
2. Deterioration of deflection characteristics under repeated loading is not considerable.

Fig. 19.16. Load-deflection diagrams of rigid frames under alternate loading.

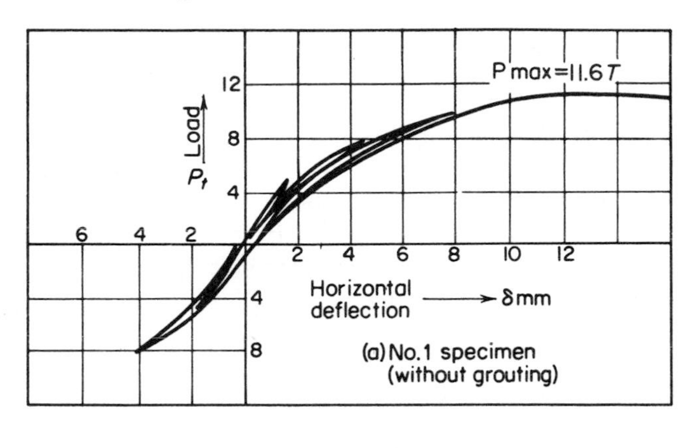

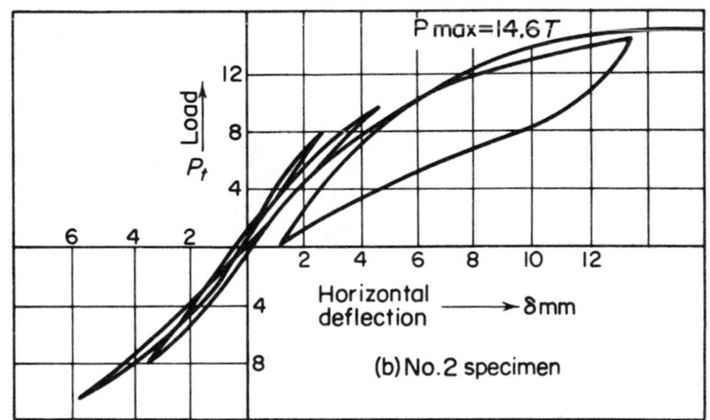

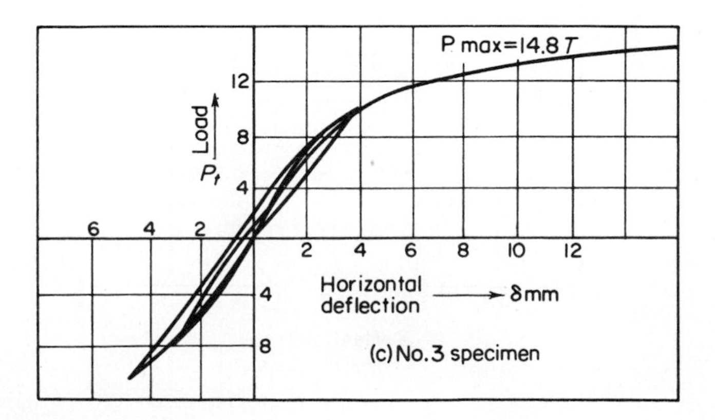

3. Ample ductilities are guaranteed.
4. Lengthening of the period of vibration of the frame, under high loads, was small.

From this limited number of tests we may draw a tentative conclusion that the yield resistance and maximum deformation capacity of prestressed concrete frames under alternate loading can be predicted by the principles of engineering mechanics.

Similarly, the behavior of prestressed concrete columns under combined axial load and moment can be predicted by the basic principles of mechanics, taking into account the effects of prestressing, shrinkage, and creep in concrete. This has been indicated by several series of carefully conducted tests published in the ASCE and ACI journals (Arnoi, 1967; Zia, 1966).

19.6 PARTIAL PRESTRESS AND NONPRESTRESSED REINFORCEMENTS

The earthquake resistance and energy absorption capacities of prestressed members can be greatly increased by the proper use of nonprestressed reinforcements.

In order to understand the design of partially prestressed beams, it is necessary to study the behavior of such beams with varying amounts of reinforcement and subjected to varying amounts of prestress. The difference in behavior of over-reinforced and under-reinforced beams is seen by comparing curves *a* and *b* of Fig. 19.17. The difference in behavior of over-prestressed and under-prestressed beams is seen by comparing curves *a*, *b*, *c*, and *d* in Fig. 19.18.

When a section is over-reinforced (Fig. 19.17) it will fail by compression in concrete before the steel is stressed

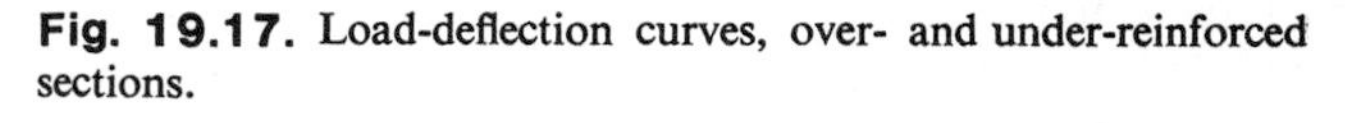
Fig. 19.17. Load-deflection curves, over- and under-reinforced sections.

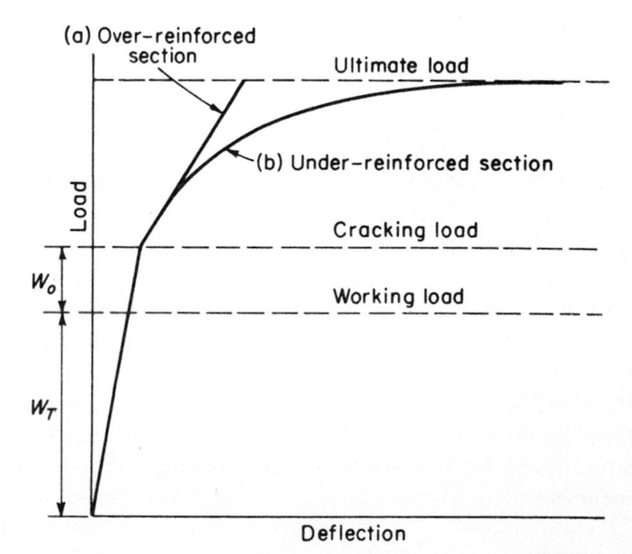

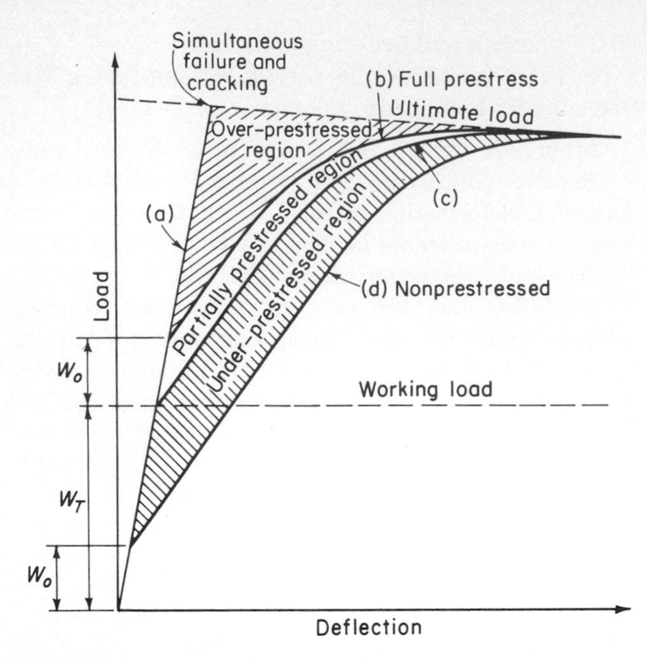

Fig. 19.18. Load-deflection curves for varying degrees of prestress (for under-reinforced sections of bonded beams).

beyond its elastic limit. Thus, the ultimate deformation of the steel and the ultimate deflection of the beam are rather small and the failure is brittle. When seriously over-reinforced, even if the steel is not prestressed, the deflection of the beam before rupture still will be limited. When a section is under-reinforced, its deflection increases very appreciably before failure, thus giving ample warning of impending collapse. Failure starts in the excessive elongation of steel and ends in the gradual crushing of concrete on the compressive side.

In order to avoid sudden and brittle failures, and also for general economy in design, most beams are under-reinforced. When an under-reinforced section is designed for full prestressing, allowing no tension in concrete under the working loads, the load-deflection relation is given by curve *b* of Fig. 19.18. Before cracking the section will carry an additional load W_0 above the working load W_T, the magnitude of that additional load being

$$W_0 = k\frac{f'I}{c_b}$$

where k is a constant depending on the span length and end conditions, f' is the modulus of rupture, and c_b the distance from cgc to the tensile extreme fibers.

If the same under-reinforced section with the same amount of steel is given somewhat smaller prestress so that cracking is reached just at the working load, the tensile stress being equal to the modulus of rupture under working load, the load-deflection relation will be given by curve *c*, where the deflection corresponding to the cracked section starts at the working load. If the beam is not prestressed at all, but still reinforced with the same amount of steel, provided that the steel is bonded to the concrete, the beam will behave as in curve *d*. It will start cracking as soon as load W_0 is reached, although its ultimate strength may not be greatly reduced.

If the beam is over-prestressed, it will crack only after the load exceeds $W_T + W_0$, and its load-deflection curve will fall between curves *a* and *b* (Fig. 19.18). In the extreme case, when the beam is very much under-reinforced but highly over-prestressed, cracking and failure may take place simultaneously so that brittle failure occurs with sudden rupture in the steel. In principle, a partially prestressed beam may have a load–deflection curve lying anywhere between curves *b* and *d*, depending on the amount of prestress. In practice, cracking is seldom permitted for prestressed concrete under working load; hence, the actual load-deflection curve will usually fall between curves *b* and *c*, and seldom below curve *c*.

The area under the load-deflection curve is a measure of the ability of a beam to stand impact load and to absorb shocks. Hence, it is seen that both the fully and the partially prestressed beams will supply a fairly large amount of resilience, while both the over-prestressed and the under-prestressed ones will supply appreciably less. The over-prestressed beams will possess less plastic energy, while the under-prestressed ones will absorb less elastic energy.

A series of tests on 15 prestressed concrete beams with unbonded tendons strengthened by nonprestressed

Fig. 19.19. Influence of additional bonded reinforcement on load-deflection response.

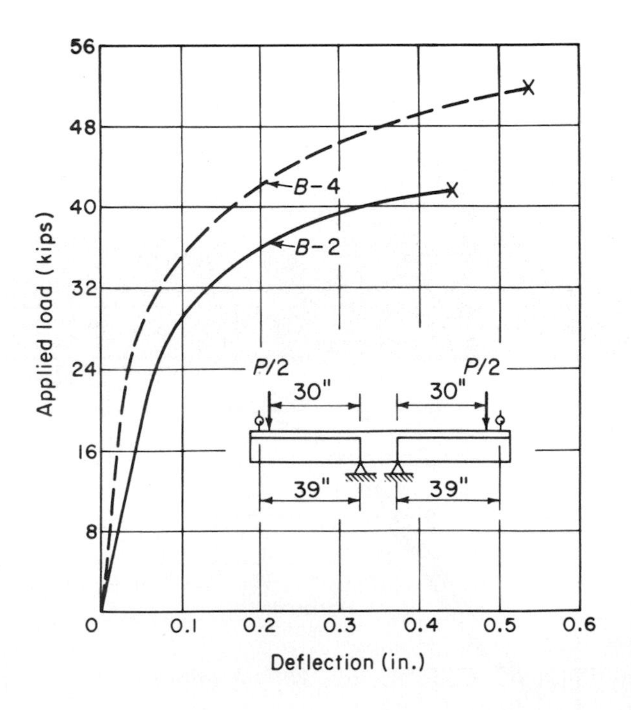

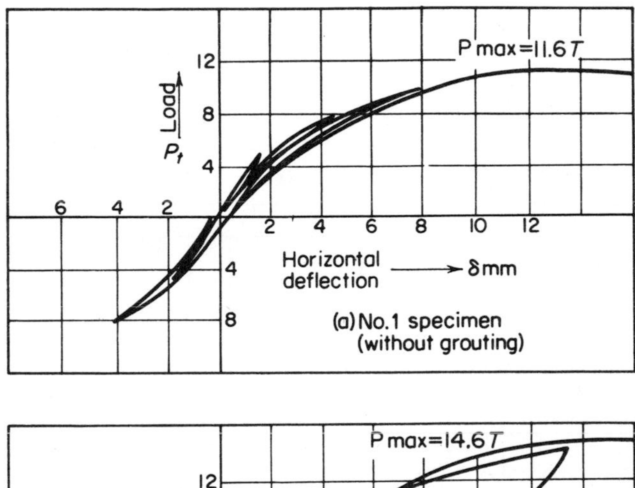

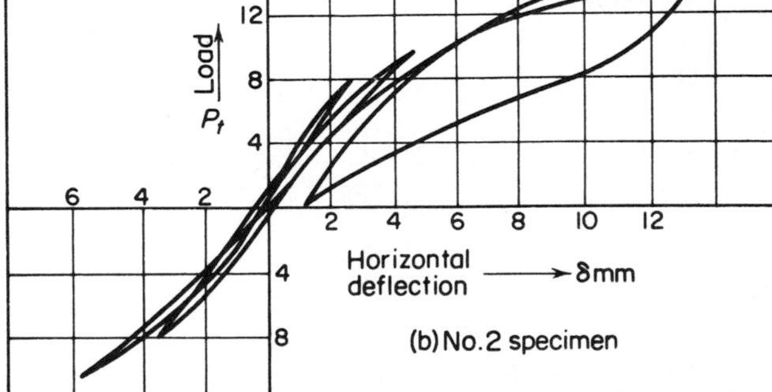

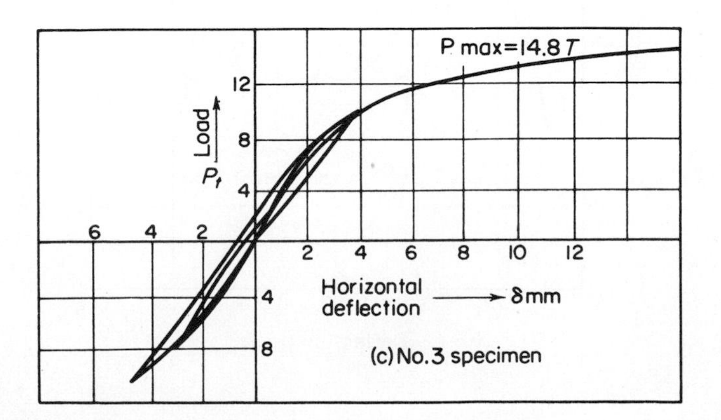

bonded reinforcement were described in a paper by Burns and Pierce (1967). Since beams containing unbonded tendons could develop some rather wide cracks, possibly resulting in a decrease of energy absorption capacity, the addition of nonprestressed reinforcement is desirable in helping to control such cracking. Figure 19.19 shows the comparative load-deflection curves for two companion beams that are identical except that nonprestressed reinforcement is added to beam B-4. Even when the nonprestressed steel acting at its yield stress was considered in the prediction of ultimate moment capacity for these beams, observed flexural strength was consistently higher than predicted.

On the other hand, an excessive amount of nonprestressed reinforcement without accompanying shear reinforcement may result in considerable reduction of ductility. Figure 19.20 shows two companion beams—CB-2 having excessive nonprestressed reinforcement producing a shear-compression type of failure that was sudden and complete. However, ductility can be restored if shear reinforcement is provided as per the ACI Code.

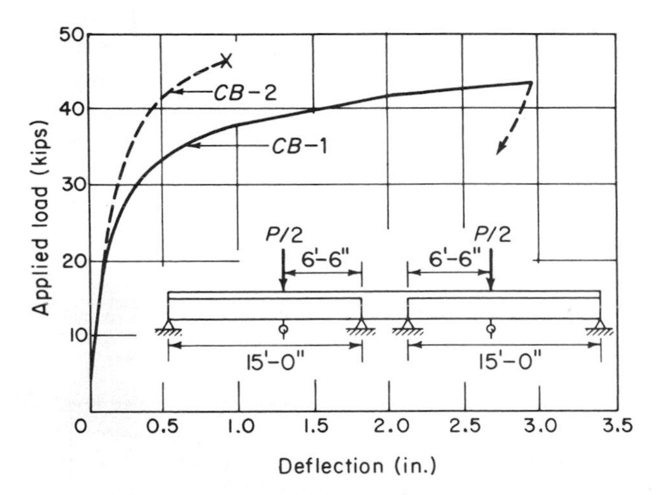

Fig. 19.20. Comparative load-deflection response for ribbed continuous beams.

In order to insure against over-reinforcement when using nonprestressed reinforcement in addition to prestressed reinforcement, the following formula is recommended as a safe measure:

$$\frac{pf_{su}}{f'_c} + \frac{p^*f_y}{f'_c} - \frac{p'f_y}{f'_c} \geq 0.30$$

where p = ratio of prestressed steel to concrete area bd; p^* = ratio of nonprestressed steel to concrete area bd; p' = ratio of compressive reinforcement to concrete area bd; f_{su} = ultimate stress in prestressed steel; f_y = yield point of nonprestressed steel; and f'_c = cylinder strength of concrete. Note that the above is in line with the 1963 ACI Code defining the limit for under-reinforced sections (ACI Code, Section 2609). Similarly, a minimum amount of steel is necessary to avoid immediate rupture at cracking, as specified in the same section of the ACI Code.

19.7 REPETITIVE LOADS

For prestressed members under the action of design live loads, the stress in steel wires is seldom increased by more than 10,000 psi from their effective prestress of about 140,000 psi. It is safe to say that, so long as the concrete has not cracked, there is little possibility of fatigue failure in steel, even though the working load is exceeded. After the cracking of concrete, high stress concentrations exist in the wires at the cracks. These high stresses may result in a partial breakage of bond between steel and concrete near the cracks. Under repeated loading, either the bond may be completely broken or the steel may be ruptured (Lin, 1963).

Numerous tests have been conducted on prestressed concrete members, giving considerable data on their fatigue strength. The results of these tests confirm the ability of the combination to stand any number of repeated loads within the working range. Failure started invariably in the wires near the section of maximum moment and often directly over the separators where the wires had a sharp change in direction.

No tests are available concerning the fatigue bond strength between high-tensile steel and concrete. However, from the results of tests on prestressed concrete beams, it seems safe to conclude that, if properly grouted, bond between the two materials can stand repeated working loads without failure. This is true because, before the cracking of concrete, bond along the length of the beam is usually low. For unbonded members, the problem of bond is clearly not an issue.

Fatigue failures at anchorage of end-anchored tendons are hardly known. When the tendons are bonded to the concrete, stress in the tendons near the end is not affected by live load. Hence there is no danger of fatigue failure, even though high localized stresses may exist in the wires at the anchorage. When unbonded, stress near the end of wires will change under live loads, but the range of stress under working loads is small so that fatigue failures are not likely.

Few data on the fatigue strength of unbonded beams are available, but there is some evidence to indicate that the use of mild steel reinforcement may greatly increase such strength. Fatigue tests on partially prestressed concrete members incorporating nontensioned wires show that such wires can stand repeated loads somewhat higher than the working loads without decreasing their eventual load-carrying capacity. This sounds quite reasonable, since the untensioned wires would be subject to the same stress range as the tensioned ones while

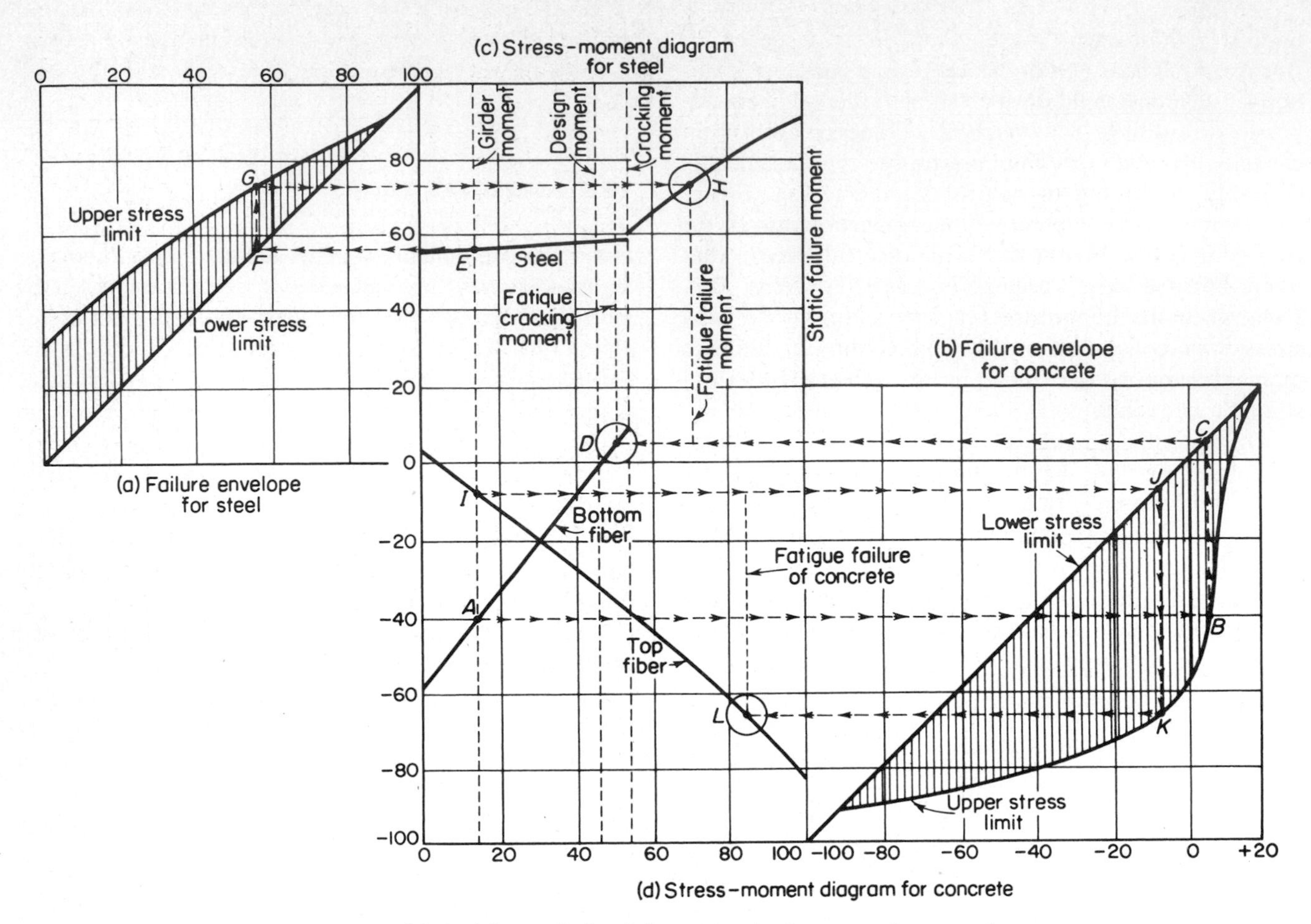

Fig. 19.21. Method for predicting fatigue strength of prestressed concrete beams.

their stress would be at a much lower level and hence there is less danger of fatigue failure.

A rational method for predicting the fatigue strength of prestressed concrete beams in bending has been developed by Professor Ekberg. It utilizes fatigue–failure envelopes for prestressing steel and concrete and relates them to the stress–moment diagram for a beam.

A typical failure envelope for prestressing steel is shown in Fig. 19.21a. This envelope indicates how the tensile stress can be increased from a given lower level to a higher level to obtain failure at 1 million load cycles. Note that all values are expressed as a percentage of the static tensile strength. Thus the steel may resist a stress range amounting to $0.27f'_s$ if the lower stress limit is zero but only a stress range of $0.18f'_s$ if the lower stress limit is increased to $0.40f'_s$. At a lower stress limit of $0.90f'_s$ or over, it takes only a negligible stress increase to fail the steel at 1 million cycles. While this fatigue envelope varies for different steels, the curve given here may be considered a typical one.

The fatigue–failure envelope for concrete is given in Fig. 19.21b. This is analogous to *a* for steel, except that it is drawn to cover both tensile and compressive stresses. This diagram indicates that if the lower stress limit is zero, a compressive stress of $0.60f'_c$ may be repeated at 1 million cycles. If the lower stress limit is $0.40f'_c$, the stress range can be $0.40f'_c$. If the compressive stress limit is $0.20f'_c$, the tensile stress limit, to produce cracking, is $0.05f'_c$.

A typical stress–moment diagram for steel is given in Fig. 19.21c, which again expresses nondimensionally both the stress and the moment by relating them to the static strength and the ultimate static moment. For example, when the external moment is 70% of the ultimate static moment, the stress in the steel is shown to be $0.80f'_s$. Similarly, Fig. 19.21d gives the concrete fiber stresses relative to the moment. It is noted that under certain loading conditions either the top or bottom fibers can be under tension rather than compression.

Combining these four portions (Fig. 19.21) it is possible to determine the fatigue-cracking moment and the fatigue-ultimate moment as limited by steel or concrete. Starting on the stress–moment diagrams at the point of dead load stress, which represent the lowest possible stress level, we can trace three paths as follows.

Steel: *E-F-G-H*
Concrete top fiber: *I-J-K-L*
Concrete bottom fiber: *A-B-C-D*

The point *H* indicates that for a maximum moment of $0.68M_{\text{ult}}$, the steel will fail in tension at 1 at million cycles. The point *L* indicates, for a maximum moment of $0.84M_{ult}$, the top fiber will fail in compression at 1 million cycles. The point D indicates that the fatigue cracking moment is $0.50M_{ult}$.

Only a few tests have been conducted on prestressed concrete members under shakedown loadings (see Section 19.4 on alternating loads), i.e., the application of complete moment reversals approaching the ultimate strength. However, such members are believed to behave in a manner similar to reinforced concrete members. In other words, if the section is not over-reinforced and if it does not fail in shear, it does possess enough ductilty to withstand a reasonable amount of shakedown loading.

19.8 DYNAMIC PROPERTIES AND DAMPING OF PRESTRESSED MEMBERS

Dynamic properties of prestressed concrete members are similar to those of reinforced concrete members, with certain exceptions. Before cracking, prestressed members are more homogeneous; after cracking, the transformed area is smaller because of the lower percentage of steel in prestressed beams. In any case, the principles of mechanics are applicable, both to prestressed and to reinforced concrete members.

A series of tests was performed at the U.S. Naval Civil Engineering Laboratory on concrete beams post-tensioned with straight unbonded bars to determine their dynamic resistance, their natural period of vibration, and their rebound characteristics. Some of the major conclusions are listed here:

1. The beams exhibited a very high capacity for recovery, with 85–90% recoverability at incipient collapse. Thus, essentially all of the energy-absorbing ability can be utilized without incurring serious permanent damage.
2. Natural periods as given by free-vibration tests agree well with the calculated period using the sonic modulus of elasticity.
3. The maximum rebound (negative deflection) for any given load curve $P(t)$ may be obtained (for loads in the elastic range) by considering an elastic beam with viscous-type damping, provided its natural period and damping factor can be predicted. The shorter the load duration, the more serious the rebound problem and the more important damping becomes.

A procedure for dynamic design for blast loadings is recommended as a result of these tests. The method suggests a preliminary design section by any conventional method using a dynamic load factor of 2.0 (100% impact), then computation of the stiffness, the natural period, and the maximum deflection and rebound under the given loading. Then the amount and location of prestress required to keep the maximum fiber stresses within the allowable limits should be designed. If the section should be modified, the process can be repeated until a satisfactory design is obtained. This elastic approach is believed to be rather conservative, unless high-tensile and compressive stresses are allowed. Since the plastic energy absorption is so much higher than the elastic energy, advantage must be taken of this reserve capacity when designing for dynamic loads. Care must be taken so that rebound effect can be taken care of by adding nonprestressed reinforcement on the tension side during rebound. For beams under high shear, the addition of stirrups also should be considered.

The damping characteristics of prestressed concrete were investigated by Professor Penzien at the University of California. Twenty concrete beams of size $6 \times 6 \times 90$ in. were post-tensioned to various stress levels to result in various stress distributions. Three significant conclusions are pointed out in the report:

1. Under steady-state conditions internal damping in prestressed concrete members may be less than 1% of critical if the initial prestress is sufficient to prevent tension cracks from developing. If tension cracks are allowed to develop, but on a microscopic scale, damping can be expected of the order of 2% of critical. If larger (visible) cracks are permitted to develop, one should expect higher damping.
2. Under transient conditions, the amount of internal damping present in prestressed concrete members depends to a great extent on the past history of loading and on the amplitude of displacements produced. For those cases where members have been dynamically loaded only a few times to a given stress level that produces considerable cracking, damping can be expected anywhere in the range of 3–6% of critical.
3. Magnitude and type of prestress in concrete members have an indirect influence on internal damping only because these parameters control the amount of cracking that can take place.

It is therefore noted that critical damping for prestressed concrete is smaller than for reinforced concrete under the design loadings because of the absence of cracking but is probably comparable to reinforced concrete in the postcracking range.

19.9 EARTHQUAKE-RESISTING COMPONENTS

While the elements in prestressed concrete buildings are basically the same as those in reinforced concrete buildings, their layouts and arrangements for earthquake resistance can be radically different. Figure 19.22 shows diagrammatically the earthquake-resisting elements in a concrete building. The vertical elements are the shear walls, the elevator shafts, and the rigid frames formed by girders and columns. The horizontal diaphragm is supplied by the floor and roof slabs, together with their supporting beams and girders.

For a conventionally reinforced concrete building, walls are economically designed to resist earthquake forces. When walls are absent or insufficient, the elevator shafts are reinforced to carry the lateral loads. In either case, the walls or the shafts serve as so-called "shear walls," acting more in shear than in bending. In a prestressed building, these elements can be vertically prestressed to serve as vertical cantilever beams fixed at the foundation and designed to resist earthquake loads in flexure rather than in shear. An outstanding example was the prestressing of the pylon to carry the earthquake forces in a 10-story garage (Ellison and Lin, 1955). These shafts, when vertically prestressed, possess high rigidity within the elastic range and resist moderately heavy earthquakes with no cracks and small deflections and consequently little damage. In case of heavy earthquake, their rigidity decreases as cracks develop so that they are able to deflect considerably, thus supplying the necessary energy absorption.

When the shafts can be prestressed to carry the major part of the lateral loads, the walls and the frames are relieved of their share and sometimes may be entirely omitted in the building. For other buildings, where the walls or frames can be prestressed or reinforced to carry the lateral loads, the shafts may not have to be prestressed.

The high ductility of a prestressed shaft is indicated by a moment-curvature relationship diagram computed for a 20 ft × 12 ft shaft section (Fig. 19.23). For this particular section, the curvature at the start of cracking is 2.4×10^{-6} rad/in. of height, while the curvature at yielding of steel is 14×10^{-6}. The curvature at ultimate load is 100×10^{-6}, more than 40 times the curvature at cracking. Although these high curvatures are obtainable only when shear failure is prevented by sufficient stirrups in the webs, they do indicate that these shafts can be designed to withstand heavy earthquakes. When the aspect ratio (height-to-width ratio) of a shaft is very low, e.g., under 2, the shear deflection of the shaft predominates and the ductility can be low; but when the aspect ratio exceeds 4, moment deflection will be more than 5 times the shear deflection, and the shaft is no longer a shear wall but essentially a cantilever beam.

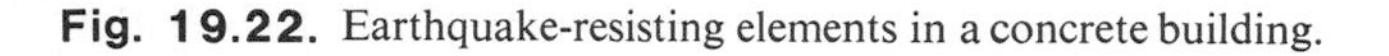

Fig. 19.22. Earthquake-resisting elements in a concrete building.

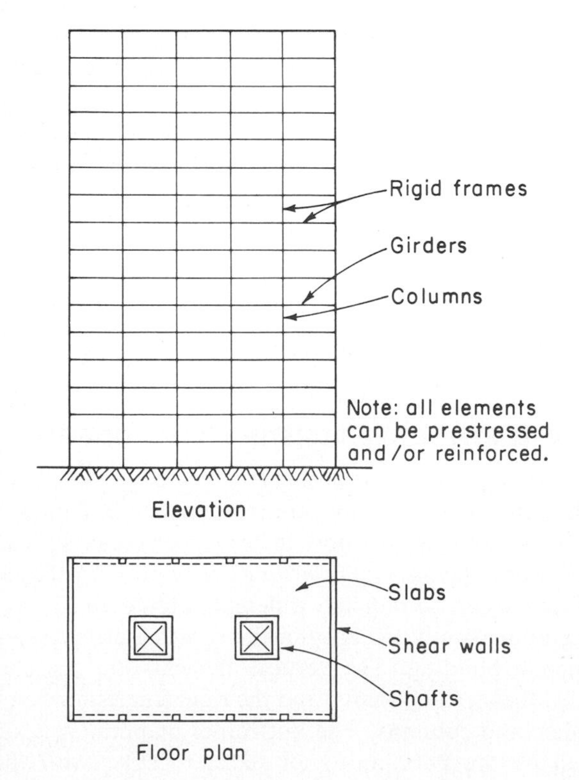

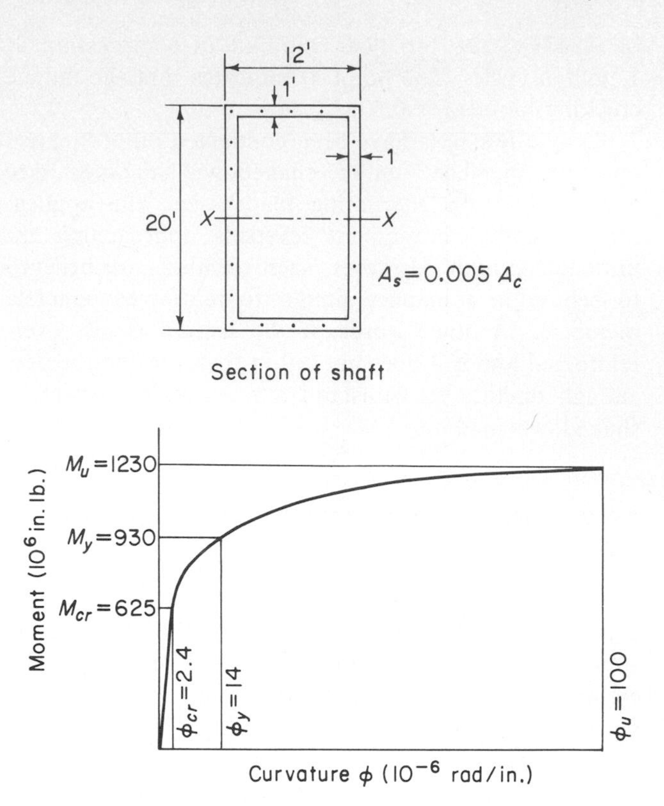

Fig. 19.23. M–ϕ relationship for a prestressed shaft section.

For a floor slab post-tensioned in place, an excellent horizontal diaphragm is provided that ties together the various vertical components of the building. If the slab is post-tensioned in two directions, the concrete, being under compression, will be able to carry high shear and moment in the horizontal plane and generally does so without cracking.

19.10 EARTHQUAKE-RESISTANT DESIGN OF MULTISTORY BUILDINGS

The seismic design of multistory buildings of prestressed concrete is essentially similar to those of conventional

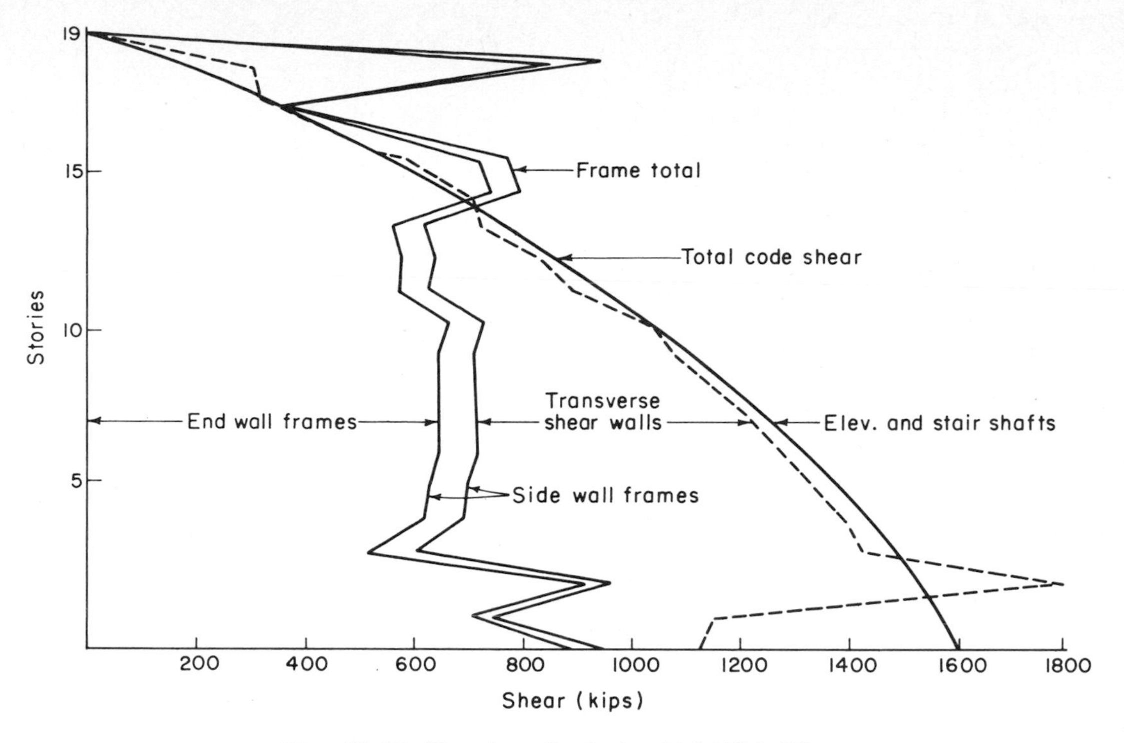

Fig. 19.24. Shear-force distribution, Nob Hill building.

reinforced concrete or steel. The main difference lies in the connections for precase members or in the post-tensioning of the vertical and the horizontal load carrying members. Furthermore, because of the frequent absence of frame action in prestressed concrete multistory buildings, their seismic resistance is more dependent on the full use of the elevator and stairway shafts. Because the nature of their resistance is sometimes different from a steel or a reinforced concrete building, careful analysis is desired to assure proper behavior and safety (Lin, 1965).

Two problems are paramount: the distribution of earthquake shear forces among the various vertical resisting elements of a building and the action of these elements when subjected to dynamic lateral forces. The shear force distribution is dependent on the elastic properties of the various resisting elements and is usually analyzed assuming rigid horizontal diaphragms.

As an example, the seismic design for the 19-story Nob Hill Apartments Building in San Francisco will be briefly described. This building is composed of 8-in. post-tensioned slabs spanning 30 ft between interior shafts and exterior walls. The earthquake force in the transverse direction is shared by the following elements (Fig. 19.24):

1. Portal action of frames, or
   a. the two end walls each acting as a rigid frame made up of wall columns and spandrel beams, and
   b. the two side walls forming a flexible frame with the post-tensioned roof and floor slabs.
2. Cantilever action of walls and shafts as vertical beams fixed at the foundation, or
   a. the transverse shear walls extending practically the full height of the building, and
   b. the elevator and stair shafts.

The load distribution was performed by computer analysis. It is noted from Fig. 19.24 that at the upper stories the cantilever deflection of the walls and shafts is so much greater than the portal deflection of the frames that the frames actually support the walls and carry much greater shear than exerted by the earthquake forces. This brings out the fact that the walls and shafts are ductile cantilevers that deflect even more than the so-called frames.

This Nob Hill building was first designed using the earthquake forces specified by the 1961 Uniform Building Code. Then the dynamic response of the structure to the north–south component of the 1940 El Centro Earthquake was computed by a digiter computer. The comparison of these static and dynamic values is shown in Fig. 19.25. Although the El Centro Earthquake produced forces and displacements of approximately five times the Code values, it was estimated that plastic deformations about twice as great as the elastic limit deflections would be produced and no irreparable damage would result from an El Centro Earthquake; this is considered to be the probable maximum that might occur at Nob Hill.

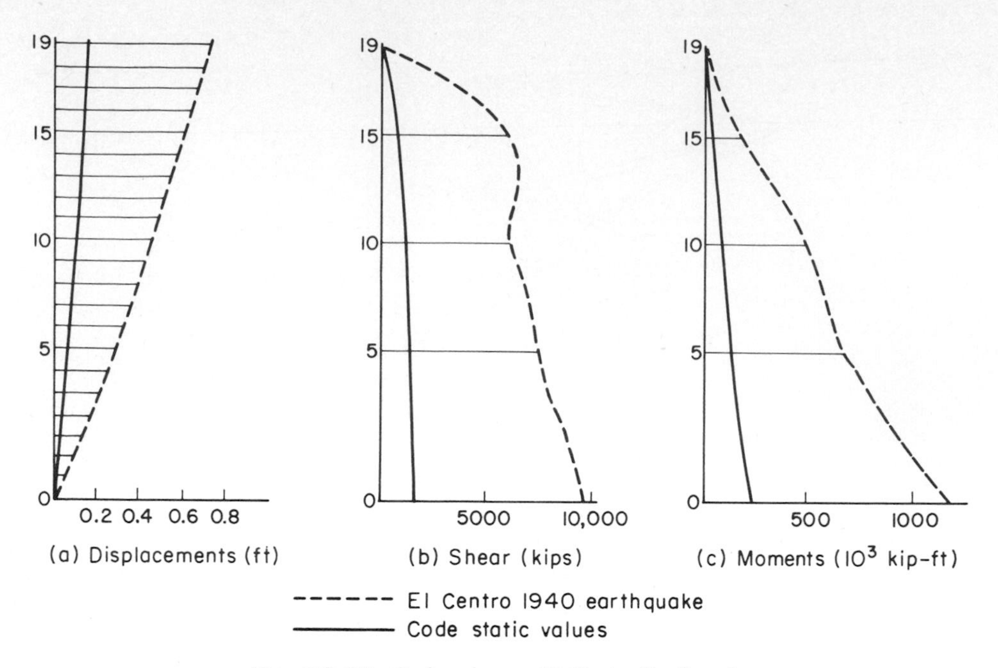

Fig. 19.25. Code values vs El Centro Earthquake.

A second example is afforded by the 15-story Apartment Building at 600 Ocean Boulevard, Long Beach, California. A plan of this building is shown in Fig. 19.26 with prestressed $7\frac{1}{2}$-in. floor slabs spanning between the columns and with the central shafts carrying the major part of earthquake resistance (Fig. 19.27). Because the slabs are rigidly connected to the columns, together they form a rigid frame that also contributes significantly to the lateral strength and rigidity of the building.

This structure is first analyzed for the forces and deflections produced by the static application of the earthquake loads specified by the 1961 Uniform Building Code and then is checked for dynamic behavior under the 1940 El Centro Earthquake (north–south component). The results are plotted in Figs. 19.28 and 19.29, comparing the code values with one-third of the El Centro values. The following are noted:

1. The lateral displacements obtained statically and dynamically are quite similar.

Fig. 19.26. Fifteen-story building, Long Beach, California.

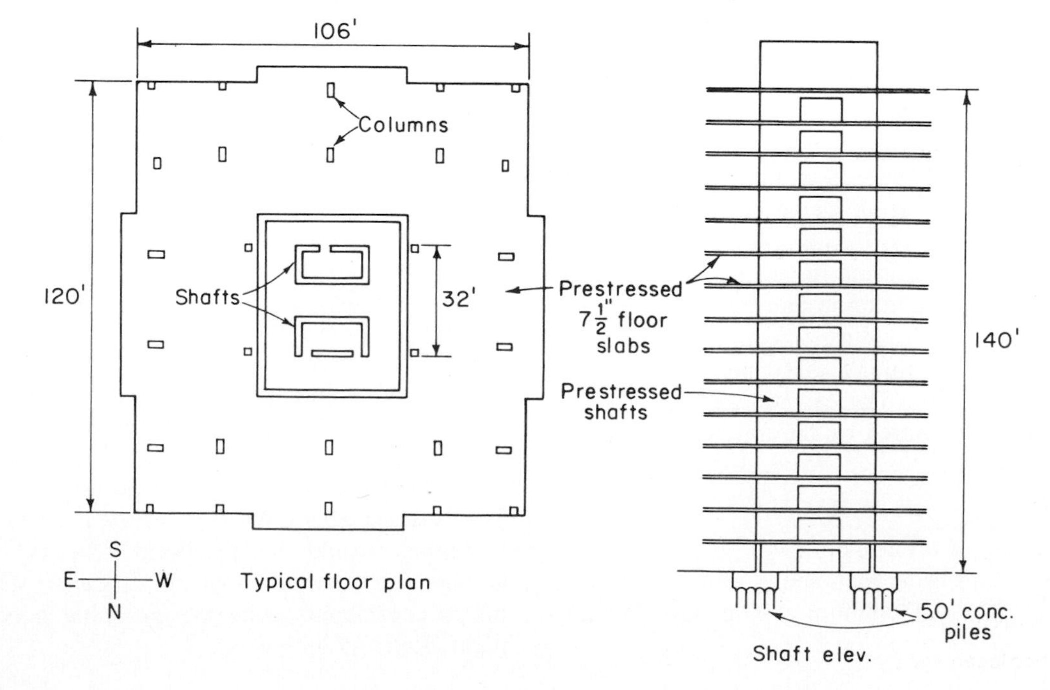

Fig. 19.27. Cable layout for 15-story building.

2. In the E–W direction (Fig. 19.28), the dynamic moments in the shafts are much higher for the upper stories.

3. In the N–S direction (Fig. 19.29), as a result of the mechanical room connecting the two shafts at the top, the dynamic moment envelope appears quite different from the static moment. Actually, when the reversal of the static moment is considered, the dynamic moments are significantly different only between the fifth and the ninth stories.

4. The shear carried by the column-slab frame is almost identical statically and dynamically, but the shear carried by the shafts is significantly higher by the dynamic analysis, especially for the upper stories and somewhat for the lower stories.

As a result of the dynamic analysis, the structure was redesigned to carry one-third El Centro Earthquake, and parts of the shafts were additionally reinforced and prestressed to satisfy these requirements.

Present knowledge of prestressed concrete enables us to design both single-story and multistory buildings for earthquake resistance in a safe and economical

Fig. 19.28. Seismic design for 15-story building (E–W direction).

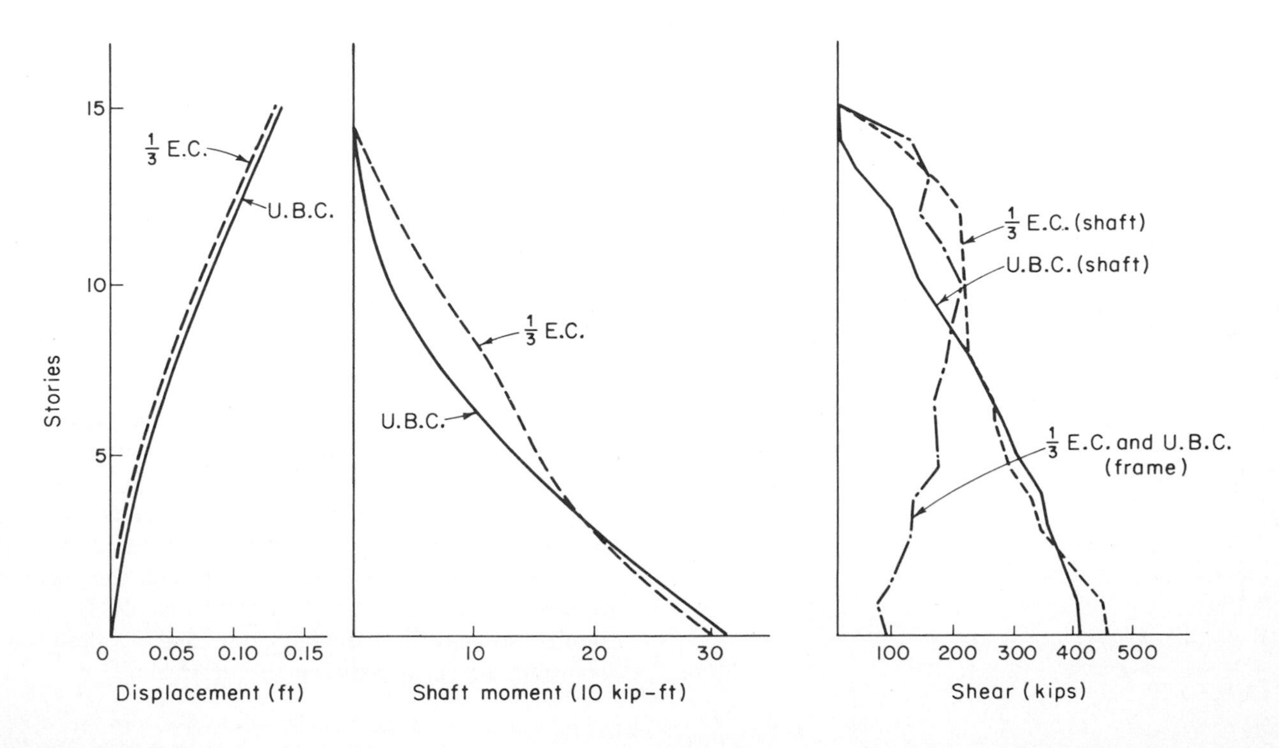

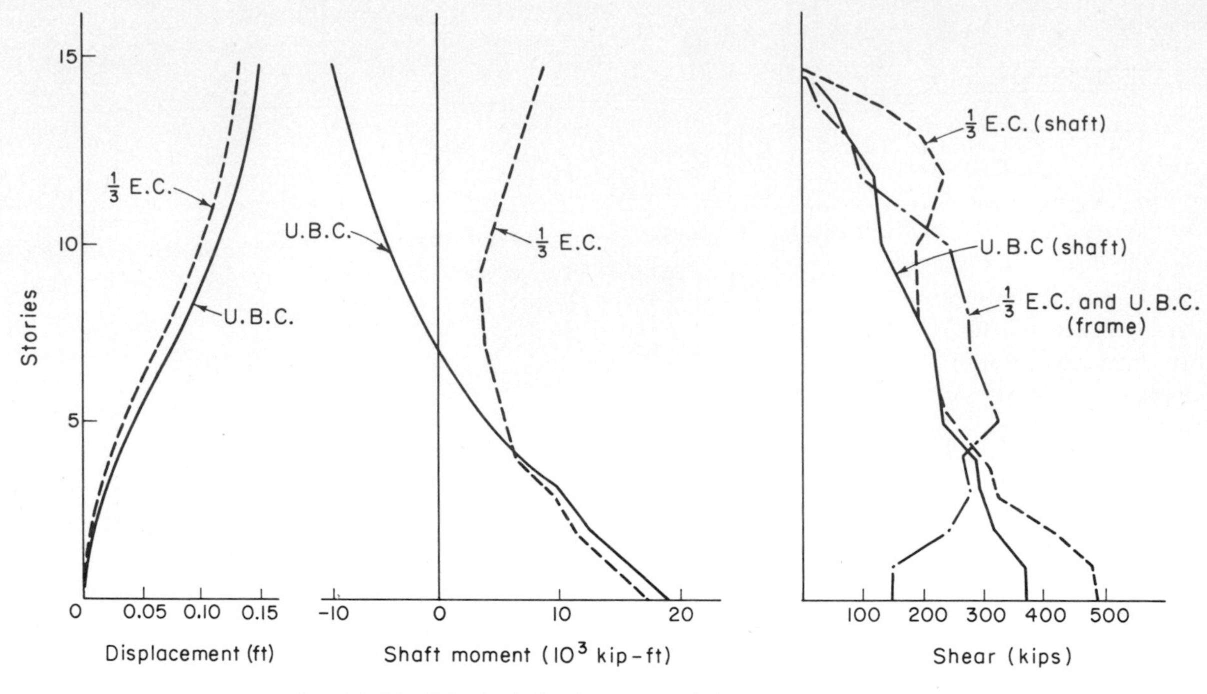

Fig. 19.29. Seismic design for 15-story building (N–S direction).

manner. Such designs can be made to satisfy both the requirements of rigidity under moderate earthquakes and the requirements of energy absorption for the resistance to extra heavy earthquakes.

19.11 PRECAST CONCRETE

The resistance of precast concrete elements against earthquakes is usually controlled by their joinery. Such joinery must supply sufficient strength to resist the maximum force developed during the earthquake and sustain the elastic or plastic deformation accompanying the movement.

Precast segments post-tensioned together to make a girder or a column (Fig. 19.30) generally form a reliable unit for earthquake resistance and can be designed similar to prestressed members under flexure

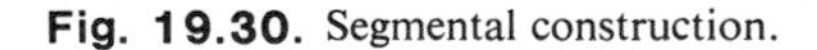

Fig. 19.30. Segmental construction.

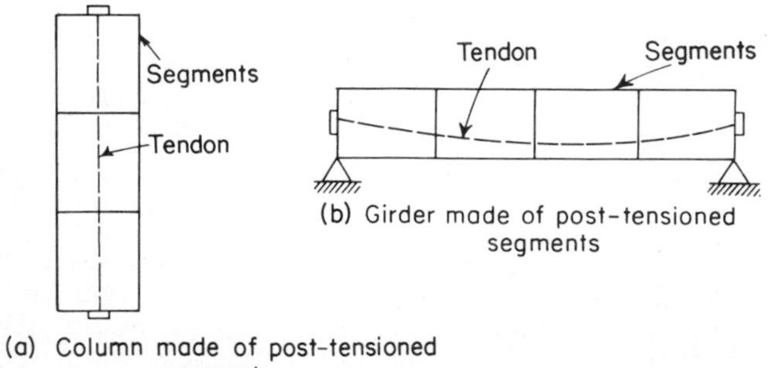

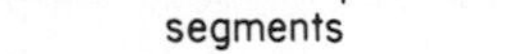

(a) Column made of post-tensioned segments

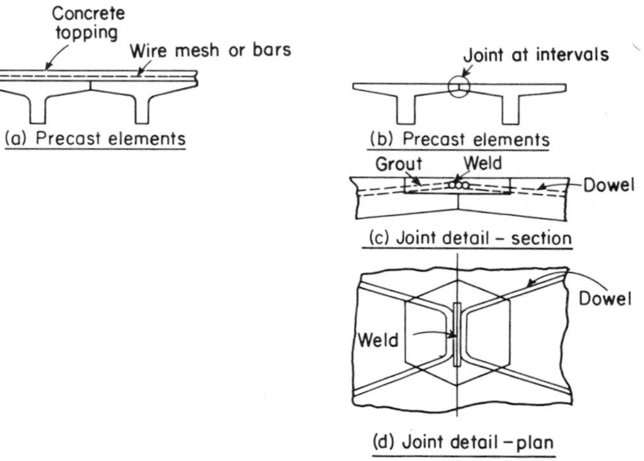

Fig. 19.31. Precast elements joined side by side.

or under combined axial load and flexure. Since the flexural ductility of the member is usually high, temporary opening up of the joints under a strong earthquake can be restored by the effect of the prestress if the tendons are not stressed beyond their yield point.

Precast floor or roof members placed side by side are connected together either by a continuous topping with reinforcement or by joints located at intervals (Fig. 19.31). Topping without reinforcement has very little ductility and may easily crack under an earthquake, thus destroying the connection between the members. Spaced joints, such as shown in Fig. 19.31, possess limited strength and ductility, but may be designed to transmit the shear between the members.

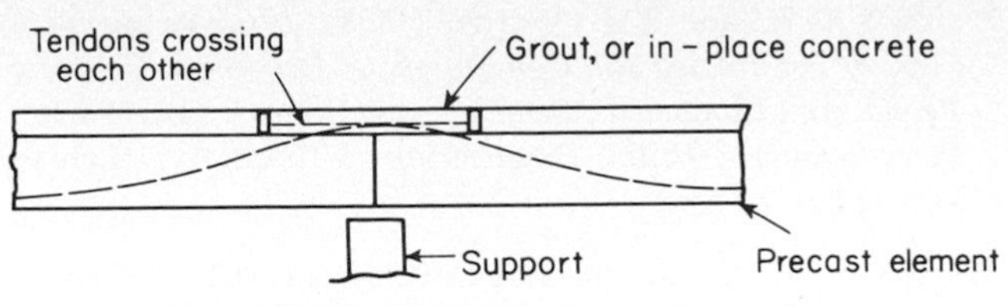

Fig. 19.32. Continuity by crossing tendons.

Often critical is the support furnished to the precast members. Failure of the supporting members or the supports themselves during an earthquake could result in dropping the precast elements. Thus, it is important to provide strength and ductility for the supporting walls or columns and for the supports thereon. Sliding supports afford no resistance against earthquakes.

Joinery for precast elements can be classified into four categories: (1) grout joints with no reinforcement across them, (2) grout joints with proper dowels or reinforcement across them, (3) plates welded to dowels or simply dowels held together by field welds, and (4) tendons crossing each other.

Grout joints with no reinforcement are not considered sufficient for earthquake resistance, except where very little force is transmitted and separating cracks are avoided. For example, plain concrete topping on precast elements can sustain a certain amount of horizontal shear (40 psi) as per the ACI Code. In almost all other cases, proper dowels must be provided. Welded joints are suitable for earthquake resistance, provided the parts to be welded together are suitably doweled into the concrete elements to develop the necessary bond.

The best way to achieve a continuous joint for earthquake resistance is by means of tendons crossing each other, either over a support (Fig. 19.32) or around the corner of the frame (Fig. 19.33). The additional provision of nonprestressed reinforcement along either side of the joint will strengthen its resistance and provide greater ductility (Fig. 19.34). Stirrups for confining the concrete at the joint will further increase its strength and ductility.

Where a floor is used to serve as a horizontal diaphragm for the transmission of earthquake forces, the precast elements must be tied together either by topping with reinforcement or by special joints at intervals. If these floor elements can be post-tensioned together, they will serve as an excellent diaphragm. Flanges of a horizontal diaphragm must be reinforced or prestressed to provide the tensile resistance under horizontal bending. The entire diaphragm must be doweled or otherwise connected to the vertical earthquake-resisting elements, which may be the stair shafts, the walls, or rigid frames.

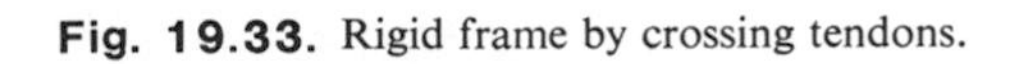
Fig. 19.33. Rigid frame by crossing tendons.

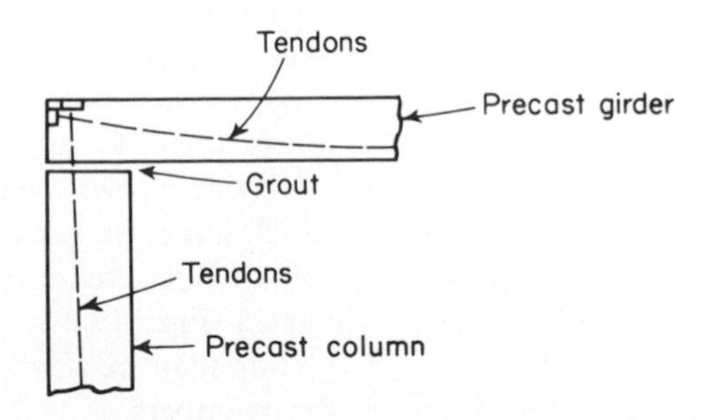

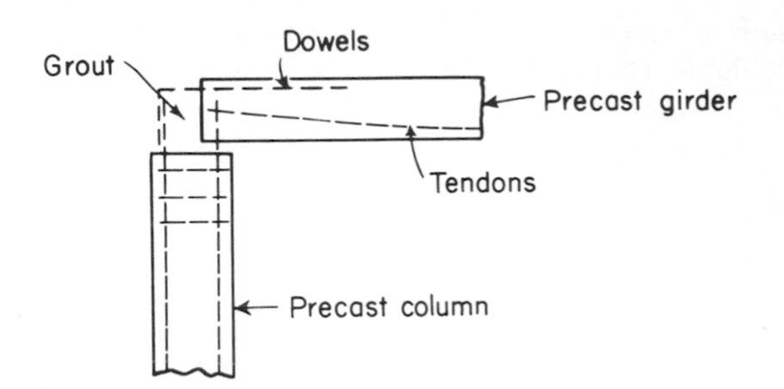

Fig. 19.34. Nonprestressed dowels for rigid frame.

The best way to form precast elements into earthquake-resisting frames or shear walls is to post-tension them together. Where this cannot be achieved, nonprestressed reinforcement across the joints can be grouted to transfer the stresses, as in conventional in-place concrete construction.

REFERENCES

Aroni, S. (May 1967). "Slender Prestressed Concrete Columns," *Preprint 467*, ASCE Structural Engineering Conference, Seattle, Washington.

"Blast Load Tests on Post-tensioned Concrete Beams," *Tech. Report 116* (May 1961). U.S. Naval Civil Engineering Laboratory, Port Hueneme, California.

Building Code Requirements for Reinforced Concrete (1963). American Concrete Institute, Detroit, Michigan.

Burns, N. H., and D. M. Pierce (October 1967). "Strength and Behavior of Prestressed Concrete Members with Unbonded Tendons," *J. Prestressed Concrete Inst.*

Caulfield, J., and W. H. Patton (1963). "Moment–Curvature Relationship of Prestressed Concrete Beams," Graduate Study Report to Division of Structural Engineering and Structural Mechanics, Berkeley: University of California.

Ellison, W. H., and T. Y. Lin (June 1955). "Parking Garage Built for $5.28 per sq ft," *Civil Engr.*, pp. 37–40.

Lin, T. Y. (1963). *Design of Prestressed Concrete Structures*, 2nd ed., New York: John Wiley.

Lin, T. Y. (October 1965). "Earthquake Resistance of Prestressed Concrete Buildings," *J. Structural Div., ASCE.*

MacGregor, J. G., M. A. Sozen, and C. P. Siess (1960). "Strength and Behavior of Prestressed Concrete Beams with Web Reinforcement," *Structural Research Series No. 201*, Urbana: University of Illinois.

Penzien, J. (January 1962). *Damping Characteristics of Prestressed Concrete*, Berkeley: Institute of Engineering Research, University of California.

Scordelis, A. C., T. Y. Lin, and H. R. May (1957). "Strength of Prestressed Concrete Beams at Transfer," *Proceedings of the World Conference on Prestressed Concrete*, San Francisco.

Uniform Building Code (1967). International Conference of Building Officials, Pasadena, California.

Walther, R. E., and R. F. Warner (1958). *Ultimate Strength Tests of Prestressed and Conventionally Reinforced Concrete Beams in Combined Bending and Shear*, Bethlehem, Pennsylvania: Fritz Engineering Laboratory, Lehigh University.

Warwaruk, J., M. Sozen, and C. P. Siess (1962). "Strength and Behavior in Flexure of Prestressed Concrete Beams," *Bulletin No. 464*, Urbana: Engineering Experiment Station, University of Illinois.

Zia, P., and F. L. Moreadith (July 1966). "Ultimate Load Capacity of Prestressed Concrete Columns," *ACI J.*, p. 767.

Chapter 20

Design of Earthquake-Resistant Structures: Towers and Chimneys

JOHN E. RINNE

Civil and Structural Engineer
Standard Oil Company of California
San Francisco, California

20.1 INTRODUCTION

Towers, such as process towers in oil refineries, and chimneys are among the simpler structures built by man and subject to earthquake ground motion. This is not to say that the response of such structures is simple. The response of even the simplest of structures to ground motion is a complex, dynamic phenomenon. There are too many unknowns to be able to predict with any degree of certainty the precise response of a particular structure to some unknown future earthquake. We have to rely upon the qualitative and, with a good seasoning of judgment, the quantitative analyses of response to assumed or recorded past earthquake ground motions. The judgment factor is largely influenced by the observation of performance of simple and complex structures in past earthquakes. The net result is that while the more or less "rigorous" analyses are very helpful in guiding the establishment of design criteria, practical design criteria at this time are simplifications of the complex dynamic phenomenon into "equivalent" static criteria applied with elastic stress limits. (The term "equivalent" is used advisedly.)

While the response of these simple structures is not simple, the structures are. They are clearly defined as cylindrical shells, sometimes tapered or step-tapered. They are unencumbered by "nonstructural" elements that might affect their response. Unlike most buildings, these structures have a single structural system upon which the safety of the structure for both normal loads and abnormal loads must rely. The dynamic characteristics of such structures can be pre-established with confidence and with such precision as is warranted. The elastic properties of the materials and the configurations are well known, as is the mass distribution. These are fundamental elements that go to define the natural period of the structure, or periods. Periods influence response to earthquake ground motion. Natural damping inherent in such structures is minimal: less than 2% of critical damping for steel structures and about 5% for reinforced concrete structures. The coupling of the structure with the supporting ground—also referred to as the soil–structure interaction—is a variable more difficult to evaluate, although much work is currently being done in this field. If a rigorous analysis is undertaken, it is important to introduce the effect of coupling. For design purposes, at the current state of the art it is sufficient and relatively conservative to assume that the structures are fixed at the base—at the top of the foundation. Then, with the fairly accurate definition of other factors influencing the dynamic characteristics, the natural periods in the fundamental and higher modes (if needed) can be calculated, as well as the mode shapes—i.e., the shape of the deflection curve in the fundamental mode, the second mode, etc. In the more rigorous evaluation of these dynamic characteristics the classical analysis involves the determination of "eigenvalues" related to the natural periods and of "eigenfunctions" related to the mode shapes. Where the manual solution of the equations that are involved would be a time-consuming and exacting task, computer programs provide these solutions in short order. Details for the rigorous computation of dynamic characteristics of specific types of structures, if not structures in general, are given in other chapters of this text.

To continue down the path of the more rigorous analyses for the moment, the response of a particualr structure to some assumed or recorded ground motion can be determined. That is, it can be determined theoretically, consistent with the assumptions and their accuracy. Of the recorded ground motions, the N–S component of the 1940 El Centro, California, Earthquake has been used so universally that it has become something of a "standard earthquake"—a reputation it does not deserve. However, it may have more general application than just at El Centro—under conditions that might be reasonably similar to those at El Centro. This record might be typical of a zone fairly close to the earthquake epicenter (30 mi at El Centro) and for a soil condition corresponding to a deep alluvium. If a rigorous analysis is made, care must be taken in the selection of an appropriate ground motion. In the case of tall, long-period towers or chimneys, it is somewhat questionable whether the El Centro record is most appropriate, because long-period structures are likely to be more responsive to more distant earthquakes. This seemed to be indicated in the 1964 Alaskan Earthquake. In any event, having selected a proper ground motion, many analyses of response of structures have been made. They have included both elastic response and elasto-plastic response. They are well documented in the technical press, and they are well appreciated for the tremendous contribution they have made to our understanding of the earthquake phenomenon.

This kind of analysis, however, is generally beyond the kind of effort that can be afforded in the design of almost all but the most critical structures. The earthquake engineer, in current practice, must resort to simplifications in design criteria and, with an understanding of his structure, apply these criteria consistently and with proper attention to details. These details cannot be stipulated in the design criteria; they must be developed as part of the application of the "art of earthquake engineering." From a practical standpoint, earthquake resistance developed consistently under simplified elastic criteria, using static "equivalent" forces to provide a simulated envelope of the dynamic forces, shears, and moments, is probably of greater importance than highly analytical solutions. This is why observed performance of structures in major earthquakes is extremely important, but to date there have been few opportunities to observe the performance of structures designed and constructed to present-day criteria. When we do have such observations, we will have a sounder basis to modify these criteria, if modification appears to be needed. Ayoji Suyehiro, the famed Japanese earthquake engineer, speaking of the Tokyo Earthquake of 1923, said: "As a practical problem, the actual fact that buildings designed under the 0.1 *g* basis resisted this earthquake fairly well is a datum more valuable than any other arguments." Despite the much more sophisticated knowledge and tools we now have at our disposal, they do little to challenge the wisdom of that early philosophy.

The most thoroughly studied and generally applied earthquake design criteria in the United States are those developed by the Structural Engineers Association of California (1967). While the primary interest of such design criteria, or codes, is the design and construction of buildings, the SEAOC code is of broader scope and has included criteria for "structures other than buildings." Towers and chimneys fall in this category.

20.2 SEAOC CODE AS APPLIED TO TOWERS AND CHIMNEYS

The SEAOC code has been written to serve as a model provision of building codes, specifically for California seismicity. As such, it specifies minimum criteria needed to assure reasonable safety and protection of life and property. If a risk analysis for some particular structure indicates that more stringent design criteria should be employed, this is outside the province of the building code. The experience and judgment of the structural engineer should influence the selection of such more stringent criteria as are believed to be justified.

Concentrating attention on the SEAOC code as applied to towers and chimneys, Figs. 20.1, 20.2, and 20.3 provide the basic criteria that define forces, shears, and moments. Figure 20.1 defines the base shear, or the total design lateral force, transferred from the supporting ground into a generalized structure. The SEAOC code formula does not include the seismicity factor Z, since for California seismicity it is unity. The International Conference of Building Officials, in adapting the SEAOC code for different zones of seismicity, introduced the Z factor in the Uniform Building Code. Z is 1.0, 0.5, 0.25, 0 in UBC zones 3, 2, 1, and 0, respectively. SEAOC makes no zonal distinctions in California. UBC recognizes areas of its zones 3 and 2 categories. Not all are in accord with the drastic variations in design criteria represented by the Z factors, particularly in view of the very short historical record of earthquakes. Much is being done, and much more needs to be done, to define the nature, the frequency, and the intensityof ground motions to be expected in the United States, as well as in the world in general. All of this comes under the heading of "seismicity."

Returning to Fig. 20.1, note that the "earthquake coefficient" is defined as the multiple of three parameters: Z (seismicity) $\times$ K (structure coefficient) $\times$ C (flexibility coefficient). Other parameters also could have been added. For example, a risk or occupancy coefficient R, greater than or equal to unity, could reflect the results of a risk analysis for a particular structure. There also has been much thought given to the influence of the supporting soil on both the ground motion and the response of the structure, including the soil–structure interaction. This has been recognized in past earthquake codes by increasing the earthquake coefficient on weaker soils. At present it is felt that the problem is not that simple. Some kinds of structures actually may be benefited by a softer, cushioning soil. Until the total effect of the soils can be adequately and consistently defined, what might be called the S factor (for soils) is considered to be unity

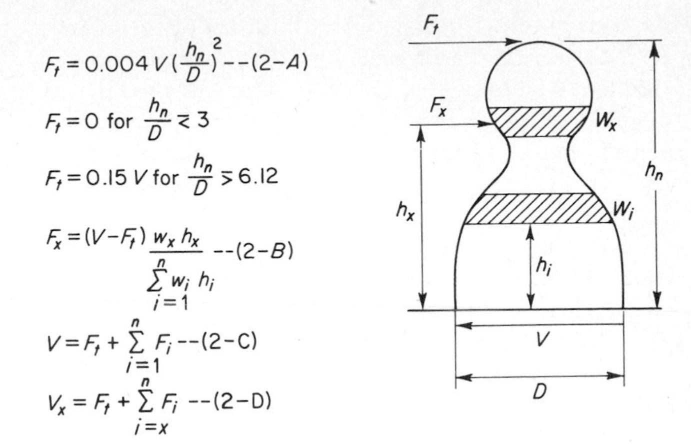

Fig. 20.2. Distribution of base shear.

Fig. 20.3. Moments.

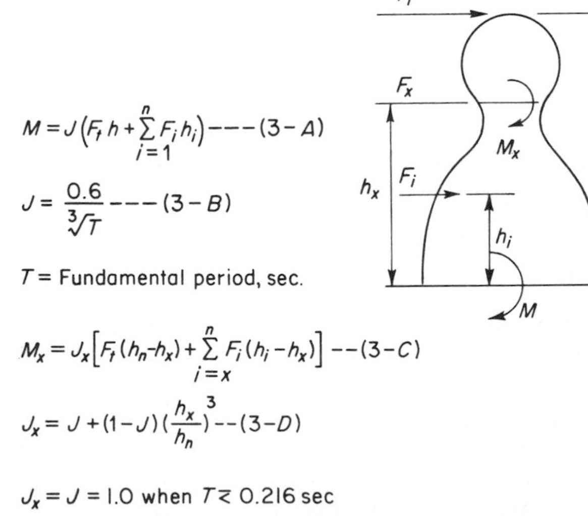

Fig. 20.1. Base shear.

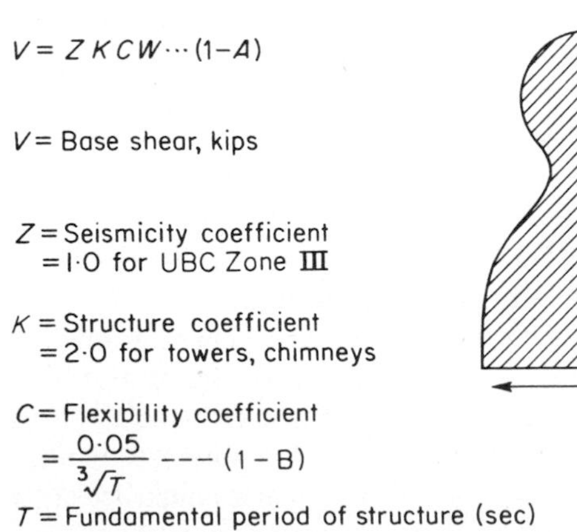

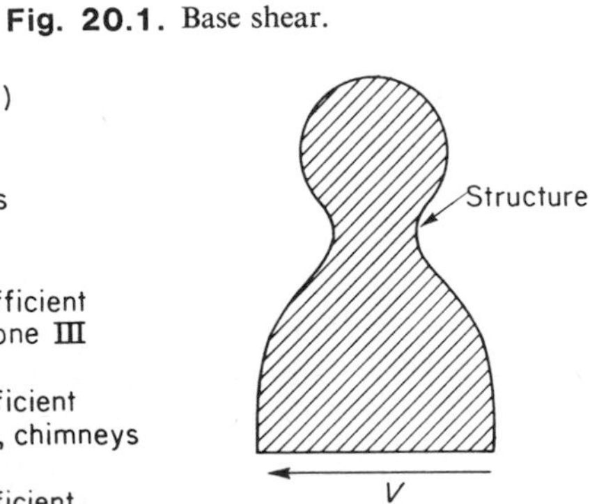

for code purposes. Isolated from the other factors that go to make up the earthquake coefficient, it is anticipated that the S factor for towers and chimneys should vary from 1.0 (rock or firm soils) upward (for progressively softer soils). Until criteria are established for giving direction to this S factor, it must be left to the analysis of the particular structure.

Note that the coefficient K is 2.0 for the type of structure considered here. This compares with corresponding coefficients for buildings that range from $K = 0.67$ to $K = 1.33$. The reasons given for the higher coefficient for structures other than buildings are essentially the differences between buildings and these other structures that were discussed earlier, explaining why towers and chimneys, specifically, are considered to be "simple structures." Briefly recapping, these are lack of nonstructural or noncomputed lateral resisting elements; a single structural system for both vertical and lateral loads; low inherent damping; and well-defined dynamic characteristics. Essentially, it is a judgment factor (of the SEAOC Seismology Committee responsible for the SEAOC code) that has set the K-values. Relatively, it would seem that the coefficients for these other structures are sufficiently high, compared to the building coefficients. In fact, a fair case could be built up to question the logic of the disparity of 1.5–3.0 times the building coefficients for these other structures. For elevated, cross-braced water towers, $K = 3.0$, with the maximum value of $KC = 0.25$ and the minimum value of $KC = 0.12$. These higher values are justified on the basis of the undesirable structural characteristics of this form of structure, especially under dynamic loading, and the higher risk factor needed to assure continuity of service of these essential components of fire-fighting systems. This is the only instance where the specified K-value is recognized to incorporate some risk factor.

The flexibility coefficient C is empirically related to the fundamental period of natural vibration of the structure by formula 1-B of Fig. 20.1. This brings up the need to have methods by which this fundamental period T can be predetermined. For buildings, some empirical formulas for T are given in the SEAOC code. For other structures, it is required that T be calculated by recognized methods. The rigorous analytical method discussed earlier certainly would yield the fundamental period with considerable accuracy. It should be pointed out, however, that if the purpose of calculating T is only for the determination of the coefficient C, then great accuracy in T is not needed. C, being related to the cube root of T, is not very sensitive to considerable error in T. An inaccuracy in T of 30% only affects C less than 10%. The value of T can be determined with better accuracy than 30% without great difficulty. Where conservative assumptions are made, such as assuming fixity at the base, this will tend to make the calculated period less than the true period. This will only increase C slightly, hence the design base shear V. Since the calculation of the fundamental period is basic to the application of the SEAOC code approach to earthquake-resistant design, a significant part of this chapter will be devoted to summarizing tools available by which this can be done—with specific reference to towers and chimneys.

In formula 1-A of Fig. 20.1, the weight W to which the combined dimensionless coefficients ZKC apply is defined as the combined dead and live load, or normal operating load, that is most likely to exist at the time of an earthquake. In the case of elevated water tanks, this is considered to be a full tank. In the case of chimneys, there is little likelihood that there would be anything more than dead load involved. In the case of process towers, there frequently will be some liquid live load as well as solids live load, plus the dead load of the equipment and supported appurtenances. The loading condition should be analyzed for each particular case. By way of comparison, the weight W for buildings is defined as dead weight including fixed equipment, except that 25% of storage area live load is to be added.

To illustrate the application of formula 1-A, if $Z = 1.0$, $T = 1.0$, and $K = 2.0$,

$$V = ZKCW = 1.0 \times 2.0 \times 0.05 \times W = 0.10W$$

Formula 1-A of Fig. 20.1 defines the design base shear V. Figure 20.2 indicates how this shear is distributed up through the structure as equivalent lateral static forces. First, if the height-to-width ratio of the structure is greater than 3, then part of the base shear, up to $0.15V$ maximum, is arbitrarily assigned to the top of the structure. The purpose of this is to increase the shear in the upper parts of the structure to account for the increasing participation of the higher modes of vibration in the response of taller, more flexible structures. It also recognizes that the fundamental mode deflection curve departs from the assumed straight line as the structure deflection is due more to bending action rather than shear deflection. This tends to cause greater shears in the top of the structure than would result from the so-called "triangular distribution" resulting from the application of formula 2-B of Fig. 20.2 with $F_t = 0$.

The formula 2-A of Fig. 20.2, for F_t, is such that for height-to-width ratio of 6.12 or greater, $F_t = 0.15V$. When the ratio is 3, $F_t = 0.036V$, at which point, and for lower ratios of height to width, the assignment of some part of the base shear to the top has little significance and F_t can be assumed to be zero. Most tower and chimney structures have height-to-width ratios greater than 6.12; hence, generally, the distribution of the base shear will be $0.15V$ as F_t at the top, with the remaining $0.85V$ distributed in accordance with the triangular distribution formula 2-B of Fig. 20.2 In applying this formula, the structure arbitrarily can be divided into convenient increments, depending somewhat upon the weight distribution but usually not more than 10 sections.

The transverse design shear at any horizontal plane in the structure is the cumulative sum of the lateral forces above that plane, resulting from the distribution of the base shear by the criteria given by formulas 2-A and 2-B of Fig. 20.2 Formula 2-C states this for any plane x above the base, giving V_x. Formula 2-D provides the corresponding value at the base V, which of course should check the base shear that previously has been distributed and now is reaccumulated.

Having determined the base shear (Fig. 20.1) and the distribution of that base shear (Fig. 20.2), the remaining basic design criteria relate to the overturning moment on the structure as a whole and at any horizontal plane x. If all of the lateral forces acted simultaneously in one direction, which is equivalent to saying that the action or response is entirely in the fundamental mode, then the overturning moment at any plane would be the moment of the lateral forces, in absolute value, above and about that plane. The formulas for J and J_x, 3-B and 3-D of Fig. 20.3, then should be unity. Actually, the design base shear represents the sum of the modal shears significant in the response of the strucure. Similarly, the base overturning moment should be the sum of the modal base moments. The base overturning moments in the different modes for a uniform towerlike structure are related to the modal shears and the structure height h_n as follows:

$$M_1 = 0.729 V_1 h_n$$
$$M_2 = 0.209 V_2 h_n$$
$$M_3 = 0.127 V_3 h_n$$

These are for structures deflecting in bending. The constants for structures deflecting in shear are substantially the same for the second and higher modes, being 0.212 and 0.127 for the second and third modes. In the first mode it is 0.636. Modes higher than the third contribute little toward base overturning. The base overturning moment corresponding to the statically distributed forces per the criteria of Fig. 20.2 would be:

$$M = 0.72 V h_n$$

So, if the design base moment were to be considered the moment of the static lateral forces, this is tantamount to saying that the base moment should correspond to the total design lateral force, or base shear, applied to the structure in the first or fundamental mode. This would be very conservative. It is clear that if any significant part of the design base shear is assignable to higher modes, the base moment will be reduced.

The commentary supporting the SEAOC code (1967) outlines the reasoning that results in a reduced base moment by a factor J as given by formula 3-A in Fig. 20.3. Formula 3-B relates J to the fundamental period, and a minimum value of J for structures other than buildings is 0.45. At any other horizontal plane x, the overturning moment is given by formula 3-C of Fig. 20.3, with J_x defined by formula 3-D. $J = 1.0$ for $T = 0.216$. For this T, and lesser values of T, $J = J_x = 1.0$.

20.3 FUNDAMENTAL PERIOD *T* DETERMINATIONS

The primary need for determining the fundamental period of vibration of a structure under design is its use in formula 1-B in determining the flexibility coefficient C. T also enters into the determination of J by formula 3-B. In both cases it is the cube root of T that is required. As mentioned earlier, this permits considerable inaccuracy in T without significantly affecting the values of either C or J.

While some approximate empirical formulas are given in the SEAOC code for pre-establishing the fundamental period T for buildings, for structures other than buildings, this period must be properly substantiated by technical data. This option is available to the earthquake engineer designing a building—in lieu of using the empirical formulas given in the code. It is rather the exception than the rule that this option is exercised for building design.

It is the purpose here to list and compare some of the methods and formulas available to compute the fundamental periods of towers and chimneys.

20.3.1 Fundamental Periods of Reinforced Concrete Chimneys

The American Concrete Institute's Committee 505 (1968), after studying the periods of a number of reinforced concrete chimneys, evolved the following empirical formula. While dimensionally inconsistent, as is the SEAOC formula for building periods, it nevertheless gives results very consistent with other methods, as shown in Table 20.2.

$$T = \frac{1.8H^2}{(3D_o - D_H)E^{1/2}} \tag{20.1}$$

where T = fundamental period in seconds
H = height in feet
D_o = outside diameter at the base in feet
D_H = outside diameter at the top in feet
E = modulus of elasticity of concrete in pounds per square inch.
$E = 10 \times$ 28-day compressive strength of the concrete

G. W. Housner and W. O. Keightley (1963) have developed the following formula for the period of a tapered cantilever. It is not only dimensionally consistent, but it has been extended in the referenced work to provide the second and third mode periods as well as the fundamental. These are important where dynamic analyses are to be made, involving the modal response to

Table 20.1. COEFFICIENTS FOR DETERMINING PERIOD OF VIBRATION OF FREE-STANDING CYLINDRICAL SHELLS HAVING VARYING CROSS SECTIONS AND MASS DISTRIBUTION*

$\frac{h_x}{H}$	α	β	γ	$\frac{h_x}{H}$	α	β	γ
1.00	2.103	8.347	1.000000	0.50	0.1094	0.9863	0.95573
0.99	2.021	8.121	1.000000	0.49	0.0998	0.9210	0.95143
0.98	1.941	7.898	1.000000	0.48	0.0909	0.8584	0.94683
0.97	1.863	7.678	1.000000	0.47	0.0826	0.7987	0.94189
0.96	1.787	7.461	1.000000	0.46	0.0749	0.7418	0.93661
0.95	1.714	7.248	0.999999	0.45	0.0678	0.6876	0.93097
0.94	1.642	7.037	0.999998	0.44	0.0612	0.6361	0.92495
0.93	1.573	6.830	0.999997	0.43	0.0551	0.5872	0.91854
0.92	1.506	6.626	0.999994	0.42	0.0494	0.5409	0.91173
0.91	1.440	6.425	0.999989	0.41	0.0442	0.4971	0.90448
0.90	1.377	6.227	0.999982	0.40	0.0395	0.4557	0.89679
0.89	1.316	6.032	0.999971	0.39	0.0351	0.4167	0.88864
0.88	1.256	5.840	0.999956	0.38	0.0311	0.3801	0.88001
0.87	1.199	5.652	0.999934	0.37	0.0275	0.3456	0.87088
0.86	1.143	5.467	0.999905	0.36	0.0242	0.3134	0.86123
0.85	1.090	5.285	0.999867	0.35	0.0212	0.2833	0.85105
0.84	1.038	5.106	0.999817	0.34	0.0185	0.2552	0.84032
0.83	0.988	4.930	0.999754	0.33	0.0161	0.2291	0.82901
0.82	0.939	4.758	0.999674	0.32	0.0140	0.2050	0.81710
0.81	0.892	4.589	0.999576	0.31	0.0120	0.1826	0.80459
0.80	0.847	4.424	0.999455	0.30	0.010293	0.16200	0.7914
0.79	0.804	4.261	0.999309	0.29	0.008769	0.14308	0.7776
0.78	0.762	4.102	0.999133	0.28	0.007426	0.12576	0.7632
0.77	0.722	3.946	0.998923	0.27	0.006249	0.10997	0.7480
0.76	0.683	3.794	0.998676	0.26	0.005222	0.09564	0.7321
0.75	0.646	3.645	0.998385	0.25	0.004332	0.08267	0.7155
0.74	0.610	3.499	0.998047	0.24	0.003564	0.07101	0.6981
0.73	0.576	3.356	0.997656	0.23	0.002907	0.06056	0.6800
0 72	0.543	3.217	0.997205	0.22	0.002349	0.05126	0.6610
0.71	0.512	3.081	0.996689	0.21	0.001878	0.04303	0.6413
0.70	0.481	2.949	0.996101	0.20	0.001485	0.03579	0.6207
0.69	0.453	2.820	0.995434	0.19	0.001159	0.02948	0.5992
0.68	0.425	2.694	0.994681	0.18	0.000893	0.02400	0.5769
0.67	0.399	2.571	0.993834	0.17	0.000677	0.01931	0.5536
0.66	0.374	2.452	0.992885	0.16	0.000504	0.01531	0.5295
0.65	0.3497	2.3365	0.99183	0.15	0.000368	0.01196	0.5044
0.64	0.3269	2.2240	0.99065	0.14	0.000263	0.00917	0.4783
0.63	0.3052	2.1148	0.98934	0.13	0.000183	0.00689	0.4512
0.62	0.2846	2.0089	0.98789	0.12	0.000124	0.00506	0.4231
0.61	0.2650	1.9062	0.98630	0.11	0.000081	0.00361	0.3940
0.60	0.2464	1.8068	0.98455	0.10	0.000051	0.00249	0.3639
0.59	0.2288	1.7107	0.98262	0.09	0.000030	0.00165	0.3327
0.58	0.2122	1.6177	0.98052	0.08	0.000017	0.00104	0.3003
0.57	0.1965	1.5279	0.97823	0.07	0.000009	0.00062	0.2669
0.56	0.1816	1.4413	0.97573	0.06	0.000004	0.00034	0.2323
0.55	0.1676	1.3579	0.97301	0.05	0.000002	0.00016	0.1966
0.54	1.1545	1.2775	0.97007	0.04	0.000001	0.00007	0.1597
0.53	0.1421	1.2002	0.96688	0.03	0.000000	0.00002	0.1216
0.52	0.1305	1.1259	0.96344	0.02	0.000000	0.00000	0.0823
0.51	0.1196	1.0547	0.95973	0.01	0.000000	0.00000	0.0418
				0.	0.	0.	0.

*Mitchell formula: $T = \left(\frac{H}{100}\right)^2 \sqrt{\frac{\Sigma\, w\, \Delta\alpha + (1/H)\,\Sigma\, P\beta}{\Sigma\, \bar{E} D^3 t\, \Delta\gamma}}$

some given ground motion. Since our interest here is primarily in the fundamental mode, the formula and related curves are given only for this mode, with some minor changes in notation to provide some degree of consistency in the nomenclature used here.

$$T = 2\pi\left[\frac{A_o pH^4}{\Omega_1^2 EI_o}\right]^{1/2} \tag{20.2}$$

where T = fundamental period in seconds
A_o = cross-sectional area of the shell at the base in square feet
P = mass density in slugs per cubic foot
H = height in feet
E = modulus of elasticity in pounds per square foot
I_o = moment of inertia of the base cross section in feet⁴
Ω_1^2 = constant given by Fig. 20.4 depending upon the ratios r_1 and r_2
r_1 = ratio of the mean diameters, top to bottom, $= (D_H - t_H)/(D_o - t_o)$
r_2 = ratio of shell thicknesses, top to bottom, $= t_H/t_o$
t_H, t_o = shell thicknesses, top and bottom, in feet

Warren W. Mitchell in an unpublished work (1962) developed a form of solution of the familiar Rayleigh principle of equating potential and kinetic energies in a vibrating system, which is especially useful in calculating fundamental periods of cylindrical, tapered-cylindrical, and step-tapered-cylindrical structures common to refinery-type vessels.

$$T = \left[\frac{H}{100}\right]^2 \sqrt{\frac{\sum w\,\Delta\alpha + (1/H)\sum P\beta}{\sum \bar{E}D^3 t\,\Delta\gamma}} \tag{20.3}$$

where T = fundamental period in seconds
H = height in feet
w = weight per foot of height over a uniform, or assumed uniform, section of the structure in pounds per foot
P = concentrated loads that may be attached to the structure at any level, which add mass but do not contribute to the stiffness of the structure
$\bar{E}$ = Modulus of elasticity in pounds per square inch over 10^6
D = diameter of each section in feet
t = thickness of each section of shell in inches
$\Delta\alpha$ = difference in the α values between the top and bottom of each section
β = constant at the section at which concentrated loads P are attached
$\Delta\gamma$ = difference in the γ values between the top and bottom of each section
$\sum$ = summation of the products of the quantities shown over the height of the structure

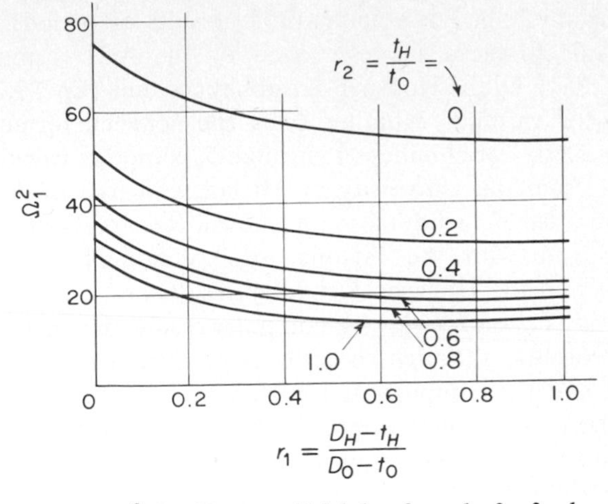

Fig. 20.4. Ω_1^2 for Housner–Keightley formula for fundamental period of tapered cantilevers.

Table 20.1 has the values for α, β, and γ. In applying this method to the 450-ft chimney, item 6 of Table 20.2, the chimney was arbitrarily divided into 10 sections. The result is in excellent agreement with those obtained by two other methods given above. An illustration of the use of the Mitchell method to a step-tapered tower is given later.

Table 20.2. FUNDAMENTAL PERIODS OF REINFORCED CONCRETE TAPERED CHIMNEYS

	Dimensions			Period in seconds				
				University of Michigan computer				
Chimney number	Height, ft	Top O.D., ft	Bottom O.D., ft	Strong axis	Weak axis	ACI formula	Housner and Keightley	Mitchell
1	299.67	9.92	26.67	1.09	1.12	1.23	1.29	
3	343.55	16.50	33.00	1.48	1.53	1.38	1.40	
3	352.5	17.17	27.42	1.83	2.12	1.84	1.84	
4	417.0	21.17	37.58	1.47	1.51	1.76	1.51	
5	534.0	18.67	35.03	2.26	2.39	3.17	2.68	
6	450.0	17.75	40.0	[Not available]		2.02	2.05	2.02

Table 20.2 shows comparative results of periods of several chimneys as determined by the ACI formula (Eq. 20.1), by the Housner–Keightley formula (Eq. 20.2), and for chimney Number 6 by the Mitchell formula (Eq. 20.3). For chimneys 1 through 5, computer calculations from the University of Michigan (a part of ACI Committee 505 study) also are shown. Results should be compared with the "strong axis" computer results. Note that the Housner–Keightley results are more consistently in accord with the computer results than are the ACI results, although the only large disparity between ACI and the computer is for chimney Number 5. The results for chimney Number 6 from the three formulas given here are in almost exact accord.

To illustrate the application of the SEAOC criteria to a chimney, the shear and moment curves for the design of the 450-ft chimney Number 6 in Table 20.2 will be developed. The basic dimensions of this chimney are as follows:

$$h_n = 450 \text{ ft}$$
$$D_o = 40.00 \text{ ft}$$
$$D_H = 17.75 \text{ ft}$$
$$t_o = 24 \text{ in.}$$
$$t_H = 7 \text{ in.}$$
$$\text{total weight } W = 10{,}814 \text{ kips}$$
$$\text{fundamental period } T = 2.02 \text{ sec}$$

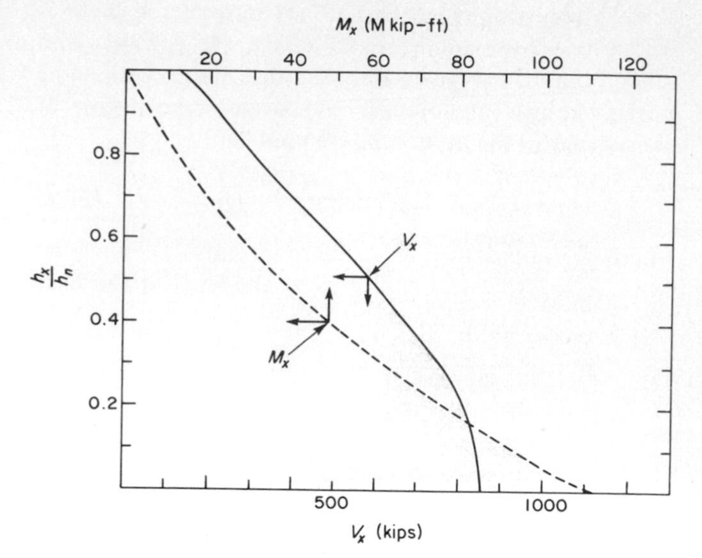

Fig. 20.5. Earthquake design shear and moments in a 450-ft reinforced concrete chimney.

$$\text{combined earthquake coefficient} = ZKC = 1.0 \times 2.0 \times 0.05/2.02^{1/3} = 0.0786$$

$$\text{base shear } V = ZKCW = 0.0786 \times 10814 = 852 \text{ kips}$$

Since H/D_o is greater than 6.12, $F_t = 0.15$ and $V = 127$

Table 20.3. EARTHQUAKE DESIGN SHEARS AND MOMENTS IN 450-FT CHIMNEY

$\frac{h}{h_n}$	$\frac{h_x}{h_n}$	w_x, kips	$\frac{w_x h_x}{h_n}$	$\frac{w_x h_x}{\sum wh}$	$(V - F_t)\left(\frac{w_x h_x}{\sum wh}\right)$	V_x, kips	$\left(\frac{h_x}{h_n}\right)^3$	J_x	M_x, kip ft
1.00						$127 = F_t$			
	0.95	341	324	0.083	60.2	187.2	0.852	0.922	2630
0.90									
	0.85	472	401	0.106	76.8	264.0	0.614	0.797	9000
0.80									
	0.75	607	455	0.117	84.8	348.8	0.421	0.695	16100
0.70									
	0.65	761	495	0.126	91.4	440.2	0.274	0.618	24000
0.60									
	0.55	937	515	0.132	95.7	535.9	0.166	0.561	32900
0.50									
	0.45	1120	504	0.129	93.5	629.4	0.091	0.522	43200
0.40									
	0.35	1310	458	0.117	84.8	714.2	0.043	0.497	55300
0.30									
	0.25	1526	381	0.098	71.1	785.3	0.0156	0.482	69100
0.20									
	0.15	1720	258	0.066	47.9	833.2	0.0033	0.476	85200
0.10									
	0.05	2020	101	0.026	18.8	852.0	0.00012	0.474	102000
0								0.474	113000
		10814 W	3892 $\sum \frac{w_x h_x}{h_n}$	1.000	$725.0 = \sum F_x$ $127.0 = F_t$ $852.0 = V$				

$$J = \frac{0.60}{\sqrt{T}} = \frac{0.60}{3\sqrt{2.02}} = 0.474 \qquad J_x = J + (1 - J)\left(\frac{h_x}{h_n}\right)^3 = 0.474 + 0.526\left(\frac{h_x}{h_n}\right)^3$$

kips. The remainder of the shear, 852 − 127 = 725 kips, will be distributed in accordance with formula 2-B of Fig. 20.2. This is done by dividing the chimney into 10 equal height sections and performing the calculation shown in Table 20.3. Summing the resulting forces progressively from the top to the bottom provides the shears at all sections. These are plotted in Fig. 20.5.

The computation of $J = 0.474$ and of the J_x values at different heights in the chimney, in accordance with the formulas 3-B and 3-D, respectively, of Fig. 20.3, are also indicated in Table 20.3. The resulting moment curve has been plotted on Fig. 20.5 also.

20.3.2 Fundamental Period of Steel Chimneys

For cantilevered structures of uniform section, Eq. 20.4 applies to materials generally:

$$T = 1.79\left(\frac{wH^4}{EIg}\right)^{1/2} \tag{20.4}$$

For steel, this can be manipulated into the following form:

$$T = 765 \times 10^{-8}\left(\frac{H}{D}\right)^2\left(\frac{12wD}{t}\right)^{1/2} \tag{20.5}$$

where T = fundamental period in seconds
H = height in feet
D = diameter in feet
w = weight per unit of height in pounds per foot
t = shell thickness in inches
E = modulus of elasticity in pounds per square foot
I = moment of inertia of the cross section in feet^4
g = acceleration of gravity = 32.2 ft/sec^2

If the base of the steel chimney is flared, the period of such a chimney can best be calculated by

$$T = 2\pi\left(\frac{0.80\delta}{g}\right)^{1/2} \tag{20.6}$$

where δ = the calculated deflection in feet at the top of the chimney due to 100% of its weight applied as a lateral load.

If the chimney is lined, the weight per foot must include the weight of the lining, and the added stiffness due to the lining must be determined as an equivalent thickness of steel. The modified thickness of steel then should be used in the appropriate formula given above. This has a good correlation with measured periods of lined stacks. Gunite lining, e.g., can reduce the period of a steel chimney on the order of 15%.

Once the period is determined, the calculation of base shear, distributed forces, shear, and moment diagrams is essentially the same as that outlined in the illustrative problem for the reinforced concrete chimney. The properties of the steel at the operating temperature must guide the allowable stresses. The operating temperature of the steel will depend upon linings or insulations that may be applied to the chimney. Whatever that temperature may be, it must be recognized that the yield point of normal carbon steel decreases about 1500 lb/in.^{2} for each 100°F rise in temperature. Buckling stresses and the need for stiffening of the shell also need investigation. Without stiffening, the thickness of the shell should be equal to or greater than that given by the following:

$$t = \frac{Ds_y}{0.24E'} \tag{20.7}$$

where t = shell thickness in inches
s_y = yield strength of the steel at the operating temperature in psi
E' = modulus of elasticity at the operating temperature in psi
D = diameter of the shell in inches

20.3.3 Fundamental Period of Step-Tapered Towers

It was mentioned earlier, in discussing methods for calculating the fundamental period of reinforced concrete chimneys, that the Mitchell formula (Eq. 20.3) has general application to uniform, tapered, or step-tapered cylindrical structures. To illustrate its use, and the use of the constants in Table 20.1, the period of a process vessel not uncommon in refinery practice has been calculated in Fig. 20.6. Having found the period to be T = 1.10 sec, the steps that follow in applying the SEAOC code criteria are essentially those outlined in some detail in the illustration of the 450-ft reinforced concrete chimney. The basic difference would be to judiciously divide the height to have sections terminate at the break points in the elevation, i.e., where the diameter or where the unit weight changes. For the purpose of calculating the period, these are the only divisions of structure that need to be made. However, for purposes of distributing the base shear, further subdivision is justified, up to perhaps 10 or more.

It is also recommended that a J factor of 1.0 be used for such structures, since it seems reasonable to expect that such a structure would vibrate primarily, if not exclusively, in the fundamental mode.

20.4 WIND AND EARTHQUAKE

While it is not the purpose here to define wind forces (they are quite adequately defined in recent reports—

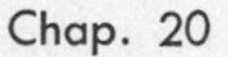

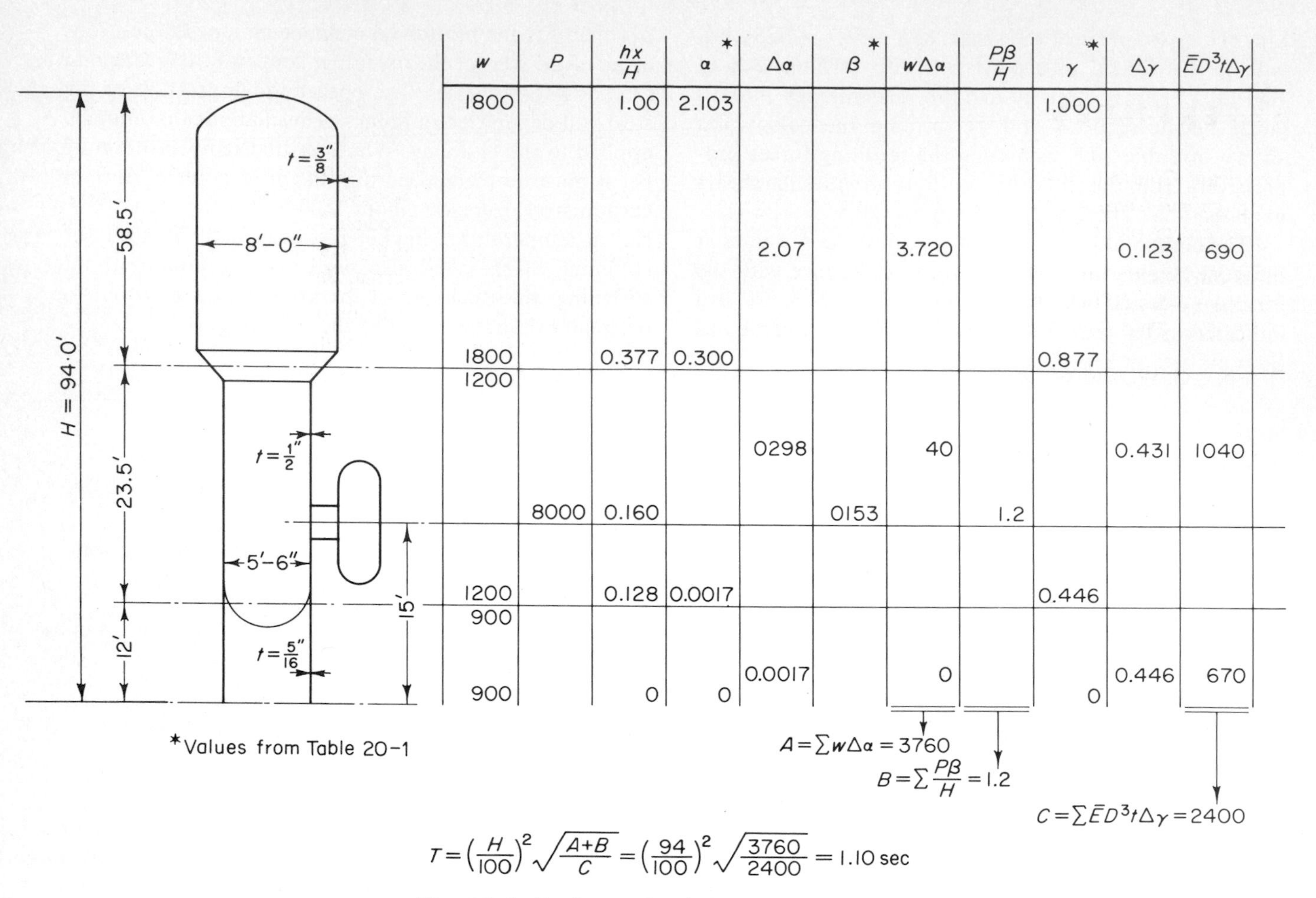

$$T=\left(\frac{H}{100}\right)^2\sqrt{\frac{A+B}{C}}=\left(\frac{94}{100}\right)^2\sqrt{\frac{3760}{2400}}=1.10\ \text{sec}$$

Fig. 20.6. Fundamental period of step-tapered tower.

ASCE Task Committee on Wind Forces, 1961, and in the Uniform Building Code), it is important to stress the need to consider wind criteria concurrently with considerations of earthquake design criteria. While these are not additive, both sets of criteria need to be investigated in detail far enough to determine the controlling or governing lateral force criteria. It is possible, e.g., for the earthquake design shear to be larger than wind shear but to have the wind overturning moment larger than the reduced earthquake overturning moment. It is not enough that the basis for selecting the lateral force design criteria be merely on the base shears.

20.5 DETAILS

Minimum design criteria, such as those given in codes, do not in themselves assure an adequately earthquake-resistant structure. They can help, but they have to be applied with good judgment and with careful attention to details. In fact, these areas of judgment and details may be considerably more important than the earthquake coefficient selected for the design. Where failures have occurred, they resulted from some shortcoming in the application of the criteria, not because of the criteria.

Generally, on tall towers of steel it is important to avoid buckling of the shell or skirt holding up the tower. This has been discussed, but it is also important to recognize that forces, shears, and moment in a major earthquake will exceed the elastic design criteria used. It is to be expected that some part of the structure necessarily will have to yield. The most logical place to have that yielding is in the anchor bolts. In order to permit this, the anchor bolts need to be long enough to allow some yielding under the extreme tensile load placed on them by the overturning effect of the lateral forces. They should be attached to the vessel through full-ring stiffeners to avoid buckling of the skirt. These details were used in the refinery structures on the Kenai Peninsula. There was $\frac{1}{8}$ in. yield in the anchor bolts, but no significant damage was sustained in the Alaska Earthquake of March 27, 1964. This experience gave considerable assurance that, properly applied, the criteria now specified are reasonable and adequate to resist major earthquakes.

REFERENCES

American Concrete Institute (1968). *Specification for the Design and Construction of Reinforced Concrete Chimneys*, standard prepared by ACI Committee 307 (formerly 505).

American Society of Civil Engineers, Task Committee on Wind Forces (1961). "Wind Forces on Structures," *Trans. Am. Soc. Civil Eng.*, **126**, 1124–1198.

Housner, G. W., and W. O. Keightley (1963). "Vibrations of Linearly Tapered Cantilever Beams," *Trans. Am. Soc. Civil Eng.*, **128**, 1020–1048.

Mitchell, W. W. (1962). "Determination of the Period of Vibration of Multi-Diameter Columns by the Method Based on Rayleigh's Principle," an unpublished work prepared for the Engineering Department of the Standard Oil Company of California, San Francisco.

Structural Engineers Association of California (1967). *Recommended Lateral Force Requirements and Commentary*, San Francisco.

INDEX